Plant Physiology

Overleaf: Global carbon and oxygen factory: a photosynthetic leaf. Scanning electron micrograph of transverse section of a *Euphorbia* leaf, ×650. (Original photograph courtesy Dr. Graham Walker, DSIR, New Zealand.)

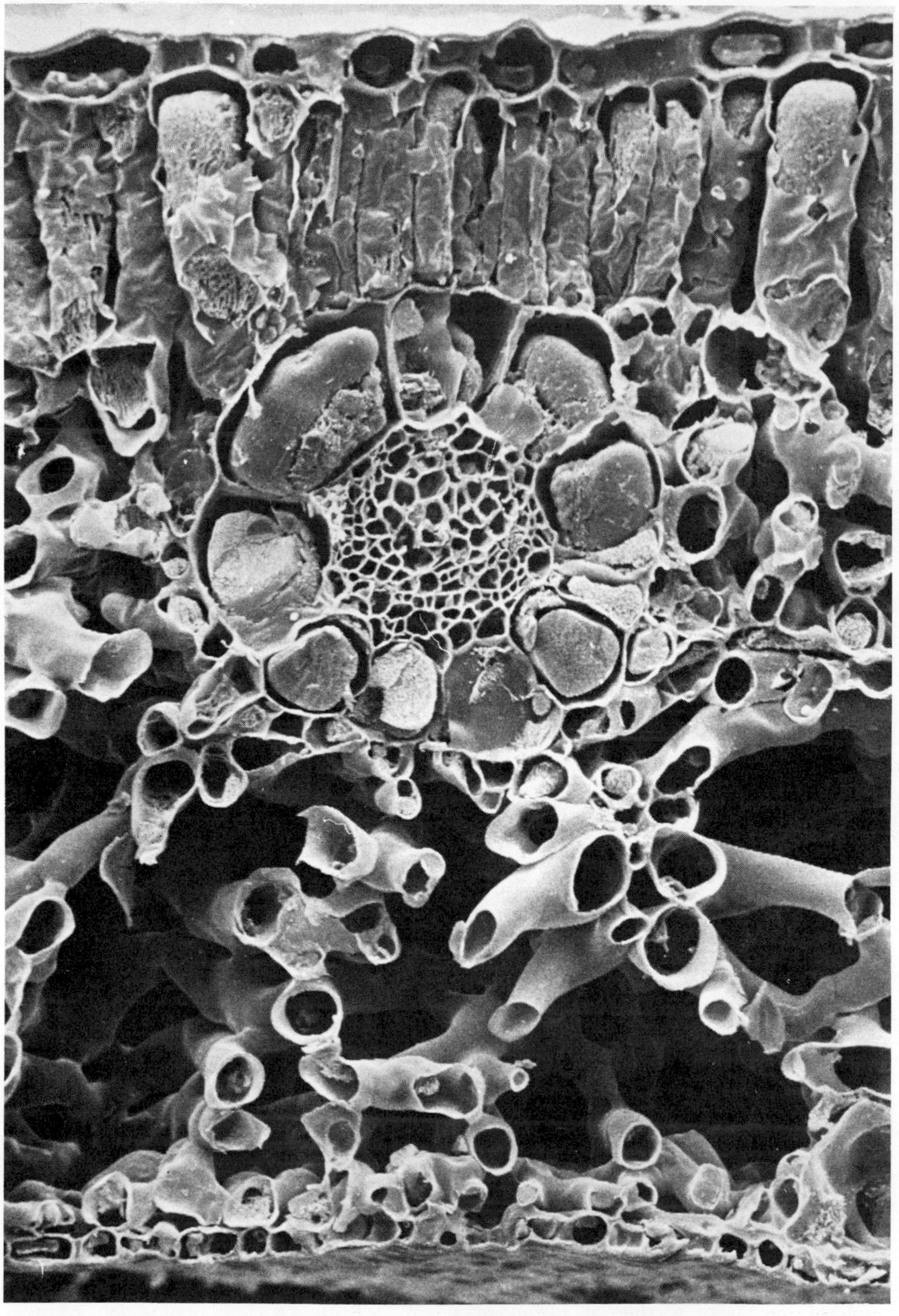

Plant Physiology

SECOND EDITION

R. G. S. Bidwell

Professor of Biology, Queen's University, Kingston, Ontario, Canada

MACMILLAN PUBLISHING CO., INC.

New York

Collier Macmillan Publishers

London

Macmillan Publishing Co., Inc.
866 Third Avenue, New York, New York 10022

Collier Macmillan Canada, Ltd.

Library of Congress Cataloging in Publication Data

Bidwell, Roger Crafton Shelford, (date)
Plant physiology.

(The Macmillan biology series)
Includes bibliographies and index.
1. Plant physiology.
QK711.2.B54 1979 581.1 78-6504
ISBN 0-02-309430-3

Printing: 1 2 3 4 5 6 7 8 Year: 9 0 1 2 3 4 5

DEDICATION

To my wife, Shirley M. R. Bidwell,
without whose encouragement, help, and forbearance
this book (and its predecessor) could not have been written.

Preface

This book is an introductory text for undergraduate students in plant physiology or experimental botany. It presents current ideas about how plants function in narrative form, uncluttered (as far as possible) with complex metabolic pathways or biochemical mechanisms. These are detailed separately in early chapters (Section II) and recalled in simplified form, or by reference, as required in the subsequent narrative. Summaries are introduced wherever the material covered in the preceding paragraphs or chapter is sufficiently complex or controversial to need review. This second edition is largely rewritten and updated. Some irrelevant material has been deleted, and a number of important topics have been clarified and revised to align them with modern developments in plant physiology.

I have tried to present the subject in a sufficiently interesting style that students can read the text easily and get an overview of large areas of plant physiology, even though the length of their course or the interest of the instructor requires that they concentrate in depth only on part of the book. As a result, this text should be useful for either one-semester or two-semester courses. The whole field of plant physiology, together with the necessary biochemical background, is treated in sufficient depth that the book will also be useful as a general background or reference text for senior and graduate students. Moreover, the book is organized so that it can easily be used for half- or quarter-courses in plant physiology, experimental botany, plant biochemistry, or plant development, and the material in it can be rearranged or taken out of sequence to suit the preference of individual instructors.

Section I is primarily review, but students should be sure that they are familiar with this essential background before proceeding. It also serves as a concise reference for the classification and occurrence of biological chemicals, plant and cell structure, water relations, and related topics. Section II covers the basic processes of the cellular metabolism of plants. Section III deals with the nutrition and metabolism of whole plants and their relationships with internal and external nutrient sources. The processes of plant development and behavior are summarized in Section IV, and the physiology of specialized groups of organisms and of special situations and plant relationships is described in Sections V and VI.

This book contains three departures from the traditional format of plant physiology texts. First, I have included (in Section III, Chapter 15) an overview of the carbon nutrition of plants that puts the overall processes of photosynthesis, photorespiration, dark respiration, and their associated nitrogen metabolism into one perspective: that of the plant. It can be read and understood by students who have not concentrated on the details of metabolism presented in Section II. The second new approach is in Section IV, in which the story of plant development is based on the life history of the developing plant rather than on the chemicals and mechanisms that control it. The mechanistic point of view is developed in Chapter 23 in the form of a summary to Section IV. The third point is the inclusion, in Sections V and VI, of discussions about the physiology of special and important organisms and associations such as seaweeds, disease, symbionts, and the factors affecting plant distribution. Plants are becoming ever more important to man, and the relationships between plants and man, particularly when they depend upon physiological factors, are an important part of the study of plant physiology.

It is impossible in a text at this level to document all the ideas and facts that must be presented. On the other hand, a list of facts, even if strung together in an acceptable narrative, is not sufficient for effective study. I have therefore provided experimental background, by describing techniques and actual experiments, for some of the more important pieces of evidence that are presented. I hope that this will make the text interesting and readable and, at the same time, stimulate the student's interest in the experimental basis of plant physiology.

I have not used references in the text. Each chapter is followed by a reading list including monographs, articles, and books. I feel that the instructor's list of essential reading will be far more valuable than any list of references I could provide, beside being up-to-date. The names of a number of prominent scientists associated with specific facts or ideas have been mentioned in the text. I have stressed the importance of searching for stimulating reading in such publications as F. C. Steward's *Treatise* on *Plant Physiology* and *Annual Reviews of Plant Physiology*. These and recent papers in the literature will provide excellent source material for class discussions or seminars.

Writing a textbook on plant physiology is a considerable task. No one plant physiologist can hope to know all about it, so I have relied heavily on the suggestions and criticisms of my colleagues and students to keep the text as accurate as possible. I wish to acknowledge the help of the several reviewers who read the manuscript of this book. For this revision I am deeply indebted to the very many colleagues and students who sent me their ideas, corrections, and suggestions for improvements. I am particularly indebted to A. T. Jagendorf, W. T. Jackson, and I. A. Tamas who exhaustively reviewed the previous edition. Many of the improvements and corrections are due to their detailed criticisms. Ultimately, of course, responsibility for what has been included or omitted is my own.

Producing a textbook is also a big task. I am most grateful to Mrs. Judy Bollen for typing much of the manuscript, and for her help in collecting and organizing illustrations and accessory material. I should also like to express my gratitude to members of the Macmillan Publishing Co., Inc., for their thoughtful assistance, forbearance, and encouragement, and particularly to Woody Chapman and Pat Larson whose help and support were indispensable during the writing and production of this book.

R. G. S. B.

Brief Contents

SECTION **V**

Physiology of Special Organisms

SECTION **VI**

Physiology of Plant Distribution and Communities

Detailed Contents

SECTION IV

The Developing Plant—Plant Behavior

Plant Physiology

SECTION I

Introduction and Background

1 Introduction

Plant Physiology

Plants germinate, grow, develop, mature, reproduce, and die. Plant physiology is the study of these processes, of how and why each plant behaves in its own peculiar way. It is the study of the organization and operation of the processes in the plant that order its development and behavior. Every plant is a product of its genetic information modified by its environment, and every part or organ of a plant is further modified by the physiological state, or internal environment, of the plant of which it is a part. Plant physiology concerns itself with the interplay of all these factors in the life of the plant.

Plants and Animals

Why should we study plant physiology as distinct from animal physiology or cell physiology? Plants and animals have evolved a basically different pattern or habit of life. Animals are mechanistic and determinate in growth and life—they are constructed like machines, whereas plants grow and operate on an architectural basis. Animals, by and large, are motile, must seek food, and have to fit into a narrow set of size limits in order to operate successfully. They are like automobiles—if they grow too big, or asymmetrically, they cannot operate successfully. Most plants are stationary and manufacture their own food, relying on what they can get within the limits of their immediate environment. They are not limited by size or mechanical considerations—they are built like houses; new rooms can be added continuously (within the confines of structural strength) without problems arising. The plant can grow and develop throughout its life, parts can be abandoned and left to die, new parts can be added here and there as required. The animal must maintain its mechanical integrity; the plant is under no such constraint. But the animal can move through its environment to find and get things it needs to live and grow. The plant can grow through its environment, but only slowly and to a limited extent. Plants must rely on what is immediately available to them.

As a result of this architectural pattern of growth and nonmotile habit of life, plants face a number of special problems in food getting and survival that have been solved in a variety of different ways. Plants must withstand not only the predictable changes in environment but also unpredictable variations in weather and climate. For example, plants that live in a desert and lack a specific water-storing device can only flourish during and immediately after a rain. Organic nutrition, a major concern for animals, is a small problem for most plants because they are **carbon autotrophic;** that is, they manufacture all required carbon compounds for themselves from carbon dioxide (CO_2). However, plants have only limited access to limited supplies of inorganic nutrients present in the soil, and a specialized and highly conservative metabolism of nitrogen, phosphorus, potassium, and other important inorganic elements has evolved in them. Animals are conservative of carbon and cycle it within their bodies, whereas they are quite wasteful of nitrogen. Plants behave in exactly the opposite way, conserving nitrogen and using carbon freely.

Being rooted in one spot, plants have a special problem with water. Like animals, they must rely on gas exchange to live. However, they require gas exchange for their major nutritional activity, photosynthesis, as well as for respiration. An automatic consequence of efficient gas exchange is the loss of water vapor. Plants need sunlight to live. The sun, by heating them, increases water loss by evaporation. Elaborate compromises have evolved in plants that enable them to conserve water and at the same time carry on efficient gas exchange of oxygen and carbon dioxide.

Another result of the nonmechanical design of plants is that they have no pumps and no closed circulation system. Plants sometimes achieve great mass and grow very tall, and, like animals, they must transport foodstuff, regulating substances, and waste materials throughout their bodies. In addition, they must transport enormous quantities of water, often to great heights. Lacking mechanical pumps, plants have a variety of chemical and physical–chemical devices for moving fluids, and an elaborate system of plumbing in the form of specialized tissues that permit the directed movement of their nutritional traffic.

Plants differ fundamentally from animals in the process of development. Animals do most of their developing in a highly coordinated way during a brief part of their life span and, having developed, achieve a more or less fixed steady state that may last for a long time. Plants continue to develop throughout their lifetime, and the various parts develop, mature, and die to a considerable extent independently of each other. Obviously, the system of controls that regulates development in plants is quite different from that in animals. Animals produce and respond to a number of highly specific hormones that affect specified tissues in specific ways. Plants react to a few generalized hormones. These can each affect all, or nearly all, the tissues in the plant and may work in one direction or another, depending on a variety of environmental and internal conditions of the affected tissue. The control of plant development is thus the result of a composite of hormonal, environmental, and nutritional factors that interact with each other and make possible the expression of the genetic characteristics of the plant in the most effective way under various environmental circumstances.

Finally, plants differ from animals in having no nervous system. Animals, operating mechanically, need nerves for constant precise control of movement. This, over a long span of time, has permitted the evolution of a brain. Plants have no need for nerves. Their behavior is largely expressed in their patterns of growth and development, processes that are controlled by biochemical and physiological integration rather than by the integration of nerves and thought.

Characteristics of Plants and Plant Life That Lead to Specialized Physiology

1. Plants are largely nonmotile and can only penetrate and exploit a limited volume of their environment.
2. Carbon autotrophy permits a profligate carbon metabolism.
3. The dependency on soil supplies of minerals results in a highly conservative mineral nutrition, particularly of nitrogen.
4. As plants evolved to a terrestrial habit, various devices have also evolved that protect them from water loss and permit them to carry on gas exchange without breathing.
5. The terrestrial habit also required the evolution of elaborate devices for obtaining and transporting water.
6. The system of structural support that has evolved in plants—rigid cell walls instead of a specialized skeleton—has created a number of problems. In spite of their massive cell walls, plant cells have a far greater degree of interconnection than animal cells. The cell contents of adjacent cells are often in intimate contact through cytoplasmic connections.
7. Plants face serious seasonal problems, and time-measuring devices have evolved so that their activities fit into the pattern of seasonal changes in their environment. Reproduction, seed production, dormancy, germination, leaf fall, and so on are all seasonally determined.
8. Because they are immobile, plants cannot hide or seek protection from the elements. Special means of protection from excesses of wind, drought, cold, heat, and light have evolved.
9. The requirements for reproduction in nonmotile organisms have resulted in the evolution of a variety of highly specialized reproductive structures and devices. The control and operation of these make up one of the most fascinating chapters in the book of plant life.
10. The problems of control and regulation in a continuously developing organism require a complex of physiological-biochemical mechanisms that have developed to a very high level in plants.
11. Evolution and adaptation of organisms take place in a physiological and biochemical sense as well as through changes in anatomy and morphology. Plants rely heavily on successful physiological or biochemical mechanisms for survival and consequently a varied and elaborate physiology has evolved.
12. Plants have no organized nervous system but rely mainly on biochemical means of communication between their parts. This results in still further elaboration of physiological mechanisms in plants that have no counterpart in animals.

All of these special problems of plant life are conveniently studied from the biophysical or biochemical point of view as if they were mechanistic devices. However, we must not lose sight of the more important aspect of these problems—the physiological organization of all these processes within the functional whole of the plant and within the relationship of the plant to its environment and to the community in which it lives. The basis of a plant's success is the ability to compete in its environment, and this ability depends largely on its physiological evolution and adaptation to the environment.

Evolution

Plants have evolved continuously, and they are still evolving. The obvious aspects of evolution, those that we see and classify, are the evolution of shape and form and habits of growth and life. Evolution of the biochemical and physiological processes in plants also occurs. It is probable that much of the evolution of biochemical processes took place very early. Even the most primitive plants have fully competent systems of photosynthesis, respiration, protein synthesis, and so on. However the physiological mechanisms that control the form and pattern of development must have evolved in parallel with those that coordinate the operation of all the parts of the plant. These processes are still evolving. There is some recent indication that, although the basic processes of photosynthesis and respiration have not changed for a long period of time, some of the subsidiary processes and the biochemical interrelationships between processes may be undergoing considerable modification as modern plants evolve in response to their constantly changing environment. Successful plants are those that can operate most effectively under the widest range of environmental conditions. Since the coming of man, success in plants has come to include maximum useful productivity for man, under conditions of cultivation as well as in the natural environment.

Applied Botany and Economics

Physiological concepts, and therefore physiologists, are frequently called upon to solve problems of economic botany and agriculture. Applied biology was at one time considered rather a second-grade subject by many so-called "pure" biologists, who looked down on research efforts directed toward a solution of economic problems. However, the tremendous success of such applied studies, the number of basic principles that have emerged from studies that began as strictly applied projects, and the recent increase in the sociological and economic importance of biological problems, particularly the specter of food shortages and starvation that haunts much of today's world, has raised applied biology to its present prominent level. Many basic research problems in applied biology are now being studied, and many more need to be tackled.

One of the most important aspects of plant science in the development of economic botany has been genetics. Selection and breeding have resulted in the development of nearly all the modern crop plants that show such desirable characteristics as high production, taste- and eye-appeal, resistance to disease, and adaptation to a wide range of climatic and environmental conditions. However, these general traits, resistance to disease, ability to withstand stringent conditions, and primary productivity, as well as specific traits, such as fruit-set, desirable characteristics of the fruit, early ripening, and so on, are all expressions of physiological processes. Previous breeding experiments were conducted by seeking desirable traits and attempting to select for them and improve them by selective breeding. The success of such programs relied on the happy coincidence of the experimenter noticing a valuable trait that was, in fact, capable of selection and development. Many of the greatest successes were lucky chances or the occasional good shot that came from thousands of "try it and see" experiments. But as the study of plant physiology progresses, it is gradually becoming possible to pick out the specific physiological or biochemical mechanisms that will result in the desired improvements and to determine what part of the genetic complement of the plant is responsible. When these new techniques are sufficiently refined, it will become possible to conduct breeding programs on a wholly new level, by selecting the best possible combinations of genetic traits to produce the most nearly perfect organism for any given set of conditions and circumstances. Typical programs now underway on crop plants are aimed at producing varieties that have high rates of photosynthesis combined with low rates of respiration (not just high metabolic rates), that produce fruit of the most useful sort, that convert the largest proportion of their assimilated carbon into fruit or seed instead of into leaves and stems, that are hardy, grow fast, and are resistant to disease.

Besides these specific traits, the overall strategy of plants can be modified to improve their usefulness to man. Their height and shape, the orientation of their leaves to the sun, the timing of their reproduction, their dependence on nutrition are all capable of modification and improvement. Modern techniques of cell physiology permit the growth of plants from single cells and even, to a limited extent as yet, the manipulation of genetic material in somatic cells. Experimental progress such as this requires the widest application of the most intimate biochemical, physiological, and genetic knowledge of plants.

There is no question that the combination of plant physiology with genetics is most important, and perhaps vitally necessary, to the future of man. The food of the world, now in spite of the green revolution in dangerously short supply, is derived from plants. The forestry industry has recently developed from a blind plundering of natural resources into an effective, self-regulating program of agriculture. Tree-breeding programs, always handicapped by the long generation time of trees, have been enormously helped by the elucidation of physiological mechanisms that permit the experimenter to force trees to sexual maturity in 2 to 4 years instead of 10 to 15 years. Marine agriculture will become a major industry, requiring entirely new insight into the habits and growth patterns of marine algae.

Plants are the major agents that clean up our increasingly polluted world—in the air, on land, and in the water. Plants also suffer from pollution. Programs of urban renewal, of city and of total environment planning, must include plants; they are essential for ground cover and protection, for water concentration, for food, beauty, and for the regeneration of the atmosphere. Plants comprise about 99 percent of the biomass of our world. They recycle somewhere in the neighborhood of 0.1 percent of the total available carbon in the biosphere every year. They are the source of many of our medicines and drugs. They have a tremendous impact on weather and weather systems by changing temperatures, evaporating water, and producing large quantities of assorted volatile chemicals. They are among the major allergenic agents in the world, as hayfever sufferers are well aware. In nearly every way, mankind depends on plants.

The understanding, development and control of plants will become more and more important as time goes on. Only complete comprehension of the physiology, biochemistry, and genetics of plants will enable us to conduct successfully the tremendous program of economic biology that will be required for the comfort, and even for the survival, of man.

Additional Reading

Janick, J., R. W. Schery, F. W. Woods, and **V. W. Ruttan:** *Plant Science.* W. H. Freeman & Co., San Francisco, 1974.

Handler, P. (ed.): *Biology and the Future of Man.* Oxford University Press, New York, 1970.

Plant Agriculture (Selections from *Scientific American*). W. H. Freeman & Co., San Francisco, 1970.

Tippo, O., and **W. L. Stern:** *Humanistic Botany.* W. W. Norton and Co., Inc., New York, 1977.

2 Chemical Background

Plant physiology is based on physical and chemical as well as biological concepts and uses the terms and systems of measurement of these sciences. Plant physiology is itself an exact science and, therefore, must be described and presented in precise terms. Many students will be familiar with the material and terms presented in this chapter. However, do not confuse familiarity with knowledge! You may not be as certain as you think. Read and make sure that you know precisely what you are reading and thinking about.

Solutions

All living matter depends on water. Protoplasm is dissolved or dispersed in water, nearly all the materials in the cell are transported in water, and almost all biological reactions take place in aqueous solution. The peculiar physical, chemical, and electrical properties of solutions and dispersions of material in water provide the basis for the chemistry and physics of living material. It is important to understand clearly the properties of solutions.

A solution consists of at least two components: the **solute** which is dispersed in molecular form throughout the **solvent.** Most natural solutions are complex in that they contain more than one solute. Within wide limits, however, the various solutes in a complex solution behave independently of each other, each one acting in the solution as if the others were not there. Exceptions to this behavior will be noted later. However, this is rather an important characteristic in the behavior of gas-water solutions, as we shall see.

Solutions of biological importance include gas in gas (so-called "mixtures" of gases are really solutions and behave as such), gas in liquid, liquid in liquid, and solid in liquid. Solutes may be nearly insoluble, slightly soluble, or very soluble. The solution is said to be **saturated** at the concentration above which no more solute can be dissolved in that solvent, that is, when layers or droplets form in a liquid-liquid solution or when crystallization or precipitation occurs in a solid-liquid solution. Unstable **supersaturated** solutions may occur, frequently as the result of gentle cooling of a saturated solution, but they are extremely susceptible to shock and the excess solute is apt to precipitate or crystallize out suddenly.

Water is probably the nearest thing to a universal solvent, and most chemicals will dissolve in water to some extent. For most compounds solubility increases with increasing temperature of the solution; however, the solubility of some compounds, such as salt, is relatively unaffected by temperature, whereas that of other compounds, including the calcium salts of some important acids, is decreased by increasing temperature.

Solutions of Gas. Solutions of gas in water are biologically of great importance. Most biological gas exchange requires that the gas be dissolved in the cell solution before it reaches the reaction sites, and gas released as the result of reaction is usually liberated into a solution before escaping from the cell as gas.

Gases of biological importance are either slightly soluble, such as oxygen (O_2), hydrogen (H_2), and nitrogen (N_2), or very soluble, such as carbon dioxide (CO_2), ammonia (NH_3), sulfur dioxide (SO_2), and sulfur trioxide (SO_3). Gases such as oxygen are only soluble to the extent of about 0.001% under normal conditions, whereas the extremely soluble gases such as CO_2 may be 100 times more soluble. The solubility of both types of gases is inversely proportional to the temperature, a most important fact for aquatic or submerged organisms that must depend on dissolved O_2 for respiration. The solubility of a gas in water is directly proportional to the pressure of that gas over the water. It follows that the solubilities of various gases in a mixture are proportional to their partial pressures. Thus the amount of O_2 dissolving in a given amount of water would be the same from air (21% O_2) at atmospheric pressure as from pure O_2 at 21% of atmospheric pressure. The amount of CO_2 dissolving in water from air at atmospheric pressure would be the same as from pure CO_2 at about 0.23 mm Hg pressure (0.03 atm)!

The extreme solubility of such gases as CO_2 and NH_3 results from their combination with water to make an acid or base in solution, according to the reactions

$$CO_2 + H_2O \rightleftharpoons H_2CO_3 \rightleftharpoons H^+ + HCO_3^-$$
$$NH_3 + H_2O \rightleftharpoons NH_4OH \rightleftharpoons NH_4^+ + OH^-$$

The state of the dissolved products is of course affected by the acidity of the solution. This is particularly true for CO_2, which can exist in several states in solution depending on the concentration of cations

$$CO_2 \rightleftharpoons H_2CO_3 \rightleftharpoons HCO_3^- \rightleftharpoons CO_3^{2-} \rightleftharpoons NaCO_3^- \rightleftharpoons NaHCO_3 \rightleftharpoons Na_2CO_3$$

Each form exists in solution as a separate entity, and the balance between them is determined by the acidity. Thus the total amount of CO_2 that will go into solution in a given volume of water is proportional not only to the partial pressure of CO_2 in the gas phase and to the temperature but also to the acidity or basicity of the water.

One of the consequences of the extreme solubility of CO_2 in water is that it will diffuse through water very rapidly. CO_2 will also diffuse through a semisolid like agar, which is composed mostly of water, as if it were water. This fact is often unrecognized, but attempts to prevent gas exchange by plugging vessels with materials like agar will be unsuccessful. CO_2 will go through almost as if the plug were not there!

Concentrations. The terms used to express the concentrations of solutions are simple and have an exact meaning. They must be used exactly. It is particularly important to specify clearly the units used and to indicate their meaning if any ambiguity exists. Pay attention to the abbreviation of terms—abbreviations are frequently misused, a common source of error in exact work.

The most common description of the concentration of a solution is the number of **moles** of a substance in exactly 1 liter of the solution at 20° C. (1 mole of any substance = 6.023×10^{23} molecules. This is Avogadro's number, and it is important to remember it. It will reappear in the discussion of energetics and photosynthesis.) The **molarity** of the solution thus expresses the number of moles per liter of solution. (*Note:* not moles per liter of solvent.) A solution containing 2 moles/liter is called 2 *M*. It is important to note that the abbreviation *M* means molar, not moles. It refers to a concentration, not an amount. This is a frequent error that must be avoided. Solutions containing a fraction of a mole per liter are nearly always expressed as decimals (for example, 0.1 or 0.05 *M* means 0.1 or 0.05 moles/liter) rather than as common fractions. It is important to note that solutions expressed in terms of molarity can be arithmetically diluted. If a 1 *M* solution is diluted to twice or ten times its volume, a 0.5 *M* or 0.1 *M* solution results.

A second, often confused, description of the concentration of a solution is made by expressing the number of moles of the solute per 1000 g of solvent; this is called the **molality** of the solution. Molality is used in situations where, for physical or chemical reasons, it is desirable to express the ratio of solute to solvent molecules, for example, in discussion of osmotic pressure. Thus 2 moles of a substance dissolved in 1 kg of water (1 liter at 20° C) makes a 2 molal (*m*) solution. Note that 1 mole of a solute dissolved in 1 kg of water will not have a volume of 1 liter—the volume may be up to 30% greater or it may be less, depending on the solute. For this reason it is not possible to make a 0.5 *m* solution from a 1 *m* solution by diluting it with an equal volume of water. It must be diluted by a volume of water equal to the volume of water in the solution.

Solution strength may be expressed as follows:

1. *Weight per unit weight.* Example: 20% (w/w) sugar in water means 20 g of sugar + 80 g of water, or 20 g of sugar per 100 g of solution.
2. *Weight per unit volume.* Example: 20% (w/v) sugar in water means 20 g of sugar per 100 ml of solution. *Note:* this has about the same concentration as a 19% w/w solution because sugar in solution takes up less room than dry sugar crystals. Thus, the solution takes up less room than the combined volumes of its components.
3. *Volume per unit volume* (often used for solutions of liquid in liquid). Example: 80% (v/v) alcohol in water means 80 ml of alcohol in 100 ml of solution. *Note:* this would be quite different from 80 ml of alcohol mixed with 20 ml of water or from 80 g of alcohol mixed with 20 g of water, because of the difference in specific gravity of the two liquids and because the volume of the solution is somewhat less than that of the two liquids separately.

Precise solutions of compounds like alcohol are often used for the differential precipitation of components in a mixture for their isolation or purification. You can see the tremendous importance of specifying the units of solutions correctly!

Concentrations of gas in solution may be expressed in the conventional way, by molarity or by percentages of weights. They are often expressed in rather sloppy units such as volumes of gas per volume of solution, for example 20 microliters per liter

(μliter/liter) or 20 ppm (parts per million) or as percentages of saturation. These terms must of course be referred to the conditions of temperature and barometric pressure. Since these conditions may be hard to duplicate, such terms should be avoided. Mixtures (that is, solutions) of gases are usually expressed in terms of the percentage of one component in the whole on a v/v basis.

Electrolytes are substances that dissociate or form ions in solution. Solutions of electrolytes are described in the usual way, but it must be remembered that many of the important properties of solutions, such as freezing-point depression and osmotic potential (see Chapter 3), do not depend on the concentration of molecules in solution but on the concentration of particles or ions. As a result, the behavior of a solution will depend not only on its concentration but on the electrolytic strength, or the tendency to ionize, of the solute.

Acids and Bases

Solutions of acids and bases are often expressed in terms of **normality.** A 1 normal solution (abbreviated 1 *N*) of acid contains sufficient acid to provide one equivalent or 1 mole of hydrogen ion (H^+) per liter of solution at 20° C. Similarly a 1 *N* base solution contains sufficient base to provide 1 equivalent or 1 mole of hydroxl ion (OH^-) per liter of solution at 20° C. Note that the acid or base need not be completely ionized; therefore, the term *normal* refers to the concentration of the total available (that is, both associated and dissociated) H^+ or OH^- ions, not their actual concentration in solution. It is evident that for an acid having one replaceable H^+ ion, a molar solution is also a normal solution. However, a 1 *N* solution of $Ca(OH)_2$ (calcium hydroxide), which contains 2 OH^- ions per molecule, will be only 0.5 *M*, and a 1 *N* solution of H_3PO_4 (phosphoric acid), containing 3 H^+ ions per molecule, will be only 0.33 *M*.

The actual concentration of H^+ ions in solution, as opposed to the total available H^+, is nearly always expressed in terms of pH. The **pH** is simply the negative logarithm of the H^+ ion concentration, expressed as its normality. It has been determined that the product of the normalities of H^+ and OH^- in pure water at 20° C is 10^{-14}. Because the concentration of H^+ must equal the concentration of OH^- at neutrality, the concentration of each ion at neutrality is $\sqrt{10^{-14}}$, or 10^{-7} *N*. The negative log of 10^{-7} is 7, so pH 7 represents neutrality. The pH of a 1 *N* solution of H^+ would be 0 (the log of $1 = 0$) and the pH of a 1 *N* solution of OH^- would be 14 (since the corresponding normality of H^+ in the solution would be 10^{-14}). A most important consequence of this system of measurement is that each unit decrease in pH indicates a tenfold increase in H^+ ion content. Thus pH 6 is ten times more acid than pH 7, and pH 1 is 100,000 times more acid than pH 6. Not only this, but the change from pH 3 to pH 2 represents a 10,000 times greater change in acidity than the change from pH 7 to pH 6.

Buffers

The action of buffers is best illustrated by the following reactions of hydrochloric acid (HCl) and sodium hydroxide (NaOH) in water (H_2O):

To 9 ml of:	*pH*	*Add 1 ml of:*	*Resulting pH*
A. The normal pH changes expected:			
1. H_2O	7	*N* HCl	1
2. H_2O	7	*N* NaOH	13
B. Sodium acetate solution is buffered against acid and acetic acid is buffered against base; the pH changes only slightly.			
3. *N* sodium acetate	8	*N* HCl	7
4. *N* acetic acid	3	*N* NaOH	4
C. Sodium acetate and acetic acid mixture is buffered against the action of both acid and base:			
5. *N* sodium acetate + *N* acetic acid	5	*N* HCl	4.7
6. *N* sodium acetate + *N* acetic acid	5	*N* NaOH	5.3
D. Sodium chloride solution has no buffer capacity:			
7. *N* sodium chloride	7	*N* HCl	1
8. *N* sodium chloride	7	*N* NaOH	13

The action of a buffer salt depends upon the fact that it is the strongly dissociating salt of a weakly dissociating acid. That is, acetic acid is a weakly dissociating acid and sodium acetate is a strongly dissociating salt. The buffer action of the salt against acid (example 3 above) is thus explained by the following reaction.

$$\underbrace{Na^+ + \text{acetate}^-}_{\text{buffer salt}} + \underbrace{H^+ + Cl^-}_{\text{added strong acid}} \longrightarrow \underbrace{Na^+ + Cl^-}_{\text{salt}} + \underbrace{\text{H-acetate}}_{\text{weak acid}}$$

The salt of the weak acid, acetic acid, is dissociated in solution. The addition of H^+ ions causes the immediate formation of acetic acid, which does not lower the pH much because it is weakly dissociating. In the same way, a solution of the weak acid is buffered against the strong base by the reaction

$$\underbrace{\text{H-acetate}}_{\text{buffer acid}} + \underbrace{Na^+ + OH^-}_{\text{added strong base}} \longrightarrow \underbrace{Na^+ + \text{acetate}^-}_{\text{salt}} + \underbrace{HOH}_{\text{water}}$$

A mixture of the weak acid and its salt will be buffered against the action of both acids and bases. Naturally occurring buffers, usually based on bicarbonate, phosphate, or weakly dissociating organic acids, have the ability to maintain pH within extremely narrow limits. The pH of a buffer depends on the degree of ionization of the acid component and the relative amounts of the acid and salt. Modern organic buffers such as TRIS [tris(hydroxymethyl)aminoethane] or TES [N-tris(hydroxymethyl)methyl-2-aminoethane sulfonic acid] have a wide range of pH control and are frequently used to stabilize pH of such delicate systems as cell-free preparations, chloroplast suspensions.

Colloids

Solutions, in which small molecules are distributed homogenously, graduate imperceptibly into colloids as the molecule size increases, or as clumps or aggregates of molecules **(micelles)** increase in size, and colloids graduate into **suspensions** or **mixtures,** in which the inhomogeneity is so great that the phases separate spontaneously. The limits between solution, colloid, and mixture are artificial. Colloidal micelles generally fall within the range of 0.001 to 0.1 μ (micron) in diameter, and are normally **ultramicroscopic** (that is, invisible in the light microscope, though they may be visible in the electron microscope). The particles are so small that colloids are stable and do not settle out under normal conditions, although larger micelles can be sedimented in high-speed or ultracentrifuges. However, colloidal particles are much larger than solute particles in a true solution.

Colloids are best described by their properties:

1. The micelles are large enough to scatter light (the **Tyndall effect,** which renders a beam of light visible in a colloid), whereas a beam of light is invisible in a true solution.
2. Colloidal particles will not pass a natural membrane filter, as true solutions will.
3. Colloidal particles are larger than those of a true solution and so diffuse more slowly.
4. Colloidal particles interact with the solvent or each other, often drastically altering the properties of a solution at low concentration (for example, the solution may gel).

Colloids have a number of special properties that arise from the fact that the total surface area of the micelles is very large. If a sphere of solid matter only 1 cm in diameter were dispersed into a 1% colloid with average micelle size, the total surface area of all of the micelles in only 100 ml of the resulting colloid would be approximately that of a football field. This means that a huge total area is available for the generation of surface-charge forces of attraction or repulsion. Since most chemical reactions occur at interfaces, the reactivity of colloids may be very great.

Some of the important properties of colloids are the result of **adsorption.** The surfaces of colloidal micelles may be electrically charged, either positively or negatively. They thus tend to attract clouds of ions or polar molecules of opposite charge. In **hydrophobic** colloids the particles have no affinity for water and stay in the colloidal state largely as the result of the mutual electrostatic repulsion of their like charges. **Hydrophilic** colloids, on the other hand, are strongly attractive to water and are surrounded by layers of oriented water molecules. Water, as shown by the structural formula

$$-O\begin{matrix} \diagup H \\ \diagdown H \end{matrix} \quad +$$

is highly polar, electropositive at one end and electronegative at the other. The layers of polarized water molecules further protect the micelles of a hydrophilic colloid from coalescing and precipitating. Most biological colloids, such as protoplasm, appear to be electronegative hydrophilic colloids; they are also characterized by weak chemical attraction between the micelles, which helps to stabilize them.

Two important results of the nature of colloids are their susceptibility to precipitation or coagulation and their ability to form gels. Coagulation is the irreversible destruction of the stability of a colloid. Any agent that robs the micelles of their protective charges or their layers of polarized molecules will cause coagulation. If the charges are neutralized by the addition of electrolytes, the particles may coalesce. The pH at which no charge difference exists between the surface of the micelle and its surrounding medium is called the **isoelectric point.** At this pH, most hydrophobic or weakly hydrophilic sols will precipitate. Strongly hydrophilic sols will not coagulate so easily; they must have their adsorbed layer of polarized water molecules removed by a dehydrating agent such as alcohol.

The formation of a gel is a quite different process from coagulation and does not result in the destruction of the colloid. A gel is merely a different state of the colloid. In a gel the micelles are associated in some manner sufficiently stable to give the colloid some degree of solidity, but not closely enough to cause flocculation or coagulation. Many gels are reversible. The usual concept of a gel supposes a more or less tangled mass of fibrillar micelles that hold the liquid phase in their interstices. The individual micelles would be loosely attached at points of intersection by weak chemical forces. The amounts of water such a structure could hold are very great—a protein gel such as gelatin may hold up to 100 times the weight of its solid phase, and a carbohydrate gel such as agar may bind up to 700 times its weight in water.

The intensity with which a gel may absorb water is astonishing—protoplasm may resist the removal of water with a force of several hundred atmospheres (1 atm = 15 lb/in.^{2}). The imbibition of water by a growing root is sufficient to split a rock or paving stone; the pioneers had a trick of splitting rock by drilling holes and fitting them with wooden plugs, which were then wetted. The water-holding power of polysaccharide gels in marine seaweeds enables them to withstand prolonged exposure to drying conditions between tides, though they are exposed to extremes of wind and sun. Colloids in certain clay soils will hold water with such intensity that they become waterlogged; many plants cannot grow in such wet clay because of the inability of oxygen to enter in sufficient amounts for normal root metabolism. The water that is held in a gel acts largely as free water, and a gel is, in fact, merely a more or less rigid colloid. Substances that would dissolve in a colloidal soil will also dissolve in and diffuse through a gel. However, some of the water in a gel, those molecules adsorbed directly onto the micelles by electrostatic forces, are not free and cannot diffuse or take part in chemical reactions or act as solvent molecules.

Chemical Bonds

Electrovalent or Ionic Bonds. These are formed between elements at opposite ends of the periodic table. Usually one partner of the bond has a nearly complete outer shell of electrons and the other has only one or two electrons in its outer shell. Each atom achieves a more stable state when the atom having a few outer electrons loses one or more to the atom having a nearly complete shell. For example, a reaction between sodium and chlorine can be diagrammed

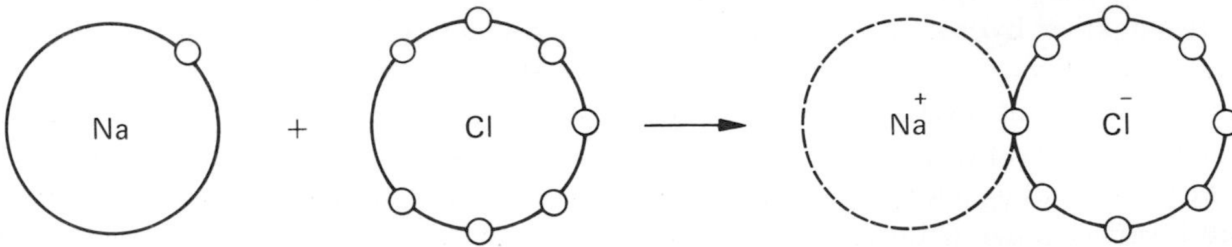

Sodium has one electron in its outer shell, chlorine has seven. In the ionic bond, sodium has given up its electron, becoming positively charged and of stable configuration, whereas chlorine has accepted the electron, becoming negatively charged and also stable. The chemical bond is itself due to electrostatic attraction between the oppositely charged atoms in the molecule. Monovalent elements transfer one electron in such a chemical union; divalent elements transfer two electrons; and so on.

Ionic bonds are nondirectional—that is, there is no fixed angle of attachment between the two partners. However, each bond has a characteristic bond length that reflects the configuration of maximum stability. Ionic bonds are usually readily ionized as follows

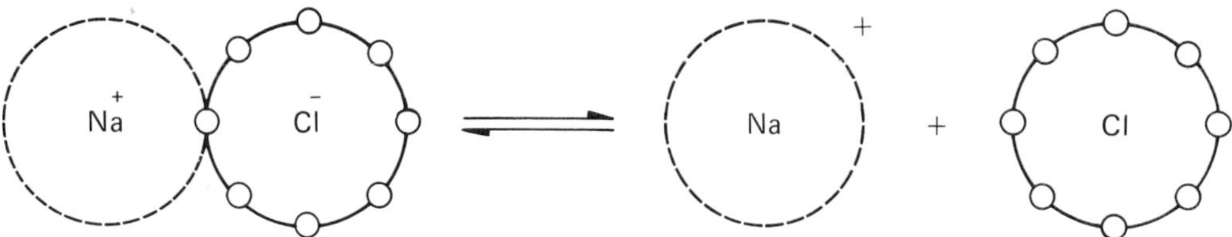

Covalent Bonds. These are formed by sharing electrons to complete outer electron shells in both partners without either atom becoming charged. The bonding of organic molecules is largely covalent. The combination of carbon and hydrogen to make methane illustrates this type of bond.

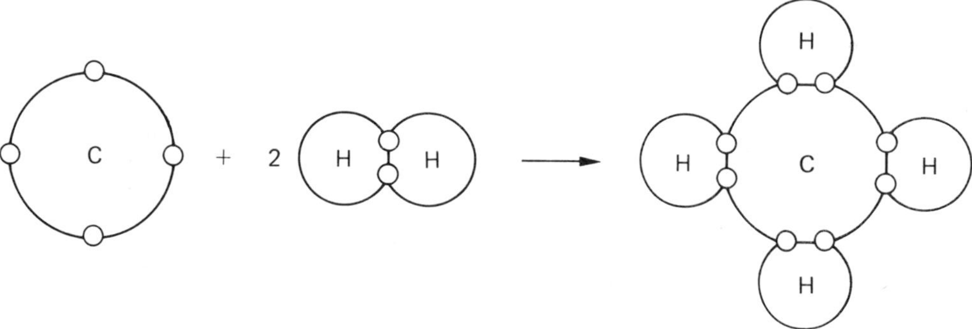

Such bonds do not ionize readily, and this is the reason for the great stability of carbon-to-carbon or carbon-to-hydrogen bonds in organic molecules.

Covalent bonds are highly directional—the member atoms are free to rotate on the axis of the bond, but the angle of the bond is relatively fixed. In some complex molecules, particularly ring structures, the bonds may be "bent" or forced from their natural angle. Such compounds are in a strained configuration and are more reactive than a corresponding unstrained configuration.

A special type of covalent bond, called the **coordinate bond,** is formed when one molecule donates two electrons to another to form a shared pair. Protons or hydrogen ions seldom exist free in solution but tend to combine coordinately with water molecules to form hydronium ions, as follows

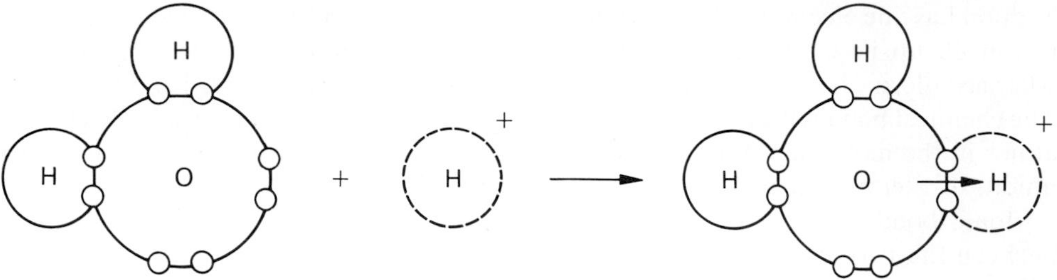

The arrow indicates the type of bond and the compound that provides the shared electron pair.

Hydrogen Bonds. Certain atoms like oxygen and nitrogen and the halogens are strongly electronegative, and have a strong electrostatic attraction for protons. Such atoms may form bonds of low or intermediate strength with hydrogen atoms on adjacent groups or molecules; these act at a considerable distance, as shown

```
H—O    O-----------H    H
   \  //            \  /
    C                 N
    |                 |
    R                 R′
```

In this example, an oxygen atom in an acid is held by a hydrogen bond to the nitrogen atom of an amine. Water is frequently held to organic molecules by hydrogen bonds, and many features of the behavior of water—for example, its high freezing and boiling points—are due to hydrogen bonding between water molecules. This type of bonding is critically important in the synthesis and maintenance of structure in such molecules as proteins and nucleic acids.

Weak Forces. A number of weak intermolecular forces may act to hold molecules in loose association. These are collectively known as the van der Waals forces, after their discoverer. Van der Waals found it necessary to postulate attractive forces between molecules of gas to explain their deviation in behavior from that predicted for an ideal gas. These forces vary in intensity, but are relatively weak when compared with ionic, covalent, or even hydrogen bonds. They arise from the slight polarization of normally nonpolar groups resulting from irregular or asymmetrical movement of electrons.

Oxidation and Reduction

Oxidation means losing an electron; reduction means gaining an electron. The electron may be accompanied by a proton (H^+), and oxygen may be added or removed from a compound. Several characteristic reactions are illustrated below (in each example oxidation is left to right; reduction is right to left).

$$\underset{\text{ethane}}{CH_3{-}CH_3} \rightleftharpoons \underset{\text{ethylene}}{CH_2{=}CH_2} + H_2 \qquad (2\text{-}1)$$

$$\underset{\text{ferrous iron}}{Fe^{2+}} \rightleftharpoons \underset{\text{ferric iron}}{Fe^{3+}} + e^- \qquad (2\text{-}2)$$

$$CH_3{-}CH_2OH + \tfrac{1}{2}O_2 \rightleftharpoons CH_3{-}CHO + H_2O \qquad (2\text{-}3)$$

ethanol acetaldehyde

$$CH_3{-}CHO + \tfrac{1}{2}O_2 \rightleftharpoons CH_3{-}COOH \qquad (2\text{-}4)$$

acetaldehyde acetic acid

(In Example 2-3 oxygen is reduced to water as the organic compound is oxidized, and vice versa.)

Many biological oxidation-reduction reactions involve electron pairs, but some are accomplished by the transfer of only one electron. Most biological oxidations or reductions are catalyzed by specific enzymes.

Some Organic Chemicals

Organic chemicals are essentially structures built upon a backbone of carbon atoms, ranging in number from one to many. They are most frequently and easily classified and referred to by the number of carbon atoms they contain. The carbon atoms may be joined together with one, two, or three covalent bonds, called single, double, and triple bonds, respectively. These are represented as

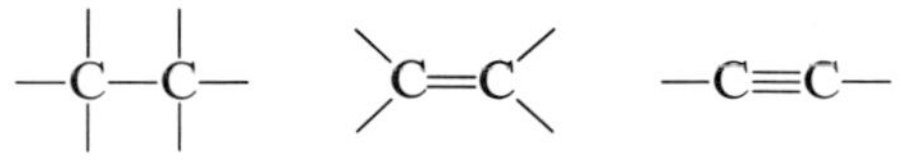

$$-\overset{|}{\underset{|}{C}}-\overset{|}{\underset{|}{C}}- \qquad \rangle C{=}C\langle \qquad -C{\equiv}C-$$

Saturated compounds have a maximum number of hydrogens; that is, they have only single bonds. **Unsaturated** compounds contain one or more double or triple bonds. An unsaturated compound has been oxidized; a saturated one has been reduced. The special properties of organic molecules depend upon their size, the number and type of substituted atoms or groups that they bear, and the degree of unsaturation in the molecule. Some important basic structures are summarized in Figure 2-1. Major groups of organic chemicals are summarized below.

Alkanes are simple organic molecules, completely saturated.

Methane: CH_4
Ethane: $CH_3{-}CH_3$
Propane: $CH_3{-}CH_2{-}CH_3$
Butane: $CH_3{-}CH_2{-}CH_2{-}CH_3$ or $CH_3(CH_2)_2CH_3$

Isomers of these compounds exist (that is, compounds having the same empirical formula but a different structure).

n-Butane: $CH_3{-}CH_2{-}CH_2{-}CH_3$
Isobutane: $CH_3{-}\underset{\underset{CH_3}{|}}{CH}{-}CH_3$ or $CH_3CH(CH_3)CH_3$

Alcohols have an —OH group substituted for a hydrogen, with the general formula ROH, where R stands for any organic radical. **Primary** alcohols have the OH group at

[*Text continued on page 20.*]

Figure 2-1. Summary of important biological chemical structures: some important low molecular weight carbon compounds.

Organic acid (carboxylic acid)	$R—C(=O)OH$, $R—COOH$
Aldehyde	$R—C(H)=O$, $R—CHO$
Ketone	$R—C(=O)—R'$, $R—CO—R'$
Alcohol	$R—CH_2—OH$, $R—CH_2OH$
Amine	$R—CH_2—NH_2$, $R—CH_2NH_2$
Amino-	$—NH_2$
Amino acid	$R—CHNH_2—COOH$
Amide	$R—C(=O)NH_2$, $R—CONH_2$
Peptide	$H—N(H)—C(R)(H)—C(=O)—N(H)—C(R)(H)—COOH$
Imine	$R—C(H)=NH$, $R—CHNH$
Imino-	$=NH$
Thiol (SH)	$R—CH_2—SH$, $R—CH_2SH$
Ether	$R—CH_2—O—CH_2—R'$, $R—CH_2—O—CH_2—R'$
Ester	$R—C(=O)—O—R'$, $R—COO—R'$
One carbon, C_1	
Methane (methyl-)	CH_4 $(—CH_3)$
Methanol	CH_3OH
Formaldehyde	$H_2C=O$

Figure 2-1. (continued)

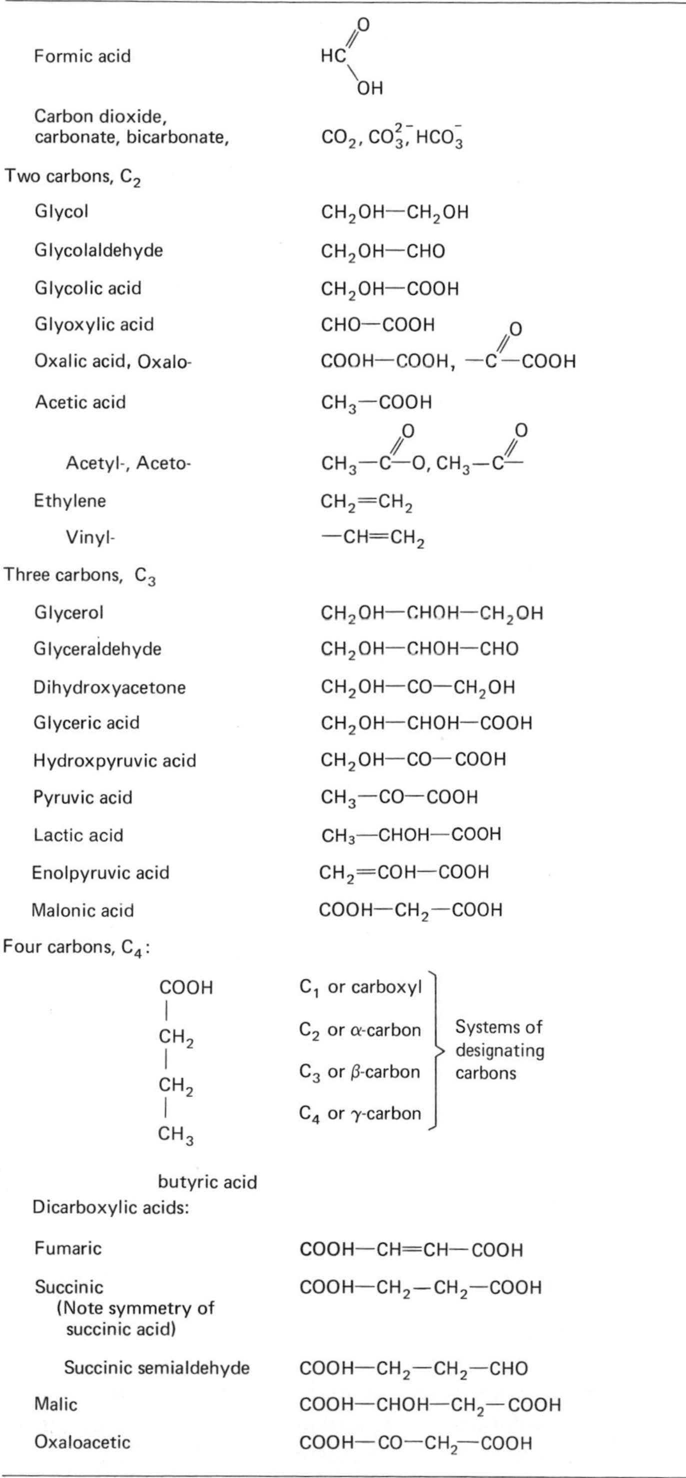

Formic acid	$HC(=O)OH$
Carbon dioxide, carbonate, bicarbonate,	CO_2, CO_3^{2-}, HCO_3^-
Two carbons, C_2	
Glycol	$CH_2OH—CH_2OH$
Glycolaldehyde	$CH_2OH—CHO$
Glycolic acid	$CH_2OH—COOH$
Glyoxylic acid	$CHO—COOH$
Oxalic acid, Oxalo-	$COOH—COOH$, $—C(=O)—COOH$
Acetic acid	$CH_3—COOH$
Acetyl-, Aceto-	$CH_3—C(=O)—O$, $CH_3—C(=O)—$
Ethylene	$CH_2═CH_2$
Vinyl-	$—CH═CH_2$
Three carbons, C_3	
Glycerol	$CH_2OH—CHOH—CH_2OH$
Glyceraldehyde	$CH_2OH—CHOH—CHO$
Dihydroxyacetone	$CH_2OH—CO—CH_2OH$
Glyceric acid	$CH_2OH—CHOH—COOH$
Hydroxpyruvic acid	$CH_2OH—CO—COOH$
Pyruvic acid	$CH_3—CO—COOH$
Lactic acid	$CH_3—CHOH—COOH$
Enolpyruvic acid	$CH_2═COH—COOH$
Malonic acid	$COOH—CH_2—COOH$
Four carbons, C_4:	

butyric acid	Systems of designating carbons
COOH	C_1 or carboxyl
CH_2	C_2 or α-carbon
CH_2	C_3 or β-carbon
CH_3	C_4 or γ-carbon

Dicarboxylic acids:	
Fumaric	$COOH—CH═CH—COOH$
Succinic (Note symmetry of succinic acid)	$COOH—CH_2—CH_2—COOH$
Succinic semialdehyde	$COOH—CH_2—CH_2—CHO$
Malic	$COOH—CHOH—CH_2—COOH$
Oxaloacetic	$COOH—CO—CH_2—COOH$

Figure 2-1. (continued)

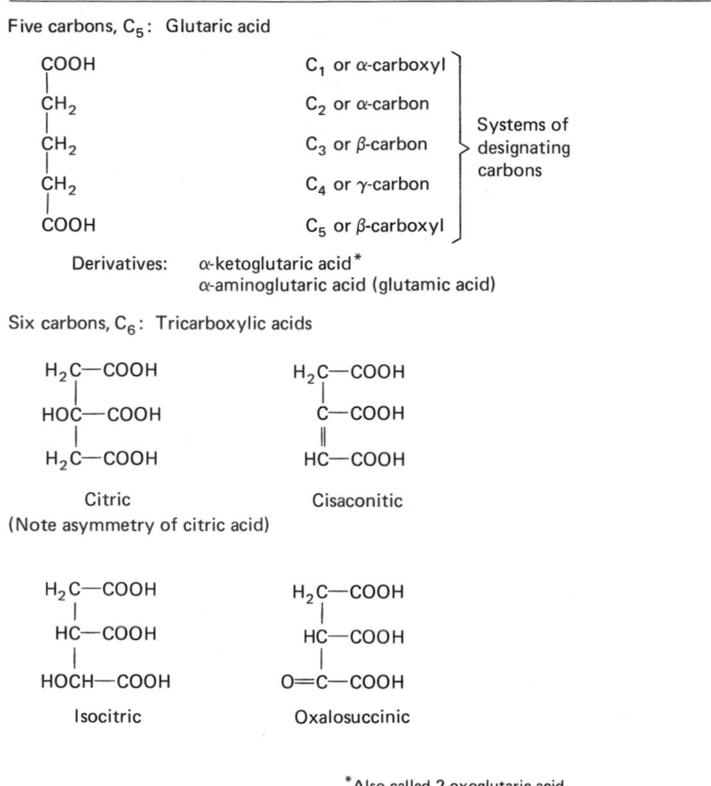

the end of a carbon chain, **secondary** or **tertiary** alcohols have the OH group attached to a carbon that has bonds to two or three other carbon atoms.

Primary or *n*-propanol: $CH_3—CH_2—CH_2—OH$ (central C bearing H above and H below)

Secondary propanol or isopropanol: $CH_3—CH(CH_3)—OH$ (C bearing H above and CH_3 below)

Tertiary butanol: $CH_3—C(CH_3)_2—OH$ (C bearing CH_3 above and CH_3 below)

Thiols are substituted alcohols in which sulfur replaces oxygen. Thiols, or —SH compounds, occur frequently and are important in biological molecules because they are readily oxidized to the S—S (disulfide bridge) form.

$$RSH + R'SH \rightleftharpoons RS—SR'$$

Ethers are alcohols in which the H of an OH group is replaced by another organic radical.

$$CH_3{-}CH_2{-}O{-}CH_2{-}CH_3$$

diethyl ether

Aldehydes result from the oxidation of alcohols or the reduction of acids. They contain an oxygen atom held by a double bond to a terminal carbon.

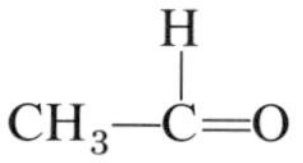

$$CH_3{-}\overset{\overset{\displaystyle H}{|}}{C}{=}O$$

acetaldehyde

Ketones are similar to aldehydes, but the =O is bonded to a secondary carbon, that is, one having two bonds to other carbon atoms.

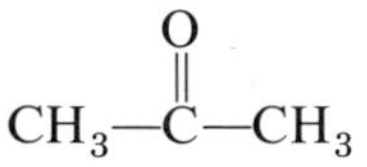

$$CH_3{-}\overset{\overset{\displaystyle O}{\|}}{C}{-}CH_3$$

dimethyl ketone

Acids have a **carboxyl** group, which may result from the oxidation of an aldehyde. This is the most highly oxidized state possible for a carbon atom.

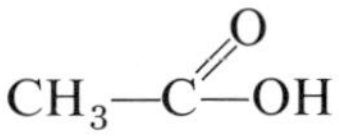

$$CH_3{-}\overset{\overset{\displaystyle O}{/\!/}}{C}{-}OH$$

acetic acid

The H attached to the oxygen atom is the ionizable or acidic hydrogen.

Organic acids may form salts exactly like mineral acids.

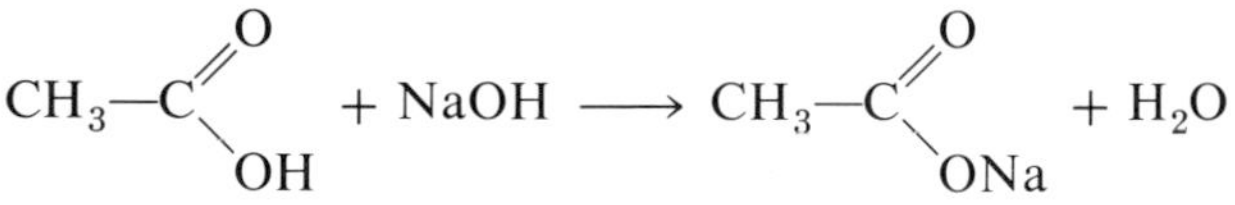

$$CH_3{-}C\begin{matrix}{/\!/}O\\ \backslash OH\end{matrix} + NaOH \longrightarrow CH_3{-}C\begin{matrix}{/\!/}O\\ \backslash ONa\end{matrix} + H_2O$$

They also form a number of substituted compounds such as esters and amides.

Esters are formed by the reaction of an alcohol with an acid.

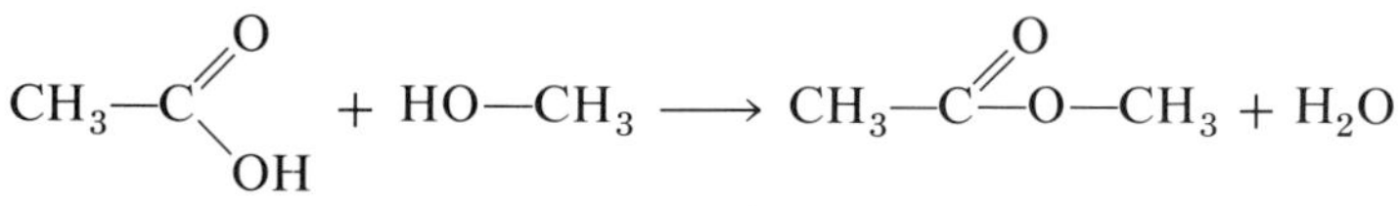

$$\underset{\text{acetic acid}}{CH_3{-}C\begin{matrix}{/\!/}O\\ \backslash OH\end{matrix}} + \underset{\text{methanol}}{HO{-}CH_3} \longrightarrow \underset{\text{methyl acetate}}{CH_3{-}\overset{\overset{\displaystyle O}{/\!/}}{C}{-}O{-}CH_3} + H_2O$$

It should be noted that, in spite of its appearance, this is not an acid–base reaction to form a salt. Esters do not ionize appreciably, and the reaction mechanism is different from the formation of a salt.

Amines have an **amino** (—NH_2) group substituted for a hydrogen.

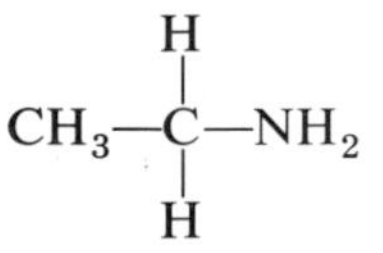

$$CH_3{-}\underset{\underset{\displaystyle H}{|}}{\overset{\overset{\displaystyle H}{|}}{C}}{-}NH_2$$

ethylamine

Amines may be primary, secondary, or tertiary, like alcohols.

Amides are formed by the substitution of an amino group for the OH of a carboxyl group.

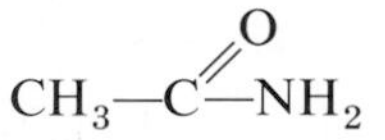

$$CH_3-C(=O)-NH_2$$

acetamide

Fats are an important class of compounds that all have the triple alcohol (trihydroxy) glycerol for a backbone. Each alcohol group forms an ester linkage with an acid, usually a straight chain unsubstituted acid called a fatty acid.

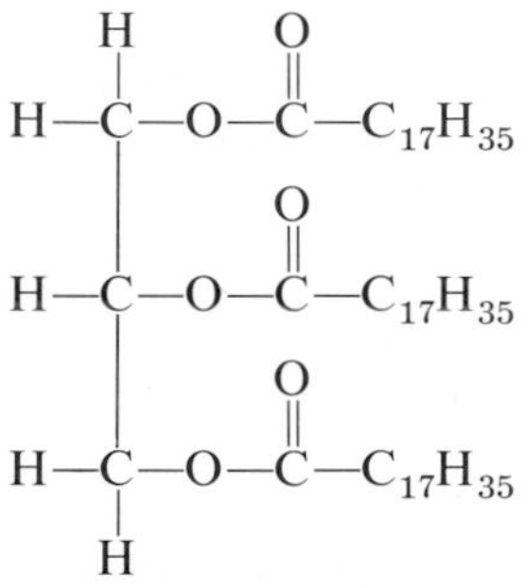

$$\begin{array}{l} H-\overset{\displaystyle H}{\underset{|}{\overset{|}{C}}}-O-\overset{\displaystyle O}{\overset{\|}{C}}-C_{17}H_{35} \\ H-\underset{|}{\overset{|}{C}}-O-\overset{\displaystyle O}{\overset{\|}{C}}-C_{17}H_{35} \\ H-\underset{\displaystyle H}{\underset{|}{\overset{|}{C}}}-O-\overset{\displaystyle O}{\overset{\|}{C}}-C_{17}H_{35} \end{array}$$

glycerol tristearate
a fat

The resulting triglyceride derives its most important property, its hydrophobic or lipophilic character, from the long-chain unsubstituted fatty acid residues. The length of the chain and its degree of unsaturation (the number of double bonds) further affects the chemical behavior of individual fatty acids and fats. A number of possible substituted glycerides is known. The most important group is the phospholipids, which have a molecule of phosphate esterified to one of the alcohol groups on the glycerol as shown

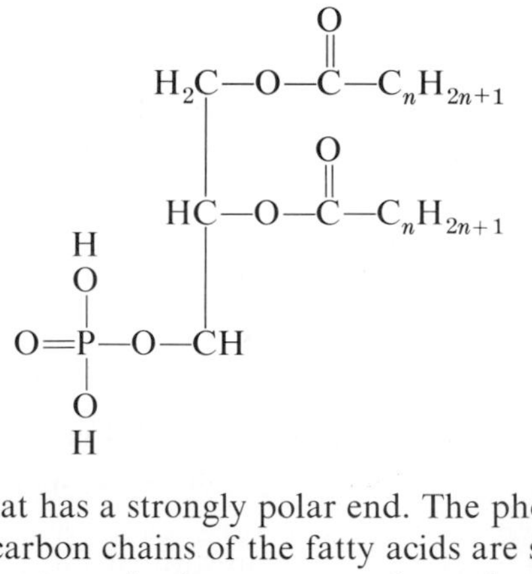

$$\begin{array}{l} H_2C-O-\overset{\displaystyle O}{\overset{\|}{C}}-C_nH_{2n+1} \\ HC-O-\overset{\displaystyle O}{\overset{\|}{C}}-C_nH_{2n+1} \\ O=P(OH)(OH)-O-CH \end{array}$$

This creates a molecule that has a strongly polar end. The phosphate group is strongly hydrophilic, whereas the carbon chains of the fatty acids are strongly lipophilic. These molecules are of great importance in the structure of membranes, as we shall later see.

Cyclic compounds may be fully saturated or unsaturated in various degrees. **Benzene,** a six-carbon ring, is a most important unsaturated molecule that forms the backbone of many compounds of biological importance. It is conventionally represented as

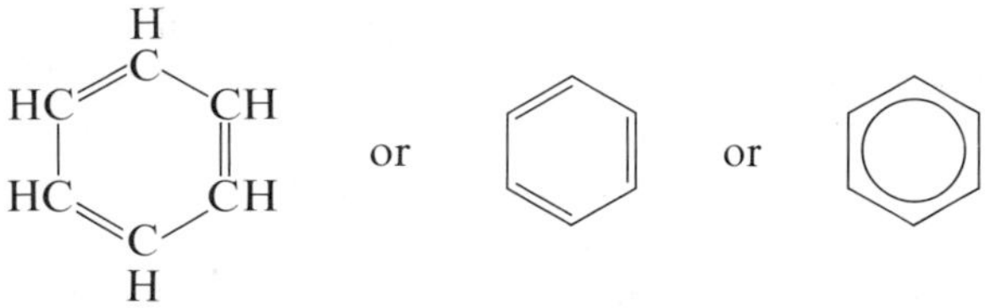

H
HC CH
HC CH
H

or

or

Substitutions may be made at any of the carbons. If two or more substitutions are made, they are usually numbered from the position occupied by the most reactive group.

Phenols are an important group of compounds having one or more OH groups substituted in single or multiple benzene rings.

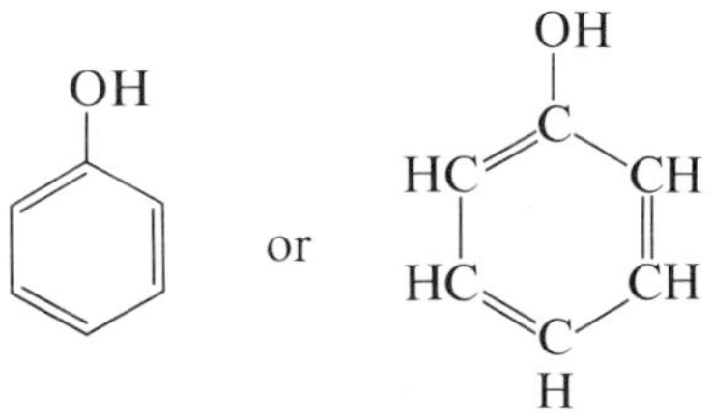

OH

or

OH
HC CH
HC CH
H

phenol

Heterocyclic compounds are ring structures that have one or more nitrogen or oxygen atoms in the ring. Some important heterocyclic rings are

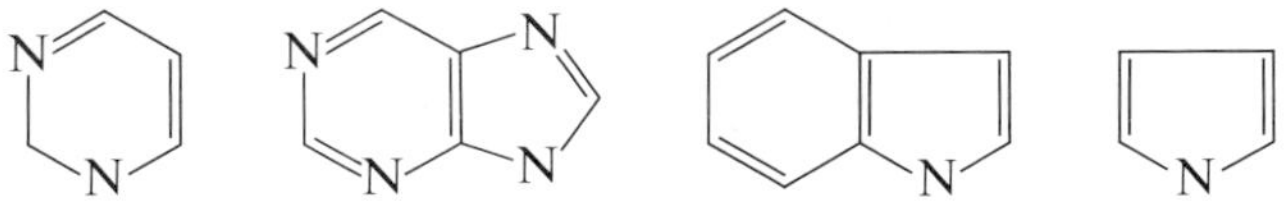

pyrimidine　purine　indole　pyrrole

Carbohydrates

Sugars have the general formula $(CH_2O)_n$. There are many possible structures having this formula, but only a limited number of them are of biological importance. In addition, there are several sugars whose formulas differ by the loss of an oxygen molecule or a water molecule or by the addition of groups such as sulfate or phosphate.

The simplest sugars of biological importance are the trioses.

$$CH_2OH—{}^*CHOH—CHO \quad CH_2OH—CO—CH_2OH$$

glyceraldehyde　dihydroxyacetone

Both have the same empirical formula; however, it may be seen that glyceraldehyde is an **aldose,** a sugar having an aldehyde group or terminal =O, whereas dihydroxyacetone is a **ketose,** having a ketone group, or a secondary =O.

Sterioisomers. Many organic molecules are optically active; that is, they can cause rotation of the plane of polarity of polarized light. Optical activity is conferred by the

* Optically active carbon.

existence in the molecule of a carbon atom that has four different groups attached to it. You can see that glyceraldehyde has one such carbon, the middle or second carbon, which is marked with an asterisk. It is therefore optically active, whereas dihydroxyacetone is not. A further consequence of the possession of such a carbon is that there are two possible spatial configurations for glyceraldehyde. Both configurations have the same formula as written above, but they are not interconvertible without chemical reaction. This is because the covalent bonds that join the atoms of this molecule together, although they can rotate, cannot be bent very much. The formula for the bond structure of a carbon atom is not correctly represented by

```
      1
      |
  4—( )—2
      |
      3
```

which suggests that all the bonds are in one plane at 90° from each other. In fact, the bonds are three-dimensional, and the angle between any two bonds is 109°. The result is that for the structure represented above two possible configurations exist, and you can see that they cannot be interconverted without breaking and remaking bonds.

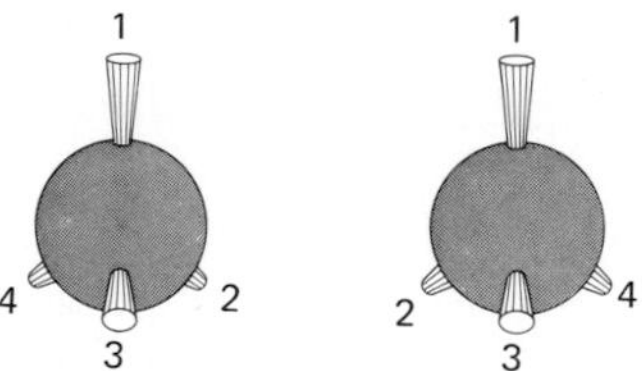

The simplest sugar that has optically active isomers is glyceraldehyde. One form rotates polarized light to the left and is called **levorotatory;** the prefix L- is added to the name to identify this compound. The other form of glyceraldehyde rotates light to the right and is called **dextrorotatory;** this compound is identified by the prefix D-.

These isomers are conventionally written

```
  CHO          CHO
   |*          *|
  HCOH        HOCH
   |            |
  CH2OH        CH2OH
D-glyceraldehyde   L-glyceraldehyde
```

D-Glyceraldehyde is the biologically important form. L-Glyceric acid and L-sugars in general are relatively unimportant biologically, are seldom formed, and are not usually metabolized by biological systems.

Sugars with four carbons, called **tetroses** and often abbreviated C_4 sugars, have two asymmetric carbons in the aldose series and one in the ketose. Carbons in an organic molecule are usually numbered from the active end of the molecule; thus, a tetrose would be represented as follows, with the carbons numbered

$$\underset{4}{CH_2OH}—\underset{3}{CHOH}—\underset{2}{CHOH}—\underset{1}{CHO}$$

Conventionally, D-tetroses and all D-sugars have the same configuration as D-glyceraldehyde in the second-last carbon; that is, the OH is written to the right

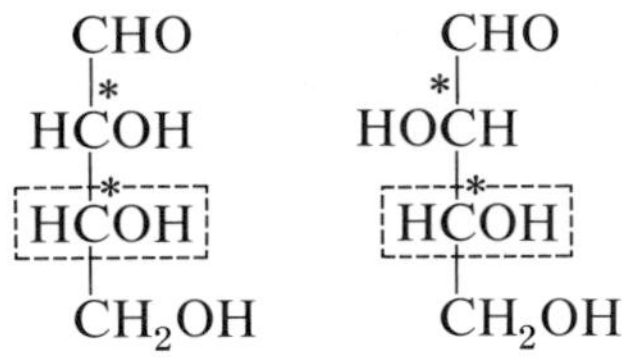

CHO | * HCOH | * HCOH | CH_2OH — D-erythrose

CHO | * HOCH | * HCOH | CH_2OH — D-threose

Unfortunately, although they are both D-sugars, they are not both dextrorotatory. D-Threose rotates light to the right (+), but D-erythrose, because its optical activity results also from the steric configuration of C-2, is levorotatory (−). Although most biologically important sugars have the D form, many are levorotatory. The symbols D- and L- thus refer to the structure rather than the optical activity of a compound.

As there are two D-tetroses in the aldose series, there are four D-pentoses, eight D-hexoses, and sixteen D-heptoses. Most of these are rarely found in nature. The structural formulas and relationships of the biologically important aldoses and ketoses are shown in Figures 2-2 and 2-3.

Lactones. The sugars in Figures 2-2 and 2-3 are represented by straight chain formulas. However, sugars containing five or more carbons normally exist in a ring structure. Two common basic structures are the **furan** (five-member) and **pyran** (six-member) rings.

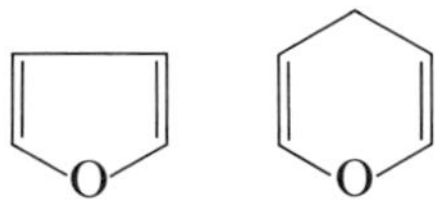

furan   pyran

The ring structures, or lactones, are called **furanose** or **pyranose,** respectively. The formulas in the Haworth or steriochemical configuration for glucose and fructose are represented as shown in the following structural formulas.

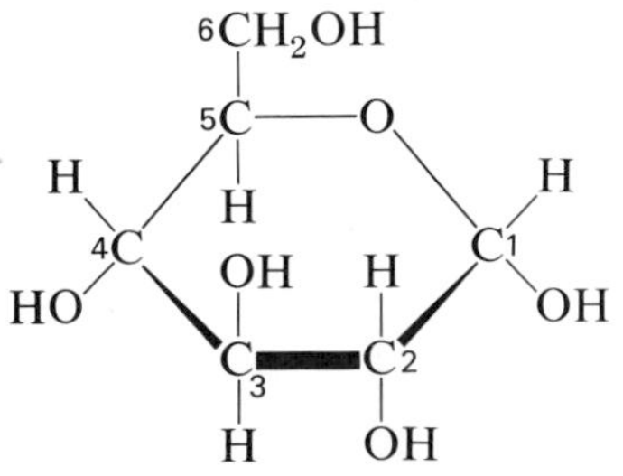

α-D-glucopyranose 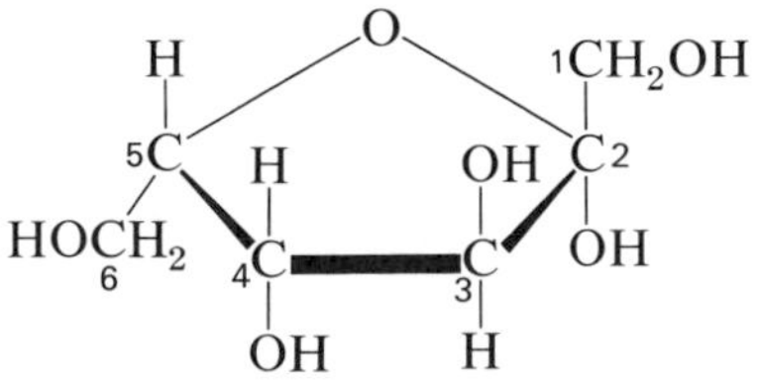   β-D-fructofuranose

Disaccharides and Polysaccharides. Glucose and fructose are joined with an anhydro linkage (by the elimination of water) to form the important **disaccharide** sucrose. The formula is shown at the top of the next page.

sucrose

α-D-glucopyranose-1 : 2-β-D-fructofuranoside

Figure 2-2. Important aldose sugars. The rotation of each sugar is shown by (+) for dextrorotatory and (−) for levorotatory compounds.

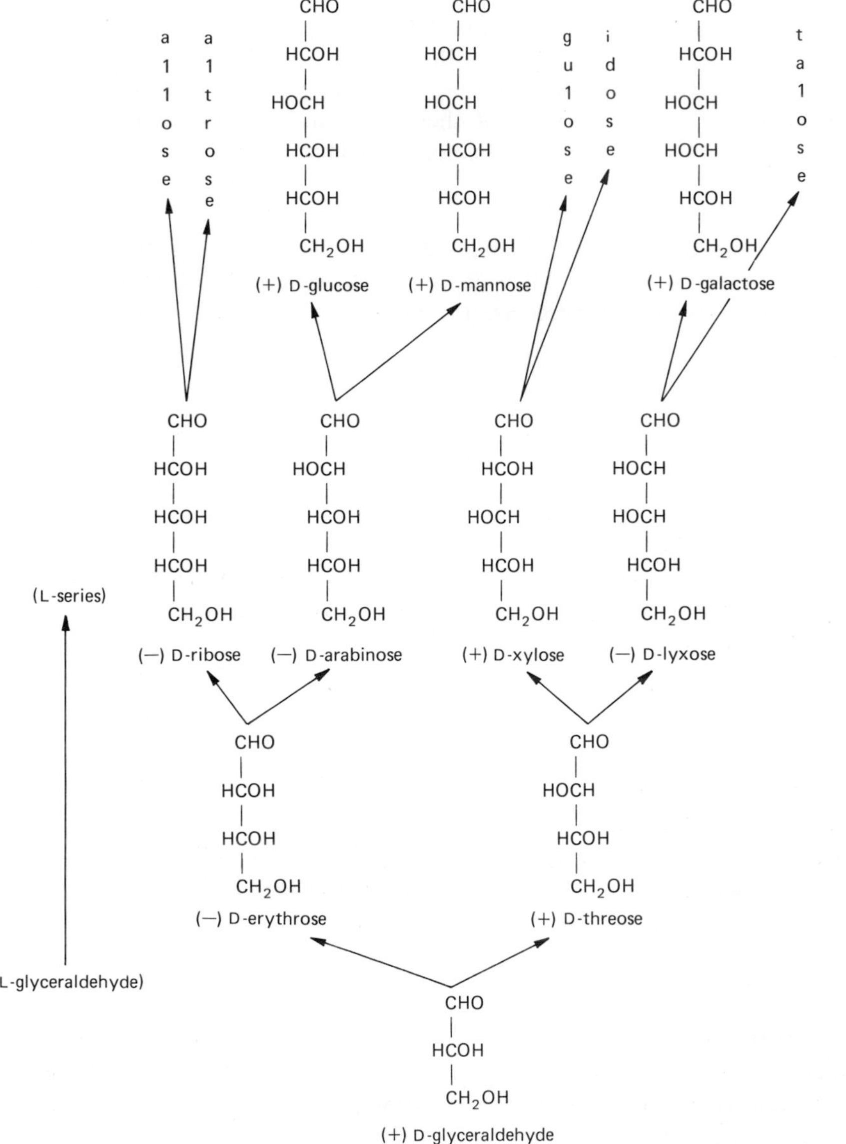

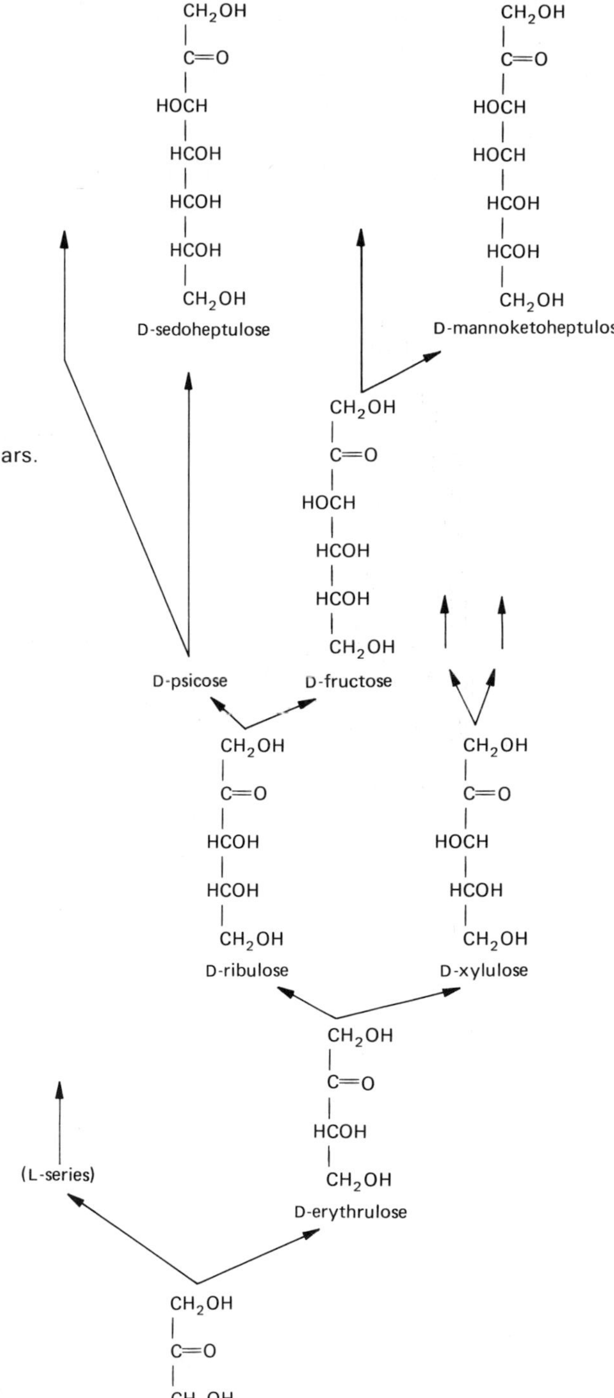

Figure 2-3. Important ketose sugars.

Note that the lactone ring structure adds another asymmetric carbon previously not present in glucose at C-1 and to fructose at C-2. The two configurations possible (in glucose, of the H and OH groups; in fructose of the OH and CH_2OH groups) are called α and β. In sucrose, glucose is in the α form and fructose is in the β form. The linkage is between C-1 of glucose and C-2 of fructose, so the resulting molecule can be described as α-D-glucopyranose-1:2-β-D-fructofuranoside.

Other important disaccharides are maltose (α-D-glucopyranose-1:4-D-glucopyrano-

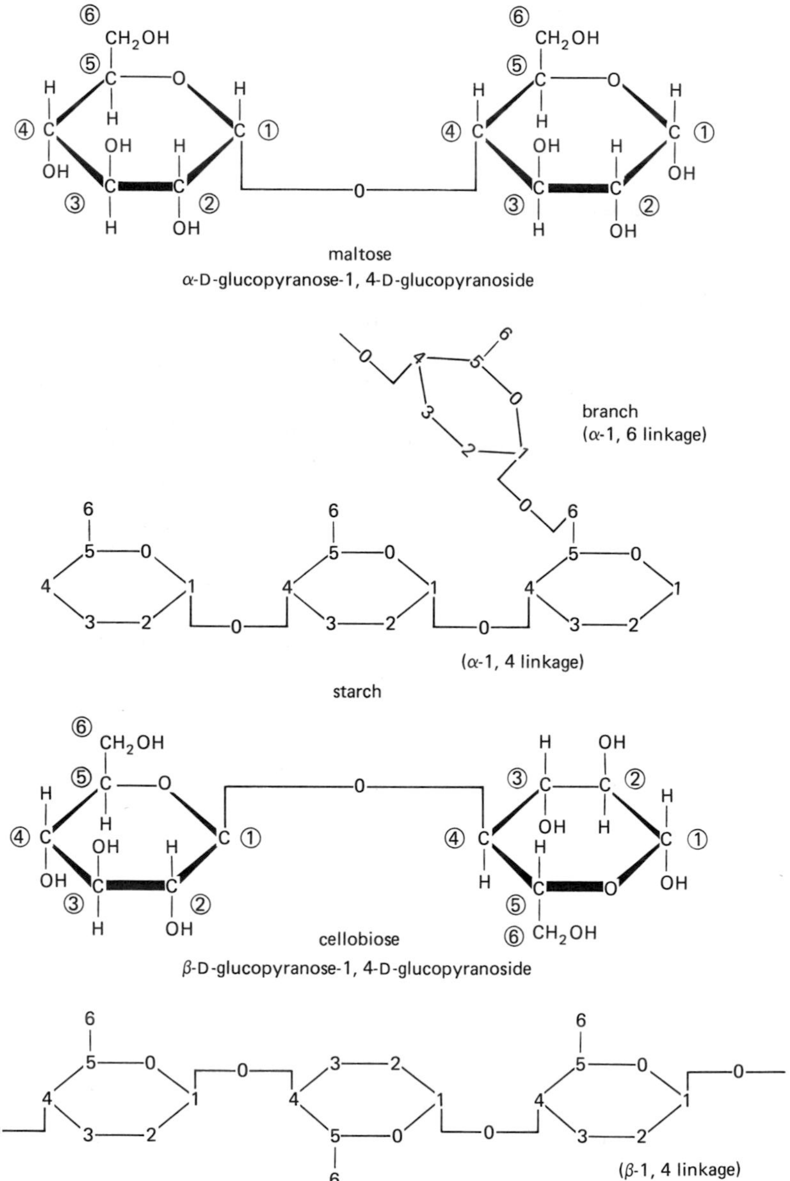

Figure 2-4. Structures of important disaccharides and their relationship to polysaccharides.

Figure 2-5. Structures of common polyhydric alcohols and uronic acids.

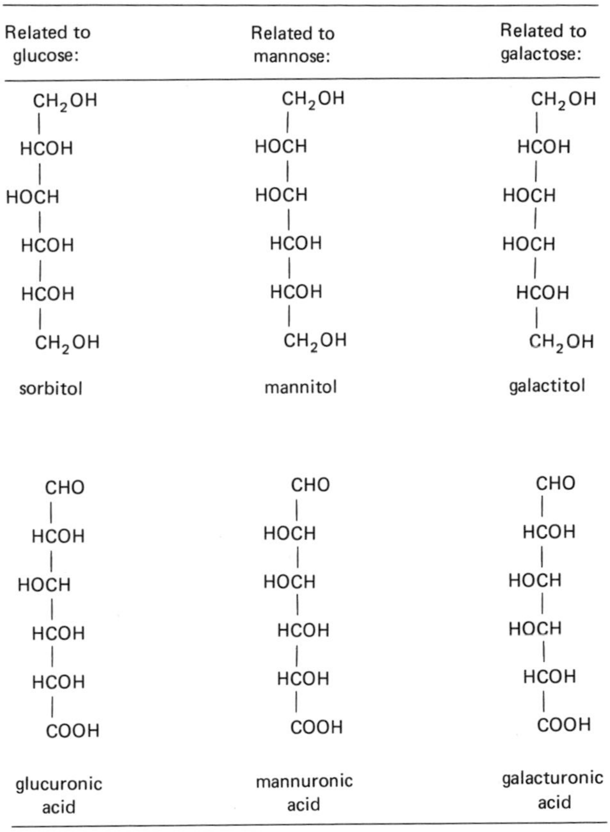

Related to glucose:	Related to mannose:	Related to galactose:
CH_2OH	CH_2OH	CH_2OH
HCOH	HOCH	HCOH
HOCH	HOCH	HOCH
HCOH	HCOH	HOCH
HCOH	HCOH	HCOH
CH_2OH	CH_2OH	CH_2OH
sorbitol	mannitol	galactitol
CHO	CHO	CHO
HCOH	HOCH	HCOH
HOCH	HOCH	HOCH
HCOH	HCOH	HOCH
HCOH	HCOH	HCOH
COOH	COOH	COOH
glucuronic acid	mannuronic acid	galacturonic acid

side) and cellobiose (β-D-glucopyranose-1:4-D-glucopyranoside). These relate to the most important plant polysaccharides, starch and cellulose, as shown in Figure 2-4. The important distinction between these two polysaccharides is as follows. Starch, a metabolically active reserve carbohydrate, is made of long chains of glucose units joined by α-1:4 linkages and with occasional side chains joined by α-1:6 linkages. Cellulose, a metabolically inactive structural carbohydrate, is made of long chains of glucose units joined by β-1:4 linkages. Other disaccharides and oligosaccharides of biological importance are listed in Table 2-1; polysaccharides are summarized in Table 2-2.

Sugar Alcohols, Uronic Acids, and Sugar Acids. The aldehyde or ketone oxygen of sugars may be reduced to an alcohol, producing polyhydric alcohols as shown in Figure 2-5. Polyhydric alcohols frequently occur in nature as storage sugars, particularly in primitive plants and the fruits of higher plants. Uronic acids are derived from hexoses by the oxidation of the sixth carbon to an acid. Some important uronic acids, found in structural and mucilaginous compounds in plants, are shown in Figure 2-5. The oxidation of the first carbon of glucose produces the sugar acid, gluconic acid. The phosphate derivative of gluconic acid is important in respiratory metabolism. The relationship of glucose to its reduced and oxidized forms is shown in Figure 2-6, see page 31.

Table 2-1. A list of biologically important oligosaccharides

DISACCHARIDES	TRISACCHARIDES
Glucose + glucose	Melezitose (glucose-1:3-fructose-2:1-glucose) turanose + glucose
Maltose (α-glucose-1:4-glucose)	Gentianose (glucose-1:6-glucose-1:2-fructose) gentiobiose + fructose or sucrose + glucose
Cellobiose (β-glucose-1:4-glucose)	Raffinose (galactose-1:6-glucose-1:2-fructose) melibiose + fructose or sucrose + galactose
Gentiobiose (β-glucose-1:6-glucose)	TETRASACCHARIDE
Trehalose (α-glucose-1:1-glucose)	Stachyose (galactose:galactose:glucose: fructose) raffinose + galactose
Galactose + glucose	PENTASACCHARIDE
Lactose (β-galactose-1:4-glucose)	Verbascose (galactose:galactose:galactose:glucose:fructose) stachyose + galactose
Melibiose (α-galactose-1:6-glucose)	DI-, TRI-, AND POLYFRUCTOSE ANHYDRIDES
Glucose + fructose	
Sucrose (α-glucopyranose-1:2-β-fructo-furanose)	
Turanose (α-glucopyranose-1:3-β-fructo-pyranose)	

Table 2-2. Some important polysaccharides

Type and examples	*Use (and source)*
Homopolysaccharides (one sugar)	
Pentosan	
araban, xylan	Structural
Hexosans	
glucans	
starch (α-1:4- and α-1:6-glucose)	Storage
cellulose (β-1:4-glucose)	Structural
fructosans	
inulin (1:2-fructofuranose)	Storage
galactosans	Structural
mannans	Storage (nuts)
laminarin (β-1:3-glucose)	Storage (seaweeds)
Heteropolysaccharides (more than one sugar)	
Pentosans, e.g., araboxylan	
Hexosans, e.g., galactomannan	
Mixed, e.g., galactoaraban	
Uronic acids	
Polyuronides	
pectic acid (1:4-D-galacturonic acid)	Structural
alginic acid (guluronic and mannuronic acids)	Structural and storage (seaweeds)
Sugars plus uronic acids	
Plant gums and mucilages	
Other sugar derivatives	
Chitin (2-acetylaminoglucose)	Structural (fungi)
Fucoidin (sulfated fucose)	Structural (seaweed)
Agar and carrageenin (sulfated galactose)	Storage and structural (seaweed)

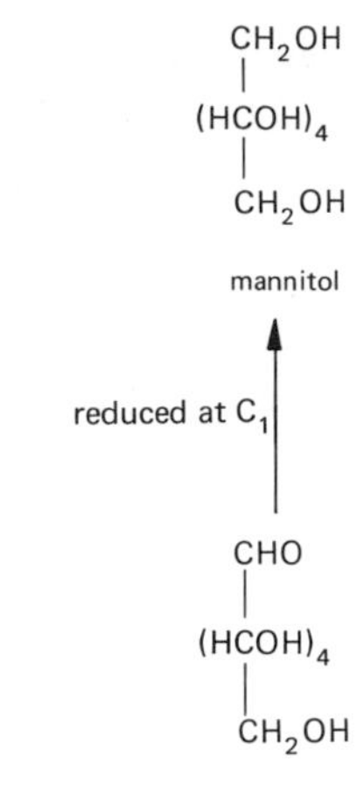

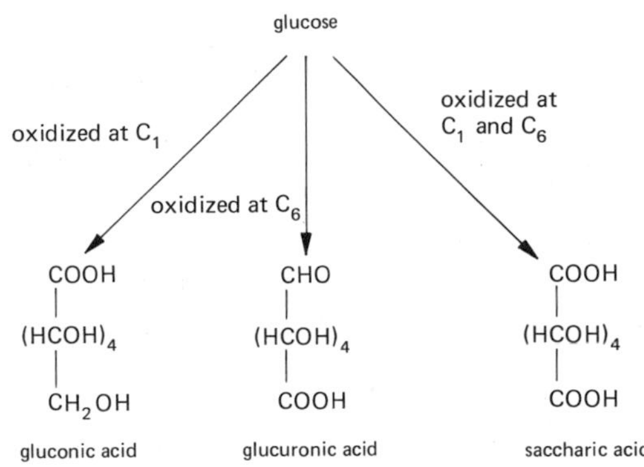

Figure 2-6. Oxidation or reduction products of glucose.

Amino Acids, Peptides, and Proteins

Although carbohydrates form a major group of reserve and structural polymers, by far the most important group of biological polymers is the proteins. These molecules are responsible for nearly all of the properties of life as we know it. There are a few readily distinguishable classes of proteins, but each class contains an enormous number of individual compounds of high molecular weight and great chemical complexity. All proteins have the same basic structure—they are polymers composed of large numbers of individual **amino acids** joined together by peptide bonds, as shown

Two amino acids form a peptide bond by the loss of water

N-terminal end

C-terminal end

Three amino acids form a tripeptide

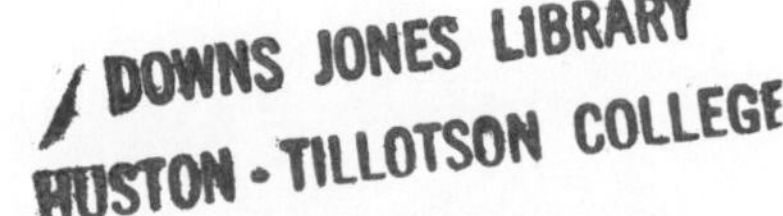

The R groups may range from H in the simplest amino acid, glycine, to one of the many more or less complex organic radicals. Twenty or twenty-one amino acids are known to occur in proteins, as well as two amides, glutamine and asparagine. Plants contain many additional amino acids that are found free but are not known to occur in proteins. A summary of the important amino acids in an approximate structural relationship is shown in Figure 2-7. The 23 amino acids and amides that commonly occur in plant proteins are marked with an asterisk.

All amino acids behave both as acids and as bases, because the amino group is a strong base and the carboxyl group is a strong acid.

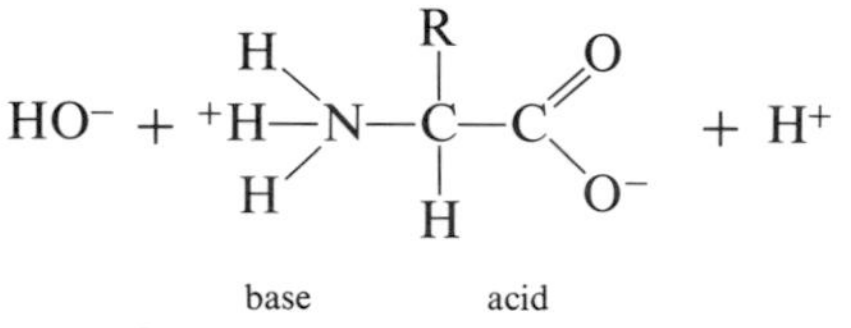

$$HO^- + {}^+H{-}NH_2{-}CHR{-}COO^- + H^+$$

base acid

Some amino acids, such as arginine or lysine, are more strongly basic because they contain extra amino (NH_2) groups; some, such as aspartic acid and glutamic acid, are more strongly acidic because of extra carboxyl (COOH) groups. The pH at which the positive charge exactly neutralizes the negative charge is called the **isoelectric point,** or **pK,** of the acid. Proteins also exhibit a pronounced isoelectric point. At the pK, proteins are very sensitive to precipitation because they do not have the protection from agglomeration provided by the presence of electrical charge.

Figure 2-7. Structure of amino acids commonly found in plants. Those known to occur in proteins are marked with an asterisk (*).

*Glycine	NH_2–CH_2—COOH
*Alanine	CH_3—CH(NH_2)—COOH
*Serine	CH_2OH—CH(NH_2)—COOH
*Phenylalanine	C_6H_5—CH_2—CH(NH_2)—COOH
*Tyrosine	HO—C_6H_4—CH_2—CH(NH_2)—COOH
Dihydroxyphenylalanine (dopa)	(HO)(HO)—C_6H_3—CH_2—CH(NH_2)—COOH

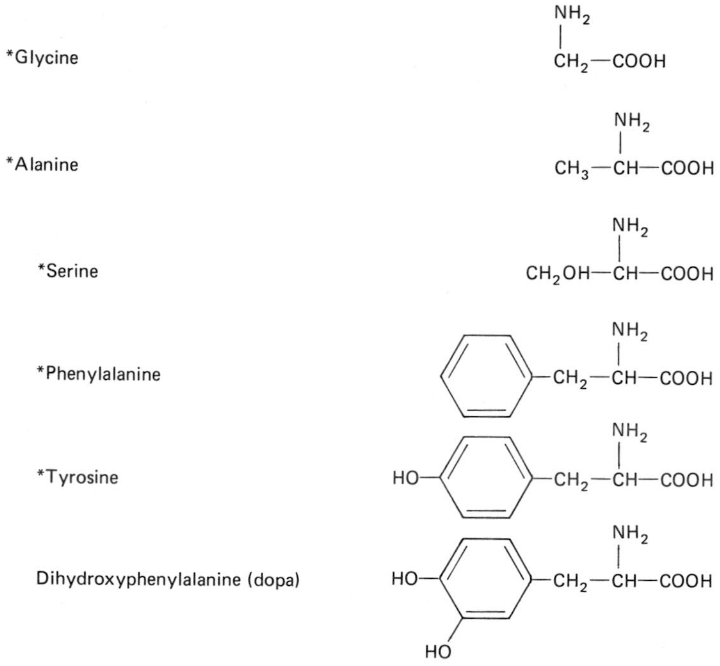

Figure 2-7. (continued)

*Tryptophan	$CH_2{-}CH(NH_2){-}COOH$ attached to indole ring (N–H)
*Histidine (Note relationship with arginine: N, N, N, N diagram)	$CH_2{-}CH(NH_2){-}COOH$ attached to imidazole ring (C, N, H–C, C–H, N–H)
*Cysteine	$HS{-}CH_2{-}CH(NH_2){-}COOH$
*Cystine	$CH_2{-}CH(NH_2){-}COOH$ / S / S / $CH_2{-}CH(NH_2){-}COOH$
α-Aminobutyric acid	$CH_3{-}CH_2{-}CH(NH_2){-}COOH$
*Methionine	$CH_2{-}CH_2{-}CH(NH_2){-}COOH$, $S{-}CH_3$
Ethionine	$CH_2{-}CH_2{-}CH(NH_2){-}COOH$, $S{-}CH_2{-}CH_3$
Homoserine	$CH_2OH{-}CH_2{-}CH(NH_2){-}COOH$
*Threonine	$CH_3{-}CHOH{-}CH(NH_2){-}COOH$
*Aspartic acid	$COOH{-}CH_2{-}CH(NH_2){-}COOH$

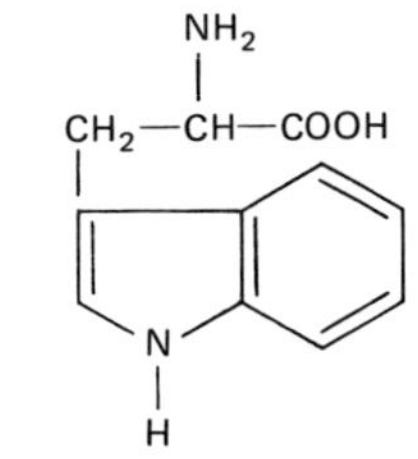

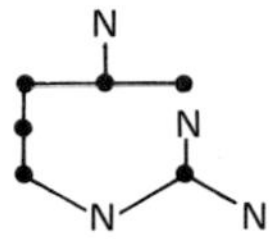

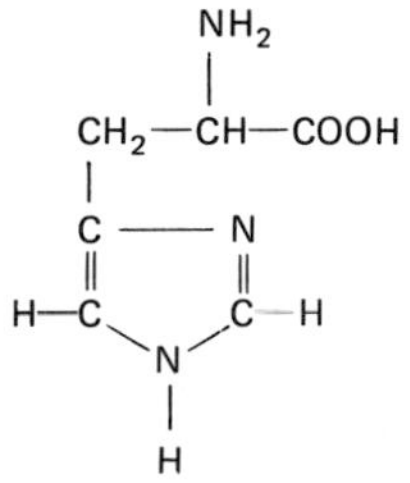

Figure 2-7. (continued)

*Asparagine	$CONH_2-CH_2-CH(NH_2)-COOH$
β-Alanine	$COOH-CH_2-CH_2(NH_2)$
Azetidine carboxylic acid	ring: CH_2–CH_2–CH(COOH)–NH– (four-membered ring)
*Glutamic acid (Various γ-hydroxy, γ-methyl, γ-methyline derivatives)	$COOH-CH_2-CH_2-CH(NH_2)-COOH$
*Glutamine	$CONH_2-CH_2-CH_2-CH(NH_2)-COOH$
γ-Aminobutyric acid	$COOH-CH_2-CH_2-CH_2(NH_2)$
*Proline	ring: CH_2-CH_2, CH_2, $CH-COOH$, NH
*Hydroxyproline	ring: $HO-CH-CH_2$, CH_2, $CH-COOH$, NH
(Note relationship with ornithine, arginine, and citrulline)	
*Valine	$(CH_3)_2CH-CH(NH_2)-COOH$
*Leucine	$(CH_3)_2CH-CH_2-CH(NH_2)-COOH$
Norvaline	$CH_3-CH_2-CH_2-CH(NH_2)-COOH$
Ornithine	$CH_2(NH_2)-CH_2-CH_2-CH(NH_2)-COOH$

Figure 2-7. (continued)

*Arginine

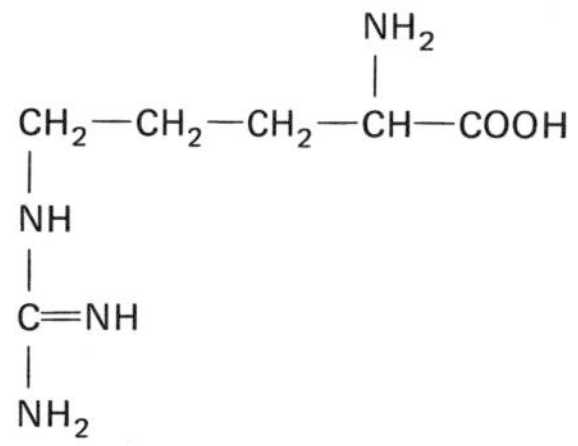

```
                                NH2
                                |
CH2—CH2—CH2—CH—COOH
|
NH
|
C=NH
|
NH2
```

Citrulline

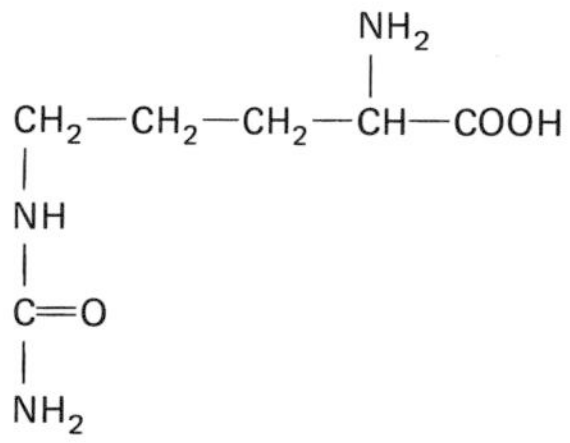

```
                                NH2
                                |
CH2—CH2—CH2—CH—COOH
|
NH
|
C=O
|
NH2
```

*Isoleucine

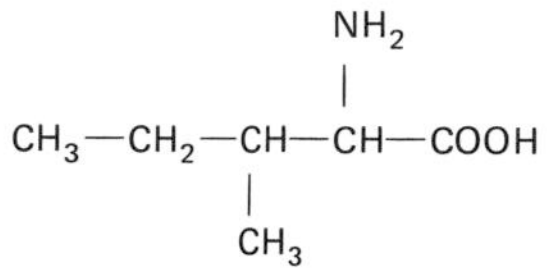

```
                          NH2
                          |
CH3—CH2—CH—CH—COOH
                |
                CH3
```

Norleucine

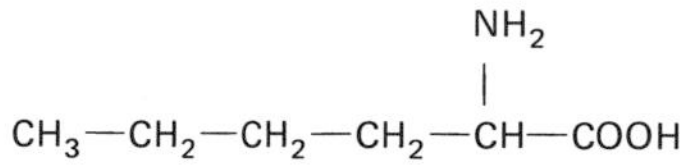

```
                                        NH2
                                        |
CH3—CH2—CH2—CH2—CH—COOH
```

*Lysine

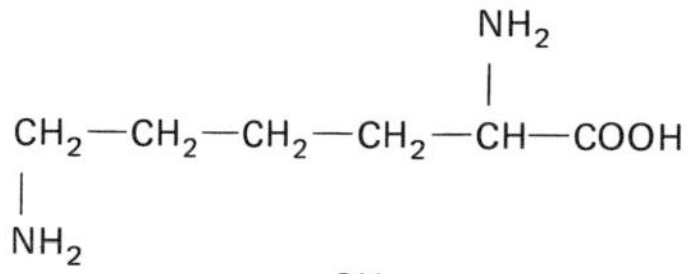

```
                                        NH2
                                        |
CH2—CH2—CH2—CH2—CH—COOH
|
NH2
```

Pipecolic acid

```
           CH2
         /     \
     CH2         CH2
      |           |
     CH2         CH—COOH
         \     /
           NH
```

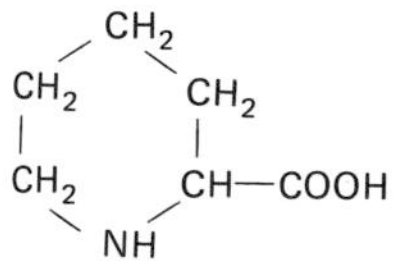

The individuality of each protein molecule is conferred partly by its **primary structure,** which is defined as the specific amino acid monomers (or residues) of which the protein is composed and the order in which these amino acids are arranged. A polymer like a protein molecule tends to twist and fold upon itself. Since proteins have regularly repeating NH_2 groups and COOH groups, they tend to form internal hydrogen bonds. Because of the configuration of the bonds and the shape and size of the monomers, hydrogen bonds tend to form between the oxygen of one residue and the nitrogen of the fourth next residue along the chain. The result is that the chain tends to form into an α helix, highly stabilized by hydrogen bonds, as shown in Figure 2-8. This configuration is called the **secondary structure** of the protein.

The helical form tends to be relatively inelastic and stiff so that proteins having extensive helices usually assume a fibrous form. In addition, the polypeptide chain may

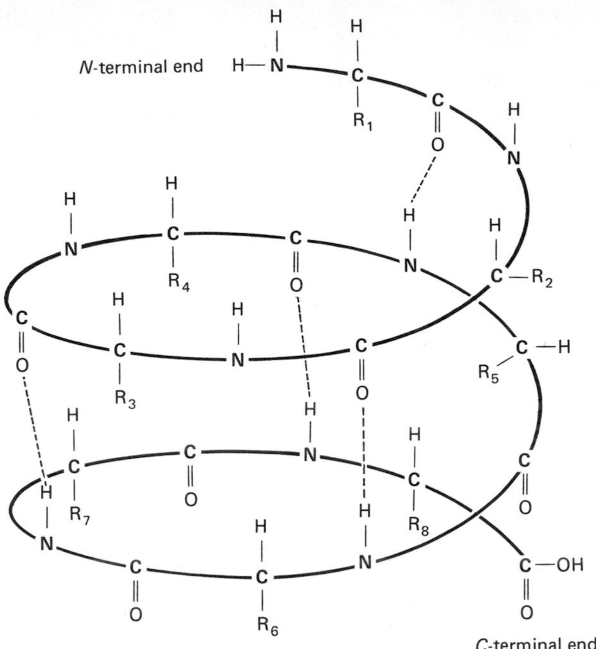

Figure 2-8. Diagram of a protein α helix, showing hydrogen bonds (dotted lines). This is really an octapeptide (containing eight amino acids). A protein molecule would contain many more.

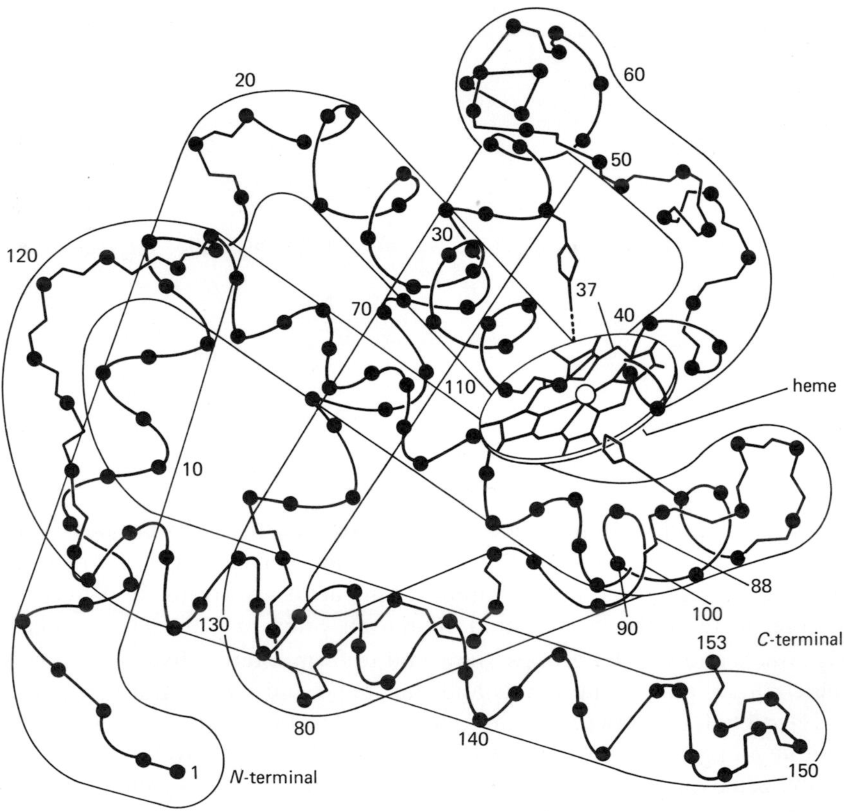

Figure 2-9. Structure of myoglobin, a protein molecule containing 153 amino acids and a heme group.

be folded and compressed, rather like an untidy tangle of thread. Nevertheless, these compressed structures are not haphazard; they are very precisely formed and held together by a number of different types of forces, from weak van der Waals forces through hydrogen bonds to covalent bonds such as disulfide (—S—S—) bridges. Most such linkages involve the side chains of the amino acids, and constitute what is called the **tertiary structure** of the protein molecule. The primary structure of a number of proteins is known. Secondary and tertiary structures are more difficult to determine and have been elucidated for only a few proteins. An example of a protein whose complete structure is known is myoglobin, shown in Figure 2-9.

Nucleic Acids

Nucleic acids are the last major group of biological chemicals that we shall examine. These are polymers of **nucleotides** in the same way that proteins are polymers of amino acids. Each nucleotide consists of a base and a sugar esterified with a molecule of phosphoric acid. The base-sugar combination, without the phosphoric acid, is called a **nucleoside.** Two sugars occur, ribose and deoxyribose (lacking an oxygen atom), and the nucleotides containing them are called ribonucleotides and deoxyribonucleotides, respectively. Ribonucleic acid (RNA) is a polymer of ribonucleotides (containing the sugar ribose), whereas deoxyribonucleic acid (DNA) is a polymer of deoxyribonucleotides (containing the sugar deoxyribose). The organic bases are adenine, uracil, cytosine, and guanine in RNA; whereas adenine, thymine, cytosine, and guanine occur in DNA. The monomers are linked together through ester bonds with the phosphate groups from the C-5 of one sugar to the C-3 of the next. The structure of ribonucleotides and deoxyribonucleotides and the linkage of their polymers are shown in Figure 2-10.

The ribonucleotides are important in the major energy transfer reactions in cellular metabolism. A second and third phosphate group may be esterified onto the first phosphate group attached to the C-5 of the ribose moiety of the nucleotide. A larger amount of energy is required for the synthesis of these phosphate esters, particularly the third. Conversely, much energy is released on hydrolysis of these bonds. Thus, energy from cellular oxidation reactions may be stored in these compounds and transported elsewhere in the cell, to be released later for synthetic reactions or to do work (see Chapter 5). The most important of the ribonucleotides is the **adenosine** series: adenosine monophosphate (AMP), adenosine diphosphate (ADP), and adenosine triphosphate (ATP), shown in Figure 2-11. The other ribonucleotides and deoxyribonucleotides also form di- or triphosphate esters. The formation of RNA and DNA takes place from the union of the triple esters of the nucleotides; two phosphate groups are eliminated in the condensation of each nucleotide.

DNA is the bearer of genetic information in the cell. Two parallel strands of DNA are quite tightly linked together, by hydrogen bonds between amino and carbonyl (C=O) groups on adjacent bases, in such a way that a purine always links with a pyrimidine. Adenine always bonds to thymine and cytosine to guanine. These pairs of bases are called **complementary;** they link only with each other because their molecular size and structure permit an exact fit only between complementary bases. The structure so formed then coils into a double helix due to the formation of hydrogen bonds, as shown in Figure 2-12. The double helix can be precisely self-duplicating because the

Figure 2-10. Structure of nucleic acid components.

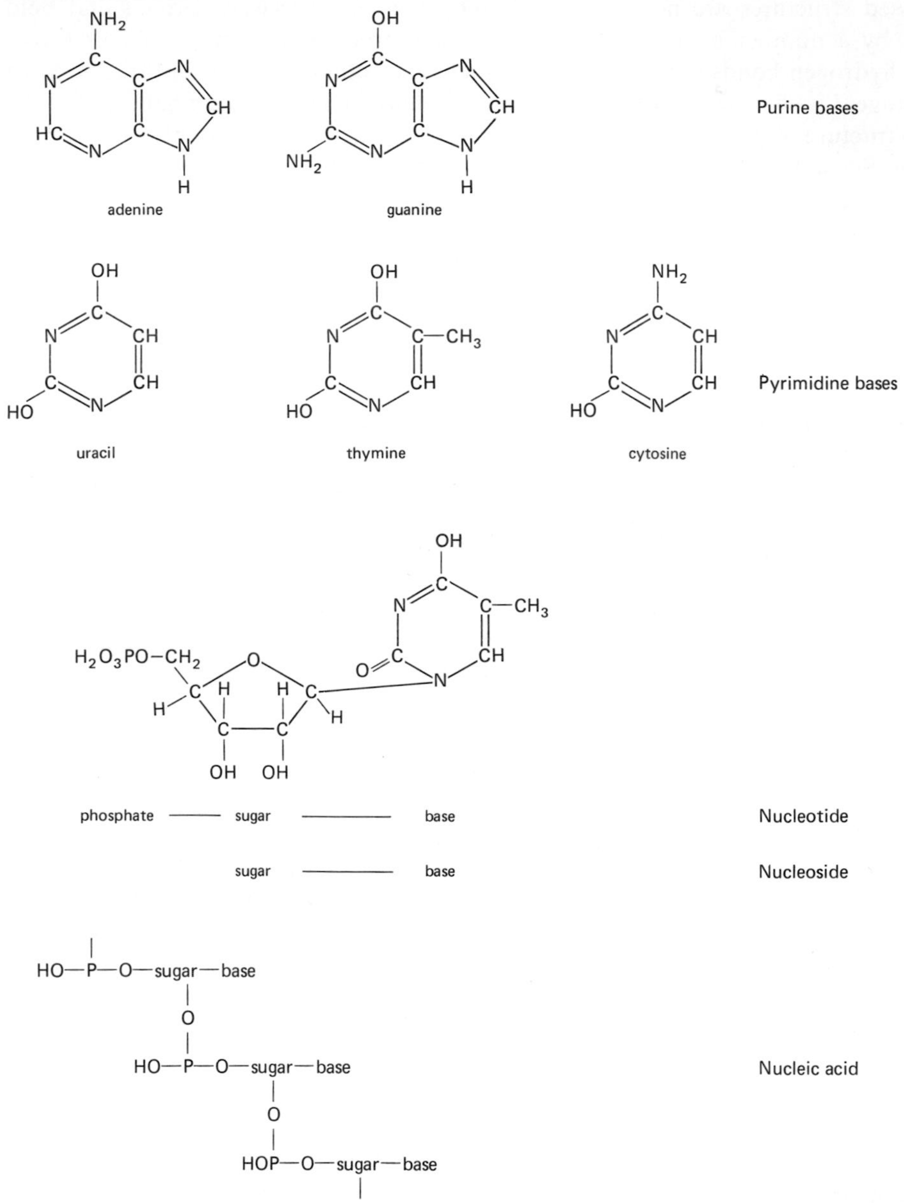

bases always pair in the same way, since only complementary bases can pair. Thus when two new strands are formed, each using half of an original double helix as a template, two precise copies of the original double helix are formed, as shown in Figure 2-13. This process occurs during cell division.

The information contained in the DNA is "read" by the synthesis of a strand of RNA, which is complementary to that part of the DNA strand forming its template (except that adenine in the DNA will be complementary with uracil in the RNA instead

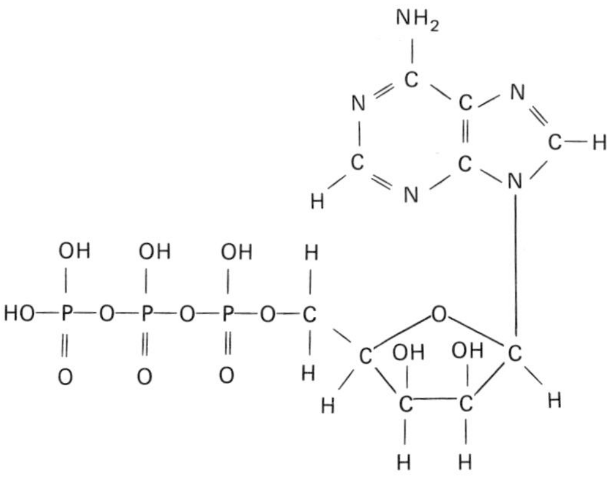

Figure 2-11. Adenosine triphosphate.

of with thymine). This process is shown in Figure 2-14. The RNA so formed is called **messenger RNA (mRNA).** The genetic information is contained in triplets of nucleotides, or **codons,** each of which codes for a particular amino acid in the manner shown in Figure 2-15.

The details are as follows: The codons are read by low molecular weight **soluble RNA (sRNA),** also known as **transfer RNA (tRNA).** Each tRNA has a triplet of bases

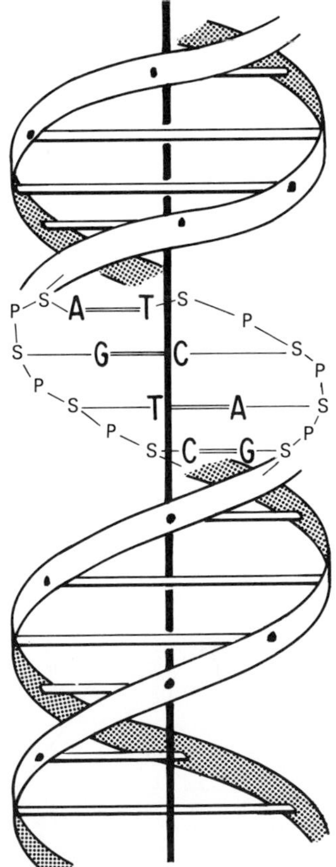

Figure 2-12. The double helix of DNA.

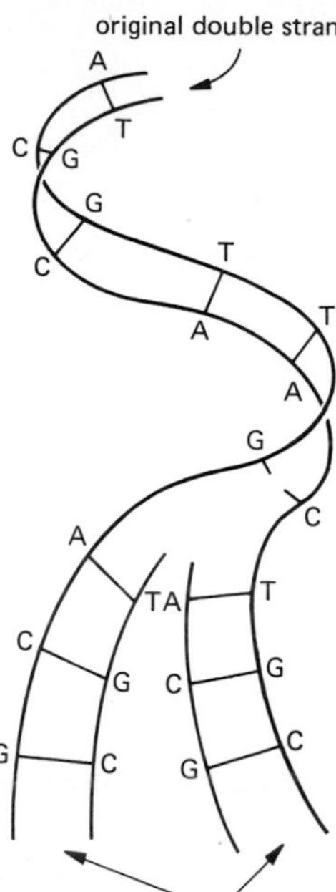

Figure 2-13. Replication of DNA.

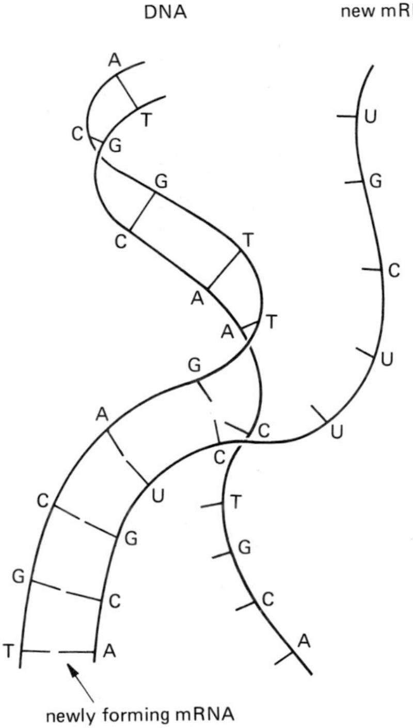

Figure 2-14. Transcription of DNA.

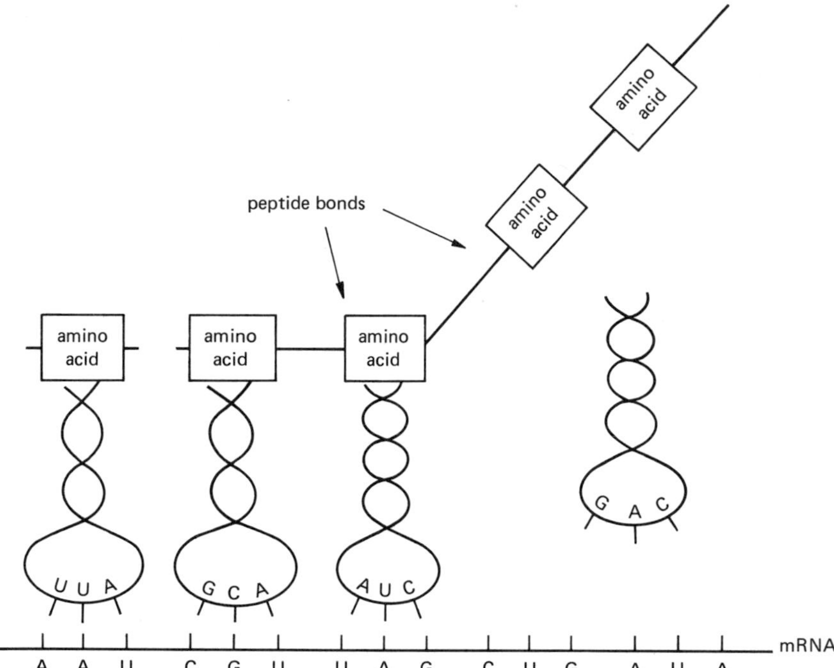

Figure 2-15. Amino acids are lined up to be joined into proteins by tRNA molecules that "read" the codons of the mRNA.

that is exactly complementary to a codon. Each specific tRNA, having a specific codon, is able to combine with the specific amino acid required by the codon. Thus the required amino acids are lined up in the sequence dictated by the mRNA, where they are successively linked together chemically and released from the tRNAs. The resulting peptide has precisely the sequence of amino acids required by the genetic information of the original DNA, as transmitted by the mRNA and as read by the tRNA. Although the amino acids, the tRNAs, and the enzymes for making amino acid–tRNA complexes are free in the cytoplasm, the enzymes and apparatus for the actual assembly of proteins are contained in minute particles called ribosomes (see page 54). It is postulated that the ribosomes move along the strands of mRNA, reading the information (that is, lining up amino acid–tRNA complexes as required) and forming the polypeptide or protein, which is then liberated. tRNAs, having delivered their amino acids to the peptide chain, move out again to complex with another molecule of their specific amino acid, and are then ready to move in again and deliver another amino acid to the peptide chain as directed by the mRNA.

This outline of protein synthesis has been very brief and is merely intended as a foundation for the later discussions of genetic mechanisms controlling metabolism and development. Students who have difficulty understanding this aspect of biochemistry should read one of the many excellent introductory texts on cell or molecular biology.

Additional Reading

The material covered in this chapter is dealt with in greater depth in up-to-date texts on organic chemistry or biochemistry.

Bronk, J. R.: *Chemical Biology*. Macmillan Publishing Co., Inc., New York, 1973.

The Molecular Basis of Life (readings from *Scientific American*). W. H. Freeman & Co., San Francisco, 1968.

Watson, J. D.: *The Molecular Biology of the Gene*. Benjamin, Menlo Park, Calif., 1970.

White, E. H.: *Chemical Background for the Biological Sciences*. Prentice-Hall, Englewood Cliffs, N.J., 1970. *Foundations of Modern Biology* series.

3 The Cell

The Cell Theory

Most biological systems are made up of units called cells. Each cell is a complete living entity; the cell is the smallest "biounit" capable of sustained independent existence. This concept, one of the basic unifying doctrines of biology, was elaborated by Schlieden and Schwann in 1839, over 150 years after the first clear recognition of the cellular nature of higher plants by Robert Hooke. The only exceptions to this concept, the viruses, are not capable of independent existence over much of their life span; viruses must be associated with and exist in cells in order to reproduce. Thus viruses are not, in the broadest sense, true living organisms.

Cells of the simplest organisms are capable of all the activities and reactions of life. In more complex organisms cells may become highly specialized, capable only of specific activities. It is thus necessary for cells to be able to communicate with each other, so that the activities of groups of specialized cells may be coordinated and the products of a metabolic process from one group of cells may be transferred to another group for further metabolism. Each cell in a multicellular organism carries initially, and perhaps for its whole life span, the totality of the genetic information in the organism. Obviously it cannot draw on or use all this information at once; therefore, the cell must have some system of selection and some external instructions to enable it to select appropriate information.

It is evident that the organism influences the developmental pattern of individual cells or, in other words, that the behavior of a cell is influenced by other cells. This is an important and basic fact in the study and understanding of plant physiology. To understand the behavior of an organism, one must know intimately the details of the behavior and capabilities of its component cells. Conversely, the study of the activities and reactions of a single cell or its parts is a barren pursuit unless the cell is studied in the perspective of the whole organism of which it is a part and which controls and directs its behavior and development.

Cells, and the organisms made of them, can be divided into two types: the **prokaryotic** (*pro,* before; *karyotic,* having a nucleus; therefore, cells having no organized nucleus or nuclear membrane) and the **eukaryotic** (*eu,* good; therefore, cells having a well-defined nucleus separated from the cytoplasm by a membrane). Bacteria and blue-green algae are prokaryotes; all higher organisms are eukaryotes. Prokaryotic cells are small, in the vicinity of 1 μ in diameter, and show comparatively little structural organization. Eukaryotic cells are usually much larger (10–100 μ or larger) and have a complex internal structure including many organelles of specific function. Although cells may have become more and more complex with evolutionary time, it is likely that they are as simple and as uncomplicated as they can be consistent with the complexity of behavior required of them. Biological systems tend often to follow the principle of Occam's razor: the simplest and most straightforward way of doing something is the most likely. Expensive (in terms of energy or material) and complicated ways of doing things are not as likely to survive evolution as simpler means to the same end. This thought (it really does not deserve to be dignified as a "principle") has been useful in the study and understanding of the complex intercellular and intracellular control systems of multicellular organisms.

The Cell and Its Parts

An electron micrograph of a cell is shown in Figure 3-1, together with diagrams of a typical plant cell showing its various parts. These diagrams should be referred to throughout the subsequent discussion. Plant cells characteristically (but not always) contain plastids, and they usually have a well-defined cell wall, which may be variously thickened. Apart from this, and the absence of a centrosome body in cells of spermatophytes, plant cells do not differ dramatically from animal cells.

The relative size of the parts of cells is shown in Table 3-1. The figures given here are only approximate, but serve to give an idea of the size relationship of a cell and its parts. The units of measurement used are shown in Table 3-2.

Cell Wall. The cell wall does not act as a physiological boundary; its main function is mechanical—supporting the cell and multicellular structures and preventing the outer membranes from bursting as a result of the hydrostatic pressures that develop inside the cell. In addition, plants lack direct mechanisms, such as phagocytosis, for dealing with invading pathogenic organisms that must gain frequent entry through wounds and natural openings. It is probable that the strong resistance of plants to infection from such invaders is due in large part to their relatively impenetrable cell walls.

Table 3-1. Approximate size relationship of cells and their parts

	Approximate usual size range
Cell	10 μ–10 mm
Nucleus	5–30 μ
Chloroplast	2–6 μ
Mitochondrion	0.5–5 μ
Peroxisome	1 μ
Ribosome	250 Å
Endoplasmic reticulum	200 Å
Unit membrane	75 Å
Protein molecule	20–100 Å

Table 3-2. Units by which the size of cells and subcellular particles are measured

1 millimeter (mm)	= 10^3 microns (μ)
	= 10^6 millimicrons (mμ) or nanometers (nm)
	= 10^7 Ångstrom units (Å)
1 μ	= 10^3 mμ (nm)
	= 10^4 Å
1 mμ (nm)	= 10 Å

The thinnest, simplest cell wall in plants is the wall enclosing the protoplasm of meristematic cells, the **primary wall.** Following cell division, a primary wall is formed between the resulting two cells. This develops from the **cell plate,** shown in Figure 3-2, a thin structure formed by the coalescence of many cytoplasmic vesicles that derive from the Golgi apparatus (see pages 54–55). The contents of these vesicles, which form the **middle lamella** of the newly forming wall, consist largely of pectin and calcium pectate, the ground substances of the middle lamella. Strands of cytoplasm called **plasmodesmata** usually penetrate the cell wall via a number of small holes or openings so that the cell contents of adjacent cells, may, in fact, be in intimate contact. It has also been suggested that elements of the endoplasmic reticulum penetrate the cell wall or are continuous with the plasmalemma, but this is not definitely known.

As soon as the cell plate is formed, cellulose begins to be deposited as long thin rods or microfibrils to form the primary wall, which is largely made up of cellulose. At first these microfibrils are usually **isotropic** (having no preferred or ordered orientation) in primary walls, as shown in Figure 3-3A. However, cell walls continue to grow, both in area and thickness, as cells elongate or divide. Cell wall growth causes the microfibrils to slip over one another and to assume an orientation more in line with the direction of growth. Subsequently formed microfibrils are usually deposited in a highly oriented way, as shown in Figure 3-3B.

Cellulose microfibrils undoubtedly form the main elements of strength or rigidity in cells, preventing the cell from swelling or bursting because of the pressure of its contents, but they are not the only constituents of cell walls. The cellulose microfibrils are embedded in an amorphous gel consisting largely of polymers other than starch or cellulose, primarily pectins and hemicellulose together with a small amount of protein. As the cell matures, the wall thickens and additional layers of oriented cellulose microfibrils are laid down until the cell becomes rigid and inelastic. In many cells massive deposition of almost pure cellulose (as in cotton) or cellulose mixed with other components such as lignins and hemicelluloses (as in tracheids or vessels) takes place.

Membranes. All the properties of living cells depend to some extent upon the properties of their membranes. A cell is surrounded by a membrane that separates it from its environment and enables it to control the entry and egress of substances selectively. Moreover, virtually all of the subcellular organelles are made of, or surrounded by, membranes or pieces of membrane, and much of the cellular enzymic machinery is mounted on or associated with membranes in one form or another. Thus, the physiology of membranes is of central importance in plant physiology.

The structure of membranes has long been a matter of controversy because it is difficult to conceive of a barrier that permits the passage of some substances and not of others and, further, allows only one-way passage of some materials. Various suggestions have been made, including sieves of different types and mosaics of lipophilic and hydrophilic compounds. It now seems probable that some of the properties of membranes result from their chemical nature, some result from the existence of pores that may perhaps act as differential sieves, and still other properties result from the existence of highly specialized transport mechanisms that actively move certain substances directionally across cell barriers.

The basic appearance of membranes has been found to be so constant from organism to organism and from organelle to organelle that the "unit membrane" concept has been put forward by J. D. Robertson and his coworkers. Membranes give the appear-

A

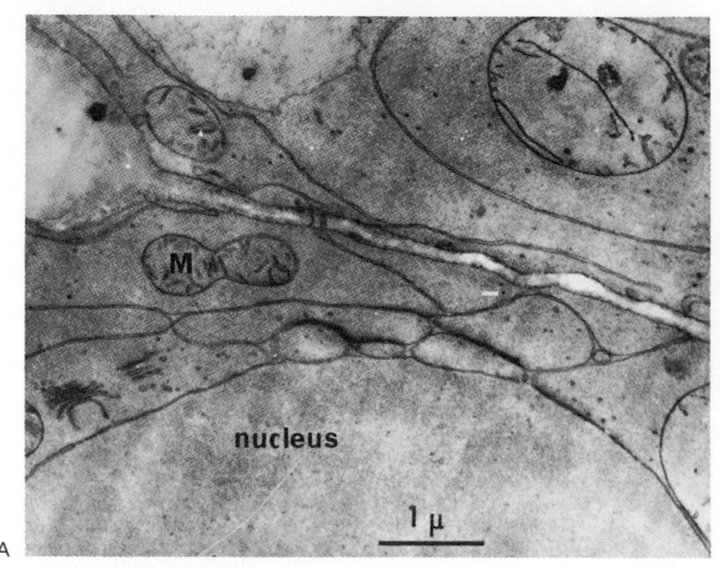

C

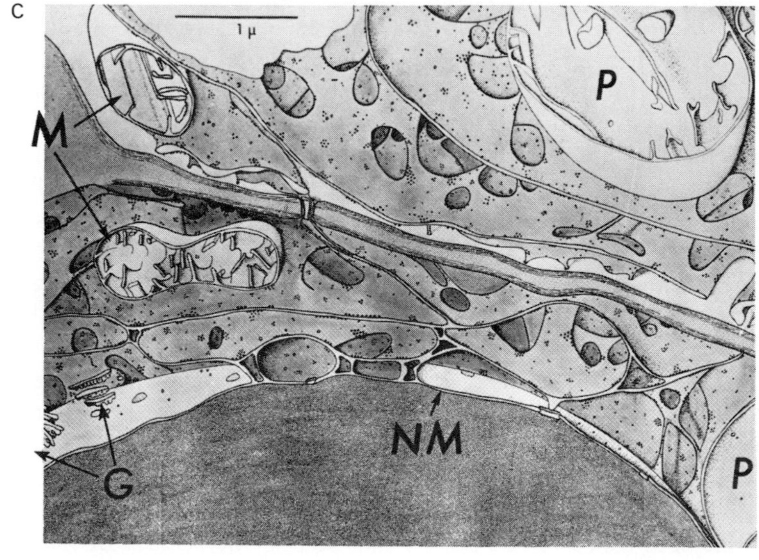

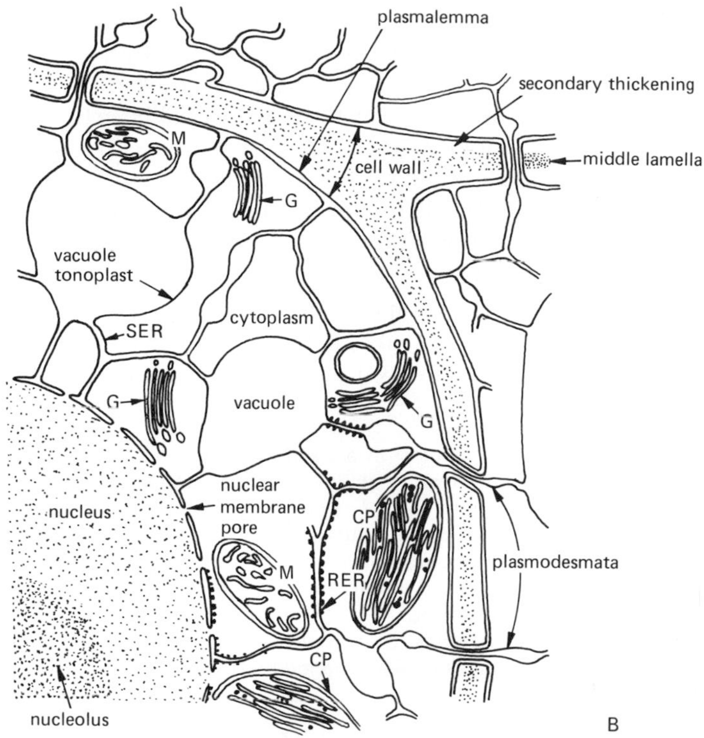

Figure 3-1. The plant cell.

A. Electron micrograph of a root cell from *Vicia*.

B. Diagram showing the ultrastructure of a cell. Code: CP, chloroplast; SER, smooth endoplasmic reticulum; RER, rough endoplasmic reticulum; G, Golgi apparatus, M, mitochondrion.

C. Three-dimensional representation of parts of two plant cells. (Compare with Figure 3-1A, from which this diagram was drawn.) The ER forms much-branched, fenestrated sheets throughout the cells and passes from one cell to another via the plasmodesmata. Sections through mitochondria (M), plastids (P), and Golgi bodies (G) are shown and both surface and transverse views of the nuclear membrane (NM).

[**A** and **C** from F. A. L. Clowes and B. E. Juniper: *Plant Cells*. Blackwell Scientific Publications Ltd., Oxford, 1968. Used with permission.]

ance of a three-layered structure in electron micrographs (see Figure 3-4), the two outer layers being largely protein and the inner layer lipid. Diagrams to illustrate current concepts of the molecular arrangement in unit membranes, as described by H. Davson and J. F. Danielli, are shown in Figure 3-5. The structure illustrated would be extremely stable, being held together by polar forces and by such attractive forces as hydrogen bonds and van der Waals forces. Yet it is flexible enough to allow for the fact that membranes are not static structures—they are able to move, to be disrupted or broken and to re-form, to bud or form vesicles by throwing folds and then pinching them off.

More recent investigations have shown that the unit membrane hypothesis is not truly universal. Some membranes appear to be composed of globules of phospholipid coated with protein rather than regular protein-coated lipid bilayers. Some membranes are clearly asymmetrical, having a thicker protein layer on one side than the other or

Figure 3-2. Electron micrograph of a cell plate forming in a young leaf of wheat (*Triticum*). It is thought that a number of the small Golgi-derived vesicles fuse to form the larger vesicles (cp), and these finally coalesce to create the cell plate itself. The cell plate is developing toward the left of the picture. On the extreme left a number of microtubules (t) appear to be "guiding" small Golgi vesicles into position (arrow). These microtubules are relatively straight, whereas those already embedded in the wall have become sinuous (S), and there is no sign of microtubules at all in the older section of the wall to the right.

[From F. A. L. Clowes and B. E. Juniper: *Plant Cells*. Blackwell Scientific Publications Ltd., Oxford, 1968. Used with permission. Micrograph originally supplied by Dr. J. D. Pickett-Heaps and Dr. D. H. Northcote.]

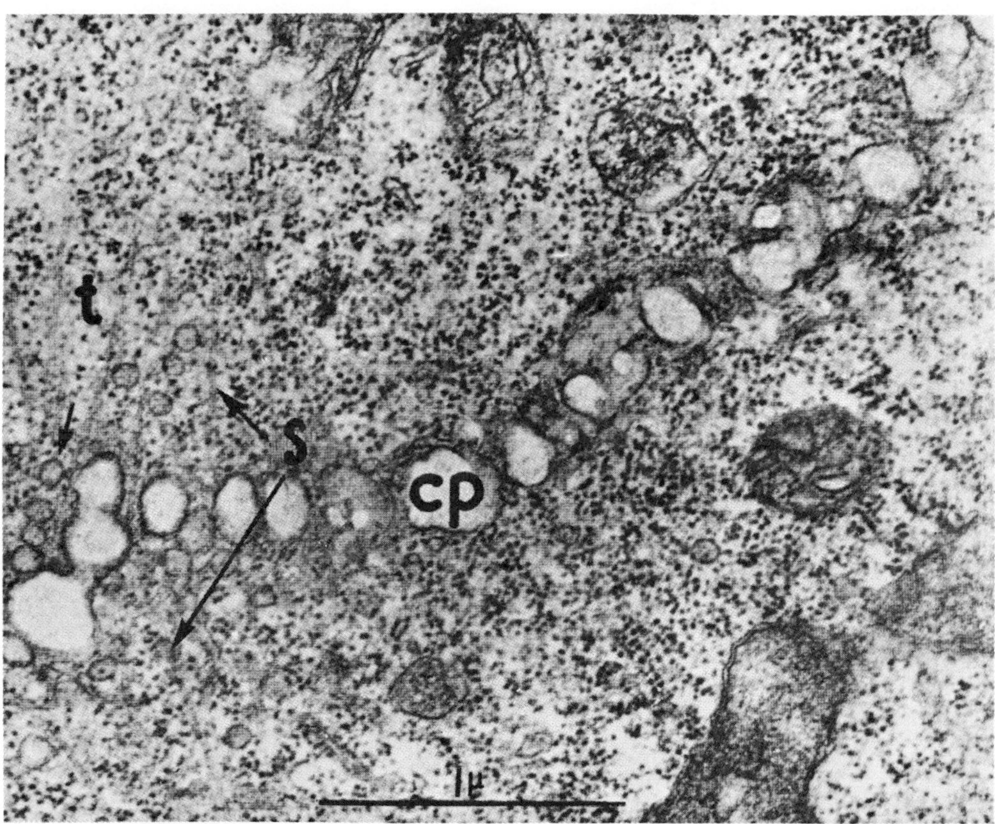

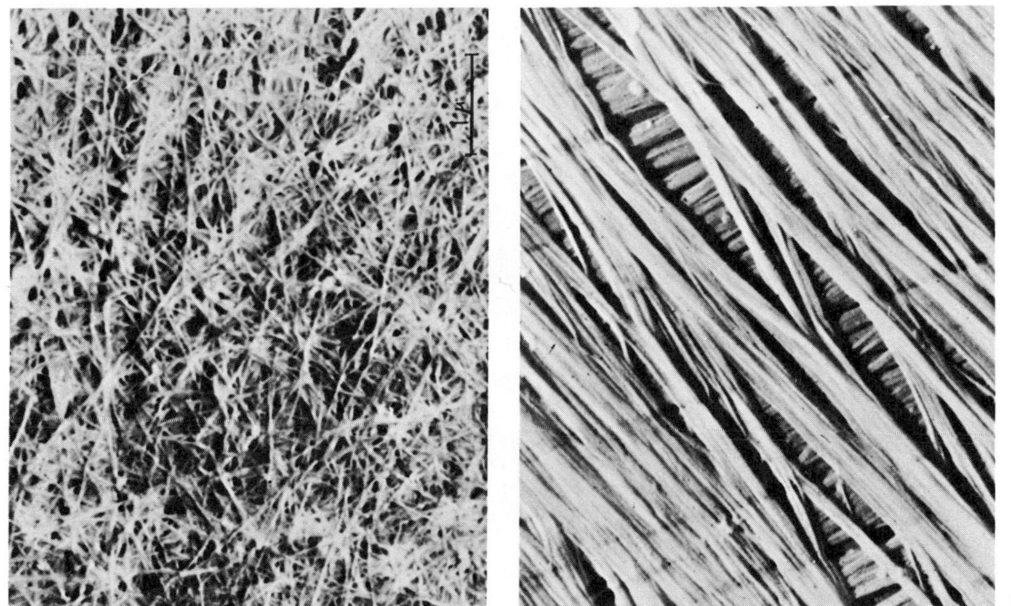

Figure 3-3. Arrangement of microfibrils in the cell walls of *Valonia,* X12,000.
A. Isotropic or randomly arranged cellulose microfibrils of the primary cell wall.
B. Parallel (anisotropic) cellulose microfibrils of secondary wall.

[From J. Bonner and J. E. Varner: *Plant Biochemistry*. Academic Press, New York, 1965. Used with permission. Photographs courtesy Dr. K. Mühlethaler, Zurich, Switzerland.]

appearing to have protein globules rather than layers. It has been suggested that protein molecules may penetrate the lipid bilayer at intervals, essentially forming hydrophilic "pores," or that proteins or polypeptide chains may be interspersed within the lipid layer.

In Chapters 5 and 12 we shall examine how membranes are involved in electron- and proton-transfer reactions leading to ATP synthesis and in ion transport. These reactions require that many proteins—enzymes of the electron transport chain and others—are specifically oriented in membranes and at their inner or outer surfaces. The American biochemist E. Racker described a model of membrane structure, shown in Figure 3-6, that satisfies these requirements. This structure can be related in general to the arrangement shown in Figure 3-5, but differs from it in showing the internal location of certain protein molecules. However, the exact structure of various membranes may differ in details that probably relate to their function.

Most subcellular structures, such as nucleus, mitochondria, or chloroplasts, are surrounded by double membranes; however, the outermost living layer of the cytoplasm, the cell membrane or **plasmalemma,** as well as the inner membrane lining the vacuole, the **tonoplast,** consist of single unit membranes. These two membranes largely control the exchange of materials between the cytoplasm and the extracytoplasmic space outside the cell and in the vacuole. They mark the boundaries of the living material of the cell. This is not to say that the cytoplasm cannot exert its influence beyond the limits of its boundary membranes. Evidently it does, since it is able to modify vacuolar contents and to conduct the synthesis of cell walls which lie outside the plasmalemma.

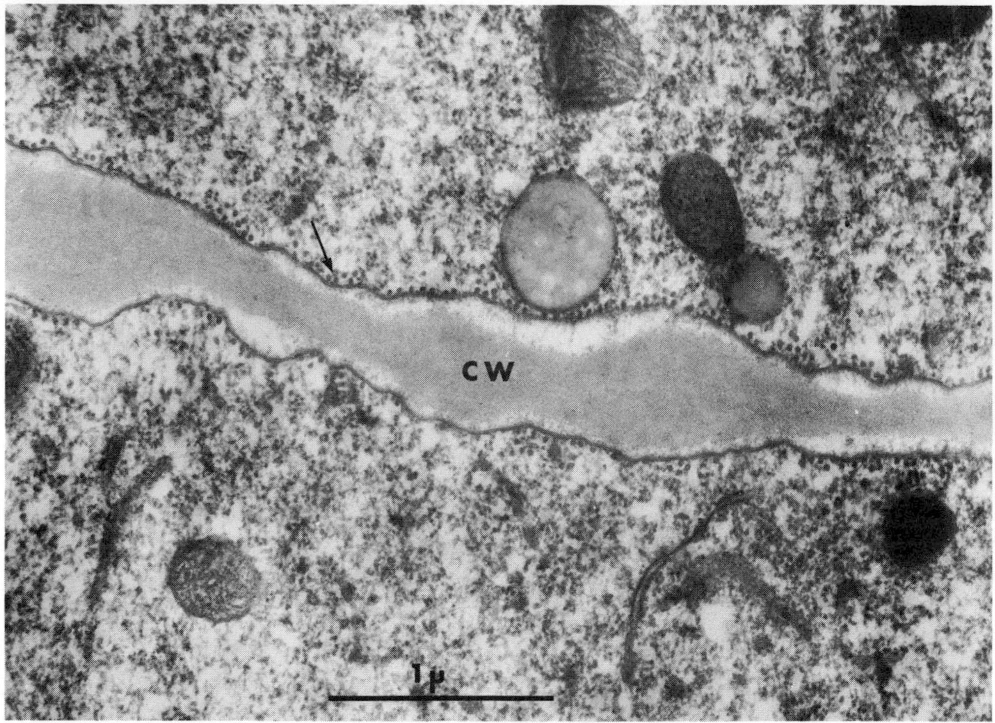

Figure 3-4. Electron micrograph of a cell wall (*Alium*) in cross section showing double layer of the plasma membrane (*note: not* a double membrane), and microtubules aligned along the membrane. [Electron micrograph courtesy Dr. A. K. Bal, Dept. of Biology, Memorial University of Newfoundland.]

Nucleus. The largest and most prominent organized inclusion in most cells is the nucleus. This structure contains a large part of the cell's genetic material, the DNA strands that are present in protein complexes forming the nucleoproteins. These are usually present as strands of chromatin (Figure 3-7), but during cell division they form into distinct chromosomes, or rather chromosome pairs, since they undergo replication prior to the start of division (see Figure 3-8). The nucleus usually contains from one to four **nucleoli** (Figure 3-7), densely staining spherical bodies that appear to be RNA reserves, presumably used during the decoding of the DNA message of the chromatin. The nucleus is surrounded by a double membrane that has many small pores (Figure 3-1A). It has been suggested, although not unequivocally proved, that large molecules such as ribonucleoproteins may pass through the pores of the nuclear membrane, thus permitting egress of informational material from the nucleus to the cytoplasm. The nuclear membrane is apparently continuous with the endoplasmic reticulum, which in turn may be connected with the plasmalemma (Figure 3-9) and thus provide a pathway between the nucleus and the surroundings of the cell. The significance of such pathways has not been fully explored.

Endoplasmic Reticulum. The endoplasmic reticulum (ER) is a network of membranes that ramifies throughout the cytoplasm of most metabolically active cells. It consists of double unit membranes, at times separating to form small vesicles. Much of

Figure 3-5. Two interpretations of the structure of cell membranes.

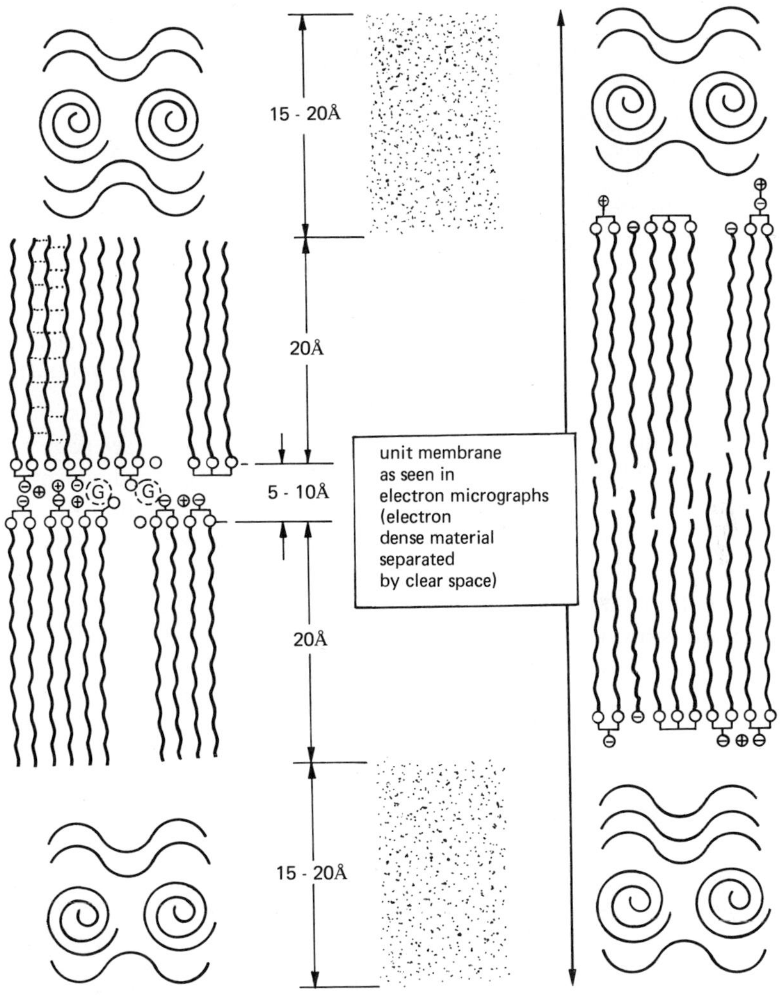

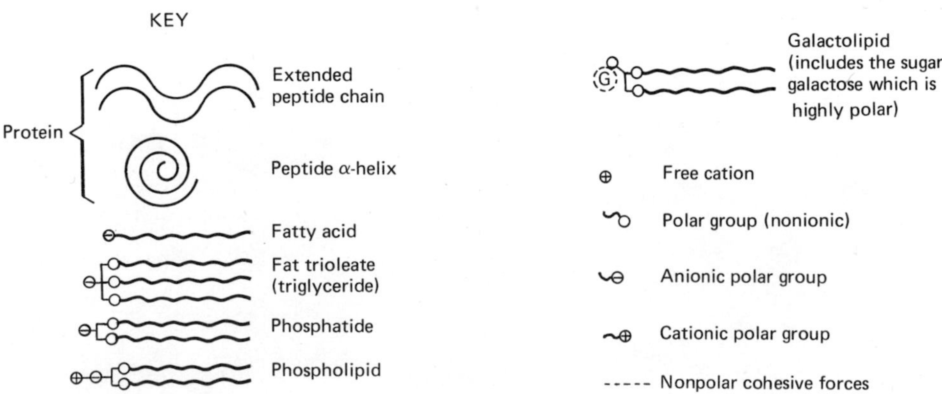

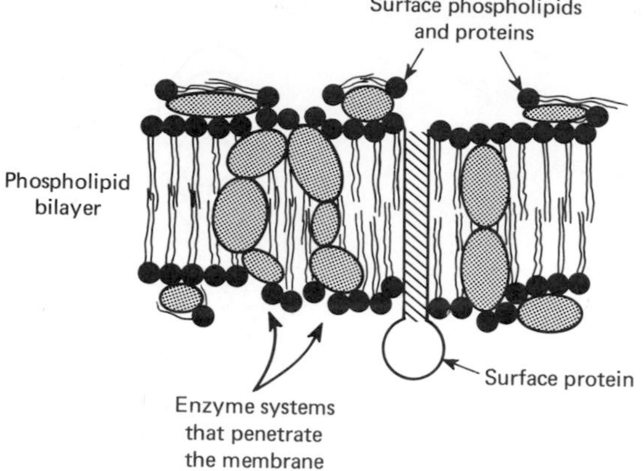

Figure 3-6. Model of membrane structure. [After E. Racker: *A New Look at Mechanisms in Bioenergetics*. Academic Press, New York, 1976.]

the endoplasmic reticulum has large numbers of ribosomes (see page 54), either attached to or associated with its outer side. This is called **rough endoplasmic reticulum;** when without ribosomes it is called **smooth endoplasmic reticulum.** Because the ribosomes are the sites of protein synthesis in the cell, it is likely that this association of ribosomes and endoplasmic reticulum is directly concerned with protein synthesis. The endoplasmic reticulum may play a part in assembling the subunits for protein synthesis and distributing the products.

Figure 3-7. Electron micrograph of a cell from the apical meristem of wheat (*Triticum*), ×10,000. The nucleus, which nearly fills the cell, contains chromatin and a large nucleolus. [Preparation by Dr. J. Mahon; micrograph courtesy Mrs. E. Paton.]

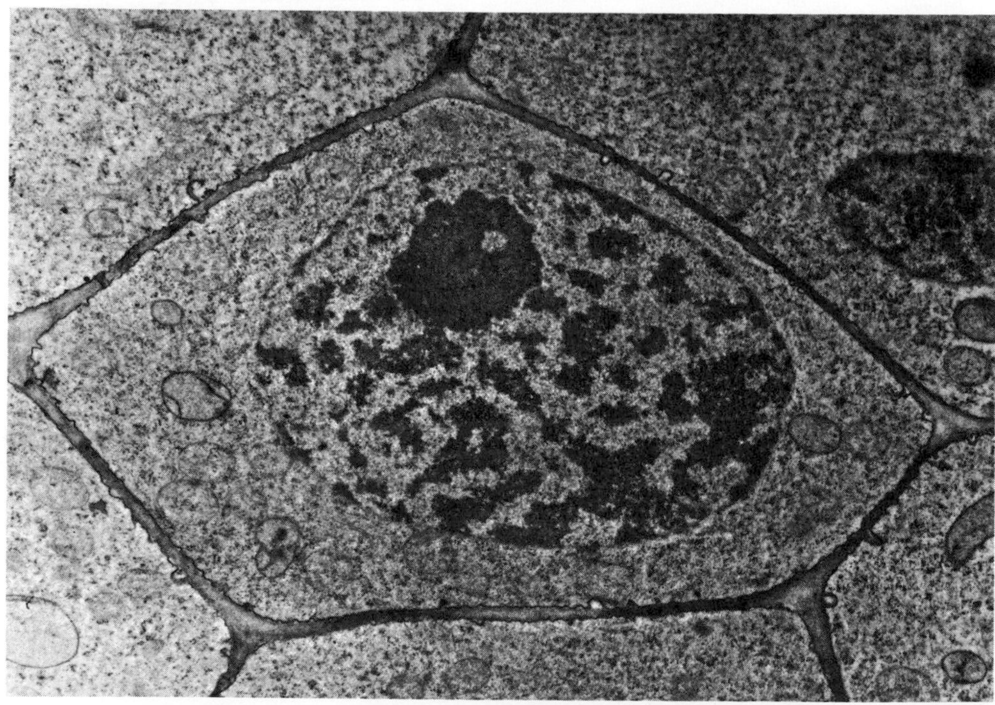

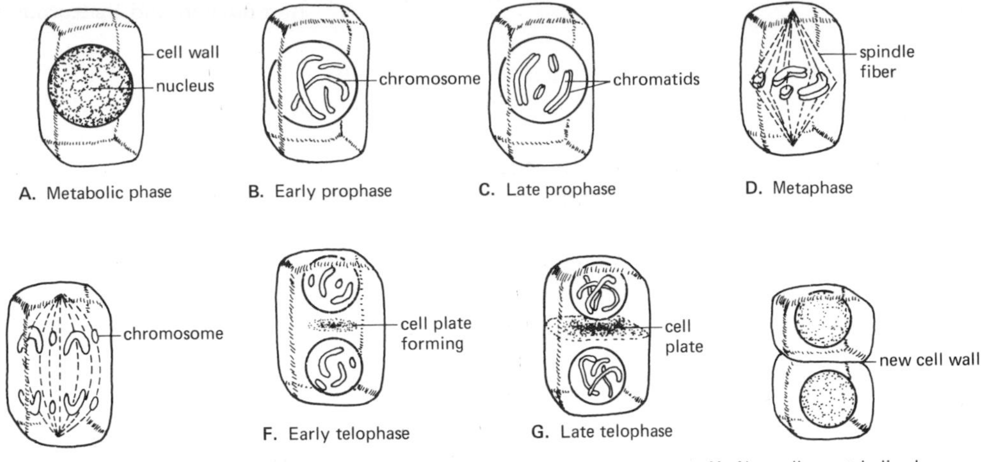

Figure 3-8. Diagram of mitosis in a plant cell. [From W. H. Muller: *Botany: A Functional Approach,* 3rd ed. Macmillan Publishing Co., Inc., New York, 1974.]

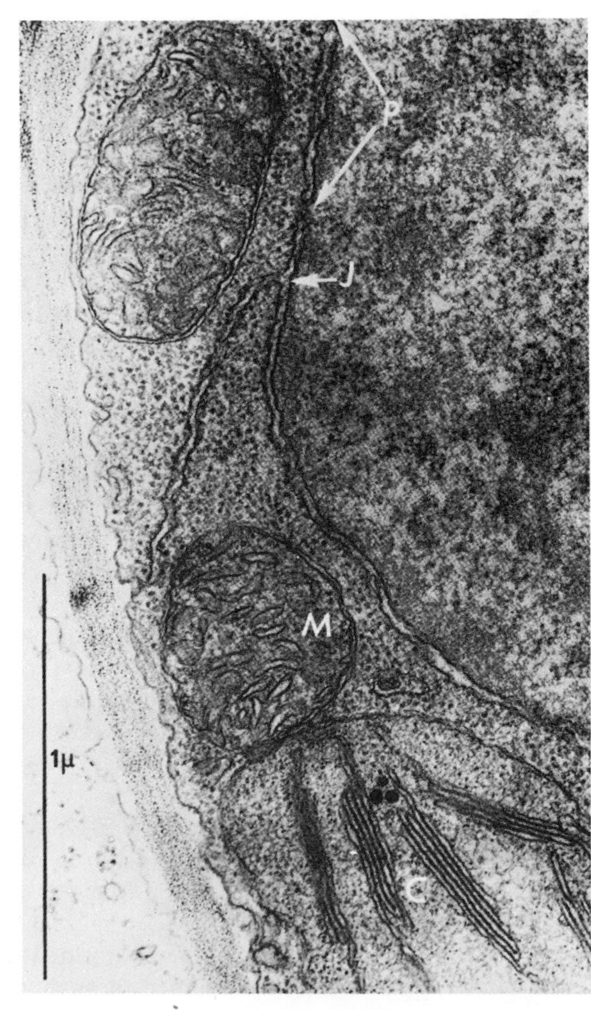

Figure 3-9. Electron micrograph of part of a cell from a spinach leaf. The nuclear membrane with pores (P) is joined to the rough endoplasmic reticulum (J). The cell also contains mitochondria (M) and a developing chloroplast (C). [From F. A. L. Clowes and B. E. Juniper: *Plant Cells*. Blackwell Scientific Publications Ltd., Oxford, 1968. Used with permission.]

It has been suggested that the endoplasmic reticulum is continuous with (and perhaps through) the plasmalemma and the nuclear envelope. If this is true, then all parts of the cell, even the nucleus, may be in close contact with a cavity system that is continuous with the outside and with other cells. The connections between the endoplasmic reticulum and the nuclear envelope are now well established (see Figure 3-9), but the association of the nuclear membrane with the plasmalemma and with the contents of adjacent cells rests on evidence that is much less clear. The possibilities of such a system are enormous, but it is not now known to what extent the endoplasmic reticulum does function as a system for material- or information-transfer between cells. A more recent interpretation suggests that the endoplasmic reticulum is not actually continuous with the plasmalemma but buds off vesicles that can associate with, and thus pass their contents through, the cell membrane. It would thus appear that there is continuity in time between the endoplasmic reticulum and the outside of the cell, but not necessarily a direct pathway from the nuclear envelope to the outside of the cell.

The formation of new cell walls during cell division begins with the alignment of microtubules (page 59) followed by the lining up of vesicles, derived apparently from dictyosomes or the endoplasmic reticulum. These vesicles contain carbohydrate material that is used to form the middle lamella of the wall. Thus the endoplasmic reticulum takes an active part in cell wall formation.

A consequence of the ramification of the endoplasmic reticulum and its continuity with the nuclear envelope is that the cytoplasm of the cell may be divided into small compartments. The divisions are not absolute because the endoplasmic reticulum is a living membrane and can be disrupted or broken and re-formed, and because there may be holes or crannies between the compartments. However, such compartments may be extremely important in the maintenance of a variety of different metabolic systems within one cell. The divisions prevent the metabolites of one system from interfering with those of another, either by the oversupply of an unwanted metabolite or by the draining of some necessary intermediate.

Golgi Apparatus and Dictyosomes. Dictyosomes are saucer-shaped bodies made up of several layers of flat vesicles or **cisternae** (singular, cisterna) composed of unit membranes (see Figures 3-1 and 3-10). One or many dictyosomes constitute the Golgi apparatus of the cell. The edges of these cisternae are frequently seen to be ballooned or swollen and evidently give rise to vesicles by pinching them off from the edges. The Golgi apparatus is primarily associated with cell wall formation and is extremely important as a major transport system of materials to the outside of the cell via the vesicles formed from the cisternae of the dictyosomes. The synthesis of cell wall polysaccharides is initiated in the endoplasmic reticulum, but is completed in the dictyosomes and then transported to the site of cell wall synthesis by vesicles derived from the dictyosomes. The main role of the Golgi apparatus thus appears to be secretion and polysaccharide synthesis.

Ribosomes. Ribosomes are the small (150–250 Å diameter) bodies that contain the machinery for protein synthesis in the cell (see page 39). They may be scattered free throughout the cytoplasm, but are often found associated with the endoplasmic reticulum, giving it a rough appearance (see Figures 3-1, 3-4, and 3-9). Protein synthesis occurs on the ribosomes, and the newly formed proteins may be liberated into the cytoplasm or passed in some manner through the membrane of the endoplasmic

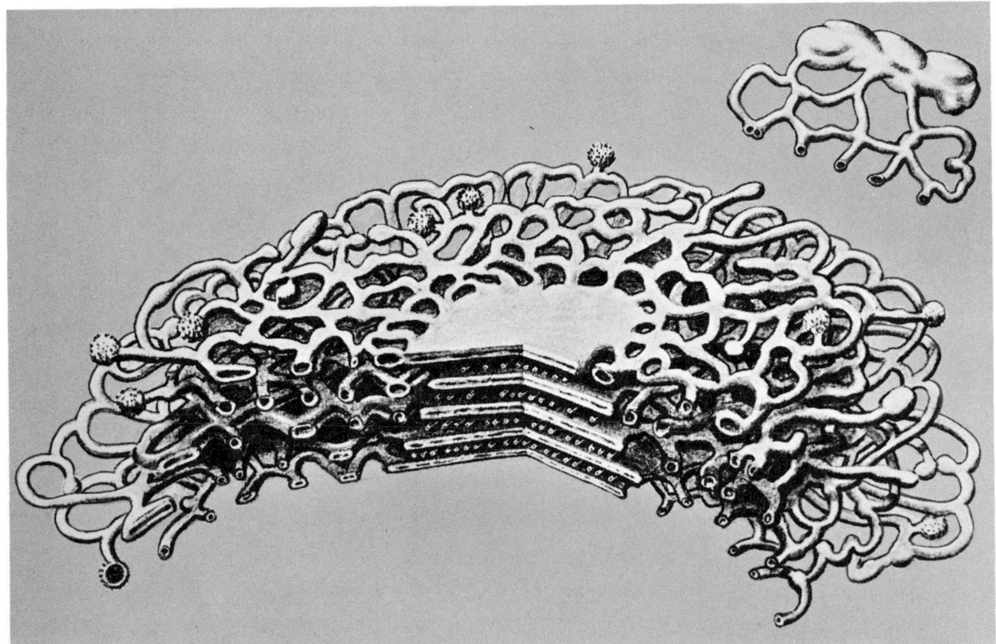

Figure 3-10. Diagrammatic interpretation of a plant dictyosome or Golgi apparatus. Insert shows a forming secretion vesicle. [From H. H. Mollenhauer and D. J. Morré: Golgi apparatus and plant secretion. *Ann. Rev. Plant Physiol.*, **17**:27–46 (1966). Used with permission. Photograph courtesy Dr. H. H. Mollenhauer.]

reticulum. It has been found that larger organelles, such as chloroplasts, mitochondria, and the nucleus, have ribosomes inside them associated with their internal membranes.

Mitochondria. These larger, usually oval, structures (about 4–7 μ long and 0.5–1 μ in diameter) contain much of the cell's metabolic machinery. They are present in large numbers in metabolically active cells but are not abundant in senescent or resting cells. The mitochondrion is made of what appears to be a normal double unit membrane. The inner layer of this membrane is deeply folded inward to form the **cristae,** transverse membranes that lie more or less crosswise in the mitochondria (see Figure 3-11). The membranes appear to have small knoblike structures about 70 Å in diameter, called F_1-ATPase, attached to their inner surface by stalks about 30 Å long, called F_0 particles. These structures are concerned with the synthesis of ATP, the cell's energy-mobilization compound. Mitochondria provide the energy, through the controlled breakdown of respiratory substrates, for the synthesis of a large part of the cell's ATP, which is in turn used to drive energy-requiring syntheses and reactions. These processes will be discussed in detail in Chapter 5.

Mitochondria have a certain degree of autonomy; they contain DNA. Plant mitochondria have sufficiently long strands of double helix to account for the informational programming of a considerable portion of their structure. However, it has been suggested that the DNA in mitochondria may be concerned only with the synthesis of some (possibly structural) proteins and that the majority of enzymic proteins is more probably programmed by nuclear DNA.

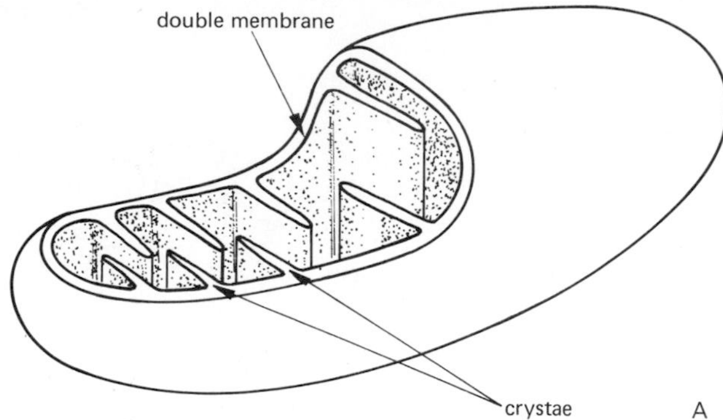

Figure 3-11. The mitochondrion. **A.** Diagram of a mitochondrion. **B.** Diagram of the structure of the mitochondrial membrane. **C.** Electron micrograph of a mitochondrion. [Original photograph courtesy of Dr. G. P. Morris, Queen's University, Kingston, Ontario, Canada.]

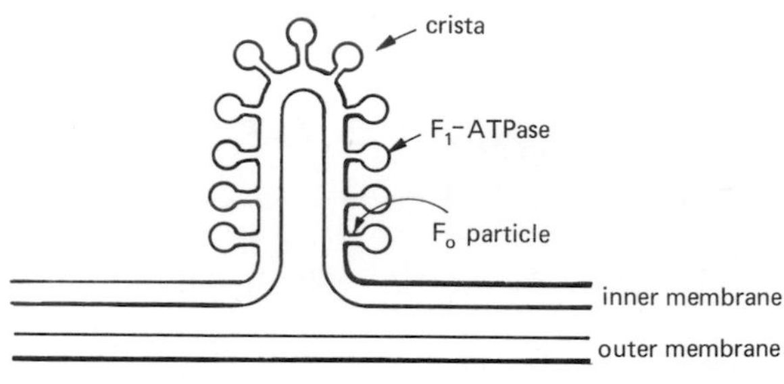

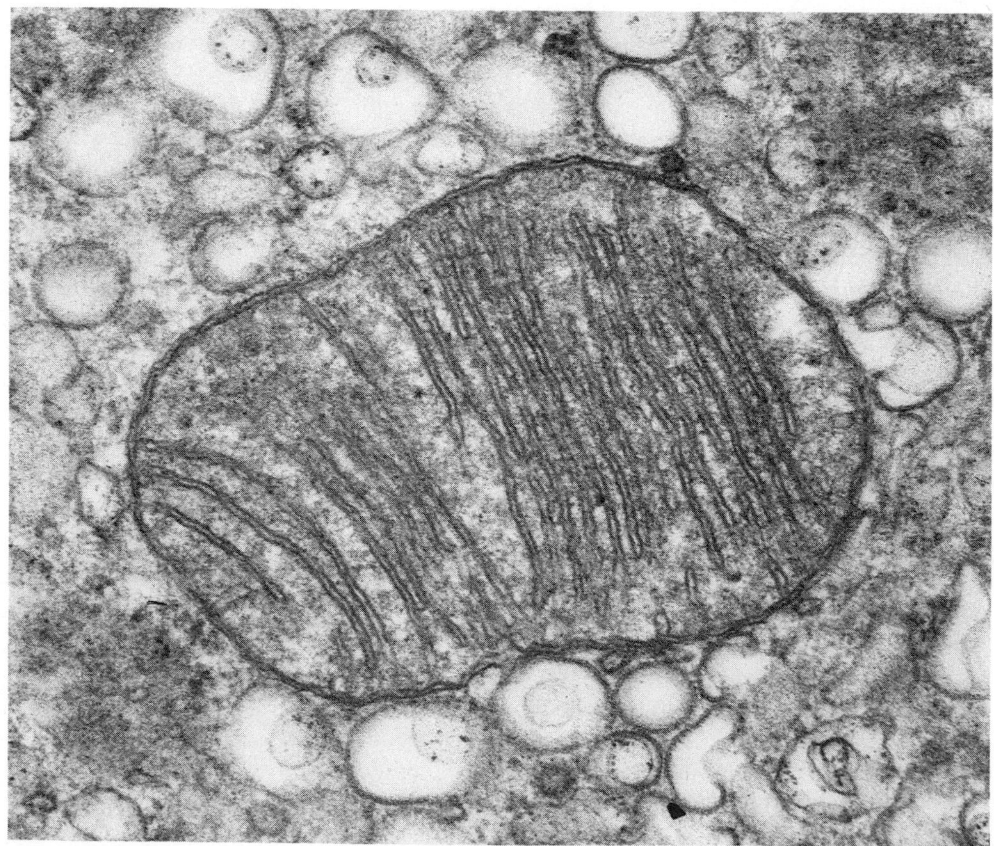

Plastids. These structures are present in many plant cells. Most familiar are **chloroplasts,** which contain the photosynthetic pigments, mainly chlorophylls, and carry on photosynthesis. **Leucoplasts** are colorless, often the site of starch granule development, and in such instances they are called **amyloplasts.** Leucoplasts and chloroplasts are interconvertible, and the exact nature of the plastid may depend on the presence or absence of light. **Chromoplasts** are specialized plastids, often angular or irregular in shape, that contain pigments other than chlorophyll and are not involved in photosynthesis. The characteristic red color of rowan (mountain ash) berries and tomatoes, for example, is due to chromoplasts that contain carotene.

Chloroplasts are usually 3–6 μ in diameter and may be very numerous in photosynthetic cells. However, a variety of spherical or disc-shaped chloroplasts are known, and some algal cells contain only one or two giant chloroplasts which may nearly fill the cell and which may assume a range of shapes from oval through star-shaped to the spiral ribbons found in *Spirogyra.*

The internal structure of chloroplasts is highly complex, as shown in Figure 3-12. Large numbers of flattened, sacklike structures called **thylakoids** (*thylakos,* a pouch) intersperse the **stroma** or ground substance. Each thylakoid is bounded by a single membrane, but because of the flatness of these structures they appear as double-membrane layers or **lamellae.** At more or less frequent intervals densely packed stacks of thylakoids, called **grana,** occur. The granal thylakoids are connected or continuous with intergranal, or stromal, thylakoids at one or more points at their edges. Chlorophyll molecules and the machinery for trapping light energy are located in the thylakoids, primarily in the grana. The enzymes which catalyze the carbon reactions of photosynthesis are largely present in the stroma.

Chloroplasts originate from minute ameboid bodies called **proplastids,** which begin to develop their internal structure by invagination of the inner layer of their double outer membrane to form vesicles. These coalesce and form a structure called the **prolamellar body.** In light, plastid development continues with the formation of thylakoids from the prolamellar body. These fuse to form grana and become green. In darkness, however, the lamellar system does not develop, and no chlorophyll is formed. Light is necessary not only for the synthesis of chlorophyll but also for the elaboration of the internal structure of the chloroplast.

Chloroplasts contain a substantial amount of DNA and are evidently capable of programming the synthesis of some of their own structural components. Chloroplast division appears to be a common phenomenon, and the chloroplasts in photosynthetic cells probably originate from the division of existing chloroplasts as well as from proplastids. The question whether chloroplasts ever arise de novo has not been finally settled. However, the consensus of current opinion is that plastids can only arise from proplastids or other plastids. If the chloroplasts are eliminated from a culture of *Euglena* cells by chemical means, that culture never regains its chloroplasts or its capacity for photosynthesis.

Glyoxysomes and Peroxisomes. These are recently discovered microscopic bodies that have now been found in many plant cells. They are electron-dense, usually approximately spherical, about 1 μ diameter, and bounded by a single membrane. They appear to be essentially "packaged units" of enzymes concerned with a specific sequence of reactions, much as mitochondria are concerned with Krebs cycle oxidation and ATP synthesis and chloroplasts with photosynthesis. Glyoxysomes contain the enzymic

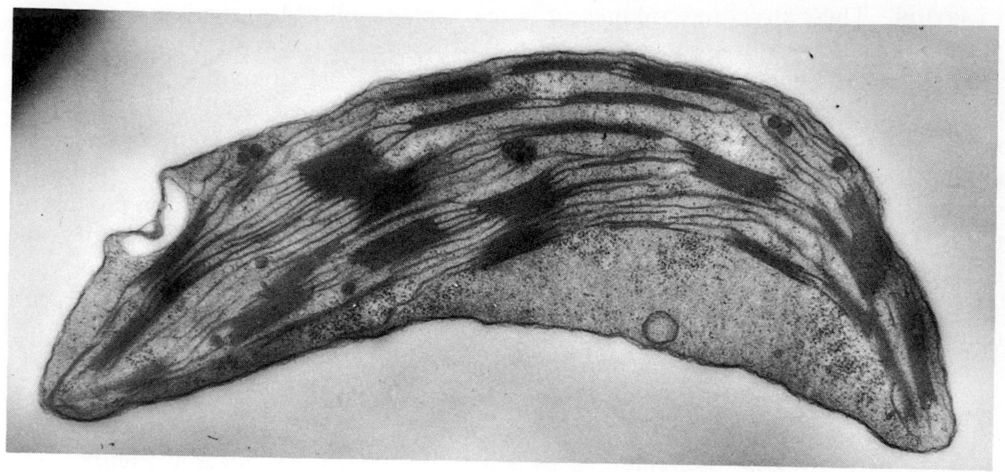

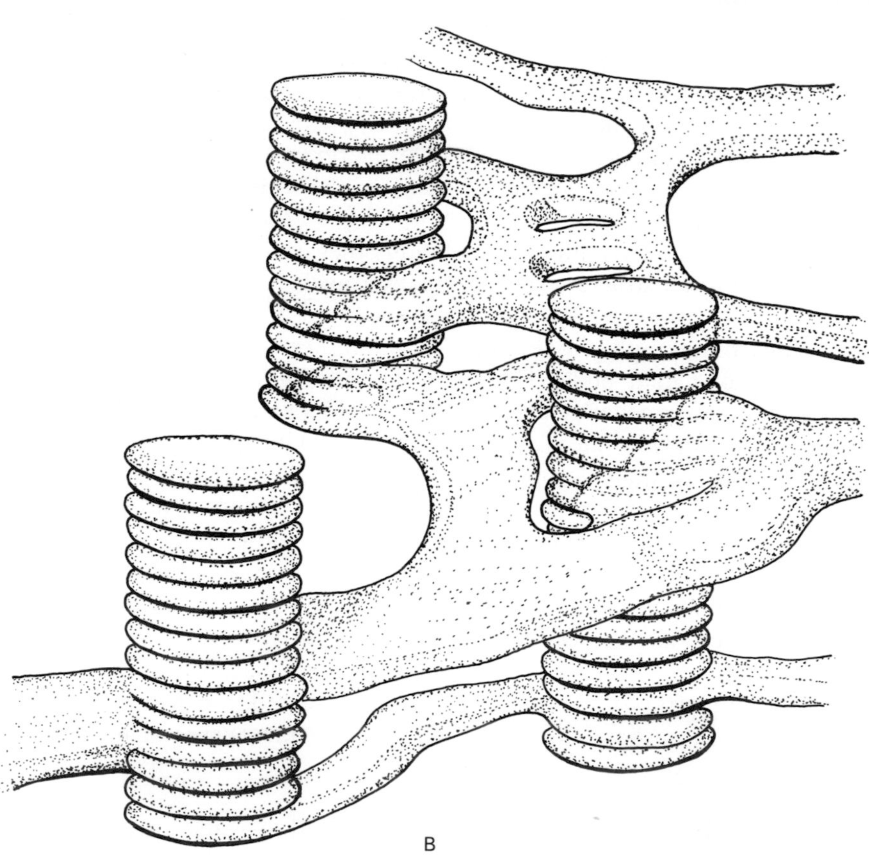

Figure 3-12. The chloroplast. **A.** Electron photomicrograph of a spinach chloroplast. [Courtesy Dr. B. F. Grant. Micrograph courtesy Mrs. E. Paton.] **B.** Diagram of grana and fret structure in a chloroplast.

machinery of the glyoxylate pathway of fat metabolism, which is important in the conversion of fats to sugars (see Chapter 6, pages 122 and 125). This process occurs primarily during the germination of fat-storing seeds, and glyoxysomes are found in large numbers in the cells of seeds such as castor beans. Glyoxysomes appear to be formed from vesicles derived from the cell's ER.

Peroxisomes are essentially similar to glyoxysomes but contain principally the enzymic machinery for the oxidation of glycolate produced in photosynthesis, and for other reactions that are part of the process of photorespiration (see Chapters 7 and 15, pages 175 and 349). They also contain the enzyme, catalase, which breaks down the poisonous substance, hydrogen peroxide, formed during the oxidation of glycolate. Peroxisomes, which occur principally in the leaves of higher plants, are probably formed during the development of the plant from glyoxysomes.

Other Subcellular Structures. In addition to the organelles described previously, various other internal structures may sometimes be discerned in cells. A **centrosome** is present in the cells of primitive plants; it is associated with the mechanism of cell division and probably also with flagellae or cilia that are present on the motile generative cells of more primitive plants. Most cells contain **microtubules,** about 200–300 Å in diameter, that are associated with cell wall synthesis and the transport of materials (see Figures 3-2 and 3-4). Microtubules are involved in the movement or alignment of cell components such as chromosomes during cell division and the vesicles containing polysaccharides that are derived from the Golgi apparatus and are destined for cell wall synthesis.

The electron microscope also reveals a number of ultramicroscopic bodies or granules whose function is unknown. Some of these may be aggregates of proteins and enzymes that conduct organized sequences of metabolic reactions, similar to glyoxysomes and peroxisomes.

The Vacuole. The vacuole is physiologically important to the cell for two reasons. It affords a storage place for materials not immediately required, and it provides a dumping ground for cellular wastes and other noxious substances that plants, lacking an excretory system, must store internally. Hydrolytic or destructive enzymes are secreted into the vacuole; there, these enzymes degrade waste material into simple substances that may be reabsorbed by the cytoplasm for reuse. Second, the vacuole functions as the water reserve in the cell; it maintains the cell's structure and rigidity by acting as an internal balloon, which, by exerting pressure on the cell wall, prevents it from distorting or collapsing. The mechanism for this will be discussed later in this chapter, beginning on page 65.

The vacuolar membrane, the **tonoplast,** is obviously a most important part of the membrane system of the plant. This typical unit membrane may be involved in the secretion of substances into the vacuole. In addition, vesicles of the endoplasmic reticulum or the Golgi apparatus are apparently able to coalesce with the tonoplast and thus eject their contents directly into the vacuole.

The vacuole may contain a range of dissolved substances: sugars, salts, acids, nitrogenous compounds, such complex compounds as alkaloids, glycosides, and the anthocyanin pigments. Small droplets or emulsions of fats, oils, and other water immiscible substances, as well as tannins, assorted polysaccharides and proteins may also be found. Crystalline deposits are also common in vacuoles of mature cells, crystals of calcium oxalate being the most usual. The presence of this complex and highly poisonous chemical dumping ground in plant cells is one of the main reasons why plant biochemistry has lagged behind animal biochemistry. The isolation of proteins, enzymes, and subcellular organelles or particles is complicated by the fact that, on disruption of the cell, the highly sensitive proteins and organelles are polluted by the vacuole, which contains many powerful precipitating or denaturing elements. The pH of vacuoles is often very different from that of cytoplasm (which is usually 6.8–8.0) and may range from as low as 0.9 to as high as 9 or 10, although alkaline values are rare. Acidic cell sap is common, usually as the result of moderate concentrations of organic acids such as citric, oxalic, or tartaric. Some microscopic algae maintain 1 *N* sulfuric acid in their vacuoles, but this is an extreme case!

Immature or actively dividing cells do not have the large, prominent, single vacuole characteristic of mature cells; instead they contain several very small vacuoles scattered through the cytoplasm. As the cell develops and matures, the small vacuoles coalesce and expand until, in the majority of fully developed cells, the large central vacuole occupies 80 to 90 percent of the total cell volume.

Water and Cells

This discussion will cover the basic mechanisms of water movement in cells and lay the groundwork for the study of water metabolism and physiology in Section III.

WATER POTENTIAL. To move requires energy. Water, like all other substances, does not move against an energy gradient; it must move down an energy gradient, giving up energy as it moves. So long as energy can be lost as a result of water movement, movement will continue. Equilibrium can only be reached when further movement does not result in any further loss of energy. This means that water always moves toward the region of lowest energy in a system. It is necessary to understand the nature of the energy and energy gradients involved in order to calculate the forces by which water moves.

Free energy is defined as the energy available (without change in temperature) to do work. The **chemical potential** of a substance under any condition (that is, whether pure, in solution, or as a member of a complex system) is the free energy per mole of that substance. Chemical potential thus measures the energy with which a substance will react or move.

Water potential is the chemical potential of water and is a measure of the energy available for reaction or movement. Under normal biological conditions the water potential is usually high enough not to limit the rates of reaction involving water (for example, in hydrolytic reactions). However, water movement depends on its potential, because the net movement of water is always from a region of higher potential to a region of lower potential. The symbol for water potential is ψ,* and it has traditionally been measured in atmospheres (atm), bars, or dynes per square centimeter (dynes/cm^2): 1 bar = 10^6 dynes/cm^2 = 29.53 in. Hg = 0.985 atm; 1 atm = 14.69 lb/in.^{2} = 1.01 bars.

A difference in water potential ($\Delta\psi$) between two regions, A and B, having water

*ψ is the Greek letter psi. A good way to remember this symbol is to recall that psi also means pounds per square inch, a measure of pressure.

potentials ψ_A and ψ_B, would be expressed $\Delta\psi = \psi_A - \psi_B$. If ψ_A is greater than ψ_B, $\Delta\psi$ is positive and water will move from A to B. If the value of $\Delta\psi$ from the equation is negative, water will move from B to A. This restates the principle given above: water moves from a region of higher potential to a region of lower potential.

The potential of pure water is, by definition, zero.* The presence of any substance dissolved in water *lowers its potential,* so that the *water potential of a solution is less than zero.* This definition only holds at atmospheric pressure. Raising or lowering the pressure around a system automatically raises or lowers the water potential by exactly the same amount.

Diffusion Molecules of gas, or of a solute in solution, are continuously in motion and tend to assume a uniform distribution throughout all the available space. Thus molecules move from a region of higher potential to one of lower potential, the process being called diffusion. Thus, for example, in an imperfectly mixed solution water molecules would diffuse down the ψ gradient from the region of more dilute solution (where the water molecules have higher ψ) to regions of more concentrated solution (where they have lower ψ). Similarly, the solute molecules would also diffuse down their concentration gradients (that is, from an area of higher concentration to an area of greater dilution) until the solution was perfectly uniform throughout.

The rates of diffusion are proportional to the kinetic energy of the molecules (their temperature), their size (diffusion rate is proportional to the square root of the molecular weight), the density of the medium through which they pass, and the gradient of concentration over which they diffuse. When uniform distribution of molecules occurs, a **dynamic equilibrium** is established and net movement of molecules ceases (although there is continuous random movement or diffusion of molecules within the framework of the equilibrium).

Differentially Permeable Membranes. Many biological membranes, particularly the plasmalemma, tonoplast, and the membranes surrounding subcellular organelles, exhibit the property of **differential permeability.** That is, because of their physical or chemical nature, water molecules pass readily through these membranes whereas molecules of substances dissolved in the water either cannot penetrate or do so more slowly than water molecules. A membrane that is almost totally impermeable to solute molecules while it is permeable to the solvent is called a **semipermeable membrane.** Most biological membranes, however, are differentially permeable rather than semipermeable.

Osmosis. Suppose a beaker or container is separated into two parts by a differentially permeable membrane, as shown in Figure 3-13A. If pure water is put on one side of the membrane and a sugar solution on the other, the water potential (ψ) on the side containing pure water will be higher than that on the other side. Sugar cannot diffuse across the membrane, but water can. Water will diffuse from the side having higher ψ (pure water) to the side having lower ψ (sugar solution) as shown in Figure 3-13B. This diffusion of water across a differentially permeable membrane from a region of higher

*This applies only to *free* water—that is, water molecules that are not bound or associated by physical or chemical forces with other substances (like the water of hydration or water in colloids). See Imbibition (page 67).

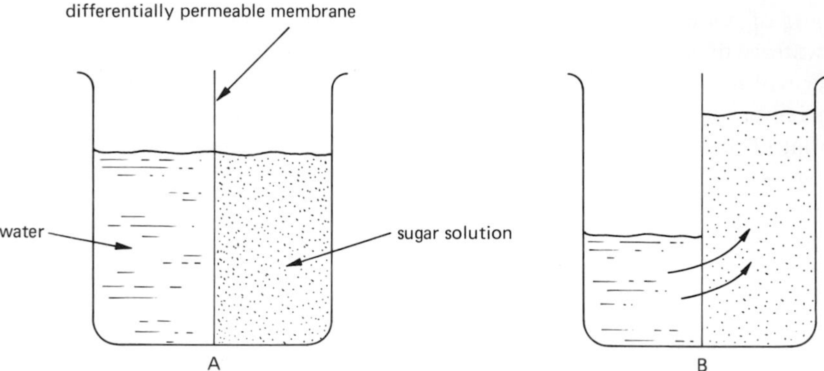

Figure 3-13. Water movement as a result of osmosis. Water diffuses down a potential gradient from a region of high potential (pure water, $\psi\pi = 0$) to a solution where the potential is lower (has a negative value).

potential (pure water or weak solution) to one of lower potential (a more concentrated solution) is called **osmosis.**

Osmotic Potential and Pressure Potential. Since water diffusing down a potential gradient is losing energy, it can be made to do work. In Figure 3-13B this is suggested by the fact that osmosis results in the transfer of water to the more concentrated solution, raising the water level in that compartment of the beaker.

Suppose now that a solution of sugar is enclosed within a sack or "cell" made of an artificial differentially permeable membrane (it could be made from a thin film of such substances as cellophane, collodion, and so on). The artificial cell is then placed in a beaker of pure water, as shown in Figure 3-14A. Water diffuses into the cell by osmosis until the cell becomes swollen or **turgid,** and the distended walls of the cell exert a pressure on the cell contents, as shown in Figure 3-14B. The pressure exerted on the liquid by the walls of a turgid cell is called **turgor pressure.** Water is now entering the cell by osmosis against a pressure gradient, so it is doing work. The potential with which pure water will diffuse toward a solution is the **osmotic potential** of that solution, called ψ_{π}. Since water diffuses from high potential (zero in pure water) to lower potential, the

Figure 3-14. Water moves by osmosis into an artificial "cell" containing sugar solution **(A)** until the cell is swollen so that its walls exert pressure on its contents, squeezing water out. At equilibrium **(B)** the pressure of water entering by osmosis is equal to the pressure squeezing water out.

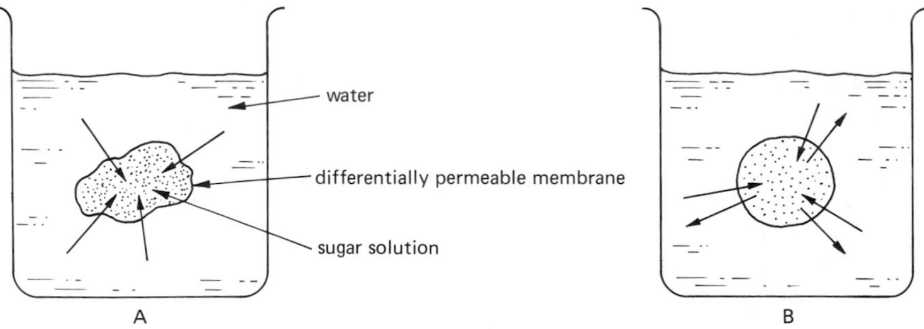

osmotic potential of a solution is always negative. The osmotic potential is thus a measure of the actual pressure that can be generated in a cell by water diffusing in by osmosis.

The result of the turgor pressure generated in the cell (Figure 3-14B) is that water is literally being squeezed out of the cell. This is another way of saying that water diffuses out of the cell down a pressure gradient. The water in the cell thus has a **pressure potential** that is positive and higher than the pressure potential of the water outside. The symbol for pressure potential is ψ_P. The pressure potential for water at atmospheric pressure is, by definition, zero. Thus values of ψ_P can range from negative to very high positive values.

When the system in Figure 3-14B is at equilibrium, the water potential is the same in all parts of the system, thus

$$\psi \text{ (outside)} = \psi \text{ (inside)}$$

But the water potential has two components, osmotic potential (ψ_π) and pressure potential (ψ_P), so at equilibrium

$$\psi_\pi \text{ (outside)} + \psi_P \text{ (outside)} = \psi_\pi \text{ (inside)} + \psi_P \text{ (inside)}$$

The external solution in Figure 3-14 is pure water at atmospheric pressure; therefore, both ψ_π and ψ_P are zero. Thus, at equilibrium

$$-\psi_\pi \text{ (inside)} = \psi_P \text{ (inside)}$$

This is another way of restating the fact that the turgor pressure developed in the cell under these conditions is numerically equal, but opposite in sign, to the osmotic potential of the fluid contained in it. If the external fluid is not water but a solution (having ψ_π less than zero), then the turgor pressure will be measured by the difference between the osmotic potential of the solutions inside and outside the cell

$$\psi_P = \Delta\psi_\pi \quad (= \psi_\pi \text{ outside} - \psi_\pi \text{ inside})$$

We are now in a position to see how these properties are measured, and how these concepts can be applied to the study of water in cells.

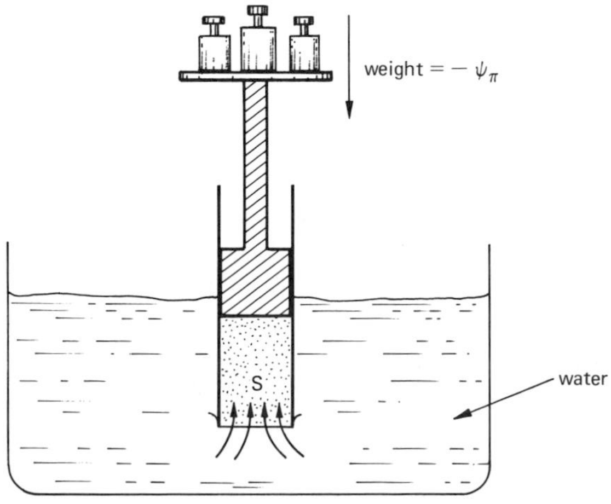

Figure 3-15. Simple apparatus for measuring the osmotic potential ($\psi\pi$) of solution S. It consists of a cylinder, having a watertight sliding piston and a differentially permeable (or semipermeable) membrane at the end, immersed in pure water ($\psi\pi = 0$).

Measuring ψ_π. The osmotic potential of a solution can be measured in an **osmometer** (a device that measures the pressure developed by osmosis) as shown in Figure 3-15. Pfeffer, using such a system, early determined that the pressure developed (P) is proportional to the solute concentration. Concentration can be expressed as $1/V$, where V is the volume of solution containing a given amount of solute.

$$P = \frac{k_1}{V} \quad k_1 = \text{a constant}$$

Van't Hoff observed that the system is sensitive to temperature T (expressed in degrees absolute) because the kinetic energy of molecules is proportional to temperature.

$$P = k_2 T \quad k_2 = \text{a constant}$$

Since these equations describe solute molecules free to diffuse as if they were a gas, the constants k_1 and k_2 can be replaced by the gas constant R, and the formulas may be combined

$$P = \frac{RT}{V}$$

P is the pressure developed in an osmometer. Because this is equal and opposite to osmotic potential at equilibrium

$$\psi_\pi = -\frac{RT}{V}$$

This relationship describes the ψ_π of a solution as if the solute molecules were gaseous. One mole of gas occupies 22.4 liters at standard conditions of temperature and pressure (STP = 1 atm, 0° C). But in a 1 M solution, 1 mole of solute occupies 1.0 liter. Thus, if it were a gas, the pressure of the solute would be 22.4 atm; therefore, the osmotic potential of a 1 M solution should be 22.4 atm. In fact, this relationship only holds for dilute solutions because other complicating factors affect the osmotic potential of concentrated solutions. Further, the osmotic potential is not proportional to molarity because, as the concentration of solute increases, the concentration of solvent decreases. This is why molality (m), which describes the relative proportions of solute and solvent, is often used to describe osmotic solutions. Finally, we have assumed that the solute is not ionized, that is, there is only one particle per molecule. The relationship developed above refers to the number of particles in solution, not to the number of molecules. Thus, a substance that undergoes complete ionization into two ions has an osmotic potential twice that of a nonionizing substance, and a salt having three ions, such as sodium sulfate (Na_2SO_4), would have three times the osmotic potential if it ionized completely.

The actual osmotic potential of a solution can be measured in an osmometer, but it is also possible to measure it by indirect means. If an unknown solution is placed in a closed chamber under controlled conditions, its water potential will come into equilibrium with the water potential of the air above it in the chamber. The relative humidity (RH) of the air is measured with an extremely sensitive hygrometer. The water potential of the solution can then be determined from the formula

$$\psi \text{ bars} = -10.7 \log 100/\text{RH}$$

Figure 3-16. A. Cell in hypotonic solution: ψ_π outside $>$ ψ_π inside; water diffuses in. **B.** Cell in isotonic solution: ψ_π outside $=$ ψ_π inside; no water movement. **C.** Cell in hypertonic solution; the cell is plasmolyzed: ψ_π outside $<$ ψ_π inside; water diffuses out.

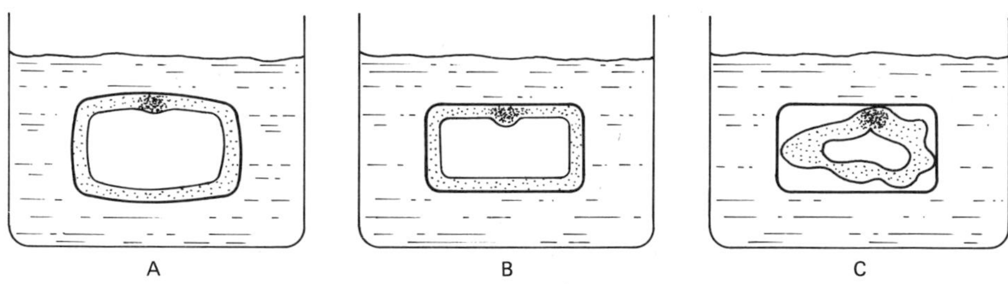

If the experiment is done at atmospheric pressure, $\psi_P = 0$ and $\psi = \psi_\pi$. A second method is to determine the freezing point depression of a solution. A 1 *m* solution of a non-ionized substance has a theoretical osmotic potential of 22.4 atm, and its freezing point is 1.86° C below that of pure water. Therefore

$$\psi_\pi = -22.4 \times \frac{\text{observed freezing point depression}}{1.86}$$

Water Potential in Cells. The concepts we have developed with an artificial cell containing a sugar solution can be transferred directly to a real cell, as shown in Figure 3-16. The membranes that surround the cell are differentially permeable, and osmosis takes place across them. If the cell is placed in a dilute solution or pure water, whose ψ_π is very high (that is, approaching zero), water will diffuse in and the cell will become turgid, as shown in Figure 3-16A. The external solution, whose concentration of solutes is less than that of the cell sap, is said to be **hypotonic** (*hypo,* less than). If the cell is placed in a solution whose ψ_π is equal to that of the cell sap, an **isotonic** solution (*iso,* the same), then no net water diffusion takes place and the cell is **flaccid** or lacks turgor (Figure 3-16B). If the external solution is more concentrated than the cell sap, or **hypertonic** (*hyper,* more than), its ψ_π is lower than that of the cell sap and water will diffuse out. Since the cell wall is relatively rigid, the protoplasm will pull away from the wall as it shrinks and the cell will become **plasmolyzed,** as shown in Figure 3-16C. Plasmolysis does not necessarily do permanent damage to the cell. If the cell is again placed in a hypotonic solution, it will quickly regain its lost water and turgor by osmosis. If the period and severity of plasmolysis are not too great, the cell will probably not be damaged.

We have seen that the water potential of the cell has two components, osmotic and pressure potentials, such that

$$\psi = \psi_\pi + \psi_P$$

When a cell is placed in water or a solution and comes to equilibrium, the water potential of the cell (ψ inside) is equal to the water potential outside (ψ outside).

$$\psi_\pi \text{ (inside)} + \psi_P \text{ (inside)} = \psi \text{ (inside)} = \psi \text{ (outside)}$$

ψ (outside) is also the sum of ψ_π (outside) and ψ_P (outside). At atmospheric pressure, since $\psi_P = 0$, then ψ (outside) $= \psi_\pi$ (outside). Hence at equilibrium

$$\psi_\pi \text{ (inside)} + \psi_P \text{ (inside)} = \psi_\pi \text{ (outside)}$$

This may be restated as

$$\psi_\pi \text{ (inside)} = \psi_\pi \text{ (outside)} - \psi_P \text{ (inside)}$$

which gives the osmotic potential of the cell sap in terms that can be measured.

The osmotic potential of cell sap may be measured by the freezing point depression or relative humidity method. However, simple direct methods are possible using the above relationship. A method often used is to make a graded series of solutions of known concentration and known osmotic potential (sucrose or mannitol are often used for this purpose). Small pieces of tissue are placed in each solution and examined microscopically after they have had time to reach equilibrium. As the solutions become stronger, the cells will be less and less turgid until some of them show signs of plasmolysis (that is, **incipient plasmolysis**). The solution in which 50 percent of the cells* show some indication of plasmolysis has approximately the same osmotic potential as the cell sap, since at plasmolysis ψ_P (inside) = 0, and

$$\psi_\pi \text{ (inside)} = \psi_\pi \text{ (outside)}$$

The turgor pressure (ψ_P) of cells can now be measured using similar techniques. Pieces of tissue of carefully measured length or weight are placed in graded solutions as before, and the change in size or weight is measured after the tissue reaches equilibrium. In the solution in which no size change takes place, the water potential of the cell is equal to that of the solution (since no water moved in or out), thus

$$\psi \text{ (outside)} = \psi_\pi \text{ (inside)} + \psi_P \text{ (inside)}$$

Since the ψ_π (inside) is known (having been determined previously), ψ_P (inside) can be calculated from the relationship

$$\psi_P \text{ (inside)} = \psi_\pi \text{ (outside)} - \psi_\pi \text{ (inside)}$$

The foregoing treatment did not include the fact that the cell wall is not entirely rigid, but elastic, so that the volume of the cells will increase as the turgor increases. The resulting relationship between ψ, ψ_π, ψ_P, and cell volume is illustrated in Figure 3-17. It may be seen that the osmotic potential of the cell sap, -12 bars in the flaccid cell, increases by dilution with water to about -8 bars as the cell expands to about 1.5 times its flaccid size. In the fully turgid cell the turgor pressure (or pressure potential) equals minus the osmotic potential, 8 bars, and the water potential of the cell sap rises from -12 bars (equaling the osmotic potential) in the flaccid cell to 0 in the turgid cell.

Movement of Water Between Cells. Water moves in and out of a cell because of differences in water potential ($\Delta\psi$) between the cell and its surrounding solution. Similarly, water can move from cell to cell by diffusing down a water potential gradient between the two cells. Thus the direction of movement of water and the force with which it will move are dependent on the water potential in each cell and, consequently, on the difference in water potential between them.

This may best be illustrated by an example: Cell A has a pressure potential (turgor pressure) of 5 bars and contains sap with an osmotic potential of -12 bars. Cell B has a

*This will give an approximation of the average or mean of all the cells; some will have higher or lower ψ_π than the average.

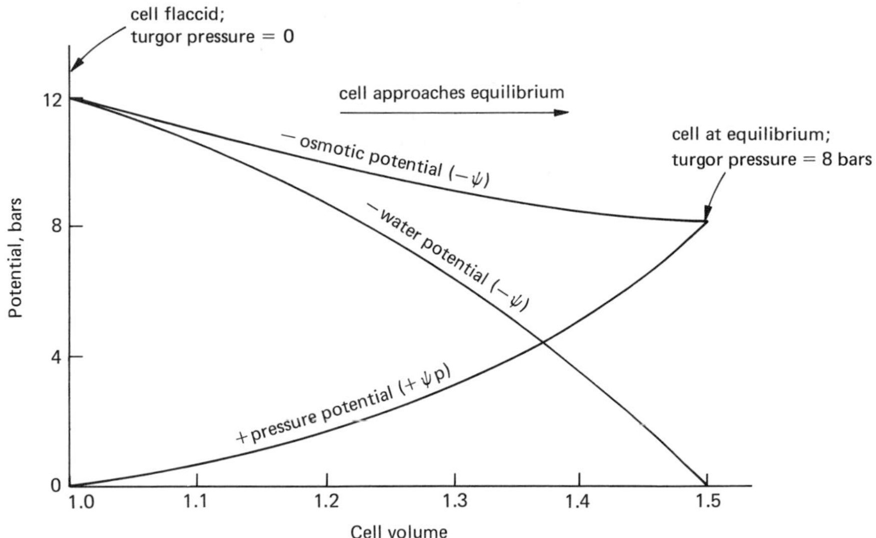

Figure 3-17. Water potential changes in an initially flaccid cell as it comes to equilibrium after being placed in pure water. Note that the osmotic and water potentials are negative values, whereas the pressure potential is positive.

pressure potential of 3 bars and an internal solution whose osmotic potential is −6 bars. If these two cells are in direct contact, which way will water move and with what force? The water potential in each cell can be described as follows

$$\begin{aligned} \text{A:} \quad & \psi = 5 - 12 && = -7 \text{ bars} \\ \text{B:} \quad & \psi = 3 - 6 && = -3 \text{ bars} \\ \text{A} - \text{B:} \quad & \Delta\psi = -7 - (-3) && = -4 \text{ bars} \end{aligned}$$

Water will move from cell B to cell A (toward the lower, or more negative, water potential) with a force of 4 bars.

The value of $\Delta\psi$ is of importance because it is directly proportional to the rate at which water will move between cells. Water moves in direct proportion to the driving force, $\Delta\psi$, and to the area of the membrane through which it is moving; it moves in inverse proportion to the resistance of the membrane. The factors of membrane area and resistance are approximately constant for a given cell; consequently, the rate (hence also the amount in a given time) of water movement is dependent on the difference in water potential, $\Delta\psi$, on either side of the membrane.

Imbibition. The process of imbibition is actively involved in water uptake under certain circumstances. This is movement of water from an area of high potential to an area of low potential, but without the assistance of a differentially permeable membrane. In addition, forces of attraction, usually either chemical or electrostatic, are involved in imbibition. Solvents are usually imbibed only into materials with which they have an affinity; water into proteins, for example, and acetone into rubber. The pressures generated by imbibition, caused by swelling of the imbibant, may be very great—imbibition pressure in a germinating seed splits the seed coat, and a seed wedged in a rock crevice may split the rock with the pressure of its imbibition of water. Imbibition of water by colloidal materials in cells helps them to withstand severe conditions of drought because of the tenacity with which imbibed water is held.

Since water moves under the influence of imbibition, the water potential (ψ) must be affected by these forces. The term **matric potential,** written ψ_M, is used to account for all the forces causing imbibition or holding water in a matrix of any sort. Thus the potential of water in a matrix (for example, in a colloid, in soil, or held in any way by surface-acting forces or imbibition) can be defined

$$\psi = \psi_\pi + \psi_P + \psi_M$$

THE OLD APPROACH TO OSMOSIS AND WATER MOVEMENT. Until recently osmosis was often explained on the basis of water diffusion from a region of high water concentration (for example, pure water) to one of lower water concentration (for example, a solution). However this is not correct because some solutions occupy a smaller volume than the same weight of pure water. In addition, by the old concept a solution in a cell or an osmometer was considered as if it *sucked* water into the cell by a force that was thought of as a negative pressure. A number of terms, common in the older literature, were developed to describe these concepts. These have now been largely abandoned in favor of the present terminology, which is based on more satisfactory thermodynamic concepts. The older terms are given here so that you can understand them if you come across them in your reading.

Term used in this book	*Equivalent older term*
Water potential (ψ)	Diffusion pressure deficit; suction pressure
Pressure potential (ψ_P)	Turgor pressure; wall pressure
Osmotic potential (ψ_π)	Osmotic pressure; osmotic concentration (these are positive terms, equal but opposite in sign to ψ_π).

Growth of Cells

The growth of plants will be considered later in Section IV, but the basic mechanisms of cell growth should be considered briefly here. The growth of plants takes place by three basic events that may occur simultaneously: by cell division, by cell enlargement, and by cell differentiation, as illustrated in Figure 3-18.

Cell division (Figure 3-18A) involves the duplication of nuclear DNA, the pairing and duplication of chromosomes, and the separation of two daughter nuclei. During telophase a number of vesicles, probably derived from the endoplasmic reticulum and the Golgi apparatus, line up across the cell in the area of the spindle and coalesce to form the cell plate, the beginning of the new common wall. The contents of these vesicles are used to make the pectic substances of the middle lamella, which eventually reaches across the cell, completing the separation of two new cells. Cellulose is now deposited in regular patterns of microfibrils, their synthesis and deposition being perhaps mediated by vesicles from the Golgi apparatus or from microtubules. During and subsequent to this process the daughter cells usually enlarge, so that each achieves the size of the

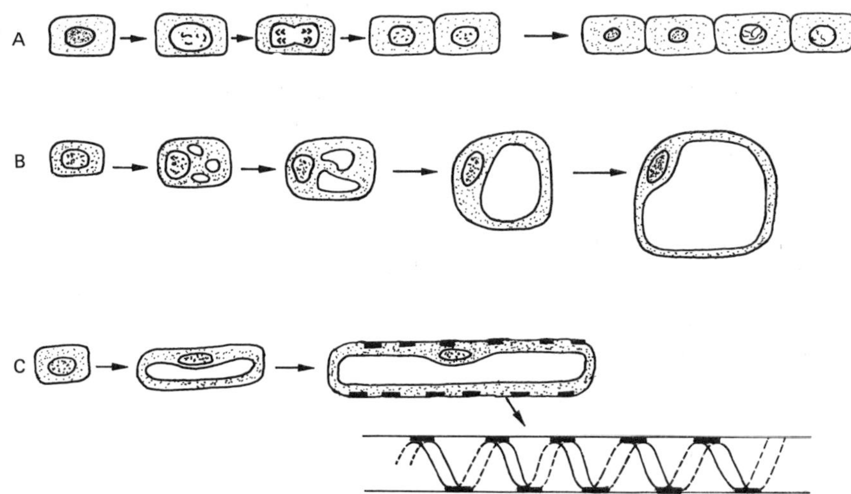

Figure 3-18. Diagrams to illustrate plant cell growth. **A.** By cell division. **B.** By cell enlargement. **C.** By cell differentiation.

original cell, by the stretching of the existing cell wall and the deposition of new material.

A cell may enlarge in a general way (Figure 3-18B) without major changes in its shape and characteristics, except that as it matures it usually develops a large vacuole and the proportion of cytoplasm decreases greatly. This type of cell is usually called **parenchyma,** and it is relatively undifferentiated. The complexity of the ultrastructure may decrease also. As the cell becomes more sluggish with age, it may lose most of its mitochondria and much of its other microcomponents. It may become highly specialized, as the photosynthetic cells of the palisade layer of the leaf (see Chapter 4), and its ultrastructure will usually reflect this specialization—in the case of the palisade cell, a great proliferation of chloroplasts.

Alternatively, the cell may grow, with or without cell division, in a highly specialized way. The illustration in Figure 3-18C represents diagrammatically the growth of a vessel element. Here the growth is in one direction only and involves the modification and differentiation of the cell into a highly distinct morphological entity. The basic processes are similar: stretching of the cell wall, deposition of many layers of oriented cellulose microfibrils, loss of much of the subcellular complexity, and development of a large vacuole.

An amazing fact about cells is that they all initially appear to have unlimited capability for growth and differentiation. All the cells in a plant are initially capable of growing in all the ways characteristic of that plant. Yet cells in different positions in the plant, though endowed with identical information from their common genetic origin, use this information in different ways to produce the multitude of different cell types in a mature plant. Evidently the cells differentiate as a result of their position in the plant, for this is the only attribute that distinguishes them from their sister cells at their formation. This capacity to recognize and react to their location in the plant is the basis of **organization,** which is one of the most impressive properties of the living organism. The concept of organization will form a central theme in the discussion of growth and differentiation in Section IV.

Additional Reading

Current cytology texts will cover the material in this chapter in greater detail. Articles on the structure and function of various subcellular organelles appear frequently in *Scientific American,* the *Annual Reviews of Plant Physiology and Biochemistry,* and as monographs.

Clowes, F. A. L., and **B. E. Juniper:** *Plant Cells.* Blackwell Scientific Publications, Oxford, 1968.

Ledbetter, M. C., and **K. C. Porter:** *Introduction to the Fine Structure of Plant Cells.* Springer–Verlag, New York, 1970.

The Living Cell (Readings from *Scientific American*). W. H. Freeman & Co., 1965.

Markham, R., R. W. Horne, and **R. M. Hicks** (eds.): The electron microscopy and composition of biological membranes and envelopes. *Phil. Trans. Royal Soc. London,* **B268:** 1–159 (1974). See particularly **W. W. Franke:** Structure and biochemistry of the nuclear envelope (pp. 67–93); **L. F. LaCour** and **B. Wells:** Nuclear pores at prophase of meiosis in plants (pp. 95–100); and **D. H. Northcote:** Membrane systems of plant cells (pp. 119–28).

Preston, R. D.: *The Physical Biology of Plant Cell Walls.* Chapman & Hall, London, 1974.

Pridham, J. B. (ed.): *Plant Cell Organelles.* Academic Press, New York, 1968.

Racker, E.: *A New Look at Mechanisms in Bioenergetics.* Academic Press, New York, 1976.

4 Structure and Growth of Familiar Higher Plants

In order that the later discussion of the growth, biochemistry, and physiological processes of plants does not suffer from the use of unfamiliar terms and concepts, a very brief description of the growth and form of typical plants and their parts will be presented here without any consideration of why things happen the way they do. The analysis of plant growth and development and the factors that control it will be considered in more detail in Section IV.

Germination

The seed is a resting structure. It is usually extremely dehydrated, largely composed of storage tissue, and surrounded by an essentially impervious cover. Metabolic processes are suspended or take place very slowly; the seed is in a state of suspended animation, mainly due to lack of water and oxygen. The process of germination is the absorption of water, the reactivation of metabolism, and the initiation of growth. A few seed coats are so impervious to water that they require extreme conditions for germination. The Kentucky Coffee Tree seed (*Gymnocladus dioica*) must be heavily scored with a file or treated with strong sulfuric acid before it will germinate, and such seeds require prolonged exposure to the weather, the action of fungi or bacteria in the soil, or even such drastic measures as exposure to a forest fire before they will germinate naturally. However, the majority of seeds begin to germinate as soon as they are wetted, providing the conditions of temperature, light, and cold pretreatment (see Chapter 22) are right.

The seed contains an embryo: one end of this embryo, the **radicle,** will form the root of the plant; the other end, the **plumule,** will form the stem and leaves. The embryo also has **cotyledons** or seed leaves (one in monocots, two in dicots, and many in gymnosperms), which may be small and occupy only a small part of the seed, as in most monocots, or may be large enough almost completely to fill the seed, as in beans and many other dicots. The seed initially contains substantial endosperm, the nutritive tissue for the embryo. In some seeds much of the endosperm may remain after germination, when it supports the nutrition of the developing embryo. In this case the cotyledons remain in the seed and function largely as absorbing organs, as in most monocots. In other seeds, particularly gymnosperms and many dicots, the process of endosperm absorption is completed before the seed is shed, and all the nutritional reserves of the

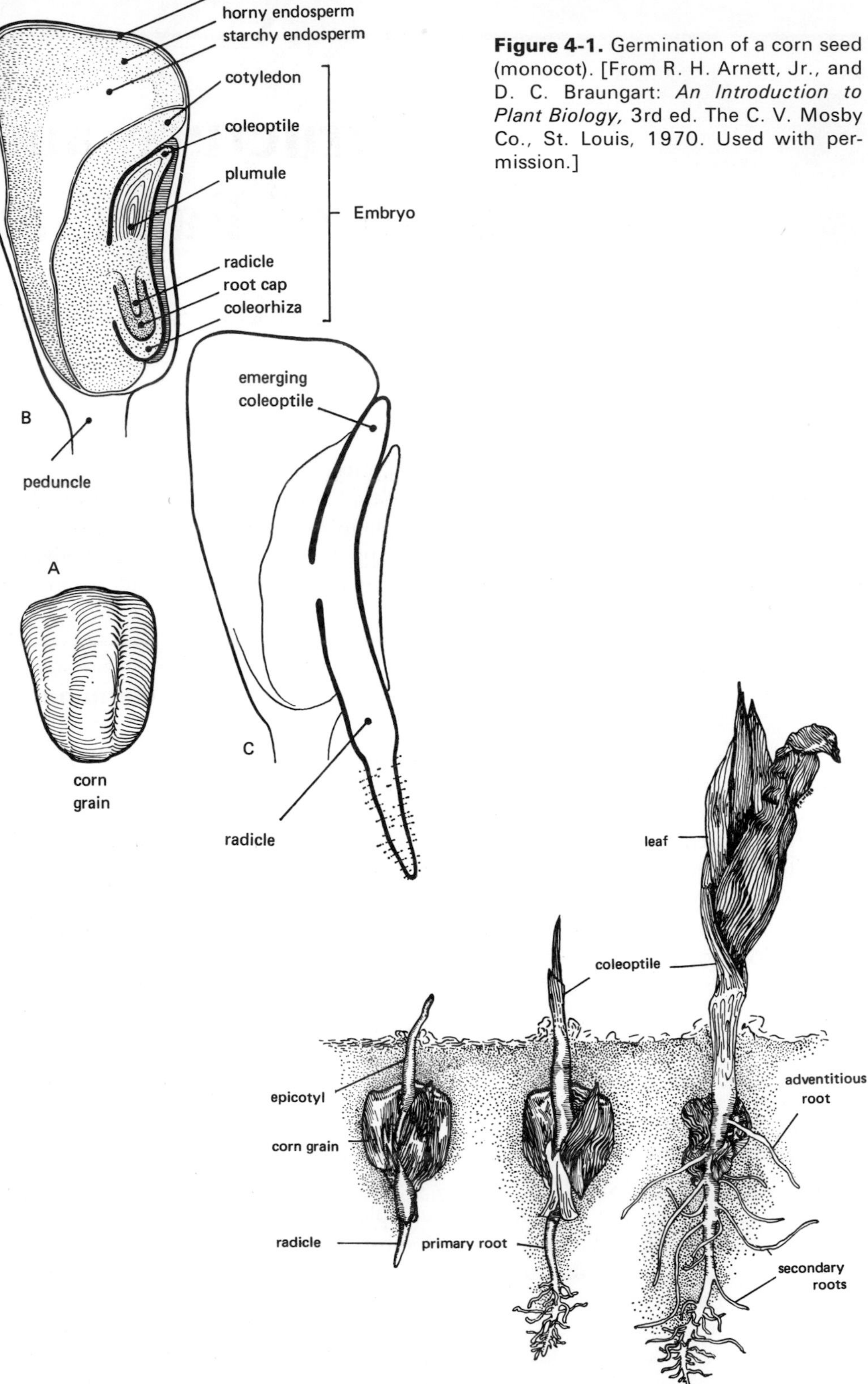

Figure 4-1. Germination of a corn seed (monocot). [From R. H. Arnett, Jr., and D. C. Braungart: *An Introduction to Plant Biology,* 3rd ed. The C. V. Mosby Co., St. Louis, 1970. Used with permission.]

seed are present in the cotyledons. In this case, the cotyledons may stay in the seed during germination, or they may be carried aloft by the growth of the embryo and subsequently they may develop into more or less normal and functional leaves.

The germination of a monocot seedling, corn (*Zea mays*), is shown in Figure 4-1. The radicle grows downward through the split seed coat to produce the **primary root,** and the shoot, encased in its protective sheath, the **coleoptile,** grows upward. When the coleoptile reaches the surface of the ground, its growth stops and the newly developing leaves of the plumule push through its top and continue to grow. The root system develops with the occasional formation of branch or **secondary roots** from the primary root, and in many monocots a strong system of **adventitious roots** may grow from the lower portion of the stem. The part of the embryo and seedling situated between the cotyledons and the radicle is called the **hypocotyl** (*hypo,* below the cotyledons), and the plumule and stem above the cotyledons are called the **epicotyl** (*epi,* above).

The germination of a typical dicot, the garden bean (*Phaseolus vulgaris*), is shown in Figure 4-2. The process is similar, except that the cotyledons are carried above the soil by a considerable extension of the hypocotyl, and instead of remaining inside the seed they turn green and become somewhat leaflike. However, as the food reserves are used they wither and finally drop off, usually about the time the primary leaves of the seedling reach the stage when their photosynthetic mechanism is fully developed and the seedling has become self-sufficient. In some seeds, for example, the snapdragon (*Antirrhinum*), the cotyledons become fully developed normal leaves that carry on photosynthesis and function throughout much of the plant's life span. In others, as for example the peach seed (*Prunus persica*) whose germination is shown in Figure 4-3, the

Figure 4-2. Germination of a bean seed (dicot). [From R. H. Arnett, Jr., and D. C. Braungart: *An Introduction to Plant Biology,* 3rd ed. The C. V. Mosby Co., St. Louis, 1970. Used with permission.]

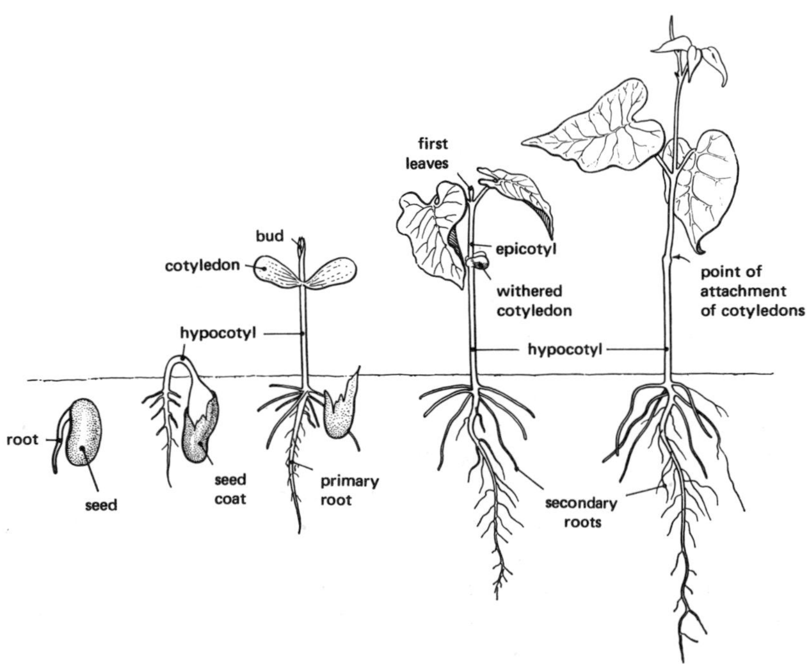

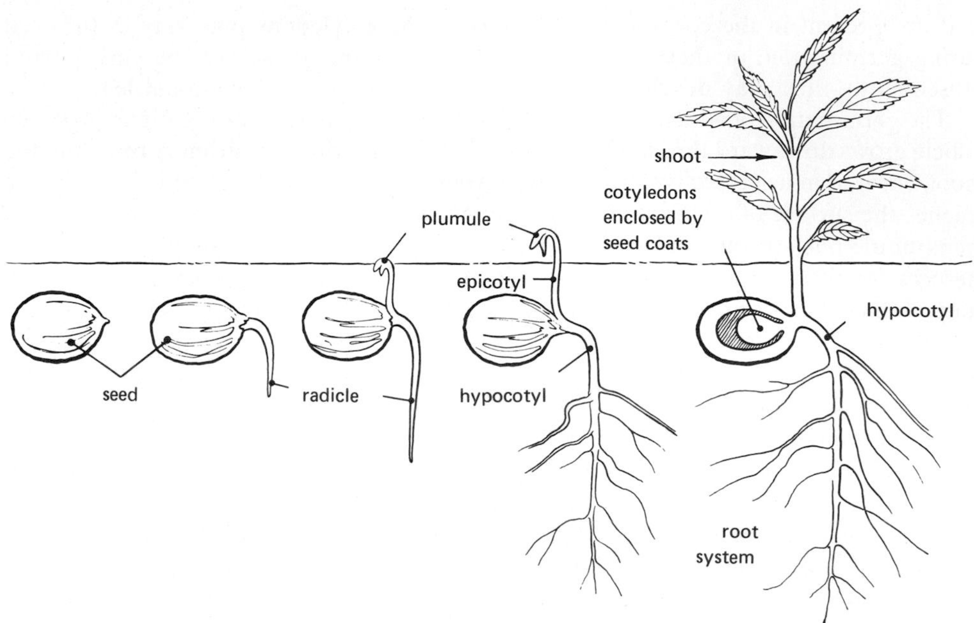

Figure 4-3. Germination of a peach seed (dicot). [From R. H. Arnett, Jr., and D. C. Braungart: *An Introduction to Plant Biology,* 3rd ed. The C. V. Mosby Co., St. Louis, 1970. Used with permission.]

cotyledons remain in the seed during and after germination. The plumule of dicots is not protected by a coleoptile. Instead, the plumule pushes through the soil in a "crook" form, called the plumule hook (Figures 4-2, 4-3). In this way the delicate newly forming leaves of the plumule are not damaged.

The Stem

The shoot apex is a dome-shaped structure, the meristem, usually surrounded by leaves, scales, or branches. The apical meristem contains a relatively small number of cells that give rise, by division, to all the other cells in the aerial portion of the plant. It may be differentiated into areas of more and less intense cell division; however, this kind of differentiation is much more pronounced in roots and is discussed in that section.

Most apical meristems contain two main zones: a **tunica,** one to several layers of cells organized in rows normal to the surface of the meristem, and a **corpus,** a body of cells less tidily arranged, beneath the tunica. The tunica cells usually divide in a plane that is perpendicular to the surface of the meristem, whereas the corpus cells divide in many different planes. The tunica usually gives rise to epidermal tissue and the corpus to the bulk of internal tissue of the stem and leaves.

Zones of cell division, elongation, and maturation occur in the stem tip, but they are not clearly separated. This is because the meristem produces not only the stem but also the leaves and branches of the shoot by the outgrowth of tissue from the rim of the apical meristem. These leaves grow rapidly ahead of the apex and enfold it. Differentiation of vascular tissue occurs first in the leaf buds, forming **leaf traces.** Below them, in the zone of elongation of the stem, a ring of **provascular strands** forms within the stem. The leaf traces differentiate downward and the provascular strands differentiate upward and ultimately establish connections. As the stem matures, the provascular strands develop into **vascular bundles** that are composed of the main conducting elements of the stem (Figure 4-4).

Dicot and monocot stems have a number of structures and cell types in common, but have certain differences in the arrangement of their tissues (Figure 4-5). Both have an outer layer of **epidermis,** usually covered on the outside with waxy **cuticle.** The main cell type of the ground material is **parenchyma,** large, thin-walled, relatively undifferentiated cells. Outside the vascular bundles is the **cortex,** usually composed of smaller, more differentiated parenchyma, and inside is the **pith,** composed of somewhat larger, thinner-walled parenchyma cells. The vascular bundles of monocots are scattered throughout the parenchyma, whereas those of dicots are arranged in a ring (Figure 4-5). Each vascular bundle contains **xylem** cells toward the center and **phloem** toward the outside. Xylem is primarily composed of dead, thick-walled conducting cells, either **vessels** (large cells with no crosswalls, forming tubelike pipes that run lengthwise through the stem) or **tracheids** (much smaller in diameter, having end walls, and usually heavier secondary thickening). The xylem may also contain **fibers** (similar to tracheids but with longer, narrower tips) that serve primarily for structural support, and strands or sheets of parenchyma cells penetrate the xylem.

Figure 4-4. Longitudinal section of a stem tip. [From R. H. Arnett, Jr., and D. C. Braungart: *An Introduction to Plant Biology,* 3rd ed. The C. V. Mosby Co., St. Louis, 1970. Redrawn from C. L. Wilson and W. E. Loomis: *Botany,* rev. ed.: 1957. The Dryden Press. Used with permission.]

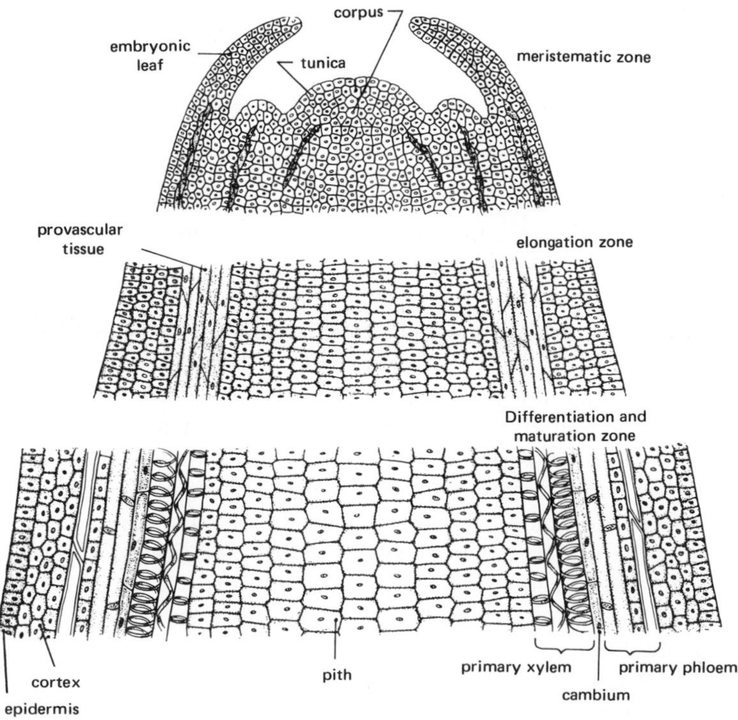

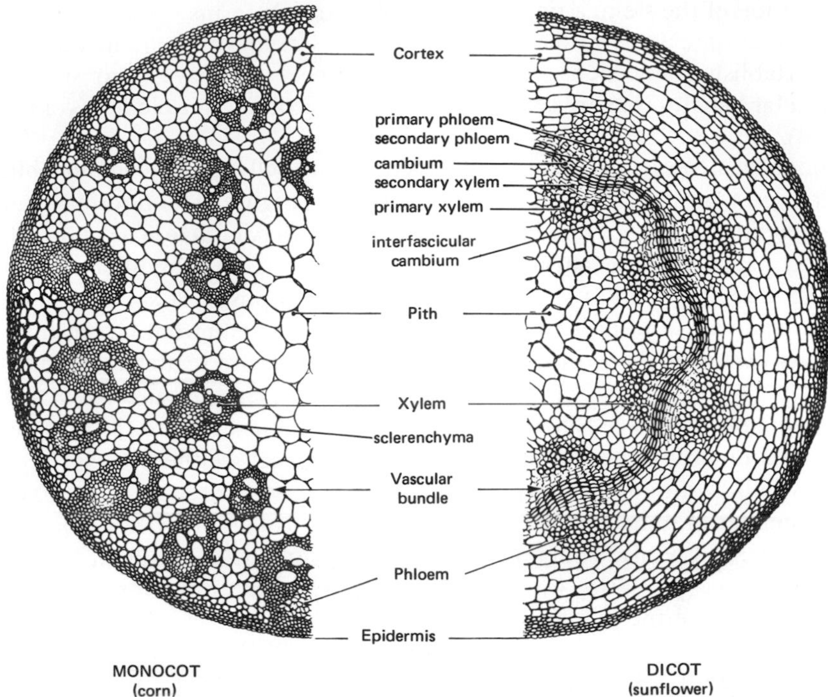

Figure 4-5. Cross section of a monocot and a dicot stem. [From R. H. Arnett, Jr., and D. C. Braungart: *An Introduction to Plant Biology,* 3rd ed. The C. V. Mosby Co., St. Louis, 1970. Used with permission.]

The phloem is composed mainly of large-diameter, thin-walled cells with characteristic sievelike end plates called **sieve elements,** lined up, end to end to make **sieve tubes.** They are associated with small parenchyma cells called companion cells. Vessels and tracheids die as they mature and lose their cell contents, but phloem cells, as well as the nonspecialized parenchyma cells of the cortex and pith, stay alive and retain some of their structural integrity. Sieve elements may lose their nuclei and undergo extensive modifications in structure (see Chapter 13), but they stay alive and apparently able to metabolize.

Vascular bundles are also frequently surrounded either partly or entirely by fiber cells, and the whole stem may have strands or a ring of heavily thickened, modified parenchyma cells called **collenchyma** (green and living) or **schlerenchyma** (dead cells). Patches of such thickened cells are often found in the phloem.

The main difference between monocot and dicot stems is in the organization of the bundles, and the existence of meristematic tissue in dicot bundles (Figure 4-5). Monocots have bundles scattered throughout the parenchyma, each containing xylem toward the inside and phloem toward the outside. The first-formed xylem, called **protoxylem** is nearest the center and later xylem, called **metaxylem,** is nearest the phloem. No cell division takes place once the bundles are formed. Secondary thickening of monocot stems is rare, and, when it occurs, new bundles are formed. A large part of the maturation and differentiation of the tissue takes place before elongation in the monocot stem.

Dicot stems are more complex and almost invariably capable of secondary growth. Initially the bundles are arranged in a circle around a central core of pith. The xylem

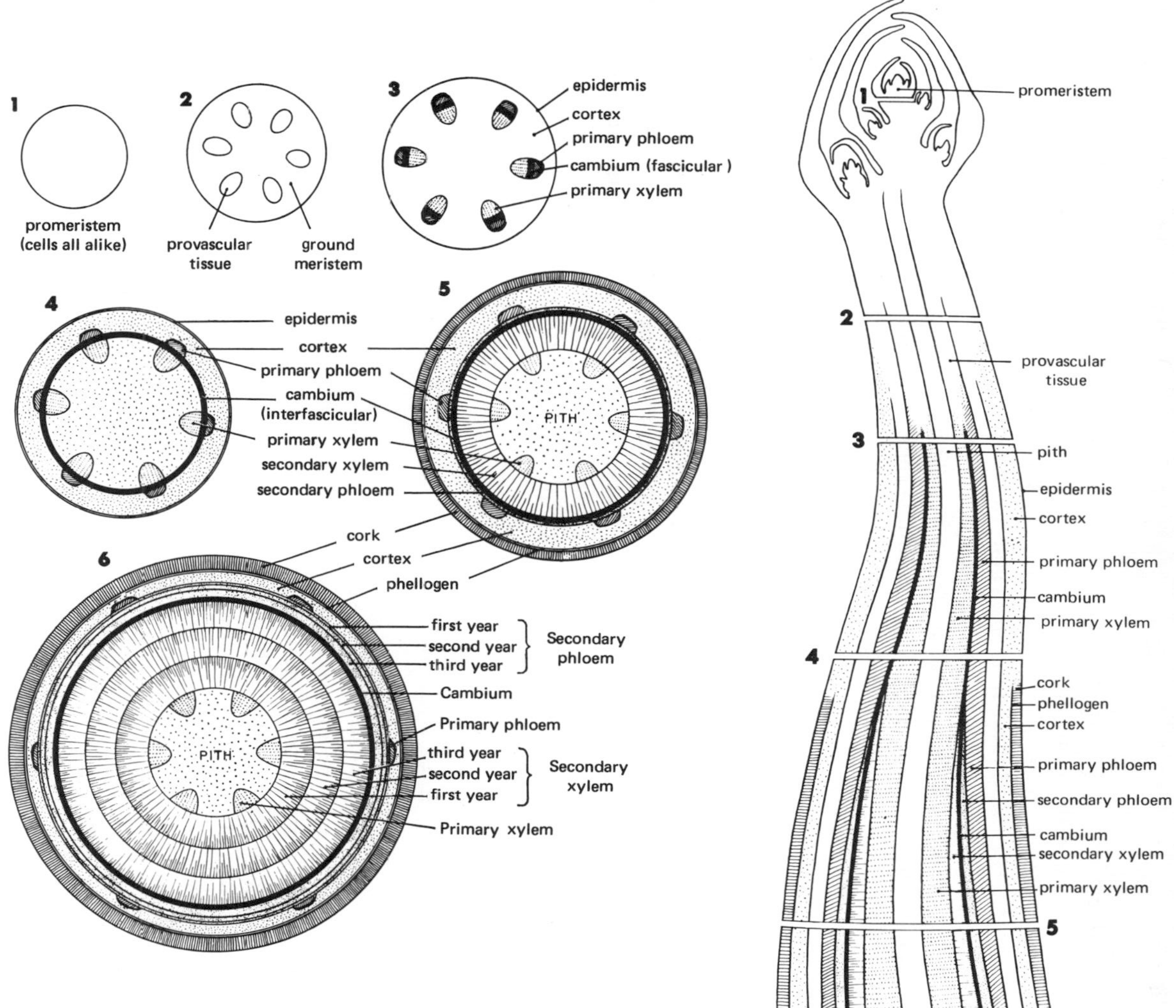

Figure 4-6. Left: Cross sections of a woody stem with secondary growth. These cross sections can be related to the longitudinal section at the right. [From R. H. Arnett, Jr., and D. C. Braungart: *An Introduction to Plant Biology,* 3rd ed. The C. V. Mosby Co., St. Louis, 1970. Used with permission.]

and phloem are separated by a layer of cells capable of division, called the **cambium.** Secondary growth occurs from this cambium by divisions that are tangential to the circumference of the stem, giving rise to new phloem cells to the outside and new xylem cells to the inside. Later, **interfascicular** cambium (*inter fascicle,* between bundles) develops by rejuvenation of parenchyma cells between the bundles. Thus a complete circle of cambium is formed, which forms a circle of xylem to the inside and a circle of phloem to the outside (Figure 4-5). The whole central section of the stem, including the phloem and everything inside it, is called the **stele.** The outer cortex, and later the outer layers of phloem, give rise periodically to cork cambium or **phellogen,** which produces the cork cells **(phellem)** that mainly constitute the bark. As the dicot stem enlarges in diameter, older bark sloughs off and new bark is formed from cork and the crushed layers of old phloem.

Perennial (woody) dicots may continue to expand for a prolonged period of time by secondary growth (Figure 4-6). Secondary xylem is deposited in annual rings containing larger-celled spring wood, which often contains the majority of vessels in woody angiosperms or hardwoods, and smaller-celled summer wood. The perennial dicot stem seldom retains more than a year or two's growth of phloem; the older phloem dies and sloughs off as the stem enlarges. The bases of branches are surrounded by new wood, forming knots in the wood.

Roots

A growing root, whether primary, secondary, or adventitious, can be roughly divided into three regions: the **meristematic region** where cell multiplication takes place, the **region of elongation and differentiation** where cell division continues to a lesser extent, and the **region of maturation** (Figure 4-7). The tip of the root is protected by a **root cap.** The meristem often contains a reserve of slowly dividing embryonic cells, the **quiescent center.** Most of the cell division resulting in root growth and the regeneration of the root cap take place around the periphery of the quiescent center, which may be involved in organizing tissue formation in the growing root. Columns of cells produced from the embryonic region expand longitudinally to produce the characteristic structure of the root. Some cells (for example, vessel elements) elongate far more than do others (for example, cortex or epidermis), which must therefore grow by further divisions. The regions of division, elongation, and maturation tend to overlap. Maturation of the cells involves the formation of **root hairs** on some epidermal cells, the differentiation of stele cells, the thickening of the walls of conducting vessels, and the differentiation of the cortex into various regions.

The generalized structures of a monocot and a dicot root are shown in Figure 4-8. The central part of the root is the stele, containing the conducting tissues xylem and phloem, occasionally with a central core of pith. The xylem and phloem cells are essentially identical with those found in the corresponding stem tissues. The tissues outside the stele are primarily the cortex, made up of parenchyma cells, and the epidermis.

In monocot roots, alternating strands of xylem and phloem form a ring of conducting tissue around the central core of the pith (Figure 4-8). Unlike monocot stems, the protoxylem is outside; maturation proceeds toward the inside (metaxylem). Outside the stele, the cortex is bounded externally by the epidermis and internally by the **endoder-**

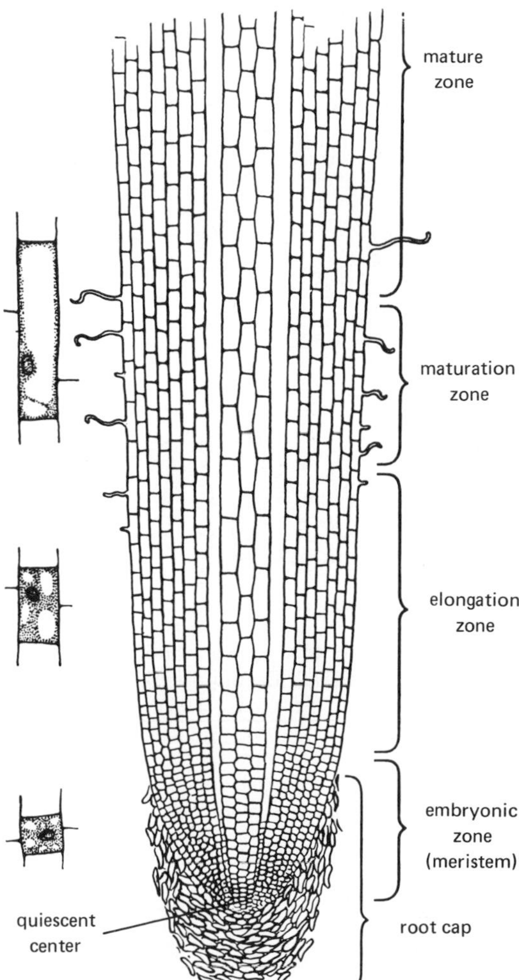

Figure 4-7. Diagram of a root and root cap. [From R. H. Arnett, Jr., and D. C. Braungart: *An Introduction to Plant Biology,* 3rd ed. The C. V. Mosby Co., St. Louis, 1970. Used with permission.]

mis. The endodermis is important in the process of water absorption and transport because its transverse walls are heavily suberized so that water cannot leak past the endodermis through intercellular spaces but must pass through the cells (see Chapter 11). In some older monocot roots considerable wall thickening occurs in the outer cortical layers to form lignified parenchyma, a supporting tissue, but in general little growth in diameter occurs in monocot roots. Some persistent monocot roots may develop secondary tissue by forming new bundles of conducting tissues in the cortex but not by the addition of new cells to already existing primary xylem or phloem.

The arrangement is similar in dicot roots, except that there is no pith and the primary xylem forms a solid core that is star-shaped in cross section (Figure 4-8). The primary phloem lies between the points of the xylem star. Outside the phloem is a layer of cells, the **pericycle,** that retain their meristematic activity. The pericycle is important because cells in this layer give rise to branch roots, as shown in Figure 4-9, a process that occurs more commonly in dicots than in monocots. Cell division in the pericycle forms a new root primordium that grows out through the cortex, either mechanically forcing its way through or enzymatically digesting the cortex cells ahead of it. Tissues at the base of

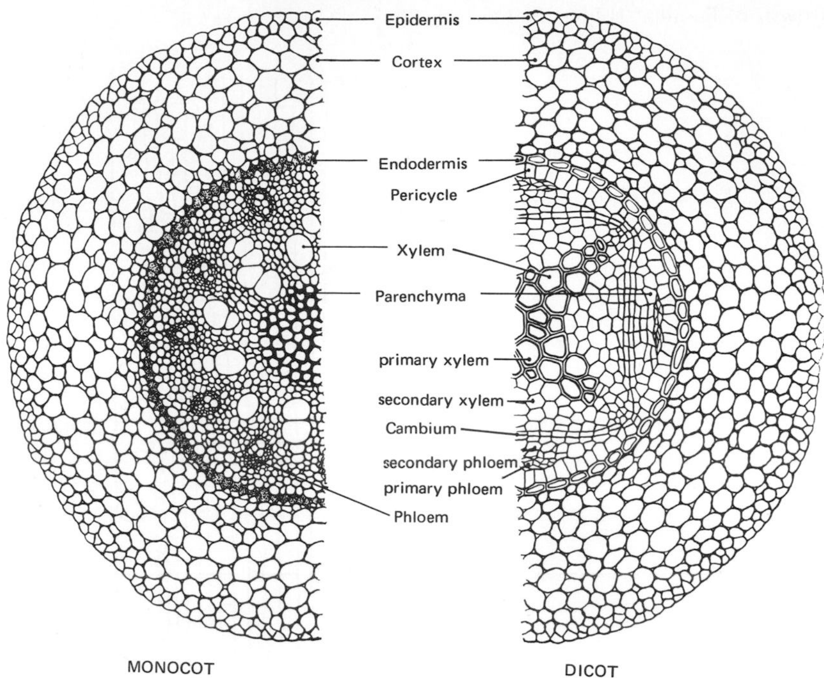

Figure 4-8. Diagram of cross sections of a monocot and a dicot root. [From R. H. Arnett, Jr., and D. C. Braungart: *An Introduction to Plant Biology,* 3rd ed. The C. V. Mosby Co., St. Louis, 1970. Used with permission.]

Figure 4-9. Origin of a branch root. [From R. H. Arnett, Jr., and D. C. Braungart: *An Introduction to Plant Biology,* 3rd ed. The C. V. Mosby Co., St. Louis, 1970. Used with permission.]

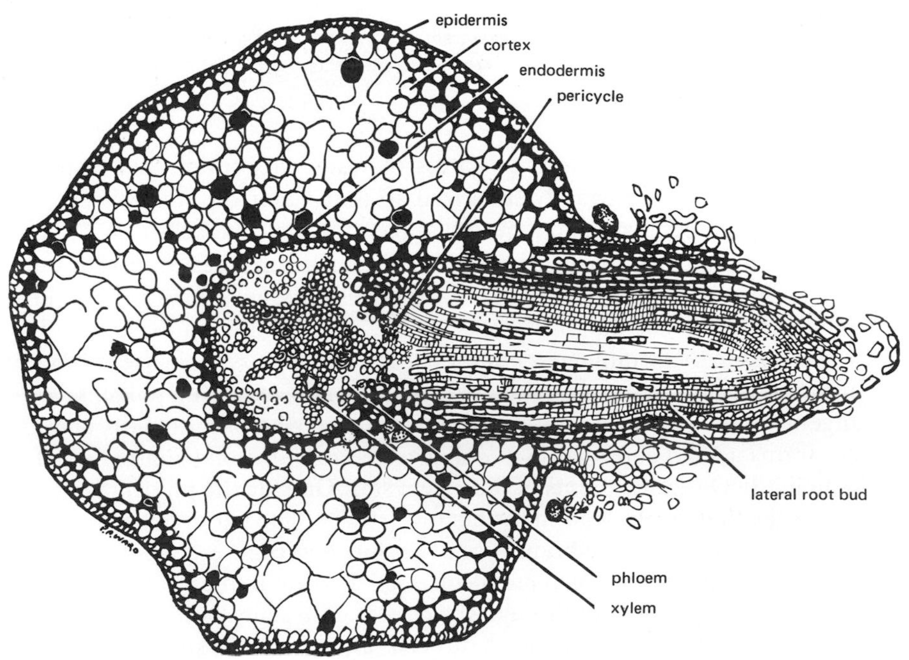

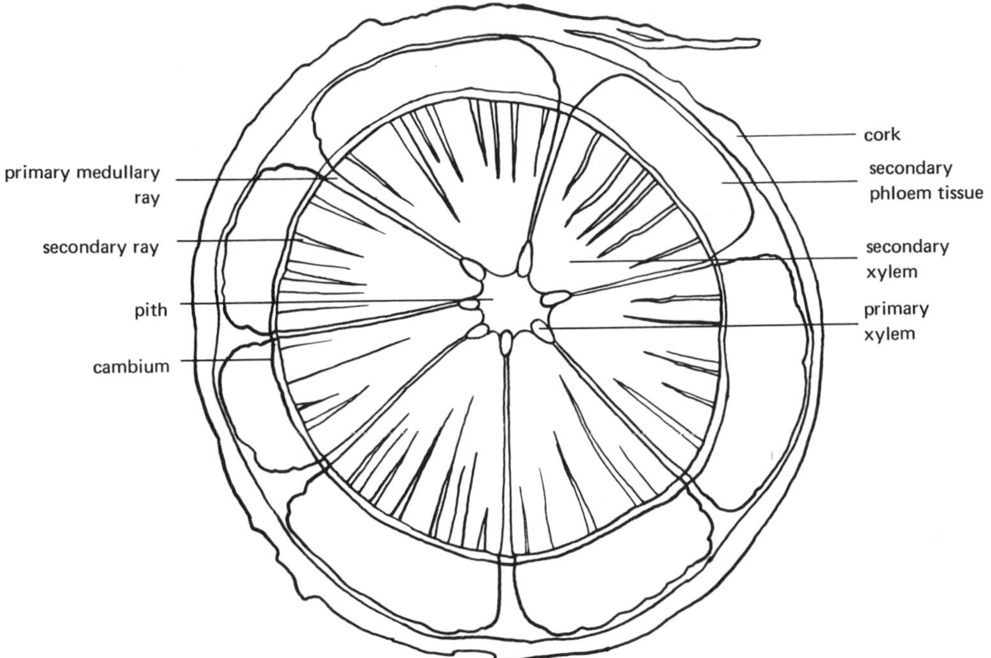

Figure 4-10. Secondary growth in a dicot root. Diagram of a transverse section through an old root of *Tilia europaea*. [From A. C. Shaw, S. K. Lazell, and G. N. Foster: *Photomicrographs of the Flowering Plant*. Longmans, Green and Co. Ltd., London, 1965. Used with permission.]

the branch root form vascular connections with the stele of the main root. Secondary growth occurs in dicot roots initially from the formation of a cambium around the xylem star, which produces new phloem to the outside and xylem to the inside (Figure 4-10). In many roots, particularly the swollen storage roots of plants like beets and turnips, additional layers of cambium may form in the phloem or in the cortex giving rise to massive secondary thickening. The outer layers of cortex slough off and cork or bark is generated by a cork cambium arising in the phloem. In older roots, secondary or branch roots may arise from meristems that develop in the phloem.

Leaf Structure

Leaves are basically stems with lateral extensions. They are usually largely preformed in buds, and a considerable part of the visible growth is expansion of cells rather than cell multiplication. Dicot leaves normally grow in length as a result of the activity of a terminal meristem, and the lateral extension of the blade is accomplished by marginal meristems on each side of the leaf.

In monocot leaves the primary meristem is at the base of the leaf, just above the **ligule** or point of attachment of the leaf. That is why grass can be (and must be!) cut frequently—the leaves continue to grow from the base. The leaf's network of veins, usually parallel in monocots and either pinnately or palmately branched in dicots, may be closed or open (that is, enclosing or not enclosing islands of parenchyma tissue). The veins are continuous with the vascular structure of the stem via the leaf trace. Veins are

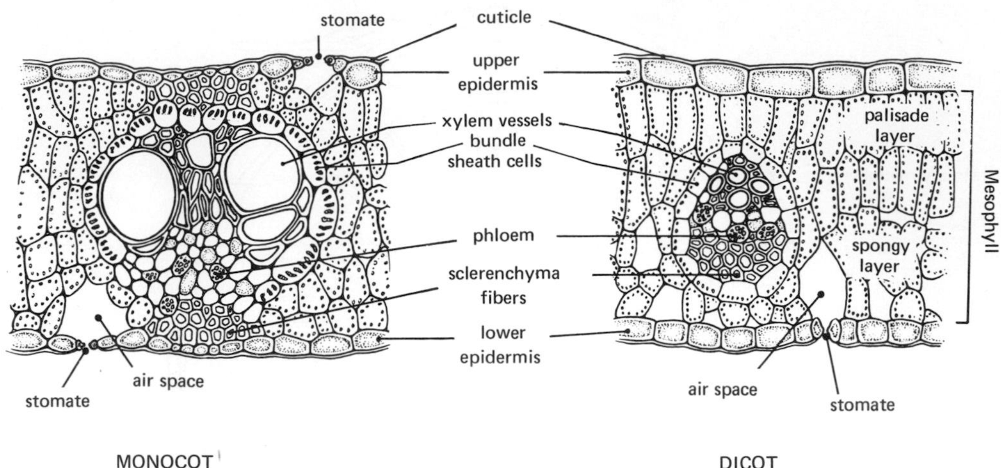

Figure 4-11. Diagram of the cross section of a monocot and a dicot leaf. [From R. H. Arnett, Jr., and D. C. Braungart: *An Introduction to Plant Biology,* 3rd ed. The C. V. Mosby Co., St. Louis, 1970. Used with permission.]

usually surrounded by a more or less developed **bundle sheath,** which may be very important in photosynthesis of some plants (see Chapters 7 and 14) and often contains masses of lignified fibers or sclerenchyma that act as stiffening (Figure 4-11).

The lamina or blade of the leaf is composed largely of parenchyma, which is the major photosynthetic tissue and contains many chloroplasts, together with an upper and lower epidermis. Epidermal cells are protected by a suberized or waxy **cuticle** and usually do not contain chloroplasts. The parenchyma in dicot leaves is arranged in two tissues—a **palisade layer,** one or two cells thick in tightly packed array, and a layer of **spongy parenchyma** that has large air spaces ramifying through it (Figure 4-11). The monocot leaf lacks a well-defined palisade layer; it is largely made up of spongy parenchyma with extensive air spaces.

The internal air spaces of the leaf are directly connected with the outside air through small pores or **stomata** (singular, **stoma,** or **stomate**). Surrounding each stoma are two cells, the **guard cells,** which open and close the stoma by their expansion and contraction. Unlike the epidermal cells, guard cells contain chloroplasts. The function and operation of stomata will be considered in detail in Chapter 14 (page 327).

Leaves exhibit a bewildering variety of form and may be much influenced in their development by environmental factors such as light, carbon dioxide content, availability of water, submergence, the age of the plant, and so on. In addition, leaves may be modified in many ways to form tendrils, thorns, insect traps, spines, and so on.

Flowers and Fruit

Flowering marks the termination of growth of the stalk or stem on which the flower is born, since flowering results from a modification of the terminal meristem. A flower is essentially a stem tip with crowded appendages, the longitudinal axis being much shortened and the appendages modified in characteristic ways to produce sepals, petals, stamens, and carpels. A fantastic number of variations on the basic structure occur, but

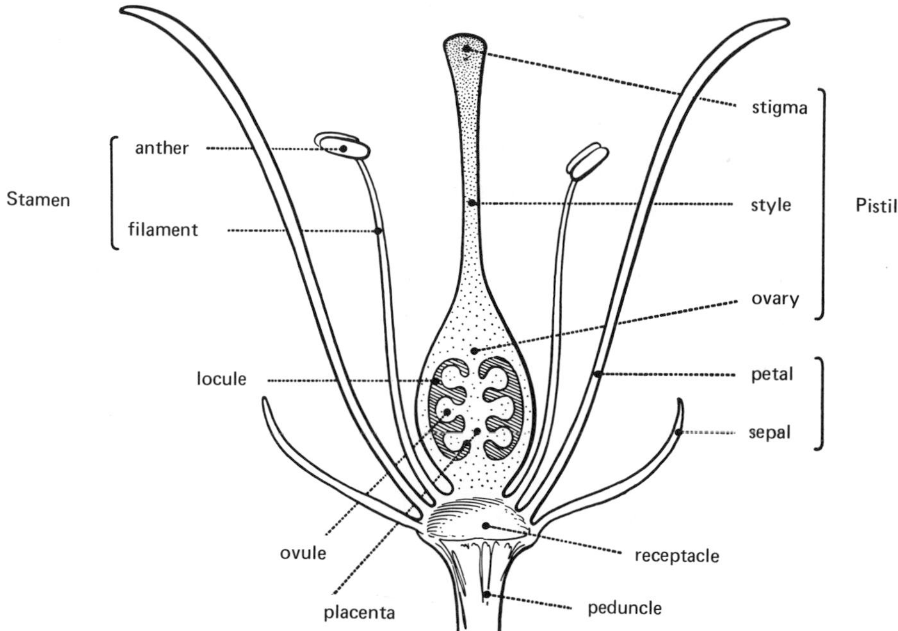

Figure 4-12. Anatomy of a flower. [From R. H. Arnett, Jr., and D. C. Braungart: *An Introduction to Plant Biology,* 3rd ed. The C. V. Mosby Co., St. Louis, 1970. Used with permission.]

the basic plan is simple, as shown in Figure 4-12. All the structures shown in Figure 4-12 are parts of the **sporophyte** (2N) generation of the plant.

Within the ovule a **megaspore mother cell** undergoes reduction division to produce an egg nucleus, usually two **polar nuclei** and usually five other nuclei, that may migrate into the opposite end of the ovule **(antipodal nuclei)** or remain near the **micropylar** end **(synergids),** as seen in Figure 4-13. These eight nuclei represent the much diminished **female gametophyte** generation of the angiosperm plant.

Meanwhile in the anther the **microspore mother cell** has undergone reduction division to produce four pollen grains, which develop into the male gametophyte. On germination, nuclear divisions take place that result in the production of a vegetative **tube nucleus,** which may be concerned with the growth of the pollen tube down the style to the ovary, and a **generative nucleus,** which later divides to produce two **sperm nuclei** (Figure 4-14).

When the pollen tube grows to and penetrates the ovary, the sperm nuclei are discharged into the female gametophyte. One of them unites with the egg nucleus to produce the diploid zygote that develops into the embryo, the new sporophyte generation. The other unites with the two polar nuclei to produce a usually triploid tissue that develops into the endosperm. The synergids and antipodal nuclei usually degenerate.

Development of a variety of different tissues in the flower and in its supporting stem (receptacle) subsequent to fertilization results in the formation of a fruit, which contains the seed or seeds. The carpels may develop, as in the tomato (a berry), and various layers of the ovary wall or its surrounding tissues may become skinlike or stony, as in the peach or pear. In addition the stem tip or base of the floral parts may be involved in fruit formation, as in the strawberry (which has a number of small dry true fruits buried in

A

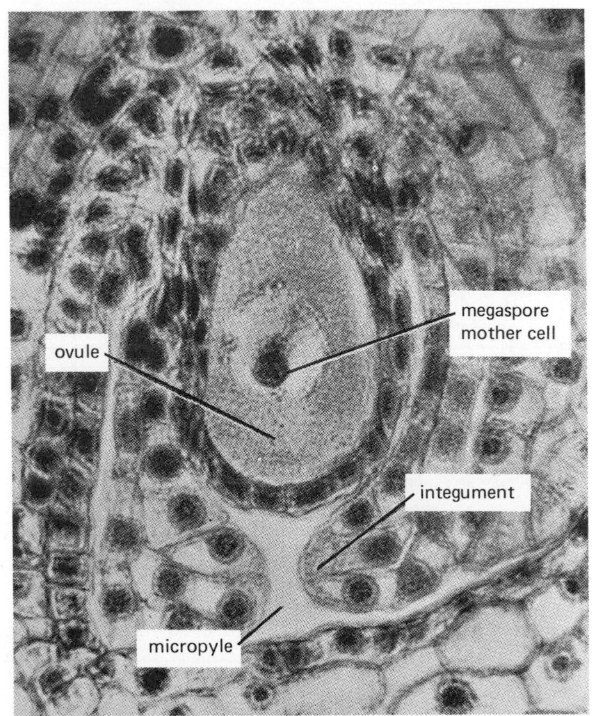

B

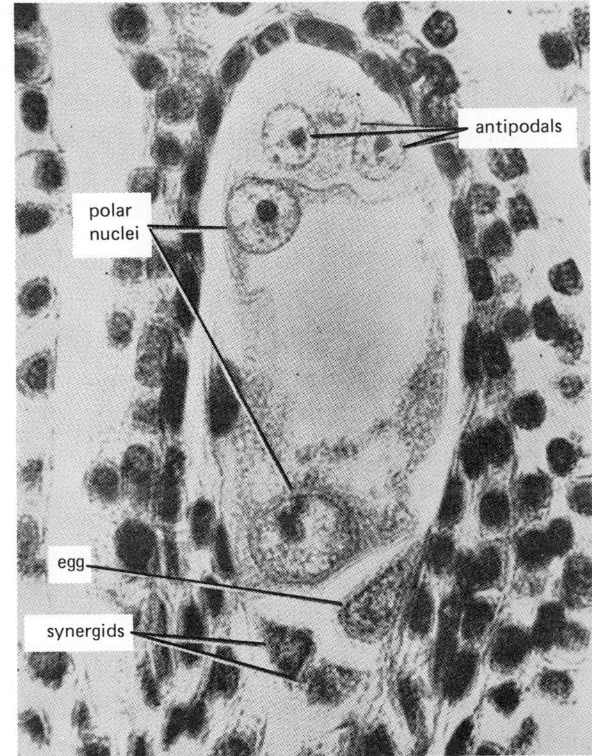

Figure 4-13. Cross section of a lily ovary showing megaspore mother cell (**A**) and later eight-nucleate stage (**B**). [From R. H. Arnett, Jr., and D. C. Braungart: *An Introduction to Plant Biology,* 3rd ed. The C. V. Mosby Co., St. Louis, 1970. Courtesy George H. Conant, Triarch Inc., Ripon, Wis. Used with permission.]

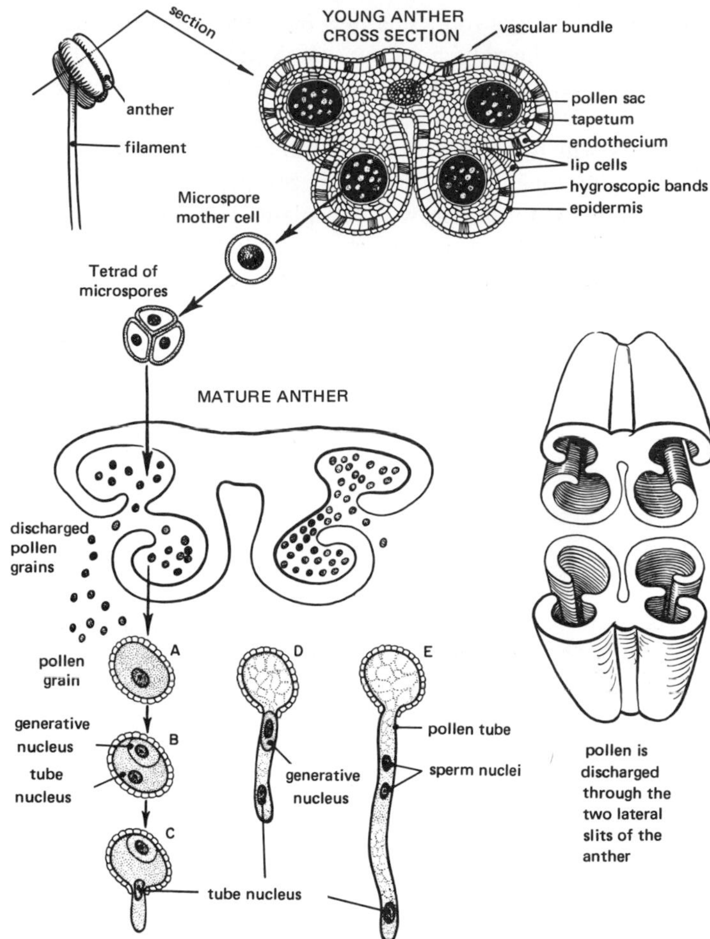

Fig. 4-14. Microspore (pollen grain) and male gametophyte development. [From R. H. Arnett, Jr., and D. C. Braungart: *An Introduction to Plant Biology,* 3rd ed. The C. V. Mosby Co., St. Louis, 1970. Used with permission.]

the surface of an accessory fruit formed from the stem tip), the apple (the ovary is surrounded by a fleshy development of the receptacle), or the banana (the banana skin is developed from the floral tube).

Meristems: Patterns of Growth

The main meristems of the plant are at the tips of the stems, roots, and all growing branch organs, and these grow continuously throughout the life of the plant or its organs. Equally important are the cambiums of the root and stem, particularly in perennial plants, which accomplish growth in girth through the annual increments of phloem and xylem. New areas of meristematic activity or cambiums are formed at intervals, usually in the phloem, that produce and regenerate the epidermal covering or bark of the stem and root. The new meristems that initiate branch roots are formed in the pericycle or the phloem of the root. Many organs formed from basic stem structures, such as leaves, petals, fruits, and so on, develop secondary meristems that operate independently of the primary meristem. Thus the blade of a leaf is formed from lateral meristems, which accomplish the lateral growth of the leaf after much of the cell division leading to its longitudinal growth is completed.

Much of the formal growth of a plant is accomplished by regulation of the relative activities of the various meristems in the plant. It is this regulation, together with the determination of the kind of tissue to be made by individual meristems, that determines the developmental pattern of the plant. This is why so much importance is attached to the biochemical and physiological activities of meristematic tissues, and why so much research has been directed toward understanding the pattern of metabolism and the control mechanisms of meristematic tissues.

Additional Reading

General textbooks on plant anatomy will cover this material in greater detail.

Esau, K.: *Vascular Differentiation in Plants.* Holt, Rinehart & Winston, New York, 1965.
Salisbury, F. B., and **R. V. Parke:** *Vascular Plants: Form and Function.* Wadsworth, Belmont, Calif., 1965.
Steward, F. C.: *About Plants.* Addison-Wesley Publishing Co. Inc., Reading, Mass., 1966.

SECTION II

Plant Metabolism

5 Energy Metabolism

Oxidation and Reduction Reactions

It is most important to understand chemical reactions. Not only are all organisms made of chemicals that must be synthesized and metabolized by chemical reactions, but all the energy used to make these chemicals and to do work in the cell is acquired, stored, metabolized, and used through the acquisition of chemicals and the operation of chemical reactions. The most common of the energy-metabolizing reactions are oxidation and reduction reactions. **Oxidation** of a compound (or of a bond in a compound) is accomplished by removing one or (usually) two electrons. **Reduction** is brought about by adding electrons. In effect, oxidation and reduction are the loss or gain of electrons, respectively. Evidently, since electrons and other charged particles cannot exist independently, when one substance is oxidized another must be reduced. A reaction in which electrons are transferred from one molecule to another, during which one compound is oxidized and the other reduced, is called a **redox** reaction.

Organic compounds tend to lose or gain electrons. A compound that tends to lose electrons (transfer them to another compound) is a reducing agent; one that tends to attract them is an oxidizing agent. Oxygen is a powerful oxidizer. In the reaction

$$\underset{\text{formaldehyde}}{\mathrm{H\cdot \underset{H}{\dot{C}}:O}} + O_2 \longrightarrow CO_2 + H_2O$$

electrons are transferred from formaldehyde to oxygen; formaldehyde is oxidized to carbon dioxide, and oxygen is reduced to water. In this reaction hydrogen ions (H^+) accompany the electrons (e^-) and electroneutrality is maintained. In the reaction

$$\underset{\text{acetaldehyde}}{CH_3\text{—}CHO} + \tfrac{1}{2}O_2 \longrightarrow \underset{\text{acetic acid}}{CH_3\text{—}COOH}$$

the oxygen molecule is reduced as it is incorporated into the oxidized organic molecule.

Redox reactions need not involve oxygen, and the majority of biological redox reactions do not. Covalent bond systems are capable of gaining or losing electrons and hydrogen ions, as in the reaction

$$\underset{\text{ethane}}{CH_3\text{—}CH_3} + \underset{\text{a disulfide}}{R\text{—}S\text{—}S\text{—}R} \rightleftharpoons \underset{\text{ethylene}}{CH_2{=}CH_2} + \underset{\text{a thiol}}{2\,RSH}$$

where ethane is oxidized and the disulfide is reduced. This type of reaction may be generalized as follows

$$\underset{\text{reduced A}}{AH_2} + \underset{\text{oxidized B}}{B} \rightleftharpoons \underset{\text{oxidized A}}{A} + \underset{\text{reduced B}}{BH_2}$$

where A and B are metabolites. Again, hydrogen ions accompany the electrons. If a molecule contains an atom that may undergo a valence change (that is, an oxidation or reduction) this may be accomplished by the addition or loss of an electron without the transport of hydrogen ions as follows

$$\underset{\substack{\text{oxidized organic}\\ \text{iron compound}}}{R\text{—}Fe^{3+}} + e^- \rightleftharpoons \underset{\substack{\text{reduced organic}\\ \text{iron compound}}}{R\text{—}Fe^{2+}}$$

An oxidant has a certain affinity for electrons (called its oxidizing power), whereas a reductant has a much lower affinity for them (its tendency to lose electrons is its reducing power). Thus, electrons lose potential energy as they pass from a reductant to an oxidant. If a powerful oxidant and a powerful reductant are allowed to react, the redox reaction goes very readily with a large release of energy

$$\underset{\text{reductant}}{AH_2} + \underset{\text{oxidant}}{B} \longrightarrow A + BH_2 + \text{energy}$$

and the reaction is termed **exergonic.** A reaction that requires or absorbs energy (for example, the reversal of the preceding reaction) is called **endergonic** and will not proceed spontaneously.

If two oxidants (or two reductants) of equal potential are mixed, no net reaction will take place because neither compound can oxidize or reduce the other. One must have a higher reducing or oxidizing potential than the other for a reaction to proceed. Later in this chapter we shall look at the question of quantifying and measuring redox potentials, the effects of concentrations of reactants, and the rate and extent to which reactions will go under the influence of such potentials and concentration gradients.

Hydrolysis Reactions

The splitting of a covalent bond by the introduction of water, called **hydrolysis,** usually results in a large loss of energy as follows

$$\text{sucrose} + H_2O \longrightarrow \text{glucose} + \text{fructose} + \text{energy}$$

Conversely, the removal of water to make an anhydride bond usually requires energy

$$\text{energy} + \underset{\text{methanol}}{CH_2OH} + \underset{\text{acetic acid}}{HO\text{—}\overset{\overset{\displaystyle O}{\|}}{C}\text{—}CH_3} \longrightarrow \underset{\text{methyl acetate}}{CH_2\text{—}O\text{—}\overset{\overset{\displaystyle O}{\|}}{C}\text{—}CH_3} + H_2O$$

It follows that anhydride bonds may contain substantial amounts of energy, and a compound containing such a bond may have a much greater potential energy than the products of its hydrolysis. This characteristic, together with the wide range of potential energies available in reducing and oxidizing compounds, provides the chemical means by which organisms are able to extract energy from their environment, store it, and use it for syntheses and to do work.

Production of ATP

A biological system requires energy for its construction and maintenance. All compounds contain potential energy stored in their bond structure that may be released when the bonds are broken. Biological systems obtain energy by controlled oxidative breaking of the bonds in fuel molecules and the utilization of the resulting energy to make new chemical bonds or to do useful work. Uncontrolled oxidation of a carbon-to-carbon bond with the direct transfer of electrons to oxygen (resulting in the formation of water) liberates all the energy in the bond as heat, which is normally useless to biological systems. Furthermore the amount of energy released is far too great for any biological system to handle—in effect, it gets too hot to hold.

Biological systems circumvent this difficulty by transferring the electrons in a series of small steps; each step is a redox reaction that liberates an amount of energy small enough to be successfully trapped by the synthesis of new bonds or used to do work. Thus electrons are passed from the original fuel molecule, which is to be oxidized, to an **electron carrier** molecule in the oxidized state, which thus becomes reduced. This carrier in turn passes the electrons along to another molecule, which passes them to another, until they are finally passed to oxygen with the formation of water. This series of electron carriers is called an **electron transport chain.**

Each member of the electron transport chain is reduced when it accepts electrons and oxidized when it passes them on. Each member of the chain is a weaker reductant than the previous one—that is, it can be reduced by preceding members of the chain, but it reduces succeeding ones. Thus, as electrons pass from member to member of the electron transport chain, energy is lost at each stage of the transfer. Part of this energy is conserved by being used to manufacture new bonds in specialized compounds that can subsequently be used to drive other reactions. These bonds are called **high energy bonds.** One of the most important of these is the phosphate-to-phosphate anhydride bond in adenosine triphosphate (ATP), shown in Figure 5-1. This compound is formed by an endergonic reaction as follows:

$$\text{ADP} + \text{Pi} + \text{energy} \longrightarrow \text{ATP} + H_2O$$

Figure 5-1. Structure of adenosine triphosphate (ATP).

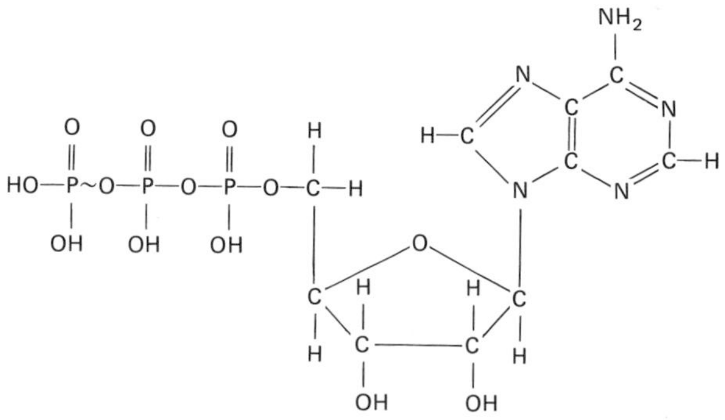

where **Pi** stands for **inorganic phosphate.** High energy bonds are often written with a squiggle (~). Thus, **~P** stands for a **high energy phosphate bond,** as, for example, in the terminal phosphate bond of ATP (A—P—P ~ P).

An electron transport reaction proceeds by an exergonic reaction as follows

$$\text{reduced A} + \text{oxidized B} \longrightarrow \text{oxidized A} + \text{reduced B} + \text{energy}$$

These reactions may be coupled together

$$\text{reduced A} + \text{oxidized B} + \text{ADP} + \text{Pi} \longrightarrow \text{oxidized A} + \text{reduced B} + \text{ATP} + H_2O$$

so that the exergonic reaction pushes or drives the endergonic one. In electron transport reactions, the synthesis of ATP usually occurs in **coupled** reactions, that is, where one reaction cannot proceed without the other. Thus the oxidation of A cannot proceed without the production of ATP. If no adenosine diphosphate (ADP) or Pi is available, the oxidation of A cannot take place. Since ATP is required to drive many synthetic reactions in the cell, being converted back to ADP + Pi in the process, cellular oxidation may be controlled by the need for ATP synthesis. If no synthetic reactions are in progress, no ATP is used, no ADP + Pi is formed, and oxidative reactions cannot proceed. This mechanism guards against the wasteful oxidation of cellular reserves. Mechanisms of coupling will be considered later (see ATP Synthesis and Group Transfer Reactions, pages 99 and 102).

An Electron Transport Chain

The majority of the oxidative reactions that yield energy in the cell are coupled to a well-defined electron transport system that has been found to operate in most animal and plant tissues in one form or another. Some variation in the nature of certain of the members occurs among groups of organisms, but the general outline is so widespread among living systems that it can be considered one of the basic reaction sequences of living organisms. A generalized scheme is shown in Figure 5-2.

The substance to be oxidized (the **substrate**), **AH₂,** reacts first with a pyridine nucleotide, usually **nicotinamide adenine dinucleotide (NAD⁺)** but sometimes with **nicotinamide adenine dinucleotide phosphate (NADP⁺).*** The structure of these nucle-

*NAD is sometimes called diphosphopyridine nucleotide (DPN) or coenzyme I. NADP is sometimes called triphosphopyridine nucleotide (TPN) or coenzyme II.

Figure 5-2. Outline of an electron transport chain. Two electrons are transferred at each step. AH_2 is the substrate being oxidized to A. $\frac{1}{2}O_2$ is the ultimate electron acceptor, being reduced to H_2O. NAD^+ ($NADH_2^+$) and FAD ($FADH_2$) are nicotinamide adenine dinucleotide and flavin adenine dinucleotide (oxidized or reduced). UQ is ubiquinone, Cyt a, b, and so on, are cytochrome pigments. (Fe^{3+}) and (Fe^{2+}) are an unknown iron-containing enzyme in the oxidized or reduced form.

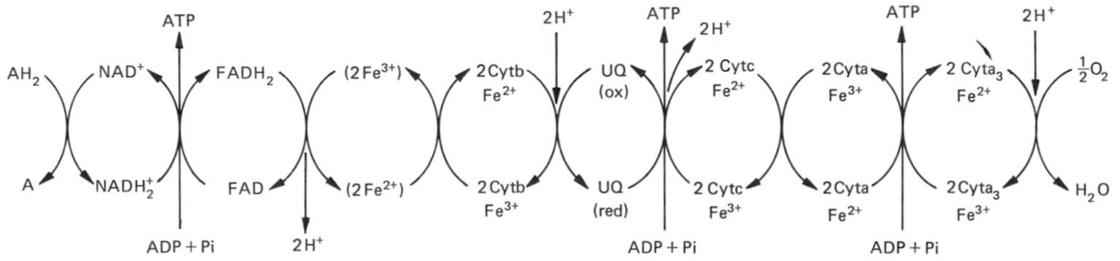

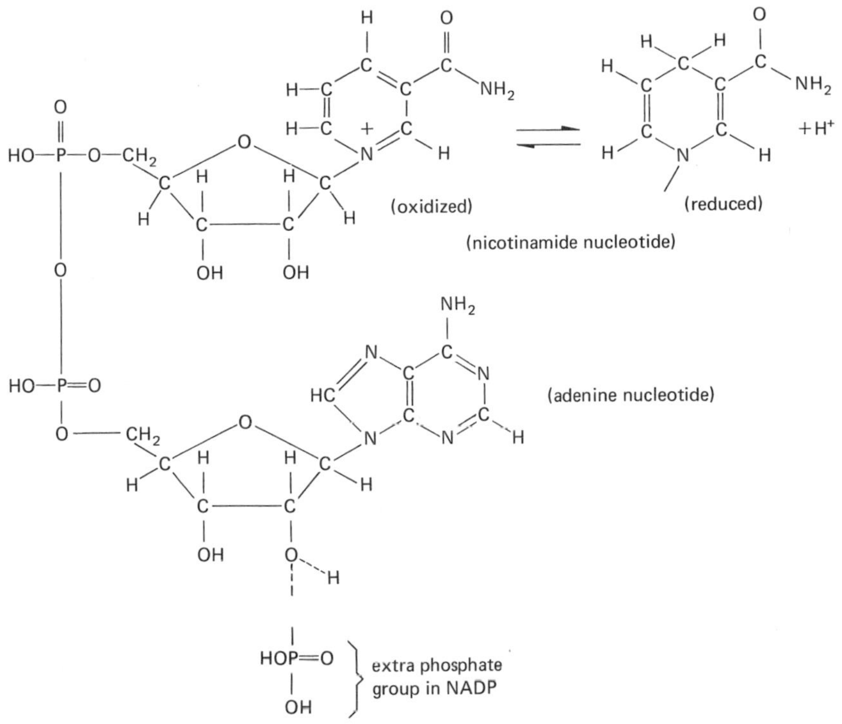

(oxidized) (reduced)

(nicotinamide nucleotide)

(adenine nucleotide)

extra phosphate group in NADP

Figure 5-3. Structure of nicotinamide adenine dinucleotide (NAD^+) and nicotinamide adenine dinucleotide phosphate ($NADP^+$). Reduction to NADH + H^+ or NADPH + H^+ takes place as shown at the nicotinamide portion of the molecule.

otides is shown in Figure 5-3. Two electrons and two H^+ ions are transferred to NAD^+, reducing it to NADH + H^+, sometimes written $NADH_2^+$ or $NADH_2$. $NADH_2^+$ then transfers two electrons and two H^+ ions to a flavin enzyme, either **flavin mononucleotide (FMN)** or **flavin adenine dinucleotide (FAD),** thus reducing it (Figure 5-4). The energy required to reduce FAD is somewhat less than the energy liberated by oxidation of $NADH_2^+$, and the excess is used to synthesize a molecule of ATP. $FADH_2$ in turn reduces an enzyme that has not been well characterized, but contains a **nonheme iron** coupled with —SH groups (not shown in Figure 5-4). This in turn reduces two molecules of the iron-porphyrin electron-transfer enzyme, **cytochrome b** (see Figure 5-5 for a diagram of a typical cytochrome). The reduction and oxidation of cytochromes are accomplished by the addition or removal of one electron at the iron portion of the molecule, converting it from a valence of +2 to +3 and back. Cytochrome b reduces a phenolic compound to its corresponding quinone, **ubiquinone** (Figure 5-6); hydrogen ions must be added at this point as well as the electrons. The hydrogen ions are not necessarily the same ones that left the chain at the oxidation of $FADH_2$—the system is aqueous, and a certain amount of H^+ is always present. Electrons from ubiquinone in turn reduce **cytochrome c;** two hydrogen ions again leave the electron transport chain. Sufficient energy is liberated at this point to synthesize a second molecule of ATP for each two electrons transferred. Cytochrome c reduces **cytochrome a,** which in turn reduces **cytochrome a_3,** a third ATP being generated for each two electrons transferred at this point.

Cytochrome a_3 is the only known member of the electron transport chain that can react with molecular oxygen. Cytochrome a and a_3 form a molecular association called

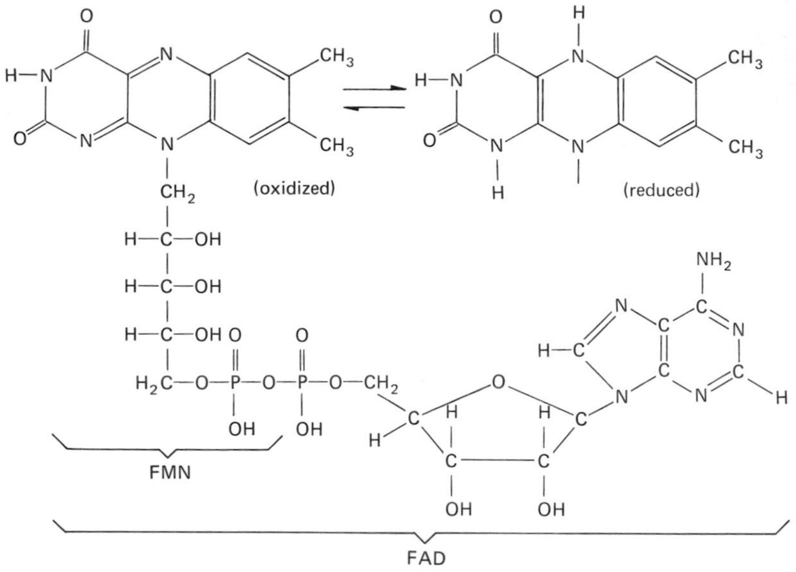

Figure 5-4. Structure of flavin mononucleotide (FMN) and flavin adenine dinucleotide (FAD). Reduction to $FMNH_2$ or $FADH_2$ takes place at the flavin portion of the molecule, as shown.

cytochrome oxidase that has not yet been chemically separated. The two enzymes appear to operate independently, but experiments have shown that they may modify each other's chemical behavior. In addition to the iron atom present in each, these two cytochromes are characterized by the presence of a copper atom in each; the copper atoms appear also to be involved in the electron transport process. The exact mechanism of reaction of the cytochrome a–a_3 complex with oxygen is not yet known. Two electrons

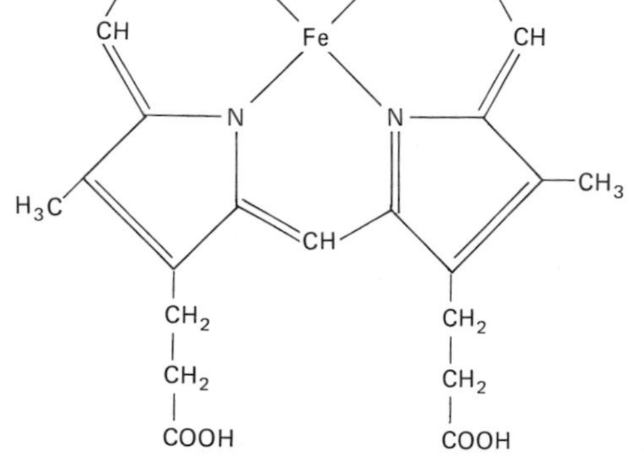

Figure 5-5. Structure of the porphyrin ring of cytochrome c, a typical cytochrome. The porphyrin is probably attached to its protein through the SH groups and by interaction of the iron atom with reactive groups in the protein.

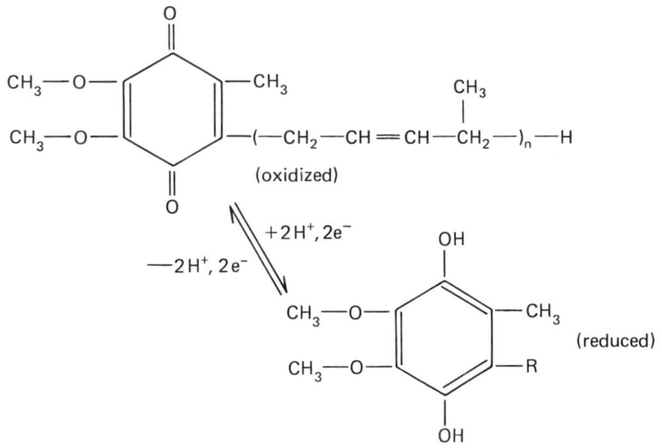

Figure 5-6. Ubiquinone. The quinone becomes a phenol when reduced. The side chain (R) is composed of 6–10 isoprenoid units (see Chapter 9).

are transferred to one atom of oxygen ($\frac{1}{2}$ O_2) together with 2 H^+ to make H_2O. This completes the transfer of two electrons from the high energy level they occupied in the fuel molecule, AH_2, to the low energy level they occupy in water. Much of the energy released by the oxidation of the fuel molecule is conserved in the three ATP molecules synthesized along the electron transfer process.

The electron transport system outlined operates in the mitochondria. Similar reactions are central to the energy trapping and storing reactions of photosynthesis. The modifications of these systems, their control and operation, and their relationship to the overall processes of metabolism, as well as the experimental evidence on which these ideas are based will be considered in the next chapter. It must be emphasized that there are many alternative routes by which electrons may pass from substrates to oxygen, but the cytochrome electron transport chain is the only one that is capable of synthesizing ATP. All other systems waste the energy derived from the oxidation of the substrate or use it directly for the coupled reduction of other substrates. The cytochrome chain is thus most important in the overall energy metabolism of the cell.

Measuring Energy Changes

The most convenient method of measuring the energy available in a bond is to measure the **standard free energy** (that is, the amount of energy available to do work) on hydrolysis of the bond. The chemical reaction

$$A + B \rightleftharpoons C + D$$

can do useful work. This is measured by the *change* in standard free energy for the reaction, ΔG^0, which can be derived from

$$\Delta G^0 = -RT \ln K$$

where R = the gas constant (1.99 cal/°C/mole); T = the absolute temperature; and K = the equilibrium constant of the reaction when the reactants are at unit activity (essentially, molar concentration). The value ΔG^0 is useful for comparing reactions, but the actual energy available is dependent on the concentration (indicated by square brackets) of the reactants and can be determined from the expression

$$\Delta G = \Delta G^0 + RT \ln \frac{[C][D]}{[A][B]}$$

ΔG thus measures the actual free energy available under any given set of conditions other than the standard ones (that is, molar concentration of the reactants). The standard free energy of hydrolysis of a number of important compounds is shown in Table 5-1. The relationship between the equilibrium constant of a reaction and the free energy available or required is shown in Table 5-2.

To calculate free energy changes for oxidation-reduction reactions we need a measure of the tendency of substances to donate or accept electrons. This is found by measuring the electric potential of the compound against that of a hydrogen electrode at unit activity (pH 0). The standard oxidation-reduction potential, E_0, is measured potentiometrically using the equation

$$E = E_0 - \frac{RT}{nF} \ln \frac{[\text{reduced}]}{[\text{oxidized}]}$$

when E = the observed potential in volts, R = the gas constant, T = the absolute temperature, n = the number of electrons transferred, and F = the faraday constant (23,000 cal/v). Thus $E = E_0$ when the reactants are at unit or equal concentration. The more commonly used value is the standard oxidation-reduction potential at pH 7 (instead of pH 0), designated E_0'. The E_0' values of a number of important biological compounds are given in Table 5-3.

The work that can be done by an oxidation-reduction reaction in terms of change of free energy, ΔG, can be calculated from the relationship

$$\Delta G = -nF \Delta E_0$$

Table 5-1. Standard free energy of hydrolysis at pH 7 ($-\Delta G^0$) for some biologically important compounds

Compound	$-\Delta G^0$, *cal/mole*
Acetyl-coenzyme A	10,500
ATP	7,400
Phosphates (ester link)	3,000
Sugar(aldose)-1-phosphates	5,000
Phosphoenolpyruvate	13,000
Glutamine (amide)	3,400
Glycoside	3,000
Sucrose	6,570
Uridine diphosphate glucose	7,600

SOURCE: H. R. Mahler and E. H. Cordes: *Biological Chemistry,* 2nd ed. Copyright 1966, 1967 by Henry R. Mahler and Eugene H. Cordes. By permission of Harper & Row, Publishers, Inc.

Table 5-2. Relationship between equilibrium constant of a reaction (K) and the standard free energy change in the reaction

K	ΔG^0, *cal/mole*	*Type of reaction*
0.001	4089	Endergonic
0.01	2726	Endergonic
0.1	1363	Endergonic
1	0	
10	−1363	Exergonic
100	−2726	Exergonic
1000	−4089	Exergonic

Table 5-3. Standard oxidation-reduction potentials (E'_0) of a number of biologically important compounds

Reaction	E'_0, v	*Reaction*	E'_0, v
O_2/H_2O	0.815	Oxaloacetate/malate	−0.17
Fe^{3+}/Fe^{2+}	0.77	Acetaldehyde/ethanol	−0.20
Cyt a, Fe^{3+}/Fe^{2+}	0.29	Riboflavin, ox/red	−0.21
Cyt c, Fe^{3+}/Fe^{2+}	0.22	Glutathione, ox/red	−0.23
Cyt b_5, Fe^{3+}/Fe^{2+}	0.02	Lipoic acid, ox/red	−0.29
Ubiquinone, ox/red	0.10	$NAD^+/NADH_2^+$	−0.32
Dehydroascorbic acid/ascorbic acid	0.08	H^+/H_2	−0.42
Fumarate/succinate	0.03	Succinate/α-ketoglutarate	−0.67
$FMN/FMNH_2$	−0.12	Acetate + CO_2/pyruvate	−0.70

SOURCE: H. R. Mahler and E. H. Cordes: *Biological Chemistry,* 2nd ed. Copyright 1966, 1967 by Henry R. Mahler and Eugene H. Cordes. By permission of Harper & Row, Publishers, Inc.

where n = the number of electrons transferred and F = the faraday constant. Thus it may be seen that the synthesis of a molecule of ATP via a two-electron transfer oxidation-reduction reaction such as cytochrome b to cytochrome c would require a ΔE_0 of 0.161 v:

$$-7400 = -2 \times 23{,}000 \times \Delta E_0$$
$$\Delta E_0 = 0.161 \text{ v}$$

In fact, it may be seen in Table 5-3 that the $\Delta E'_0$ of the cytochrome b–cytochrome c electron transfer is approximately 0.2 v. This is more than enough energy to make a molecule of ATP. The residue of energy is not conserved, but is used to push the equilibrium over toward the synthesis of ATP. In other words, it is used to "make the reaction go"—to make it go more rapidly and more nearly to completion.

This is a most important concept in biochemistry. As was pointed out earlier, the free energy available from a reaction depends on the concentration of the reactants and products, as well as on the reaction constant. The higher the concentration of reactants, the faster the reaction will go; the higher the concentration of products, the slower the reaction will go. A reaction that has a large equilibrium constant (see Table 5-2) will go nearly to completion in spite of the fact that the concentration of products will be high and the concentration of reactants will be low. A series of reactions can form a sequence in which one or more of the products of one reaction form the reactants of the next. In such a sequence, if one of the reactions has a large equilibrium constant, that is, it is a strong exergonic reaction, then it will tend to drive the whole sequence of reactions. Thus, the apparently wasted energy in the powerful exergonic reaction in the sequence is in fact being used to drive the whole sequence of reactions.

Many biological reaction sequences contain such an exergonic reaction, often the hydrolysis of a high energy or intermediate energy phosphate bond by a phosphatase enzyme. This energy-liberating reaction thus serves to keep the whole reaction sequence going in a forward direction, and prevents it from coming to equilibrium with a large amount of unreacted substrate still present. These considerations make it clear that the direction in which a reaction goes is governed not only by its equilibrium constant, but also by the concentrations of the reactants and products. It is thus possible for an

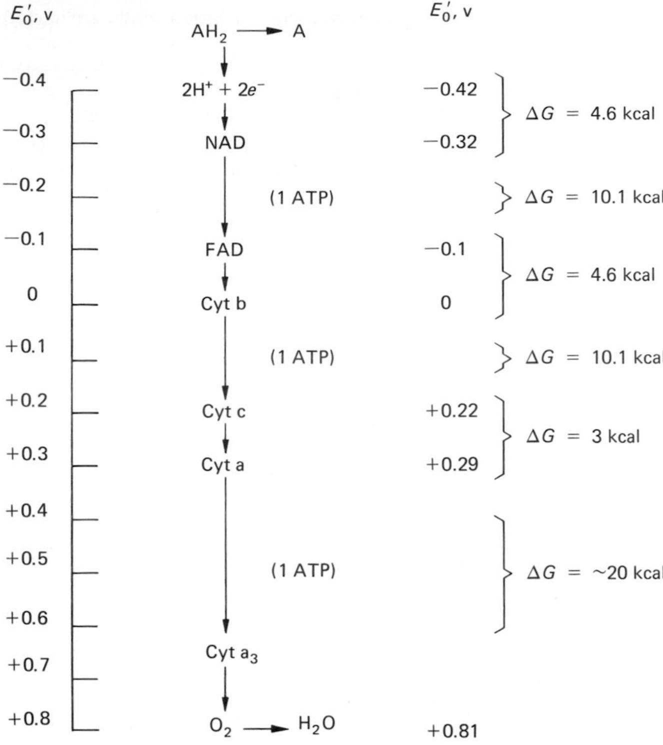

Figure 5-7. Approximate energy levels of intermediates in the electron transport chain. Overall free energy change from E_0' − 0.42 v to + 0.81 v is about 56 kcal/mole for the transfer of 2 electrons. The synthesis of ATP requires about 7.5 kcal/mole.

apparently unfavorable reaction to be used in a biosynthetic process, in spite of its large energy requirement, by coupling it with another energy-liberating reaction.

An energy model of the electron transport system described earlier in this chapter is presented in Figure 5-7. This places the components of the system in perspective, showing the points at which energy is used to make ATP and where energy is wasted or used to make a reaction run against unfavorable concentration gradients. It should be evident that the exact presentation of reduction-oxidation potentials or changes in free energy is not possible because the concentration of reactants in biological systems varies from situation to situation. Thus, the model in Figure 5-7 is only an approximation.

High Energy Compounds

Certain molecules such as ATP are important in driving many of the metabolic and synthetic reactions of biological systems. Since these molecules essentially provide the energy to make reactions go, they have been classified by F. Lipmann as high energy compounds. These compounds are characterized by a large negative free energy of hydrolysis; that is, on hydrolysis they yield large amounts of energy. Compounds that yield only small amounts of energy are known as **low energy compounds.** As a general guide, high energy compounds usually yield at least 7000 cal/mole or have an E_0' value of +0.3 v or lower.

There are several major types of high energy compounds, the most common of which contain high energy phosphate bonds, often abbreviated ~P. The most important of these are phosphoric acid anhydrides (P-P) such as ATP, carboxylic-phosphoric anhydrides such as acetyl phosphate, (acetyl-P), and enol phosphates such as phosphoenolpyruvate (PEP). Essentially these bonds are unstable; their hydrolysis, by the introduction of a molecule of water, results in the formation of a far more stable product or products with a resultant loss of energy. Other high energy bonds of importance are the thiol esters, of which the most important is acetyl-coenzyme A (acetyl-CoA). Certain amino acid esters can be classified as high energy compounds, as can such compounds as S-adenosylmethionine (a methyl group donor) or uridine diphosphate glucose (a glucosyl donor). The ΔG^0 of each of these compounds is listed in Table 5-1. Many electron donors, such as $NADH_2^+$, $NADPH_2^+$, and lipoic acid, that have low E_0' values (Table 5-3) can clearly be classified as high energy compounds. All these compounds are important in the energy metabolism of the plant.

Mechanism of ATP Synthesis

How is the metabolic energy that is released in the electron transport chain coupled to the formation of ATP? This is a question that has baffled scientists for decades, and the answer is not yet clear. Several theories have been put forward, and one of these, the **chemiosmotic theory,** has found widespread acceptance. However, it must be emphasized that the details are not yet known, some equivocal data are not yet understood, and no theory of ATP synthesis has as yet been proved.

The **chemical hypothesis** of ATP synthesis involves the formation of a high energy bond with a hypothetical protein intermediate while a member of the respiratory chain is reduced

$$AH_2 + \text{enzyme} + \text{protein} \longrightarrow A \sim \text{protein} + \text{enzyme-}H_2$$

The energy in the protein-substrate bond is then used to synthesize ATP

$$A \sim \text{protein} + \text{ADP} + \text{Pi} \longrightarrow A + \text{protein} + \text{ATP}$$

The chemiosmotic hypothesis, put forward by the British biochemist P. Mitchell (now working in the United States), is a modification of this. The respiratory chain is used to separate charges in the reaction

$$\underset{\text{hydrogen}}{H} \longrightarrow \underset{\text{hydrogen ion}}{H^+} + \text{electron}^-$$

The two charged species are separated on opposite sides of the mitochondrial membrane by the electron carrier enzymes, which, according to the hypothesis, are so arranged in the inner mitochondrial membrane that they transport hydrogen to the outside and electrons to the inside. As a consequence, hydrogen ions, separated from electrons in the electron transfer reactions, are passed to the outside of the inner mitochondrial membrane. Furthermore, the hydrogen atoms transported through the membrane must be derived from water by the reactions

$$H_2O \longrightarrow OH^- + H^+$$
$$H^+ + e^- \text{ (from electron carrier)} \longrightarrow H$$

As a result, hydroxyl ions accumulate on the inside of the inner membrane. This situation, and the way in which the various electron transport components are thought to participate, is shown in the upper part of Figure 5-8.

It must be remembered that the reactions forming ATP from ADP and Pi involve the removal of a water molecule. The situation created by the spatially organized electron and hydrogen ion transfer reactions in the preceding equations generates a powerful chemical and potential gradient, since hydrogen ions are on the outside of the inner membrane of the mitochondria while hydroxyl ions are on the inside, and the outside of this membrane becomes positively charged while the inside becomes negatively charged. This potential tends to pull the components of water together very strongly. However, the gradient cannot collapse with the simple formation of water and the liberation of heat because the inner mitochondrial membrane is essentially impermeable to hydrogen or hydroxyl ions. There are, however, pathways through which hydrogen ions may penetrate the membranes, and these are via the stalked protrusions of the inner mitochondrial membrane that contain the enzyme ATPase.

ATPase, like most enzymes, is capable of catalyzing the reaction either forward or backward according to existing conditions, and so it can not only hydrolyze ATP but also synthesize it. It is hypothesized that the F_1-ATPase particles (see page 56) are so arranged that hydrogen ions can enter via the F_0 particles, or stalks, only when ADP and Pi are present. Under the influence of the strong potential gradient, two hypothetical intermediates, X and I (at least one of which is thought to be an active site on the F_1-ATPase) form an anhydride bond that acts to remove oxygen from hydroxyl groups of the Pi. The oxygen is used to form water with hydrogen ions entering from the outside. Under the influence of the same gradient, hydrogen ions from ADP and from the hydroxyl groups of the Pi leave the F_1-ATPase on the inside, where they form water by combining with hydroxyl groups derived from the electron transport process previously described. The ADP and Pi radicals so formed unite to generate ATP. This sequence is shown diagramatically in the lower half of Figure 5-8.

It is evident that the action of **uncoupling agents,** chemicals that allow electron transport to proceed (often at greatly accelerated rates) without the concomitant formation of ATP, can be readily explained by the chemiosmotic hypothesis. Their action is thought to result from their effect on membranes: they render the membranes more permeable to hydrogen ions, which can leak across the membrane and so collapse the gradient with the direct formation of water and the resulting loss of energy. Similarly, the coupling of electron transport to the transport of charged particles (for example, K^+, Ca^{2+}) across membranes is easily explained by this hypothesis. So long as the ion can permeate the membrane, it will diffuse across down the electrochemical gradient generated by the hydrogen ion gradient across the membrane. This coupling will be examined in greater detail in Chapter 12 where ion transport is considered.

It must be remembered that this scheme is still hypothetical. However, it seems to fit the experimental data better than alternative hypotheses, and many biochemists now believe that this, or some similar scheme, gives the best account now possible for ATP synthesis. Alternative ideas include the high energy intermediate or **chemical-coupling** hypothesis and the **paired moving charge** hypothesis. The chemical-coupling hypothesis requires high energy intermediates that have never been isolated, and it is difficult to

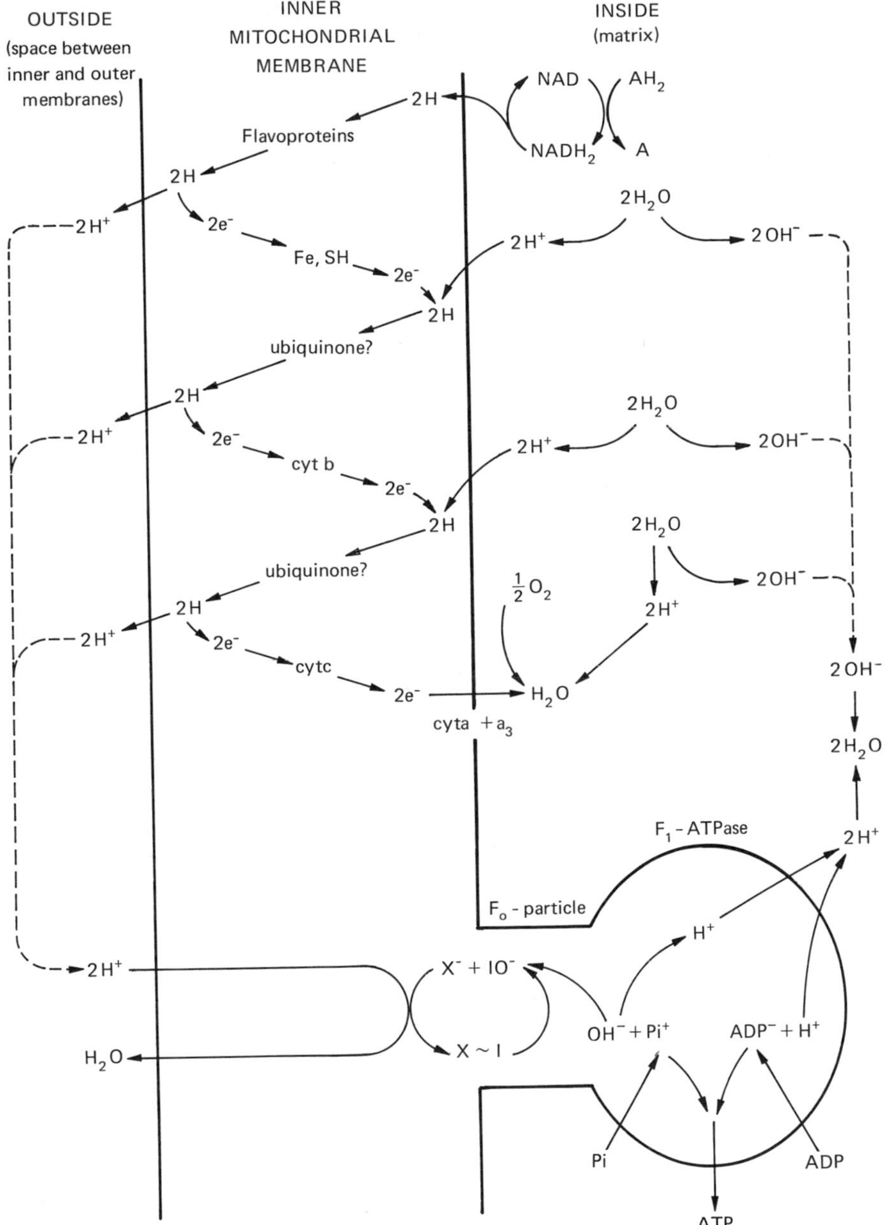

Figure 5-8. Schematic diagram of mitochondrial membrane showing how the electron transfer system, by alternately carrying protons + electrons, or electrons only, could generate a proton-hydroxyl gradient across the membrane (upper part of diagram). The lower part of the diagram shows one hypothetical way in which the gradient could be discharged through the ATPase particles to make ATP. Other mechanisms are also possible.

explain the action of uncouplers through this hypothesis. The paired moving charge hypothesis requires that electrons move across membranes via specified channels under the influence of the electrochemical gradient formed by electron carriers, and that positively charged ions, encapsulated in special proteinaceous molecules called **ionophores,** move in parallel channels or "tunnels" under the influence of coulombic interactions between the negatively charged electrons and the positively charged ionophores. Again, the required structures are hypothetical, and the evidence supporting this hypothesis is not very strong.

Among the strongest evidence supporting the chemiosmotic hypothesis is the fact that the required pH gradient can be demonstrated, and, conversely, if a pH gradient is applied to isolated mitochondria or chloroplasts, then ATP synthesis takes place. The chemiosmotic hypothesis also requires intact membranes and the complete isolation of the internal and external spaces around the chemiosmotically active membrane, criteria that must be met if mitochondrial or chloroplast phosphorylation is to take place. The weight of biochemical evidence now seems to lie in favor of the relatively simple and elegant chemiosmotic hypothesis.

Group Transfer Reactions

It is evident that high energy compounds may react energetically with a common component of their environment, usually water. The formation of many chemical bonds in the synthesis of biological molecules and structural elements requires the elimination of water, as for example in the synthesis of sucrose

$$\text{glucose} + \text{fructose} \longrightarrow \text{sucrose} + H_2O - 6600 \text{ cal}$$

This reaction, as written, cannot be conducted by biological systems. Only certain reactions can eliminate water and form anhydride bonds; the most common of these is the synthesis of ATP

$$\text{ADP} + \text{Pi} \longrightarrow \text{ATP} + H_2O - 7200 \text{ cal}$$

the energy being supplied by electron transport reactions. Once energy is present in an anhydride bond, it may be conserved by the transfer of a group without the intervention of water. Thus energy may be transferred to another molecule with the transfer of the group.

$$\text{ATP} + \text{glucose} \longrightarrow \text{glucose-1-phosphate} + \text{ADP} + 1200 \text{ cal}$$

In this case, the phosphate group is transferred. The glucose-1-phosphate ester now has most of the energy from the high energy phosphate bond of ATP. (Not quite all the energy—1200 cal is lost, but this serves to make the reaction go.)

An enzyme from the organism *Pseudomonas,* sucrose phosphorylase, can catalyze the next group transfer (in this case it is the glycosyl group that is transferred)

$$\text{glucose-1-phosphate} + \text{fructose} \longrightarrow \text{sucrose} + \text{Pi} - 600 \text{ cal}$$

and much of the original energy is still stored but has now been transferred to the anhydride bond present in sucrose. In this way, the hydrolysis of ATP has been coupled to the synthesis of sucrose.

The synthesis of sucrose in higher plants actually takes place by an even more interesting sequence of group-transfer reactions involving the nucleotides UDP, UTP, and a glycosyl-substituted nucleotide, **uridine diphosphate glucose** (UDPG) in the following sequence

$$\text{UDP} + \text{ATP} \longrightarrow \text{UTP} + \text{ADP}$$

ATP is resynthesized by electron transport reactions elsewhere in the cell. The UTP then reacts with glucose-1-phosphate (G-1-P) to generate the glucosyl-transferring molecule UDPG, and **pyrophosphate,** which is a molecule consisting of two phosphates joined by an anhydryl bond (P ~ P), written PPi.

$$\text{UTP} + \text{G-1-P} \longrightarrow \text{UDPG} + \text{PPi}$$

$$\text{PPi} + H_2O \longrightarrow 2\ \text{Pi} + 8000\ \text{cal}$$

The pyrophosphate is hydrolyzed to inorganic phosphate with the release of a large amount of energy. This reaction tends to drive the whole sequence forward. UDPG then transfers the glucose moiety to a molecule of fructose to generate sucrose

$$\text{UDPG} + \text{F} \longrightarrow \text{sucrose} + \text{UDP}$$

Alternatively, UDPG may react with fructose-6-phosphate (F-6-P) to make sucrose phosphate

$$\text{UDPG} + \text{F-6-P} \longrightarrow \text{sucrose phosphate}$$

$$\text{sucrose phosphate} + H_2O \longrightarrow \text{sucrose} + \text{Pi} + 3000\ \text{cal}$$

The hydrolysis of sucrose phosphate yields further energy, which makes possible the accumulation of high concentrations of sucrose. This sequence of reactions, involving both group-transfer reactions and exergonic reactions, effectively permits the synthesis and concentration of very large amounts of sucrose such as are found in sugarcanes and sugarbeets, and in the cells of photosynthetic plants. Once again, however, through several intermediate stages, the hydrolysis of ATP (and of G-1-P) has been coupled to the synthesis of sucrose.

The principle of energy conservation through group-transfer reactions is very important. Once a high energy bond is hydrolyzed, its energy is wasted. Sometimes, however, it is necessary to hydrolyze such bonds directly to force reactions to go. Thus, the energy of hydrolysis of a high energy bond can be used to make a reaction go in spite of a large concentration of the product. Cells contain a number of hydrolytic enzymes that do this. However, the cell must have effective means of preventing the indiscriminate action of such enzymes as **phosphatase** (which hydrolyzes phosphates) or **ATPase** (which attacks ATP) from wastefully destroying all available substrate. Such enzymes are usually secreted in special bodies or compartments of the cell and only released when needed. Most phosphatases are highly specific; they are only present in those organelles where they are required for metabolic purposes but are excluded from those organelles where synthetic reactions require the maintenance of effective concentrations of phosphorylated substrates.

The "Energy Charge" Concept and Metabolic Control

Cells contain a finite amount of energy-storing compounds, particularly the adenosine phosphates (AMP, ADP, ATP), that can be present as either high or low energy compounds. Thus a cell can bc said to be "fully charged" when all its adenylate is present as ATP. Similarly, when all ATP is hydrolyzed to AMP, the cell is "fully discharged." These energy states are analogous to the states of an electrolytic battery, that can be charged or discharged.

The level of charge in a cell can be calculated from the expression

$$\text{percent charge} = \frac{[\text{ATP}] + \frac{1}{2}[\text{ADP}]}{[\text{ATP}] + [\text{ADP}] + [\text{AMP}]} \times 100$$

This gives a value representing the energy status of a cell compared with its fully charged condition. The approximate relationship of amounts of ATP, ADP, and AMP, to the percent charge of a cell is shown in Figure 5-9.

Cells are normally at about 80 percent of full charge. This level is maintained by a mechanism called **feedback control.** Feedback means that some product of a reaction sequence influences one of the reactions leading to its production so that a constant level of the product can be maintained. A household thermostat is a feedback controller. The amount of heat produced (by the furnace) is influenced by the amount of heat present (as indicated by temperature); more or less is added as needed.

The feedback mechanisms that maintain energy-charge balance in cells are as follows. Certain reactions that synthesize ATP are influenced positively by the concentration of AMP (that is, the formation of ATP is increased by increasing AMP concentration). Some ATP-utilizing reactions are influenced positively by the amount of ATP and negatively by the amount of AMP present (that is, increasing ATP speeds its utilization, and increasing AMP slows its production). Thus control is achieved not only by the absolute amounts of ATP or AMP present but by their relative concentrations. Other controls of this sort will be discussed in more detail in Chapter 6.

The operation of the feedback system may be seen from the solid-line graphs in Figure 5-10. When the ATP/AMP ratio is very low, the cell's energy budget is low and the cell is discharged. ATP-synthesizing reactions then run at high rate, ATP-utilizing reactions are slowed, and the cell becomes charged. When the charge level approaches 80 percent, the ATP-synthesizing reactions slow down and ATP-utilizing reactions speed up until a balance is achieved somewhere about 20 percent short of full charge.

In addition, however, the reactions generating or using ATP may have other functions. One of the most important of these is the provision of intermediates for the syntheses of cell components. It is thus important that some secondary controls be incorporated into this charge-balancing mechanism, or synthetic reactions might be completely cut off in a fully charged cell. Thus, many such reactions are also sensitive to

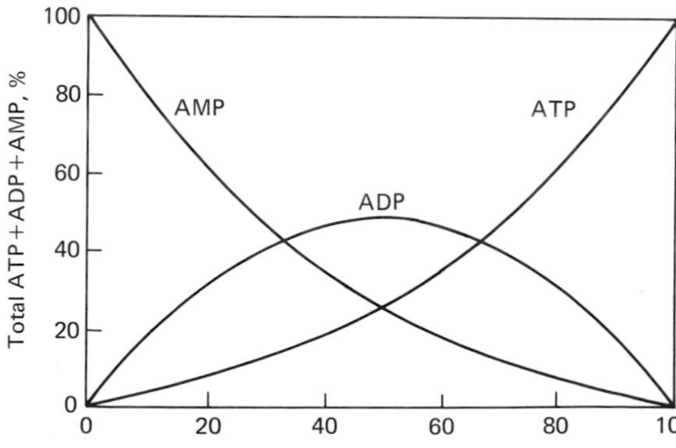

Figure 5-9. Relationship between concentrations of ATP, ADP, and AMP (as percent of total adenylate) to the percent charge of a cell.

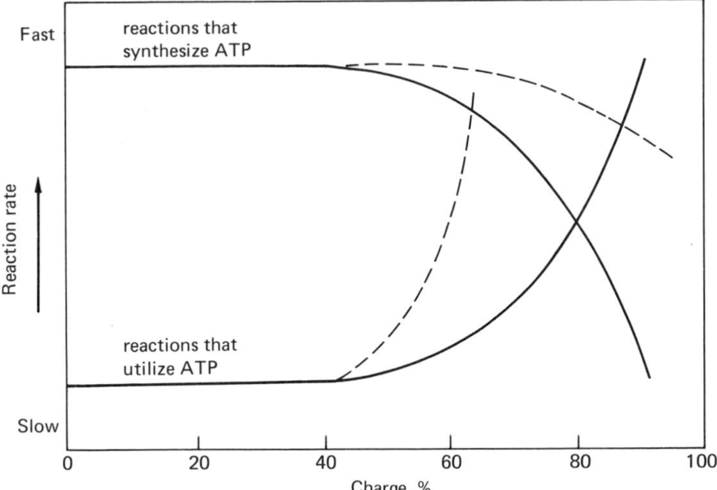

Figure 5-10. Maintenance of charge at about 80 percent by feedback-controlled reactions. Dotted lines show the situation where reactions also lead to needed intermediates, resulting in a tendency to "over-" or "undercharge."

the concentrations of intermediates resulting from their operation (often several reaction steps removed). The effect of this sensitivity is to modify the behavior of "charging" or "discharging" reactions as shown by the dotted lines in Figure 5-10, leading to a tendency toward "overcharging" or "undercharging." This is presumably the reason why charge balance is normally maintained at about 80 percent—it is sufficient for emergency needs, yet not so high as to prohibit some flexibility of operation.

The importance of the concept of feedback control in maintaining a specific condition in a dynamic system cannot be overemphasized. It is the most important single device for regulating and controlling the activities of cells and organisms and for maintaining constant appropriate conditions within the organism in the presence of continually shifting external or environmental factors. It is the prime means by which plants are protected from being entirely at the mercy of the equilibrium constants of their reactions and the concentrations of their metabolites. Feedback controls prevent the wasteful oxidation of all available substrates or the accidental overproduction of unwanted metabolites. They are essential for the maintenance and balance of all the metabolic activities of organisms.

Enzyme Action

Chemical reactions run in the direction that liberates energy. However, most reactions will not run spontaneously, even in a forward direction, without some initial input of energy. For example, although wood burns with the liberation of large amounts of energy, it will not do so spontaneously but must be ignited. Before molecules can react together, a certain amount of energy must be introduced to activate them; this energy requirement is called the **energy of activation.** The energy input is necessary to make the molecules more reactive, perhaps by bringing molecules into closer association or by putting them under a stress or strain of some sort. Certain substances, called **catalysts,** which are not themselves consumed in reactions, have the effect of reducing the energy of activation and thus making molecules more reactive. We have already mentioned **enzymes;** these are special protein molecules in cells and organisms that act as biological catalysts. Enzymes function by reducing the energy of activation of molecules, thus enabling reactions that are thermodynamically possible to go.

The mechanism of enzyme action is illustrated diagramatically in Figure 5-11. The structure of each enzyme is so arranged that it can bond (by hydrogen bonds, ionic forces, and weak intermolecular forces) with the substrate. In so doing, the substrate becomes activated, perhaps by being held in closer conjunction with another substrate, or by being placed under a strain (that is, by molecular distortion). The substrate reacts, and the product is released from the enzyme surface, as shown in Figure 5-11A. The enzyme remains unchanged and free to mediate the reaction of more molecules of substrate. Many enzymes are reversible, that is, they will mediate a reaction either in a forward or a backward direction, provided that this is thermodynamically possible. It must be recognized that an enzyme does not change the direction of a reaction—only its rate.

Figure 5-11. Diagrammatic representation of enzyme action or inhibition. In fact, the "fit" between enzyme and substrate is not a geometric one, as shown, but the result of many points of interaction of weak attractive forces, hydrogen bonds, and the like. It should be remembered that the relative sizes of enzyme and substrate or inhibitor molecules is probably not as shown; the enzyme may be hundreds of times larger.

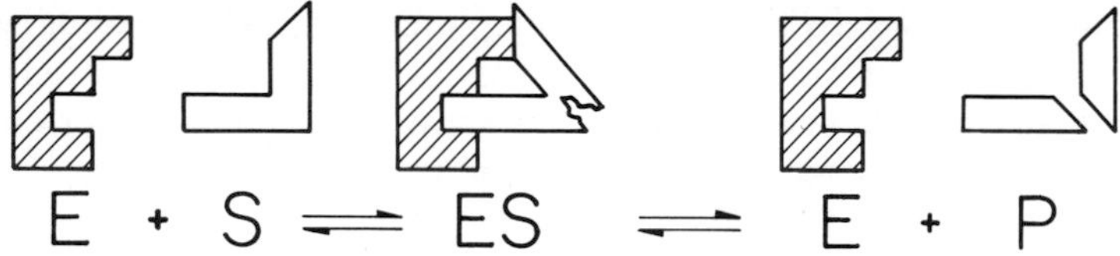

A. Enzyme action. E = enzyme, S = substrate, P = product.

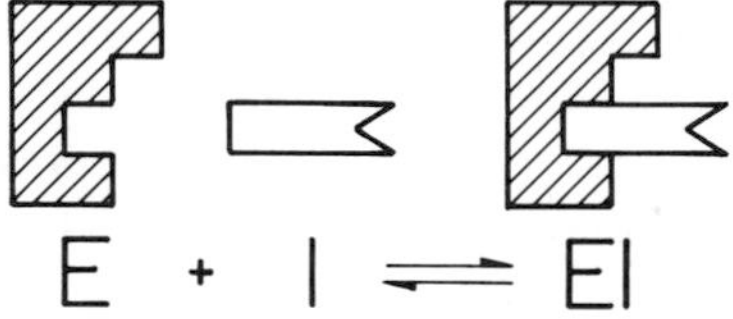

B. Inhibition by competition for the reactive site. The more strongly E binds with I, the less easily EI will dissociate, and the more strongly inhibitory (or toxic) I will be. I = inhibitor.

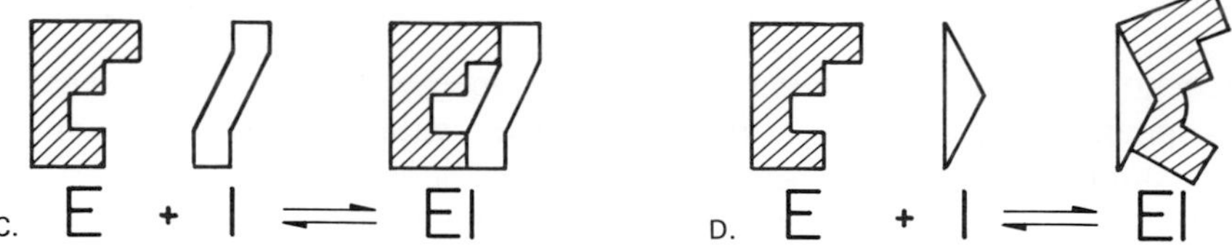

C and D. Inhibition by inactivation of the enzyme, not involving the reactive site. In C the reactive site is masked, and in D it is distorted by an allosteric inhibitor.

Enzymes may be inhibited by a poison that combines with the reactive site of the enzyme and thus competes with the substrate (Figure 5-11B). In this case, if the enzyme-inhibitor complex can dissociate, the inhibition can be overcome by increasing the concentration of the substrate. Alternatively, the inhibitor may complex with some other site on the enzyme molecule in such a way that the enzyme is prevented from combining with the substrate (Figures 5-11C and D). Because the inhibitor and substrate are not competing for the same reaction site, this type of inhibition cannot be relieved by the addition of more substrate. There are also certain substances that activate enzymes, that is, make them more effective. This is the basis of **allosteric effects,** in which a molecule other than the substrate reacts with a special site on the enzyme, separate from the reactive site, and causes a **conformational change** (that is, a change in the shape or tertiary structure of the enzyme), that either activates or inhibits the enzyme. Many allosteric effects are involved in the feedback control of metabolism. For example, the end product of a reaction sequence involving several steps and several enzymes may allosterically inhibit an earlier step in its own production, so that the rate of synthesis of the end product is controlled by the amount present. Some examples of this important mechanism will be discussed in the next chapter.

Additional Reading

Lehninger, A. L.: *Biochemistry.* Worth Publisher, Inc., New York, 1970, Chaps. 8, 13, 14, 17.
Lehninger, A. L.: *Bioenergetics.* W. A. Benjamin, Inc., Menlo Park, Calif., 1971.
Peusner, L.: *Concepts in Bioenergetics.* Prentice-Hall, Inc., Englewood Cliffs, N.J., 1974.
Westley, J.: *Enzyme Catalysis.* Harper & Row, New York, 1969.

6 Respiration

Introduction

So far we have looked only at the flow of energy. However, organisms also have mass, and they both acquire mass in their synthetic reactions and lose it in respiration. Furthermore, the reactions whereby energy is transformed and used are chemical. The flow of materials, in syntheses and respiration, is just as important as the flow of energy. In this chapter we shall study the overall process of respiration as it occurs in the cells and organs of plants. Our main concern will be with the sources of carbon, the intermediary metabolism, and the control systems that regulate respiration. Later (particularly in Chapters 15 and 21) we shall examine in more detail the relationships between respiration and other metabolic systems, and the patterns of respiration in the developing plant.

The primary process of respiration is the mobilization of organic compounds and their controlled oxidation to release energy for the maintenance and development of the plant. We shall consider first the carbon reactions that are summarized in the equation

$$C_6H_{12}O_6 + 6\ O_2 \longrightarrow 6\ CO_2 + 6\ H_2O + \text{energy}$$

representing the oxidation of a molecule of hexose. The carbon reactions of respiration involve two distinct processes. The first, **glycolysis,** is a series of reactions constituting the **Embden–Meyerhoff–Parnass (EMP)** pathway (named for three of the principal scientists whose work led to its elucidation), which also forms the basis of anaerobic respiration or **fermentation.** The EMP pathway converts a molecule of hexose to two molecules of pyruvic acid. These are then decarboxylated, and the remaining two-carbon fragments are completely oxidized by the second of the two main processes, the **tricarboxylic acid** or **citric acid cycle,** also called the **Krebs cycle** for the famous British biochemist Sir Hans Krebs who first demonstrated the reactions. We shall also examine an important pathway of hexose catabolism that bypasses the EMP pathway, the **hexose-monophosphate shunt** or **pentose shunt.**

Glycolysis

Reactions. The reactions of the EMP pathway of glycolysis are outlined in Figure 6-1 together with the enzymes that catalyze each reaction. The first step uses ATP to

Figure 6-1. Reactions and enzymes of the Embden-Meyerhoff-Parnass (EMP) pathway of glycolysis.

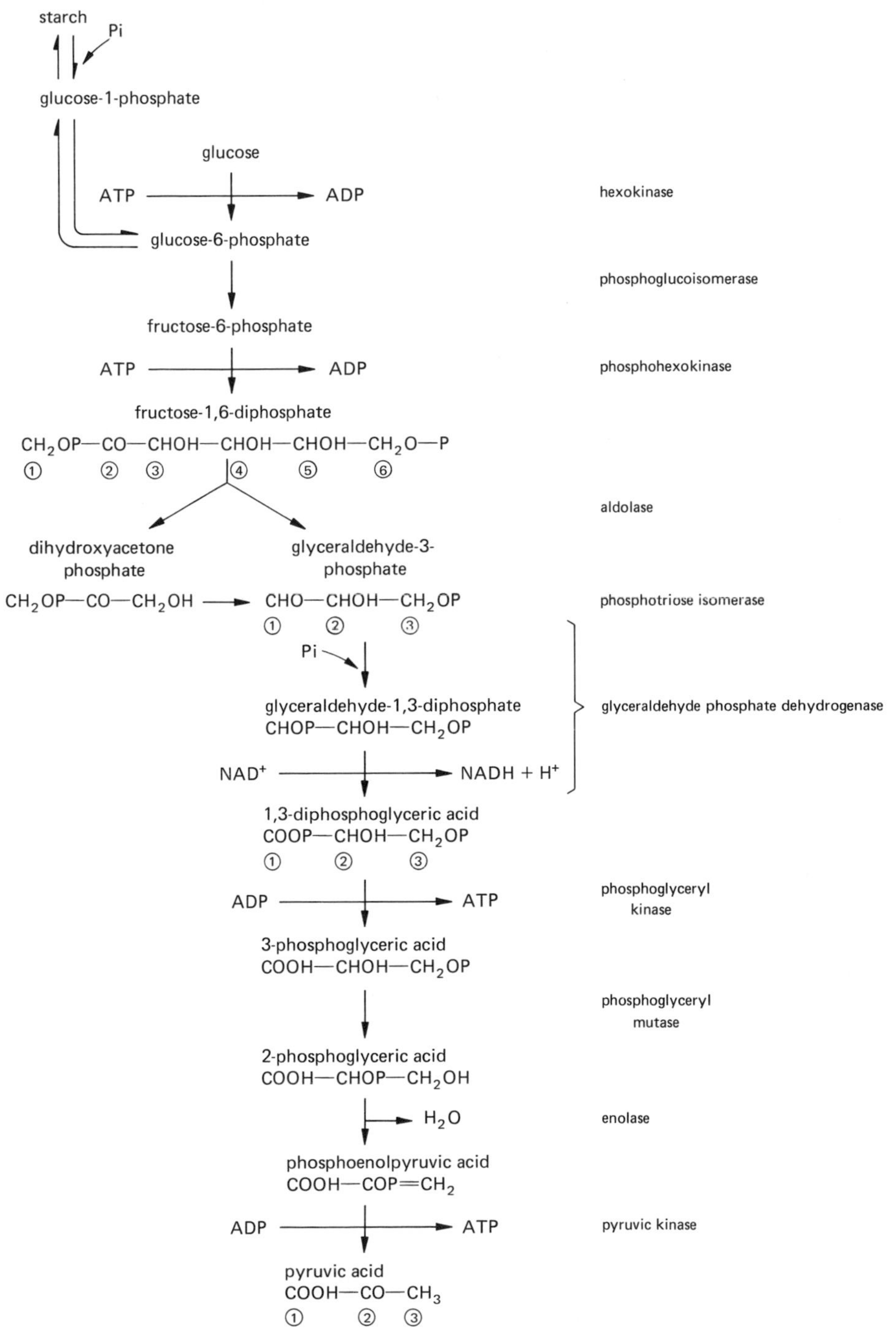

phosphorylate hexose, a hexokinase reaction (**kinases** are enzymes that add a phosphate group; **hexo**kinase phosphorylates a hexose). The resulting glucose-6-phosphate (G-6-P) is converted into its isomer, fructose-6-phosphate (F-6-P), by phosphoglucoisomerase (an **isomerase** alters the structure of a compound without changing its formula). A second molecule of phosphate is next introduced from ATP by the enzyme phosopho-hexokinase.

The fructose diphosphate (FDP) so formed now undergoes a split catalyzed by **aldolase** (an enzyme that catalyzes reactions between or produces aldehyde-alcohol compounds) to the ketotriose, dihydroxyacetone phosphate (DHAP), which contains C_1, C_2, and C_3 of the original hexose, and the aldotriose glyceraldehyde-3-phosphate (GAP), which contains C_4, C_5, and C_6. This and the subsequent steps of glycolysis are illustrated in greater detail in Figure 6-1 to show the relationship between the individual carbon atoms of the intermediates.

The two trioses are interconverted and exist in equilibrium through the action of the enzyme phosphotriose isomerase. DHAP is converted into GAP, and this compound is then oxidized by glyceraldehydephosphate dehydrogenase to 1,3-diphosphoglyceric acid (the **dehydrogenases** remove hydrogen from compounds, thereby oxidizing them). In this reaction part of the energy of the oxidation is used to reduce NAD^+ to $NADH + H^+$ (oxidized and reduced NAD and NADP will be written NAD–NADH and NADP–NADPH in future). The remainder of the energy of oxidation is conserved by the esterification of inorganic phosphate on C_1 of the GAP molecule to form a high energy acyl phosphate.

In the next reaction this phosphate group is transferred to ADP to generate ATP, catalyzed by phosphoglyceryl kinase. The resulting 3-phosphoglyceric acid (PGA) is converted to 2-phosphoglyceric acid by phosphoglyceromutase (a **mutase** changes the position of esterified phosphate), and this is converted, by the removal of a molecule of water, to phosphoenolpyruvate (PEP) by **enolase,** an enzyme which mediates coversion to and from the enol form. Enols have a double bond (-ene) and an adjacent alcohol group (-ol). The conversion of PEP to pyruvate by pyruvate kinase involves the transfer of the phosphate group to ADP making ATP. The energy for this transfer is derived from the conversion of the highly reactive and unstable PEP to the more stable pyruvic acid.

Energy Balance. The energy balance of glycolysis is readily determined. In the initial conversion of glucose to FDP, two molecules of ATP are consumed; subsequently, however, two are generated directly in the oxidation of two molecules of glyceraldehyde diphosphate and two more are generated in the conversion of two molecules of PEP to pyruvate. This direct syntheses of ATP is called **substrate phosphorylation.** The net balance is thus two molecules of ATP synthesized by substrate phosphorylation for each molecule of glucose converted to pyruvate. In addition, during the oxidation of two molecules of GAP to PGA, two molecules of NAD are reduced to NADH. The reoxidation of each molecule of NADH by oxygen via the electron transport chain generates three molecules of ATP, which adds up to six more molecules of ATP per molecule of glucose. Thus the overall net production of glycolysis, in terms of high energy intermediates per mole of glucose catalyzed, is 2 moles of ATP + 2 moles of NADH, or 8 moles of ATP. This represents only about 60 kcal/mole of glucose, or about 10 percent of the total energy available in glucose. Some of the energy liberated in the conversion of glucose to pyruvate is lost as heat in the process, but a greater proportion of the energy of glucose is still locked in the pyruvate molecules, and is released in the oxidative reactions of the Krebs cycle.

Krebs Cycle

Formation of Acetyl-Coenzyme A. The two molecules of pyruvate resulting from glycolysis of a molecule of hexose next undergo a series of reactions that converts them into a derivative of acetic acid, acetyl-coenzyme A **(acetyl-CoA),** in which form they enter the Krebs cycle. The group transfer agent, **coenzyme A (CoA),** participates in a number of important reactions, including the decarboxylation of pyruvate and α-keto-glutarate in oxidative metabolism and the oxidation of fats to acetate. CoA is made up of a molecule of the vitamin pantothenic acid and a molecule of ATP. Its active group—the linkage that serves to transfer groups such as acetyl radicals—is the SH group, which can be oxidized and reduced. The structure of CoA is shown in Figure 6-2.

The reaction sequence that leads to the formation of acetyl-CoA is outlined in Figure 6-3. In the first step pyruvate reacts with thiamine pyrophosphate **(TPP** or **cocarboxyl-ase)** to form an acetaldehyde-TPP complex and CO_2. The complex reacts with the cofactor α-lipoic acid in the oxidized state to form acetyl-lipoic acid complex, releasing TPP. The acetyl-lipoic acid complex reacts with CoA to form acetyl-CoA and reduced lipoic acid. Note that the acetaldehyde has been oxidized to acetate and the lipoic acid has been reduced by this reaction. The lipoic acid is reoxidized by NAD, and the NADH so formed is reoxidized by the cytochrome electron transport system, resulting in the generation of three molecules of ATP per molecule of pyruvate oxidized. The structures of TPP and α-lipoic acid are shown below; complexes form at the points marked with arrows. Oxidized α-lipoic acid has an S-S bond (dotted line); in the reduced form hydrogens are added to the sulfur atoms.

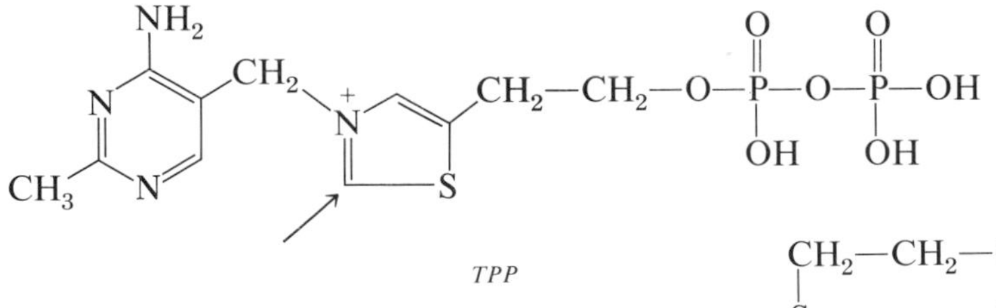

NH_2, N, N, CH_3, CH_2, $\overset{+}{N}$, S, $CH_2—CH_2—O—P(=O)(OH)—O—P(=O)(OH)—OH$

TPP

$CH_2—CH_2—CH—(CH_2)_4—COOH$ with S–H on the first and third carbons (S - - - - S)

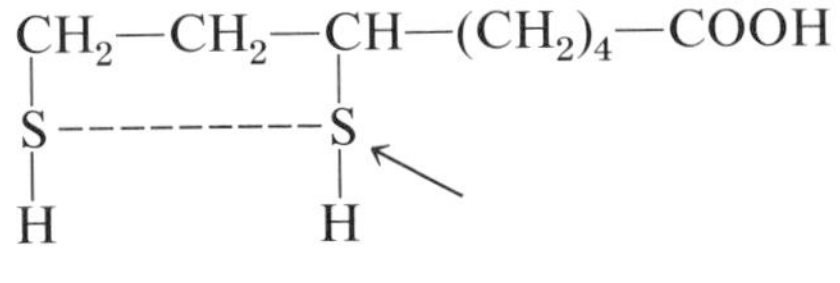

α-lipoic acid

Reactions of the Cycle. Acetyl-CoA is the fuel of the Krebs cycle, the oxidative system that completes the conversion of carbon from respiratory substrates to CO_2. The need for a cycle instead of a direct oxidation is twofold. First, the direct oxidation of acetate to CO_2 would have to proceed via one-carbon compounds, and these are extremely reactive and, so to speak, difficult to hold. Thus the acetate, instead of being directly oxidized, is attached to a "handle." The resulting larger molecule is oxidized step by step to the size of the original "handle," which can then accept a new acetate molecule to be oxidized, and so on. The second advantage of a cycle is that a number of more complex intermediates are made during the operation of the cycle, and these

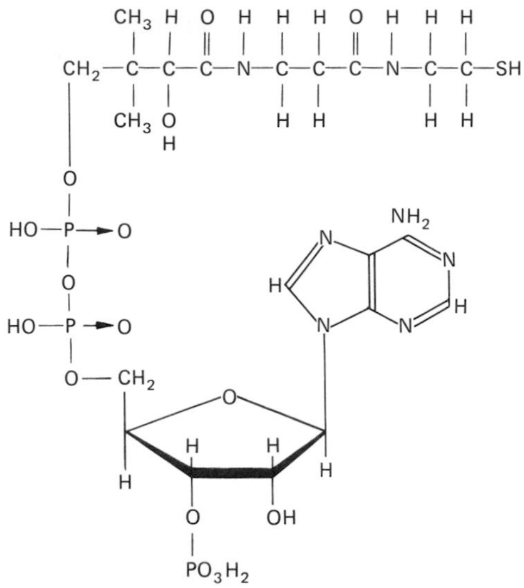

Figure 6-2. Structure of coenzyme A (CoA). The two basic parts of the molecule are derived from pantothenic acid (top part) and ATP.

intermediates may serve as starting points for the synthesis of other cell components. This function of the respiratory system will be described in some detail on page 121.

The reactions of the Krebs cycle are outlined in Figure 6-4. Details of the relationships of carbon atoms are shown in Figure 6-20 (page 142) in the discussion on ^{14}C tracer investigations. The first step in the cycle is the addition of acetate from acetyl-CoA to oxaloacetate making citrate, the first tricarboxylic acid of the cycle. The next series of reactions shifts the OH group from the middle carbon of citrate to the next carbon to make isocitrate, which can then be oxidized to oxalosuccinate. This shift is necessary because the carbonyl group must adjoin a carboxyl for the subsequent reactions. Next,

Figure 6-3. Conversion of pyruvate to acetyl-CoA (pyruvic dehydrogenase).

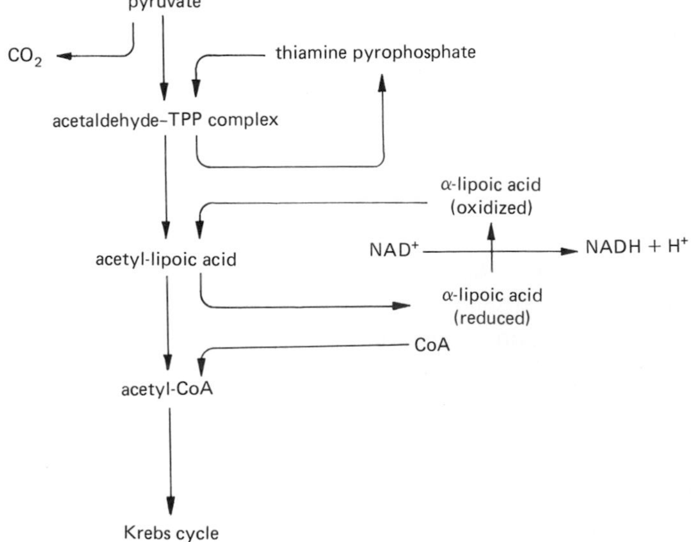

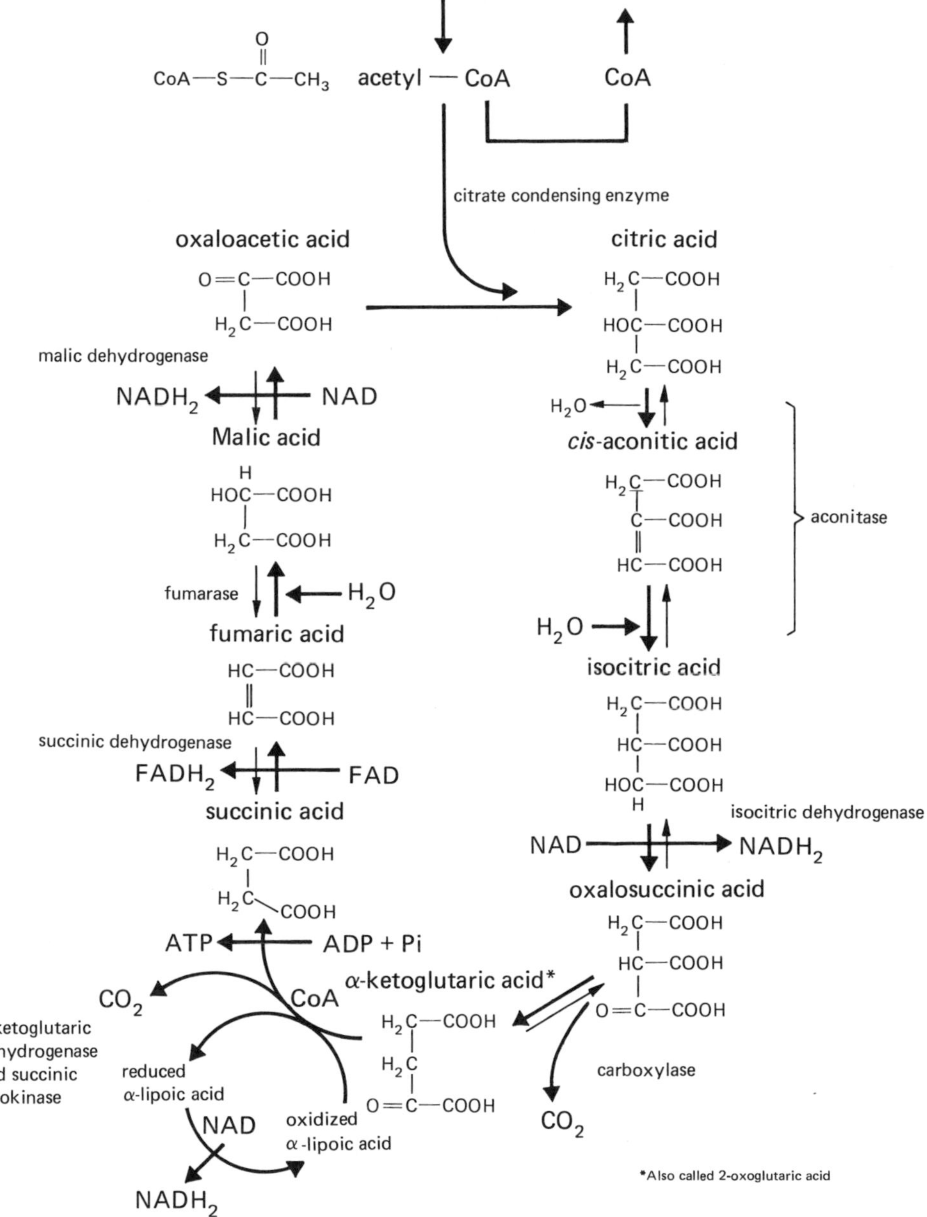

Figure 6-4. Krebs cycle. Reversible reactions are shown by double arrows; heavy arrows indicate the direction of normal cycle operation.

the central carboxyl group is removed leaving the five-carbon α-ketoglutarate, which is oxidatively decarboxylated to the four-carbon succinate. The reactions of the four-carbon dicarboxylic acids complete the oxidative steps of the cycle and regenerate the starting four-carbon acid, oxaloacetate.

The individual reactions are as follows. The addition of acetate to the carbonyl carbon of oxaloacetate is carried out by the **citrate condensing enzyme.** This is an important reaction type and may be used to create long chain and branched chain compounds by the reaction of acetyl-CoA with a variety of carbonyl compounds. The citrate so formed now undergoes the removal and replacement of a molecule of water, which shifts the OH from C_3 to C_4 of the molecule. Both reactions, the removal of water to make *cis*-aconitic acid and its replacement to make isocitric acid, are catalyzed by **aconitase.** This step is the site of inhibition by fluoroacetic acid, a compound found free in large quantities in the South African plant "Gibflaar" (*Dichapetalum cymosum*). Fluoroacetate itself is not inhibitory, but it forms fluoroacetyl-CoA, which reacts with oxaloacetate to form fluorocitrate. This analog of citrate is a competitive inhibitor of aconitase, and blocks the cycle at this point. Another important fact of this reaction is that citrate behaves as an asymmetric molecule in the aconitase reaction because of its three-point attachment to the enzyme (the three carboxyls form an asymmetric pattern, as shown in Figure 6-5). Thus the subsequent oxidation is at the opposite end of the molecule from that formed by the newly added acetate carbons. This has important consequences in the investigation of the cycle and its associated metabolic pathways using radioactive tracers, as we shall see later (page 141).

Isocitrate is oxidized to the keto acid oxalosuccinate by isocitric dehydrogenase, which transfers two electrons and two H^+ to NAD; the NADH formed is reoxidized via the electron transport system. Oxalosuccinate is decarboxylated by a **carboxylase** (an enzyme that adds or removes carboxyl groups) to α-ketoglutaric acid* and CO_2. α-Ketoglutaric acid is then oxidatively decarboxylated via a nonreversible reaction to succinic acid and CO_2. This reaction is essentially similar to the decarboxylation of pyruvate. It requires TPP and oxidized lipoic acid to form succinyl–CoA; the reaction is catalyzed by α-ketoglutaric acid dehydrogenase. The reduced lipoic acid so formed reduces NAD and becomes reoxidized in the process. Succinyl–CoA is converted by

*Also called 2-oxoglutaric acid.

Figure 6-5. Stereospecificity of aconitase. C_1 and C_2 of citric acid are derived from acetyl-CoA and C_3-D_6 from oxaloacetate.
A. Shows a molecule of citrate correctly oriented for a three-point attachment, through the three carboxyl groups, to the enzyme surface.
B. The arrows show the active site.
C. Shows a molecule of citrate incorrectly oriented.

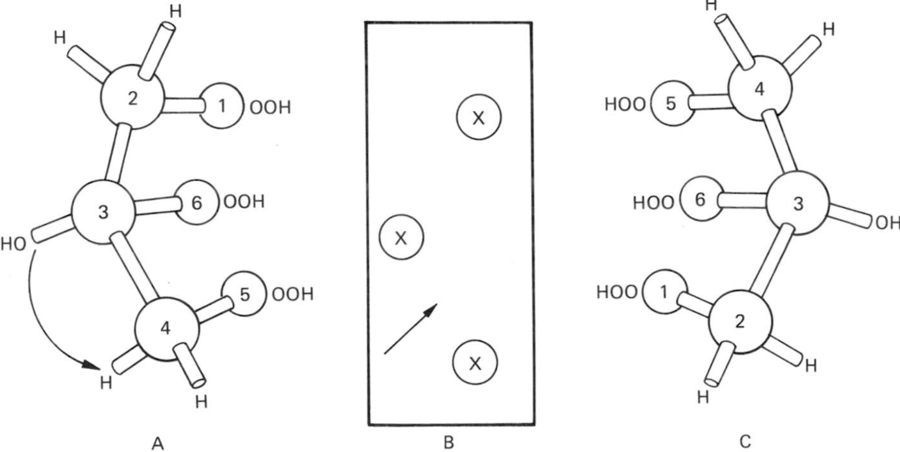

succinic thiokinase to succinic acid and CoA. In the **thiokinase** reaction the energy of the thioester bond of succinyl–CoA is used to convert ADP + Pi into ATP.

The oxidation of succinate to fumarate by succinic dehydrogenase differs from other oxidations in the cycle in that two H^+ and two electrons are transferred directly to flavin adenine dinucleotide (FAD)—the coenzyme of succinic dehydrogenase—rather than through NAD (see also Figure 6-21, page 144). The $FADH_2$ so formed reacts with the cytochrome system in the usual way; however, it produces only two molecules of ATP in the transfer of electrons to oxygen. This step in the cycle is strongly inhibited by malonic acid, a three-carbon analog of succinic acid that inhibits the enzyme succinic dehydrogenase by binding to it but not reacting.

Fumarate is converted to malate by the addition of water about the double bond, catalyzed by fumarase. This step in the Krebs cycle does not liberate any energy, but it prepares the four-carbon acid for a subsequent energy-releasing oxidation (malic acid to oxaloacetic acid). The reactions of succinate and fumarate, unlike those of citrate and the other asymmetric members of the cycle, are symmetrical; that is, the enzymes involved cannot distinguish between the two ends of the molecule. Carbons derived from either end of the original carbon structure of succinyl-CoA become indistinguishable or mixed in subsequent reactions. In the final step of the cycle, malic acid is oxidized by malic dehydrogenase to oxaloacetic acid, NAD being reduced in the process. This completes the reactions of the Krebs cycle.

Energy Balance. For each molecule of pyruvate oxidized to acetyl-CoA and for the three NAD-linked oxidations of the cycle, one pair of electrons and one pair of H^+ ions are carried to oxygen via the electron transport chain, producing three ATP molecules in the process, for a total of 12 ATP. In addition, the FAD-linked oxidation of succinate generates two more ATP, and the regeneration of CoA from succinyl-CoA generates one ATP. Thus the total ATP synthesis for one turn of the cycle (the oxidation of one molecule of pyruvate to CO_2 and H_2O) is 15 ATP, or 30 ATP per molecule of glucose. Glycolysis, it will be recalled, generates an additional eight molecules of ATP per molecule of glucose, bringing to 38 the total molecules of ATP that can be generated in the complete combustion of a molecule of glucose to CO_2 and H_2O. The total energy balance is thus strongly in favor of catabolism—the energy recovered as ATP represents only about one half of the total energy of combustion of glucose. The remainder of the energy is lost as heat and used to operate the system, that is, to maintain a favorable balance of intermediates so that reactions will proceed forward at effective rates.

Pentose Shunt

Reactions. This pathway, also known as the **hexose-monophosphate shunt** or the **direct oxidation** pathway of glucose catabolism, is a sequence of reactions that essentially converts glucose into triose phosphate and CO_2. Only one molecule of CO_2 is produced per glucose molecule; the rest of the carbons undergo a complex reorganization. The cycle is shown in Figure 6-6, together with the enzymes responsible for the reactions. Several of these are similar to, or identical with, enzymes of the glycolytic sequence.

Two extremely important enzymes that occur here and also in the pathway of carbon reduction of photosynthesis (the Calvin cycle, described in Chapter 7) are **transketolase** and **transaldolase.** Transketolase transfers the first two carbons from a ketose-P to an aldose-P, producing a new ketose-P that has two carbons more than the receiving aldose and a new aldose that has two carbons less than the donating ketose.

ketopentose-P		aldopentose-P		ketoheptose-P		aldotriose-P
				H_2COH		
H_2COH		HC=O		C=O		
C=O		HCOH		HCOH		HC=O
HCOH	+	HCOH	$\xrightarrow[\text{reaction}]{\text{transketolase}}$	HCOH	+	HCOH
HCOH		HCOH		HCOH		H_2COP
H_2COP		H_2COP		HCOH		
				H_2COP		

There are two transketolase reactions in the cycle. One of them converts xylulose-5-phosphate (Xu-5-P) and ribose-5-phosphate (R-5-P) into sedoheptulose-7-phosphate (S-7-P) and glyceraldehyde-3-phosphate (GAP), and the second converts Xu-5-P and erythrose-4-phosphate (E-4-P) into fructose-6-phosphate (F-6-P) and GAP.

Transaldolase transfers the top three carbons from a ketose-P to an aldose-P, producing a new ketose and new shorter aldose.

ketoheptose-P		aldotriose-P		ketohexose-P		aldotetrose-P
H_2COH				H_2COH		
C=O				C=O		HC=O
HCOH				HCOH		HCOH
HCOH	+	HC=O	$\xrightarrow[\text{reaction}]{\text{transaldolase}}$	HCOH	+	HCOH
HCOH		HCOH		HCOH		H_2COP
HCOH		H_2COP		H_2COP		
H_2COP						

The transaldolase reaction in the pentose shunt converts S-7-P and Xu-5-P into F-6-P and E-4-P. The net result of the transketolase and transaldolase reactions is the conversion of three C_5 sugars into two C_6 and one C_3 sugars. These intricate reactions have been set out so that they can be clearly seen in the lower central portion of Figure 6-6.

Another new enzyme is ribulose phosphate epimerase (**epimerases** change the configuration, that is, the plane of symmetry, of compounds), which alters ribulose-5-phosphate (Ru-5-P) at C_3, converting it to xylulose-5-phosphate (Xu-5-P). Phosphoribose isomerase converts Ru-5-P into its isomer, ribose-5-phosphate (R-5-P). Two oxidative steps take place: the oxidation of G-6-P to 6-phosphogluconate by G-6-P dehydrogenase, and the oxidative decarboxylation of phosphogluconate to Ru-5-P by 6-phosphogluconic acid dehydrogenase. Both these dehydrogenases are linked to the coenzyme NADP, which in turn can reduce NAD via a transhydrogenase. NADH can then reduce the electron transport chain and produce ATP. Alternatively, NADPH may be used as the reducing agent in a number of synthetic reactions, such as fat synthesis.

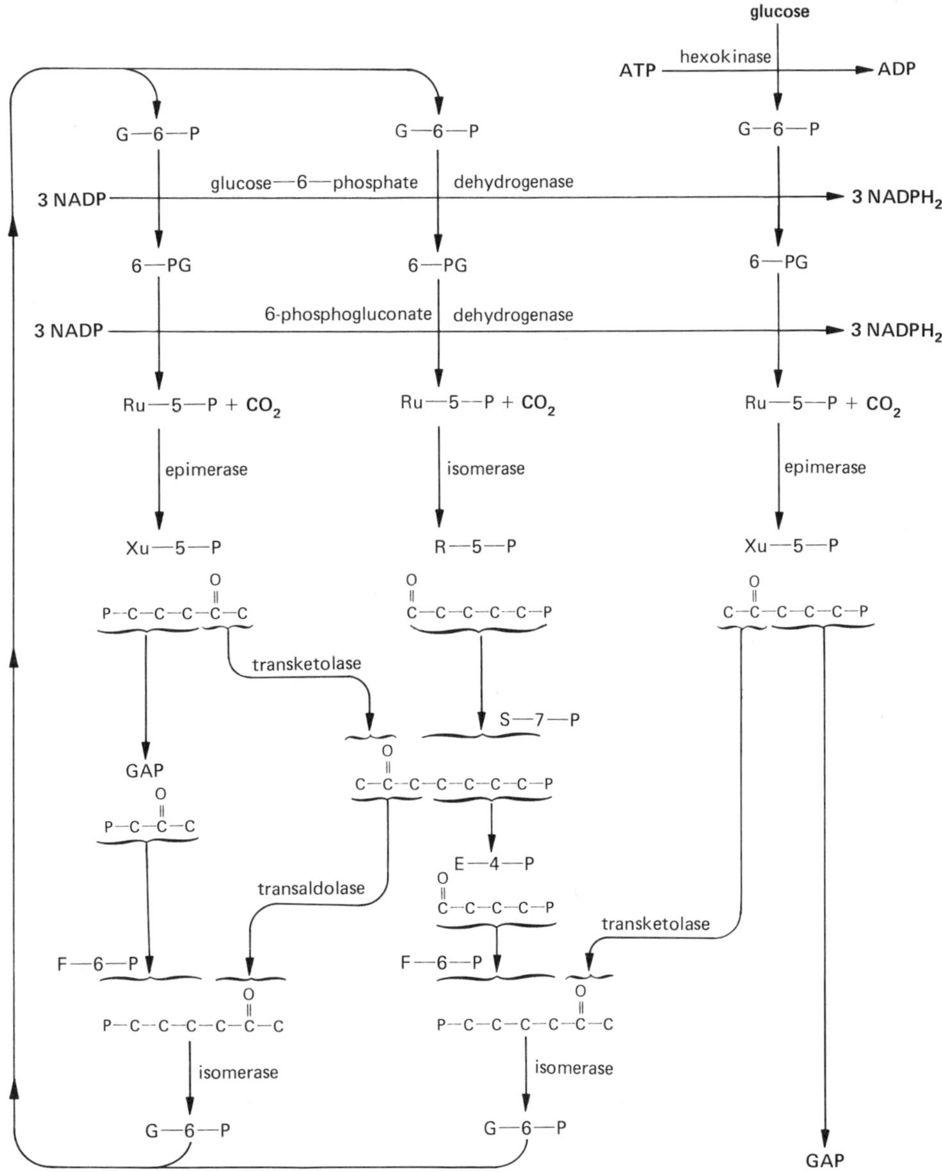

Figure 6-6. Pentose shunt. All the reactions are reversible, except the phosphorylation of glucose by hexokinase. As written here, one turn of the cycle converts a molecule of glucose to glyceraldehyde-phosphate plus 3 CO_2.

The products of the oxidation are CO_2 and triose phosphate. It is possible for two molecules of triose phosphate to combine via the aldolase reaction (Figure 6-1) after the isomerization of one molecule of GAP, the product of the cycle, to DHAP. The resulting FDP can be converted to G-6-P which may then reenter the cycle, so that the complete oxidation of hexose to CO_2 can take place. Alternatively, and probably much more commonly, the GAP produced can enter the triose phosphate pool of the cell and be oxidized to pyruvate, thence to be oxidized by the Krebs cycle.

Energy Balance. Two molecules of NADP are reduced for each molecule of CO_2 produced from glucose, resulting in the formation of six molecules of ATP, or 36 ATP per molecule of glucose oxidized. One ATP is needed to phosphorylate the glucose initially; therefore, the net gain is 35 ATP per glucose, making this pathway of oxidation slightly less efficient than glycolysis and the Krebs cycle (38 ATP per glucose). If the triose phosphate produced by the shunt enters the glycolytic pathway, the energy recovered is somewhat higher: 18 ATP are produced for the production of three CO_2, less one for the initial phosphorylation of glucose; the subsequent oxidation of triose phosphate to pyruvate produces five more ATP, and the oxidation of pyruvate produces 15 more ATP, for a total of 37 molecules of ATP per molecule of glucose oxidized.

Fermentation

In the absence of oxygen the oxidative reactions of the Krebs cycle cannot take place, and organisms deriving their energy from the catabolism of glucose must rely exclusively on the energy liberated in glycolysis. However, a further problem arises. The NADH formed during the oxidation of GAP cannot be reoxidized by oxygen; hence,

Figure 6-7. Pathways of fermentation, which lead to the reoxidation of NADH.

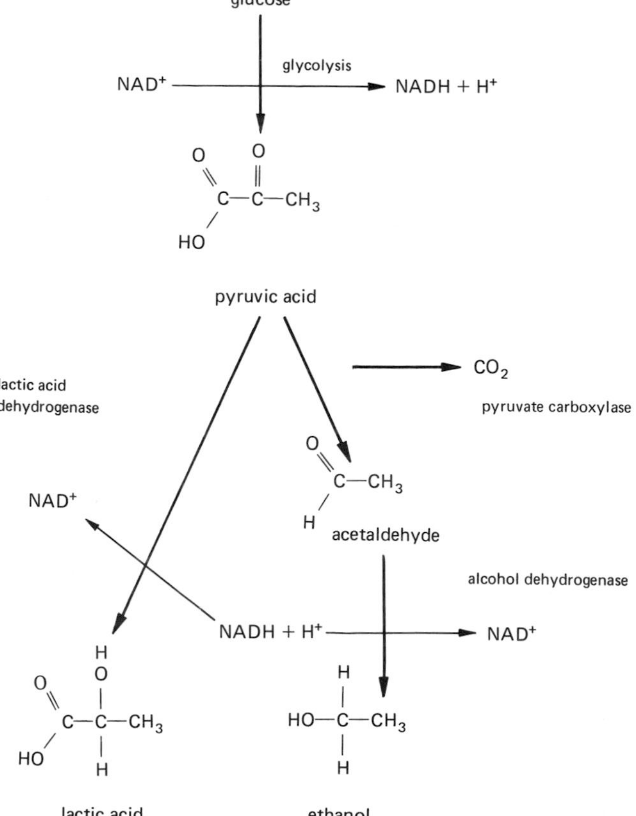

some other system is required to produce the continuous supply of NAD required for the operation of glycolysis. This problem has been solved in various organisms in two important ways.

One solution is the reduction of pyruvate to lactate, catalyzed by lactic acid dehydrogenase, which converts NADH to NAD in the process. This is the terminal step in animal systems, which often conduct glycolysis and oxidative metabolism in distantly separated sites. The reaction is characteristic of some plant systems and of certain bacteria, for example *Lactobacillus,* which sours milk by the production of lactic acid.

The second solution is the decarboxylation of pyruvate to acetaldehyde, catalyzed by pyruvic carboxylase, and the subsequent reduction of acetaldehyde to ethanol by alcohol dehydrogenase, with the reoxidation of NADH. These reactions are outlined in Figure 6-7. This is the pathway commonly found in yeasts—the alcoholic fermentation of sugars is a very important reaction (sociologically as well as physiologically!). It is probable that bulky organs such as potato tubers, which may be short of oxygen at their centers because of the long diffusion path, may conduct alcoholic fermentation as do germinating seeds before the disruption of the seed coat permits the entry of oxygen.

As the result of the necessity of reoxidizing NADH, the energy production from the conversion of one molecule of glucose to two molecules of lactate or of ethanol + CO_2 is very small; only two molecules of ATP (about 15 kcal/mole) are produced, and this is only about 2.5 percent of the total energy present in the glucose molecule. Thus the recovery of energy from glucose by fermentation is extremely low, and organisms that rely on fermentation alone for the production of energy must consume very large quantities of sugar in the process.

Localization of Pathways

Cells may be fractionated to separate their subcellular organelles, usually by disruption of the cells or tissue followed by careful centrifugation. Since organelles vary somewhat in their densities, they can be spun out singly or in groups, the heaviest first, by carefully controlling the speed of the centrifuge or the density of the suspending liquid. Alternatively, the organelles can be separated on a density gradient. Liquids of decreasing density are carefully layered from the bottom in a centrifuge tube, and the cell debris is layered on top of this. The whole gradient is then centrifuged carefully so that the layers do not mix. Various components of the cell are spun through the lighter density layers of the gradient until they reach the layer whose density is the same as their own; here they stop. It is possible to determine the location of specific enzymes in the various layers of the gradient. This makes it possible to find out which enzymes are associated with specific particles or organelles in the cell and which are free or soluble in the cytoplasm.

It has been shown that the enzymes of the glycolytic sequence and the pentose shunt are largely soluble; that is, they do not associate with any particle but are present either free in the cytoplasm or only loosely associated with cytoplasmic membranes such as the endoplasmic reticulum. The enzymes of the Krebs cycle, on the other hand, are largely in the mitochondria, as are the enzymes of oxidative phosphorylation. This means that the products of glycolysis must be able to enter or leave as necessary. On the other hand, it is unlikely that the mitochondrial membrane is freely permeable to intermediates of the cycle, otherwise it would be impossible to maintain them inside at the concentrations necessary for the operation of the cycle. Thus a certain number of transport enzymes are necessary to move compounds into and out of the mitochondria as required. These enzymes themselves may need ATP to drive them. They are discussed further in Chapter 12.

The mitochondrion thus constitutes a fully self-contained system that conducts, among others, the basic energetic reactions of the cell. Many of the enzyme systems within the mitochondria are closely associated in a structural way. As we saw in Chapter 5, the enzymes of the electron transport system are tightly bound and highly structured within the inner, folded layer of the mitochondrial membrane, and in close physical association with the ATP-synthesizing enzymes. Such close binding and association are necessary for the efficient transfer of intermediates, electrons, and energy. Other groups of enzymes are also known to be structurally bound together to form **multienzyme complexes.** The group of enzymes that accomplish the decarboxylation of pyruvic acid has been shown to constitute such a complex in bacteria, as have the similar enzymes that decarboxylate α-ketoglutarate in plants, and the enzymes associated with fat synthesis (Chapter 9). Certain other groups of enzymes are also closely bound in organelles (for example, the enzymes of the glyoxylate cycle which are found in glyoxysomes in seeds).

This fact has tended to become a principle, and some physiologists have been led to the mistake of assuming that all such reaction mechanisms are associated with organelles or organized structures in the cell. For example, the reactions of nitrogen fixation were at one time thought to be organized in **nitrosomes,** and the primary reactions of photosynthesis in **quantasomes,** neither of which turned out to be real entities. Generalization is an important process in the development of science. However, these examples should show that it is also an extremely dangerous process and must be undertaken with great care.

Mobilization of Substrates

Major substrates of plant respiration (see Figure 6-8) are starch and related polysaccharides, stored soluble sugars such as sucrose, fats, and proteins. Under certain circumstances low molecular weight compounds, such as organic acids or simple sugars, may accumulate. These may serve as respiratory substrates, but they simply enter respiratory metabolism at appropriate points and will not be further considered here (see Figure 6-8).

Starch is frequently a major substrate of respiration and is usually degraded by the phosphorylase reaction to glucose-1-phosphate (G-1-P).

$$\underset{\text{starch}}{\text{—G—G—G—G}} + \text{Pi} \xrightarrow{\text{phosphorylase}} \text{—G—G—G} + \text{G-1-P}$$

G-1-P can be converted by the enzyme phosphoglucomutase to G-6-P and so enter the glycolytic sequence. However, starch phosphorylase can only attack α-1:4-glycosidic links, and the 1:6 links of amylopectin must be disrupted by the so-called R enzyme (amylo-1,6-glucosidase), which yields molecules of free glucose. Alternative systems for the degradation of starch are α-amylase and β-amylase, both of which yield the disaccharide maltose. Maltose is hydrolyzed to glucose by the widely distributed enzyme maltase. α-Amylase attacks internal 1:4 links in the starch molecule, breaking the chain

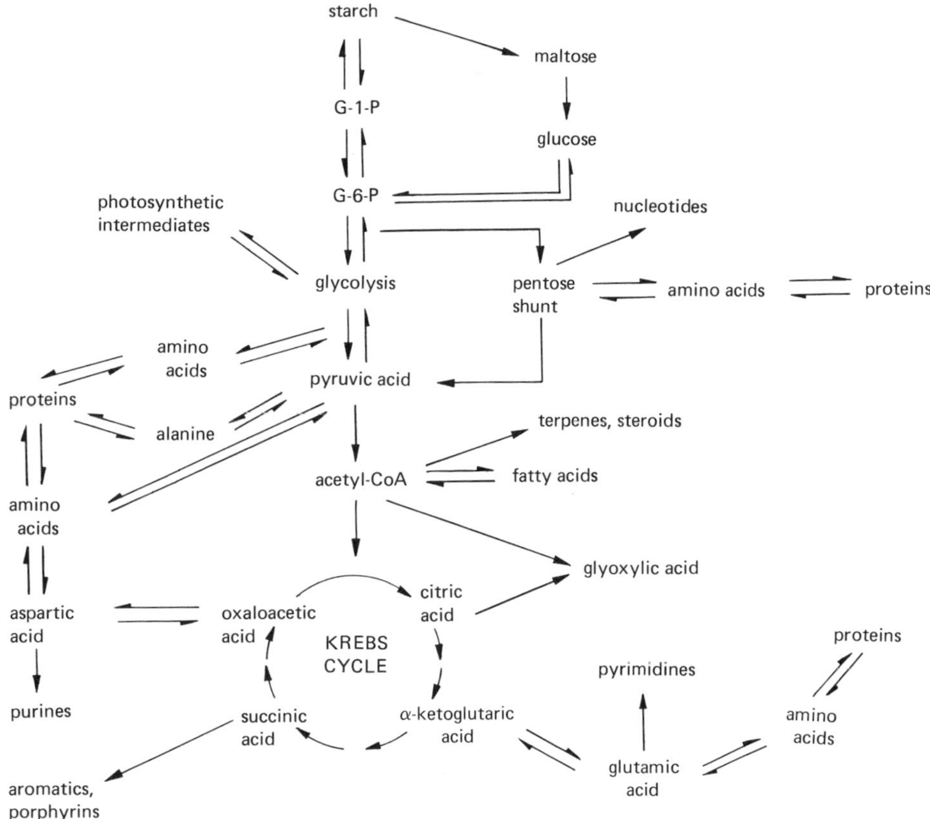

Figure 6-8. Relationships between intermediates of respiratory metabolism and other metabolic sequences.

into small fragments. β-Amylase only attacks the subterminal link in a chain, liberating the terminal two hexose units as maltose.

Naturally, odd-numbered groups of glucose units are left over after amylase attack; these always contain three hexose residues, never one. If branching 1:6 links were present, a number of dextrins containing up to seven glucose units (up to two each side of the 1:6 linked residue) may be left over. These are called **limit dextrins,** and may serve as primers for the synthesis of new molecules of starch (see Chapter 9). The amylase degradation of starch seems to be the system most commonly used by seeds for the mobilization of their reserves; leaves and starch-storing structures such as potato tubers mobilize starch largely by the phosphorylase reaction.

The entry of sucrose into respiratory metabolism probably occurs largely via the hydrolytic enzyme invertase, which is nearly universally distributed in plant tissues. Invertase hydrolyzes sucrose directly to an equimolar mixture of glucose and fructose called **invert sugars** (sucrose is dextrorotatory, but the mixture of glucose + fructose is levorotatory because of the strong levorotation of fructose; thus on hydrolysis the direction of rotation is inverted). The bond energy of the sucrose glycosidic bond is wasted in this reaction. The enzyme sucrose phosphorylase, which converts sucrose to G-1-P and fructose in *Pseudomonas* (see page 102), is not found in higher plants.

Little work has been done on the in vitro degradation of other polysaccharides in plants, but hydrolytic enzymes have been reported that degrade compounds such as inulin (polyfructosan) to disaccharides or monosaccharides; these can be converted into G-6-P or F-6-P and so enter the glycolytic pathway.

Other substrates of respiration are fats and proteins, though their use is not so general as that of sucrose or starch. Fats are degraded by the process known as β-oxidation, which cuts off molecules of acetic acid from the acidic end of fatty acids in the form of acetyl-CoA, and this can enter the Krebs cycle directly. The reaction mechanism is complex and all of its steps have not yet been demonstrated in plants; however, its operation is probably as outlined in Figure 6-9. ATP is produced in the reoxidation of FADH and NADH, which are reduced by the fatty acid oxidation. The glycerol left after the oxidation of the fatty acid residues of fat is phosphorylated by an appropriate kinase to form glyceryl phosphate; this can be oxidized to DHAP, in which form its carbon can enter directly into the glycolytic pathway. The degradation of fats for respiration is probably not a general phenomenon but occurs during the germination of certain types of fat-storing seeds (see Chapter 17).

Proteins are frequently used as substrates of respiration in plants. This may occur under conditions of starvation or during germination of seeds whose major storage reserve is protein. In addition some tissues or organs undergo a continuous breakdown of resynthesis of protein. This may take place during growth, when there is a continuous change in the complement of enzymes from those that conduct the reactions of growth and development to those that conduct the reactions characteristic of the mature tissue. Protein turnover is also characteristic of the continuing metabolism of certain tissues and tissue cultures. This process and its metabolic relationships will be considered in more

Figure 6-9. Cycle of fatty acid oxidation (β oxidation) producing acetyl-CoA.

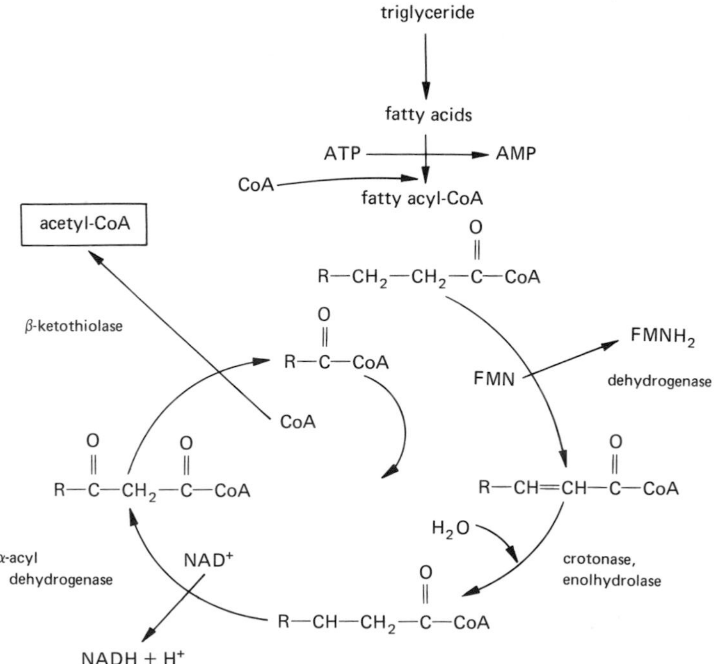

detail in Chapters 8 and 15. In this discussion it is important to stress that a number of intermediates of respiratory pathways, particularly of the Krebs cycle, form the carbon skeletons of important amino acids, and by reverse reactions these are the points of entry of protein carbon into respiratory metabolism. The most important of these are the α-keto acids pyruvic acid, α-ketoglutaric acid and oxaloacetic acid. These α-keto compounds are able, by the process of transamination (see Chapter 8), to form the amino acids alanine, glutamic acid, and aspartic acid, all important in the synthesis of proteins and as precursors of other amino acids. These relationships are shown in Figure 6-8.

Carboxylation Reactions

One of the consequences of a cyclic reaction sequence like the Krebs cycle is that for every molecule of acetate oxidized one molecule of acceptor is required and only one is regenerated. The rate of the reactions depends in part on the concentration of the intermediates; there must be a pool of oxaloacetate ready for the initiation of the cycle reactions. If the concentration of intermediates in the cycle falls to a low level, the cycle will slow down or stop. Every time a molecule of oxaloacetate or α-ketoglutarate or any other intermediate of the cycle is removed from the cycle, one less molecule of oxaloacetate is available for subsequent operation of the cycle.

Certain reactions are thus necessary to replenish the intermediates of the cycle when they are drained for synthetic reactions. In these reactions members of the cycle, or compounds feeding the cycle, are converted directly into new molecules of some member of the cycle. Such reactions are called **anaplerotic**—they do not contribute to the energy pool of the cell, but they serve to regenerate intermediates of the cycle that may have been depleted. Carboxylation reactions that convert a three-carbon acid of the glycolytic pathway into a four-carbon acid of the Krebs cycle are important anaplerotic reactions. However, these carboxylations have a number of other functions in plants, including the synthesis of intermediates in embryonic tissues or tissues that do not have a photosynthetic CO_2-fixing mechanism. These will be discussed in Chapter 15.

The most important of the carboxylation reactions is catalyzed by PEP carboxylase and results in the formation of oxaloacetate.

$$\text{PEP} + CO_2\ (+\ \text{ADP}) \xrightleftharpoons[]{\substack{\text{PEP carboxylase} \\ \text{(or PEP carboxykinase)}}} \text{oxaloacetate} + \text{Pi}\ (\text{or} + \text{ATP})$$

This reaction may also be catalyzed by the phosphorylating enzyme PEP carboxykinase, in which case it results in the synthesis of $\sim$ P in the form of ATP and proceeds somewhat slowly.

At least two enzymes are known which might carboxylate pyruvate, making a four-carbon acid. One of these is malic enzyme, which catalyzes the reaction between pyruvate and CO_2 to form malate.

$$\text{pyruvate} + CO_2 + \text{NADPH} \rightleftharpoons \text{malate} + \text{NADP}$$

However, this may not be a very important anaplerotic reaction because it goes more readily in the reverse (decarboxylation) direction. Also, this reaction is NADP specific and so is probably not mitochondrial. An NAD-specific malic enzyme is present in mitochondria, but its function is almost exclusively to decarboxylate malate.

An enzyme found in animal tissue and also in *Pseudomonas* uses the energy of ATP hydrolysis to speed a reaction that also carboxylates pyruvate but produces oxaloacetate.

$$\text{pyruvate} + CO_2 + \text{ATP} \longrightarrow \text{oxaloacetate} + \text{ADP} + \text{Pi}$$

All of these reactions regenerate intermediates of the Krebs cycle by carboxylating pyruvate (or PEP) instead of decarboxylating it and thus bypassing the oxidative reactions of the cycle. These reactions may also be important in synthetic metabolism under certain circumstances, particularly during the early growth of seedlings and in the dark metabolism of roots and even of green tissues in the absence of light.

A rather specialized sequence of reactions leading to the carboxylation of phosphoenol pyruvate (PEP) is found in the photosynthetic tissues of certain plants. In this reaction sequence energy from the hydrolysis of ATP and pyrophosphate is used to drive the carboxylation by forming the substrate, PEP, by a strongly exergonic reaction.

$$\text{pyruvate} + \text{Pi} + \text{ATP} \xrightarrow[\text{dikinase}]{\text{pyruvate, phosphate}} \text{PEP} + \text{AMP} + \text{PPi}$$

$$\text{PPi} \xrightarrow{\text{pyrophosphatase}} 2\ \text{Pi}$$

$$\text{AMP} + \text{ATP} \xrightarrow{\text{adenylate kinase}} 2\ \text{ADP}$$

$$(\text{sum : pyruvate} + \text{Pi} + 2\ \text{ATP} \longrightarrow \text{PEP} + 2\ \text{ADP} + 2\ \text{Pi})$$

$$\text{PEP} + HCO_3^- \xrightarrow{\text{PEP carboxylase}} \text{oxaloacetate} + \text{Pi}$$

This reaction sequence, in which the formation of the substrate for carboxylation is catalyzed by the enzyme pyruvate, phosphate dikinase, will be discussed at greater length in Chapter 7, page 177.

Glyoxylate Cycle

Another important anaplerotic pathway is the glyoxylate cycle, an internal cycle that enables a molecule of citrate and a molecule of acetyl-CoA to be converted ultimately into two molecules of oxaloacetate by the reactions outlined in Figure 6-10. The important step in this cycle is the cleavage of isocitrate (formed by the normal operation of the citrate condensing enzyme) into a molecule of succinate and a molecule of glyoxylate by the enzyme isocitratase. The succinate is converted to oxaloacetate by the normal pathways of the Krebs cycle. The glyoxylate forms the substrate of a condensation reaction with acetyl-CoA essentially similar to the reaction between oxaloacetate and acetyl-CoA. This reaction is catalyzed by the enzyme malate synthetase. A molecule of malate is formed by this reaction, which can then be oxidized to form a second molecule of oxaloacetate.

Although this cycle is capable of increasing the concentration of oxaloacetate, there is little evidence for its widespread operation in such a capacity. It seems to be much more important as a means whereby the fat that is stored in a number of seeds, such as castor beans, can be mobilized and converted into sugars, which are then suitable for transport to the growing part of the embryo. Acetyl-CoA is formed by the β-oxidation of fats, and this is converted to oxaloacetic acid by the glyoxylate cycle in glyoxysomes. The oxaloacetic acid is decarboxylated by the reverse operation of a PEP carboxylase,

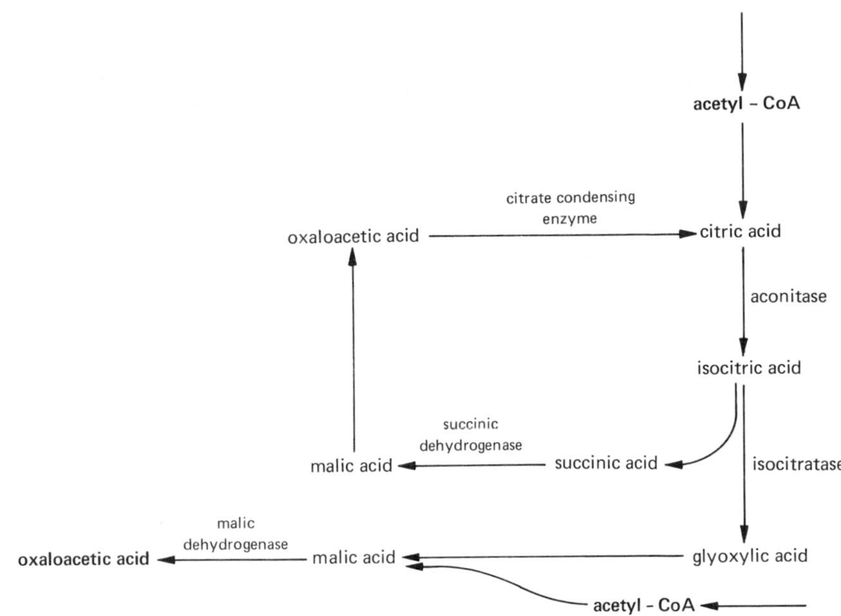

Figure 6-10. Glyoxylate cycle: 2 acetyl-CoA ⟶ 1 C_4 acid.

producing CO_2 and PEP. PEP is then reduced to form PGA, and, by reversal of the EMP pathway, fructose-1, 6-diphosphate is formed. This is converted to fructose-6-phosphate by a phosphorylase (not by the backward operation of phosphohexokinase) which in turn can be converted to glucose-6-phosphate. Glucose-6-phosphate may be converted to glucose-1-phosphate, and this, with fructose-6-phosphate, forms the substrates for sucrose synthesis. Thus, the carbon that was stored as fat in the seeds can be mobilized and converted into sugars, the usual transport form of carbon in plants. The production of sugars from PEP is frequently called **gluconeogenesis.** A curious fact emerges from a consideration of the reactions of gluconeogenesis. It is most important that the phosphohexokinase that phosphorylates fructose-6-phosphate and the phosphatase that dephosphorylates it be kept separate or closely regulated in the cell. Otherwise, these two enzymes would couple to form a cyclic reaction that would function essentially as an effective ATPase, a situation that would not be advantageous to the plant!

Control of Respiration

Pasteur Effect. Long ago Pasteur noted that the pathway of yeast metabolism could be affected by oxygen: low oxygen favored fermentation, whereas high oxygen inhibited fermentation and stimulated oxidative respiration as well as promoting the use of carbon from sugars for synthetic reactions. This was the first recognition of a control system for metabolism; plant physiologists and biochemists are still arguing about how it works! One probable mechanism is a regulation of the ratio of ATP/ADP by oxygen. In the absence of oxygen, oxidative metabolism cannot take place and the main avenue for the synthesis of ATP is cut off. The continuing metabolism of the cell uses up available ATP and a large amount of ADP and Pi is produced, which stimulates fermentation. Because the energy liberated (and ATP formed) by fermentation is limited, much larger amounts of substrate are utilized to maintain the same rate of synthetic reactions.

Experiments with dinitrophenol (DNP), a chemical that **uncouples** oxidative phosphorylation, confirm this view. (**Uncoupling** means separating the reactions of the cytochrome system from those that make ATP so that no ATP is made during electron transfer.) Although DNP does not affect glycolytic reactions directly, it greatly stimulates glycolysis by reducing the amounts of ATP and permitting a pileup of ADP. The American physiologist H. Beevers showed that DNP can cause a substantial switch of respiration toward fermentation even in the presence of oxygen.

In the absence of oxygen the Krebs cycle cannot operate, so the intermediates of the Krebs cycle are not available for synthetic reactions. Pyruvate is also unavailable because its reduction to lactate or ethanol is required to reoxidize the NADH produced in triose phosphate oxidation. In addition, many synthetic reactions using these intermediates require oxidative metabolism. Hence, under anaerobic conditions, little or no sugar carbon can be diverted to cellular synthesis, whereas the presence of oxygen would permit these reactions to take place.

Feedback and Allosteric Control. Respiration is an exergonic (or exothermic) process; that is, it liberates energy. Thus, if no controls were exerted, respiration would run at full speed continuously until all supplies of substrate were used up. In fact a number of control mechanisms are at work, interlocking or integrating the various phases of cellular metabolism. We shall consider two main types of control: allosteric effects and feedback control. **Allosteric** effects are exerted when a small molecule, often not directly related to the reaction, may promote or inhibit a specific enzyme eaction by combining with a secondary, or allosteric, site on the enzyme. **Feedback** mechanisms involve the inhibition or (less frequently) the stimulation of a reaction by one of its ultimate products, often as the result of an allosteric effect. **Feedforward** control may also occur, in which a reactant can affect a subsequent step in its own metabolism. This situation is not so common.

The points of control in metabolic pathways are not fortuitous. Usually an early and a late reaction in a metabolic sequence are strongly exothermic (that is, they have a strong negative free energy change) and are thus nearly irreversible. This prevents the pileup of substrates or intermediates, and the enzymes in question are called **pacemakers.** The pacemakers are the obvious sites that require regulation, as are enzymes that mediate reactions at the site of major branches in a metabolic chain. Some of the control points in respiration are summarized in Figure 6-11.

The first pacemaker in glycolysis is phosphofructokinase, which phosphorylates F-6-P to FDP using ATP as a P donor. This enzyme is allosterically inhibited by ATP at high concentration and is activated by ADP and Pi. Inhibition by ATP prevents the reaction from "running away" in the presence of high concentrations of the substrate, F-6-P, when the demand for ATP is low (that is, there is an excess of ATP present). The enzyme is also inhibited by citrate, which tends to pile up in the presence of excess Krebs cycle activity. As might be expected, a reaction at the lower end of the glycolytic sequence, the pyruvate kinase conversion of PEP to pyruvate, is also inhibited by ATP.

The Krebs cycle enzymes are similarly under internal control. The enzyme aconitase is inhibited by ATP and NADH and enhanced by ADP; a preparation from *Neurospora*

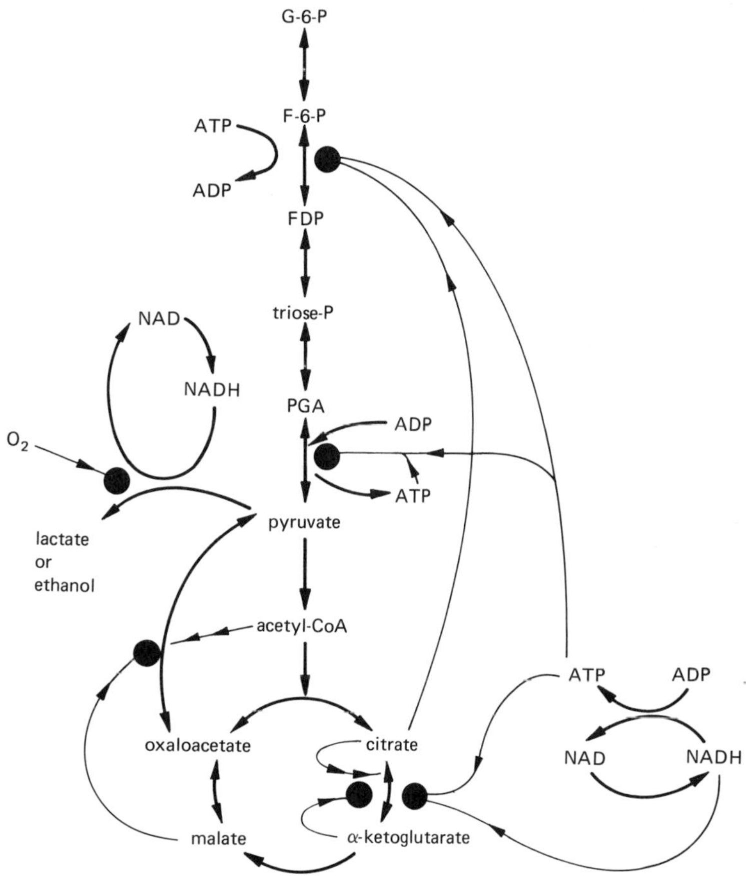

Figure 6-11. Generalized scheme of respiration showing some of the sites of feedback control. The levels of ATP, ADP, NADH, and NAD all directly affect any reaction in which these compounds take part (not all shown, to avoid confusion). —▸—▸ stimulates; —▸● inhibits.

is inhibited by α-ketoglutarate and enhanced by citrate. As might be expected, the anaplerotic reactions are also under allosteric control. Pyruvate carboxylase and PEP carboxylase are sometimes (but not always) activated by acetyl-CoA, a compound that would accumulate if the Krebs cycle were slowed down by the removal of intermediates which required replacement. PEP carboxylase is under the control of a product of its own reaction: the carboxylation is inhibited by malate, which is readily formed from oxaloacetate resulting from the carboxylation of PEP.

Cofactor Control. The important reactions of respiration are subject to a direct precursor-product control through the ratios of NADH/NAD and ATP/ADP. Glycolysis and fermentation are substantially regulated by the requirements of the cell for ATP and NADH—the reduction reactions of fermentation (pyruvate ⟶ lactate or acetaldehyde ⟶ ethanol) are dictated by the absolute requirement for NAD, which cannot be formed from NADH through the electron transport chain in the absence of oxygen. Because mitochondrial oxidation of NADH via the electron transport chain is tightly coupled to ATP formation (that is, ADP and Pi are required substrates), all the oxidation reactions of the Krebs cycle are controlled by the cell's requirement for ADP or NADH, which supply the driving force for synthetic reactions, growth, salt uptake, and so on. In effect, the rate of respiration is controlled by the energy charge level of respiring cells, as described in Chapter 5 (page 103). The addition of DNP, which uncouples oxidative phosphorylation, often causes dramatic increase in CO_2 production because the whole oxidative process is liberated from the restraints imposed by the requirement for ADP and Pi as obligatory substrates of reactions. The not infrequent failure of DNP to stimulate CO_2 production may be due to the fact that respiration is also under the control of other systems, as discussed above.

Side Reactions. Inevitably the rate of respiration will be affected by cellular requirements for intermediates of syntheses. Thus, draining intermediates will affect the reactions that generate them by mass action, increasing the overall rate of respiration. Since the further metabolism of diverted intermediates (for example, for amino acid and protein synthesis) also requires ATP or reduced cofactors, a further increase in the rate of respiration will occur. Thus it may be seen that respiration is closely integrated with the energy- and substrate-requiring activities of the cell. All the metabolic activities of cells and organisms are so integrated. Without such integration, metabolic chaos would prevail in the cell's complex network of intermediary reactions.

Other Respiratory Systems and Oxidases

Phenol Oxidases. Several enzymes that oxidize phenols to quinones are known. Two of the most important are monophenol oxidase (tyrosinase) and polyphenol oxidase (catechol oxidase). These enzymes participate in the characteristic "wound reaction" of plants and contribute to wound respiration by converting phenols released in the wound to quinones, which are toxic to microorganisms, thus helping in the prevention of infection. The brown color that often develops rapidly on wounding (for example, when a potato tuber or apple fruit is cut or bruised) is a result of this reaction. It is evident from the rapid reaction that occurs on wounding that the enzyme and its substrate, which both appear to be soluble, are maintained apart from each other in the normal cell, sequestered in different parts or compartments of the cell.

Phenol oxidase may be coupled to the oxidation of cellular components as follows

AH$_2$ → A; phenol → quinone (coupled)
quinone → phenol; $\frac{1}{2}$ O_2 → H_2O (phenol oxidase)

Though this reaction may be active in senescence, it is not normally important in respiration. It is possible for the oxidation and reduction of phenol to be coupled to substrate oxidation by NADP

AH$_2$ → A; NADP → NADPH (coupled)
NADPH → NADP; quinone → phenol (coupled)
phenol → quinone; $\frac{1}{2}$ O_2 → H_2O (coupled)

The phenol oxidases are involved in the chemical manipulation of precursors of lignin synthesis and other such chemical components of the cell. Quinones such as coenzyme Q (ubiquinone) are important in the electron transport chains of respiration and photosynthesis (Chapter 7). However, in these circumstances they do not react directly with oxygen as in the phenol oxidase reaction because this would effectively short-circuit the electron transport chain and prevent its proper operation as an ATP-synthesizing system.

Ascorbic Acid Oxidase. Ascorbic acid, or vitamin C, is a common component of plants. It may be oxidized to dehydroascorbic acid by the enzyme ascorbic acid oxidase as shown in Figure 6-12. Ascorbic acid oxidase appears to exist both as a cell wall bound and as a free enzyme. This oxidase is associated with a number of **redox enzymes** (that is, reducing one substrate as they oxidize another) as the terminal oxidase—that is, the enzyme transferring electrons to oxygen. It is linked to certain dehydrogenases via the SH component **glutathione** in the following sequence

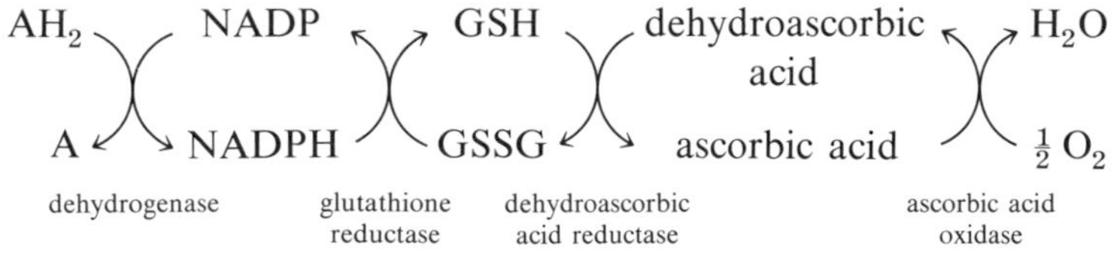

AH_2 / A	NADP / NADPH	GSH / GSSG	dehydroascorbic acid / ascorbic acid	H_2O / $\frac{1}{2}O_2$
dehydrogenase	glutathione reductase	dehydroascorbic acid reductase		ascorbic acid oxidase

where GSH is reduced glutathione and GSSG is oxidized glutathione. This may be an important reaction in the production of oxidized intermediates for cellular synthesis, for example, for the oxidation of sugars to acids or the decarboxylation of amino acids.

Catalase and Peroxidases. These enzymes use hydrogen peroxide (H_2O_2) as a substrate. Catalase normally acts only to destroy H_2O_2, whereas peroxidases oxidize various substrates as well

$$H_2O + H_2O_2 \xrightarrow{\text{catalase}} 2\ H_2O + O_2$$

$$H_2A + H_2O_2 \xrightarrow{\text{peroxidase}} 2\ H_2O + A$$

It was previously thought that catalase acted only as a detoxifying agent for enzymes, such as glycolic acid oxidase, that produce H_2O_2 (see below), but it has now been shown to have peroxidase activity as well. Peroxidation of the hormone indoleacetic acid is thought to be catalyzed by catalase, and this may have an important regulatory effect by

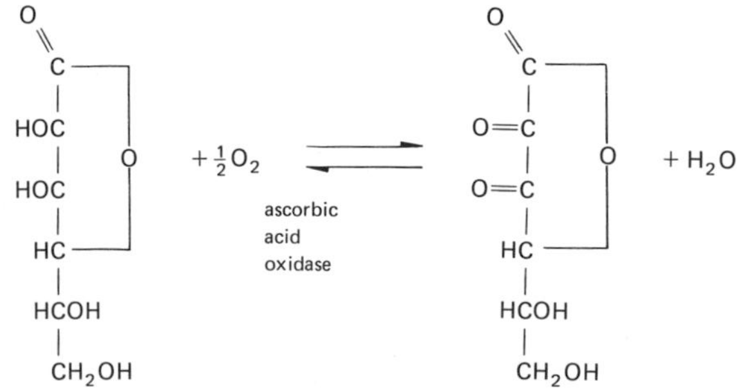

Figure 6-12. Action of ascorbic acid oxidase.

controlling IAA content. In addition NADH or NADPH can be oxidized by peroxidase, which may be important in the control of cellular metabolism. Like phenol oxidase, these enzymes may be involved in lignin biosynthesis.

Glycolic Acid Oxidase. This extremely important enzyme catalyzes the conversion of glycolic acid to glyoxylate. The product of the oxidation is not H_2O but H_2O_2, which requires the presence of catalase for its conversion into H_2O. Glycolic acid oxidase may be linked to the oxidation of certain substrates, for example, ethanol, via glyoxylate reductase, as follows

$$\begin{array}{ccccccccc} \text{ethanol} & & \text{NAD} & & \text{glycolate} & & O_2 & & \\ & \times & & \times & & \times & & & \\ \text{acetaldehyde} & & \text{NADH} & & \text{glyoxylate} & & H_2O_2 & \xrightarrow{\text{catalase}} & H_2O + \tfrac{1}{2}O_2 \\ & \text{ethanol dehydrogenase} & & \text{glyoxylate reductase} & & \text{glycolic acid oxidase} & & & \end{array}$$

Glycolic acid oxidase may be important in the synthesis of glycine, which can be derived from glyoxylate. It is probably important in **photorespiration,** the process of CO_2 release by photosynthetic tissue in light. Photorespiration does not yield usable energy as does dark respiration and is a quite different process. It will be discussed in Chapters 7 and 15.

Participation of Other Oxidases in Respiration. The oxidases described above probably do not participate in respiration except in ancillary reactions. First, none of them has been shown to be coupled with the cytochrome electron transport system, the only effective way the cell can generate ATP. Second, cytochrome oxidase has a very high affinity for oxygen (that is, it reacts very readily), whereas the other oxidases do not. Third, studies with poisons generally support the view that respiration proceeds largely via cytochrome oxidase, which is poisoned by cyanide (CN) or by carbon monoxide (CO), the CO inhibition being reversed by light. The possible terminal oxidases are shown in Table 6-1, which indicates some of their relevant properties.

Plants live in an atmosphere containing an abundance of O_2 and contain a number of powerful reducing compounds. It is now known that the chemical interaction of these compounds may produce oxygen radicals or "excited" forms of oxygen, which would be extremely toxic because of their powerful oxidizing capacity. It is possible that some of the oxidases mentioned previously may be important as scavengers that rid the cell of excess oxygen, particularly in the excited form. It seems likely that these enzymes, in fact, perform a number of functions in cells.

Table 6-1. Some properties of terminal oxidases

Enzyme	*Affinity for O_2*	*Coupled with ATP synthesis*	*CN sensitivity*	*CO sensitivity*	*Light reverses CO effect*
Phenol oxidase	medium	—	+	+	—
Ascorbic acid oxidase	low	—	+	—	
Glycolic acid oxidase	very low	—	—	—	
Cytochrome oxidase	very high	++	+	+	+
Cytochrome bypass	high	+	—	—	

"Alternative" Respiration

Although some CN- or CO-insensitive respiration may take place in plant tissues (in certain tissues respiration is almost completely insensitive to CN), the process of oxidative phosphorylation is much affected. The efficiency of a preparation or a tissue in making ATP via oxidative reactions may be expressed as the ratio of molecules of phosphate esterified per atom of oxygen consumed; this is called the P/O ration. When a tissue having CN-insensitive respiration, such as the *Arum* spadix, is poisoned by CN, the overall rate of respiration is not much affected while phosphorylation declines substantially; therefore, the P/O ratio decreases. Data from the spectrometric study of cytochromes (see page 143) show that an alternative pathway exists. Electrons are bypassed from cytochrome directly to oxygen either via an autooxidizable cytochrome b (cytochrome b_7) or else via a special autooxidizable cytochrome a. In either event, electrons are "short circuited" to oxygen and miss the phosphorylating site at the CN-sensitive cytochrome a_3 (cytochrome oxidase). The alternative hypothesis that glycolic acid oxidase participates in CN-insensitive respiration is not attractive because of the enzyme's low affinity for oxygen and the fact that it is not coupled with ATP synthesis.

Recent data suggest that alternative, CN-insensitive respiration is common in germinating seeds and storage tissues, such as potatoes, and also during the climacteric in aging leaves or ripening fruits. The significance of this pathway of respiration is not clear. It has been suggested that cyanide is a normal byproduct of metabolism, and since it cannot escape from tissues with an impervious cover, such as seeds, or from very massive tissues, such as potato tubers, its concentration may build up to a level where it would inhibit the normal electron transport chain. The alternative, cyanide-insensitive pathway would obviate this problem. However, the evidence is by no means conclusive.

Factors Affecting Respiration of Tissues

The physiology of respiration as a process is much affected by the relationship of respiration to other metabolic processes in the plant, to its overall requirements in growth and development, to the availability of suitable substrates, and to the physical and physiological situation of the plant. Many of these interrelations will be dealt with in detail in succeeding chapters. At this point it is not appropriate to examine the whole physiological interrelationship of respiration. Instead we wish to deal briefly with the basic responses of the respiratory mechanism to internal and external environmental conditions.

Respiratory Quotient and Substrates of Respiration. Early in the study of respiration it was recognized that compounds at different stages of oxidation could serve as substrates of respiration. It was thought that some indication of the kind of substrate undergoing oxidation might be obtained from a measurement of the amount of CO_2 produced for each molecule of O_2 consumed. This ratio, expressed as

$$\frac{\text{moles of } CO_2 \text{ produced}}{\text{moles of } O_2 \text{ absorbed}}$$

is called the **respiratory quotient** or **RQ.** When carbohydrate is completely oxidized by the general reaction

$$C_6H_{12}O_6 + 6\ O_2 \longrightarrow 6\ CO_2 + 6\ H_2O$$

The RQ is 6 CO_2/6 O_2 or 1.0. When fats, proteins, or other highly reduced compounds are oxidized, the RQ is less than 1. For example, in the oxidation of glycerol trioleate

$$\underset{\text{glycerol trioleate}}{C_{57}H_{104}O_6} + 83\ O_2 \longrightarrow 57\ CO_2 + 52\ H_2O$$

The RQ is $\frac{57}{83} = 0.69$. When compounds that are already partially oxidized, such as organic acids, serve as substrates of respiration, the RQ is above 1, as in the oxidation of citric acid

$$\underset{\text{citric acid}}{C_6H_8O_7} + 4\tfrac{1}{2}\ O_2 \longrightarrow 6\ CO_2 + 4\ H_2O$$

where the RQ is $6/4\frac{1}{2} = 1.33$. Thus the RQ may give some indication of the class of compounds being oxidized or, at least, of the state of oxidation of the substrate.

However, a number of other conditions may result in larger changes in the RQ. For example, the occurrence of fermentation in a tissue causes an abnormally high RQ, whereas the partial oxidation of a substrate, which might absorb O_2 and result in the liberation of substantial amounts of energy but no CO_2, would result in an unusually low RQ. Similarly the retention of O_2 or CO_2 in a bulky tissue could give misleading experimental values. In the other direction, the inability of the tissue to absorb O_2 (as in a germinating seed) might result in fermentation, with its characteristic high RQ.

Thus, although the chemical composition of the substrate may sometimes determine the RQ, the value is just as likely to be a reflection of the process rather than an indication of the substrate. For this reason the RQ is now of major importance in physiological studies of respiration, except in carefully controlled and properly understood situations.

Age and Tissue Type. By and large, young tissues respire more strongly than old, developing tissues more than mature, and tissues undergoing other metabolic or energetic activities (such as salt uptake or water uptake) respire more than resting tissues. This is the natural consequence of the fact that respiration is the process that liberates energy for all the cell's other activities.

However, certain conditions modify this concept. The first is that respiratory substrates change as tissues mature, and the overall process as well as the efficiency of respiration changes with development. For example, the respiration rate of seedlings usually rises rapidly during germination to a peak during the most rapid period of seedling growth, then it falls as the initially formed tissues mature (Figure 6-13).

Furthermore when the respiration rate of a single organ such as a leaf is measured, it follows a characteristic curve during its development (Figure 6-14). Respiration is at its highest during leaf growth, then it falls to a steady state during the period of the maturity of the leaf. There is often a **climacteric,** or a brief rise to a new high level, signaling the onset of irreversible processes of degeneration that marks the senescence and death of the organ. Generally the P/O ratio (atoms of phosphorus esterified per atom of oxygen absorbed) declines markedly during and after the climacteric rise, indicating a breakdown of the integrated system of energy transfer in the mature leaf.

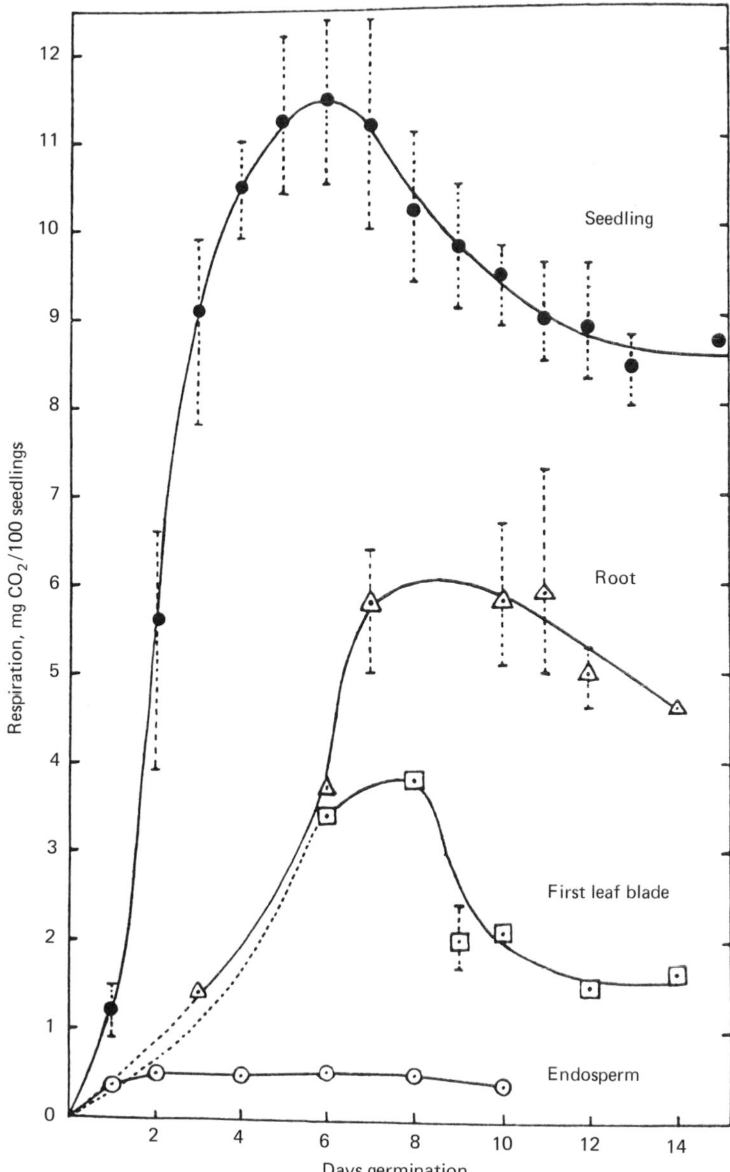

Figure 6-13. The respiration of germinating barley seedlings. [From B. F. Folkes, A. J. Willis, and E. W. Yemm: The respiration of barley plants. VIII, the metabolism of nitrogen and respiration of seedlings. *New Phytol.*, **51**:317–41 (1952). Used with permission.]

Late in the period of senescence proteins begin to break down and form the substrate of respiration. There may be a final brief period of high carbon dioxide production as organization collapses and the cells die, although this may also be due in part to the rapid multiplication of endogenous or invading microorganisms.

Different tissue types have different respiration rates depending on their metabolic activity, the relative mass of their nonmetabolic or structural components, and their

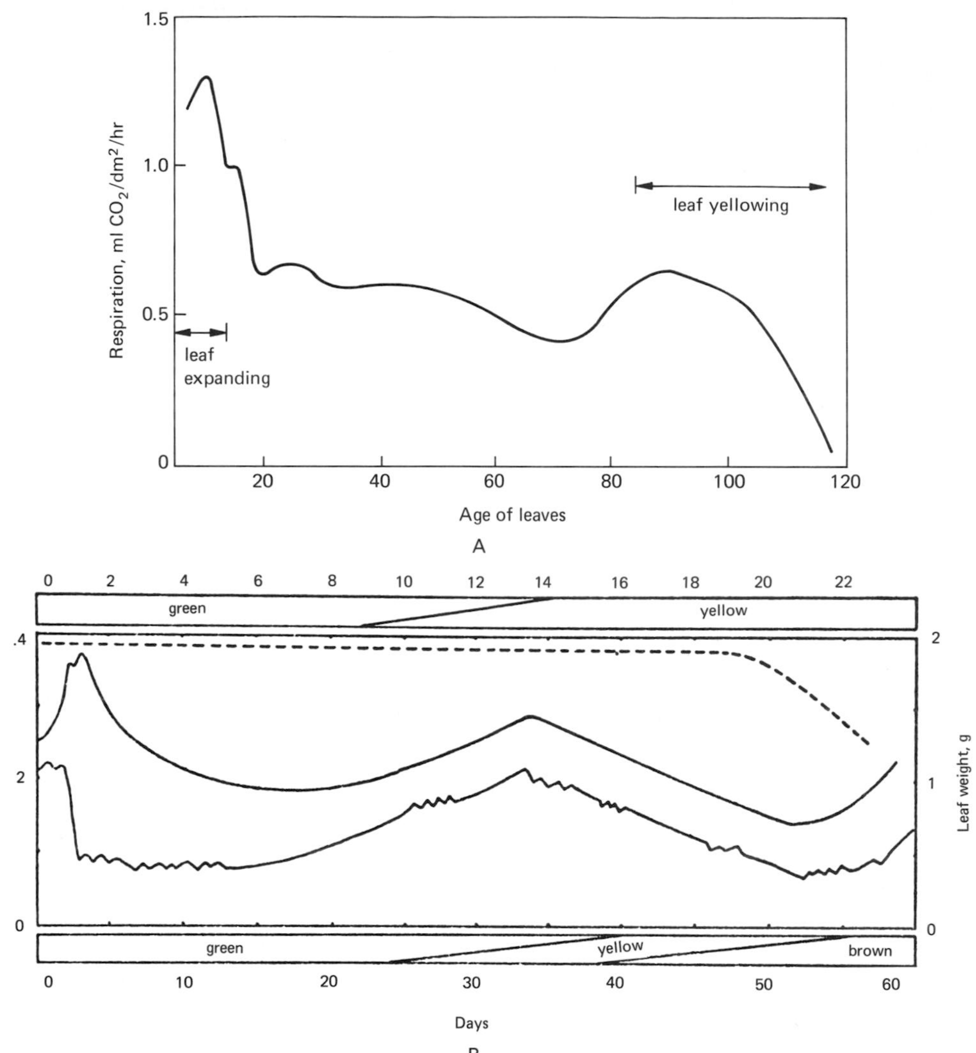

Figure 6-14. Respiration of leaves.

A. The respiration of strawberry leaves attached to the plant. CO_2 production was measured in an air stream at 24.5° C. The leaves were periodically sealed into a respiration chamber over a period of 3 months. [From S. E. Arney: The respiration of strawberry leaves. *New Phytol.,* **46:**68–96 (1947). Used with permission.]

B. Respiration drift-curves for *Tropaeolum majus* (upper solid curve and time scale) and *Prunus laurocerasus* (lower solid curve and time scale). The broken line gives the leaf weight for *Tropaeolum* with ordinates on the right. [From W. O. James: *Plant Respiration*. The Clarendon Press, Oxford, 1953. Used with permission.]

accessibility to oxygen. A summary of respiration rates of various tissues is given in Table 6-2. Respiration rates of metabolically inactive tissues, such as scales, stems, older leaves and roots, or bulky tissues such as fruits in the resting stage, are lower than the rates of actively growing or metabolizing tissues. Rates for tissues having a mass of dead or nonmetabolic material, such as woody stems, may be very low. However, the rate per cell or per unit of protein of some components in such tissues as stems (that is, phloem

Table 6-2. Rates of respiration of some tissues

Tissue	*Respiration rate, μmoles O_2/hr* Per g fresh wt.	*Per g dry wt.*
Man		
resting	10	
running	200	
Mouse		
resting	100	
running	900	
Kidney		900
Brain		600
Bacteria		10,000
Barley seed	0.003	
Wheat seedling	65	
Wheat leaf		
5 day	22	
13 day	8	
Healthy laurel leaf	9	
Starved laurel leaf	1.3	
Barley root	50	
Carrot root	1	
Potato tuber	0.3	
Undeveloped apple fruit	10	
Mature apple fruit	0.5	
Whole potato plant	5	
Pea seed		0.005
Barley seedling		70
Tomato root tip		300
Beet slices		50
Sunflower plant		60
Aroid spadix		2,000

companion cells, cambium, or parenchyma) may be very high. The respiration of bulky organs like potatoes may be much affected by the diffusion rate of oxygen. Dormant seeds may carry on a very slow gas exchange, but it is unlikely that this is really slow respiration. More probably it represents a process of autooxidation and decay rather than organized metabolism.

Temperature. Respiration, like other enzymic processes, is affected by temperature. Within certain limits the rate of enzyme reactions approximately doubles for every 10° C rise in temperature. This is quantitatively expressed by the value Q_{10}, given by the expression

$$Q_{10} = \frac{\text{rate at } (t + 10)^\circ \text{ C}}{\text{rate at } t^\circ \text{ C}}$$

Q_{10} values for respiration are usually between 2 and 3 at temperatures between 0 and 20° C. Above this temperature there is often a decrease in Q_{10} that is probably caused by limitation of oxygen due to the reduced solubility and slow diffusion of this gas. As the temperature increases above 35° C there may be a progressively more rapid breakdown of respiration due to the destruction of enzymes by heat and the breakdown of the

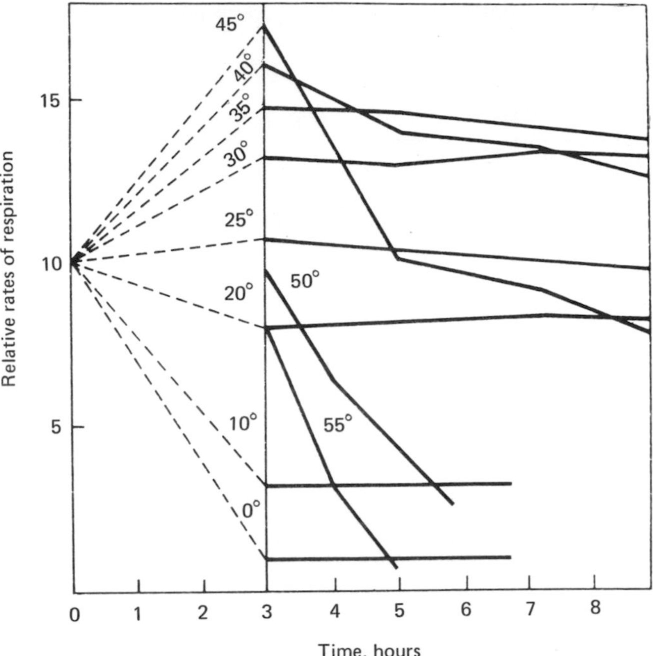

Figure 6-15. The effect of temperature on the rate of respiration of 4-day-old pea seedlings (*Pisum sativum*). At time 0 temperatures were changed from 25° C to those shown. [From R. M. Devlin: *Plant Physiology*. Published by Van Nostrand Reinhold Co. Copyright 1966, 1969 by Litton Educational Publishing, Inc., New York, 1966. Used with permission.]

respiratory mechanism (Figure 6-15). When leaf temperature is raised very quickly, the respiration rate increases rapidly until a brief dramatic rise (the **climacteric**) marks the breakdown of cellular organization and the flooding of oxidative enzymes with substrates. Then the enzymes themselves are inactivated by heat (Figure 6-16). If the temperature is raised more slowly, heat inactivation precedes tissue breakdown, and a shortage of substrates develops that prevents the dramatic climacteric (Figure 6-16).

This brief summary indicates the complex interrelationships that exist between various internal factors affecting respiration response to temperature.

Oxygen. We have considered the effects of oxygen on respiration and fermentation, and the presence of oxygen is clearly essential for oxidative metabolism. However, the cytochrome oxidative system has a high affinity for oxygen; therefore, it is saturated even at very low oxygen partial pressures. Data for a soybean leaf that illustrate this fact and also show that, unlike dark respiration, photorespiration is strongly affected by oxygen concentration are presented in Figure 6-17. On the other hand, oxygen must diffuse to sites of oxidation. Because it is rather insoluble and diffuses slowly as a result, the rates of respiration, particularly in bulky tissues, may frequently be limited by oxygen supply. This is clearly shown in Figure 6-18 where the respiration rate for intact germinating pea seeds is proportional to oxygen concentration (that is, to its diffusion rate) over the whole range 0—100 percent, whereas respiration of seeds with the testa (seed coat) removed is saturated at 20 percent oxygen.

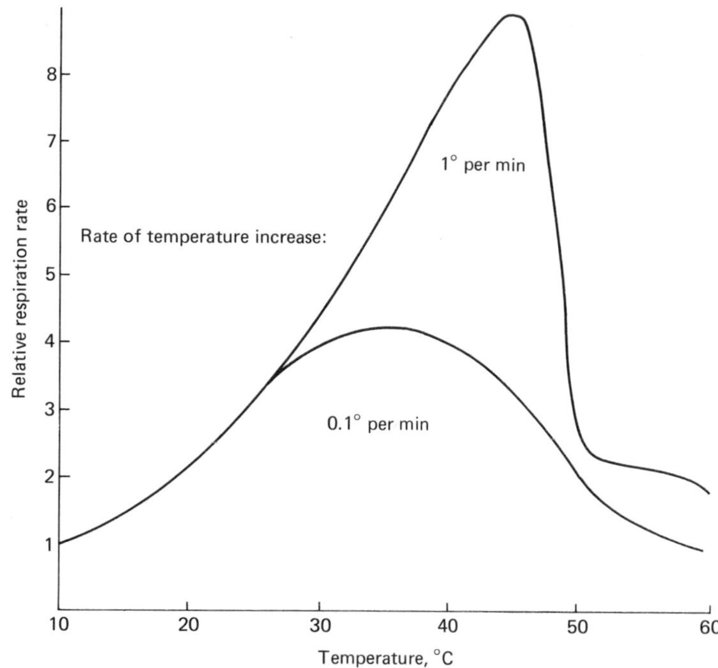

Figure 6-16. The effect of increasing temperature on the rate of respiration of wheat leaves. [Data from undergraduate Plant Physiology Laboratory, University of Toronto, 1962.]

Figure 6-17. Effect of oxygen on the rate of photorespiration (PR) and dark respiration (R_D) in detached soybean leaves. [From M. L. Forrester, G. Krotkov, and C. D. Nelson: *Plant Physiol.*, **41:**422–27 (1966). Used with permission.]

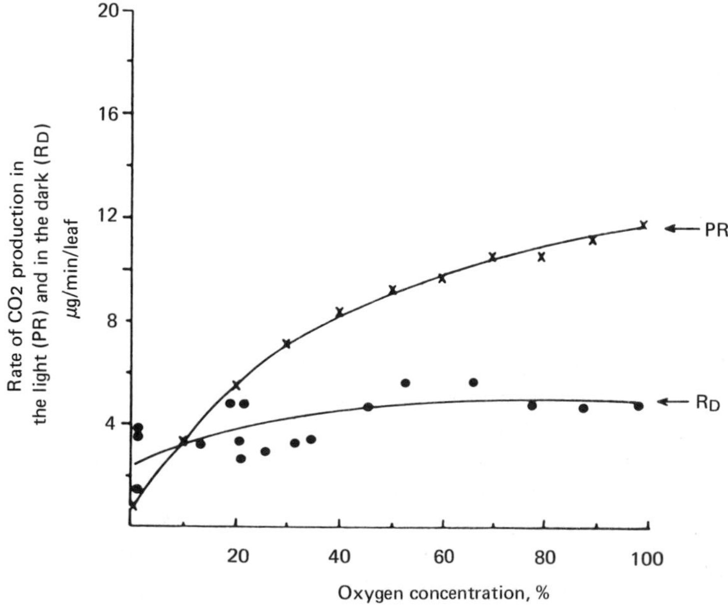

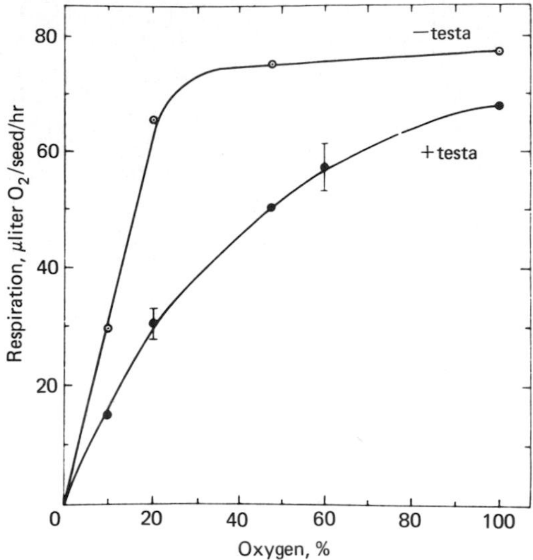

Figure 6-18. The effect of oxygen concentration on the respiration of pea seeds. Removing the testa allowed oxygen to diffuse readily into the seed, which it could not do if the testa was intact. [From E. W. Yemm: In F. C. Steward (ed.): *Plant Physiology*: *A Treatise,* Vol. IVA. Academic Press, New York, 1965, pp. 231–310. Used with permission.]

Carbon Dioxide. As an end product of the reaction, high concentrations of carbon dioxide might be expected to inhibit respiration. It does in fact inhibit respiration somewhat, but only in concentrations that greatly exceed those normally found in air. The mechanism for this inhibition is not clear; in fact, several different mechanisms may be involved. The anaerobic respiration (that is, carbon dioxide production by fermentation) of germinating pea seeds is inhibited about 50 percent by 50 percent carbon dioxide in air. This may relate to a mechanism for maintaining dormancy in seeds; however, its mode of action is not clear. Carbon dioxide has an inhibitory effect on succinoxidase, but this would only affect aerobic respiration, which may not be very important at the start of germination. The decarboxylation steps in respiration are all reactions that yield a substantial amount of energy, so it is unlikely that even a relatively high concentration of end product (carbon dioxide) would have much affect on the reaction rates. Carbon dioxide does, however, have a strong effect on stomata (see Chapter 14). High concentrations of carbon dioxide generally cause stomata to close, and the inhibitory effect that has been observed on leaf respiration may well be due to this effect.

Salts. When roots are absorbing salts, the respiration rate rises. This rise has been linked to the fact that energy is expended in absorbing salts or ions, and the required energy is supplied by increased respiration. This phenomenon, called **salt respiration,** will be discussed in Chapter 12.

Wounding and Mechanical Stimulus. It has long been known that mechanical stimulation of leaf tissue results in increased respiration for a short time—usually a few minutes to an hour. The kind of stimulation seems important. Compression or tension seems to have little effect, bending has more, and shearing stress stimulates respiration most. The mechanisms are unknown. Sound waves have been said to stimulate respiration, among other processes, but no rigorous proof has been offered. Actual wounding or disruption of tissues greatly stimulates respiration for three reasons. The first is the rapid oxidation of phenolic compounds that takes place when the organized separation of these substrates from their oxidases is disrupted. Second, the normal processes of glycolysis and oxidative catabolism are increased as the disruption of the cell or cells results in greatly increased accessibility of substrates to the enzymic machinery of respiration. Third, the usual consequence of wounding is the reversion of certain cells to the meristematic state, followed by callus formation and the "healing" or repair of the wound. Such actively growing cells and tissues have much higher rates of respiration than resting or mature tissue.

The Study and Measurement of Respiration

Respiration has been studied in many ways, using many techniques. The study of respiration is basic to the study of tissue metabolism and biochemistry. It would be far beyond the scope of this book to attempt even an outline of all the experimental work that lies behind the account of respiration given here, but it is possible to examine briefly some of the important experimental approaches and techniques that have been used.

Measurement of Rates. The easiest and most effective method for measuring respiration is to measure the gaseous product or substrate—carbon dioxide or oxygen. Quantitative measurement of gas volumes by manometry has been a standard technique since the development of the **Warburg apparatus** by Otto Warburg. This apparatus consists of a set of accurately thermostated chambers in which samples of plant material are placed. The chambers are fitted with sensitive manometers, and the whole apparatus is shaken continuously. The pressure change resulting from net gas exchange can be measured over a period of time in each manometer, and then alkali may be admitted to a side arm so that the carbon dioxide, which was produced in respiration, is absorbed. The resulting pressure change is due to carbon dioxide produced, and the difference between the pressure change resulting from net gas exchange and that from carbon dioxide absorption is the result of oxygen uptake.

This sensitive apparatus and the more recently developed **respirometer,** which works on the same principle but has a constant pressure adjusted by a micrometer-operated volume-changing device instead of a manometer, have produced much of the quantitative data on tissue respiration. The obvious drawback of this apparatus is that the necessity for small, sealed chambers precludes the use of large samples or attached organs such as leaves or whole plants. For less sensitive measurements, carbon dioxide can be absorbed in alkali solution and determined gravimetrically or by titration.

Several recently developed instruments allow sensitive *continuous* measurement of rates of oxygen uptake or carbon dioxide production. Carbon dioxide can be measured with very great sensitivity in a flowing gas stream by the infrared gas analyzer, which detects carbon dioxide by its absorption of infrared light rays. The graphs showing the changes in leaf respiration with rapid temperature rise presented in Figure 6-16 were made using an infrared carbon dioxide analyzer on a sample of two or three wheat leaves, indicating the sensitivity and adaptability of this technique.

Both carbon dioxide and oxygen concentration in either air or liquid can be measured using polarographic devices having electrodes that basically resemble pH electrodes. These methods permit continuous, sensitive measurement of gas exchange by all types of tissue under almost any experimental situation and have proved immensely valuable in modern plant physiology. A specific example of the value of the extremely

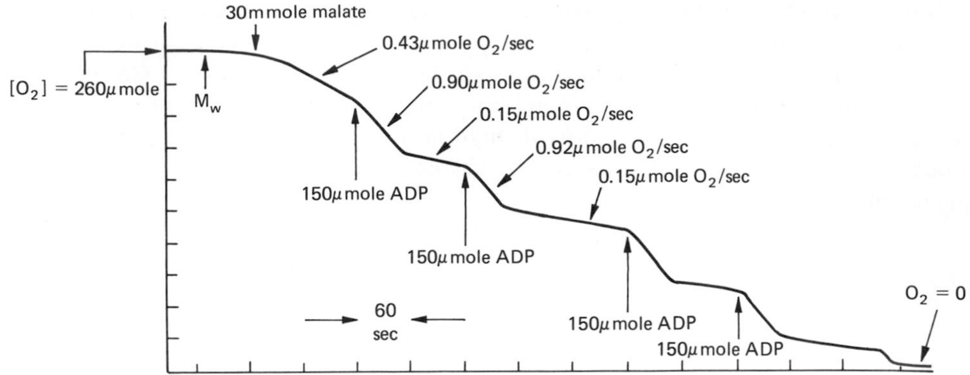

Figure 6-19. Polarographic (oxygen electrode) trace of oxygen utilization by isolated mung bean mitochondria. The steady state oxygen utilization by the mitochondria is shown following additions of malate and ADP to the suspension. The ADP control of oxidation rate is strikingly evident. The ADP/O ratio, equivalent to the P/O ratio, calculated from the trace is about 3. [Data of W. Bonner, Jr., from J. Bonner and J. E. Varner: *Plant Biochemistry*. Academic Press, New York, 1965. Used with permission.]

sensitive polarographic method of oxygen uptake measurement is shown in Figure 6-19, from the work of W. D. Bonner, Jr.

Mitochondria of mung bean are suspended in the appropriate medium, and the rate of oxygen uptake is very small. A carbon substrate is added (malic acid), and the rate of oxidation increases. Then ADP is added (Pi is present in the medium), and immediately the rate of oxidation increases greatly until the ADP is all converted to ATP. The addition of ADP may be repeated, and each time respiration is stimulated until the ADP is used up. Careful examination of the data indicates that about 3 μmoles of ADP are used for each microatom* of oxygen consumed. This means that the P/O ratio of this preparation is 3, which is the theoretical value assuming three ATP produced for each pair of electrons transported via the electron transport system from malate to oxygen.

The presentation of quantitative data for respiration has always been a problem to physiologists. In many instances, as in the experiments reported in Figures 6-16 to 6-19, the comparison of rates under different conditions (for example, oxygen tension), or at different times, is important. So long as the same sample is used for all measurements in a series, no careful measurement of absolute tissue quantities is required.

For comparative purposes, however, some measure of tissue amounts must be made. Fresh weight is often used, but it may be much influenced by water content. Dry weight obviates this difficulty, but the tissue is destroyed and it may not always be feasible to measure dry weight. Further, dry weight represents only total dry matter, not total metabolizing matter. Two tissues having the same number of metabolizing cells of the same activity may have widely different contents of fibrous or sclerenchyma tissue, which would invalidate a dry weight comparison. Comparison has been made on the basis of total nitrogen, total protein, or nucleic acid content in attempts to get around this problem, but it is evident that such practices are open to the same sort of criticism. Moreover, since they presuppose relationships between respiration and these quantities

*A microatom is the number of micrograms equivalent to the atomic weight of the substance, just as a micromole is the number of micrograms equivalent to its molecular weight: 2 μatoms of O = 1 μmole O_2

that may not exist, they could be misleading. Expressions of reaction rates on the basis of specific enzyme or reactant concentration have been useful in estimating the activity of specific systems, but the formidable amount of secondary analysis required precludes the frequent use of this method. Probably the most useful conventions are the simplest: per unit fresh weight or per unit organ (for example, per leaf) when sequential measurement of the same sample is possible.

Understanding Pathways. Before the discovery of radioactive tracers many biochemical techniques were used to determine pathways of carbon reaction, including (1) determining that the required enzymes were present and operating at the required speed, (2) studying reconstituted or partial systems, and (3) using inhibitors to block specific points in the reaction sequence. However, these techniques seldom give a definitive answer. Enzymes may be difficult to isolate, and their activities may change or be lost in the process. Reconstituted systems may or may not be identical with natural ones; the degree of association of enzymes, or their degree of coupling, may be quite different in vivo and in vitro. A number of specific inhibitors can be used to determine if specific reactions are operating. For example, the oxidation of glyceraldehyde-phosphate to PGA is blocked by idoacetamide, arsenite and thiol (SH) group inhibitors. The conversion of PGA to PEP is powerfully inhibited by fluoride. The Krebs cycle is blocked at the aconitase step by fluoroacetate, and at succinic dehydrogenase by malonate. However, inhibitors are seldom completely specific, and the interpretation of results obtained with their use is very difficult. Thus, even by these techniques it could not easily be proved that the pathway under investigation was necessarily the only one or the normal one.

The discovery and use of tracers have made it possible to study reactions much more precisely by (1) kinetic studies of the passage of labeled (radioactive) carbon through the steps of the reaction sequence, (2) the use of specifically labeled intermediates to determine the exact patterns and quantities of carbon distribution in metabolism, and (3) the use of isotopes to clarify the mechanism of individual reactions in a sequence.

An example of the use of specifically labeled intermediates is the use of radioactive pyruvic acid to examine the reactions of the Krebs cycle leading through citric acid to α-ketoglutarate. The reaction mechanism and labeling patterns are shown in Figure 6-20. It will be seen that pyruvate labeled in its first carbon (pyruvate-1-^{14}C) cannot pass this label to citric acid or α-ketoglutarate when entering the cycle by decarboxylation because the C-1 of pyruvate is converted to CO_2 in the synthesis of acetyl–CoA. However, pyruvate carbon may enter the Krebs cycle after its carboxylation, and this would lead to ^{14}C-labeled citric acid from pyruvate-1-^{14}C. On the other hand, ^{14}C in C-2 or C-3 of pyruvate will enter the cycle as a result of either the decarboxylation or the carboxylation of the pyruvate. A comparison of the amount of radioactivity entering the cycle from pyruvate-1-^{14}C or pyruvate-2-^{14}C will therefore provide evidence showing the relative amounts of pyruvate carbon entering the cycle by these two pathways; that is, the relative importance of the anaplerotic carboxylation reaction.

Further information can be had from this system. The experimental evidence shows that when pyruvate-2-^{14}C is used, all the radioactivity is found in the β-carboxyl (C-5) of α-ketoglutarate. This clearly demonstrates the steric specificity of the enzyme aconitase, which operates on citric acid (see Figure 6-5). If there were no steriospecificity, the radioactivity from C-2 of pyruvic acid would end up in both terminal carboxyls of citric acid and, consequently, in both carboxyls of α-ketoglutaric acid.

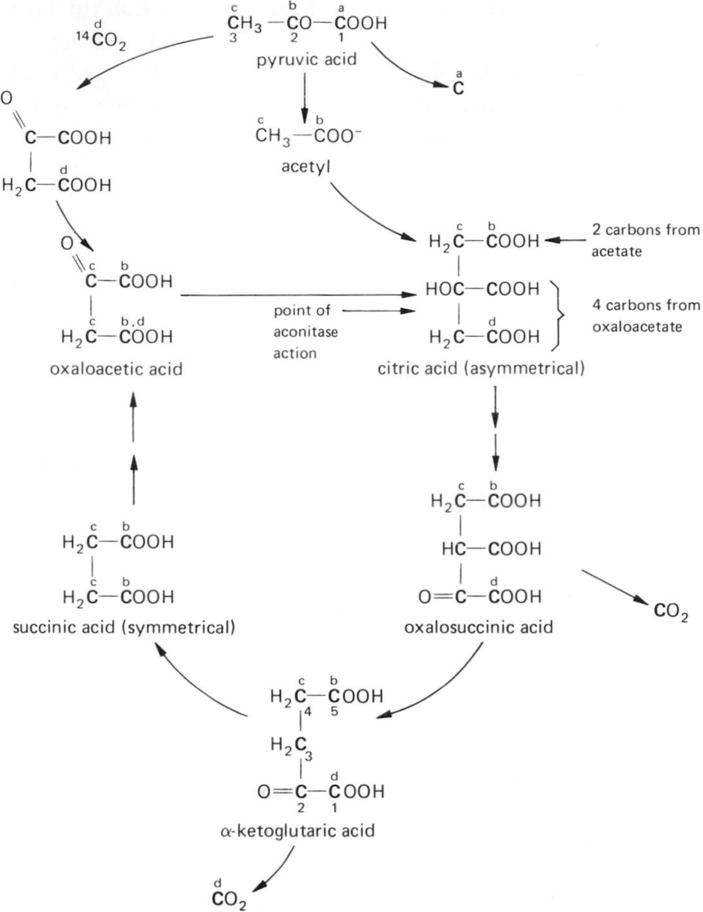

Figure 6-20. Kreps cycle (some reactants omitted at broken arrows) showing distribution of ^{14}C label in various intermediates after supply either of specifically ^{14}C-labeled pyruvate, at *a, b,* or *c,* or of $^{14}CO_2$ at *d*. Labeling for only one turn of the cycle is shown here. Further metabolism would spread the radioactive carbon. Carbon atoms of some compounds are numbered thus: $\underset{1}{C}$, $\underset{2}{C}$, and so on.

In experiments with intact tissues, a certain amount of radioactivity does get redistributed. The radioactivity in the β-carboxyl of α-ketoglutarate will be distributed between the two carboxyls of succinic acid (which is symmetrical) and the succeeding four-carbon acids. Thence, it will reenter the cycle and label the middle and lower carboxyls of citric acid, thus introducing a small amount of radioactivity into the α-carboxyl of the α-ketoglutarate formed in subsequent turns of the cycle. Certain difficulties may arise in interpreting data because the reactions of the four-carbon acids of the Krebs cycle are reversible. Thus, asymmetrically labeled molecules of oxaloacetic acid may become symmetrically labeled through back reaction to fumaric or succinic acid and subsequent reconversion back to oxaloacetic acid.

The derivation of glutamic acid from α-ketoglutaric acid in the Krebs cycle has been demonstrated in vivo by similar experiments, in which the labeling pattern of glutamic acid conforms with that predicted in α-ketoglutarate after the supply of specifically labeled intermediates such as pyruvate or acetate. The amount of redistribution of ^{14}C found in glutamic acid provides data from which it is possible to calculate how much carbon is being recycled and what percentage is being sidetracked into glutamic acid synthesis.

A very interesting modification of this technique is the determination of the pathway of glucose catabolism by the use of specifically labeled glucose. Glucose labeled either at C-1 or C-6 is supplied to two identical tissue samples, and the radioactivity of the CO_2 evolved by each sample is measured. If glucose is being catabolized via the EMP pathway, the same amount of $^{14}CO_2$ will be produced from C-1 and C-6 of glucose because both ends of the molecule are treated alike and that C-6/C-1 ratio will be 1. However, if the glucose is being catabolized via the pentose shunt pathway, the C-6/C-1 ratio will be initially very much less than 1. This is because the C-1 of glucose is released in the first reaction, whereas C-6 is not oxidized until much later, after cyclic reaction of the products of the first oxidation. If both pathways are operating, intermediate values will be obtained. In this way information about the pathways of carbon metabolism that would be impossible to get without the use of isotopes can be rapidly obtained.

Enzymology. No matter how cleverly experiments are performed with tracers to elucidate pathways of reactions, the final proof of the existence of any reaction rests on the isolation, or proof of operation, of the necessary enzymic machinery in the tissues. Many of the enzymes of respiratory metabolism have been isolated or demonstrated beyond doubt in plant systems, using the standard methods of biochemistry, and their specific requirements and cofactors have been determined. Many of these enzymes have been located in specific organelles, and, indeed, it has been shown that several of the enzymes responsible for respiratory or oxidative metabolism exist in different forms, sometimes with different cofactors, in different locations in the cell. Studies on the primary structure of enzyme proteins are beginning to indicate that some of these similarities may be the result of parallel or convergent evolution patterns—that is, the enzymes have evolved from different sources but now perform similar functions. Other enzymes doubtless represent the result of divergence in biochemical evolution; that is, two distinct enzymes may evolve from one common prototype, each now performing either similar or different functions, but under different circumstances or in different locations in the cell.

An extremely powerful tool in the study of electron transfer agents is **spectroscopy.** Each enzyme capable of being oxidized and reduced has a characteristic spectrum of light absorption, and this spectrum undergoes a substantial change when the enzyme is converted from the oxidized to the reduced form or vice versa. Thus, enzymes can be recognized and identified by their **absorption spectra.** Further, it can be demonstrated that they are taking part in specific reactions because they can be recognized by their **difference spectra.** This is the spectral change that occurs when an electron carrier is converted from the oxidized to the reduced state, or vice versa. The difference spectrum is obtained by subtracting the spectrum of the enzyme in the oxidized state from that of the same enzyme in the reduced state. In addition to this, some enzymes undergo a characteristic spectral change when poisoned (for example, by carbon monoxide), and this spectral change can be correlated with the activity of the enzyme.

It is thus possible, by obtaining difference spectra in (1) the presence or absence of substrates, (2) the presence or absence of oxygen, or (3) the presence or absence of

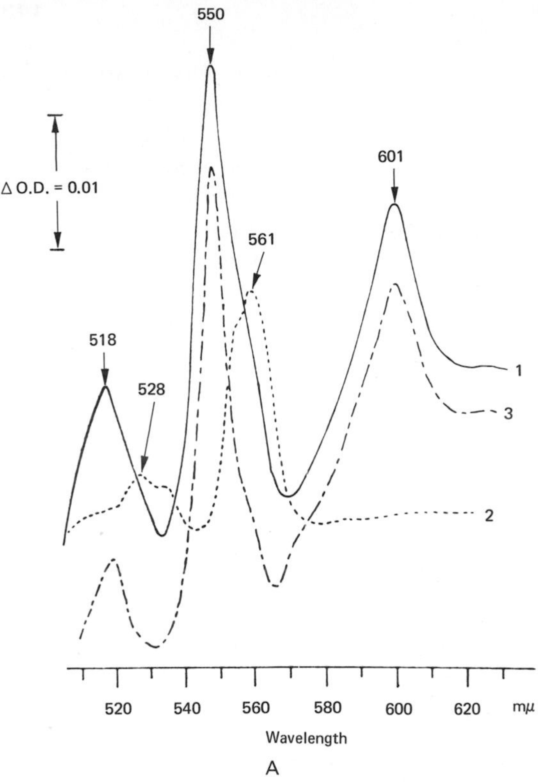

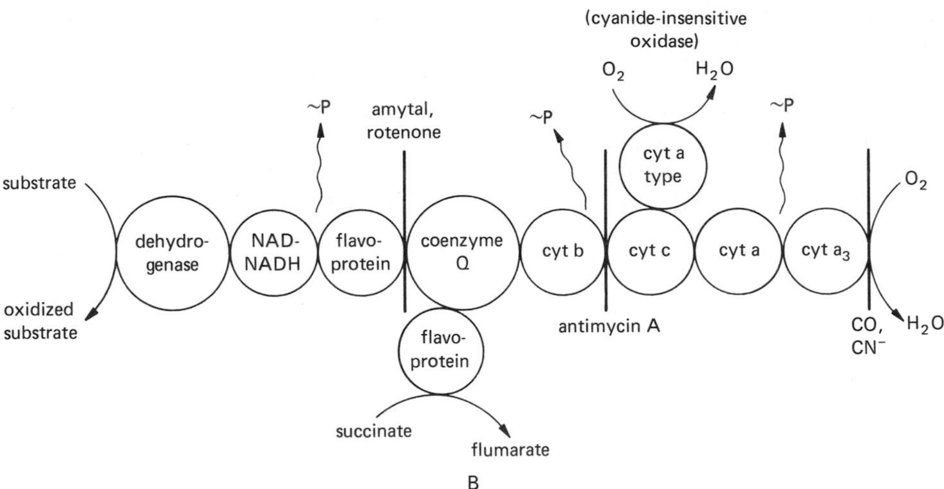

Figure 6-21. The respiratory chain in mitochondria.

A. Difference spectra of isolated wheat root mitochondria. These spectra show how the a, b, and c cytochromes can be optically separated. Curve 1: reduced minus oxidized. Curve 2: substrate + antimycin A + aeration minus oxidized. In this spectrum the a and c components are oxidized; only the b components are reduced. Curve 3: succinate + antimycin A + CN^- minus substrate + antimycin A + aeration. Here the reduced b cytochromes cancel each other out, and only the a and c components appear in the spectrum. [From J. Bonner and J. E. Varner: *Plant Biochemistry*. Academic Press, New York, 1965. Used with permission.]

B. Diagram of the electron transport chain in mitochondria, showing sites of phosphorylation (~P) and inhibitor action. The CN-insensitive terminal oxidase is not poisoned by antimycin, so must divert electrons before they reach cytochrome c.

specific inhibitors whose site of action is known, to demonstrate the participation of specific enzymes in the electron transport chain and to determine their relative positions in the chain.

An example of how this technique was used to examine the electron transport system of wheat-root mitochondria is presented in Figure 6-21A, taken from the work of W. D. Bonner, Jr. Curve 1 gives a composite difference spectrum (reduced minus oxidized) of all three cytochromes, a, b, and c. The cytochrome b spectrum is separated in curve 2. This is obtained by putting cytochrome b in either the oxidized or reduced state by adding or omitting the substrate, and by using the poison antimycin A, which blocks the transfer of electrons from cytochrome b to cytochrome c. The cytochrome a and c spectra are cancelled because they are kept continuously in the oxidized state by the presence of oxygen. In curve 3, cytochrome b is cancelled out by being kept in the reduced state by the use of antimycin A during both parts of the difference spectrum measurement. The difference spectrum of cytochromes a and c is obtained by the contrast of the cyanide-poisoned spectrum with the spectrum in the presence of oxygen.

Figure 6-21B shows a schematic representation of the mitochondrial electron transport system, with the position of some well-known inhibitors marked. The cyanide-insensitive bypass and the sites of ATP ($\sim$P) synthesis are also shown. The data in Figure 6-21A should be examined with reference to Figure 6-21B.

Additional Reading

Articles in *Annual Reviews of Plant Physiology* on cell energetics, mitochondrial function, and respiration.

Up-to-date biochemistry textbooks may be consulted for further details on respiratory pathways and enzyme mechanisms.

Beevers, H.: *Respiratory Metabolism in Plants.* Row, Peterson & Co., Evanston, Ill., 1961.

Forward, D. F.: The respiration of bulky organs. In F. C. Steward (ed.): *Plant Physiology: A Treatise,* Vol. IVA. Academic Press, New York, 1965.

Steward, F. C. (ed.): *Plant Physiology: A Treatise,* Vol. IA. Academic Press, New York, 1960.

Yemm, E. W.: The respiration of plants and their organs. In F. C. Steward (ed.): *Plant Physiology: A Treatise,* Vol. IVA. Academic Press, New York, 1965.

7 Photosynthesis

Introduction

Our approach to the study of photosynthesis depends upon how we define it. Some research scientists who work on photosynthesis have considered only the reactions actually involving light quanta, whereas others have examined the whole spectrum of reactions that result from the initial stimulus provided by light. We shall take the broader view and consider all the processes of metabolism that go on as the direct result of the light-gathering activities of the photosynthetic apparatus. It is evident that there must be interactions between the anabolic process of photosynthesis and all the other catabolic and anabolic processes in the cell, but these interactions will be considered in detail in Chapter 15.

Basically, photosynthesis is the absorption of light energy and its conversion into stable chemical potential by the synthesis of organic chemicals. It is best approached as a three-phase process:

1. The absorption of light and retention of light energy.
2. The conversion of light energy into chemical potential.
3. The stabilization and storage of the chemical potential.

In describing these phases we shall also have to examine the mechanics of photosynthesis—the physical nature of the subcellular apparatus in which the process takes place and the chemical machinery that is required.

Photosynthesis is important for a number of reasons. From man's point of view its greatest importance is its role in the production of the world's foodstuff and oxygen; therefore, photosynthesis is often studied in terms of its end products. In the overall process, however, these are secondary. The important concept is the trapping and transformation of energy.

A simple analogy is worth considering. When a quantum of light strikes an object—say, a black rock—a molecule of the rock absorbs the energy of the light quantum. This molecule becomes momentarily more energetic or "hot"; that is, an electron of the rock molecule assumes a higher energy orbital—the electron is raised to a higher energy level. The electron does not stay at the higher level long; almost immediately it falls again to its former level (or **ground state**), and the extra energy absorbed by the rock molecule is reemitted at once as heat. This process is illustrated in Figure 7-1A.

When a quantum of light strikes and is absorbed by a molecule of chlorophyll in a plant, the molecule becomes energized and an electron is raised to a higher energy level, just as in the rock. However, instead of returning to the ground state immediately with the loss of all the absorbed energy as heat, the electron is held at the higher energy level by being transferred to an appropriate electron-accepting compound. In the process the compound that receives the electron becomes reduced, and the energy that entered the chlorophyll molecule has now been trapped and converted into chemical potential in a reduced chemical bond. The initial bond so formed may be quite unstable; however, it is stabilized by a series of chemical transformations so that the energy is stored and can later be released in the reactions of respiration, as illustrated in Figure 7-1B. Life may thus be viewed as an electric current—the analogy is illustrated in Figure 7-1C. The

Figure 7-1. Process of photosynthesis and an electrical analogy (*hν* is the symbol for a quantum of light, or photon).
A. Light strikes a rock and is converted to heat.
B. Light strikes a chlorophyll molecule, its energy is converted to chemical potential, stored, and later used.
C. An electrical analogy of photosynthesis.

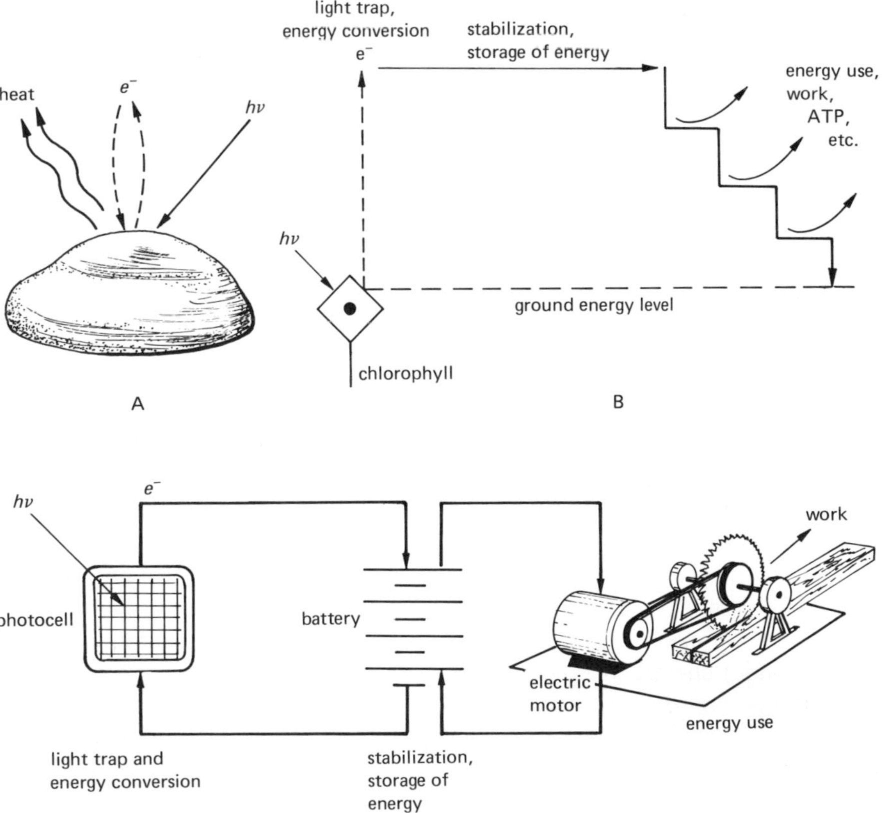

photosynthetic apparatus, electron transfer agents, and carbon chemistry are the wires, batteries, and electric motors of the living system. They exist, as do the wires and components in an electric circuit, as a means of converting and transferring energy. It must be remembered, however, that organisms are not made only of energy. The point of storing energy is that it can be used to conduct reactions, to synthesize the stuff that plants and animals are made of.

Historical Background

Research in photosynthesis presents an orderly historical development leading to the present understanding of the processes involved. Aristotle thought that light was necessary for the growth of plants, but it was Stephen Hales in 1727 who first clearly recognized that light was required in the process by which plants gain nourishment from the air. (Previously it had been thought that plants obtained their substance only from water and the soil.) During the period 1790 to 1815, experiments by Priestley, Ingen-Housz, Senebier, and de Saussure established that the green parts of plants absorb CO_2 from the atmosphere and produce O_2 when illuminated, the process being reversed in the dark. It was soon recognized that water is also absorbed and converted into organic matter in light. By the middle of the nineteenth century the German physiologist Mayer had recognized the true importance of light in photosynthesis, and indeed its importance to the whole biotic world, in supplying the energy for all biological processes through photosynthesis. By 1900 photosynthesis had been extensively studied and had been put on a quantitative basis shown in the equation

$$6\,CO_2 + 6\,H_2O \xrightarrow{\text{light energy}} C_6H_{12}O_6 + 6\,O_2$$

Much of the work at that time was directed toward understanding this reaction in terms of a chlorophyll-mediated reaction (perhaps together with other pigments) of CO_2 with H_2O leading to the formation of reduced carbon that would polymerize into the known sugars produced in photosynthesis. Unfortunately this concept was wrong. The wrong questions were being asked, hence, much of the work produced only puzzling results that could not be interpreted.

The groundwork for the correct approach was laid by the English physiologist F. Blackman in 1905 in his work on the interrelations of light and temperature effects on photosynthesis. He found that the rate of photosynthesis varied with temperature at high light intensities (that is, $Q_{10} = 2$ or 3) but was unaffected by temperature ($Q_{10} = 1$) at low light. Thus when photosynthesis was limited by light, it was temperature insensitive, indicating that the overall process was limited in rate by a temperature-insensitive (presumably nonenzymic) reaction requiring light. On the other hand, when saturated by light, the process was limited in rate by a temperature-sensitive (therefore presumably enzymic) chemical reaction.

Further experiments indicated that the temperature-sensitive portion of photosynthesis could also be limited by reducing the CO_2 concentration, whereas the temperature-insensitive portion did not require CO_2. The conclusion followed that photosynthesis consists of at least two reaction sequences—one requiring light, a nonchemical reaction that has come to be called the **light reaction,** and the other a chemical, enzymic, CO_2-requiring reaction that does not require light, called the **dark reaction** or the **Blackman reaction.**

Work with inhibitors or poisons by the German scientist O. Warburg and many others established that these two reactions are quite independent. Experiments by many scientists, particularly Warburg and later the Americans R. Emerson and W. Arnold, showed that the efficiency of photosynthesis (the amount of photosynthesis per unit of light) could be greatly increased by interrupting the light with short periods of darkness. This showed that the light process is very rapid, whereas the dark process is slower. In a brief flash of light the light reaction would build up an excess of its end product, which could be used during subsequent dark intervals by the dark reactions for the fixation of CO_2. Since the dark intervals could not be extended for more than a few seconds without loss of CO_2-fixing ability, it was concluded that the products of the light reaction are extremely labile and decompose rapidly if not used at once in the dark reaction.

In 1924 the American microbiologist C. B. van Niel proposed a mechanism accounting for the dual nature of the photosynthetic reaction based on his observations of bacterial photosynthesis. He showed that the process in certain bacteria could be represented by

$$CO_2 + 2\,H_2A \longrightarrow [CH_2O] + H_2O + A_2$$

where H_2A represents a reduced substance (H_2O in plants) that donates electrons and H^+ ions for the reduction of CO_2 to carbohydrate and H_2O. The formula $[CH_2O]$ represents a single carbon atom at the level of carbohydrate, not a specific chemical. The reaction could be divided into two parts

$$2\,H_2A \xrightarrow{\text{light}} A_2 + 4\,[H] \qquad (1)$$

$$4\,[H] + CO_2 \longrightarrow [CH_2O] + H_2O \qquad (2)$$

Van Niel suggested that, instead of reacting with CO_2 as earlier postulated, water was a substrate from which H^+ ions and electrons could be derived for the reduction of CO_2 as shown in the two equations below

$$2\,H_2O \xrightarrow{\text{light}} O_2 + 4\,[H]$$

$$4\,[H] + CO_2 \longrightarrow [CH_2O] + H_2O$$

The nature of the reductant was not known, and it was clear that the second reaction was complex and probably involved several steps. However, the basic implications were extremely important: the light reaction is the light-induced splitting, or **photolysis,** of water to produce O_2 and reducing power, and the dark reaction is the use of this reducing power to reduce CO_2 to carbohydrates and water.

The validity of this view was not finally established until 1941 when the American scientists S. Ruben and M. D. Kamen and coworkers were able to use isotopically enriched oxygen, $^{18}O_2$, to establish that all of the O_2 produced in photosynthesis was derived from water, and none from the CO_2. The three possible reactions are shown below.

$$C^{16}O_2 + H_2^{18}O \longrightarrow [CH_2^{16}O] + {}^{16,18}O_2 \qquad (1)$$

$$C^{16}O_2 + H_2^{18}O \longrightarrow [CH_2^{18}O] + {}^{16}O_2 \qquad (2)$$

$$C^{16}O_2 + 2\,H_2^{18}O \longrightarrow [CH_2^{16}O] + H_2^{16}O + \qquad (3)$$

It was found that the O_2 produced in photosynthesis with $H_2^{18}O$ has essentially the same

^{18}O content as the water used, indicating that the oxygen was derived from splitting the water as in reaction (3), not from a reaction of water with CO_2.

This concept was related to the biological system in 1937 by the work of the English biochemist R. Hill, who was the first to get a partial reaction of photosynthesis to work in isolated chloroplasts. His preparations could produce O_2 and simultaneously reduce added electron acceptors in light, a process that has been called the **Hill reaction.** Although Hill was unable to couple this reaction with the reduction of CO_2, it clearly represents the first step, or light reaction, of photosynthesis.

Much later M. Calvin's group in Berkeley, California, using $^{14}CO_2$, was able to demonstrate finally that the dark reaction of photosynthesis can indeed go on in darkness. Suspensions of algal cells were shown to be able to fix CO_2 for a brief period of time in darkness, after illumination, but the total "carryover" of reducing power into darkness was not large. In the period from 1955 to 1960 the American physiologist D. I. Arnon and his coworkers showed that there are two products of the light reaction, ATP and a reducing agent later shown to be NADPH, and that these two high energy compounds are consumed in dark reactions resulting in the reduction of CO_2. Arnon was able to show that the components of the light reaction, which lead to the reduction of NADP and the production of ATP (by **photophosphorylation**), are tightly bound in the chloroplast, but that the enzymes of the dark reaction are soluble and leak out during the isolation of chloroplasts. When these enzymes were added back, the entire process would proceed, although the reaction rates were very low. During the preceding decade, Calvin's group, working with $^{14}CO_2$, had elucidated the reactions of the carbon reduction cycle **(Calvin cycle)** in which CO_2 is fixed and reduced to carbohydrate using the NADPH and ATP generated in the light reaction.

The most important step in the elucidation of the nature of the light reaction of photosynthesis was the suggestion by R. Hill and F. Bendall in 1960 that this process could be considered as a two-step electron transport system, as outlined in Figure 7-2. In the first step electrons from H_2O are raised from the ground level to an intermediate level (resulting in the production of O_2), and in the second step they are raised to the reducing level of H_2, with the formation of NADPH. The transfer of electrons from the

Figure 7-2. Outline of the two-step electron transport system of photosynthesis.

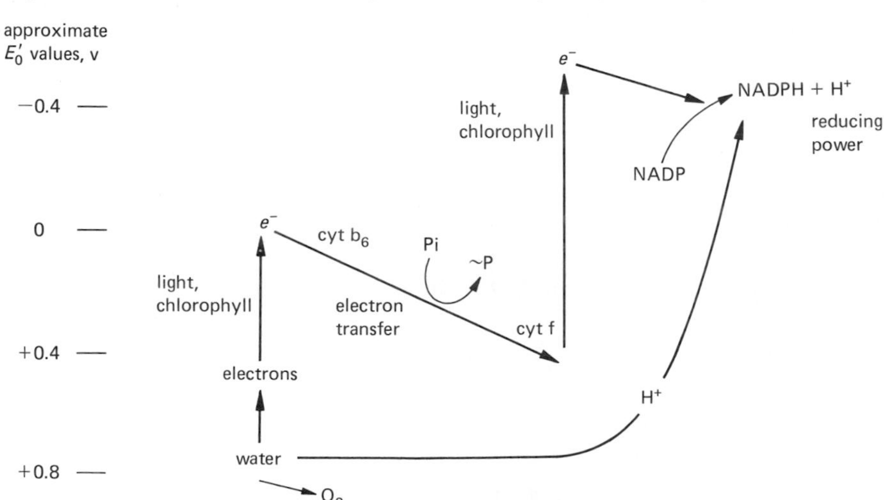

level of the top of the first step to the bottom of the second, a lowering in their potential, could take place via a conventional electron transport system involving two known cytochromes located in the chloroplast, cytochrome b_6 ($E_0' = 0.0$ v) and cytochrome f ($E_0' = +0.37$ v). Electron transfer through these components could generate ATP in the usual way (rather than viewing photosynthetic phosphorylation as a special sort of reaction, different from the oxidative phosphorylation of respiration). The approximate potentials of the electrons at different stages in the system are shown in Figure 7-2.

We can summarize the discussion to this point as follows: light energy, absorbed by chlorophyll, is used to remove electrons from water (causing the liberation of oxygen) and to raise them via a two-step transport process to the reducing level required to reduce carbon dioxide. In the process, Pi is esterified to ADP making ATP, which is used, together with the reducing power generated, to drive a cycle of chemical reactions by which carbon dioxide is fixed and reduced to carbohydrate. We shall now examine the components of the system more carefully in order to develop the details of the currently accepted model.

It should be noted now that the terms "light" and "dark" reactions are not really very satisfactory. The only true light reaction is the absorption of photons. All the electron transport reactions that follow (and that were included in the early concept of the light reaction) are actually dark reactions, since they can be chemically energized and made to run in the dark. The dark reaction is in reality a large number of reactions that do not require light. Several of them are light activated; most if not all of them do not operate in the dark under normal conditions. For these reasons we shall abandon the terms light and dark reactions and call them instead the **electron transport** and **carbon reactions** of photosynthesis.

Electron Transport Reactions

LIGHT. Since the primary reactions of photosynthesis involve the absorption of light by pigments, it is necessary to examine some of its basic properties. Light is electromagnetic energy propagated in discrete corpuscles called **quanta** or **photons.** As the energetics of chemical reactions are usually described in terms of kilocalories per mole of the chemicals (1 mole = 6.02×10^{23} molecules), light energies are usually described in terms of kilocalories per mole quantum or per **einstein** (1 mole quantum or 1 einstein = 6.02×10^{23} quanta).

The color of light is determined by the wavelength (λ) of the light radiation. At any given wavelength, all the quanta have the same energy. The energy (E) of a quantum is inversely proportional to its wavelength. Thus the energy of blue light ($\lambda = 420$ mμ) is of the order of 70 kcal/einstein, and that of red light ($\lambda = 690$ mμ) about 40 kcal/einstein. The symbol commonly used for quantum, $h\nu$, is derived from this relationship. In any wave propagation, the frequency (ν) is inversely proportional to the wavelength. Since $E \alpha 1/\lambda$, then $E \alpha \nu$. Planck's constant (h) converts this to an equation. $E = h\nu$. Thus $h\nu$, used to designate a quantum, refers to the energy content of the quantum.

The relationship between the energies of light, both as calories per mole quanta (per einstein) and as E_0' values, and the energies required to conduct certain reactions is shown in Figure 7-3. It may be seen that the energy of a red quantum is just sufficient to raise an electron from OH^- to the reducing level of H_2; a UV quantum contains nearly twice this amount of energy. Thus, there is enough energy in a quantum of light (barely

Wavelength of light, mμ	Color of light		Energy of quanta kcal/mole	E_0', v
350	Ultraviolet		80	3.5
450	Blue	visible		
550	Yellow	visible	60	2.6
650	Red	visible		
750	Infrared		40	1.7

Figure 7-3. Energy of light quanta at different wavelengths.

enough in a red quantum) to split water. This does not mean that such a reaction takes place, only that it is thermodynamically possible. In fact, the conversion of energy from one form into another invariably results in the loss of some energy to the surroundings during the process of conversion—no machine can be 100 percent efficient. As a result, the whole energy of a quantum can never be available to conduct a chemical reaction.

The proportion of light energy that is available in photosynthesis has long been a matter of controversy and is still not settled. Much attention was focused in the period from 1930 to 1950 on the quantum requirement of photosynthesis, a measure of the efficiency of the process. The initial thought was that, since 1 quantum can conduct only one molecular reaction, the number of quanta absorbed per molecule of O_2 produced might indicate the number of reaction steps involved.

In complex experiments requiring most carefully controlled conditions, Warburg found quantum requirements of 4 quanta per molecule of oxygen produced. This means 1 quantum per [H] derived from H_2O for the reduction of CO_2. From Figure 7-3, the energy required to derive one reducing equivalent from H_2O is close to 30 kcal/mole, and the energy available in quanta of red light, which Warburg used in his experiments, is only about 40 kcal/einstein. This is an efficiency of 75 percent, which is extremely high for any energy conversion machine.

Warburg's experiments could not be repeated in other laboratories, and it is now generally accepted that at least 8–12 quanta are required to reduce one molecule of CO_2 and produce one molecule of O_2. The scheme represented in Figure 7-2 satisfies the requirement of at least 8 quanta for each O_2 produced or CO_2 reduced, because it takes 4 ($e^- + H^+$) to reduce one CO_2 to carbohydrate level and each electron transferred requires 2 quanta. This formulation is not the only one that would work, as we shall later see, but it is consistent with the evidence.

Pigments. So far the only light-absorbing pigment we have mentioned is chlorophyll. Initially the green pigment of plants was recognized as the substance responsible for light absorption in photosynthesis, absorbing red and blue but not green light. However, it has long been known that there are a number of different pigments in plants of various colors and that even chlorophyll is not a simple substance but a group of related pigments. Some of the coloring matter of plants was discovered to be outside the chloroplasts, diffused through the cytoplasm or else present in special bodies, sometimes like plastids but often irregularly shaped or deeply angular, called **chromatophores.**

Chloroplasts are the site of photosynthesis; therefore, the pigments outside the

Table 7-1. A list of photosynthetic pigments

Pigment	*Where found*	*Light absorbed*
Chlorophyll a	All green plants	Red and blue-violet
b	Green plants, not red or blue-green algae or diatoms	Red and blue violet
c	Brown algae, diatoms	Red and blue-violet
d	Red algae	Red and blue-violet
Protochlorophyll	Etiolated plants	Near-red and blue-violet
Bacteriochlorophyll	Purple bacteria	Near-red and blue-violet
Bacterioviridin	Green sulphur bacteria	Near-red and blue-violet
Phycocyanin	Blue-green, red algae	Orange red
Phycoerythrin	Red, blue-green algae	Green
Carotenoids (carotenes, xanthophylls)	Most plants, bacteria	Blue, blue-green

chloroplasts (notably the blue and red **anthocyanins,** the yellow **xanthophylls,** and some of the red to orange **carotenes**) are not associated with photosynthesis. However, a number of pigments other than chlorophyll are found within chloroplasts, including some xanthophylls and carotenes, and extensive experiments have been conducted to determine whether they play a part in photosynthesis. Some of these pigments are present in special groups of plants, whereas others have nearly universal distribution. A list of the important photosynthetic pigments with some basic information about them is presented in Table 7-1. The chemical structure of some of these pigments is summarized in Figure 7-4 and 7-5.

Figure 7-4. Graphic formula of chlorophyll a. Several tautomers (with different arrangement of double bonds) are possible. The rings numbered I to IV are pyrrole rings, V is the cyclopentanone ring.

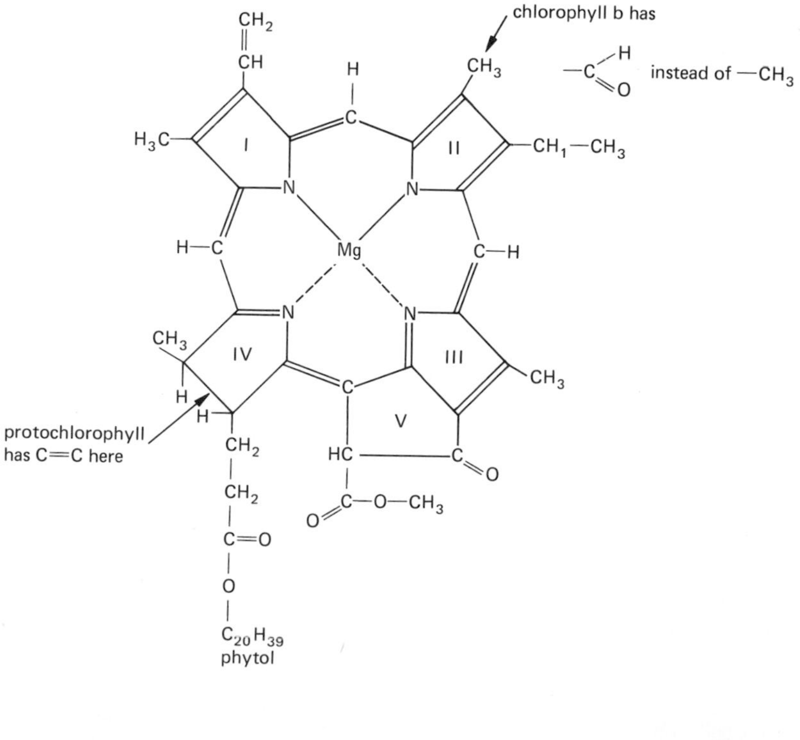

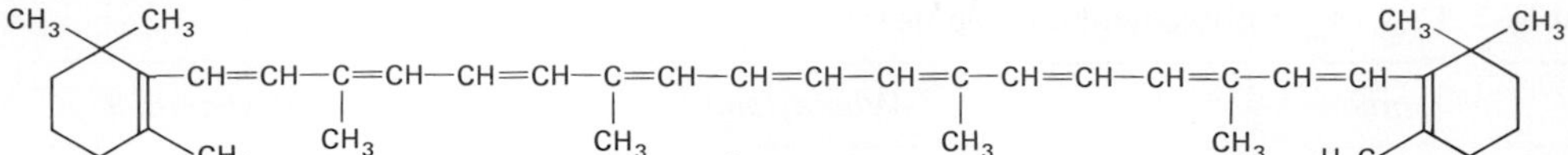

β-carotene

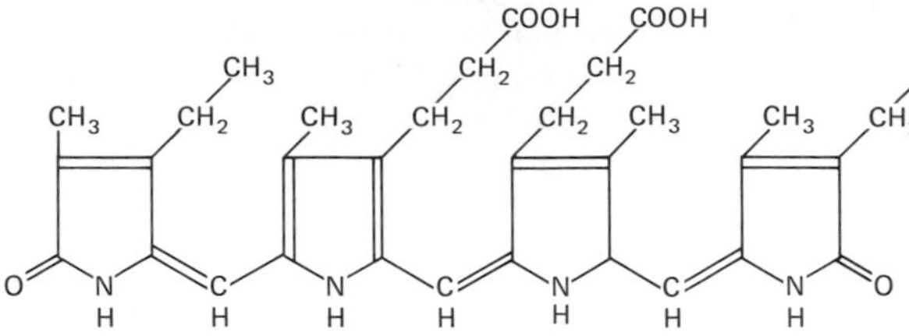

phycocyanin

Figure 7-5. Formulas of a carotene and a phycobilin pigment. Compare the **cyclic** tetrapyrrole nucleus of chlorophyll with the **linear** tetrapyrrole nucleus of the phycobilin.

Chlorophyll a is universally present in all photosynthetic plants. Chlorophyll b is present in most green plants, but its place is taken by **phycocyanin** in blue-green algae, **fucoxanthin** in brown algae, and **phycoerythrin** in red algae. Photosynthetic bacteria have a far-red absorbing chlorophyll, **bacteriochlorophyll.** There are a number of carotenoid pigments, one or more of which are present in nearly all photosynthetic organs.

It may be seen from Figure 7-4 that chlorophyll is a tetrapyrrole and bears a close resemblance to the chemical structure of heme and the cytochromes. Chlorophyll differs from these iron enzymes in that it contains an atom of magnesium, which does not appear to participate directly in electron transfer reactions as does the iron of cytochromes. Chlorophyll is characterized by a cyclopentanone ring (V), and a number of characteristic side groups at various points. The identity of the side groups provides the identity of the various chlorophylls. The most important of these is the phytol ester attached to ring IV. This provides a long chain lipophilic "tail" to the molecule that is extremely important in the orientation and anchoring of chlorophyll molecules in the chloroplast lamellae.

Chlorophyll has the potential for a number of reaction mechanisms in light absorption. It might change its energy content by resonance of its coordinate bond structure (the alternating single and double bonds can resonate by "changing places" back and forth), by the reduction of one of its double bonds, by the reduction of the quinone (=O) in ring V, or by the loss of a single electron in the double bond structure. Experiments with deuterium or tritium isotopes of hydrogen have indicated quite clearly that chlorophyll does not participate in H transfers or oxidation-reductions involving H. Chlorophyll appears to participate in energy transfer reactions both by electron transport (that is, oxidation and reduction by the gain and loss of an electron) and by resonance (a direct transfer of energy; see The Light Trap, page 159).

The chemical structure of the accessory phycobilin pigments, phycocyanin, and phycoerythrin, is similar to that of chlorophyll in that they are tetrapyrrole compounds; however, they differ in being linear instead of cyclic in structure, and they do not have a metal component (Figure 7-5). The carotenes, including fucoxanthin, are closely related to vitamin A, and are basically long chain lipophilic molecules with a more active terminal group at each end. The synthetic pathways of these compounds will be briefly considered in Chapter 9. The porphyrin pigments require light for their synthesis in most plants, hence the colorless or pale yellow appearance of dark-grown or **etiolated** leaves. The last step of chlorophyll synthesis, the reduction of protochlorophyll to chlorophyll, is accomplished at the expense of light energy absorbed by the protochlorophyll molecule itself. The pathway of chlorophyll synthesis is also described in Chapter 9.

The contribution of the various chloroplast pigments to photosynthesis has been a subject of intense experimentation for many years. With the development of modern methods of spectroscopy, it has become possible to compare the absorption spectrum of an organism (that is, the spectrum of light absorbed by the whole organism, which would depend on the presence and concentrations of all the pigments in the organism) with the action spectrum of photosynthesis in the organism. The **action spectrum** of photosynthesis measures the effectiveness of light of various wavelengths in conducting photosynthesis. It is possible to determine which pigments are present from an analysis of the absorption spectrum, and to determine their relative contribution to photosynthesis by comparing the absorption spectrum with the action spectrum.

Some curves for the marine alga *Ulva* (especially suitable for this study because it grows in uniform thin sheets) are shown in Figure 7-6, together with the absorption spectra of the major pigments. It may be seen that the action and absorption curves compare well over much of the spectrum, showing the contribution of the chlorophylls and the phycobilins to photosynthesis; however, there is a clear discrepancy in the carotene area, showing that carotenes are not as efficient photosynthetic pigments as the chlorophylls or the phycobilins in this organism. There is evidence that carotenes may be effective in some plants, but they are not universally involved in photosynthesis. Alternative evidence stresses their role in the stabilization of chlorophyll in plastids and in the prevention of autooxidation or light destruction of chlorophylls.

Pigments that are involved in photosynthesis are capable of carrying on certain reactions even after they are extracted from the chloroplasts. They can absorb light and **fluoresce**—that is, reemit the light energy absorbed as light, but necessarily of longer wavelength (that is, of lower energy). Chlorophyll solutions fluoresce dark red. If the reactions of photosynthesis are blocked by poisons, fluorescence will occur in vivo because the energy absorbed cannot be used. The fluorescence spectrum is characteristic of the pigment, so it is possible to tell which pigment is fluorescing, hence, which was activated. In addition, energy acceptors (that is, oxidized substrates capable of being reduced) can quench the fluorescence, demonstrating the ability of the pigment to transfer its energy to these acceptors instead of reemitting it as light. Chlorophyll solutions can be made to catalyze reactions when illuminated as, for example, the formation of polymers from a solution of a monomer, showing that the light energy can be used to create free radicals. The transfer of energy from molecule to molecule has been demonstrated in a mixture of chlorophyll a and chlorophyll b. When the mixture is illuminated by light that can be absorbed only by chlorophyll a, an analysis of the fluorescence spectrum shows that chlorophyll b has been made to fluoresce, indicating energy was transferred from the absorbing chlorophyll a molecules to chlorophyll b.

So far we have not considered the efficiency of different wavelengths of light in conducting photoreactions. It appears that the high energy blue light absorbed by chlorophyll is not used efficiently. The basic requirement is for a specific number of quanta, and the energy of the quanta (providing they can be absorbed by chlorophyll) is unimportant. Red quanta (40 kcal/einstein) are as effective as blue quanta (70 kcal/einstein); the extra energy of the blue quanta is wasted. Presumably if a

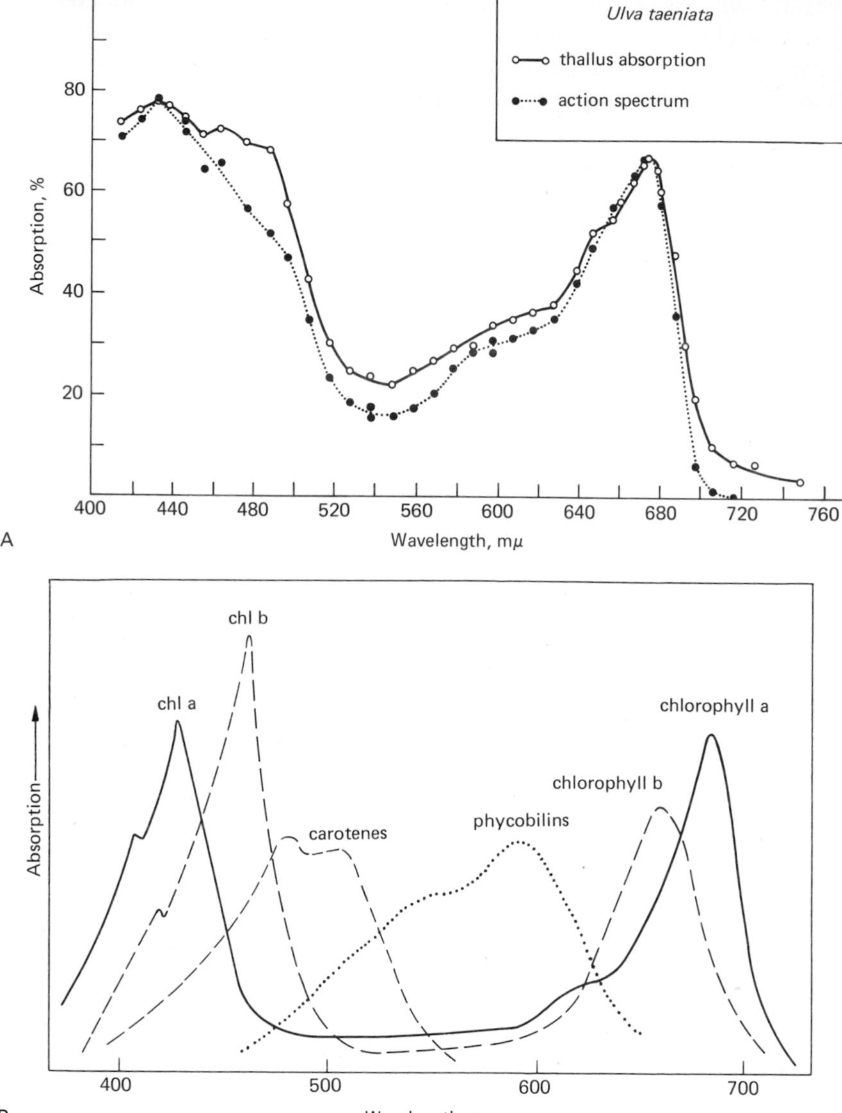

Figure 7-6. Comparison of the action and absorption spectra of an organism and the absorption spectra of important photosynthetic pigments.
A. Action spectrum and absorption spectrum of the green alga *Ulva*. Note the discrepancy between the spectra at 460 to 500 mμ. [From F. T. Haxo and L. R. Blinks: *J. Gen. Physiol.*, **43:**404 (1950). Used with permission.]
B. Absorption spectra of some photosynthetic pigments.

quantum is of the appropriate wavelength to be absorbed, it will be effective. However, an important exception to this behavior is the co called **red drop**—a decided decrease in efficiency found in many organisms at the far-red end of the absorption spectrum, usually over 685 nm.

Emerson found that the efficiency of red light at a wavelength of about 700 nm could be increased by adding shorter wavelength light (650 nm). In other words, the rate of photosynthesis in light of the two wavelengths together was greater than the added rates of photosynthesis in either alone. This is called the **Emerson effect,** after its discoverer, and has been important in shaping the thinking that has developed the current concept of the photosynthetic mechanism. Evidently, light is absorbed separately by two different pigment systems, one of longer wavelength than the other, and the satisfactory functioning of photosynthesis requires that both systems be activated. By careful spectral analysis of this effect, it has been found that the shorter wavelength system (now called system II) contains, besides some chlorophyll a, a substantial amount of chlorophyll b and accessory pigments such as phycobilins. The long wavelength system (system I) has a higher proportion of chlorophyll a and less of the other pigments. These two light-absorbing systems are correlated with the two light absorptions postulated by Hill and Bendall (Figure 7-2). We shall now consider the current position and some of the evidence on which it is based.

ELECTRON TRANSPORT. The current picture of photosynthesis as an electron transport process is summarized in Figure 7-7. Photosystems I and II are each capable of absorbing quanta of light. Each contains specialized molecules of chlorophyll that are able to lose electrons and regain them from a different source. These **reactive centers,** which will be described in a subsequent section (The Light Trap), contain specialized molecules of chlorophyll a that absorb at a longer wavelength than usual. The reactive center of photosystem II absorbs light at 680 nm, and the pigment is called P_{680}. The corresponding reaction center for photosystem I is P_{700}, which has an absorption maximum at approximately 700 nm.

Electrons are removed from hydroxyl radicals and transferred via P_{680} in photosystem II to an as yet unknown electron acceptor "Q," which has a potential (E_0') of about 0 to − 0.1 v. "Q" may in fact represent several factors or a pool of interacting quinones or quinonelike components. Electrons are then passed to **plastoquinone,** which transfers

Figure 7-7. A model of photosynthesis. Dotted line shows the path of electrons in cyclic phosphorylation.

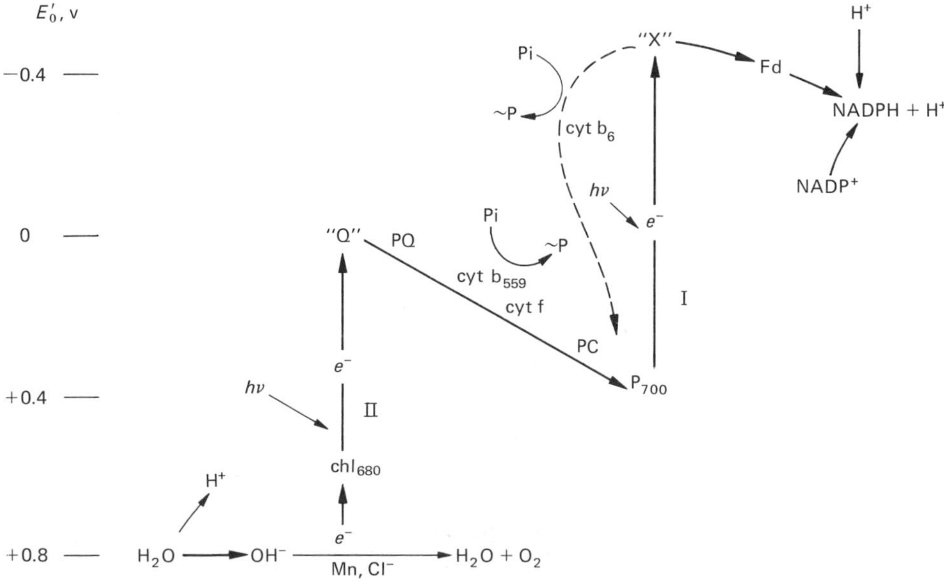

both hydrogen ions and electrons in the same way as ubiquinone in the respiratory electron transport chain. Plastoquinone passes electrons to cytochrome b_{559} and cytochrome f, which in turn pass them to **plastocyanine,** a copper enzyme that has an E_0' close to 0.4 v. A molecule of ATP is generated from ADP and Pi for each electron pair that passes down this electron transport chain in a manner that closely resembles oxidative phosphorylation. The process of **photosynthetic phosphorylation** will be described on page 162.

Photosystem I transfers its energy to P_{700}, which has an E_0' of about 0.4 v. As a result, P_{700} transfers an electron to an unidentified acceptor called "X," having a strongly negative potential ($E_0' = -0.5$ to $-$ 0.6 v). P_{700} regains its electron from plastocyanin, reoxidizing it in the process. The strong reducing factor "X" reduces **ferredoxin** (E_0' about $-$ 0.45 v), a nonheme iron enzyme widely distributed in plants and concerned with reducing reactions in nonphotosynthetic as well as photosynthetic plants (see Chapter 8, Nitrogen Fixation). Ferredoxin reduces NADP to NADPH through the flavoprotein enzyme NADP reductase.

It is possible also for electrons to pass from "X" or ferredoxin through an electron transport chain that would presumably include a hydrogen-transferring agent as well as the known cytochrome b_6 to plastocyanin, and so back to P_{700}. A phosphorylation is coupled to this electron transfer. Since the electrons in this system would pass through a cyclic path from P_{700} through ferredoxin back to P_{700}, the process, which thus is capable of generating ATP at the expense of light energy without reducing NADP, is called **cyclic phosphorylation.** The generation of ATP during the transfer of electrons from photosystem II to photosystem I is called **noncyclic phosphorylation.** A third possibility is that electrons might be transferred from ferredoxin back to oxygen, reducing it to H_2O. It is possible that this process might also involve an electron transport chain and make ATP. The experimental basis for **pseudocyclic phosphorylation,** as it is called, is not nearly so firm as that for cyclic or noncyclic phosphorylation.

Experimental Evidence. Much of this scheme has not yet been proved, and many points of dispute have arisen. Some possible alternatives will be considered later. However, it has gained wide acceptance and is supported by several types of evidence. Although the absorption spectra of system I and system II overlap, system I has a maximum much further in the red (approximately 690 nm) than system II, which has a maximum about 650–670 nm (or lower in blue-green algae). Consequently it is possible to energize system I or system II to some extent independently by using beams of monochromatic light of the appropriate wavelength. This is coupled with the analysis of the state of components of the electron transport chain by difference spectroscopy to determine whether they are oxidized or reduced. Thus, the illumination of chloroplasts with short wavelength light tends to reduce plastoquinone and the cytochromes, whereas long wavelength light tends to oxidize them.

There are a number of specific inhibitors that block the chain at specific points. The combined use of these inhibitors, selected wavelengths of activating light (**actinic** light), and differential spectroscopy has clarified much of the process. The inhibitor 3(3,4-dichlorophenol)-1,1-dimethylurea (DCMU), a selective herbicide, inhibits system II. In the presence of DCMU, when photosystem I is illuminated the cytochromes become oxidized, and they cannot be reduced by light that mainly activates system II, as would happen in the absence of DCMU. This shows that the cytochromes are normally reduced by system II and oxidized by system I and, thus, link them in the manner shown in Figure 7-7. Another approach is possible because photosystem II appears able to fluoresce, and the quenching of fluorescence by oxidized components (that is, electron acceptors) has been used to study the system.

A third, powerful approach has been the addition of artificial electron donors to reactivate poisoned systems, or the addition of components of known potential that can accept or donate electrons at specific points in the electron transport chain. For example, the proposed separation of the site of cyclic phosphorylation from that of noncyclic phosphorylation rests largely on observations that electron donors, such as phenazine methosulfate (PMS), 2,6-dichlorophenolindophenol (DPIP), or ferredoxin itself, can catalyze phosphorylation in the absence of oxygen, with system I light and in the presence of DCMU. Noncyclic phosphorylation, on the other hand, requires both light systems and the presence of oxygen, and is inhibited by DCMU.

A fourth and extremely important approach has been developed by the American physiologist R. P. Levine and his associates. They used mutants of the alga *Chlamydomonas* lacking one or another of the components of the electron transport chain. Thus, for example, a mutant lacking cytochrome f can reduce plastoquinone and cytochrome b with system II light, but it cannot reduce plastocyanin and P_{700}. System I light oxidizes plastocyanin and P_{700} but not plastoquinone and cytochrome b. A mutant lacking cytochrome f still possesses cyclic phosphorylation, but a mutant lacking plastocyanin does not; this indicates that cyclic phosphorylation transfers electrons from the top of system I back to plastocyanin (Figure 7-7).

A fifth technique for studying the electron transport system rests on the discovery that when chloroplast membranes are carefully disrupted and fragmented they can be separated by high-speed centrifugation into heavier and lighter particles. The heavier particles contain a much higher proportion of chlorophyll b, are able to liberate oxygen, and are poisoned by DCMU. The lighter particles contain a higher proportion of chlorophyll a, are able to reduce NADP and generate ATP, but do not liberate oxygen. It is thought that the heavier particles represent photosystem II and the lighter particles represent photosystem I. Data from this kind of experiment are not entirely clear, and the results seem to depend to a considerable extent on the exact technique used to disrupt chloroplasts. This may be accomplished by forcing a chloroplast suspension through a tiny orifice under the influence of high pressure (the French pressure cell) or by sonication; various detergents or surfactants that help to solubilize lipid materials may be used to assist in the disruption. Chloroplast fragments are not able to conduct the full reactions of photosynthesis. They often require added cofactors (because of the disruption or to replace cofactors that have washed out during their preparation), and they are usually unable to make ATP. However, much progress has been made in identifying these functional units with structures that can be seen in the electron microscope (see Structural Relationships, page 161).

Although a great deal of experimental evidence has been amassed on the function of the electron transport system in photosynthesis, much of it is conflicting or difficult to interpret, and it is possible that we have not yet arrived at the final definition of the system. The scheme presented in Figure 7-7 appears to be a useful working approximation.

The Light Trap. Understanding the nature of the physical reaction that traps light energy requires some consideration of the structure of the photosynthetic apparatus. Much effort has gone into the search for a **photosynthetic unit,** that is, a biochemical or biophysical unit capable of the complete reaction of photosynthesis. Several attempts have been made to identify structural components of the chloroplast with such a self-contained unit.

However, it is now realized that this was a barren concept because complete photosynthesis requires the coordination of a series of processes that are distributed throughout the organized membrane structure of the chloroplast. Small particles that can be seen in electron micrographs were called *quantasomes*, reflecting the hypothesis that they might be photosynthetic units. However, they are now known to be part of the system that synthesizes ATP, and the complete process of photosynthetic electron transport requires the coordination of areas of the underlying lamellae.

It was early recognized that if chloroplasts were broken into small fragments, the minimum pieces that could conduct the Hill reaction contained at least several hundred chlorophyll molecules, and it now appears that the complete light reaction of photosynthesis will not proceed in fragments having less than a thousand or more chlorophyll molecules. Further, studies with the inhibitor DCMU suggest that one DCMU molecule per 2000 chlorophyll molecules is required for complete inhibition. The inference is that a large number of chlorophyll molecules is associated with each **reaction center** (the site where light energy is used to transport electrons). Careful analyses have shown that P_{700}, cytochrome f, and cytochrome b are each present in the ratio of one molecule of each to several hundred chlorophyll a molecules, which supports this view. Since P_{700} is the electron-transferring pigment, it appears that the other chlorophyll molecules serve as light-gathering agents that transfer light energy to P_{700}. It would be highly uneconomical to have a complete electron transfer system associated with every chlorophyll molecule because the chlorophylls would be so screened by the enormous mass of all the associated electron transport systems that they would be unable to function as light absorbers.

The light trap of system I thus appears to consist of a large array of chlorophyll molecules that can each absorb light and pass the resultant excitation energy along from molecule to molecule to the reaction center (see Figure 7-8). Accessory pigments must also be present and involved. The energy is passed by resonance transfer between adjacent molecules, not by actual electron transfer. Somewhere in each array is a group

Figure 7-8. Possible structure and function of the light trap of photosystem I. ⟶ indicates electron transport; - - -→ indicates excitation energy transfer.

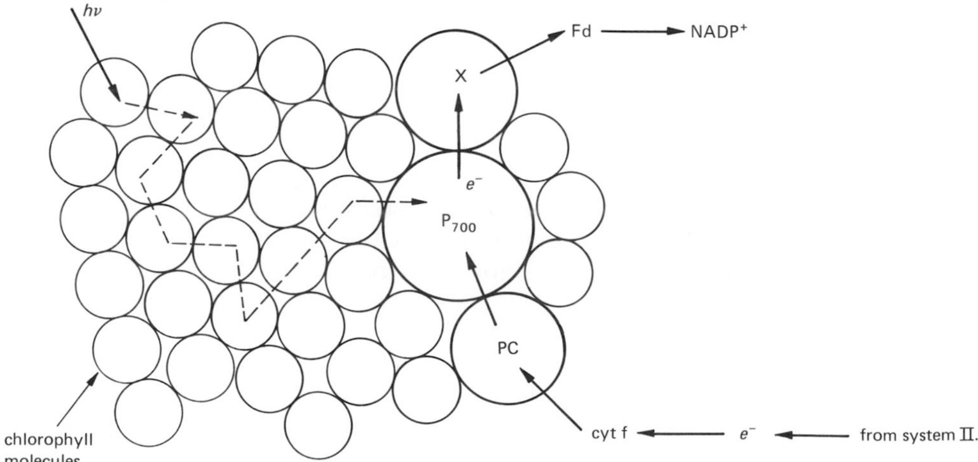

of chlorophyll molecules that, because of their physical orientation, absorb longer wavelength light and permit a small amount of the energy of excitation to be lost as heat. This serves to trap the energy of excitation, which can no longer escape to the higher energy levels of surrounding molecules.

In the center of this complex is a molecule of P_{700} (probably a chlorophyll molecule or dimer in a special association with its protein component) that is associated with cytochrome f and plastocyanin as well as with "X," the electron acceptor. The excitation passes to P_{700}, which ejects an electron to "X," reducing it, and regains an electron from plastocyanin, oxidizing it.

The light trap of photosystem II is essentially similar, except that it may contain a considerably larger number of chlorophyll b molecules and a higher proportion of accessory pigments. The reaction center of photosystem II, P_{680}, has a slightly shorter absorption maximum than that of photosystem I. Photosystem II absorbs electrons from water and passes them to "Q" and plastoquinone.

The kinetics of photosynthesis have also been studied through the measurement of **electron spin resonance (ESR),** a technique that detects changes in the paramagnetism of photochemical intermediates when unpaired electrons are formed during photochemical events. This technique has been used in efforts to identify primary electron donors and acceptors in the two photosystems. There is some indication that a tightly bound complex of ubiquinone and an iron-containing component serve as the primary acceptor of electrons in bacterial photosynthesis. The identity of the primary acceptors from photosystems I and II in plants is not yet clear. Recent experiments suggest that the electron donor in photosystem II, the mechanism that transfers electrons from water to P_{680}, may involve a complex of quinones or quinone derivatives together with manganese. However, much additional work is needed.

Release of Oxygen. The system in photosynthesis about which the least is presently known is the mechanism that produces oxygen. Four electron transport reactions must take place, involving four molecules of water, in order to generate one molecule of oxygen. Just how this is accomplished, in view of the fact that electrons are transported singly, is not clear. It has been suggested that a **water-splitting enzyme** may contain an association of four electron-transporting molecules (perhaps chlorophylls) so oriented that the removal of four electrons from water could result in the overall reaction

$$4\,H_2O \rightarrow \begin{cases} 4\,H^+ \\ 4\,OH^- \longrightarrow 4\,e^- + 2\,H_2O + O_2 \end{cases}$$

Manganese and chloride ions appear to be required for the evolution of oxygen, and the reversible oxidation-reduction of manganese may be involved in the liberation of the oxygen. CO_2 or bicarbonate is also required for oxygen production (apart from its role as substrate of photosynthetic carboxylation). The intermediates and cofactors in this reaction are still unknown.

Structural Relationships. The overall structure of the chloroplast lamellar system (see page 57) is now well defined. Recently **freeze-etch** electron microscopy has provided new details of the thylakoid membranes. In this process, preparations are frozen and then splintered with a microtome knife. The tissue tends to split along natural lines of cleavage, usually along membrane surfaces. Shadowing the preparation with metal then provides a relief picture of the inner or outer surface of the membrane. An electron micrograph prepared in this way is shown in Figure 7-9A, together with an interpretive diagram (Figure 7-9B). A number of particles of different sizes are visible on the various surfaces of the membrane. Some of these appear to be related to aggregates of enzymes or electron carriers associated with photosystems I or II, or with the ATPase-coupling factor (see the following paragraph). Figure 7-9C shows a hypothetical model, based on electron microscopic evidence and the results of subfractionation experiments, that shows the location of photosystem I particles or assemblies (capable of conducting cyclic phosphorylation only) and of photosystem I plus photosystem II assemblies, capable of noncyclic electron transport.

ATP Synthesis. The chemiosmotic hypothesis of Mitchell accords well with available evidence on photosynthetic ATP synthesis. Essentially, the formation of ATP by either cyclic or noncyclic electron transport is similar to mitochondrial ATP synthesis resulting from oxidative electron transport (see page 99). The transfer of electrons and hydrogen

Figure 7-9. Chloroplast structure.
A. Electron micrograph of freeze-fracture faces and surfaces of grana lamellae, ×100,000.
B. Diagrammatic interpretation of A. The numbers refer to (*1*) exterior thylakoid surface, (*2*) fracture plane immediately beneath exterior surface, (*3*) fracture plane beneath interior surface of thylakoid, (*4*) interior thylakoid surface. The enlargement shows a possible arrangement of chlorophyll molecules in the particles (visible in A) embedded in the thylakoid membrane.
C. Diagrammatic representation of granal structure and the distribution of photosystems I and II. [From Park, R. B., and P. V. Sane: *Ann. Rev. Plant Physiol.*, **22:**395–430 (1971), and Anderson, J. M.: *Nature,* **253:**536–37. Used with permission. Photograph kindly supplied by Dr. Park.]

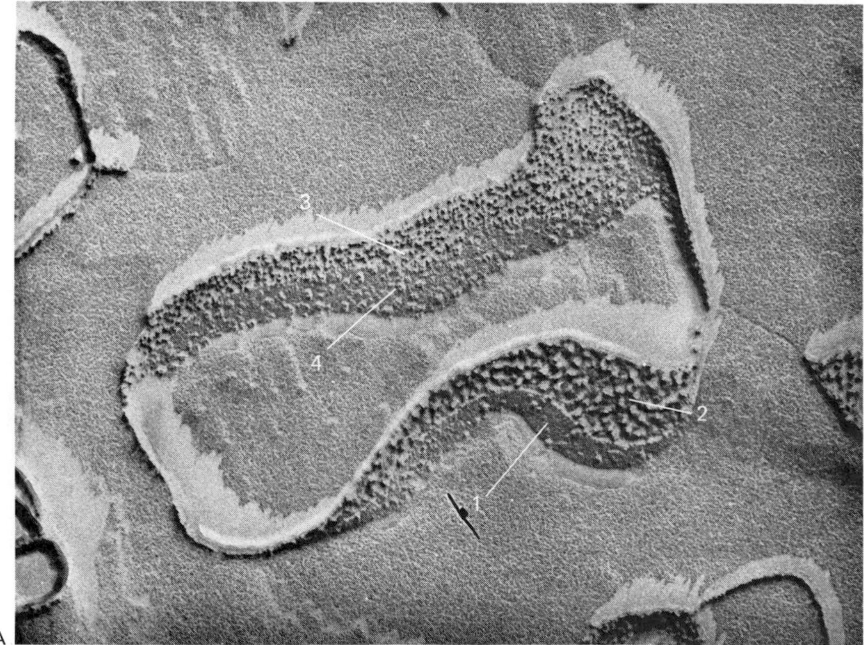

A

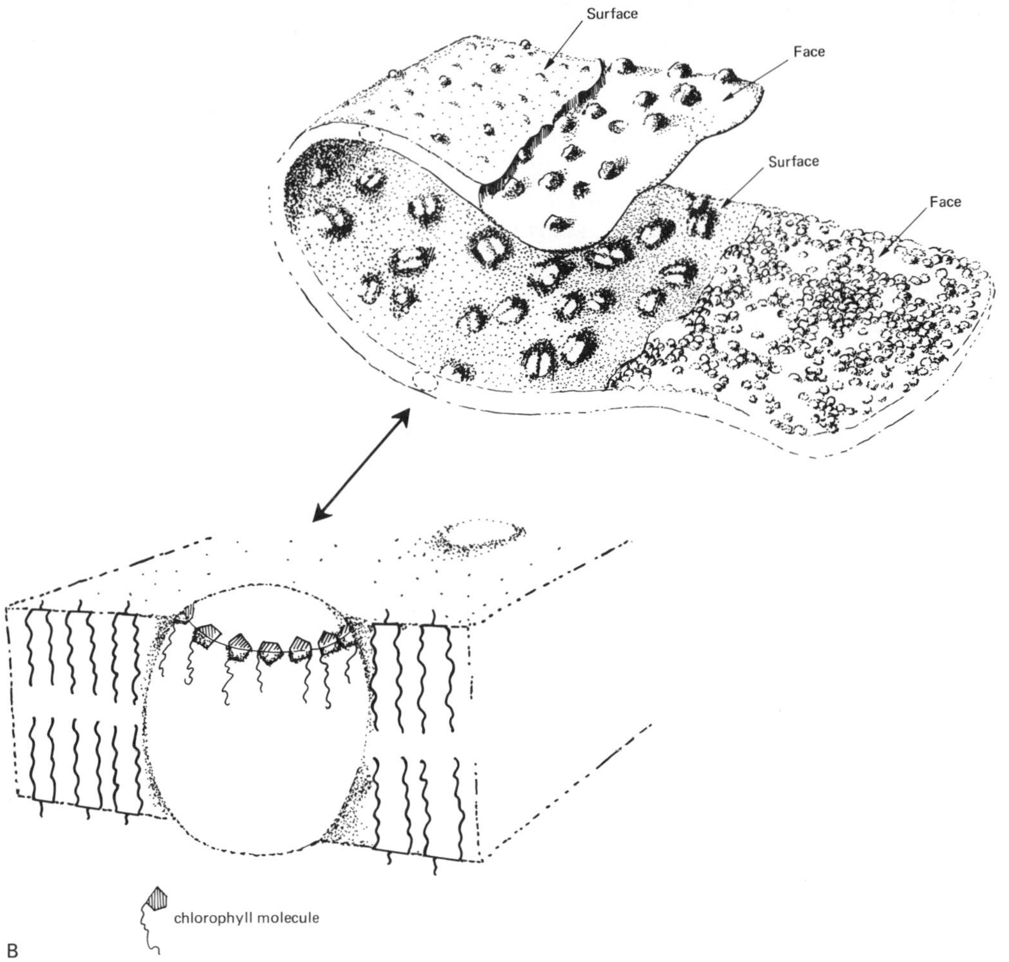

Surface
Face
Surface
Face
chlorophyll molecule

B

C

Photosystems I plus II — granal thylakoid
Photosystem I — intergranal thylakoid

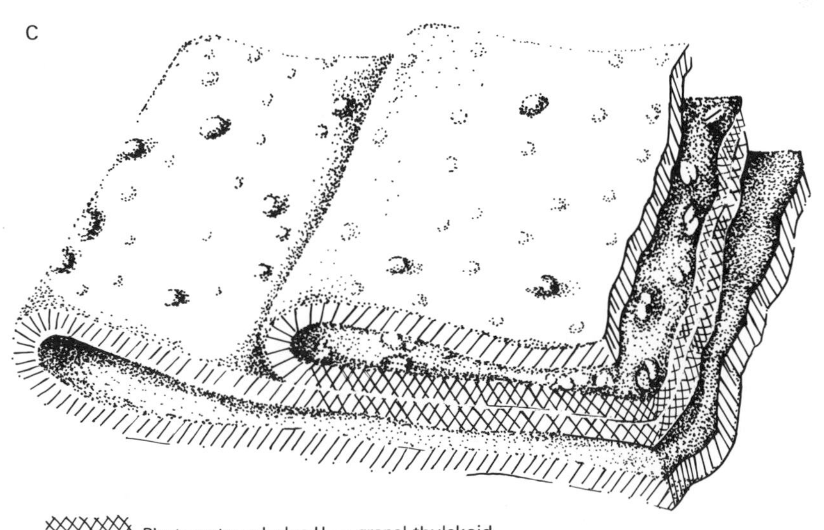

ions (protons) across the thylakoid membrane provides a separation of charges; these are allowed to come together again through the ATPase enzyme and a **coupling factor** (CF) that is identifiable as a particle protruding from the outer surface of the membrane (see Figure 7-9). Figure 7-10 shows a model system drawn from the work of Dr. A. T. Jagendorf (Cornell University) and Dr. N. E. Good (Michigan State University). The critical requirement of this scheme, apart from the spatial organization of cytochromes, is the mobility of plastoquinone, which actually transfers electrons and protons across the membrane from outside to inside. The photosystems I and II activate the charge separation by transporting electrons to the outside. The coupling factor operates in the same way as the F_0 and F_1-ATPase of mitochondria (see Figure 5-8, page 101).

Evidence supporting this view includes Jagendorf's original discovery that raising the external pH of isolated chloroplasts will cause phosphorylation to take place in the dark, and the observation that pH gradients do occur across the thylakoid membranes in light as required. It has been found that the photooxidation of hydrogen donors (water, catechol) leads to phosphorylation, whereas the oxidation of electron donors such as ferricyanide does not. Finally, uncoupling agents that permit the leakage of protons across the membrane by a path other than the coupling factor, although they do not affect electron transport, prevent photophosphorylation from accompanying electron transport.

This model suggests that two sites of phosphorylation should occur—one for each photosystem—since each photosystem causes the release of a pair of protons for each pair of electrons transferred. Recent experiments with the inhibitor dibromothymoquinone have permitted the separation of intact thylakoids with the activities of photosystems I and II, and it has been possible to demonstrate the fact that each photosystem is capable of phosphorylation. In very carefully prepared chloroplasts, ATP production approaches the theoretical values that are predicted by this model. This raises a question about the need for cyclic phosphorylation. However, it is unlikely that phosphorylation runs at 100 percent efficiency all the time, and it seems likely that photosynthetic ATP is also available to the cytoplasm (see Chapter 15). If this is so, then cyclic phosphorylation

Figure 7-10. Model of chemiosmotic interpretation of photosynthetic phosphorylation in chloroplasts, based on ideas of Dr. N. E. Good and Dr. A. T. Jagendorf. This diagram should be compared with Figure 5-8 (page 101) and Figure 7-7 (page 157).

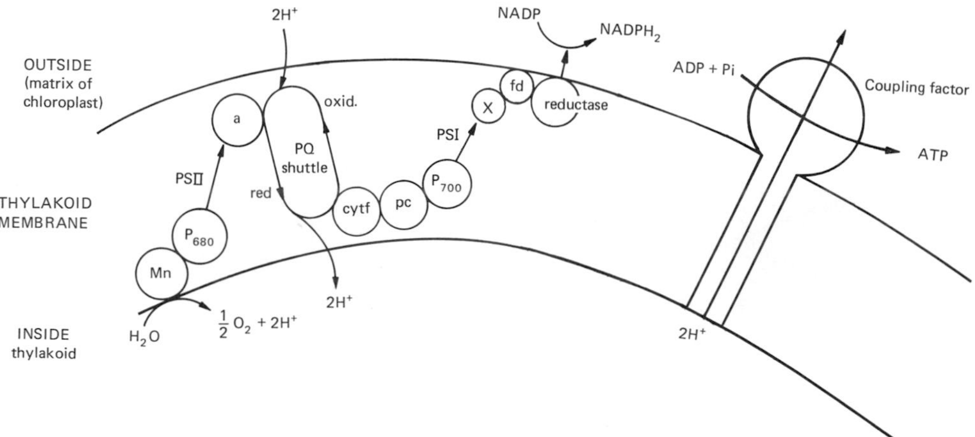

is probably of importance, since the carbon reactions of photosynthesis require at least 1.5 ATPs for each reducing equivalent (NADPH) used. This requirement is sufficient only for the production of monosaccharides, and extra ATP is required for the formation of sucrose, starch, and other compounds made in the chloroplast.

An interesting experimental confirmation of the location and participation of coupling factor follows the use of immunological techniques. It has been possible to isolate the coupling factor and, by injecting it into a rabbit, to cause the formation of an antibody to coupling factor. When the antibody is added to a chloroplast suspension, electron transport continues unhindered but the synthesis of ATP is prevented. This shows two things: that the coupling factor, which would be inactivated by the antibody, is indeed involved in phosphorylation, and that it must be located on the outside of the thylakoid membranes in order that the antigen could have access to it.

Energy Balance. We have seen that the generally accepted scheme requires four quanta of red light (40 kcal/einstein) to produce one molecule of NADPH and one or two ATPs. The energy input is thus $4 \times 40 = 160$ kcal/mole, and the recovery is 51 kcal available from the oxidation of 1 mole of NADPH plus 7–15 kcal for the ATPs, which amounts to 35 to 40 percent of the energy input. The energy losses occur in the conversion of light energy to chemical potential and in the stabilization of the high energy (excited) pigments and intermediates—that is, in "trapping" energy or the migrated electron so that it cannot fall back into thc level or position from which it came. Considering the complexity of the conversion, this is a remarkably high level of efficiency.

Carbon Reactions: The Calvin Cycle

Introduction. The light reactions result in the production of reducing power, which reduces NADP, and in the production of ATP. In the dark reactions, these agents are used to reduce CO_2. Until the discovcry of radioactive carbon, the intermediates and pathways of the synthesis of sugars could only be guessed at. When Calvin and his coworkers supplied $^{14}CO_2$ to suspensions of the alga *Chlorella,* they found that the major initial product of photosynthesis was the three-carbon acid, 3-phosphoglyceric acid (PGA). This product was isolated and chemically degraded, and it was found that most of the ^{14}C was in the carboxyl position of the molecule.

$$\text{P—O—CH}_2\text{—CHOH—}^{14}\text{COOH}$$

Further examination of the products of short-term photosynthesis in $^{14}CO_2$ revealed that fructose-1,6-diphosphate (FDP) early became radioactive, and upon degradation it was found to contain most of its radioactivity in the middle two carbons.

$$\text{P—O—CH}_2\text{—CHOH—}^{14}\text{CHOH—}^{14}\text{CHOH—CO—CH}_2\text{—O—P}$$

These observations suggested (1) that the synthesis of hexose takes place via a reversal of the reaction system of glycolysis, and (2) that a cyclic scheme is required for the regeneration of a CO_2 acceptor. With the development of paper chromatography and the combination of this powerful analytical tool with the use of radioactive carbon, the whole complex of carbon reduction and CO_2-acceptor regeneration were quickly worked out.

Radioactivity and Chromatography. These tools have had such a dramatic impact on plant physiology that it is well to examine them briefly. Radioactive elements are chemically identical with their stable isotopes and differ only in the mass of their nuclei. They decay at a rate that is proportional to their instability by disintegrating with the ejection of radiations—usually γ rays or β particles. Since these are ionizing radiations, they can be detected by various instruments that detect ionization, particularly the Geiger-Mueller counter or the scintillation counter. The Geiger-Mueller counter detects the production of ionization in a special ion chamber, the Geiger-Mueller tube, which has a thin window through which the ionizing radiations can enter. The scintillation counter detects and counts flashes of light emitted by a crystal or a special liquid on absorption of a radiation. The decay rate of isotopes is exponential, being proportional to the amount present. The stability or decay time of an isotope is measured by its **half-life,** the time required for half of any starting amount to decay. For short-lived isotopes (for example, ^{32}P half-life = 14 days) corrections must be made from day to day to allow for this decay. ^{14}C has a half-life of about 6000 years, so no correction need be applied.

Because radioactive atoms are chemically like their stable counterparts, they undergo the same chemical reactions. Except for some slight effects, which may result from the fact that ^{14}C is slightly heavier than ^{12}C, radioactive carbon atoms behave in precisely the same manner as nonradioactive ones; thus they enter all the reaction sequences and label all the compounds that are involved in the normal metabolism of carbon. If $^{14}CO_2$ is supplied to a plant in light for a period of 5 sec, those compounds synthesized during that 5 sec, and only those compounds, will become radioactive. Moreover, if a new (radioactive) carbon atom is added to an existing (nonradioactive) compound, only that atom will be radioactive. The presence and position in the molecule of the radioactive atom can be determined by chemically degrading the compound, carbon by carbon, and assaying each carbon independently for its radioactivity. Using this technique it is possible to detect the pathways of carbon in metabolism. These in turn must be correlated with the types of reactions involved and the nature of the enzymes that are present to achieve a complete understanding of the metabolic system.

Paper chromatography is a technique for separating the components of a complex mixture of organic chemicals, such as the soluble constituents of a plant. Plant tissue is extracted, usually with a solvent that kills the cells and denatures the enzymes (hot ethanol or methanol is commonly used) and the extract is concentrated. Then a few drops of the extract are applied to the corner of a sheet of filter paper, usually about 20–50 cm in each dimension, and one edge is dipped in a suitable organic solvent. The solvent moves through the paper by capillary action. A certain amount of water is present in the paper, tightly bound and immobile. Different substances move with the moving organic solvent at different rates, depending on their relative solubility in the stationary water phase, in the moving organic phase of the solvent system, and to some extent on their adsorption onto the paper. The result is that the components of the mixture are separated into a row of discrete spots.

In complex mixtures separation of all components may not be complete, so a second solvent system may be used at right angles to the first, to resolve previously unseparated spots. The components can be visualized by spraying the chromatogram with a reactive agent that produces colored derivatives (for example, ninhydrin reacts with amino acids to produce a blue or purple color), and radioactive spots can be located by **radioautography** (also called **autoradiography**). The chromatogram is placed next to a sheet of x-ray film in a holder for a period of hours or days, depending on the intensity of radioactivity. Those spots that are radioactive cause a darkening of the film over their location, so that the film records a picture of the position and intensity of radioactivity of the spots on the chromatogram. Illustrations showing typical apparatus used in paper chromatography and a chromatogram and radioautograph of a plant extract are shown in Figures 7-11 and 7-12.

Using these techniques it is possible to identify which compounds are present in a plant and, in an experiment with a labeled substrate, to determine which ones were made from that substrate and how much. Individual spots can be cut out of the paper and the compound in the spot can be degraded by microchemical techniques, yielding further information about metabolic pathways.

Figure 7-11. Paper chromatography.
A. Photograph of a chromatography cabinet in use in the author's laboratory and a rack of paper chromatograms about to be run.
B. How paper chromatograms are run in two solvents and developed. The solvents may also be run downward instead of up, by hanging the papers from troughs in the top of the cabinet.

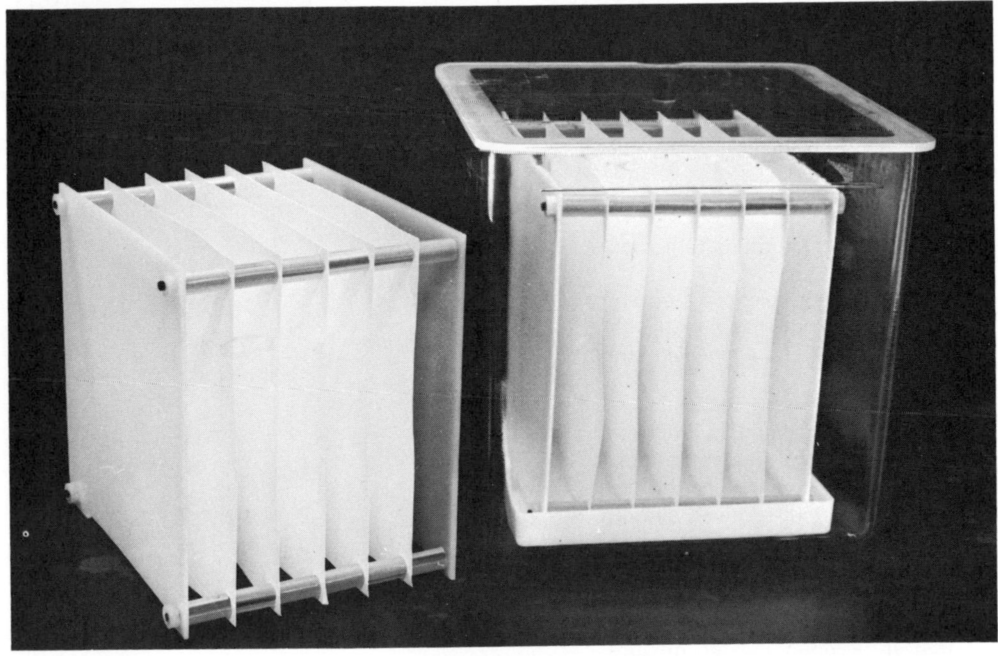

A

B

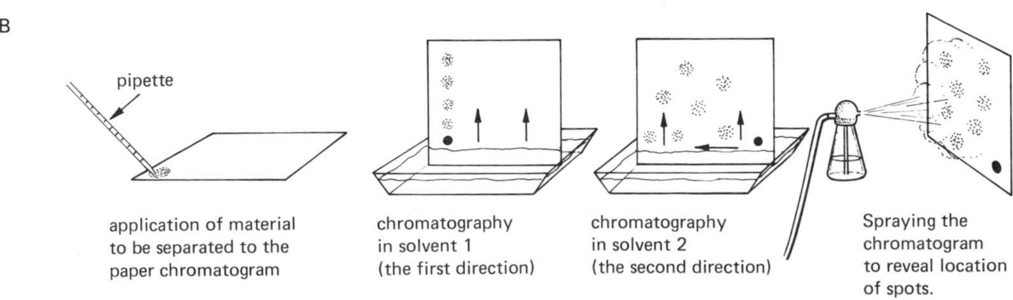

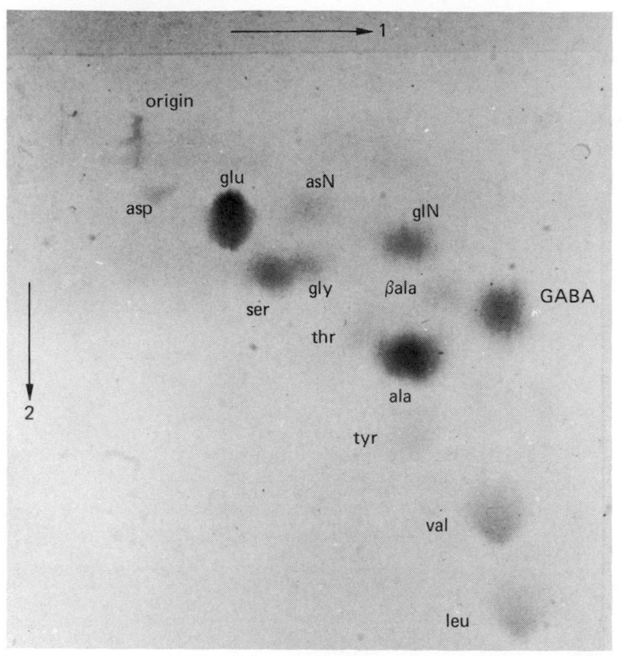

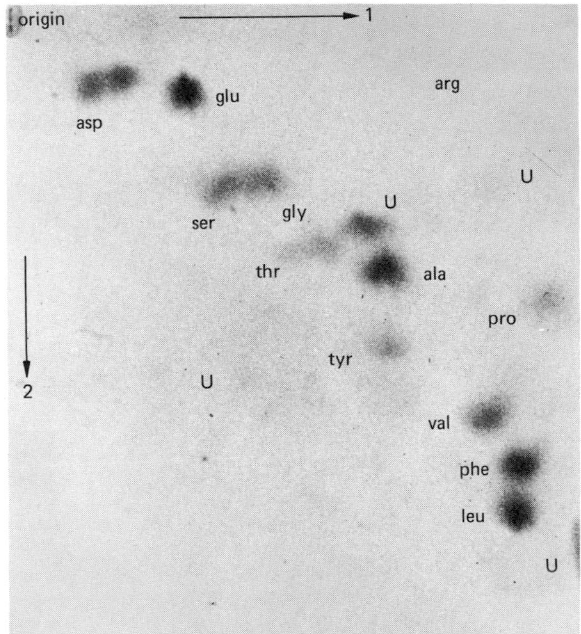

Figure 7-12. Chromatogram and radioautograph.

A. Photograph of a chromatogram of wheat-leaf extract sprayed with ninhydrin reagent to reveal amino acids.

B. Radioautograph of a chromatogram of the protein hydrolysate of wheat leaves after the supply of $^{14}CO_2$ for 1 hr in light. All the compounds that received carbon from products of photosynthesis are radioactive and can be seen as dark spots on the x-ray film. The chromatograms were run first in a phenol-water mixture (1), then in a propanol-ethyl acetate-water mixture (2). The extracts were applied at the origin. Code: ala, alanine; βala, β-alanine; arg, arginine; asp, aspartic acid; asN, asparagine; GABA, γ-aminobutyric acid; glu, glutamic acid; glN, glutamine; gly, glycine; leu, leucine; phe, phenylalanine; pro, proline; ser, serine; thr, threonine; tyr, tyrosine; U, unknown; val, valine.

The Calvin Cycle. The reactions of the carbon reduction cycle of photosynthesis and the enzymes that mediate them are shown in Figure 7-13. The reactions as outlined result in the synthesis of one molecule of triose from the fixation of three molecules of CO_2. The reactions may be summarized

$$3\,CO_2 + 9\,ATP + 6\,NADPH \longrightarrow GAP + 9\,ADP + 8\,Pi + 6\,NADP$$

To make hexose phosphate, two complete turns of the cycle are required. One of the GAP molecules so formed is converted to DHAP, and the two triose phosphates unite by the aldolase reaction. The resulting fructose diphosphate is converted to fructose-6-phosphate by phosphatase. The summary for the overall production of a molecule of hexose phosphate would then be

$$6\,CO_2 + 18\,ATP + 12\,NADPH \longrightarrow F\text{-}6\text{-}P + 18\,ADP + 17\,Pi + 12\,NADP$$

It is worthwhile to follow the reaction sequence through the cycle in Figure 7-13 to get a clear understanding of the pathways taken by each carbon atom. For convenience, the newly added molecules of CO_2 are marked with an asterisk on Figure 7-13, so that the level of labeling and the sit of radioactivity of each compound can be seen for one turn of the cycle following the addition of $^{14}CO_2$. The peculiar labeling pattern of RuBP* results from the mixing of three molecules of Ru-5-P, two of which are labeled differently from the third.

The student should determine the pattern of labeling in key compounds after the introduction of a second round of $^{14}CO_2$ and compare the results he gets with the results of an actual experiment, recorded in Table 7-2. The labeling patterns are not always found to be exactly as predicted. Certain asymmetries that have been found can be explained by rearrangements caused by reversal of the transketolase and aldolase reactions, which redistribute ^{14}C into new patterns. It should be noted that, although the

*RuBP (for ribulose bisphosphate) has recently superceded in biochemical nomenclature the older term RuDP (for ribulose diphosphate). Properly speaking this is a correct (but I think silly) distinction. A diphosphate (like ADP) has the second phosphate esterified on the first. A bisphosphate has two phosphates on separate carbons. RuBP is also sometimes called RuP_2. Other bisphosphates (FDP, SDP) should properly be so named. Since the older, more familiar nomenclature for these compounds is still widely used in the literature, it has been retained in this chapter.

Table 7-2. Pattern of radioactivity in Calvin cycle intermediates from an experiment in which $^{14}CO_2$ was supplied to a culture of *Scenedesmus obliquus* for 5.4 sec.

	Radioactivity in individual atom, percent			
Carbon atom	*PGA*	*Fructose*	*Sedoheptulose*	*Ribulose*
1	82	3	2	11
2	9	3	2	10
3	9	43	28	69
4	—	42	24	5
5	—	3	27	3
6	—	3	2	—
7	—	—	2	—

SOURCE: J. A. Bassham and M. Calvin: *The Path of Carbon in Photosynthesis.* Prentice-Hall, Inc., N.J., 1957. Used with permission.

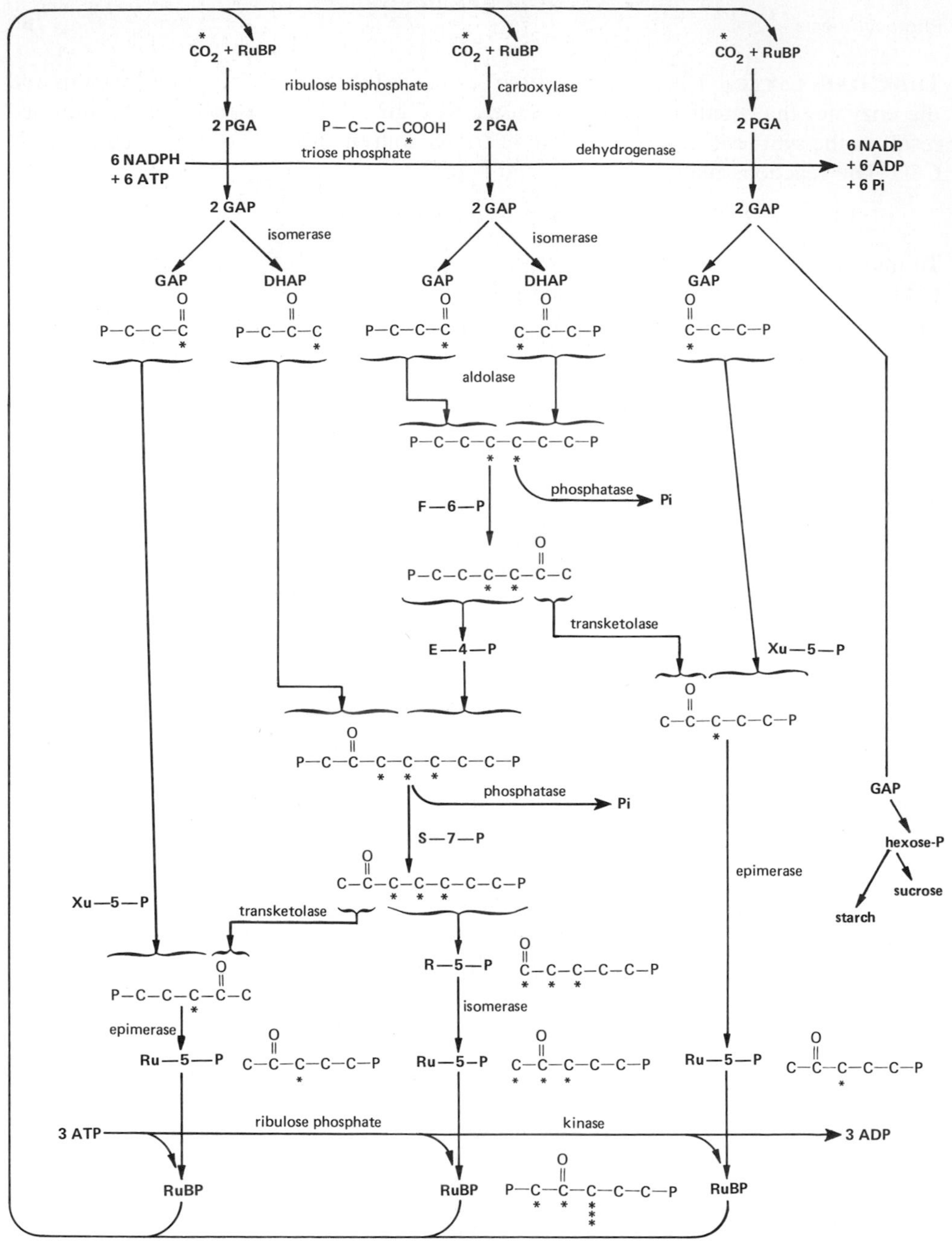

Figure 7-13. The Calvin cycle. The —OH and —H groups have been omitted for clarity; only the =O or —P groups are shown. If radioactive carbon dioxide ($^*CO_2$) goes through one turn of the cycle, the labeling patterns shown will result.

ABBREVIATIONS:

RuBP = ribulose biphosphate†
PGA = phosphoglyceric acid
GAP = glyceraldehyde phosphate
DHAP = dihydroxyacetone phosphate
FDP = fructose diphosphate†
F-6-P = fructose-6-phosphate
E-4-P = erythrose-4-phosphate
SDP = sedoheptulose diphosphate†
S-7-P = sedoheptulose-7-phosphate
R-5-P = ribose-5-phosphate
Xu-5-P = xylulose-5-phosphate
Ru-5-P = ribulose-5-phosphate

† See footnote for p. 169.

cycle as described in Figure 7-13 produces one molecule of triose for each three CO_2 molecules added, it could just as easily be arranged to produce PGA, a pentose, a hexose, or C_2 compounds, all of which are known as end products of photosynthesis.

The individual reactions and enzymes are as follows: In the first reaction a molecule of the five-carbon sugar ribulose bisphosphate (RuBP) is carboxylated. An unstable six-carbon intermediate is formed that hydrolyzes spontaneously to yield two molecules of phosphoglyceric acid (PGA) as shown

$$\begin{array}{ccccc} & & CH_2OP & & CH_2OP \\ & & | & & | \\ & & C{=}O & & CHOH \\ & & | & & | \\ {}^*CO_2 & + & CHOH & \xrightarrow{H_2O} & {}^*COOH \\ & & | & & + \\ & & CHOH & & COOH \\ & & | & & | \\ & & CH_2OP & & CHOH \\ & & & & | \\ & & & & CH_2OP \\ & & \text{RuBP} & & \text{2 PGA} \end{array}$$

This reaction was one of the earliest elucidated by Calvin's kinetic studies with $^{14}CO_2$ and paper chromatography. Cells were allowed to reach steady-state photosynthesis with $^{14}CO_2$, so that all pools of cycling intermediates were saturated with ^{14}C. Then the sequence was interrupted by darkening the cells, so that the products of the light reaction were cut off and the reducing reactions stopped. Carboxylation could continue, however, resulting in the simultaneous disappearance of the CO_2 acceptor and the appearance of the product of carboxylation. The graph in Figure 7-14 shows the results of such an experiment, which lead to the conclusion that RuBP is the CO_2 acceptor and PGA is the product of carboxylation.

Figure 7-14. Light-dark changes in concentrations of PGA and RuBP. [From J. A. Bassham and M. Calvin: *The Path of Carbon in Photosynthesis,* ©1957. By permission of Prentice-Hall, Inc., Englewood Cliffs, N.J. Original figure courtesy Dr. J. A. Bassham.]

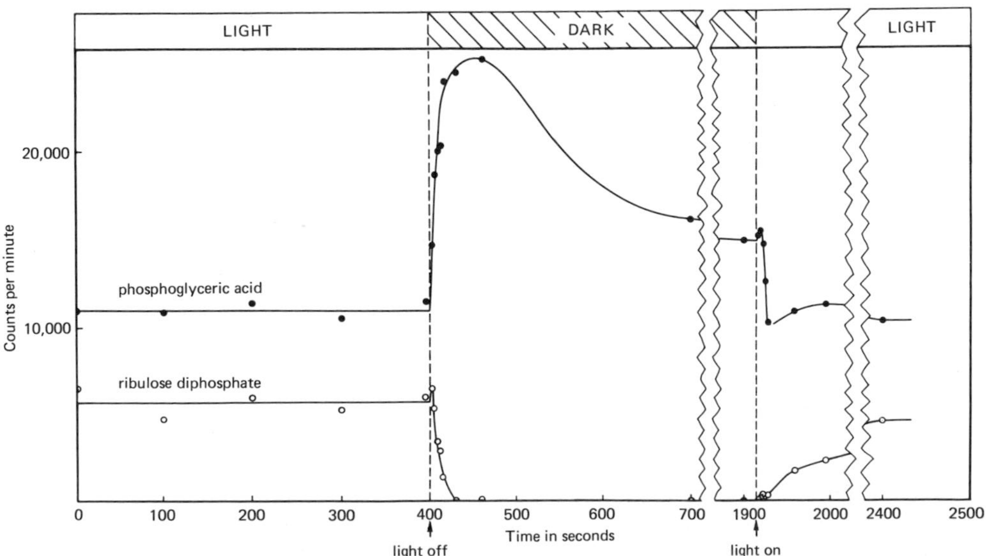

The enzyme responsible for this carboxylation, **RuBP carboxylase (RuBPcase,** also known as carboxydismutase) has been the object of much study, and many of its characteristics in vivo are now well known. The enzyme is readily isolated from leaves and appears to be the major protein fraction in photosynthetic tissues. It is possible to isolate an apparently pure protein (called **fraction I** protein) that contains RuBPcase activity, but it also contains enzymic activity responsible for several other reactions of the cycle that can only be removed with great difficulty. It has been suggested that RuBPcase and certain other enzymes of the cycle are closely associated with each other, which may account for the great efficiency of operation of this enzymic mechanism. Measurements made in vitro with bicarbonate gave very low rates of reaction and suggested that the enzyme required an extremely high concentration of substrate. However, recent experiments by the research group of the British physiologist D. A. Walker have shown that the affinity of the enzyme for its natural substrate, CO_2, is sufficiently high to account for the rate of photosynthesis in vivo. There is some evidence that this enzyme is light-activated in vivo. In addition, its synthesis requires light, and it does not appear in dark-grown leaves until they have been illuminated for three to five hr.

The PGA produced in the RuBPcase reaction is phosphorylated ("primed" for the reduction reaction) by phosphoglycerylkinase, ATP being the donor. The resulting 1,3-diphosphoglyceric acid is reduced by a NADPH-specific triose phosphate dehydrogenase to 3-phosphoglyceraldehyde (GAP). Some of the GAP is then converted to dihydroxyacetone phosphate (DHAP) by triosephosphate isomerase, and fructose diphosphate (FDP) is synthesized from the two trioses by aldolase (see page 110). The reactions from PGA to FDP are similar to the reverse of the glycolytic sequence from FDP to PGA, except that the triosephosphate dehydrogenase in chloroplasts is NADPH linked whereas that of the cytoplasm is NAD linked. Other differences may also exist. Difficulties in reconciling the low in vitro activities of some of the enzymes with the high activities required for in vivo photosynthesis have led to the suggestion that the enzymes may be activated in the chloroplast, which would further differentiate them from their glycolytic counterparts.

The conversion of FDP to fructose-6-phosphate (F-6-P) is accomplished by a phosphatase yielding inorganic phosphate. The phosphatase appears to be specific to sugar diphosphates. This is one of the three "energy wasting" steps in the cycle that ensure that the reactions will proceed in a forward direction and not back up with the massive production of intermediates or stop altogether. F-6-P next undergoes a transketolase reaction (see page 116) that removes the two top carbons as the thiamine pyrophosphate (TPP) derivative of glycolaldehyde, leaving the tetrose erythrose-4-phosphate (E-4-P). The E-4-P condenses by aldolase reaction with DHAP to form sedoheptulose diphosphate (SDP), and this is converted by a second energy-liberating step to sedoheptulose-7-phosphate (S-7-P) and Pi by a phosphatase reaction. The S-7-P undergoes a transketolase reaction in which the two top carbons are removed as TPP-glycolaldehyde, leaving the pentose ribose-5-phosphate (R-5-P). This is converted to ribulose-5-phosphate (Ru-5-P) by phosphopentose isomerase. The TPP-glycolaldehyde derived from F-6-P and F-7-P in the transketolase reaction is transferred to GAP forming xylulose-5-phosphate (Xu-5-P), which is converted to R-5-P by a phosphopentose epimerase. The R-5-P is converted to Ru-5-P by an isomerase and is phosphorylated by phosphoribulokinase, ATP being the donor, to produce ribulose bisphosphate (RuBP) and ADP (a second "priming" reaction that prepares the pentose for carboxylation). The use of ATP to make a low energy ether-phosphate link represents the third point in the cycle where energy is "wasted," providing an irreversible step to maintain speed and direction in the cycle.

Control Points. A number of control mechanisms have been suggested that may regulate the operation of the cycle. First, the cycle is a very efficient **autocatalytic** device. Since the end product (either a triose or a hexose) is also an intermediate, it is possible to rearrange the cycle so that it produces a larger number of starter molecules (RuBP) for each turn of the cycle (at the expense of the normal product). Thus, the cycle can be used to build up its own intermediates and to increase its own rate of operation. This might be necessary because certain of the intermediates (for example, PGA, triose phosphate, and glycolate) might leave the chloroplasts during periods of darkness, so that the concentration of intermediates would become very low. Under these circumstances the cycle could only operate slowly until the level of intermediates had built up again. The depletion of the cycle may also be prevented by other regulatory reactions, principally the light activation of certain of the enzymes of the cycle. When the leaf is in darkness, several of the reactants become inactive (particularly the carboxylase, the two phosphatases, and Ru-5-P kinase.) Finally, the balance of synthesis of the various possible products of the cycle (hexoses, pentoses, trioses, PGA, or C_2 compounds) is maintained by regulation, often through allosteric effects, of various reactions in the cycle by other members of the cycle or cofactors such as ATP or ADP. In these ways, balance and high-speed operation are maintained, and the cycle is able to react rapidly and easily to the demand from other parts of the cell for a variety of products that may be needed for subsequent metabolism.

Energy Balance. The reactions of the cycle can be summarized

$$6\ CO_2 + 18\ ATP + 12\ NADPH \longrightarrow C_6H_{12}O_6 + 18\ (ADP + Pi) + 12\ NADP + 6\ H_2O$$

The 18 ATP represents a total of about 140 kcal and the 12 NADPH a total of about 615 kcal. Thus the energy input is about 755 kcal. The energy recovered in hexose is about 670 kcal/mole, which represents an efficiency of nearly 90 percent. The wasted 10 percent is the energy input used to keep the cycle running. This sequence of reactions, by which plants stabilize and store the chemical potential recovered from light energy, is thus remarkably efficient. The high efficiency is largely the result of the effective positive feedback system of the Calvin cycle, in which the reaction is continuously fed by its own products and by which the concentration of intermediates can be rapidly built up by internal reactions and maintained at the proper level for maximum operation of the cycle.

Other Photosynthetic Pathways

RuBP Oxygenase. It has recently been suggested that RuBPcase is able to react with oxygen as well as with carbon dioxide. This **RuBP oxygenase** activity converts RuBP into one molecule of phosphoglycolic acid and one molecule of PGA, instead of into the two molecules of PGA that are formed when CO_2 is fixed

$$O_2 + \begin{array}{l} CH_2OP \\ | \\ C{=}O \\ | \\ CHOH \\ | \\ CHOH \\ | \\ CH_2OP \end{array} \longrightarrow \begin{array}{l} CH_2OP \\ | \\ COOH \\ \text{phosphoglycolic acid} \\ + \\ COOH \\ | \\ CHOH \\ | \\ CH_2OP \end{array}$$

RuBP PGA

Experiments show that O_2 is a competitive inhibitor of carboxylase activity, and CO_2 competitively inhibits oxygenase activity. Furthermore, the in vivo reaction using $^{18}O_2$ results in glycolate labeled in the carboxyl end. An active phosphatase converts P-glycolate to glycolate and Pi at the chloroplast surface.

Alternate pathways for the formation of glycolate have been suggested, but they are not so well documented. It is possible that thymine pyrophosphate–glycolaldehyde (the C_2 fragment of the transketolase reaction) can be oxidized to make glycolate, or that the carboxylation product of RuBPcase might be oxidized to yield CO_2 and glycolate. However, the weight of the evidence at present suggests that the oxygenation of RuBP is the main source of glycolate in photosynthesizing tissue.

Glycolate Pathway. Clearly an oxidizing step in a reducing reaction like photosynthesis is wasteful. The oxygenase reaction may, in fact, be an unavoidable consequence of the fact that RuBPcase does not distinguish very strongly between CO_2 and O_2. But all the carbon diverted by this reaction is not lost. The work of many investigators, particularly N. E. Tolbert and his colleagues at Michigan State University, has led to the formulation of the **glycolate pathway** of metabolism shown in Figure 7-15. This is a scavenging mechanism that permits the rescue of carbon that would otherwise be lost as glycolate, and the reintroduction of three quarters of it back into the chloroplast as glyceric acid.

Four main points should be noted. The first is that the glycolate pathway involves three distinct sites of metabolism: chloroplasts, peroxisomes, and mitochondria. These organelles are presumably closely associated, otherwise the diffusion path of carbon would be too long for the pathway to function effectively. Experimental evidence leading to the understanding of the involvement of various organelles came to a large extent from Tolbert's laboratory. Cell homogenates were carefully centrifuged through density gradients to separate their constituents. The various enzymes of the pathway were located in specific layers in the centrifuge tubes that coincided with the position of specific organelles.

The second point is that the reaction of glycolic acid oxidase produces H_2O_2, and this poisonously powerful oxidant is destroyed by the catalase reaction in peroxisomes. Catalase is largely confined to peroxisomes in photosynthetic leaves and forms a very convenient marker enzyme when peroxisomes are separated by centrifugation. The enzyme glycolic acid oxidase has a rather low affinity for oxygen. It is selectively poisoned by the inhibitor α-hydroxypyridinemethanesulfonic acid (HPMS); in the presence of HPMS glycolate oxidation is blocked and glycolic acid piles up in leaves.

The third point is the production of glycine and serine by the glycolate pathway. These compounds have often been found to be very early products of photosynthetic CO_2 fixation; radiochromatograms of plant extracts after a few minutes fixation of

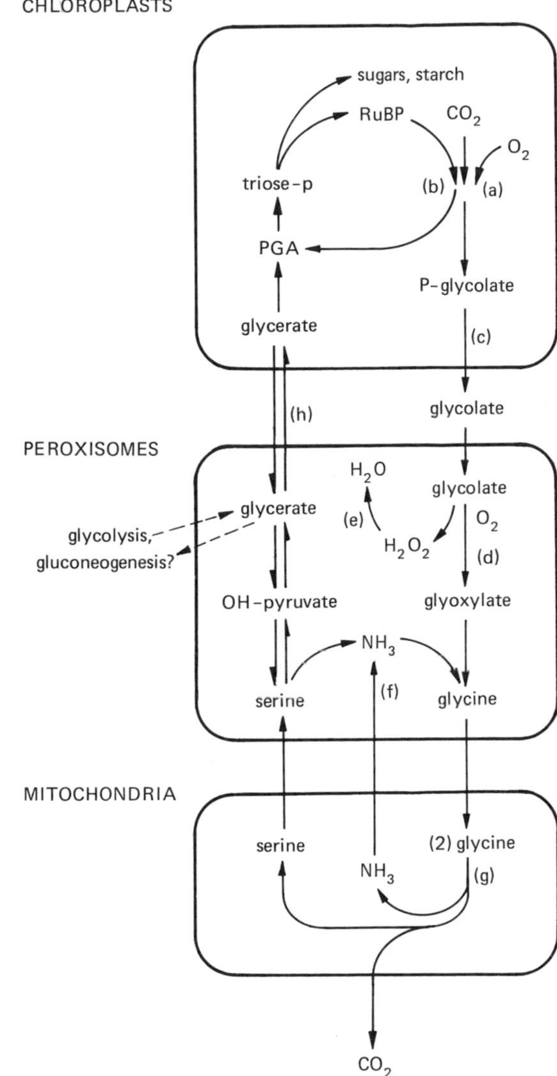

Figure 7-15. Reactions of the glycolate pathway.
(a) RuBP oxygenase } the same enzyme
(b) RuBP carboxylase }
(c) phosphatase
(d) glycolic acid oxidase
(e) catalase
(f) transaminase
(g) glycine decarboxylase and hydroxymethyl transferase
(h) return of carbon to chloroplast
Note that two molecules of glycine are required to make one serine at (g). The left over amino group is presumably returned to the peroxisomes to make more glycine at (f).

$^{14}CO_2$ often show a large amount of radioactivity in serine and glycine. However, in the absence of O_2, which prevents the metabolism of glycolate by preventing both its formation and its oxidation, radioactive glycine and serine are not formed. HPMS also prevents serine and glycine formation.

Photorespiration. The fourth point arising from the glycolate pathway is that one molecule of CO_2 is produced and one molecule of O_2 is absorbed for every two molecules of glycolate that are oxidized. The uptake of O_2 and production of CO_2 in light by photosynthesizing tissue are termed **photorespiration.** Characteristically, photorespiration is greatly inhibited by reduced O_2 concentration. This would be expected because the oxidative function of the carboxylase cannot operate in the absence of O_2, and the oxidation of glycolate requires quite a high concentration of O_2. High pH promotes the oxygenase function of RuBPcase, and raising the pH in vitro also increases photorespiration. Furthermore, CO_2 produced in photorespiration by a leaf that is fixing

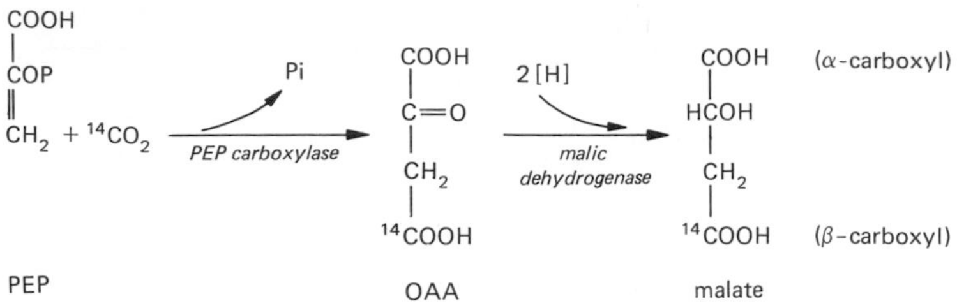

$$\begin{array}{c} COOH \\ | \\ COP \\ \| \\ CH_2 \end{array} + {}^{14}CO_2 \xrightarrow[\text{PEP carboxylase}]{\nearrow Pi} \begin{array}{c} COOH \\ | \\ C{=}O \\ | \\ CH_2 \\ | \\ {}^{14}COOH \end{array} \xrightarrow[\text{malic dehydrogenase}]{2[H]} \begin{array}{c} COOH \\ | \\ HCOH \\ | \\ CH_2 \\ | \\ {}^{14}COOH \end{array} \begin{array}{l} (\alpha\text{-carboxyl}) \\ \\ \\ \\ \\ \\ (\beta\text{-carboxyl}) \end{array}$$

PEP OAA malate

Figure 7-16. β-carboxylation reaction and reduction leading to malate labeled in the C-4 or β-carboxyl. The malic dehydrogenase may be linked to NADH or NADPH.

$^{14}CO_2$ quickly becomes radioactive as the ^{14}C spreads through the Calvin cycle intermediates and enters the glycolate pathway. Finally, specific inhibitors of glycolate oxidation, such as HPMS, also inhibit photorespiration. All these lines of experimental evidence support the view that photorespiration results from the operation of the glycolate pathway.

It is important to note, however, that the glycolate pathway is not a true respiratory metabolism because it is not associated with electron transport or energy production. In addition, several other possible sources of respiratory CO_2 in light are known, besides the possibility of a continuation of dark respiration in light. Also, oxygen may be absorbed by a variety of oxidative reactions in light. Thus, the processes known collectively as photorespiration are complex and involve the interrelationship of several metabolic systems as well as the various organelles shown in Figure 7-15. The whole problem of the interrelationship between photosynthesis, respiration, and photorespiration will be covered in depth in Chapter 15.

C_4 Photosynthesis. Kinetic experiments with leaves, algae, or isolated chloroplasts usually provide evidence of secondary carboxylation reactions. Small amounts of malic acid labeled in the β-carboxyl (C-4) are often found. This suggests that the carboxylation of phosphoenol pyruvate (PEP) to form oxaloacetic acid (OAA) may be linked to a light-dependent reduction. An example of this type of β-carboxylation is shown in Figure 7-16.

In the mid-1960s the physiologists H. P. Kortschack and C. E. Hartt, working in Hawaii, noted that in certain plants the major early product of $^{14}CO_2$ fixation was not PGA but C_4 dicarboxylic acids, and these were initially labeled in the β-carboxyl. As a result of extensive experiments by the Australian physiologists M. D. Hatch and C. R. Slack and others, a cyclic carboxylation pathway has been proposed based on the β-carboxylation reaction. An outline of this pathway is shown in Figure 7-17.

Figure 7-17. Outline of the Hatch and Slack cycle of C_4 photosynthetic carboxylation.

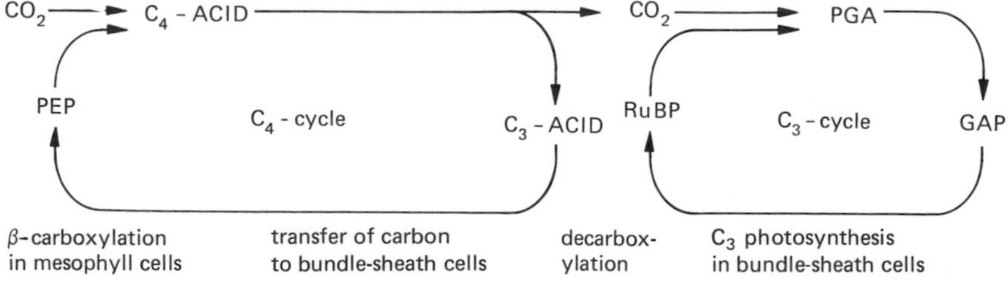

The basic reactions of C_4 photosynthesis are as follows. First there is a β-carboxylation at one site in the leaf, the **mesophyll cells.** The C_4 acid formed by the β-carboxylation of PEP is transferred to the cells surrounding the vascular bundles of the leaf, called the **bundle-sheath cells.** There it is decarboxylated, and the CO_2 so formed is fixed by the Calvin cycle (C_3 Photosynthesis). The C_3 acid formed by the decarboxylation is returned to the mesophyll cells and converted back into PEP.

The special characteristic of this photosynthetic cycle is the separation of the two carboxylations. Most plants having C_4 photosynthesis have a specialized leaf anatomy called the **Kranz** type, shown in Figure 7-18. In C_3 plants, the parenchyma cells are organized into two distinct tissues, the palisade layer and the spongy parenchyma, and there are conspicuous air spaces. In C_4 leaves the veins are closer together, and each is surrounded by a layer of bundle-sheath cells that contain large numbers of chloroplasts. These are surrounded by mesophyll cells that largely fill the leaf; air spaces are much smaller. Thus the distance for diffusion of CO_2 to the carboxylation sites is short. Furthermore, mesophyll cells are seldom more than one or two cells' distance from the bundle-sheath cells, so the transfer of acids to and fro need not be over a very long distance.

The complete story of C_4 photosynthesis, shown in Figure 7-19, is quite complicated because several assorted variations on the main pathways have been found in different plants. The first stable acid formed (oxaloacetic acid—OAA—is unstable and breaks down on isolation, so is seldom identified unless special precautions are taken to protect it from degradation) may be malate, which requires for its formation a reduction step using photosynthetically generated NADPH (reaction b), or aspartate, which requires transamination (reaction c). Malate or asparate may be transferred to bundle-sheath cells. Malate is then decarboxylated by malic enzyme, regenerating the NADPH required for its synthesis. The pyruvate remaining after the decarboxylation is returned to the bundle-sheath cells.

If aspartate is the C_4 acid, it is reconverted back to OAA in the bundle-sheath cells. In some plants it is then decarboxylated to form PEP which is converted to pyruvate, a reaction sequence involving the sequential hydrolysis and synthesis of ATP. In other plants the OAA is reduced to malate in the mitochondria of the bundle-sheath cells, using NADH. The malate is decarboxylated by mitochondrial malic enzyme that regenerates NADH. When aspartate is the mobile C_4 acid, alanine is the returning C_3 acid instead of pyruvate. This prevents the pile-up of NH_3 in bundle-sheath cells. C_4 plants are frequently classified according to the enzyme that decarboxylates the C_4 acid, as shown in Figure 7-19.

The regeneration of PEP is a curious and interesting reaction. The enzyme **pyruvate, phosphate dikinase** requires ATP and Pi, producing PEP, AMP, and pyrophosphate (PPi). The PPi is hydrolyzed to 2 Pi by a pyrophosphatase, and the AMP reacts with another molecule of ATP to make two molecules of ADP. Thus, in sum, two molecules of ATP are converted to ADP for each molecule of PEP synthesized. This overall reaction is powerfully exothermic and tends to run strongly in the direction of PEP synthesis (see also Chapter 6, page 124).

It will be noted from this discussion and from Figure 7-19 that C_4 photosynthesis may involve cytoplasmic enzymes (notably the carboxylation enzymes itself) as well as mitochondrial and chloroplast enzymes. In this way it resembles the glycolate pathway. Thus the old idea that photosynthesis is carried on exclusively in chloroplasts can no longer be considered strictly true. In this pathway, as in the glycolate pathway, the metabolic sequences call for the collaboration of all parts of various cells in different

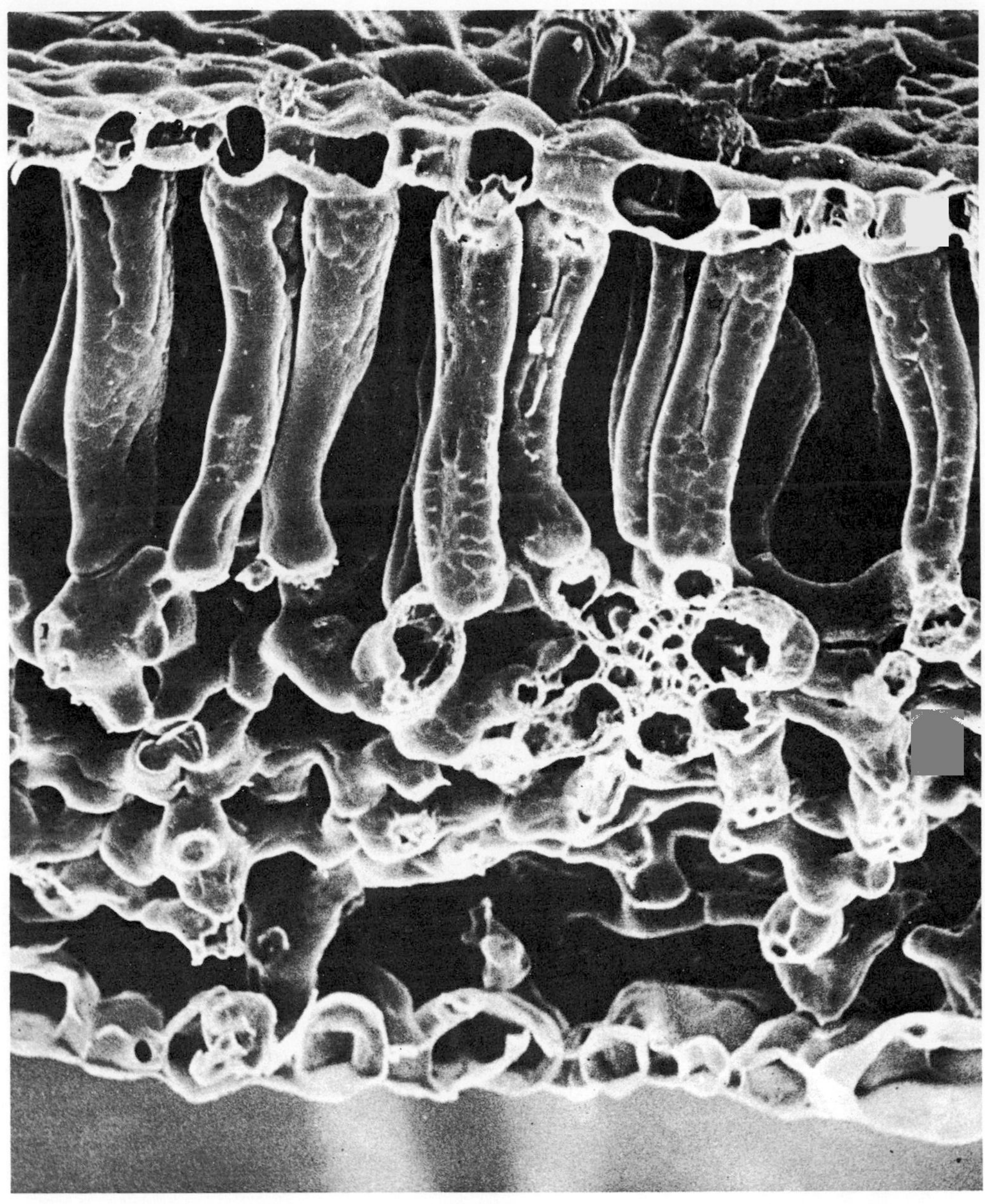

Figure 7-18. Scanning electron micrographs of the photosynthetic tissues of (**A**) a C_3 leaf (bean—*Phaseolus vulgaris*, ×420) and (**B**) a C_4 leaf (sedge—*Cyperus rotundus*, ×600). Note the conspicuous bundle-sheath cells, small air spaces, and the overall organization that characterizes the Kranz anatomy of the C_4 leaf. [Sources: *A*, from J. A. Troughton and L. A. Donaldson; *Probing Plant Structure*. McGraw-Hill Book Co., New York, 1972, used with permission. B, photograph kindly supplied by Prof. C. C. Black, University of Georgia.]

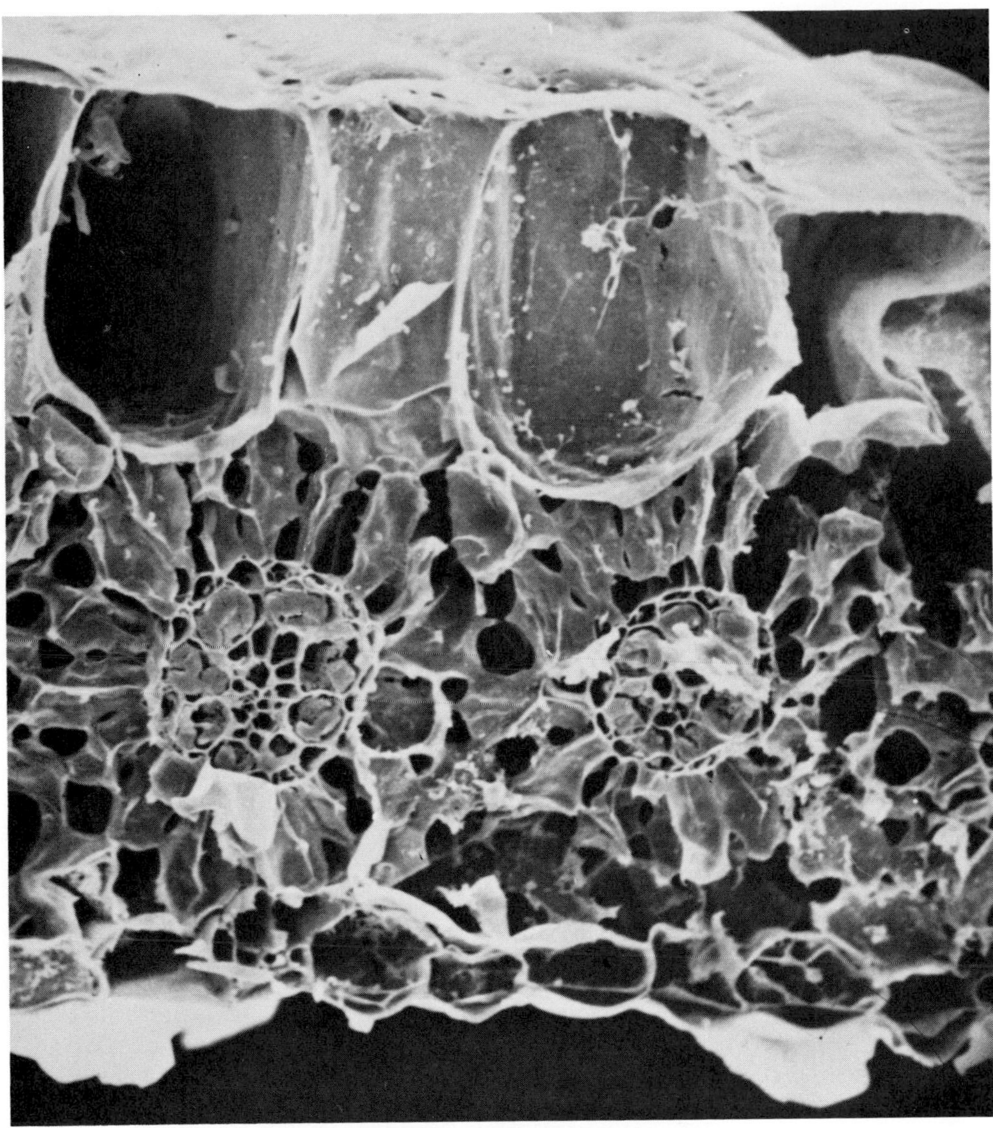

B

parts of the plant. The identification of the various enzyme activities in the organelles of either bundle-sheath or mesophyll cells has been possible by careful selective isolation of tissues, by separating the chloroplasts (bundle-sheath chloroplasts make starch while mesophyll chloroplasts do not; thus the bundle-sheath chloroplasts are denser and can be separated by differential centrifugation), and by isolating organelles from the cells of different tissues that have been carefully separated by mechanical means.

An interesting technique for demonstrating the distribution of RuBPcase has been developed by P. W. Hattersley in Australia. He first isolated chloroplast RuBPcase, then injected it into rabbits to make an RuBPcase antiserum. The rabbit antiserum was applied to leaf sections, then a sheep antiserum (raised against rabbits) tagged with a fluorescent marker was added. The fluorescent marker thus labeled RuBPcase in situ. It can be seen from Figure 7-20 that in a C_3 leaf RuBPcase is distributed in all photosynthetic cells, but in a C_4 leaf it is present almost exclusively in the bundle-sheath cells.

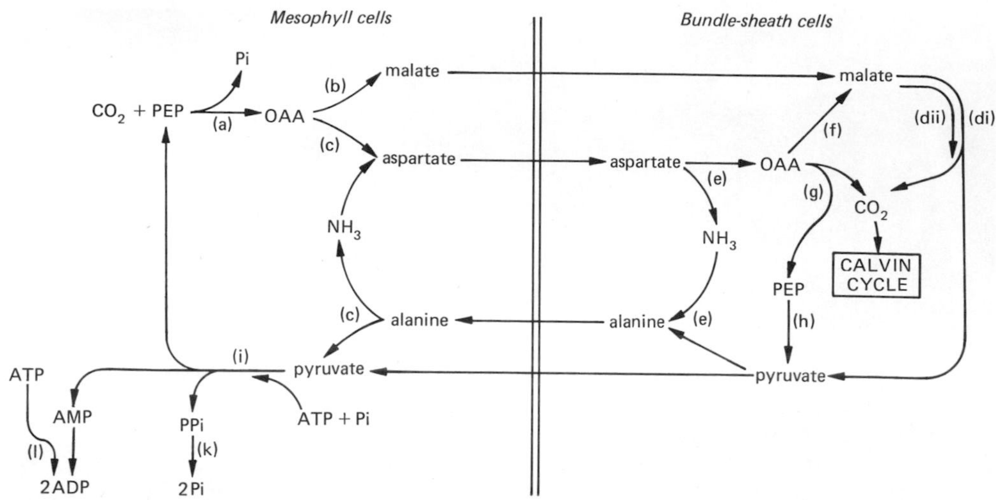

Figure 7-19. The Hatch and Slack cycle and the proposed pathways of C_4 photosynthesis.

Reaction	Site	Notes	*Light activated*
(a) PEP carboxylase	cytoplasm (probably)	substrate is bicarbonate	
(b) malate dehydrogenase	chloroplasts	NADPH ⟶ NADP	+
(c) transaminase	cytoplasm		
(d) malic enzyme	(i) chloroplasts (ii) mitochondria	NADP ⟶ NADPH NAD ⟶ NADH	++
(e) transaminase	mitochondria or cytoplasm		
(f) malic dehydrogenase	mitochondria	NADH ⟶ NAD In plants that have reaction (dii)	
(g) PEP carboxykinase	mitochondria (probably)	ATP ⟶ ADP	
(h) pyruvic kinase	cytoplasm	ADP ⟶ ATP	
(i) pyruvate, phosphate dikinase	chloroplasts	2ATP ⟶ 2ADP (i, k, l)	+ (i, k, l)
(k) pyrophosphatase	chloroplasts		
(l) adenylate kinase	chloroplasts		

Note: C_4 plants are often classified NADP-malic enzyme, or NAD-malic enzyme, or PEP carboxykinase (PCK) types, depending on their mechanism of C_4 acid decarboxylation.

In some C_4 plants, particularly those in which malate is the C_4 acid transferred, the bundle-sheath chloroplasts appear to lack organized grana or to have a much reduced grana structure, as shown in Figure 7-21. This may be associated with the fact that malate-transferring plants have a lower requirement for NADPH synthesis in the bundle-sheath chloroplasts, because the decarboxylation of malate generates NADPH. Thus, a considerable part of the NADPH requirement of the Calvin cycle may be met

---

Figure 7-20 (Opposite). Fluorescent-die staining to locate RuBPcase in photosynthetic tissue. **A.** A C_4 grass, *Digitaria brownii,* in which only bundle-sheath cells are fluorescent, ×380. **B.** A C_3 grass, *Danthonia bipartita,* in which all photosynthetic cells are fluorescent, ×240. [See P. W. Hattersley, L. Watson, and C. B. Osmond, *Aust. J. Plant Physiol.,* **4:**523–39 (1977) for details of technique. Photographs kindly supplied by Dr. Hattersley.]

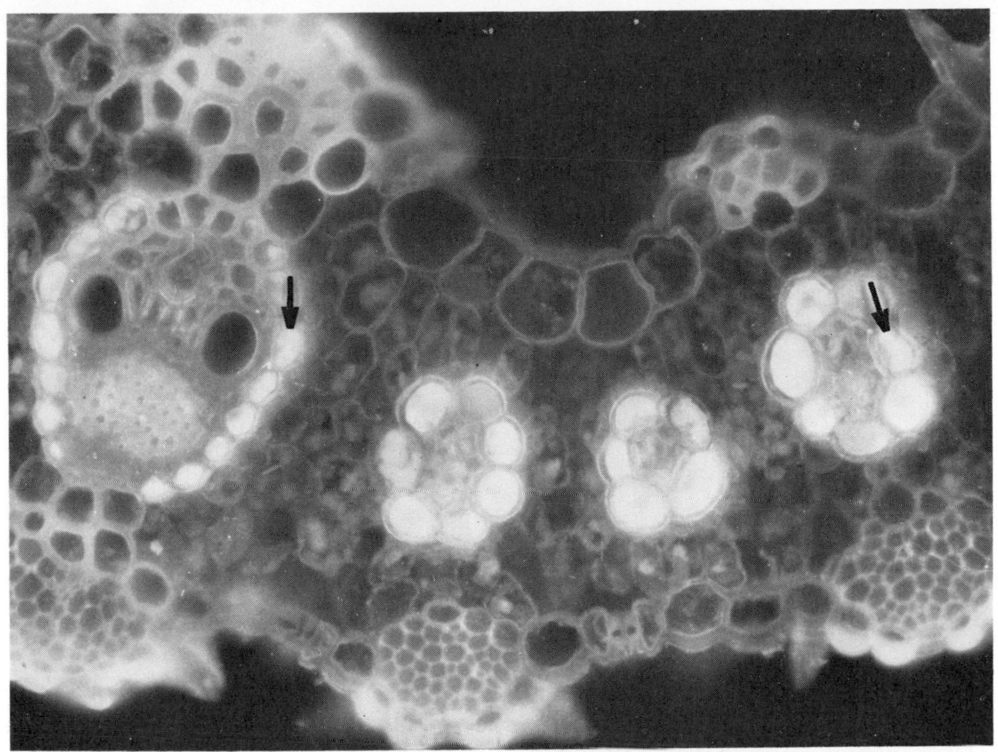

A

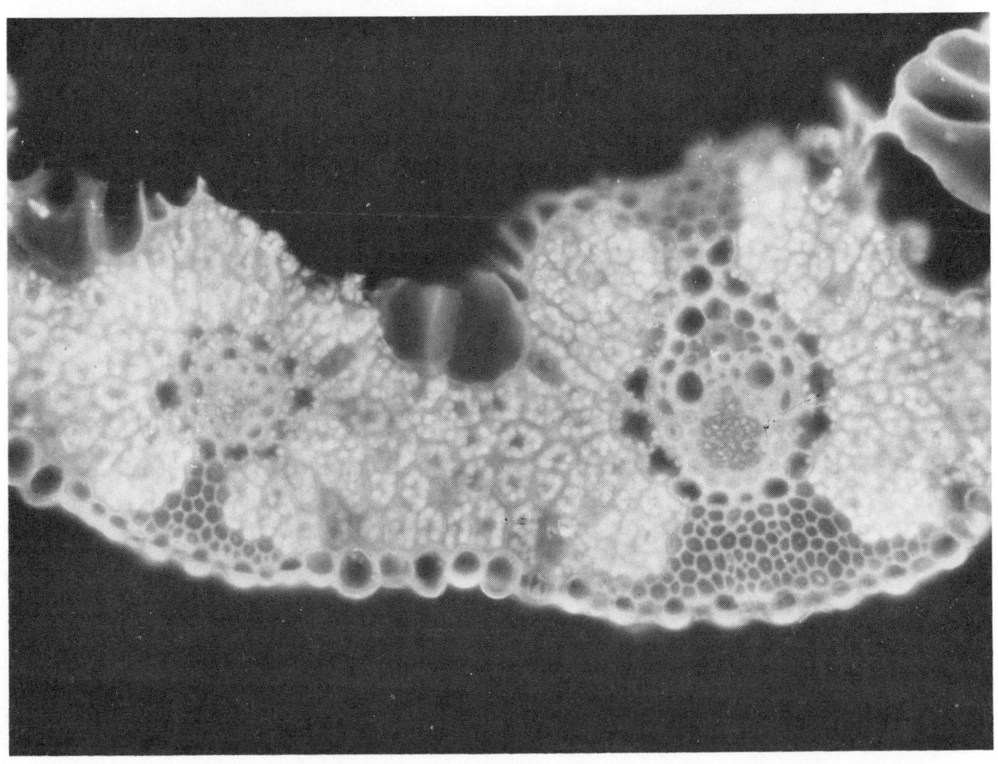

B

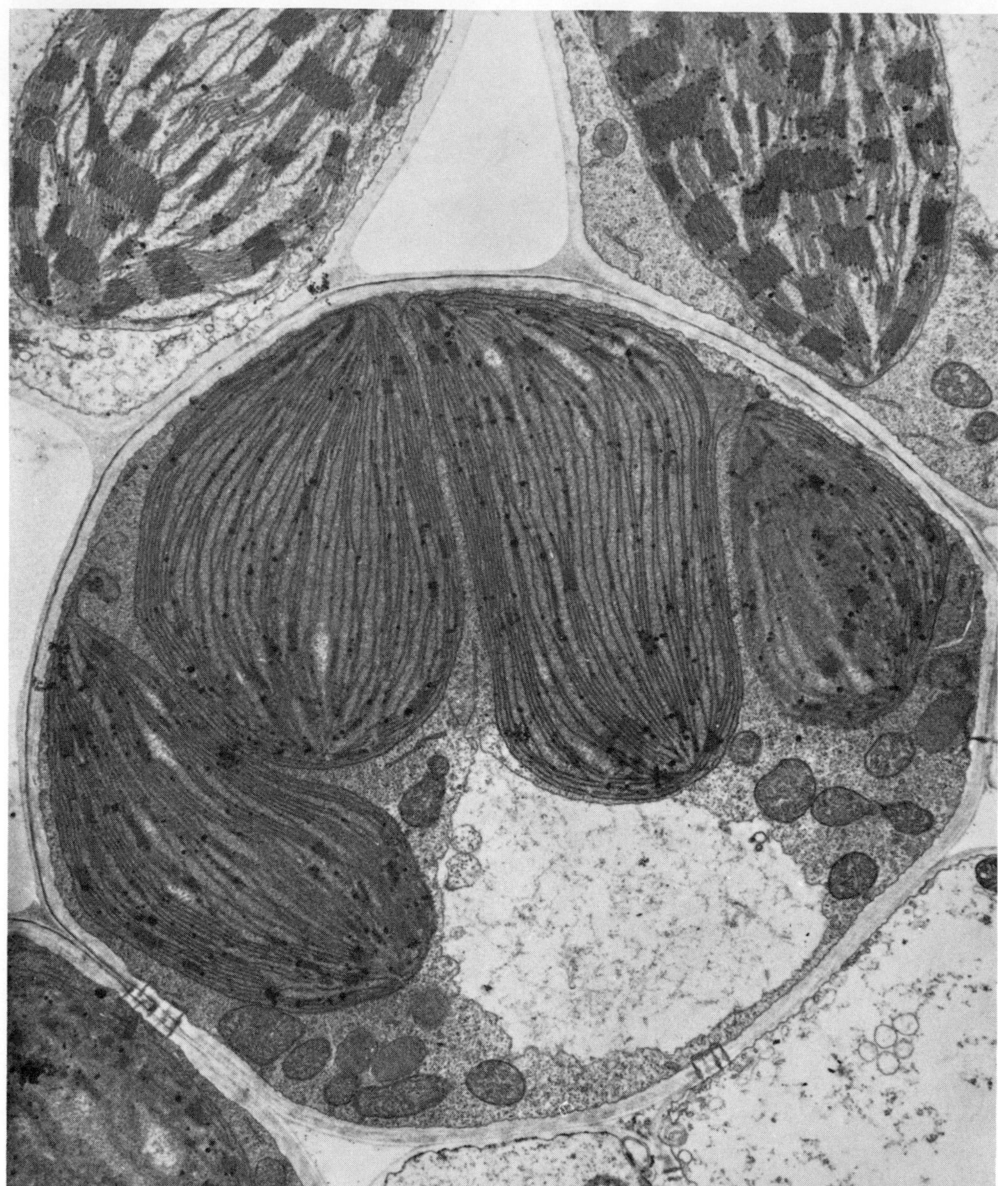

Figure 7-21. Electron micrograph of chloroplasts from crabgrass (*Digitaria sanguinaris*), ×16,000, showing agranal bundle-sheath chloroplast (center) and granal chloroplasts in two adjoining mesophyll cells (above). [Photograph kindly supplied by Prof. C. C. Black, University of Georgia.]

from this source. It will be remembered that the operation of the entire electron transport system of photosynthesis, which is necessary to generate NADPH, appears to require the apposition of two or more thylakoids, which is to say, the formation of grana. On the other hand, ATP synthesis associated with photosystem I goes perfectly well in intergranal membranes. This differentiation of chloroplast types, however, is not entirely clear. It is probable that the bundle-sheath cells of most C_4 plants are in fact capable of complete photosynthesis of the C_3 type. The C_4 cycle is primarily a device for improving the rate of photosynthesis.

Summary of C_4 Photosynthesis: Its Significance to Plants That Possess It. The story of C_4 photosynthesis is far from complete. Much remains to be clarified about the evolution of this pathway and its significance to the plants that possess it. The taxonomy of C_4 photosynthesis is interesting: C_4 plants are found in several groups of tropical grasses and sedges, and a number of dicotyledonous families also have representatives. A curious fact is that several families (and even certain genera) have members of both C_3 and C_4 types. The inference is that C_4 photosynthesis has arisen independently in several different groups of higher plants and that it is thus a recent evolutionary development.

Its significance is severalfold. Plants having C_4 photosynthesis are capable of much higher rates of photosynthesis, and, because of the high affinity of PEP carboxylase for CO_2 and the strongly exothermic reaction leading to PEP synthesis, they are able to absorb CO_2 strongly from a much lower CO_2 concentration than are C_3 plants. This means that they can maintain high rates of photosynthesis when the stomata are nearly closed, a decided advantage to plants living in a dry hot climate. No CO_2 is lost in photorespiration by C_4 plants—either they lack the metabolism of photorespiration or else any CO_2 so produced (in the bundle-sheath cells because that is the primary site of RuBPcase) is refixed by the mesophyll cells and so cannot escape.

Most of the bad agricultural weeds in the world and some of the most productive crops are C_4 plants. References to various checklists of C_4 plants will be found in R. H. Burris and C. C. Black (eds.): *CO_2 Metabolism and Plant Productivity* cited at the end of this chapter (page 191). However, some C_3 plants equal C_4 plants in productivity, and many C_4 plants are not competitive in all situations. So the C_4 syndrome does not automatically confer special advantages, nor is it always "more efficient" or "better" than C_3 photosynthesis, as is often claimed. In fact, perhaps the most important point about C_4 photosynthesis is that it is *less* efficient; that is, it uses more light energy to fix CO_2 than does C_3 photosynthesis. This means that C_4 plants, mostly tropical or of tropical origin, can use some of their excess light to run the C_4 cycle to concentrate CO_2 in the bundle-sheath cells where it can be more rapidly fixed by the C_3 cycle. In a sense, the C_4 cycle is a CO_2 supercharger. It burns a lot of fuel (the energy used to generate ATP for the pyruvate, phosphate dikinase reaction); but if the fuel is available and free (light) then its use can confer a decided advantage.

C_4 photosynthesis is not necessarily an absolute or invariable reaction sequence. Plants have been characterized as "aspartate formers" or "malate formers" according to their C_4 acid; but some C_4 plants can produce either or both to varying degrees. Also, the C_4 cycle need not operate always nor need it be the sole primary carboxylation pathway—the C_3 cycle can still fix atmospheric CO_2 and may, under proper conditions (high CO_2, low light, abundant water), be the main carboxylation system. The C_4 reaction may be used under certain conditions to store large quantities of C_4 acids that could be used as a source of CO_2 for photosynthesis at a time when, perhaps due to stomatal closure under severe water stress, atmospheric CO_2 is no longer available.

Photosynthesis used to be considered as an absolute, unvarying mechanism. It should now be apparent that this is not true. C_4 photosynthesis is merely another example of the wide variability and adaptability of plants to their variable environment. Most plants are capable of some β-carboxylation. C_4 plants have capitalized on this reaction and coupled it with an anatomical structure conferring decided advantages (though even this is not invariable) to achieve a more effective use of light energy for concentrating CO_2 at the place where it is needed: the site of CO_2 reduction.

Crassulacean Acid Metabolism. Early physiologists noted that certain succulent plants of the family *Crassulacea* increased their acid content markedly at night, decreasing it by day. Later it was found that these plants absorb CO_2 in darkness, but frequently not in light. This overall pattern of photosynthetic metabolism is called **Crassulacean acid metabolism (CAM).** It involves the synthesis of malic acid by β-carboxylation at night, and the breakdown of malic acid in daytime to liberate CO_2 for photosynthesis. CAM plants are usually succulent, possess xeric characteristics (reduced leaves, thick cuticle, sunken stomata, and so on), and live in dry arid climates. This type of metabolism permits them to carry on photosynthesis even when stomata are tightly closed during the heat and drought of the day, using CO_2 absorbed during the damper, cooler night.

The metabolic pathways of CAM are outlined in Figure 7-22. A checklist of CAM plants will be found in R. H. Burris and C. C. Black (eds.): *CO_2 Metabolism and Plant Productivity* cited at the end of this chapter (page 191). In darkness, stored carbohydrates are converted by glycolysis to PEP, which is carboxylated (PEP carboxylase) to malic acid. The malic acid is stored in the vacuole. In light, malate is decarboxylated (usually by malic enzyme, in some plants by PEP carboxykinase) to give pyruvic acid and CO_2. The CO_2 is used for normal C_3 photosynthesis. Pyruvic acid may be oxidized to CO_2, providing more CO_2 for photosynthesis, or it may be reconverted to PEP or PGA and used for sugar synthesis, or it may be reintroduced into the photosynthetic cycle. The fate of pyruvic acid is not known with certainty, and probably involves all the possible reactions that this centrally located metabolite can undergo.

CAM is not an obligate pathway. If stomata are open in daylight, CO_2 may be absorbed and fixed in the usual way, bypassing CAM. On the other hand, CAM seems to be tightly regulated either by diurnal rhythms or by allosteric feedback control of PEP carboxylase by malate. Since the carboxylation enzyme is in the cytoplasm, whereas malate is stored in the vacuole, no regulation will occur until the malate concentration becomes so high that no more can enter.

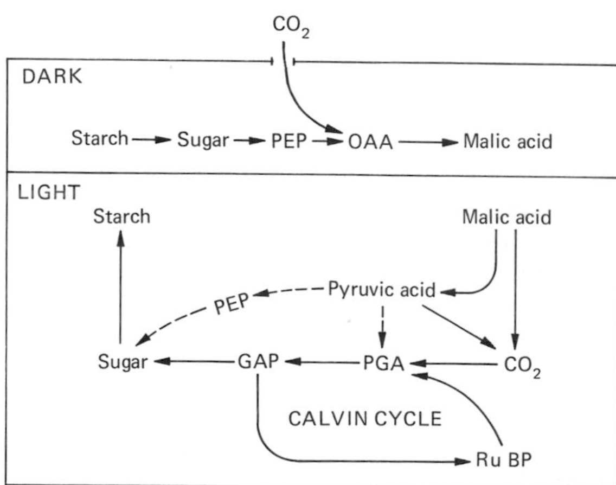

Figure 7-22. Outline of CAM. The dark carboxylation reaction is catalyzed by PEP carboxylase. Malate is decarboxylated by malic enzyme or PEP carboxykinase. Unproved pathways are shown by dotted lines.

CAM is similar in many ways to C_4 photosynthesis, except that the β-carboxylation and C_3 photosynthesis are separated in time instead of in space. Unlike C_4 photosynthesis, CAM does not confer high rates of photosynthesis on plants that possess it. In fact, CAM is very inefficient, but it does permit the continuation of photosynthesis under extremely xeric conditions.

Photosynthesis of Other Compounds. Many compounds are produced via the operation of cellular metabolic pathways (such as the Krebs cycle or glycolysis) using as substrates products of photosynthesis that leave the chloroplast in the light. A considerable proportion of the protein synthesis that goes on in a leaf is done in the chloroplast, and many of the amino acids that enter leaf protein are synthesized more or less directly from carbon recently fixed in photosynthesis. These amino acids appear to be made both in the chloroplast and in the cytoplasm, from carbon that leaves the chloroplast. It appears probable that a major proportion of the leaf's lipids are formed from glycerol and acetyl-CoA derived more or less directly from photosynthesis. The glycerol presumably arises directly from the reduction of glyceraldehyde phosphate, but the source of acetyl-CoA is not known. It may arise directly from the TPP–glycolaldehyde by reduction, or by glycolysis from PGA which has left the chloroplast.

Sucrose and Starch Formation. Not all of the recent products of photosynthesis become immediately available to the cell. A substantial part is stored as starch in the chloroplasts or amyloplasts, and starch formation has long been recognized as a principal end point of photosynthesis. Alternatively, in many plants (principally monocotolydons) carbohydrate is stored as sucrose, either in the photosynthetic cells or, as in sugarcane, in the vacuoles of special storage cells in the stem. Both sucrose and starch were first thought to be synthesized by phosphorylase reaction with G-1-P as a substrate

$$\text{G-1-P} + \underset{\text{fructose}}{\text{F}} \rightleftharpoons \underset{\text{sucrose}}{\text{GF}} + \text{Pi}$$

$$\text{G-1-P} + \text{—G—G—G} \rightleftharpoons \underset{\text{starch}}{\text{—G—G—G—G}} + \text{Pi}$$

Starch synthesis by this reaction requires the existence of a primer or starter molecule, either maltose or a glucose polymer. It is really a chain-lengthening reaction. However sucrose phosphorylase is not found in higher plants, and the equilibrium of these phosphorylase reactions is such that the reactions would be expected to run in the direction of starch or sucrose breakdown rather than synthesis. The discovery of the **uridine diphosphate glucose (UDPG)** transfer system by the Argentine biochemists L. F. Leloir and C. E. Cardini opened the way to understanding oligosaccharide and polysaccharide synthesis by glucose transfer reactions.

There are two probable pathways for the synthesis of sucrose.

$$\text{UDPG} + \text{F} \longrightarrow \underset{\text{sucrose}}{\text{GF}} + \underset{\text{uridine diphosphate}}{\text{UDP}} \quad (1)$$

$$\text{UDPG} + \text{F-6-P} \longrightarrow \underset{\text{sucrose phosphate}}{\text{GF-6-P}} + \text{UDP}$$

$$\text{GF-6-P} \longrightarrow \text{GF} + \text{Pi} \quad (2)$$

The second reaction forms sucrose phosphate, which yields sucrose on hydrolysis. This hydrolysis is strongly exothermic and thus essentially irreversible, and is probably the device that permits the accumulation of large quantities of sucrose in certain leaves (for example, sugarbeet and many monocots that do not normally make starch in their leaves). The UDPG is made via the reaction sequence

$$\text{UDP} + \text{ATP} \longrightarrow \underset{\text{uridine triphosphate}}{\text{UTP}} + \text{ADP}$$

$$\text{UTP} + \text{G1P} \longrightarrow \text{UDPG} + \underset{\text{pyrophosphate}}{\text{PPi}}$$

The hydrolysis of PPi yields 8 kcal/mole and thus can be coupled to make the reaction run strongly toward the synthesis of sucrose. The synthesis of starch is accomplished by essentially the same reaction sequence

$$\text{—G—G—G} + \text{UDPG} \longrightarrow \underset{\text{starch}}{\text{—G—G—G—G}} + \text{UDP}$$

or by a similar reaction in which the glucosyl transfer agent is adenosine diphosphate glucose (ADPG) instead of UDPG. Again, as for phosphorylase, a primer or starting molecule is required. Presumably the ATP required to synthesize UDPG or ADPG is derived in photosynthesis from cyclic phosphorylation.

The ready interconversion of F-6-P $\rightleftharpoons$ G-6-P $\rightleftharpoons$ G-1-P permits rapid synthesis of sucrose and starch without the appearance of free hexoses. This accounts for the early discovery that ^{14}C-labeled glucose supplied to leaves would rapidly give rise to labeled starch and to sucrose labeled equally in both hexose moieties, but only slowly to free labeled fructose. This also settled the difficult problem that faced early physiologists, who tried to answer the question: "Which sugar (or starch) is the first product of photosynthesis?" It now seems most likely that free hexoses found in small quantities in most photosynthetic tissues are the result of sucrose or starch hydrolysis, and are not their precursors.

It is an interesting reflection on the difficulties of plant biochemistry that, although the site of starch formation is well known to be the chloroplast, the site of sucrose synthesis is not known with certainty. This is particularly surprising because sucrose is a most important commercial biochemical (sugarcane, sugarbeets) and it is the form in which carbon is translocated in plants. Isolated higher plant chloroplasts, although they can photosynthesize as effectively as leaves if carefully prepared, seem unable to make sucrose. Experiments in which leaves are given $^{14}CO_2$ and chloroplasts are isolated in nonaqueous solvents (to prevent leakage of sucrose which is very water soluble) suggest that it is made at or outside the boundary of the chloroplast. Hexoses and sucrose seem unable to pass through the chloroplast membrane with any facility, but trioses and other low molecular weight compounds penetrate without difficulty.

There are two main possibilities: either sucrose or sucrose phosphate is made in chloroplasts and rapidly exported by an active transport mechanism (that could dephosphorylate sucrose phosphate at the chloroplast membrane), or sucrose is made in the cytoplasm adjacent to the chloroplasts from triose phosphates, which readily permeate the chloroplast membrane. Triose phosphates could easily be converted, by reverse glycolysis, to the hexose phosphate precursors of sucrose synthesis. The answer to this interesting and important problem will have to await the discovery of better biological or biochemical techniques.

Factors Affecting Photosynthesis

Photosynthesis does not require the same rigorous internal rate control as does respiration—in fact, it is obviously to the advantage of a plant for photosynthesis to proceed rapidly when conditions are right. Evidently plants must adjust their overall efficiency to the maximum light intensity they usually encounter. Plants that live in shade need a highly efficient light-gathering system because they must develop maximum fixation rates at low light intensity. Leaves situated in open places require much less efficient light traps, otherwise they would absorb too much light energy and heat up as a consequence. This level of control is not provided by biochemical adjustments but by the overall developmental pattern of the plant. Also the rate of photosynthesis achieved by a plant relates to its physiological condition—the situation under which it grew, its nutritional status, genetic factors, the state of its stomata, and so on.

Finally, various secondary pathways—carboxylation, the C_4 cycle, CAM, the glycolate pathway, as well as dark respiration—all impinge on the operation of photosynthetic pathways. Net photosynthesis, or net CO_2 assimilation, is a resultant of the gross or overall rate of photosynthetic CO_2 fixation and the loss of CO_2 by photorespiration and other respiratory pathways. Each of these metabolic systems may react to the whole range of internal and external factors in different ways. For this reason we shall consider photosynthesis in Chapter 15 in the context of the nutritional metabolism of the plant as a whole, and as it relates to other pathways of carbon metabolism in the plant. Certain specific factors that affect the photosynthetic pathways will be briefly mentioned here.

Temperature. We have mentioned earlier the interrelations between light intensity and temperature examined by Blackman. Essentially, although the light absorption reaction is not much affected, the enzymic or dark reactions are strongly dependent on temperature. In the physiological range between 5 and 25–30° C, photosynthesis usually has a Q_{10} of about 2, as would be expected. Certain organisms can continue CO_2 fixation at extraordinary extremes of temperature—some conifers at −20° C and algae that inhabit hot springs at temperatures in excess of 50° C, but in most plants photosynthesis ceases or declines sharply beyond the physiological limits mentioned above.

Much early work on photosynthesis was concerned with determining the optimum conditions for maximum rates, but the results obtained were conflicting and difficult to evaluate. This is because, as we have seen, photosynthesis is a highly complex process and the optimum for any one factor is bound to be much affected by the levels of several others. Thus the effects of CO_2 concentration, temperature, and light on photosynthesis are all interrelated, and all, to a greater or lesser extent, are dependent on a number of physiologcal or anatomical characteristics of the plant. Under field conditions, temperature in fact does not greatly influence the rates of photosynthesis over the range of 16 to 29° C, unless the light intensity is sufficiently high that the dark reactions are limiting.

Oxygen. Oxygen strongly affects photosynthesis in several ways. Certain of the photosynthetic electron carriers may transfer electrons to oxygen, and ferredoxin in particular appears to be sensitive to O_2. In bright light, high oxygen leads to irreversible damage to the photosynthetic system, probably by oxidation of pigments. Carotenes in the chloroplasts tend to protect chlorophylls from damage by **solarization,** as this is called.

The oxygenase reaction of RuBPcase provides the most important site of O_2 effect on photosynthesis. O_2 competitively and reversibly inhibits the photosynthesis of C_3 plants

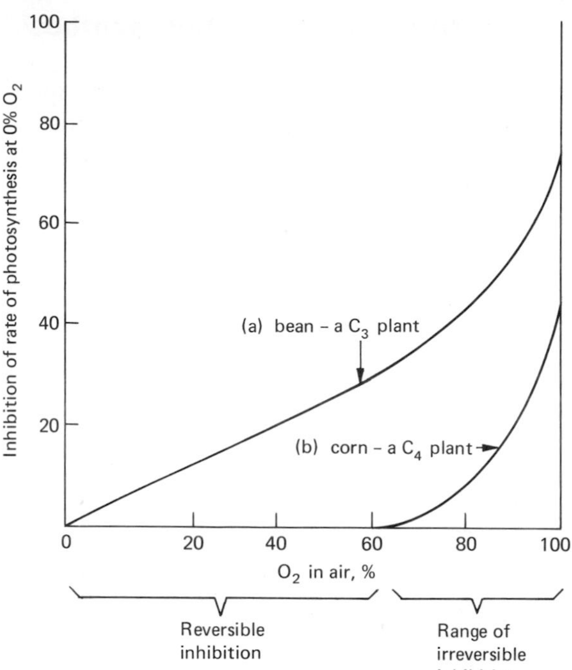

Figure 7-23. Inhibitory effects of oxygen on photosynthesis. [Data of W. B. Levin and R. G. S. Bidwell.]

over all concentrations of oxygen; at high O_2 (80 percent or over) irreversible inhibition also takes place. C_4 plants do not release CO_2 in photorespiration (either because they do not photorespire or because photorespired CO_2 is reabsorbed by the C_4 cycle), and their photosynthesis is not affected by O_2 until very high concentrations are reached that cause irreversible damage to the photosynthetic system. A comparison of the effects of oxygen on C_3 and C_4 plants is shown in Figure 7-23.

Carbon Dioxide. Under field conditions, carbon dioxide concentration is frequently the limiting factor in photosynthesis. The atmospheric concentration of about 0.033 percent (330 ppm) is well below carbon dioxide saturation for most plants; some do not saturate until a concentration of 10 to 100 times this is reached. Characteristic carbon dioxide saturation curves are shown in Figure 7-24. Evidently photosynthesis is much affected by carbon dioxide at low concentrations, but is more closely related to light intensity at higher concentrations. At reduced carbon dioxide concentrations the path of carbon may change dramatically—much higher glycolate production has been noted at low carbon dioxide concentrations, due to the increased relative level of O_2.

As CO_2 concentration is reduced the rate of photosynthesis slows until it is exactly equal to the rate of photorespiration. In C_3 plants this occurs at a CO_2 concentration of about 50 ppm. This CO_2 concentration, at which CO_2 uptake and output are equal, is called the **CO_2 compensation point** (abbreviated Γ). The CO_2 compensation point of C_4 plants, which do not release CO_2 in photorespiration, is usually very low, from 2 to 5 ppm CO_2. Characteristic CO_2 curves for C_3 and C_4 plants are shown in Figure 7-24.

Light. As one might expect of a light-dependent process, the intensity of light directly affects the rate of photosynthesis. The graphs in Figure 7-25 show characteristic light

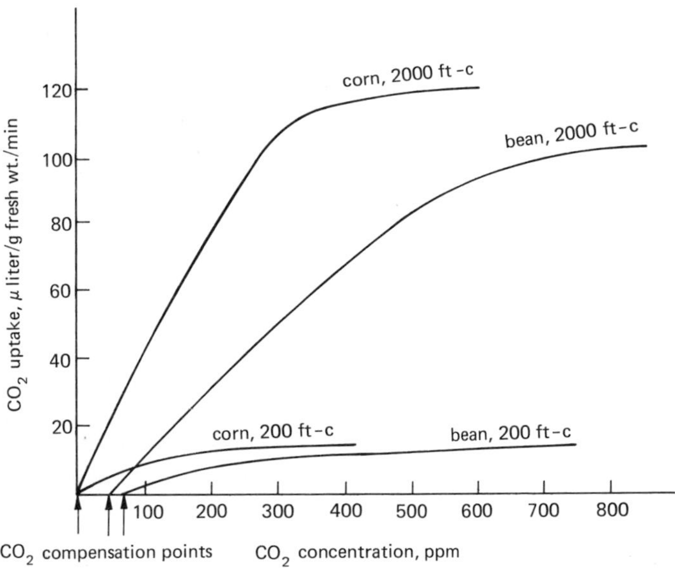

Figure 7-24. The effect of CO_2 concentration on photosynthesis of a C_3 leaf (bean, *Phaseolus vulgaris*) and a C_4 leaf (corn, *Zea mays*). [Data of R. G. S. Bidwell.]

Figure 7-25. The effect of light intensity on photosynthesis of a C_3 leaf (bean, *Phaseolus vulgaris*) and a C_4 leaf (corn, *Zea mays*). [Data of R. G. S. Bidwell.]

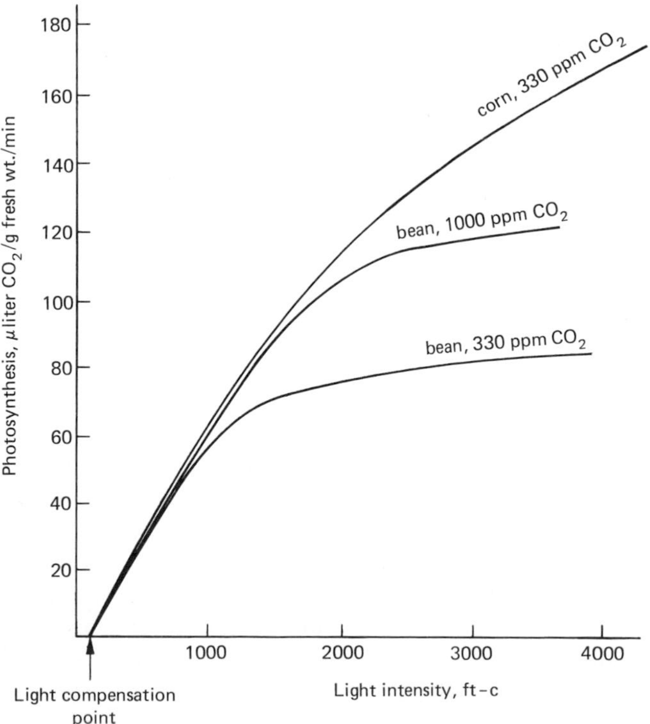

curves for C_3 and C_4 leaves. They show that photosynthesis saturates quite sharply at the intensity at which the dark reactions become limiting, as shown by the fact that high CO_2 concentration permits higher light saturation levels than low CO_2. The point at the other end of the graph, at which photosynthesis equals respiration and no net gas exchange occurs, is called the **light compensation point.** In general, the light compensation point of shade-tolerant plants is much lower than that of sun plants. It usually falls in the range of 20–100 ft-c.

The quality of light also affects photosynthesis. We have noted that far-red light is inefficient *by itself* in photosynthesis and requires additional light of shorter wavelength to be used efficiently. It has also been noted that if red light is supplemented by a relatively weak source of blue light the rate of photosynthesis may be substantially increased and the products of photosynthesis may be affected. The Russian physiologist A. Nichiporovitch observed that the production of protein is enhanced in blue light and of carbohydrate in red light. It has been suggested that flavin-linked enzymes may be primed or activated by the absorption of blue light by their yellow-colored flavin cofactors, and the synthesis of certain amino acids may thus be enhanced.

The Evolution of Photosynthesis

We cannot now say with certainty how photosynthesis evolved, but it is possible to make an intelligent guess along the lines opened up by the writings of Oparin and Haldane. The result of such speculation is useful because it places the basic processes of photosynthesis in a new perspective. We can imagine the first organisms living in an organic "chemical soup" made of compounds that were formed earlier, from an atmosphere rich in reduced carbon and nitrogen, as the result of irradiation with ultraviolet light and electrical discharges. These primitive organisms lived in an aqueous medium with a strongly reducing atmosphere, surrounded by all the substances needed for growth. Synthetic reactions would require only the absorption of the required substrates, and the energy for synthesis could be derived from the breakdown of strongly reducing compounds in coupled reactions. However the supply of required intermediates would quickly become exhausted, and enzymic systems would have evolved to make use of some of the energy present in nonrequired substrates for the synthesis of needed intermediates. In the process, an electron transport mechanism for the extraction of energy from substrates and the synthesis of high energy intermediates would evolve.

Eventually the supply of high energy electrons would become very low, and some way would be needed to make use of low energy electrons—in other words, to add energy to the system. The electron transport enzymes, by their nature, must absorb light, and presumably the next step was the development of an enzyme that would not only transfer electrons but could also make use of the light it absorbed to boost the electrons to a higher energy level. This would result in a system of photosynthesis similar to that of some photosynthetic bacteria which use electrons from donors of intermediate energy, raised to a high energy level by light absorption, for synthetic reactions. The evolution from bacterial photosynthesis to higher plant photosynthesis requires only one further step—the cooperation of two light-absorption reactions coupled by the electron transport chain so that electrons from a universal electron donor, water, could be used. In this way the photosynthetic system is freed from any external requirement for reduced electrons.

It is therefore possible that carboxylation reactions and the synthesis of organic chemicals from carbon dioxide began very early in the history of organisms, and the utilization of light energy in a primitive way for synthesis may well have been an ancient development. However it is likely that the evolution of photosynthesis as it now occurs in higher plants is comparatively recent. Thus oxygen is probably also a comparatively recent addition to the earth's atmosphere.

It has been suggested that the earth's atmosphere in Carboniferous times, when the growth of plants exceeded that in any other era, contained 5 percent carbon dioxide and only 5 percent oxygen. This, together with the prevalent warmth and high humidity of that period, represents the ideal conditions for plant growth. Plant growth in that era provides us now with coal and oil. But perhaps it also provides us with an object lesson: the Carboniferous plants used up their natural resources (CO_2) in a wild extravagance of proliferation and growth and polluted the environment for themselves forever with the waste products of their own metabolism (O_2). Neither the environmental ecology of the world nor the Kingdom of the Plants has ever recovered from this event!

Additional Reading

Articles on photosynthesis, photorespiration, and chloroplast structure and function in *Annual Reviews of Plant Physiology and Biochemistry.*

Bassham, J. A., and **M. Calvin:** *The Path of Carbon in Photosynthesis.* Prentice-Hall, Inc., Englewood Cliffs, N.J., 1957.

Burris, R. H. and **C. C. Black** (eds.): *CO_2 Metabolism and Plant Productivity.* University Park Press, Baltimore, Md., 1976.

Calvin, M., and **J. A. Bassham:** *The Photosynthesis of Carbon Compounds.* W. A. Benjamin, Inc., New York, 1962.

Energy conversion by the photosynthetic apparatus. *Brookhaven Symposia in Biology,* No. 19, 1967.

Gibbs, M. (ed.): *Structure and Function of Chloroplasts.* Springer-Verlag, New York, 1971.

Govindjee (ed.): *Bioenergetics of Photosynthesis.* Academic Press, New York, 1975.

Gregory, R. P. F.: *Biochemistry of Photosynthesis.* John Wiley & Sons Ltd., London, 1971.

Hatch, M. D., C. B. Osmond, and **R. O. Slatyer** (eds.): *Photosynthesis and Photorespiration.* John Wiley & Sons, New York, 1971.

8 Nitrogen Metabolism

Nitrogen Fixation

Although Earth's atmosphere is 80 percent nitrogen, this element is often in short supply to organisms, particularly to plants, because only certain microorganisms are capable of assimilating molecular nitrogen and converting it into forms available to plants. These microorganisms are of four principal types: symbiotic microorganisms living in the roots of certain plants, certain free-living heterotrophic soil bacteria, photosynthetic bacteria, and some photosynthetic blue-green algae. The importance of nitrogen fixation cannot be overestimated—the largest part of the organic nitrogen in the world must come from this process. A figure of 100,000,000 tons per year has been suggested. In North America, the nitrogen fixed probably outweighs the amount applied as fertilizer by a factor of 3 or 4. Not only is the market value of fixed nitrogen (as legumes) over 3 billion dollars per year in the United States, but (unlike the commercial manufacture of fertilizers) it is a nonpolluting process. Nitrogen fixation is so important that much research effort is now being expended to increase the efficiency and rates of nitrogen fixation in crops such as legumes that have this capability. In addition, some scientists are trying, by selection and by genetic modification experiments, to generate new symbiotic associations so that important field crops such as the cereals that do not now fix nitrogen can be made nitrogen autotrophic. The saving in the cost and effort of nitrogen fertilization if a nitrogen-fixing corn or wheat could be developed is inestimable.

Symbiotic Nitrogen Fixation. The first indication that plants could fix atmospheric nitrogen was obtained in 1838 by Boussingault, who showed that legumes could increase the nitrogen content of their soil. In 1886 the German physiologists H. Hellriegel and H. Wilfarth demonstrated that bacteria inhabiting the nodules of legumes were responsible for the process; plants without nodules or grown in sterile soil were unable to fix nitrogen, and could not grow in nitrogen-deficient soil. The legumes are the main group of plants that symbiotically fix nitrogen, the bacterial symbiont being a member of the genus *Rhizobium*. Certain nonleguminous plants also bear root nodules containing microorganisms that can fix nitrogen. Examples are alder (*Alnus* spp.), the bog-myrtle (*Myrica gale*), and sea-buckhorn (*Hippophaë rhamnoides*). The symbionts found in nodules of these plants are probably Actinomycetes, not bacteria.

A number of blue-green algae that fix nitrogen form symbiotic associations with higher plants, including the aquatic fern *Azolla,* some tropical herbaceous plants, and some species of cycads (a primitive coniferous plant). In the latter, a species of *Anabaena* invades roots of the plant which then reverse their geotropic polarity and grow upward to the soil surface where photosynthesis of the nitrogen-fixing symbiont can take place (see also Chapter 27, page 632).

The process of nodulation is very interesting and has been much studied. Apparently the first step is the production by the roots of a substance that attracts *Rhizobium* bacteria, which then secrete a material that appears to contain hormones causing the root hairs to curl into a crook shape. Extracellular enzymes produced by the bacteria may be involved in the invagination of partial destruction of the cell wall of the root hair. The bacteria then invade the root hair and move into the cortex in a threadlike structure that consists of mucilaginous material in which the bacteria are embedded. Cortical cells are stimulated to divide by the infection thread, and these divisions result in polyploid cells. There are also usually some tetraploid cells in the cortex of any normal diploid root. The nodule structure is formed from the repeated division of these polyploid cells. The bacteria that infect the nodule cells increase greatly in size and diversity of shape and stop dividing. The bacteroids, as they are now called, very nearly fill the infected cells. The nodule grows by the division of host cells, and it becomes effectively supplied with vascular tissue. Diagrams illustrating the process of infection are shown in Figure 8-1. The appearance of a typical well-nodulated legume is shown in Figure 8-2.

The *Rhizobium* component of legume nodules is specific at least to the genus of the legume (six different groups are recognized). For successful nodulation it is necessary to have a substantial inoculation of the bacteria. Farmers who practice crop rotation usually include a leguminous crop to maintain the fertility of the soil, and it is a common practice to inoculate the seed or soil with a culture of the appropriate *Rhizobium* at the time of planting or shortly after to ensure the maximum benefits of a nitrogen fixing crop. Nodulation and nitrogen fixation are affected by the nitrogen status of a plant—minimum levels of fixed nitrogen are necessary in the soil after germination to ensure vigorous plants; thereafter the amount of nitrogen fixed is inversely proportional to the amount of fixed nitrogen available. An appropriate level of carbohydrate nutrition, and thus of photosynthesis, is also necessary for effective nitrogen fixation, since the energy for nitrogen fixation is derived from the respiration of carbohydrates. The nitrogen fixed in nodules is rapidly converted to amino acids, a process that requires carbon skeletons resulting from respiratory activity. Organic nitrogen is transferred to the host plant via the xylem, primarily in the form of asparagine in legumes or citrulline in the alder.

Effective nodules were early found to contain a pink pigment called **leghemoglobin** which was shown to be similar to mammalian hemoglobin. This pigment appears to be involved as an oxygen carrier in the process of nitrogen fixation in nodules. It is probably important in the oxidative metabolism that provides the energy for driving nitrogen fixation, but it may also have a role in protecting the nitrogen-fixing enzyme, which is sensitive to oxygen, from the effects of atmospheric oxygen. It does not occur in free-living nitrogen fixers, and it appears to be an ameliorating rather than an essential component of active nodules.

Nonsymbiotic Nitrogen Fixation. Nitrogen is fixed in free-living microorganisms in two important ways: as a photosynthetic reduction of nitrogen by photosynthetic bacteria or blue-green algae and as a nonphotosynthetic process occurring in certain soil

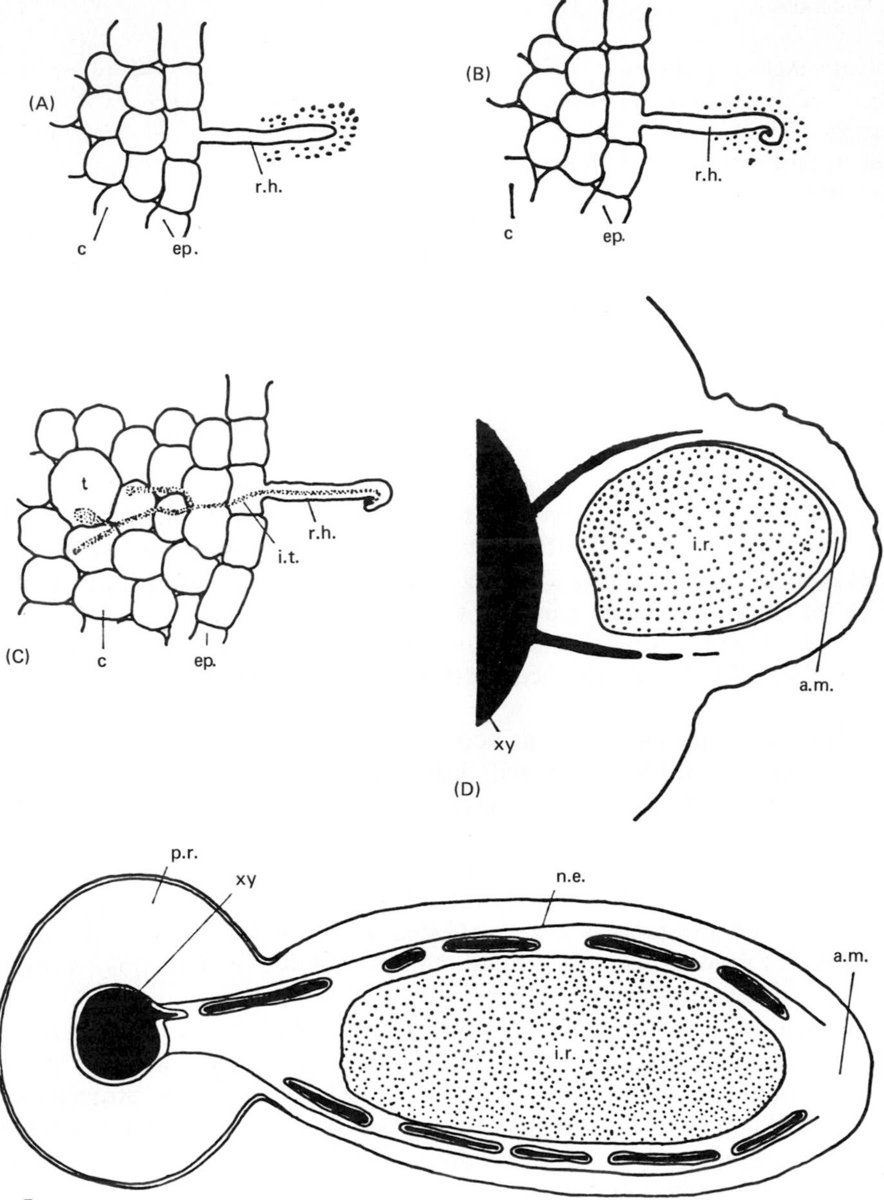

Figure 8-1. Initiation and structure of pea nodules.

A. *Rhizobia* aggregate round root hairs.

B. Root hairs curl.

C. *Rhizobia* infect root hair and move through the root hair and inner cortex until they enter a tetraploid cell. This stimulates meristematic activity.

D. A central infected region and apical meristem become distinguished.

E. Longitudinal section through nodule showing central infected region, apical meristem and nodule endodermis.

Code: a.m., apical meristem; c, cortex; ep., epidermis; i.r., infected region; i.t., infection thread; n.e., nodule endodermis; p.r., primary root; r.h., root hair; t, tetraploid cell; xy, xylem.

[From W. D. P. Stewart: *Nitrogen Fixation in Plants*. Athlone Press, London, 1966. Used with permission.]

A

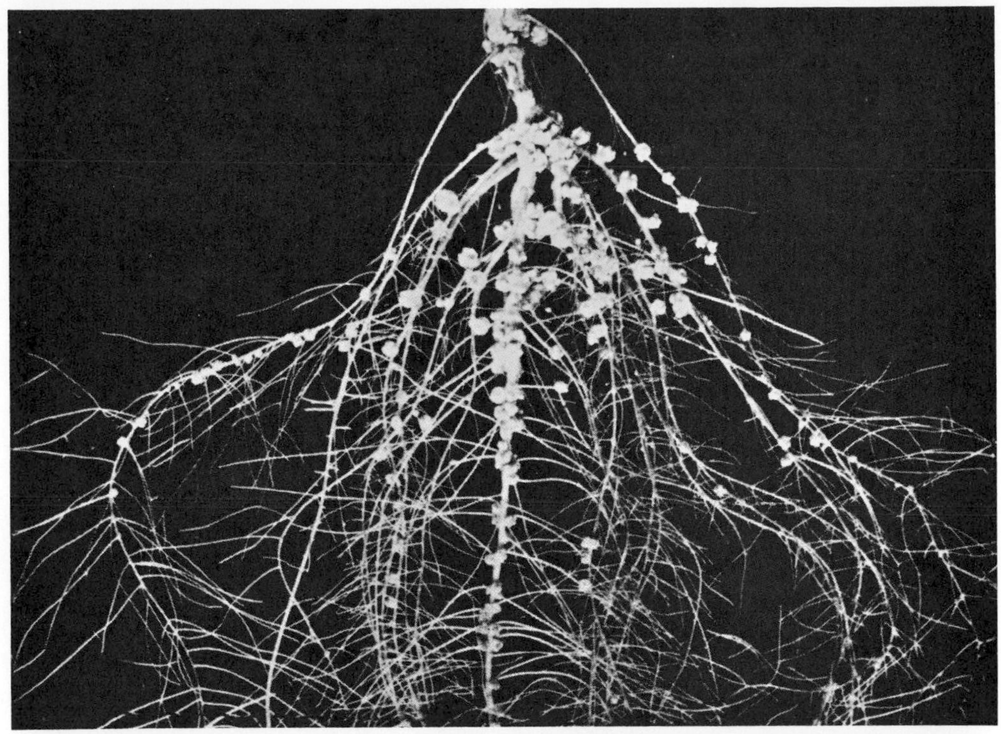

B

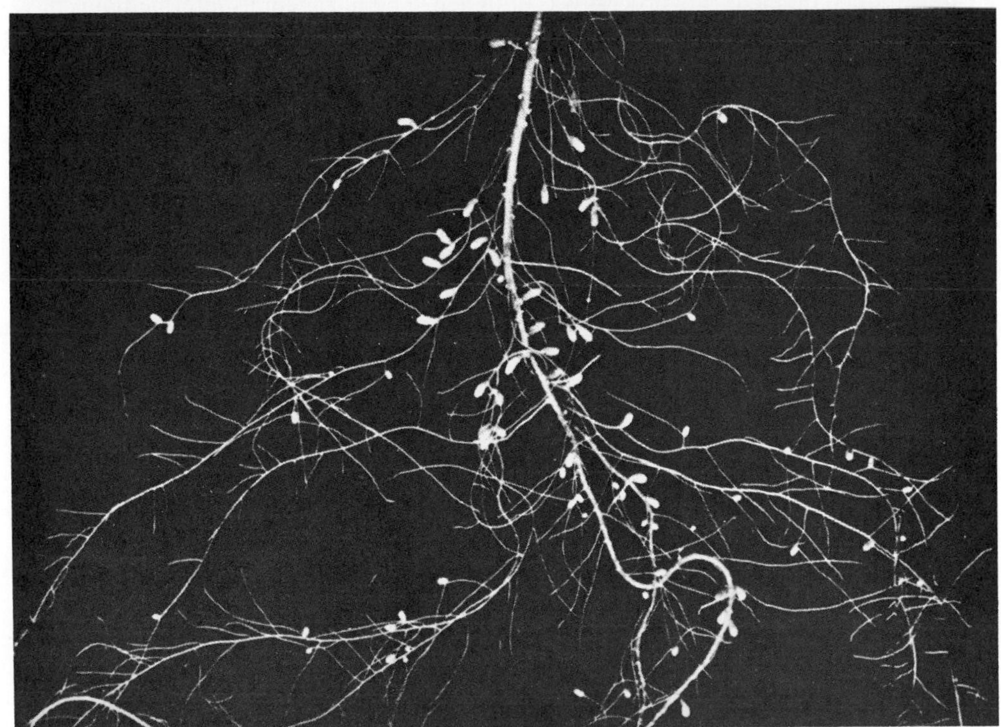

Figure 8-2. Typical appearance of nodulated roots of (**A**) soybean and (**B**) clover. [From W. D. P. Stewart: *Nitrogen Fixation in Plants*. Athlone Press, London, 1966. Used with permission.]

microorganisms. Photosynthetic blue-green algae such as *Anabaena* and *Nostoc* can fix nitrogen by a reaction essentially similar to that used to fix carbon dioxide. A number of green and purple sulfur bacteria as well as nonsulfur photosynthetic bacteria (for example, *Rhodospirillum, Rhodopseudomonas, Chlorobium, Chromatium*) can also fix nitrogen using light energy. These organisms do not derive electrons from water, as do blue-green algae, but from some reduced electron donor such as sulfide, sulfur, hydrogen, or organic compounds (see Chapter 7, page 190). In addition, many of these organisms appear able to fix nitrogen in darkness, when supplied with appropriate reduced electron donors, in a manner that resembles nitrogen fixation by nonphotosynthetic bacteria such as *Azotobacter* or *Clostridium pasteurianum*. Nitrogen fixation in the photosynthetic forms is responsive to light, whereas in nonphotosynthetic forms a supply of carbohydrates or other organic compounds is necessary to provide reducing power and to generate the required ATP. Nitrogen fixation is inhibited by oxygen, hydrogen, and certain poisons such as CO. Usually the presence of organic nitrogen compounds or NH_3 greatly reduces the fixation of molecular nitrogen.

Mechanism of Nitrogen Fixation. For a long time no progress was made in elucidating the mechanism of the nitrogen-fixing reaction or the intermediates in the process because it proved very difficult to obtain cell-free extracts that would fix nitrogen. It was not until the mid-1950s to 1960, when J. E. Carnahan and his group at the Du Pont laboratories isolated the two enzyme systems from *C. pasteurianum,* that a clearer understanding of the process was obtained. The nitrogenous intermediates have been hard to find, both because they are enzyme bound and do not exist free in any quantity in the cell and because no satisfactory radioactive isotope of nitrogen exists. The stable isotope, ^{15}N, has provided some valuable information, but the techniques for its assay are complex and less sensitive than those for radioactive isotopes.

The pathways and mechanisms of nitrogen fixation are not yet fully understood. NH_3 appears to be the final end product. It can be seen from Figure 8-3 that *Azotobacter* cultures utilize NH_3 immediately when it is supplied but require a substantial lag-phase before nitrate (NO_3^-) is utilized. This suggests that NH_3 is a naturally occurring nitrogen compound, whereas the organism has to adapt to the use of nitrate. When $^{15}N_2$ is supplied, $^{15}NH_3$ has the highest isotope content, higher than asparagine, glutamine, or amino acids.

The conversion of N_2 to NH_3 requires the addition of six electrons and six hydrogen ions per molecule of N_2 reduced. This is accomplished by the transfer of reducing power to the enzyme **nitrogenase,** which catalyzes the reduction. The source of electrons may be from respiratory electron transfer, from photosynthesis in autotrophic nitrogen fixers, or by a **phosphoroclastic** split (the insertion of a phosphate molecule) of pyruvate

$$\text{pyruvate} + \text{Pi} \longrightarrow \text{Acetyl-P} + CO_2 + 2H^+ + 2e^-$$

The transfer of electrons to the nitrogenase is not yet clearly understood in higher plants, but is thought to be via NADH and ferredoxin, a nonheme iron enzyme that also functions in photosynthetic electron transport (see Chapter 7). Alternative electron donors are known from various microorganisms, and isolated enzyme preparations can be made to work with various artificial donor systems. NADPH may be involved in photosynthetic nitrogen fixation.

The actual conversion of N_2 to NH_3 takes place on the surface of the nitrogenase, a complex enzyme that appears to contain both molybdenum and iron at its reactive site.

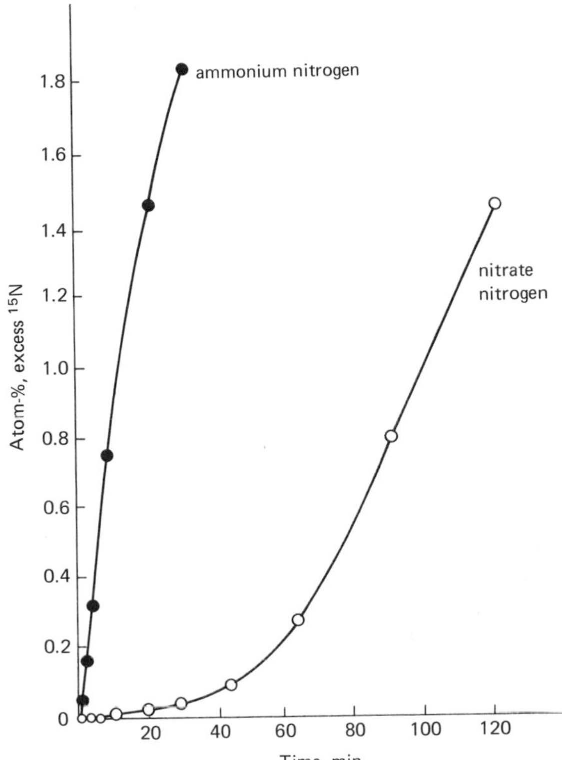

Figure 8-3. Uptake of ^{15}N-labeled ammonium or nitrate nitrogen by *Azotobacter vinelandii*. [From W. D. P. Stewart: *Nitrogen Fixation in Plants*. Athlone Press, London, 1966. Used with permission.]

N_2 apparently bonds to both metals at the reactive site and is then reduced to NH_3 via the sequence

$$N_2 \longrightarrow \underset{\text{diimide}}{N_2H_2} \longrightarrow \underset{\text{hydrazine}}{N_2H_4} \longrightarrow NH_3$$

These intermediates do not occur free, being labile and extremely poisonous, but are bound to the enzyme complex. For reasons that are not clear the reduction reaction requires the hydrolysis of ATP. The required ATP is generated by oxidative metabolism (for example, by pyruvate oxidation via the Krebs cycle). This takes place in the host tissue in nodules. The transfer of O_2 may be facilitated by leghemoglobin. A hypothetical model of the reactive site is shown in Figure 8-4, and a generalized scheme showing the reactions of nitrogen fixation is given in Figure 8-5.

The enzyme nitrogenase is capable of using a number of other substrates that contain triple bonds besides N_2 ($N\equiv N$). These include N_2O, nitriles such as $HC\equiv N$ or $CH_3—C\equiv N$, and alkynes such as acetylene ($CH\equiv CH$) which is reduced to ethylene ($CH_2=CH_2$). The latter reaction has provided a powerful field and laboratory test for N_2-reducing ability. Direct measurement of N_2 reduction is impossible. The use of ^{15}N (which remains the final proof that N_2 is being reduced) is an expensive and cumbersome technique requiring the use of a mass spectrometer, and it is not easily adapted to field use. However, acetylene reduction is easily measured with great sensitivity because ethylene can be detected by biological assays (see Chapter 23) or by the use of the gas chromatograph. This permits rapid analysis in the field as well as in the laboratory, an

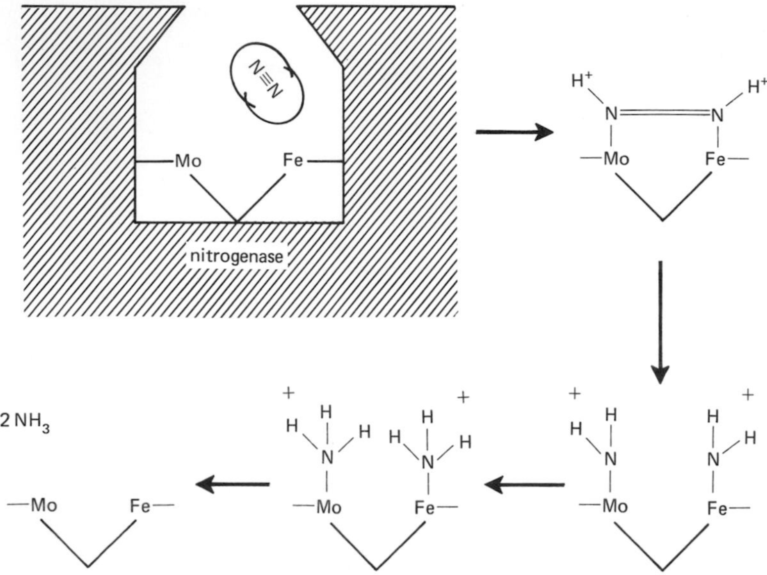

Figure 8-4. Hypothetical reaction site of nitrogenase. The reactive site is in a pocket of the enzyme surface so that only small molecules have access to it. The space between the Mo and Fe atoms is variable and accommodates the changing bond length between the two nitrogen atoms as they are reduced. Electrons may be added in *pairs* or *singly*. [Adapted from R. C. Burns and R. W. F. Hardy: *Nitrogen Fixation in Bacteria and Higher Plants*. Springer-Verlag, New York, 1975.]

Figure 8-5. Outline of the metabolism associated with nitrogen fixation.

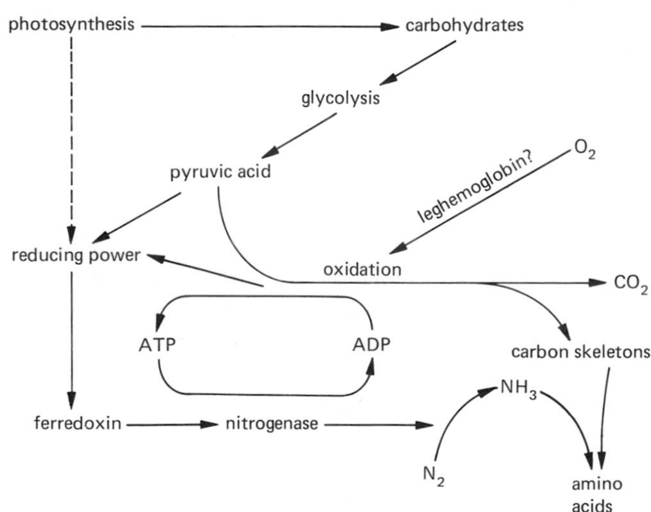

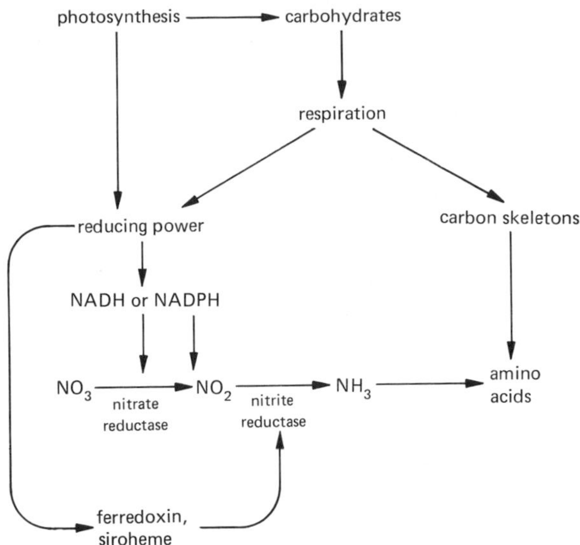

Figure 8-6. A general scheme of the metabolism associated with nitrate reduction.

important development in our never-ending hunt for crop systems that have higher productivity.

An important point brought out in Figure 8-6 is the relationship between demand for photosynthate and nitrogen fixation. If the plant fixes too much nitrogen, it will be short of carbohydrate for growth and the result will be the same as if the plant was short of nitrogen—it will be stunted. This stresses the importance of a well-regulated balance between photosynthesis, growth, and nitrogen fixation in plants.

Nitrate Reduction

Mechanism of Nitrate Reduction. NH_3 produced by nitrogen fixation or fertilization is rapidly converted to nitrate by bacterial action in the soil, and thus most nitrogen is available to the plant in the form of the highly oxidized nitrate ion (NO_3^-). Since almost all organic nitrogen compounds contain, or are made from, fully reduced nitrogen (NH_3), the plant must reduce nitrate in order to make it available for organic synthesis. As in nitrogen fixation, the study of the intermediates of nitrate reduction has been difficult because of the difficulty of isolating the enzymes from leaves and because of the poisonous nature of the intermediate states of nitrogen.

To convert NO_3^- to NH_3, a total of eight electrons must be added per molecule. Many possible intermediate compounds of nitrogen have been suggested, and the most likely series is as follows

$$\underset{\text{nitrate}}{NO_3^-} \longrightarrow \underset{\text{nitrite}}{NO_2^-} \longrightarrow \underset{\text{hyponitrite}}{N_2O_2^{2-}} \longrightarrow \underset{\text{hydroxylamine}}{NH_2OH} \longrightarrow \underset{\text{ammonia}}{NH_3}$$

where each step is accomplished by the transfer of two electrons. However, only two enzymes are required to accomplish this reduction. The first is **nitrate reductase,** which catalyzes the conversion of nitrate to nitrite (NO_2^-). Nitrite is then reduced all the way to ammonia by **nitrite reductase** without the appearance (or release) of any identifiable intermediates.

Nitrate reductase is now well documented. The reduction of nitrate proceeds at the expense of NADH or NADPH oxidation, coupled by FAD to the reductase, a molybdenum enzyme

$$\begin{array}{ccccccc} \mathrm{NADH_2^+}\ \text{or}\ \mathrm{NADPH_2^+} & & \mathrm{FAD} & \overset{2\,\mathrm{H^+}}{\uparrow} & 2\,\mathrm{Mo^{5+}} & \overset{2\,\mathrm{H^+}}{\downarrow} & \mathrm{NO_3} \\ & \times & & \times & \text{reductase} & \times & \\ \mathrm{NAD^+}\ \text{or}\ \mathrm{NADP^+} & & \mathrm{FADH_2} & & 2\,\mathrm{Mo^{6+}} & & \mathrm{NO_2 + H_2O} \end{array}$$

Nitrite reductase is less well known. This enzyme appears to be linked with ferredoxin in some tissues, but some reduction will proceed at the expense of FMN in roots. The functional metal is probably iron. Recently an iron-containing porphyrin called **siroheme** has been isolated from plant and bacterial sources. This electron transfer agent is capable of mediating the complete nitrite reductase reaction from NO_2^- to NH_3.

In addition to nitrite reductase, which converts NO_2^- all the way to NH_3, reductases that reduce either nitric oxide (NO) or hydroxylamine (NH_2OH) have been described from some plant sources. However, the association of these enzymes with the normal path of nitrate reduction in plants has not been demonstrated. These intermediates are so poisonous that it seems unlikely that they would ever occur free in the cell as components of a major metabolic pathway.

Nitrate Reduction and Metabolism. (See Figure 8-6.) The energy and reducing power for nitrate reduction must be derived from photosynthetic or oxidative metabolism, frequently the respiration of carbohydrates. In addition, when large amounts of nitrate are reduced, the resulting ammonia must be rapidly combined with carbon skeletons to form organic nitrogen compounds, otherwise toxic quantities of ammonia would accumulate. Thus the massive accumulation of nitrate may tax a plant's carbohydrate reserves and photosynthetic capacity quite substantially, resulting in either stunted or strongly vegetative growth.

Nitrate reduction is strongly stimulated by light. The reducing enzymes are often found to be NADPH linked. However, although nitrite reductase is thought to be located in chloroplasts, nitrate reductase is known to be present in the cytoplasm. In many plants CO_2 and nitrate compete for reducing power, particularly at limiting light intensities. At high light intensities the addition of nitrate to a photosynthesizing leaf or cell is likely to increase oxygen production. All these lines of evidence indicate that there is a direct utilization of photosynthetically generated reducing power for the reduction of nitrate or nitrite in photosynthetic organisms. However, this is not obligatory because many plants reduce nitrate in the roots, and many nonphotosynthetic organisms are also capable of nitrate and nitrite reduction. In addition, both nitrate and nitrite reduction may take place in photosynthetic organisms in the dark.

It has been found in some organisms that photosynthetic nitrate reduction will not proceed in the absence of CO_2. This suggests the need for concomitant carbohydrate metabolism, even if some or all of the reducing power is derived directly from photosynthetically reduced NADPH or ferredoxin. It is possible that, in addition to reduced pyridine nucleotides, nitrate or nitrite reduction also requires ATP, which might need to be derived from respiration. Evidence indicates that nitrate reduction may be strongly stimulated by blue light. Flavin pigments absorb blue light, and the possibility exists either that flavins may be photochemically reduced or that flavoprotein enzymes may require light activation for maximum efficiency of operation.

Nitrate reductase is strongly induced by light, by the presence of nitrate, and in some plants by certain hormones such as gibberellic acid and cytokinins. Unlike many enzymes, those mediating the reduction of nitrate and nitrite do not appear to be inhibited by their end product, NH_3. On the other hand, a build up of NH_3 in cells will reduce the activity of nitrate reductase by inhibiting the production of NADPH or NADH.

A scheme summarizing the process of nitrate reduction is shown in Figure 8-6, emphasizing the possible relationships between photosynthesis, respiration, and nitrate reduction.

Absorption of Nitrogen by Plants

Plants absorb nitrogen in four important forms: as nitrate, ammonium ion, organic compounds such as amino acids, and as urea. Nitrate is the most abundant available form of nitrogen and the most important source to plants. Ammonia is sometimes relatively abundant, for example where nitrogen fixation is going on, or in wet, anaerobic soils. Ammonia is toxic, however, and its uptake in large quantities may put a severe strain on the carbohydrate metabolism of the plant in the provision of carbon skeletons for its detoxification. Organic nitrogen, usually in the form of amino acids, may become available to the plant due to the death and decay of plant or animal matter. Under these circumstances plants are competing with bacteria for the nitrogen, which the bacteria will normally convert to molecular nitrogen (N_2) or nitrate in the course of their metabolism. Organic nitrogen does not normally constitute a major source of nitrogen for plants. Likewise urea is not normally important, but it has been found to be an effective fertilizer that can be applied as a foliar spray. Urea is absorbed through the leaves, thus can be cheaply applied to crops, together with pesticides. In this way the waste from fertilizer uptake by weeds or loss to soil runoff, a major problem in nitrogen fertilization today (see Chapter 30), is greatly reduced.

Inorganic Nitrogen. Nitrate is absorbed through the roots and either reduced there or transported to the leaves prior to reduction. Many plants, such as tomato, normally reduce nitrate in the roots, unless the soil is chilled when it is transported to the aerial portions of the plant. In other plants, such as grasses, nitrate is normally transported to leaves and may accumulate in large amounts there, being reduced as required. Nitrate reduction will normally proceed more rapidly in the daytime than at night because of the availability of carbon substrates and reducing power from photosynthesis.

Ammonia is used by many plants and preferentially by a few. Those plants that are able to absorb it in large amounts include many acid plants, such as *Rumex,* which are able to detoxify ammonia by forming ammonium salts of organic acids. Certain other plants, known as **amide plants** (including beet, spinach, squash) are able to form large amounts of the amides glutamine or asparagine. These compounds are formed from the corresponding dicarboxylic amino acids as shown on the next page.

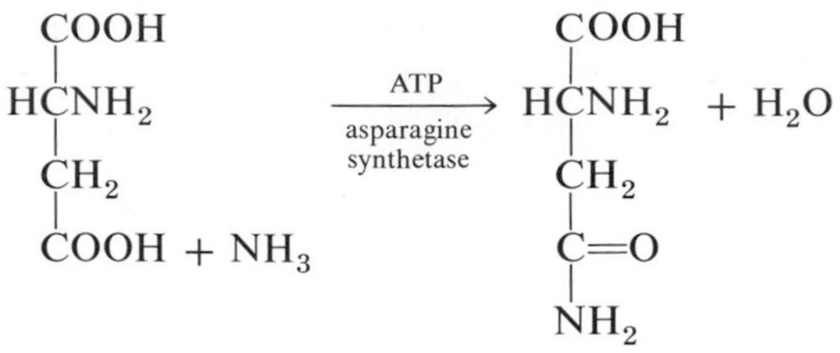

$$\begin{array}{c} \text{COOH} \\ | \\ \text{HCNH}_2 \\ | \\ \text{CH}_2 \\ | \\ \text{COOH} + \text{NH}_3 \end{array} \xrightarrow[\text{asparagine synthetase}]{\text{ATP}} \begin{array}{c} \text{COOH} \\ | \\ \text{HCNH}_2 \\ | \\ \text{CH}_2 \\ | \\ \text{C{=}O} \\ | \\ \text{NH}_2 \end{array} + H_2O$$

aspartic acid — asparagine

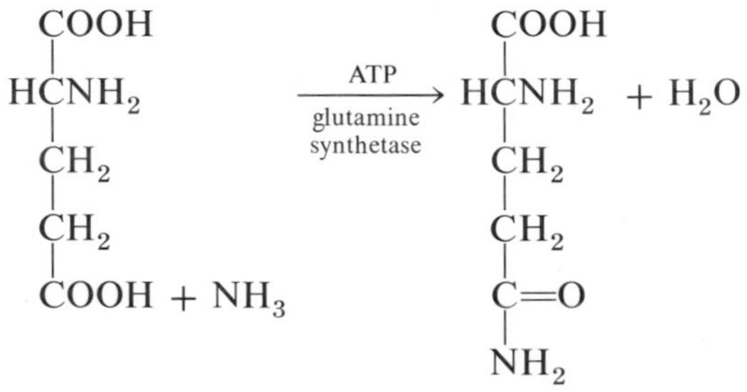

$$\begin{array}{c} \text{COOH} \\ | \\ \text{HCNH}_2 \\ | \\ \text{CH}_2 \\ | \\ \text{CH}_2 \\ | \\ \text{COOH} + \text{NH}_3 \end{array} \xrightarrow[\text{glutamine synthetase}]{\text{ATP}} \begin{array}{c} \text{COOH} \\ | \\ \text{HCNH}_2 \\ | \\ \text{CH}_2 \\ | \\ \text{CH}_2 \\ | \\ \text{C{=}O} \\ | \\ \text{NH}_2 \end{array} + H_2O$$

glutamic acid — glutamine

The details of these reactions will be described later (page 211). Swiss chard or beet plants, for example, can withstand quite high concentrations of ammonium salts, detoxifying the ammonia by the manufacture of large amounts of glutamine. Wheat leaves under similar circumstances use the carbon of available sugars for the manufacture of asparagine. Plants that do not accumulate glutamine may nevertheless use this reaction to accumulate organic nitrogen by transferring the amide nitrogen immediately to α-ketoglutaric acid to make glutamic acid. This reaction is described on page 204.

Some plants, when supplied with both ammonium (NH_4^+) and nitrate (NO_3^-) ions in liquid culture, will take either the anion or cation depending on the pH. If the culture solution is basic, the plant will absorb NH_4^+, eliminating H^+ by exchange, which thus lowers the pH by forming nitric acid (HNO_3) with the nitrate left behind. Conversely, if the pH is acidic, the plant will absorb NO_3^-, eliminating OH^- in exchange which raises the pH by forming ammonium hydroxide (NH_4OH). Seedlings and very young plants tend to absorb NH_4^+ preferentially, whereas mature plants absorb NO_3^-. This may relate to the greater abundance of carbohydrates and reducing power in the mature actively photosynthesizing plant. Certain plants such as rice, which live in water-logged, anaerobic soils, require NH_3 or reduced organic nitrogen fertilizer and cannot survive on nitrate alone.

Organic Nitrogen. It was found in the 1940s that urea applied to plants could be absorbed directly through the leaves as well as by the roots. It seems probable that the urea is directly hydrolyzed to NH_3 and CO_2 by the urease-mediated reaction

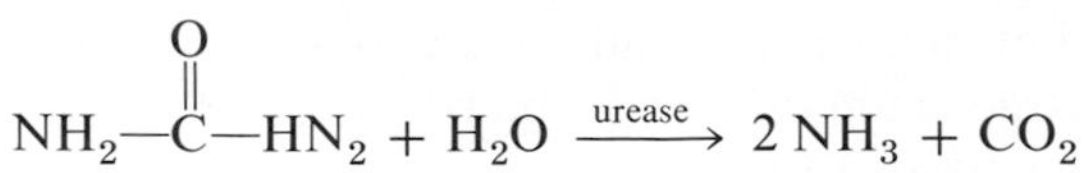

$$\underset{\text{urea}}{NH_2{-}\overset{\overset{\displaystyle O}{\|}}{C}{-}HN_2} + H_2O \xrightarrow{\text{urease}} \underset{\text{ammonia}}{2\,NH_3} + CO_2$$

Plants supplied with ^{14}C-labeled urea or $^{14}CO_2$ showed the same pattern of carbon incorporation. Urea may be incorporated directly (for example, by condensation with ornithine to form arginine). It has also been suggested that urea may be converted directly into carbamyl phosphate, a precursor of pyrimidines and the amino acid citrulline

$$\underset{\text{urea}}{NH_2{-}\overset{\overset{\displaystyle O}{\|}}{C}{-}NH_2} + Pi \longrightarrow \underset{\text{carbamyl phosphate}}{NH_2{-}\overset{\overset{\displaystyle O}{\|}}{C}{-}O{-}P} + NH_3$$

Organic nitrogen in the form of amines or amino acids can be absorbed and utilized, and many plants benefit strongly from its application. Certain amino acids may be toxic in large amounts, however. Cultured plant cells or explants often require specific organic nitrogen sources, such as asparagine, glutamine, glycine, or other amino acids. Carnivorous plants such as sundew, Venus flytrap, or pitcher plants unquestionably utilize the breakdown products of protein from the insects they catch. It was noted as long ago as 1829 that the daily nourishment of a Venus flytrap plant with strips of rump steak greatly improved its growth!

Amino Acids

Amino acids are the building blocks of proteins, and in plants they serve a number of additional functions in the regulation of metabolism and the transport and storage of nitrogen (see Chapter 2, pages 31–35 and Figure 2-7). Amino acids may be formed (1) directly from ammonia and appropriate carbon skeletons, (2) by transamination from existing amino acids, or (3) by modifications or changes to the carbon skeletons of already formed amino acids. Several common amino acids are made by more than one pathway, and many of the pathways are not completely known as yet. It appears likely that specific pathways of metabolism are associated with particular physiological conditions or activities of the plant, such as its state of nutrition, growth, or development. This matter will be further discussed later in this chapter (page 213). Many complex biochemical reactions are required for amino acid biosynthesis and metabolism; we shall consider the main ones here.

Formation of Organic Nitrogen. There are two main pathways of entry of NH_3 into the organic union. The first is via **reductive amination,** the reverse of oxidative deamination. The important reaction in this pathway is the formation of glutamic acid from the corresponding keto acid, α-ketoglutaric acid.* The reaction is catalyzed by glutamic acid dehydrogenase. The first step is the apparently spontaneous reaction of the keto acid with NH_3 to form α-iminoglutaric acid, which is then reduced to the α-amino acid under the influence of the enzyme,

*Also called 2-oxoglutaric acid.

$$\underset{\alpha\text{-ketoglutaric acid}}{\begin{array}{c}COOH\\|\\C{=}O\\|\\CH_2\\|\\CH_2\\|\\COOH\end{array}} + NH_3 \longrightarrow \underset{\alpha\text{-iminoglutaric acid}}{\begin{array}{c}COOH\\|\\C{=}NH\\|\\CH_2\\|\\CH_2\\|\\COOH\end{array}} \xrightarrow[\text{glutamic acid dehydrogenase}]{\nearrow H_2O \quad NADH_2^+} \underset{\text{glutamic acid}}{\begin{array}{c}COOH\\|\\HC{-}NH_2\\|\\CH_2\\|\\CH_2\\|\\COOH\end{array}} + NAD^+$$

This is the only reaction of its kind that has been well authenticated. Parallel reactions leading to the formation of alanine from pyruvate and of aspartate from oxaloacetate have been suggested. The Russian physiologist V. L. Kretovitch has found that certain plant tissues or homogenates respond strongly to the addition of pyruvate and NH_3 by the synthesis of alanine, but the possibility of the amination of α-ketoglutaric and its transamination with pyruvate cannot be ruled out.

The glutamic dehydrogenase pathway of amino group synthesis is not very satisfactory because it has a rather low affinity for NH_3 and its reaction rate is slow for such an important metabolic step. A newly discovered enzyme, **glutamic acid synthetase,** answers these questions, since it has a high affinity for NH_3 and reacts at the required speed. It is coupled with glutamine synthetase to make glutamic acid by the following sequence

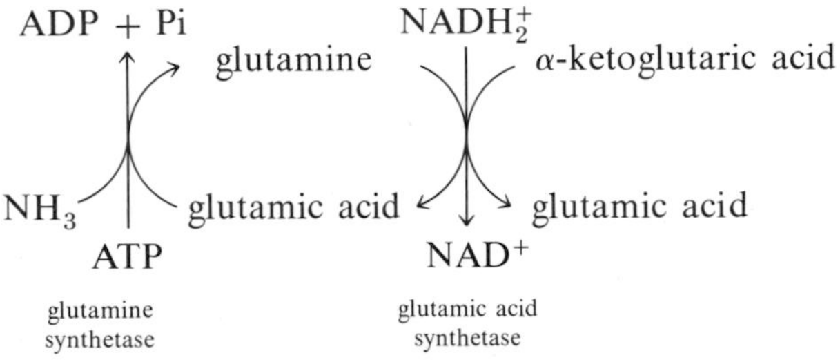

ADP + Pi ↑ glutamine NADH$_2^+$ α-ketoglutaric acid

NH_3 glutamic acid ← → glutamic acid

ATP NAD$^+$

glutamine synthetase — glutamic acid synthetase

The source of reducing power for glutamic acid synthetase may be NADH, as shown, or ferredoxin in photosynthetic tissue. The net reaction is the conversion of α-ketoglutaric acid and NH_3 into glutamic acid; only small amounts of glutamine are required:

$$NH_3 + \alpha\text{-ketoglutaric acid} + ATP + NADH_2^+ \xrightarrow{\text{(glutamine)}} \text{glutamic acid} + ADP + Pi + NAD^+$$

This sequence is associated with nitrogen fixation, and these enzymes, rather than glutamate dehydrogenase, are commonly found in nitrogen-fixing organisms.

Aspartic acid may be synthesized by the **aspartase** reaction, which adds NH_3 to the double bond of fumaric acid, in a manner analogous to the synthesis of malate by the addition of water to the double bond of fumaric acid.

$$\begin{matrix} COOH \\ | \\ CH \\ \| \\ CH \\ | \\ COOH \end{matrix} + NH_3 \xrightarrow{\text{aspartase}} \begin{matrix} COOH \\ | \\ HC{-}NH_2 \\ | \\ CH_2 \\ | \\ COOH \end{matrix}$$

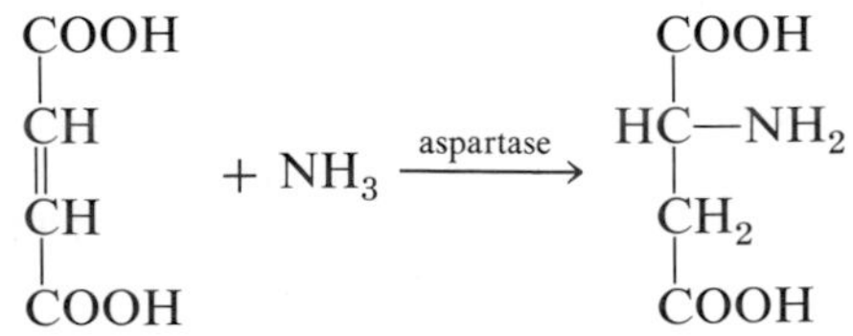

fumaric acid — aspartic acid

The reaction was discovered in bacteria. It has been found in seedlings and leaves of green plants, but it does not seem a likely major pathway for the entry of NH_3 into organic union.

Transamination. In this important reaction the amino group of one amino acid is transferred to a keto acid to form a new amino acid. Enzymes catalyzing this reaction are called **transaminases,** or **aminotransferases.** The coenzyme of this reaction is **pyridoxal phosphate,** which participates in the reaction as follows

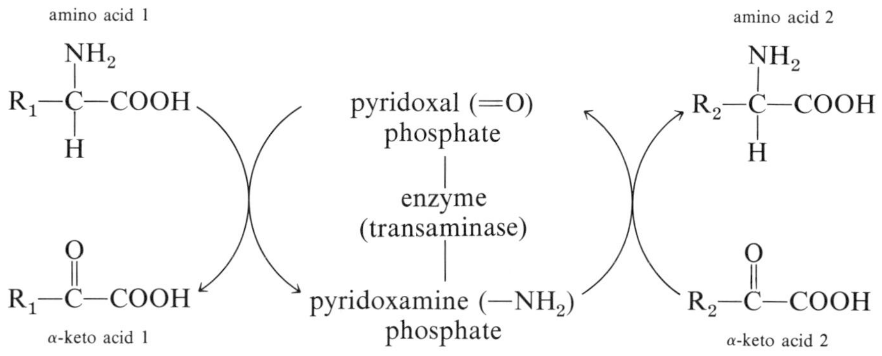

amino acid 1: $R_1-CH(NH_2)-COOH$

amino acid 2: $R_2-CH(NH_2)-COOH$

pyridoxal (=O) phosphate
|
enzyme
(transaminase)
|
pyridoxamine ($-NH_2$) phosphate

α-keto acid 1: $R_1-C(=O)-COOH$

α-keto acid 2: $R_2-C(=O)-COOH$

As a specific example

$$\underset{\text{glutamic acid}}{HOOC-CH(NH_2)-CH_2-CH_2-COOH} + \underset{\text{pyruvic acid}}{HOOC-C(=O)-CH_3} \xrightarrow{\text{alanine aminotransferase}} \underset{\alpha\text{-ketoglutaric acid}}{HOOC-C(=O)-CH_2-CH_2-COOH} + \underset{\text{alanine}}{HOOC-CH(NH_2)-CH_3}$$

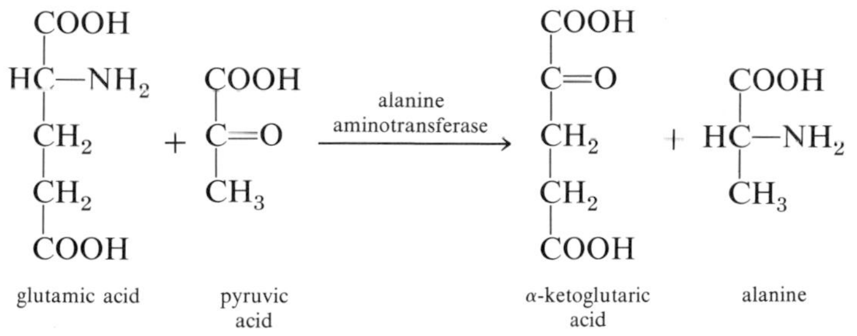

The end result is the transfer of the amino group from amino acid 1 to keto acid 2, with the formation of amino acid 2. The most widespread amino donor is glutamic acid. Glutamic acid–oxaloacetic acid (aspartic synthesizing) and glutamic acid–pyruvate (alanine synthesizing, illustrated above) transaminases are the most active in plants. Some transaminases are quite specific as to nitrogen donor or acceptor.

Since many of the enzymes have not been isolated in pure form, it is not clear precisely what their specificity is or whether nonspecific enzymes are in fact enzyme mixtures. Transaminase activity between glutamic acid and at least 17 or 18 α-keto acids is known to occur in plants, and several others involving α-, β-, or γ-keto acids are suspected. Thus it is possible that a whole range of amino acids may be made by transamination from glutamic acid. Aspartic acid and alanine are also effective amino donors in transamination, and it is probable that several others may participate in lesser degree. The nonprotein amino acid γ-aminobutyrate has been found to be active as an amino donor in transamination.

Carbon Transformations. Many amino acids are formed from preexisting amino acids by modification of their basic carbon skeleton or by the substitution of various groups on their carbon chain. The formation of certain other amino acids requires the synthesis of the appropriate α-keto acid, often by similar or parallel reactions, prior to amino acid synthesis by transamination.

Characteristic types of reactions are listed below.

1. Decarboxylation: Example

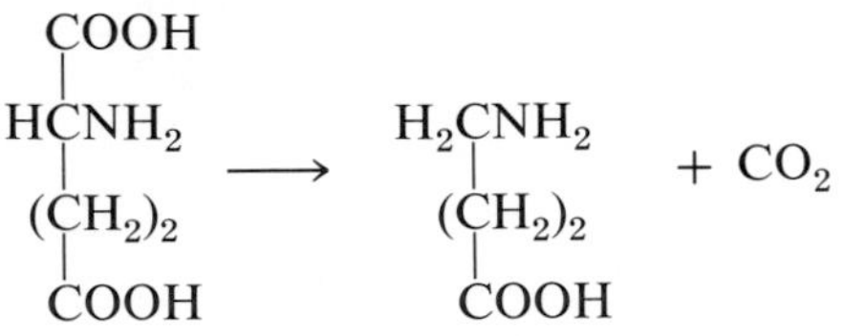

$$\begin{array}{c}COOH\\ |\\ HCNH_2\\ |\\ (CH_2)_2\\ |\\ COOH\end{array} \longrightarrow \begin{array}{c}H_2CNH_2\\ |\\ (CH_2)_2\\ |\\ COOH\end{array} + CO_2$$

glutamic acid — γ-aminobutyric acid

2. Reduction: Example

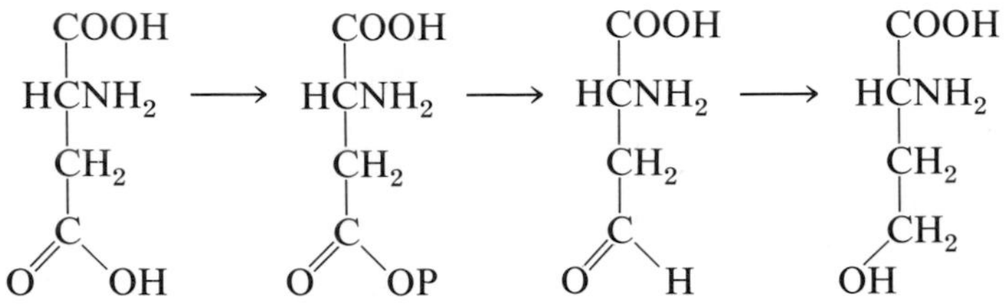

$$\begin{array}{c}COOH\\ |\\ HCNH_2\\ |\\ CH_2\\ |\\ C(=O)OH\end{array} \longrightarrow \begin{array}{c}COOH\\ |\\ HCNH_2\\ |\\ CH_2\\ |\\ C(=O)OP\end{array} \longrightarrow \begin{array}{c}COOH\\ |\\ HCNH_2\\ |\\ CH_2\\ |\\ C(=O)H\end{array} \longrightarrow \begin{array}{c}COOH\\ |\\ HCNH_2\\ |\\ CH_2\\ |\\ CH_2\\ |\\ OH\end{array}$$

aspartic acid — β-aspartyl phosphate — aspartic semialdehyde — homoserine

3. Oxidation: Example

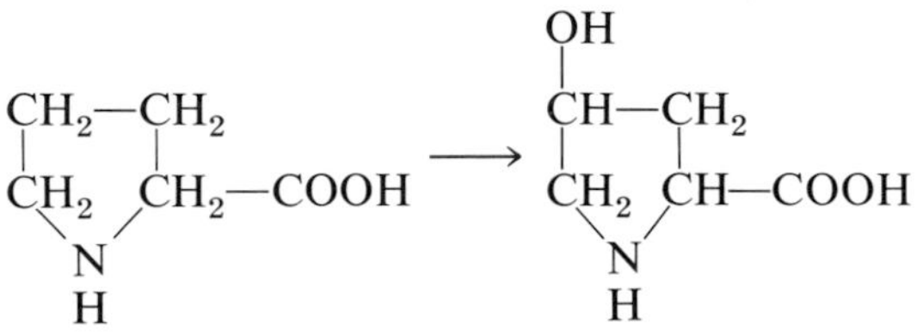

$$\begin{array}{l}CH_2—CH_2\\ |\qquad\quad |\\ CH_2\quad CH_2—COOH\\ \quad\diagdown\ \diagup\\ \quad\ N\\ \quad\ H\end{array} \longrightarrow \begin{array}{l}OH\\ |\\ CH—CH_2\\ |\qquad\quad |\\ CH_2\quad CH—COOH\\ \quad\diagdown\ \diagup\\ \quad\ N\\ \quad\ H\end{array}$$

proline — hydroxyproline

4. Substitution: Example

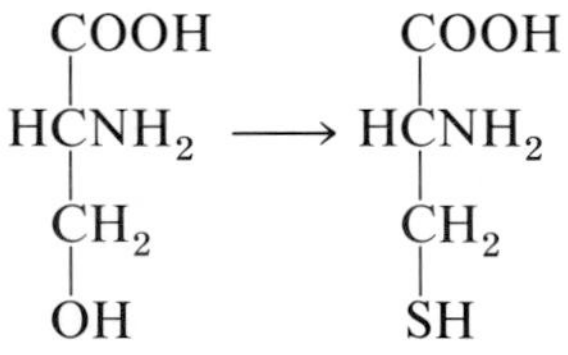

$$\begin{array}{c}COOH\\ |\\ HCNH_2\\ |\\ CH_2\\ |\\ OH\end{array} \longrightarrow \begin{array}{c}COOH\\ |\\ HCNH_2\\ |\\ CH_2\\ |\\ SH\end{array}$$

serine — cysteine

5. Internal group transfer: Example

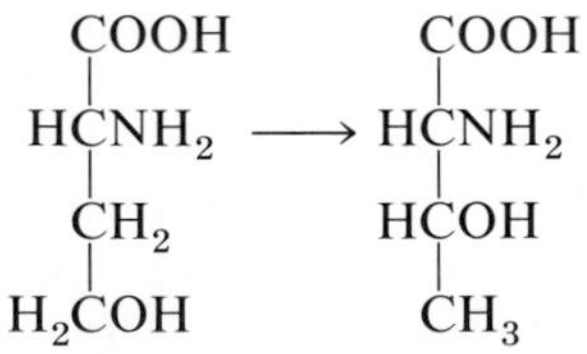

$$\begin{array}{c}COOH\\ |\\ HCNH_2\\ |\\ CH_2\\ |\\ H_2COH\end{array} \longrightarrow \begin{array}{c}COOH\\ |\\ HCNH_2\\ |\\ HCOH\\ |\\ CH_3\end{array}$$

homoserine — threonine

6. Condensation: Example

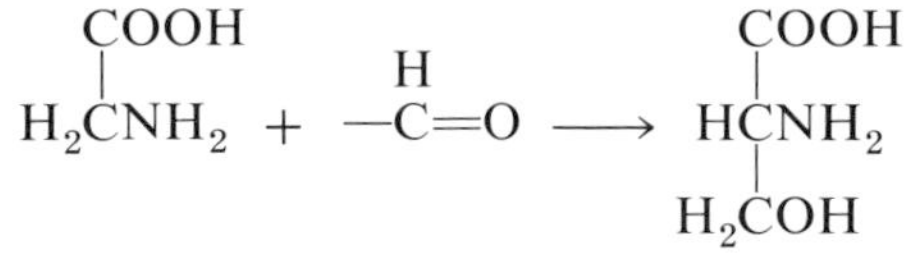

$$\begin{array}{c} COOH \\ | \\ H_2CNH_2 \end{array} + \begin{array}{c} H \\ | \\ -C{=}O \end{array} \longrightarrow \begin{array}{c} COOH \\ | \\ HCNH_2 \\ | \\ H_2COH \end{array}$$

glycine — formaldehyde derivative — serine

Condensations involving such donors as acetyl-CoA are important in the synthesis of several long chain or branched chain amino acids.

7. Ring formation: Example

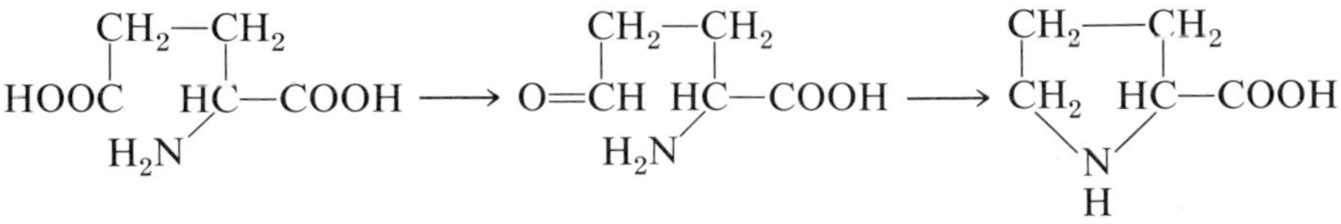

$$\begin{array}{l} CH_2{-}CH_2 \\ HOOC\quad HC{-}COOH \\ \quad H_2N \end{array} \longrightarrow \begin{array}{l} CH_2{-}CH_2 \\ O{=}CH\quad HC{-}COOH \\ \quad H_2N \end{array} \longrightarrow \begin{array}{l} CH_2{-}CH_2 \\ CH_2\quad HC{-}COOH \\ \quad\ N \\ \quad\ H \end{array}$$

glutamic acid — glutamic semialdehyde — proline

These are generalized reactions, and many parallel sequences are known that lead to other amino acids or their α-keto precursors.

The extensive early work on *Escherichia coli* by the American bacteriologist P. H. Abelson and his associates showed that there are groups or "families" of amino acids derived from one precursor of "family head" by various reactions in that organism, and subsequent work has shown this to be true also for higher plants. Typical experiments have been done by supplying one amino acid labeled with ^{14}C and determining the range of compounds that become labeled as the result. All the labeled compounds must have been derived from the fed precursor. Compounds with higher specific activities are more directly derived than compounds of lower specific activities.

Branch points in chains can be detected by **isotope competition** experiments. In this technique, a labeled precursor is fed in parallel experiments either alone or together with the unlabeled compound suspected to be at or just prior to the branch point. All those compounds derived from the branch that include the fed unlabeled compound will have their specific activities reduced (that is, they will be diluted) in the samples that received the nonradioactive competitor. This technique is illustrated in Figure 8-7.

The pathways of synthesis of many groups of amino acids have been worked out in detail, and the reactions are precisely known. Others are only partially known, or else the pathways of carbon are known without the specific reactions having been studied in vitro. Since the same amino acid may be made by more than one pathway via different intermediates, the "family relationship" sometimes becomes slightly confused. The main relationships are presented in Figure 8-8. It must be recognized that under certain circumstances and in certain organisms somewhat different patterns may be obtained. However, the ones shown here can be considered general. The importance of the "family" concept is that biochemical regulating mechanisms must exist to control the flow of carbon into the various pathways. Imbalance of this biochemical regulation system leads to serious problems in metabolism.

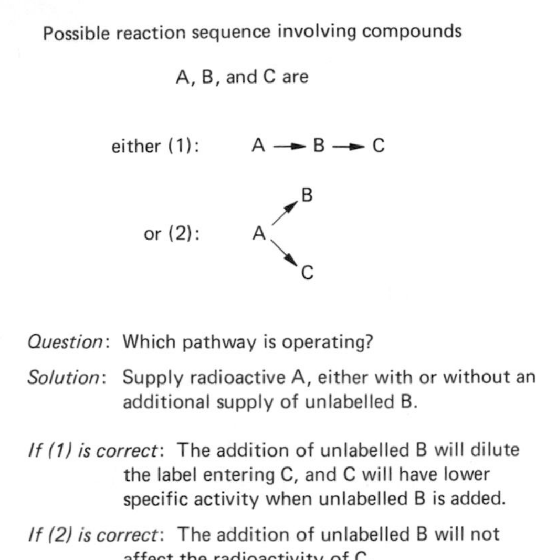

Figure 8-7. Isotope competition experiment to determine a metabolic pathway.

Some Metabolic Patterns. Amino acids are not usually involved in such major metabolic activities as the Krebs cycle or glycolysis as organic acids are because the special function of amino acids in the cell is the transformation and metabolism of nitrogen, not energy. However, there are several important biosynthetic or metabolic sequences that do involve nitrogen compounds in addition to the pathways of synthesis and degradation of amino acids themselves. The glycolate pathway of sugar synthesis, involving glycine and serine, has already been mentioned (see page 174 and Figure 7-15). It has also been suggested that aspartate as well as malate or oxaloacetate may act as the carboxyl-transporting C_4 acid in the Hatch and Slack pathway of photosynthesis (see Figure 7-19). Another metabolic system involving amino acids is the ornithine cycle, which makes urea in animals (Figure 8-9). All the reactions of this cycle are known in plants, but it is not clear whether the cycle functions, as such, in plants.

Methionine is widely involved in **transmethylation** reactions in the synthesis of methylated compounds (for example, thymine, methylated phenols, alkaloids, lignin, and so on). Tryptophan is a precursor of the growth hormone indoleacetic acid, losing its amino group by transamination and undergoing decarboxylation of the pyruvate side chain to acetate. Proline and hydroxyproline hold an interesting relationship. Hydroxyproline is not formed in the soluble state in plants and is, in fact, quite toxic. Proline is first incorporated into protein, then converted to hydroxyproline, where it is involved in the crosslinking by weak bonds that helps to establish the tertiary structure of proteins.

Amides

The two amides, glutamine and asparagine, play a central role in the nitrogen metabolism of plants. Both the amino and amido groups of glutamine and asparagine are involved in many specific and general reactions of nitrogen. The amides are major

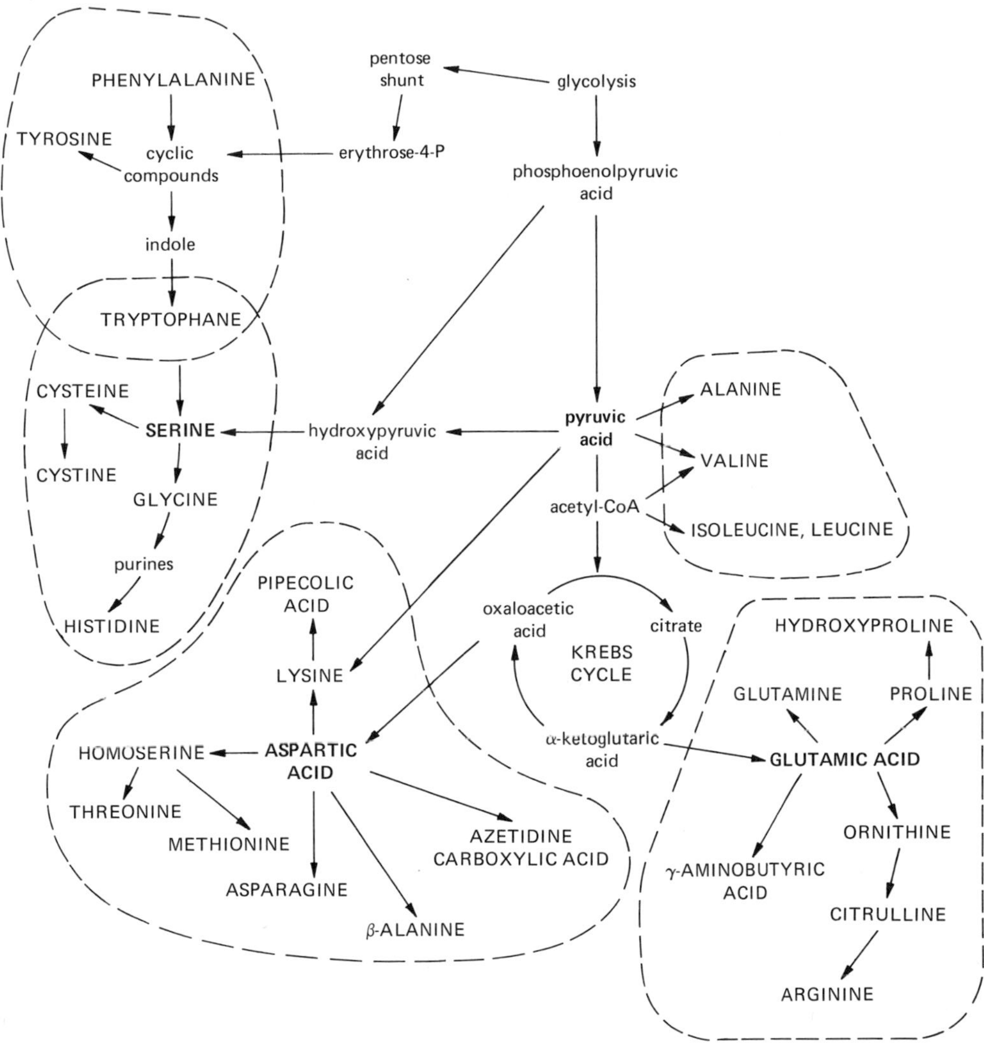

Figure 8-8. Metabolic relationships of amino acids. Major "families" are circled with dotted lines and family heads are printed in boldface type. Amino acids are printed in capitals, other compounds in lower case type.

nitrogen storage and transport compounds and may achieve extremely high concentrations under appropriate conditions. Glutamine and asparagine are homologs, apparently differing only by one carbon in chain length, yet their behavior is not that of homologs. It has been suggested that the chemical structure of asparagine is not correctly represented by the structural formula commonly used, but plausible alternatives are not supported by experimental data.

Nevertheless, glutamine and asparagine differ widely in biochemical and physiological behavior, and the reason for these differences is not clear. Asparagine is rather insoluble in water; glutamine is very soluble. Asparagine is stable and requires heating in moderately strong acid solution to hydrolyze it; glutamine is most unstable, hydro-

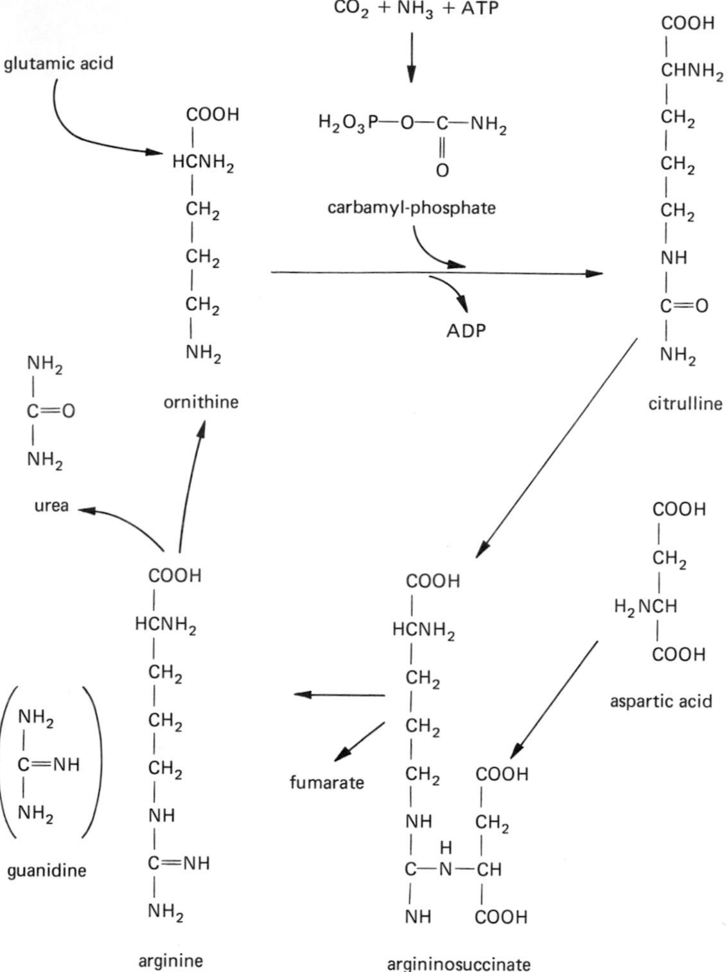

Figure 8-9. The ornithine cycle.

lyzing slowly at room temperature and being completely hydrolyzed to glutamic acid in boiling water or dilute acid solutions. The ninhydrin reaction, used to visualize amino acids on chromatograms and in their quantitative colorimetric measurement, is quite different for asparagine and glutamine. Glutamine liberates carbon dioxide and gives a purple color, like most other amino acids, whereas asparagine does not immediately liberate carbon dioxide and gives a brown color similar to that of proline and other cyclic imino acids. No satisfactory explanation is available for these differences in chemical behavior, but they may well be related to the differences in biological behavior of the two amides.

Synthesis. Glutamine is formed from glutamic acid and NH_3 by the enzyme glutamine synthetase. (**Synthetases,** as might be assumed from the name, mediate syntheses.) Energy for the synthesis is derived from the hydrolysis of ATP, and Mg^{2+}, Co^{2+}, or Mn^{2+} is required. The reaction can be represented as follows:

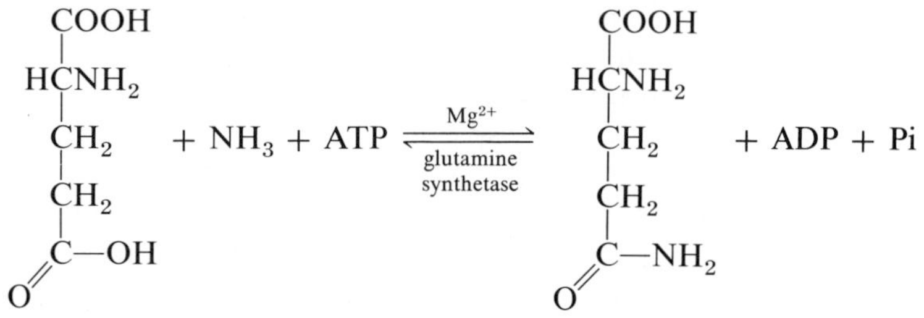

$$\begin{array}{c} COOH \\ | \\ HCNH_2 \\ | \\ CH_2 \\ | \\ CH_2 \\ | \\ O{=}C{-}OH \end{array} + NH_3 + ATP \xrightleftharpoons[\text{glutamine synthetase}]{Mg^{2+}} \begin{array}{c} COOH \\ | \\ HCNH_2 \\ | \\ CH_2 \\ | \\ CH_2 \\ | \\ O{=}C{-}NH_2 \end{array} + ADP + Pi$$

glutamic acid — glutamine

The exact mechanism of this reaction is not known. An enzyme-glutamyl derivative, synthesized via an enzyme-phosphate intermediate, has been suggested, but there is also evidence that glutamyl phosphate is involved in the reaction. The enzymic conversion of glutamine to glutamic acid requires ADP and Pi, as does the glutamyl transfer reaction that replaces the amide group on the glutamine

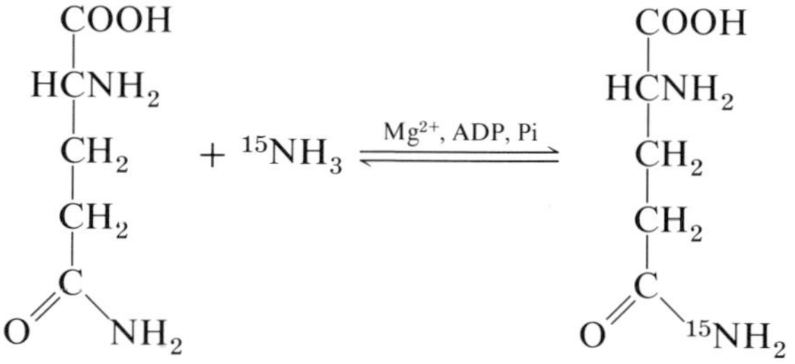

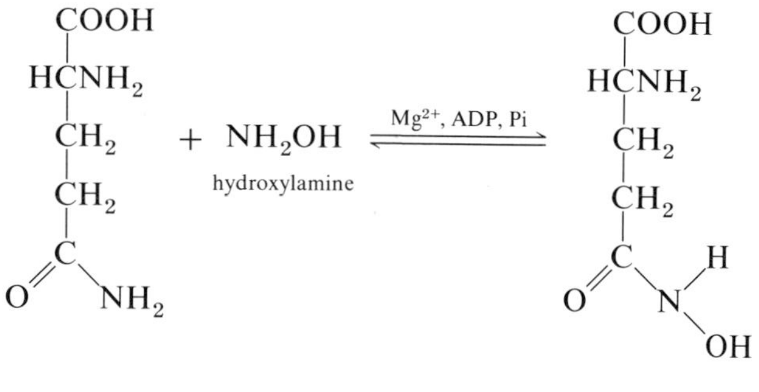

$$\begin{array}{c} COOH \\ | \\ HCNH_2 \\ | \\ CH_2 \\ | \\ CH_2 \\ | \\ O{=}C{-}NH_2 \end{array} + {}^{15}NH_3 \xrightleftharpoons{Mg^{2+},\ ADP,\ Pi} \begin{array}{c} COOH \\ | \\ HCNH_2 \\ | \\ CH_2 \\ | \\ CH_2 \\ | \\ O{=}C{-}{}^{15}NH_2 \end{array}$$

$$\begin{array}{c} COOH \\ | \\ HCNH_2 \\ | \\ CH_2 \\ | \\ CH_2 \\ | \\ O{=}C{-}NH_2 \end{array} + \underset{\text{hydroxylamine}}{NH_2OH} \xrightleftharpoons{Mg^{2+},\ ADP,\ Pi} \begin{array}{c} COOH \\ | \\ HCNH_2 \\ | \\ CH_2 \\ | \\ CH_2 \\ | \\ O{=}C{-}N(H)OH \end{array}$$

which indicates that the ATP $\rightleftharpoons$ ADP reaction is tightly coupled to the amide bond synthesis.

The corresponding enzyme that makes asparagine from aspartic acid has been hard to find and difficult to work with. Cell-free extracts have been prepared from wheat germ or lupine seedlings that catalyze the reaction

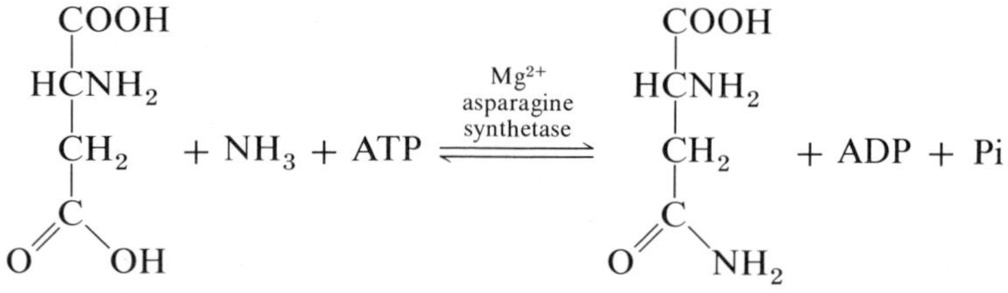

$$\begin{array}{c} COOH \\ | \\ HCNH_2 \\ | \\ CH_2 \\ | \\ O{=}C{-}OH \end{array} + NH_3 + ATP \xrightleftharpoons{\substack{Mg^{2+} \\ \text{asparagine} \\ \text{synthetase}}} \begin{array}{c} COOH \\ | \\ HCNH_2 \\ | \\ CH_2 \\ | \\ O{=}C{-}NH_2 \end{array} + ADP + Pi$$

aspartic acid — asparagine

but the enzyme activity is much lower than that for glutamine synthetase. There is substantial evidence that much asparagine in plants is formed from other pathways, both from isotope experiments utilizing specifically labeled aspartate as a substrate and from balance studies on the loss of aspartic acid and concomitant appearance of asparagine. Additional pathways have now been shown.

Several plants, particularly those having active cyanide (CN) metabolism and high cyanoglycoside content, such as flax (*Linum usitatissimum*), sorghum (*Sorghum vulgare*), clover (*Trifolium rapens*), and sweet pea (*Lathyrus odoratus*), are able to incorporate HCN into asparagine by the reaction

$$\underset{\text{cysteine}}{\begin{array}{c} COOH \\ | \\ HCNH_2 \\ | \\ CH_2 \\ | \\ SH \end{array}} + HCN \longrightarrow \underset{\beta\text{-cyanoalanine}}{\begin{array}{c} COOH \\ | \\ HCNH_2 \\ | \\ CH_2 \\ | \\ C{\equiv}N \end{array}} \xrightarrow{+H_2O} \underset{\text{asparagine}}{\begin{array}{c} COOH \\ | \\ HCNH_2 \\ | \\ CH_2 \\ | \\ C(=O){-}NH_2 \end{array}}$$

It has been suggested (but without experimental verification) that asparagine formation via a $C_1 + C_3$ condensation may be widespread.

Metabolism. In certain plants, for example, in germinating lupin seeds, asparagine is derived in large amounts directly from protein and from the amidation of aspartic acid derived from protein hydrolysis. Some asparagine may be made from aspartic acid freshly synthesized from Krebs cycle acids. There is evidence that in pea roots asparagine may be synthesized from the four-carbon acids arising from the carboxylation of pyruvate or phosphoenolpyruvate and, rarely, under special circumstances (for example, in starving wheat leaves), asparagine may be derived in large amounts from the metabolic products of endogenous sugars. However, the larger part of asparagine found in plants appears to be derived from the products of protein breakdown. Asparagine is not normally metabolized rapidly; when ^{14}C-asparagine is introduced into plants or cells it is usually metabolized relatively slowly compared with either sugars, amino acids, or glutamine.

Glutamine, like asparagine, also may be derived from glutamic acid liberated in protein breakdown but, unlike asparagine, massive synthesis of glutamine frequently takes place from carbon derived from stored carbohydrate or directly from photosynthesis. The most usual stimulus to glutamine formation is the supply of nitrogen either as ammonia or nitrate. Kretovitch has shown that the response to supplied NH_3 is the formation of glutamine in almost all plants examined, even in acid plants like *Sedum* and plants that normally contain asparagine. In contrast to asparagine, carbon from supplied glutamine rapidly enters the general metabolism of the cell, presumably via glutamic acid and α-ketoglutaric acid. Glutamine is a specific nitrogen donor in a number of important syntheses. It supplies the nitrogen atoms at positions 3 and 9 of the purine ring (see page 221) and the amide nitrogen of NAD and NADP. The amide nitrogen is also used in the synthesis of glucosamine and its derivatives, the monomers of the **chitins,** which are cell wall components of insects, many fungi, and a few higher plants. The cyclic nitrogen of the amino acids histidine and tryptophan is also supplied by glutamine.

Perhaps the most important reaction of glutamine is its involvement in the conversion of NH_3 and α-ketoglutarate to glutamic acid via the glutamine synthetase and glutamic acid synthetase reactions (page 204). In this reaction the higher energy of the amide nitrogen is used to assist in the conversion and transfer of amide nitrogen (from glutamine) into amino nitrogen (in glutamic acid). No such reaction sequence involving the amide nitrogen of asparagine has been found.

A number of compounds related to glutamine are found in plants. Derivatives of glutamine having methyl, methylene, or hydroxy groups substituted in the gamma position occur quite commonly in plants and occasionally make up a large proportion of the soluble nitrogen. These derivatives appear to act, like glutamine, as storage compounds for nitrogen. Experiments have shown, however, that they are not usually rapidly metabolized. Similar derivatives of asparagine are not known, although certain amide nitrogen-substituted asparagines have been found. The metabolic significance of these is not understood as yet.

Behavior of Glutamine and Asparagine. The roles of these two amides was once thought to be interchangeable in different species of plants. It is now clear that each performs certain special functions, although these functions may overlap in different species of plants and under different situations. As a result, our understanding of their metabolism and behavior is far from complete. Both amides are involved in nitrogen translocation. Asparagine seems to be the important compound involved in the mobilization of nitrogen stored as protein in seeds of plants such as lupine, but glutamine is far more often the major translocation compound in growing herbaceous plants and trees. Both amides may accumulate as a result of the presence of excess nitrogen; however, asparagine accumulation is usually associated with protein breakdown, whereas glutamine is more likely to be involved in the mobilization of nitrogen for protein synthesis.

This situation has led to the generalization by the Russian physiologist D. N. Prjanishnikoff that the presence of asparagine characterizes an unhealthy plant, whereas glutamine indicates a healthy plant. This generalization is borne out by the fact that asparagine accumulates in rust-infected wheat and other parasitized plants. Similarly, F. C. Steward at Cornell University has shown that if growth is affected or inhibited for various reasons in whole plants or tissue cultures (for example, by growing plants at the wrong day-length or by withholding a needed growth factor or nutrient) then asparagine predominates. When normal growth is permitted, glutamine becomes the dominant amide. Glutamine is associated with light or daytime metabolism in mint leaves, whereas asparagine is predominant in darkness (Figure 8-10).

In other work, Steward has shown that glutamine predominates in potatoes grown or developed in warm, dry environments with longer days, whereas asparagine is dominant in potatoes established in poorer climatic conditions. A comparison of English, American, hybrid, and greenhouse-developed potato cultivars is shown in Table 8-1. It may be seen that the glutamine/asparagine ratio is high in the American varieties conditioned to good weather and excellent growing conditions, whereas the ratio is much lower in the English varieties and the greenhouse variety developed under less favorable climatic conditions! The British physiologist E. W. Yemm observed that as protein breakdown occurs in starving barley leaves, glutamine rapidly accumulates. As starvation proceeds and the leaves begin to yellow, asparagine accumulation begins and exceeds that of glutamine. With metabolic disruption and eventually the death of the leaves from

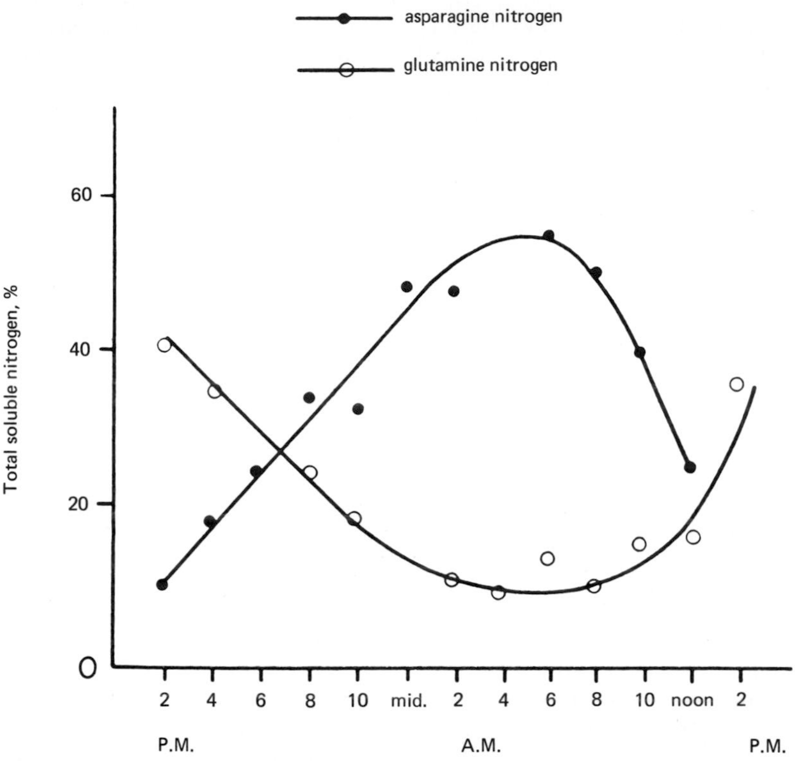

Figure 8-10. Diurnal variation in the composition of soluble nitrogen of leaves of *Mentha piperita* L. grown under short days. [From F. C. Steward (ed.): *Plant Physiology: A Treatise*, Vol. IVA. Academic Press, New York, 1965. Used with permission.]

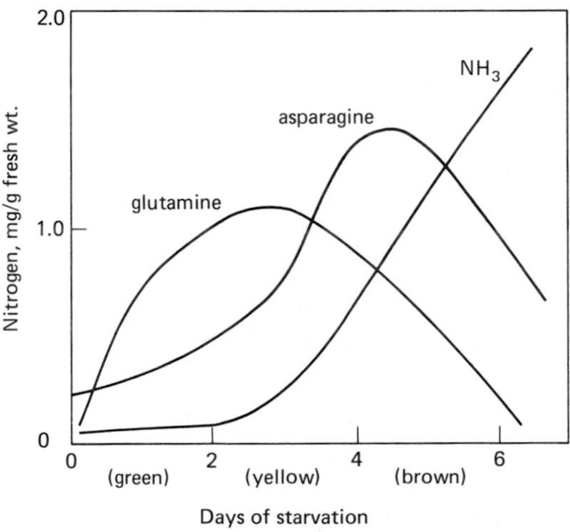

Figure 8-11. Changes in glutamine, asparagine, and NH_3 in barley leaves during starvation. [Redrawn from data of E. W. Yemm: *Proc. Roy. Soc. London*, **B123:**243 (1937).]

Table 8-1. Differences in the composition of the soluble nitrogen of various strains of potato tuber

Amino acid, μg/g fresh wt.	*American cultivar (Sebago)*	*American cultivar*	*Hybrid-British and American parentage*	*British cultivar (Katahdin)*	*Cultivar developed in greenhouse: short days, cool conditions (Houma)*
Aspartic acid	107	78	205	246	168
Glutamic acid	178	134	190	276	431
Serine	66	67	53	42	29
Glycine	28	—	—	—	48
Asparagine	138	1661	2200	2672	1083
Threonine	100	54	43	76	81
Alanine	131	55	68	72	24
Glutamine	3021	1142	858	754	91
Lysine	63	42	30	66	72
Arginine	356	159	55	110	81
Methionine	83	66	—	—	—
Proline	—	25	225	—	—
Valine	244	197	68	141	254
Leucines	93	167	25	24	196
Phenylalanine	138	203	—	—	—
Tyrosine	128	179	Trace	Trace	Trace
γ-Aminobutyric acid	300	244	225	86	240

SOURCE: F. C. Steward (ed.): *Plant Physiology: A Treatise,* Vol. IVA. Academic Press, New York, 1965. Used with permission.

starvation, first glutamine and then asparagine breaks down, liberating fee NH_3 (Figure 8-11). In some plants, such as legumes, however, asparagine has been shown to fill the role normally played by glutamine, being associated with synthetic reactions and the transfer of nitrogen in anabolic reactions such as the formation of seed proteins.

There are many similar observations indicating that the main roles of asparagine and glutamine in plants are nitrogen transport, ammonia detoxification and storage, and nitrogen mobilization for synthesis. Asparagine appears to be more frequently associated with protein breakdown or catabolic reactions, whereas glutamine is associated with anabolic reactions and growth. Glutamic acid and glutamine may also be involved in the nitrogen transfer that occurs during protein degradation and resynthesis, either in protein turnover or in the reworking of the cell's complement of structural and enzymic proteins during development.

Proteins

Proteins are the basis of all life because all the reactions of living systems are catalyzed by protein enzymes. The synthesis and structure of proteins are described in Chapter 2, pages 31–37. They are made of sequences of amino acids linked together by peptide bonds (primary structure). The polypeptide is frequently coiled in the form of an α helix stabilized by hydrogen bonds (secondary structure). This in turn may be much folded and twisted into a three-dimensional tertiary structure stabilized by a variety of weak to strong forces or bonds. Loss of tertiary structure or denaturation occurs when solvents, pH, heat, or other factors disrupt the bond system, and the protein may be precipitated or coagulated, a process that is often irreversible. Loss of enzymic properties nearly always accompanies denaturation, indicating that the active site of enzymes requires the tertiary folding of the appropriate primary and secondary structure.

Types of Proteins. The commonly used scheme of protein classification is based primarily on solubility rather than on chemical structure. However, we do not yet know enough to provide a structural basis for protein classification, and the present system is widely used.

A. Simple Proteins. These consist only of amino acids.

Albumins. Soluble in water and dilute aqueous salt solutions. A major class of proteins, many of which have enzymic activity.

Globulins. Insoluble or only slightly soluble in water but soluble in dilute aqueous salt solutions. Globulins may be salted out of aqueous solution by the addition of ammonium sulfate to one half the saturation concentration. They are important storage and enzymic proteins.

Prolamines. Insoluble in water, soluble in 50–90 percent aqueous ethanol. These are high-proline proteins, often found as storage proteins in seeds such as zein (corn), gliadin (wheat), and hordein (barley). Some enzymes (as papain) are prolamines.

Glutelins. Insoluble in neutral solvents but soluble acid or base. Major storage proteins in seeds.

Protamines. Low molecular weight proteins extremely rich in arginine. Protamines associate with nuclear proteins but may not be widespread in plants.

Histones. Characterized, like protamines, by their high content of basic amino acids. They are water soluble. Histones are usually found in the nucleus of the cell, and are associated in some manner with the nucleoproteins. They may have a specific role as nuclear gene inhibitors or regulators.

B. Conjugated Proteins. These compounds contain, beside their polypeptide chain, a different substance (called the **prosthetic group**) bound to them by salt linkages or covalent bonds. They are grouped according to the prosthetic group.

Mucoproteins. Combinations of protein and a carbohydrate, often a hexose or pentose polysaccharide. Animal mucoproteins contain hexosamine, but plant mucoproteins do not. They may be components of membranes.

Lipoproteins. Complexes of protein and a variety of lipoid types. These are the major structural proteins of membranes.

Nucleoproteins. Proteins, often protamines or histones, combined by salt linkages with nucleic acids. There is some doubt about the nature of the association, which may be an artifact of extraction.

Chromoproteins. An important and diverse group of proteins having various pigments for prosthetic groups. These are almost all important enzymes. Hemoglobin, chlorophyll-protein complexes, carotenoid proteins and flavoproteins are examples.

Metalloproteins. Many common enzymes have a metal prosthetic group more or less closely associated with the protein moiety. The molybdenum of nitrate reductase is a good example.

Protein Formation and Breakdown. The source of amino acids for protein synthesis varies from tissue to tissue and according to the physiological situation of the tissue in which protein synthesis is taking place. In developing seedlings some amino acids are translocated to the growing tips from the storage tissues in the endosperm or cotyledons. In this case they are derived either from the breakdown of proteins or from stored nitrogen and carbon from stored carbohydrate. Other amino acids are made in the leaves or at the site of protein synthesis, the carbon being largely derived from products of photosynthesis. Growing tips synthesize certain of their own amino acids from the carbon of sugar translocated from elsewhere in the plant, and they may be unable to use exogenous supplies of those particular amino acids. Still other amino acids are made exclusively in the leaves or roots and translocated to the growing root or stem tip. Leaves appear to be able to synthesize most of their own amino acids, but nevertheless some may be formed in roots from carbon exported from the leaves as sugars. This may follow from the fact that absorbed nitrogen, once reduced to NH_3, is quickly converted to amino acids, and this is the form in which nitrogen is transported back to the leaves. Within leaf cells, some amino acids are formed directly from products of photosynthesis, presumably in the chloroplasts. Others must be formed in the cytoplasm from carbon derived from products of photosynthesis that left the chloroplasts. The resulting amino acids are then returned to the chloroplasts where most of the leaf protein synthesis takes place.

Much protein synthesis takes place in meristematic or developing tissues. Mature tissue may undergo some protein turnover, but the pace of protein synthesis declines substantially on maturation. Leaves apparently continue to synthesize proteins, although at a reduced rate, until they become senescent. Detached leaves usually show no increase in their protein content, but they nevertheless continue to show protein turnover. This suggests that some factor essential for protein accretion, but not for protein synthesis, is derived from the roots. The nature of this factor or stimulus is not known, but it may be a cytokinin or some related substance. Kinetin application to detached leaves causes soluble amino acid mobilization, retards protein breakdown, and causes some synthesis.

Protein breakdown in plants is catalyzed by proteolytic enzymes of various sorts, and the peptide bond energy is not conserved (that is, no ATP or other high energy compounds are made). There is some indication that **lysosomes,** small microbodies containing proteolytic and other degradative enzymes found in animal cells, may also be present in plants. In many cells the main vacuole assumes this function. Protein degradation that occurs during turnover of proteins may not involve the same process as the rather general proteolysis that occurs in senescence or in germinating seeds. A more selective, perhaps partial, breakdown mechanism seems more likely.

Protein Turnover. The idea of protein turnover probably originated with I. P. Borodin in 1878, who suggested that protoplasmic activities such as respiration require continuous breakdown and regeneration of protein. Early analyses could not sustain or refute this idea, although it was considered by many botanists during the next 50 years. The British physiologists F. G. Gregory and P. K. Sen, following extensive analyses of the interrelationships between respiration, sugar content, and protein metabolism in barley leaves, proposed a scheme whereby a substantial part of cell respiration is sustained by amino acids derived from a protein cycle, as shown in Figure 8-12. This

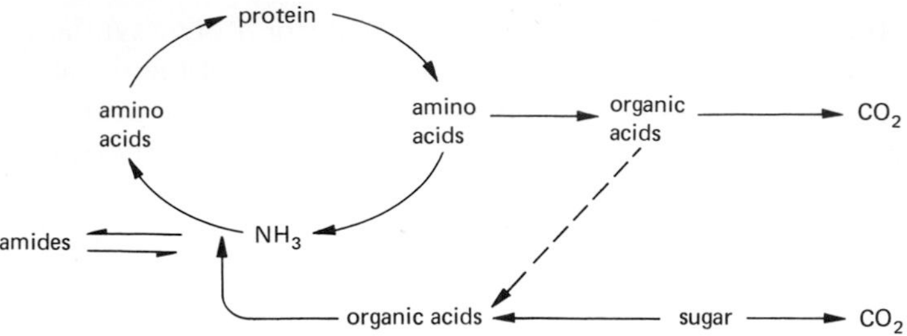

Figure 8-12. Cyclic protein metabolism as envisaged by F. G. Gregory and P. K. Sen in 1937.

scheme was proposed before the knowledge that a cycle of organic acid metabolism was involved in respiration or that the carbon skeletons of amino acids could be derived from this source, but it clearly anticipated these ideas.

The fact that protein turnover could occur was demonstrated by H. B. Vickery and his coworkers at the Connecticut Agricultural Experimental Station, who found that detached leaves could incorporate $^{15}NH_3$ into proteins in the absence of net protein synthesis. F. C. Steward and his associates, using ^{14}C-labeled substrates, showed that protein turnover occurs in cultured carrot tissues and that the pace of protein turnover is proportional to growth and respiration rate. Moreover, it was possible, as was predicted by Gregory and Sen's model, to show that amino acids derived from protein breakdown are largely oxidized to carbon dioxide, while simultaneously protein is synthesized from amino acids newly formed from sugar carbon. Protein turnover also takes place in leaves, but in these organs it seems more likely to be a reflection of the biochemical differentiation that takes place during development. Turnover is rapid in growing leaves or in cotyledons that are developing into photosynthetic organs, but it decreases greatly with maturity and ceases during senescence. The relationship with respiration seems more likely to be connected with the requirement of ATP for peptide bond synthesis and carbon skeletons for amino acid formation.

Peptides

Peptides may be formed as the result of partial (that is, faulty) synthesis of proteins or by their partial degradation. A few tissues contain substantial amounts of peptides, but these do not appear to have major physiological significance. However, certain peptides have an important physiological role as cofactors in enzyme reactions. These peptides usually have their own biosynthetic pathway unrelated to the ribosomal system or protein synthesis. The tripeptide **glutathione,** γ-glutamylcysteinylglycine, is important as a hydrogen transfer agent associated with several redox enzymes. Glutathione is formed by the condensation of glutamate and cysteine to form γ-glutamylcysteine; ATP is hydrolyzed to ADP + Pi in the process. Glycine is then added, a second molecule of ATP being used in the process. Presumably the amino acids are phosphorylated prior to the formation of the peptide bonds. Other important compounds containing peptide links are **tetrahydrofolic acid** (involved in formyl and methyl transfer reactions) and **pantothenic acid** (a part of the CoA molecule). The auxin, indoleacetic acid, may be inactivated when supplied to plants in excess, by conjugation with aspartic acid to form the inactive peptide derivative, indoleacetylaspartic acid.

Purines and Pyrimidines

The purine and pyrimidine bases, important components of the nucleic acids (Chapter 2, pages 37–41), are synthesized from simple cellular components by complex reaction sequences. The free purines and pyrimidines are not synthesized as such. The purines are built up stepwise on a molecule of ribose-5-phosphate (R-5-P), resulting in the direct formation of purine nucleotides. The basic pyrimidine structure, orotic acid, is synthesized directly, then attached to R-5-P, and the other pyrimidine nucleotides are formed from this ribotide. The synthesis of deoxyribotides is accomplished by the reduction of the corresponding ribotide. R-5-P, the amino acids glutamine, glycine, and aspartic acid, carbamylphosphate, ATP, CO_2, and derivatives of tetrahydrofolic acid (THFA) are all used to donate carbon, nitrogen, and phosphorus in stepwise syntheses of remarkable beauty and economy. The reactions leading to purines are outlined in Figure 8-13 and those leading to pyrimidines in Figure 8-14.

The R-5-P required is probably derived from pentose-shunt metabolism or possibly from Calvin cycle intermediates formed during photosynthesis. The amino acids come

Figure 8-13. Synthesis of purines, starting from ribose-5-phosphate (R5P). The end products of the synthesis are printed in capitals. The new atoms added by each reaction are circled with a dotted line.

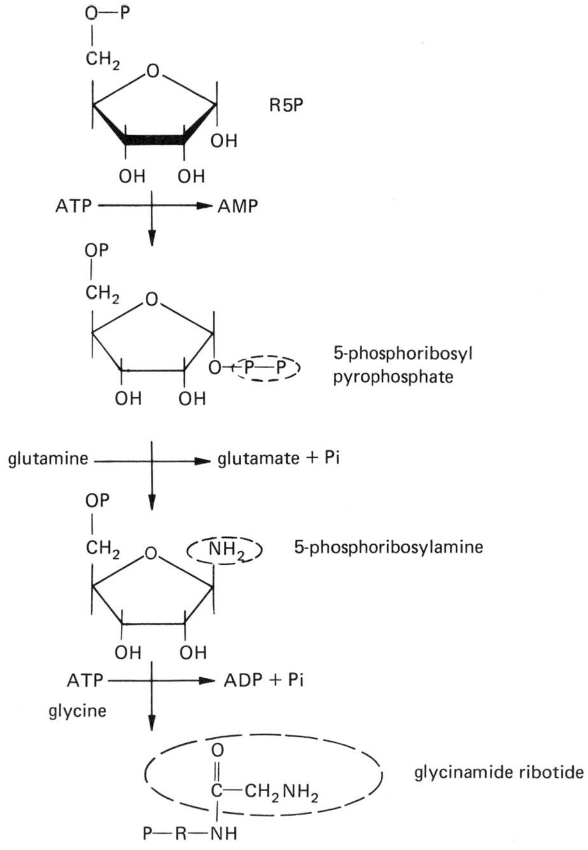

Figure 8-13. (continued)

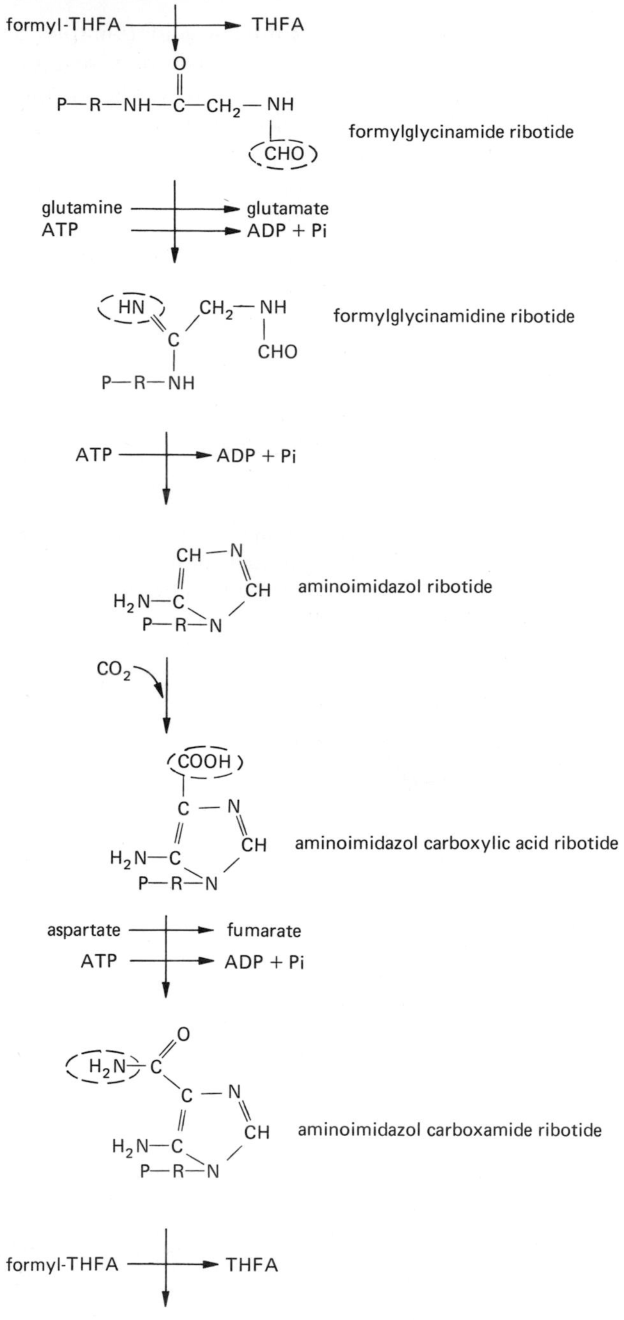

formyl-THFA → THFA

P—R—NH—C(=O)—CH$_2$—NH—(CHO) formylglycinamide ribotide

glutamine → glutamate
ATP → ADP + Pi

(HN)=C(—CH$_2$—NH—CHO)—NH—R—P formylglycinamidine ribotide

ATP → ADP + Pi

aminoimidazol ribotide

CO_2

(COOH) aminoimidazol carboxylic acid ribotide

aspartate → fumarate
ATP → ADP + Pi

(H$_2$N)—C(=O) aminoimidazol carboxamide ribotide

formyl-THFA → THFA

Figure 8-13. (continued)

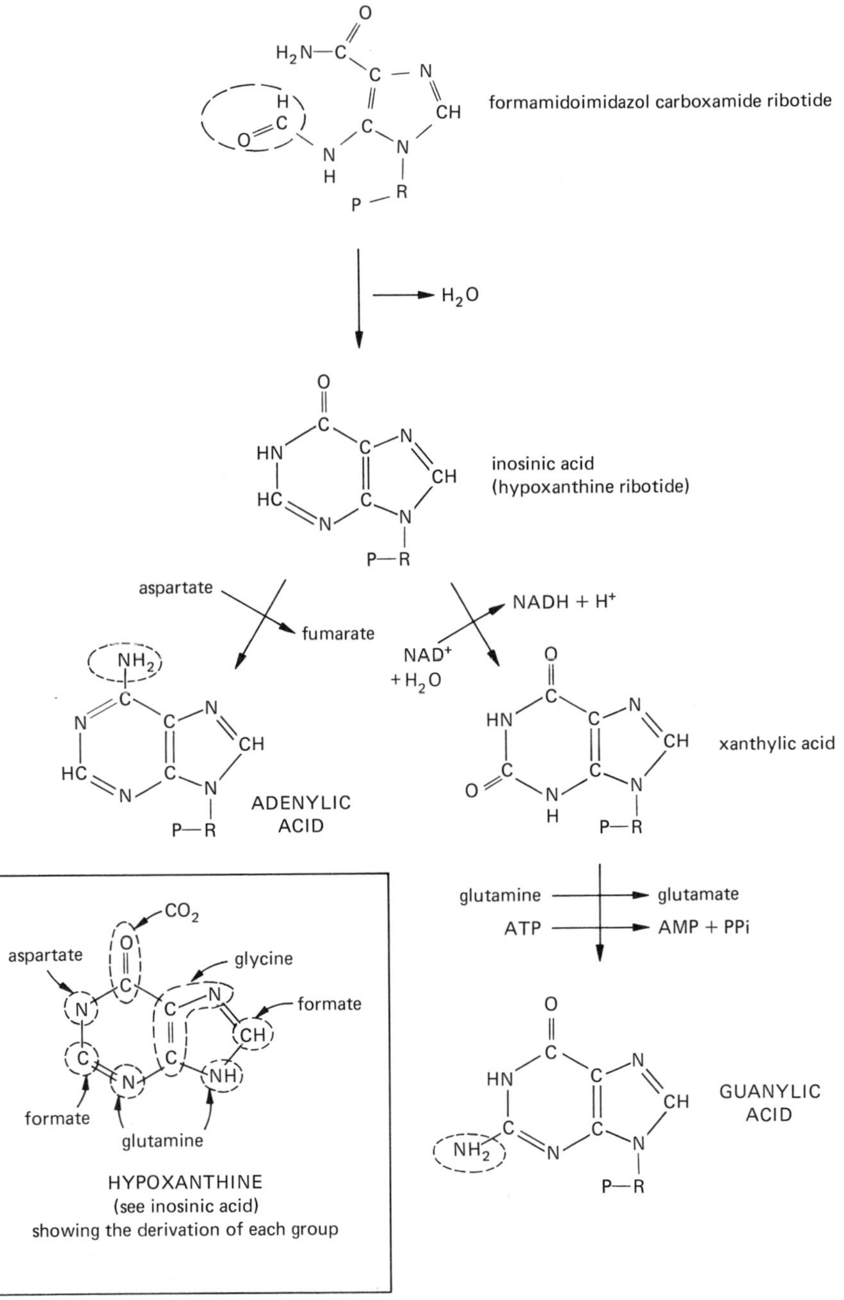

formamidoimidazol carboxamide ribotide

$\rightarrow H_2O$

inosinic acid
(hypoxanthine ribotide)

aspartate → fumarate

$NAD^+ + H_2O$ → $NADH + H^+$

ADENYLIC ACID

xanthylic acid

glutamine → glutamate
ATP → AMP + PPi

GUANYLIC ACID

CO_2
aspartate
glycine
formate
formate
glutamine

HYPOXANTHINE
(see inosinic acid)
showing the derivation of each group

Figure 8-14. Pyrimidine synthesis. The end products of this synthesis are printed in capitals.

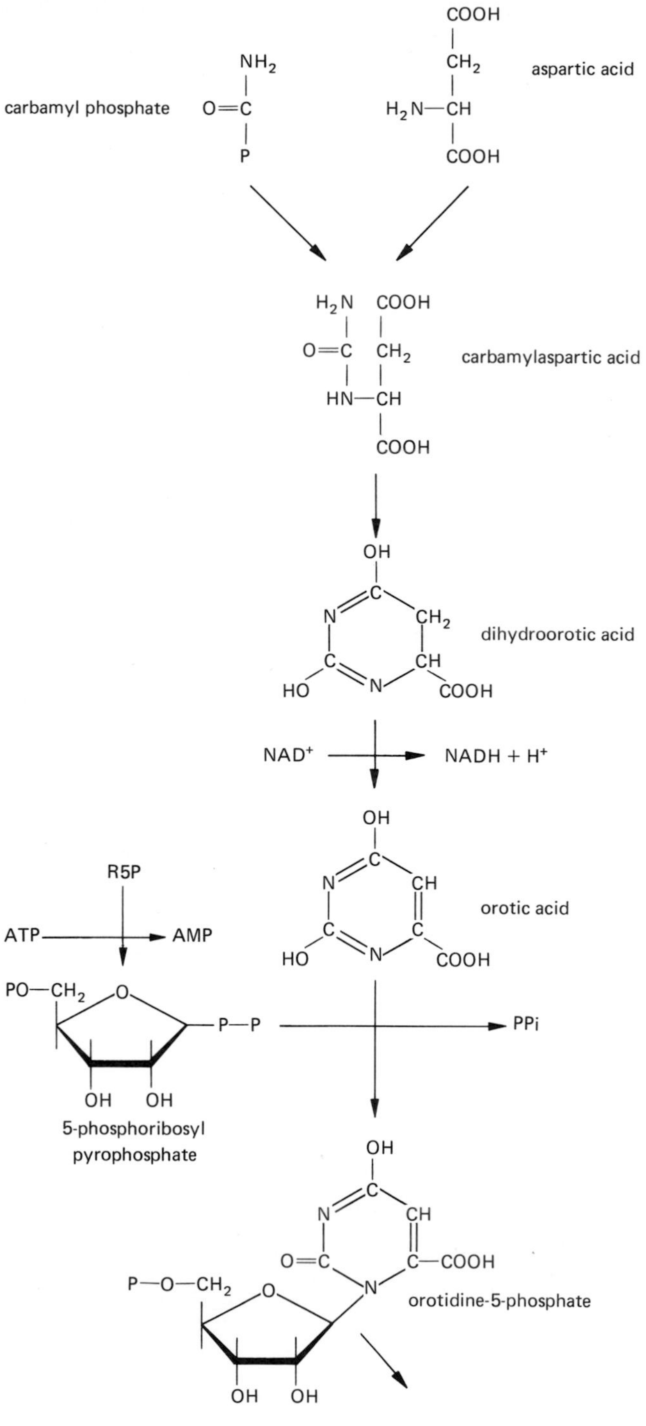

Figure 8-14. (continued)

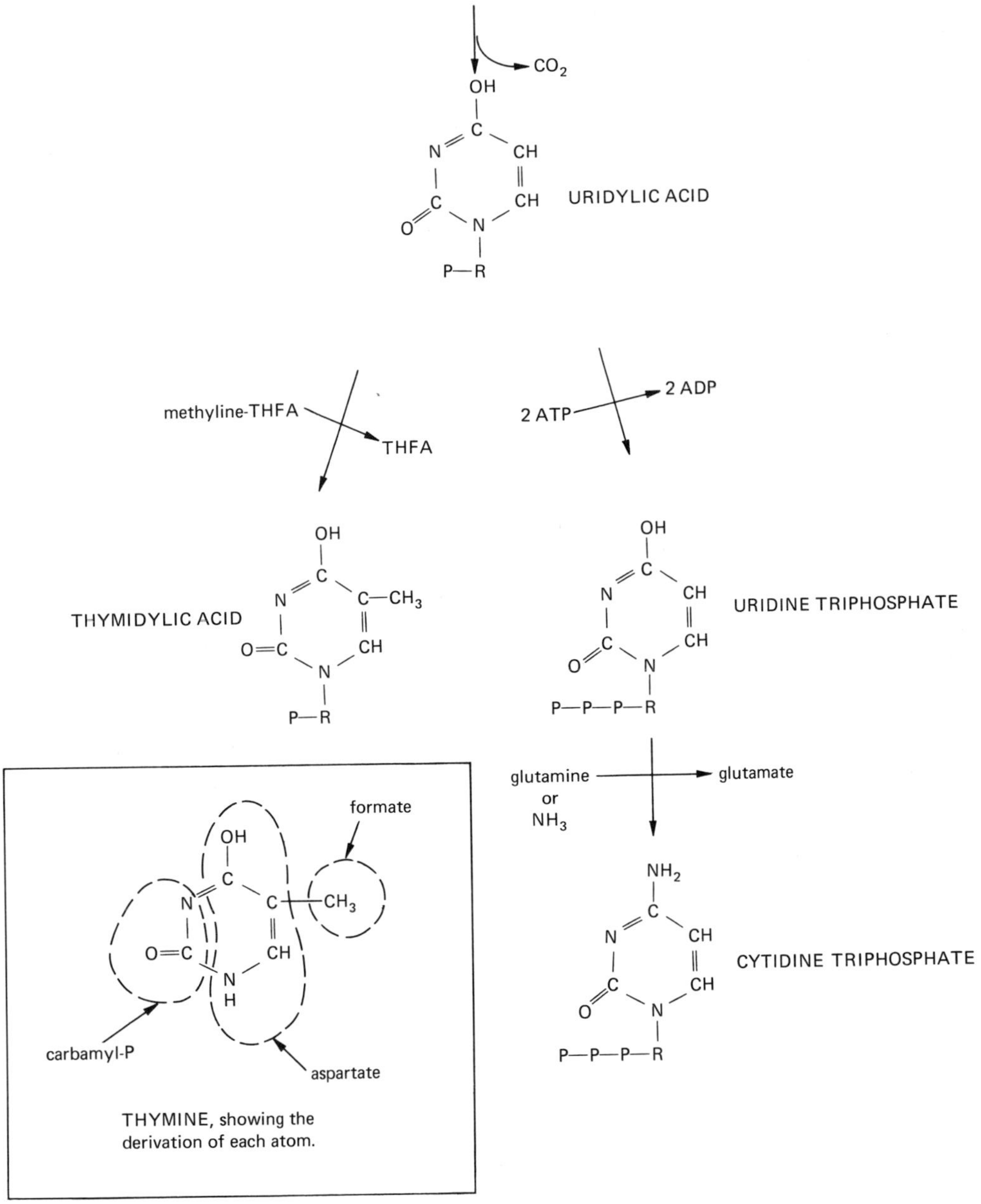

THYMINE, showing the derivation of each atom.

from the cellular pools derived from respiratory metabolism. Carbamylphosphate is probably synthesized in the reaction

$$CO_2 + NH_3 + ATP \rightleftharpoons NH_2{-}\overset{\overset{\displaystyle O}{\|}}{C}{-}O{-}P + ADP$$

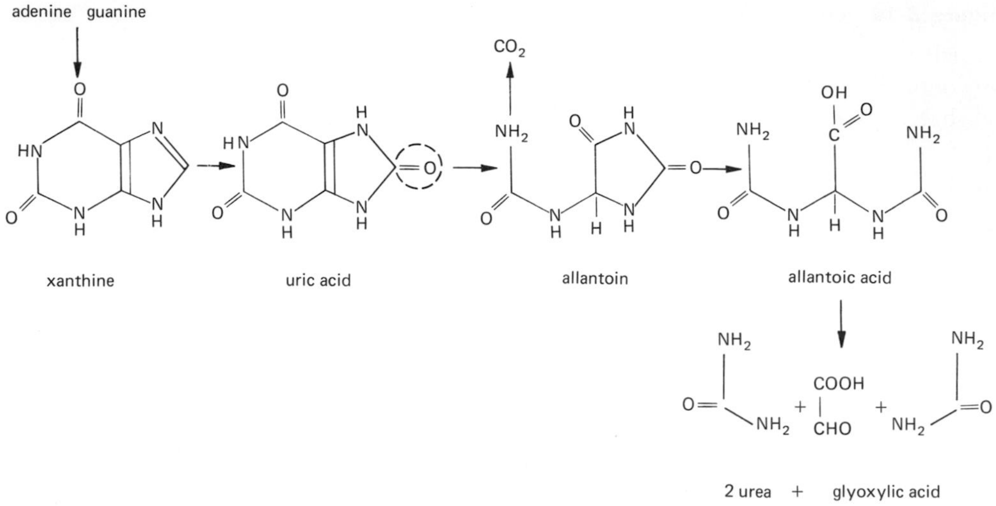

adenine guanine → xanthine → uric acid → allantoin (CO_2) → allantoic acid → 2 urea + glyoxylic acid

Figure 8-15. Purine degradation.

although it might be derived from the breakdown of citrulline

$$\begin{aligned} \text{citrulline} &\longrightarrow \text{ornithine} + \text{carbamate} \\ \text{carbamate} + \text{ATP} &\longrightarrow \text{carbamylphosphate} + \text{ADP} \end{aligned}$$

The methylene and formyl derivatives of THFA are formed from the reaction of THFA with a suitable hydroxymethyl donor such as serine

$$\text{serine} + \text{THFA} + \text{NADP} \longrightarrow \text{methylene THFA} + \text{glycine} + H_2O$$
$$\text{methylene THFA} \longrightarrow \text{formyl THFA}$$

A suitable methyl donor may form methyl THFA which can be oxidized to methylene THFA, and formic acid may also be converted directly to formyl THFA.

Figure 8-16. Pyrimidine degradation.

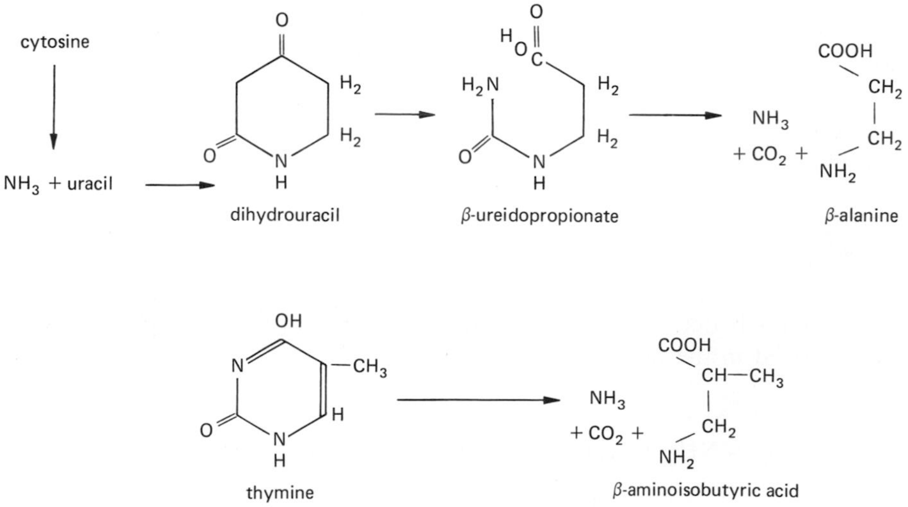

cytosine → NH_3 + uracil → dihydrouracil → β-ureidopropionate → NH_3 + CO_2 + β-alanine

thymine → NH_3 + CO_2 + β-aminoisobutyric acid

The breakdown of purines results in the formation of alantoin and alantoic acid, which are ultimately converted to urea and glyoxylic acid (Figure 8-15). Pyrimidines decompose through the hydrolytic opening of the ring to give a β-ureide, which is probably decomposed to NH_3 and CO_2, but the reaction mechanism has not been worked out in plants (Figure 8-16). Ribotides, and particularly deoxyribotides, resulting from RNA and DNA catabolism are quickly metabolized. If this were not so, their presence in cells might be prejudicial to normal metabolism.

Alkaloids

The alkaloids represent an extremely heterogeneous group of compounds having one or more nitrogen atoms, usually in a heterocyclic ring. They range in complexity from simple amines such as ricinine (Figure 8-17) to complex steroid glycosides such as solanine (also shown in Figure 8-17). Over 1000 alkaloids are known from 1200 plant species, and most plants probably contain small amounts of some alkaloidal material. The alkaloidal plants, known to contain large or striking amounts of one or more alkaloid, are scattered erratically throughout almost every group of plants, except perhaps the algae. Some alkaloids are widespread, whereas others are known from one species or genus only. A few of the better known ones, together with the plants from which they are most frequently derived, are shown in Figure 8-18.

Alkaloids are widely known for their powerful physiological effects on animals. You will probably be aware of the effects of most of the ones shown in Figure 8-18! So far, however, no general view of their function in plants has emerged. It has been suggested that they serve as a protection mechanism, but certain alkaloid-rich plants, like tobacco, are at least as susceptible—possibly more so—to insect pests as many nonalkaloidal plants. Saprophytic plants and parasites apparently flourish on alkaloidal plants. Alkaloids are unlikely to form effective nitrogen storage compounds because of their usually low nitrogen content and the small quantities stored. Alkaloidal plants are normally capable of forming asparagine, glutamine, or arginine, which are much more effective ammonia detoxifiers or storage compounds. Alkaloids are frequently found in young, actively growing parts of plants but may be localized in other tissues such as bark, roots, or leaves. They often appear to be synthesized in the roots and translocated

Figure 8-17. Structures of two plant alkaloids.

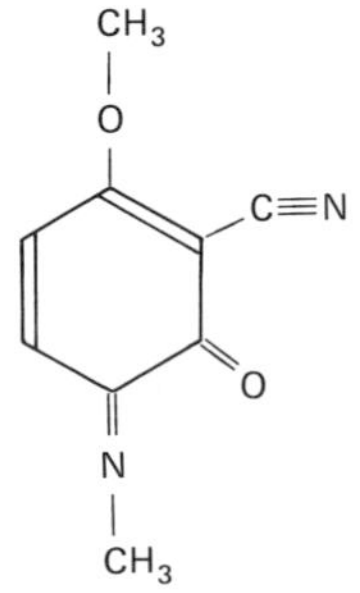

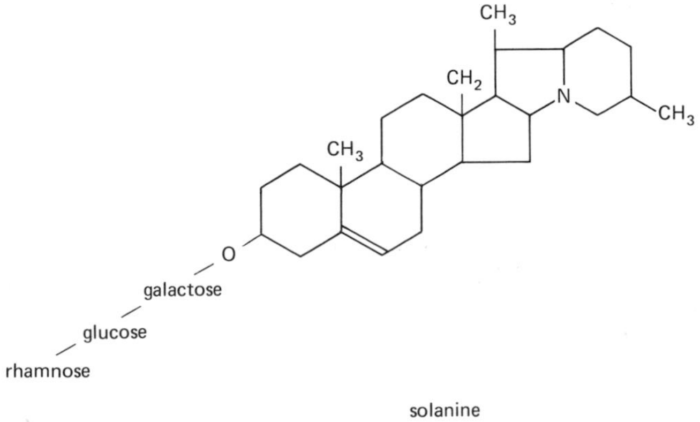

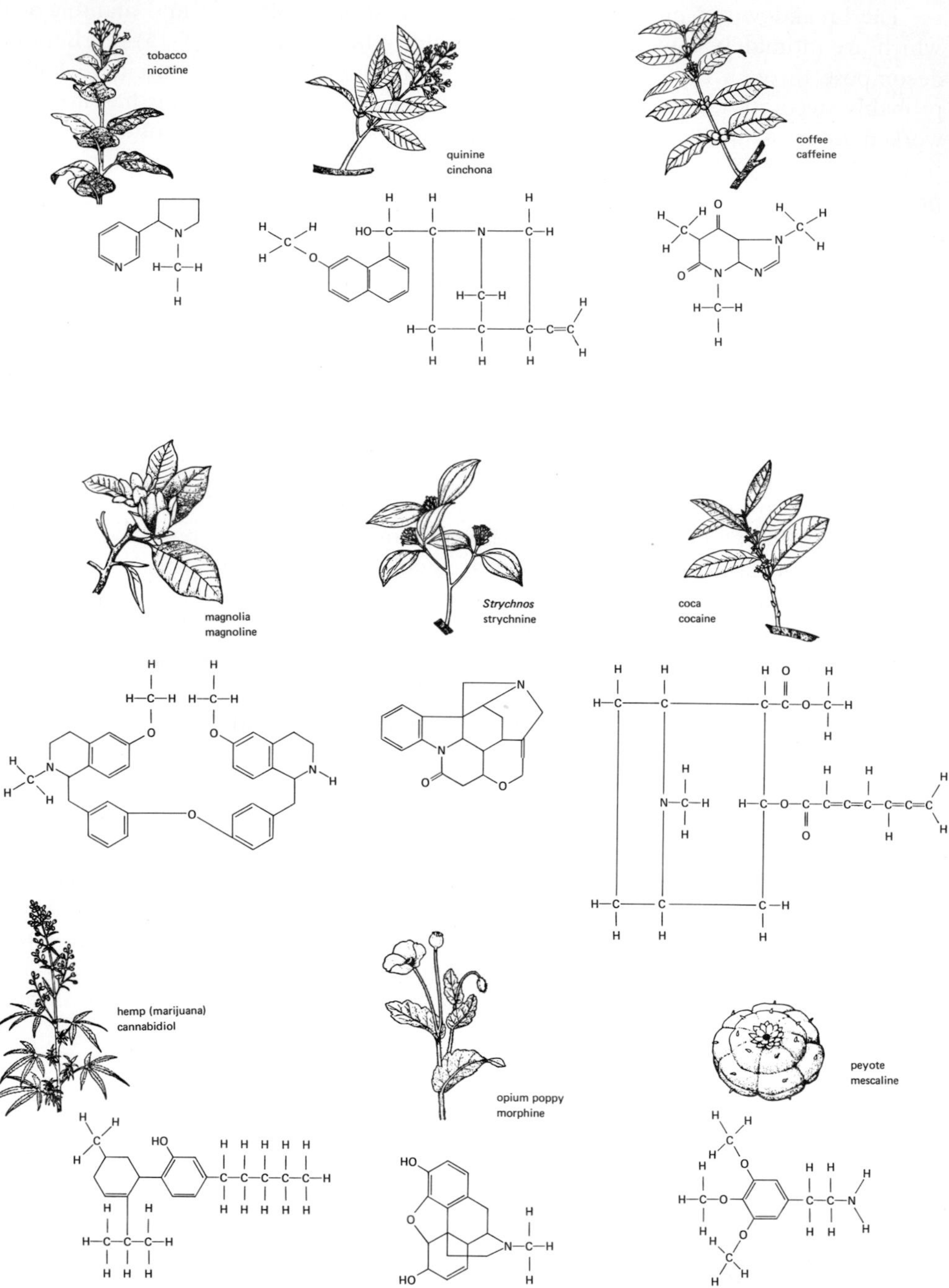

Figure 8-18. Plant alkaloids. [From P. R. Ehrlich and P. H. Raven: Butterflies and plants. *Sci. Am.*, **216**(6):106 (1967). Used with permission.]

elsewhere in the plant. Active metabolism of alkaloids has been found to occur in some plants. They may perhaps play some part in the metabolism or control of development, but no obvious relationships have been found, and their true role in plant biology is unknown at present.

Additional Reading

Articles in *Annual Review of Plant Physiology* under the heading "Nitrogen Metabolism."

Bidwell, R. G. S., and **D. J. Durzan:** Some recent aspects of nitrogen metabolism. In **P. J. Davies** (ed.): *Historical and Current Aspects of Plant Physiology,* pp. 152–225, Cornell University Press, Ithaca, N.Y., 1975.

Boulter, D., R. J. Ellis, and **A. Yarwood:** Biochemistry of protein synthesis in plants. *Biol. Rev.,* **47:**113–175 (1972).

Burns, R. C., and **R. W. F. Hardy:** *Nitrogen Fixation in Bacteria and Higher Plants.* Springer-Verlag, New York, 1975.

Chibnall, A. C.: *Protein Metabolism in the Plant.* Yale University Press, New Haven, Conn., 1939 (reprinted 1964).

Hewitt, E. J., and **C. V. Cutting** (eds.): *Recent Aspects of Nitrogen Metabolism in Plants.* Academic Press, New York, 1968.

McKee, H. S.: *Nitrogen Metabolism in Plants.* Clarendon Press, Oxford, 1962.

Nutman, P. S. (ed.): *Symbiotic Nitrogen Fixation in Plants.* Cambridge University Press, London, 1976.

Steward, F. C., and **D. J. Durzan:** Metabolism of nitrogenous compounds. In **F. C. Steward** (ed.): *Plant Physiology: A Treatise,* Vol. IVA. Academic Press, New York, 1965.

Stewart, W. P. D.: *Nitrogen Fixation in Plants.* The Athlone Press, London, 1966.

Webster, G. C.: *Nitrogen Metabolism in Plants.* Row, Peterson Co., White Plains, N.Y., 1959.

9 Polymers and Large Molecules

In this chapter we shall consider briefly the metabolism and formation of some of the important polymeric substances in the plant and of a few of the physiologically more important complex substances that have not previously been mentioned. Further information on the synthesis of organic components not covered here may be found in biochemistry textbooks. Although many of these compounds are present in plants, their metabolism is known only from studies with bacterial or animal systems and will be presented here in outline. This chapter is primarily intended to give a brief outline of some of the more important pathways for reference purposes.

Polysaccharides

Starch. Starch is usually degraded by amylase or phosphorylase reactions (Chapter 6), but its synthesis is accomplished by uridine diphosphate glucose (UDPG) or adenine diphosphate glucose (ADPG) transglycosylase. This reaction

$$\begin{matrix}\text{UDPG}\\ \text{or}\\ \text{ADPG}\end{matrix} + \text{glucose}_n \xrightarrow{\text{transglycosylase}} \begin{matrix}\text{UDP}\\ \text{or}\\ \text{ADP}\end{matrix} + \text{glucose}_{n+1}$$

discovered by Leloir's group in Argentina, makes only α-(1:4) linkages. ADPG appears to be a more effective donor in most plant material that has been tested. This reaction is allosterically stimulated by the primary product of photosynthesis, phosphoglyceric acid (PGA), providing an effective positive feedback loop that ensures rapid starch formation in light. The synthesis of UDPG or ADPG is accomplished by the reaction

$$\begin{matrix}\text{UTP}\\ \text{or}\\ \text{ATP}\end{matrix} + \text{G-1-P} \xrightarrow{\text{phosphorylase}} \begin{matrix}\text{UDPG}\\ \text{or}\\ \text{ADPG}\end{matrix} + \text{PPi}$$

The G-1-P could arise from F-6-P produced in photosynthesis or from sucrose, the usual translocation form of carbohydrate. Sucrose could also transfer glycosyl residues more directly to starch via sucrose synthetase (Chapter 7), as follows

$$\text{sucrose} + \text{UDP} \xrightarrow[\text{sucrose synthetase}]{} \text{UDPG} \xrightarrow[\text{starch synthetase}]{} \text{starch}$$
$$(\text{UDPG step also releases} \nearrow \text{fructose})$$

An important fact of this starch-synthesizing enzyme (sometimes called starch synthetase) is that it requires a primer or acceptor of at least two glucose residues, that is, maltose or a maltose oligosaccharide. The same is thought to be true of phosphorylase, although the de novo synthesis of amylose by muscle phosphorylase without primer has been reported. This report suggests that phosphorylase might be involved in starting starch synthesis. Another group-transferring enzyme is the **D-enzyme** of potato, which can transfer groups of two or more glucose units from one α-(1:4) linked chain to another. This enzyme could accomplish the synthesis of long chains if the residues remaining after transfer were removed. All these enzymes add the new glucose residue to the nonreducing end of the acceptor molecule.

The α-(1:6) linkages that make the branches found in amylopectin are synthesized by the **Q-enzyme,** or "branching factor," discovered by C. Cori. This enzyme can transfer groups of glucose residues with α-(1:4) linkage onto the 6-position of a glucose residue in another similar chain, making a branch through the newly formed α-(1:6) link. The initial branch length is at least four glucose units long. The resulting branched molecule can continue to grow at both its nonreducing ends by phosphorylase or starch synthetase activity.

The mechanism of starch synthesis in vivo is still unknown. Mixtures of Q-enzymes and phosphorylase do not produce a natural mixture of amylose and amylopectin but only amylopectin, whose degree of branching is proportional to the amount of Q-enzyme present. The relationship between Q-enzyme and UDPG- or ADPG-starch transglucosylase is not clear.

INULIN. The synthesis of this β-(2:1)-polyfructosan and the related β-(2:6) linked **levans** is not well understood. Fructose units are apparently transferred to the 1- or 6-position of sucrose, which is always present as the nonreducing terminal. UDP-fructose, recently isolated from *Dahlia* tubers, may be an intermediate. Inulin is an important storage form in roots and stems of such plants as *Dahlia* and *Helianthus tuberosum* (Jerusalem artichoke). Interestingly, these plants tend to make starch, not inulin, in their leaves. A number of plants, particularly monocots, form fructosans in their stems and leaves, usually of the levan type. Inulin-type fructosans, however, are found in the ears of cereal plants.

CELLULOSE. The recent work of W. Hassid in California shows that cellulose is made in a manner analogous to starch, but the donor of glucose units is guanosine diphosphate glucose (GDPG), forming β-(1:4) linkages. Cell-free preparations from *Lupinus albus* (white lupin) make cellulose from UDPG, but this may involve a UDPH-GDPG transfer.

OTHER POLYSACCHARIDES. A number of other polysaccharides, mostly structural (that is, cell wall components), are formed by transglycosyl reactions from the appropriate UDP or ADP derivative. These include pectins, pectic acid, hemicellulose, and a variety of xylans, arabans, and mixed polysaccharides. Pectic substances include pectic acid, an α-(1:4)-poly-D-galacturonic acid forming the middle lamella of cell walls, and pectins that are essentially similar but have the carboxyl groups masked by the formation of methyl esters.

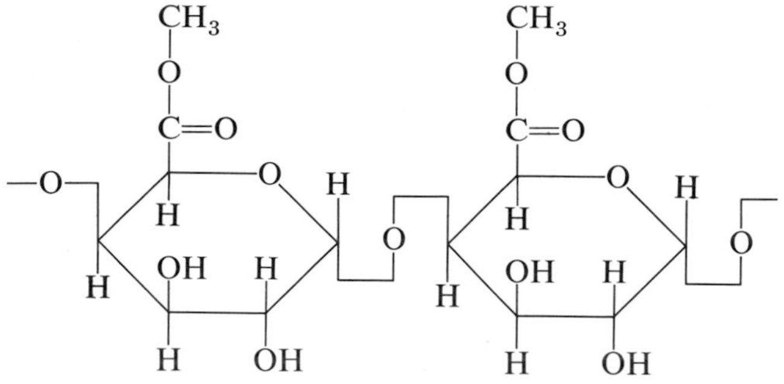

pectin

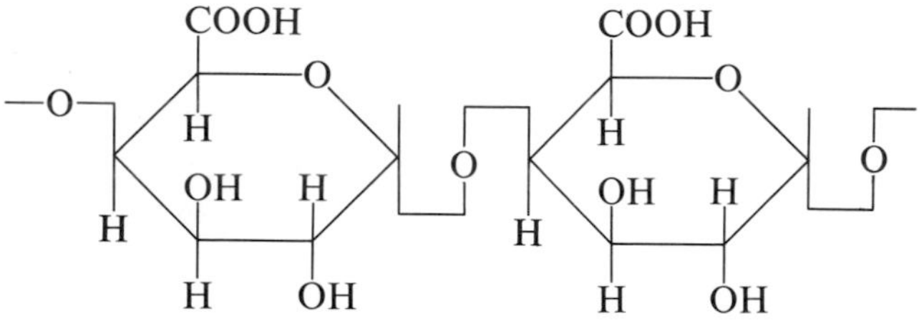

pectic acid

Other sugars frequently may be included in the structure of pectic substances. Hemicelluloses represent a poorly defined group of polysaccharides, usually composed of several different sugars and uronic acids. Details of their syntheses are lacking, but like pectic substances and the related algal polysaccharides (Chapter 2) they are probably made by transglycosyl reactions.

Lipids

The plant lipids fall into three main groups: the fats and oils, largely present as food storage; the phospholipids and glycolipids, mainly structural components of membranes; and the waxes, which form the outer protective coating or cuticle of most plants.

The synthesis of fatty acids is a fairly complex process that was first worked out for animal and microbial systems. Since most of the natural fatty acids consist of even numbers of carbon atoms and their β oxidation results in the production of acetyl-CoA, it was previously thought that they were formed by a reversal of this oxidation process (see page 122, Figure 6-9). Then it was discovered that CO_2 was not only a stimulator of fatty acid synthesis but a necessary reactant, although it is not itself converted to fat. This fact was explained by the discovery that malonyl-CoA, not acetyl-CoA, is the principal carbon donor for fatty acid synthesis. Malonyl-CoA is made from acetyl-CoA and CO_2 by a carboxylase

$$\underset{\text{acetyl-CoA}}{CH_3{-}\overset{\overset{\displaystyle O}{\|}}{C}{-}CoA} + ATP + CO_2 \xrightarrow[\text{carboxylase}]{\text{biotin, }Mg^{2+}} \underset{\text{malonyl-CoA}}{COOH{-}CH_2{-}\overset{\overset{\displaystyle O}{\|}}{C}{-}CoA} + ADP$$

This reaction is inhibited by palmitic acid, a common fatty acid, which provides an effective feedback control mechanism. The required acetyl-CoA is presumably derived from the oxidation of pyruvate. This takes place in the mitochondria, however, and acetyl-CoA does not pass the mitochondrial membrane. Presumably the acetate moiety is transferred via some carrier to cytoplasmic CoA, as occurs in animal mitochondria. Alternatively, acetyl-CoA could be synthesized from citrate, which can diffuse out of the mitochondria, by a mechanism similar to the animal reaction

$$\text{citrate} + \text{ATP} + \text{CoA} \longrightarrow \text{acetyl-CoA} + \text{oxaloacetate} + \text{ADP} + \text{Pi}$$

Malonyl-CoA donates an acetyl group in a reductive reaction, releasing CO_2 and reduced CoA; the acceptor is acetyl-CoA or the CoA derivative of an even-numbered fatty acid, as follows

$$CH_3{-}\overset{O}{\overset{\|}{C}}{-}CoA + COOH{-}CH_2{-}\overset{O}{\overset{\|}{C}}{-}CoA + 2\,NADPH \rightleftharpoons$$

$$CH_3{-}CH_2{-}CH_2{-}\overset{O}{\overset{\|}{C}}{-}CoA + CoASH + 2\,NADP + CO_2 + H_2O$$

CoASH is reduced CoA. The reaction is in fact not quite so simple as shown above. The CoA derivative of the fatty acids involved is first transferred to a special protein called the **acyl carrier protein (ACP).** After the condensation, a CoA derivative of the resulting longer chained acid is again formed. The synthesis of palmitic acid, a 16-carbon saturated fatty acid (no double bonds) would be summarized as follows:

$$\text{acetyl-CoA} + 7\,\text{malonyl-CoA} + 14\,\text{NADPH} \longrightarrow$$

$$CH_3{-}(CH_2)_{14}{-}COOH + 7\,CO_2 + 8\,CoA + 14\,NADP + 6\,H_2O$$

The intermediate in the subsequent synthesis of fat is probably the fatty acyl-CoA, not the free acid. The synthesis of unsaturated fatty acids takes place by reductive reactions, perhaps involving ferredoxin and NADPH. This process can take place in the chloroplast and it is enhanced by light.

The glycerol backbone of triglycerides (Chapter 2, page 22) is probably derived from the glycolytic intermediate dihydroxyacetone phosphate (DHAP) by reduction.

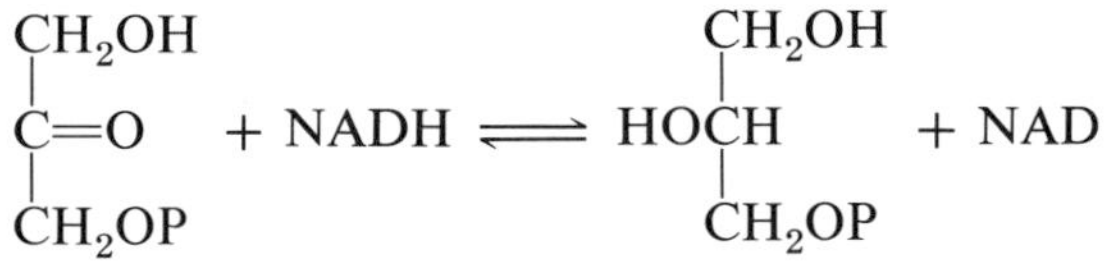

$$\underset{\text{DHAP}}{\begin{array}{c} CH_2OH \\ | \\ C{=}O \\ | \\ CH_2OP \end{array}} + NADH \rightleftharpoons \underset{\text{glycerophosphate}}{\begin{array}{c} CH_2OH \\ | \\ HOCH \\ | \\ CH_2OP \end{array}} + NAD$$

Alternatively, glycerol might be formed from some nonphosphorylated intermediate and be phosphorylated by glycerokinase, ATP being the donor.

The final step in the synthesis of a fat is the combination of three fatty acids by ester linkage with glycerol phosphate in the following two reactions.

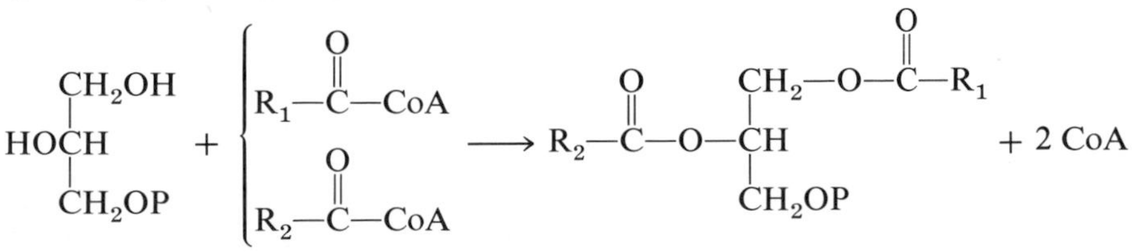

$$\underset{\text{glycerophosphate}}{\begin{array}{c} CH_2OH \\ | \\ HOCH \\ | \\ CH_2OP \end{array}} + \underset{\text{fatty acyl-CoA}}{\left\{\begin{array}{l} R_1{-}\overset{O}{\overset{\|}{C}}{-}CoA \\ R_2{-}\overset{O}{\overset{\|}{C}}{-}CoA \end{array}\right.} \longrightarrow \underset{\text{phosphatidic acid}}{\begin{array}{r} CH_2{-}O{-}\overset{O}{\overset{\|}{C}}{-}R_1 \\ R_2{-}\overset{O}{\overset{\|}{C}}{-}O{-}CH \qquad \\ CH_2OP \quad \end{array}} + 2\,CoA$$

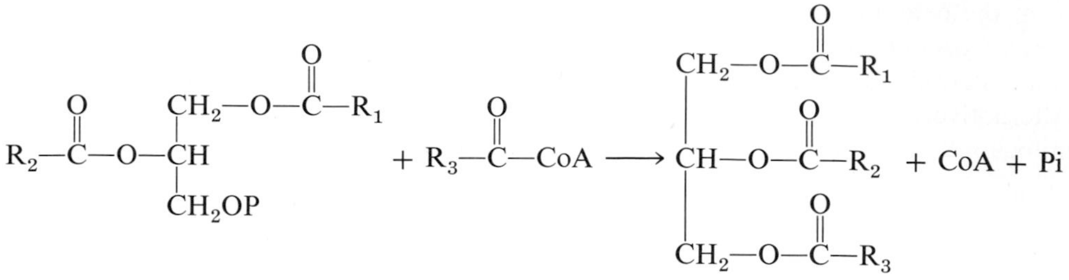

$$\begin{array}{l} CH_2\text{—}O\text{—}\overset{O}{\overset{\|}{C}}\text{—}R_1 \\ R_2\text{—}\overset{O}{\overset{\|}{C}}\text{—}O\text{—}CH \\ CH_2OP \end{array} + R_3\text{—}\overset{O}{\overset{\|}{C}}\text{—CoA} \longrightarrow \begin{array}{l} CH_2\text{—}O\text{—}\overset{O}{\overset{\|}{C}}\text{—}R_1 \\ CH\text{—}O\text{—}\overset{O}{\overset{\|}{C}}\text{—}R_2 \\ CH_2\text{—}O\text{—}\overset{O}{\overset{\|}{C}}\text{—}R_3 \end{array} + \text{CoA} + \text{Pi}$$

phosphatidic acid — fatty acyl-CoA — triglyceride

The intermediate **phosphatidic acids** are used in the formation of substituted fats. These include **lecithins, cephalins,** and phosphatidyl derivatives of glycerol and inositol, whose structures are shown in Figure 9-1. The first step in the synthesis of substituted fats is the phosphorylation of **choline** (for lecithins) or **ethanolamine** (for cephalins). The choline or ethanolamine is then transferred to cytidine triphosphate, forming a CDP-derivative, which transfers the choline or ethanolamine to the phosphatidic acid acceptor and releases cytidine monophosphate. The nucleotide is then rephosphorylated at the expense of ATP.

Other substituted fats include the glycolipids and sulfolipids, also shown in Figure 9-1. These compounds are found in chloroplasts and various actively metabolizing parts of the cell and may be important cofactors in some metabolic or synthetic reactions.

The plant waxes are long chain carbon compounds, often pure hydrocarbons, sometimes with alcohol, aldehyde, or ketone groups. Waxes are largely deposited on the

Figure 9-1. Some substituted lipids.

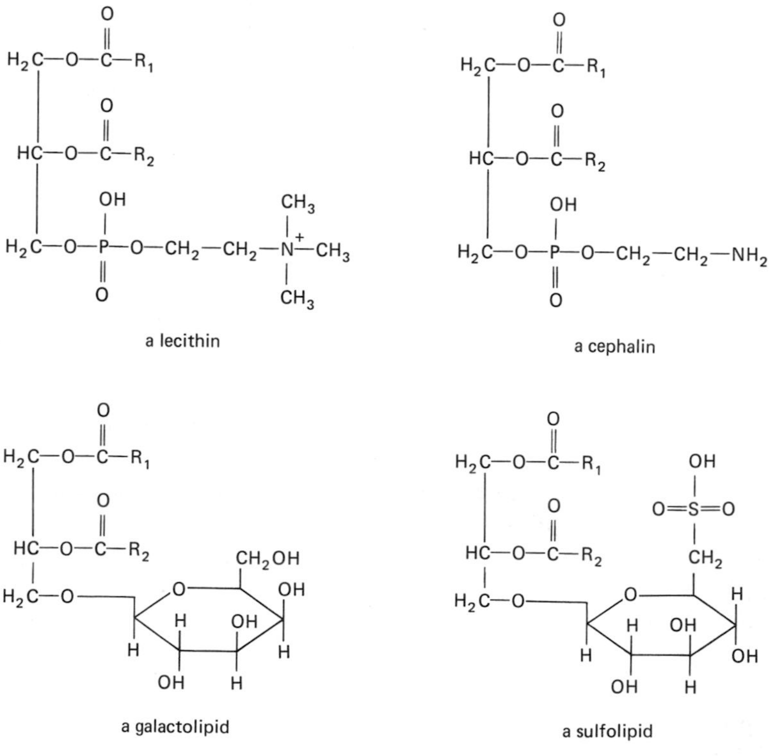

a lecithin

a cephalin

a galactolipid

a sulfolipid

outer surface of aerial portions of the plants, where they form a protection against water loss, infection, and mechanical damage to the delicate epidermal cells. Most of the plant waxes that have no oxygen or that have alcohol or ketone oxygens attached somewhere along the chain have an odd number of carbon atoms. Those that have an active group containing oxygen (for example, alcohol or carboxylic acid) at the end of the chain have even-numbered carbon chains.

Chlorophyll

The structure of chlorophyll was given earlier (Figure 7-4, page 153). The pathway of its biosynthesis was worked out in animals and bacteria and is shown in Figure 9-2. The starting materials are succinic acid, as succinyl-CoA, and glycine. These combine to form **δ-aminolevulinic acid;** two molecules of this substance condense to form **porphobilinogen,** which contains the pyrrole ring structure. Four molecules of porphobilinogen then condense to form the **porphyrinogen,** which has the basic tetrapyrrole structure of the porphyrin nucleus. Porphyrinogen undergoes modifications to make **protoporphyrin IX,** which has essentially the structure of chlorophyll but lacks the Mg atom. The Mg is introduced, the cyclopentanone ring V is formed, and the resulting **protochlorophyllide,** after conjugation with the appropriate protein in the plastid, is reduced through the action of light, which it absorbs, to **chlorophyllide a.*** The phytyl group is then esterified onto the propionic acid on ring IV, forming chlorophyll a. Chlorophyll b is apparently formed from chlorophyll a by oxidation of the methyl group on ring II to the aldehyde.

Recently it has been suggested that δ-aminolevulinic acid may be made in corn and perhaps other leaves by a much simpler reaction involving the reduction of α-ketoglutarate and the transamination of the product

$$
\begin{array}{c}
\mathrm{COOH} \\ | \\ \mathrm{CH_2} \\ | \\ \mathrm{CH_2} \\ | \\ \mathrm{C{=}O} \\ | \\ \mathrm{COOH}
\end{array}
\xrightarrow{\text{NADH} \quad \text{NAD}}
\begin{array}{c}
\mathrm{COOH} \\ | \\ \mathrm{CH_2} \\ | \\ \mathrm{CH_2} \\ | \\ \mathrm{C{=}O} \\ | \\ \mathrm{HC{=}O}
\end{array}
\xrightarrow{\text{alanine} \quad \text{pyruvate}}
\begin{array}{c}
\mathrm{COOH} \\ | \\ \mathrm{CH_2} \\ | \\ \mathrm{CH_2} \\ | \\ \mathrm{C{=}O} \\ | \\ \mathrm{H_2C{=}NH_2}
\end{array}
$$

α-ketoglutaric acid — dioxovaleric acid — δ-aminolevulinic acid

This synthesis has been shown in *Chlorella* and a photosynthetic bacterium, as well as in corn, beans, barley, and cucumber. It is probably of widespread occurrence in plants and may be of greater importance than the δ-aminolevulinic acid pathway mentioned above.

Iron is essential for the synthesis of δ-aminolevulinic acid; a shortage of iron results in a characteristic chlorosis of green tissue. The synthesis of chlorophyll from protochlorophyll or protochlorophyllide requires light in most species. The most effective wavelengths for this transformation are 450 and 650 nm, which are the absorption maxima of protochlorophyll. All the steps in chlorophyll synthesis from δ-aminolevulinic acid occur in chloroplasts, but the site of synthesis of δ-aminolevulinic acid itself is not known.

*The term **chlorophyllide** is usually applied to a chlorophyll that lacks the phytyl group. One that lacks the Mg atom is sometimes called a **chlorophyllin.**

Figure 9-2. Synthesis of chlorophyll.

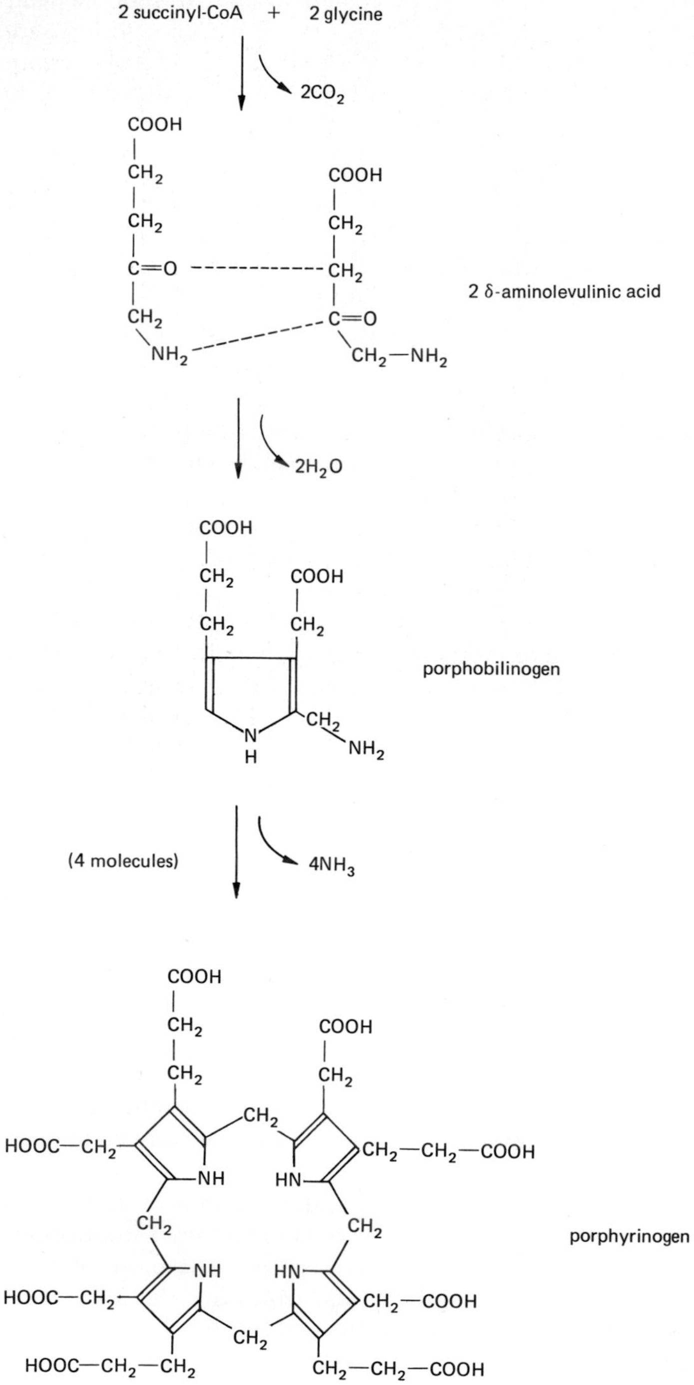

2 succinyl-CoA + 2 glycine

↓ (− $2CO_2$)

2 δ-aminolevulinic acid

COOH–CH_2–CH_2–C=O–CH_2–NH_2 COOH–CH_2–CH_2–C=O–CH_2–NH_2

↓ (− $2H_2O$)

porphobilinogen

↓ (4 molecules) (− $4NH_3$)

porphyrinogen

Figure 9-2. (continued)

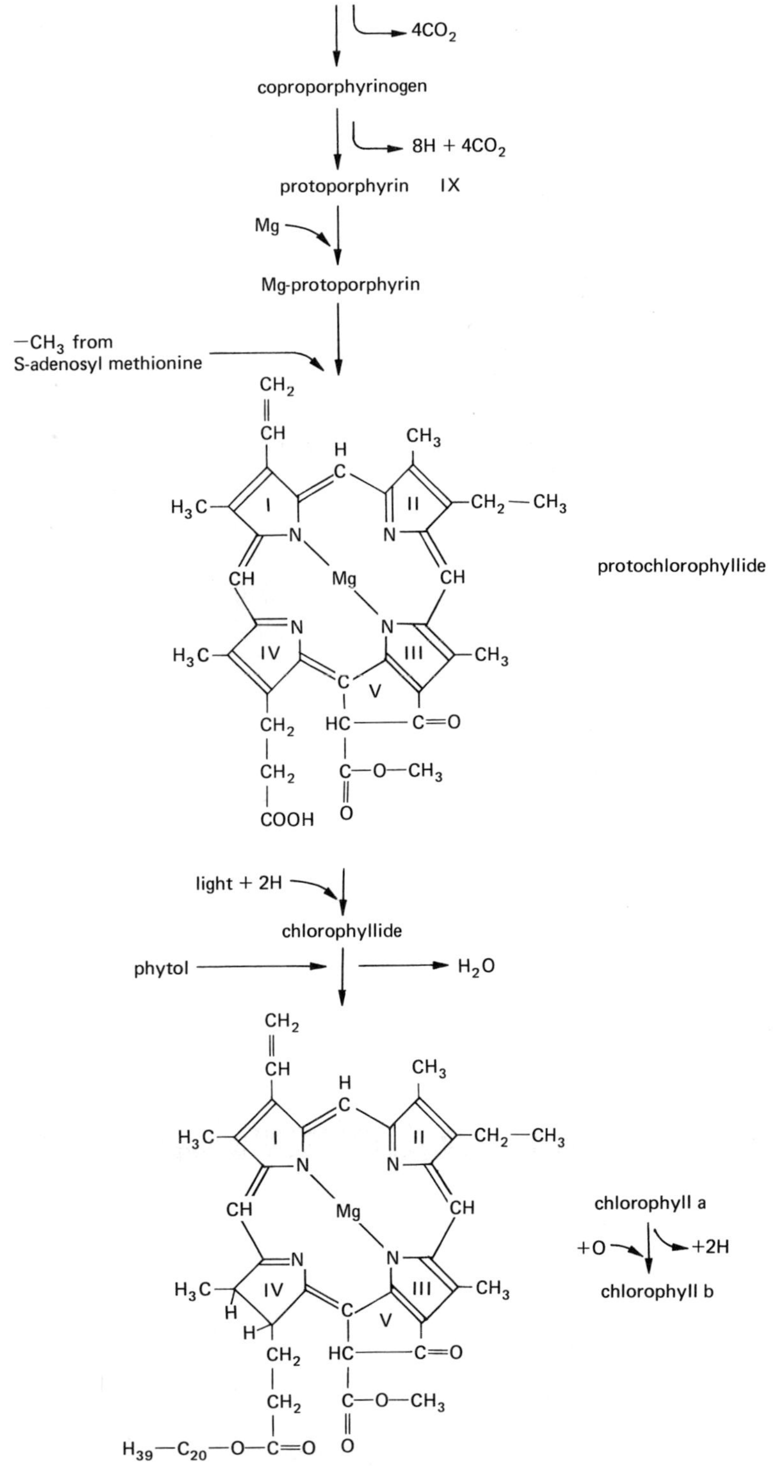

$\downarrow \rightarrow 4CO_2$

coproporphyrinogen

$\downarrow \rightarrow 8H + 4CO_2$

protoporphyrin IX

Mg $\rightarrow \downarrow$

Mg-protoporphyrin

$-CH_3$ from S-adenosyl methionine $\rightarrow \downarrow$

protochlorophyllide

light + 2H $\rightarrow \downarrow$

chlorophyllide

phytol $\rightarrow \downarrow \rightarrow H_2O$

chlorophyll a

$+O \rightarrow \downarrow \rightarrow +2H$

chlorophyll b

Isoprenoids

This very diverse group of plant constituents is made up essentially of polymers of the compound **isoprene**

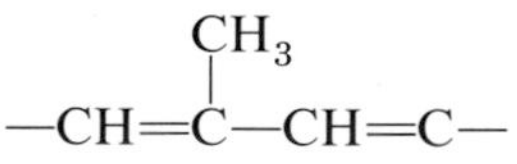

$$-CH{=}\underset{}{\overset{\displaystyle CH_3\atop |}{C}}-CH{=}C-$$

The isoprenoids include steroids, the carotene pigments and vitamin A, rubber, terpenes, essential oils, the phytol chain in chlorophyll, and some important plant hormones, the

Figure 9-3. Synthesis of terpene precursors isopentenyl pyrophosphate and dimethylallyl pyrophosphate.

$CH_3-\overset{O}{\overset{\|}{C}}-CoA + CH_3-\overset{O}{\overset{\|}{C}}-CoA$ acetyl-CoA

⇅ (→ CoA)

$CH_3-\overset{O}{\overset{\|}{C}}-CH_2-\overset{O}{\overset{\|}{C}}-CoA$ acetoacetyl-CoA

acetyl-CoA → ⇅ (→ CoA)

$CH_3-\overset{OH}{\overset{|}{C}}(CH_2-COOH)-CH_2-\overset{O}{\overset{\|}{C}}-CoA$ β-hydroxy-β-methyl glutaryl-CoA

$2\,NADPH_2^+$ → ⇅ → $2\,NADP^+$, CoA

$CH_3-\overset{OH}{\overset{|}{C}}(CH_2-COOH)-CH_2-CH_2OH$ mevalonic acid

2 ATP → ⇅ → 2 ADP

$CH_3-\overset{OH}{\overset{|}{C}}(CH_2-COOH)-CH_2-CH_2-O-P-O-P$ mevalonic acid pyrophosphate

ATP → ↓ → ADP + Pi, CO_2

$(CH_3)(CH_2{=})C-CH_2-CH_2-O-P-O-P$ isopentenyl pyrophosphate

⇅

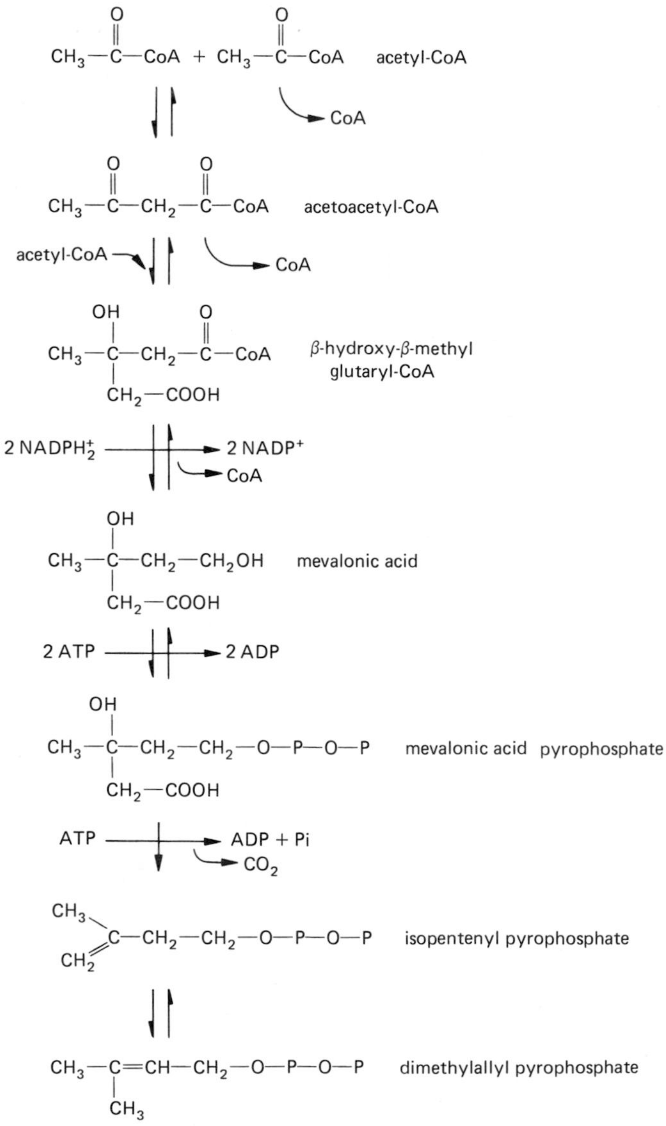

$CH_3-\underset{CH_3}{\underset{|}{C}}{=}CH-CH_2-O-P-O-P$ dimethylallyl pyrophosphate

gibberellins and abscisic acid. There are hundreds and possibly thousands of chemicals in this group, and all plants possess the ability to synthesize some of them. Certain plants are rich in isoprenoids, and a few make very large amounts of characteristic isoprenoids such as rubber. This important polymer is produced in several groups of plants, notably the Euphorbiaceae to which group the commercial rubber tree, *Hevea brasiliensis,* belongs.

Isoprene itself does not occur naturally. The basic building block of the isoprenoids is **isopentenyl pyrophosphate,** which is formed from three molecules of acetyl-CoA as shown in Figure 9-3. This compound can condense with a molecule of **dimethylallyl pyrophosphate,** an isomer derived from itself, to form **monoterpenes,** and additional isopentenol pyrophosphate groups may then be added. The structural relationship of common isoprenoids is shown in Figure 9-4.

Some members of this group are of great physiological importance. Gibberellins, powerful growth hormones, are derived from diterpenes that form intercyclic structures. Abscisic acid, which induces dormancy in plants, is also derived from isoprenoid intermediates, probably through a similar metabolic pathway. It has been suggested that part of the control mechanism of growth and dormancy is accomplished by a metabolic switch between the synthesis of gibberellic acid and abscisic acid (see also Chapters 22 and 23).

A number of animal hormones are derived from the sesquiterpenes (three isoprene units), including the insect juvenile hormone and male sex attractant. Since insects

Figure 9-4. Relationships of isoprenoid compounds. (Isoprene units are circled; *nx* = number of isoprene units.)

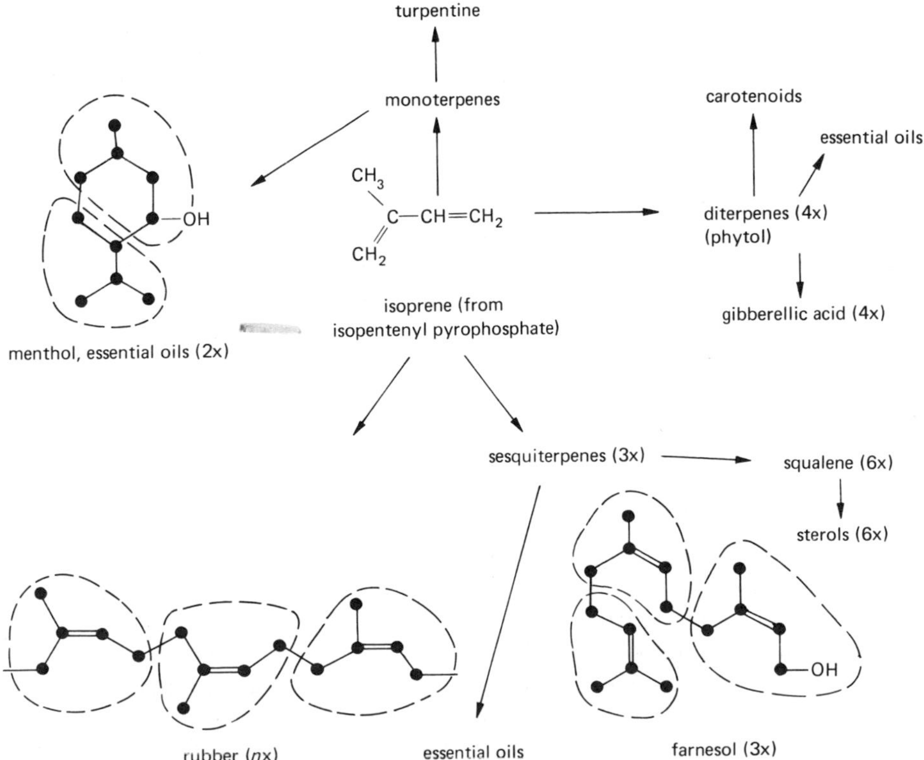

appear unable to manufacture sesquiterpenes, they presumably derive the carbon skeletons for these compounds from their plant food. Similarly vitamin A, which many animals are unable to synthesize, is derived from carotene synthesized in plants. The steroids, which often exert a powerful physiological influence in animals, are also derived from sesquiterpenes. Many of the characteristic aromatic and flavoring compounds in plants are essential oils (essence—as in a perfume). Rubber and gums are among the most valuable commercial products of plants. Chewing gum is an interesting mixture of isoprenoids, being made of a variety of rubber isomers and triterpenols, usually flavored with monoterpenols such as the essential oils from mint.

Phenols and Aromatic Compounds

All plants contain a number of chemicals of varying degrees of complexity whose basic unit is the benzene ring. The ring or rings may be more or less reduced, and many substitution groups are possible. Many important natural and synthetic hormones are substituted phenols or their derivatives. Certain amino acids, the prosthetic groups of some enzymes, and the structural material lignin are phenolic substances. Most phenolic substances are derived from intermediates of respiratory metabolism via the shikimic acid pathway, described below.

Aromatic Amino Acids, Indoleacetic Acid. Phenylalanine, tyrosine, and tryptophan are formed via the **shikimic acid pathway** illustrated in Figure 9-5. The starting compounds are phosphoenol pyruvate, derived from glycolysis, and erythrose-4-phosphate, which could be derived either from photosynthetic metabolism or from the pentose shunt in respiration. The first major stable intermediate having a benzene ring is shikimic acid, which gives its name to this pathway. An additional molecule of pyruvic acid is attached at C_3 to form chorismic acid. The pyruvyl side chain is transferred by an internal shift to C-1, forming prephenic acid. This is a major branch point, leading via phenylpyruvic acid to phenylalanine or via *p*-hydroxyphenylpyruvic acid to tyrosine.

Chorismic acid provides the branch point for indole synthesis, leading to tryptophan and the indole nucleus for the important plant growth substance, indoleacetic acid (see Figure 9-5). The side chain is removed and nitrogen is added to make anthranilic acid. A phosphoribosyl pyrophosphate (PRPP) derivative (see purine synthesis, page 219) is formed, which is converted to the glyceryl phosphate derivative of indol, and this derivative is converted to tryptophan by reaction with serine. Tryptophan may be converted by two possible pathways to indoleacetic acid, as shown in Figure 9-6. An interesting side branch leads to serotonin, an important factor in the transmission of nerve impulses in animals.

Simple Phenols and Lignin. A number of simple phenols are derived from intermediates of the shikimic acid pathway or from phenylalanine or tyrosine. These include cinnamic, coumaric, caffeic, ferulic, protocatechuic, chlorogenic, and quinic acids, as shown in Figure 9-7. They are widely distributed in plants, but their functions are not well known. Some have antibacterial or antifungal properties and may be involved in the resistance of certain plants to disease. In addition, many related compounds called **coumarins,** having a double ring structure (Figure 9-7), are found in plants. These compounds or their derivatives are often extremely toxic to animals, for example,

Figure 9-5. Aromatic amino acid biosynthesis. End products of the synthetic pathways are printed in capitals.

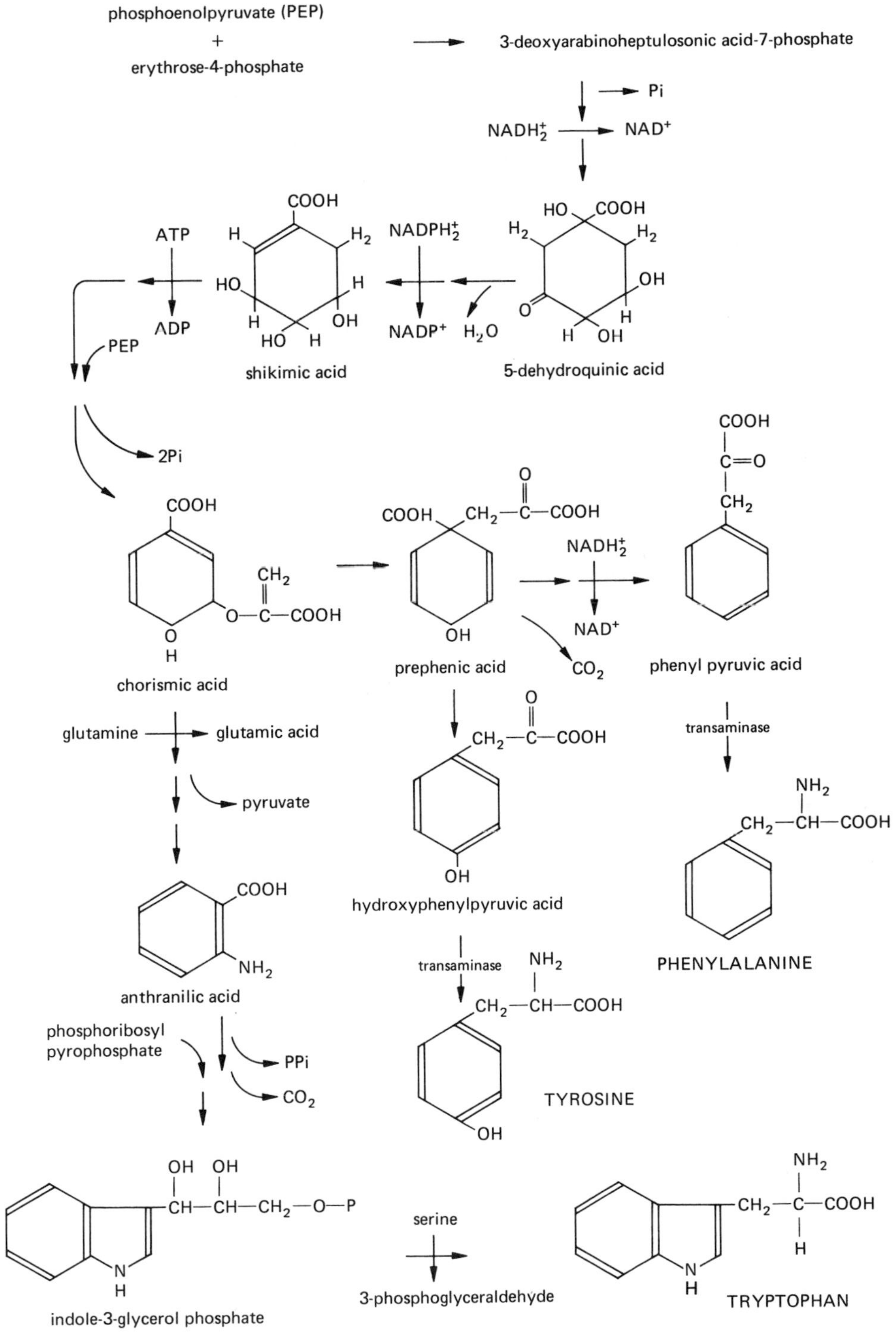

phosphoenolpyruvate (PEP) + erythrose-4-phosphate → 3-deoxyarabinoheptulosonic acid-7-phosphate

→ Pi

$NADH_2^+$ → NAD^+

5-dehydroquinic acid

$NADPH_2^+$ → $NADP^+$; H_2O

shikimic acid

ATP → ADP

PEP

→ 2Pi

chorismic acid

prephenic acid

$NADH_2^+$ → NAD^+; CO_2

phenyl pyruvic acid

transaminase

PHENYLALANINE

hydroxyphenylpyruvic acid

transaminase

TYROSINE

glutamine → glutamic acid

→ pyruvate

anthranilic acid

phosphoribosyl pyrophosphate

→ PPi

→ CO_2

indole-3-glycerol phosphate

serine → 3-phosphoglyceraldehyde

TRYPTOPHAN

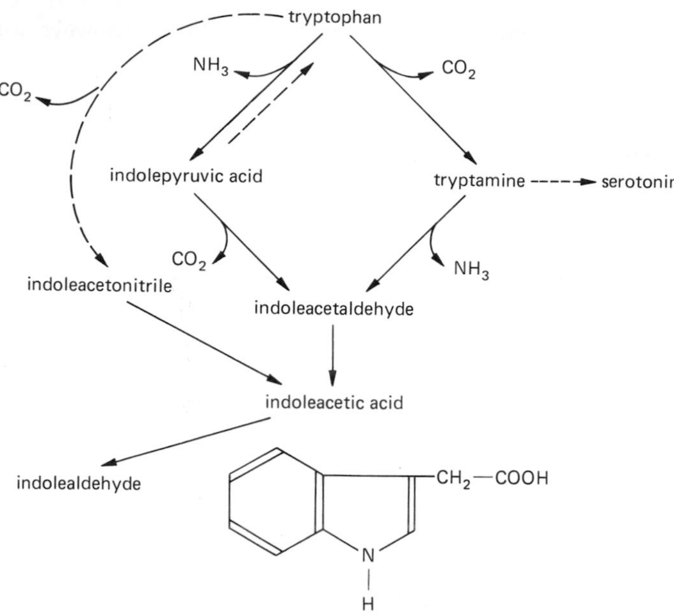

Figure 9-6. Indole transformations. The tryptamine and indole-pyruvate pathways occur in many plants. The indoleacetonitrile pathway is restricted to *Brassicaceae* and related groups. [From illustrations courtesy Dr. F. Wightman, Carleton University, Ottawa, Canada.]

dicoumarol formed from coumarin in clover during storage. The coumarins may be formed in plants in response to invasion by parasites, and thus make the plant resistant to the invasion.

An interesting enzyme in the metabolism of phenolics is **phenylalanine ammonia-lyase (PAL),** which catalyzes the deamination of phenylalanine to cinnamic acid (Figure 9-7), an important precursor of flavonoid compounds (shown in Figure 9-10). This important enzyme catalyzes the branching reaction that leads from the shikimic acid

Figure 9-7. Simple phenols and derivatives.

Simple Phenols

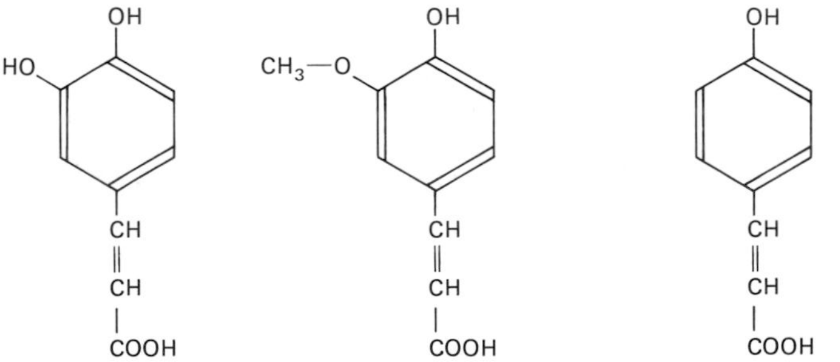

OH, HO, CH, CH, COOH — caffeic acid

OH, CH_3—O, CH, CH, COOH — ferulic acid

OH, CH, CH, COOH — *p*-coumaric acid

Figure 9-7. (continued)

chlorogenic acid

trans-cinnamic acid

protocatachuic acid

Coumarins

scopoletin

coumarin

Lignin Monomers

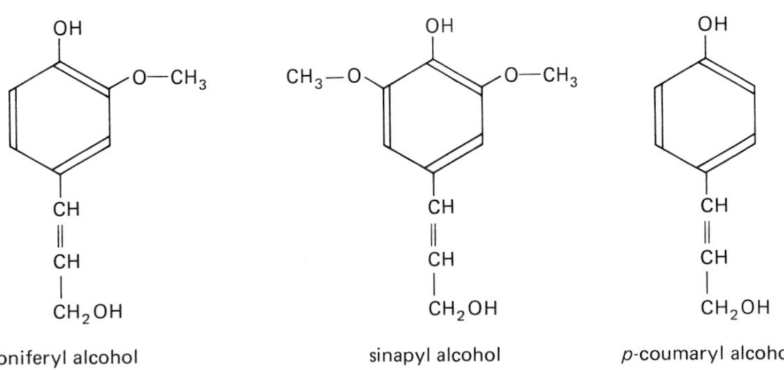

coniferyl alcohol

sinapyl alcohol

p-coumaryl alcohol

pathway (Figure 9-5) to a wide range of secondary products such as lignins, phenols, and coumarins, as well as flavones and anthocyanins. The interest lies in the fact that the activity of this enzyme is affected by a wide variety of external and internal factors. It varies with the developmental state of the plant. It is stimulated by wounding, by infection, and by the growth-regulating substance ethylene (see page 395). These factors are probably related: wounding and infection often stimulate ethylene formation in tissues. The phenolic compounds formed as the result of PAL stimulation include powerful bactericidal compounds and the precursors of lignins needed for the repair of wounds. Certain other hormones (notably IAA) inhibit PAL, and its activity is increased by high carbohydrate levels.

The most studied facet of PAL is its activation by light, as shown in Figure 9-8. Various light treatments, in addition to the blue light shown in Figure 9-8, activate the enzyme in a wide range of tissues. The lag phase is at present unexplained, but may relate to a response of phytochrome, a pigment involved in photomorphogenesis (see Chapter 20). There are conflicting data and viewpoints about the nature of the stimulation. Inhibitors of protein synthesis prevent the increase of PAL, and this suggests that light stimulates its synthesis. However, the enzyme appears to be in a state of continuous synthesis and degradation (**turnover,** see page 217), and the light effect may be the inhibition of its breakdown rather than the stimulation of its synthesis. Experiments in which tissue was incubated with heavy water (D_2O) during activation showed no more D incorporation into the enzyme in light-treated tissue than in dark controls. The light stimulation thus appeared to be the activation of existing enzyme or the prevention of its destruction. This is further supported by the rapid loss of activity after a time even if light treatment is continued, as seen in Figure 9-8. Inhibitors of protein synthesis applied at peak stimulation prevent subsequent loss, which suggests that protein metabolism is involved in the loss of activity as well as its increase. For example, the protein-synthesis inhibitors could equally be inhibiting the production of activators or inactivators of PAL. These ideas are summarized in Figure 9-9.

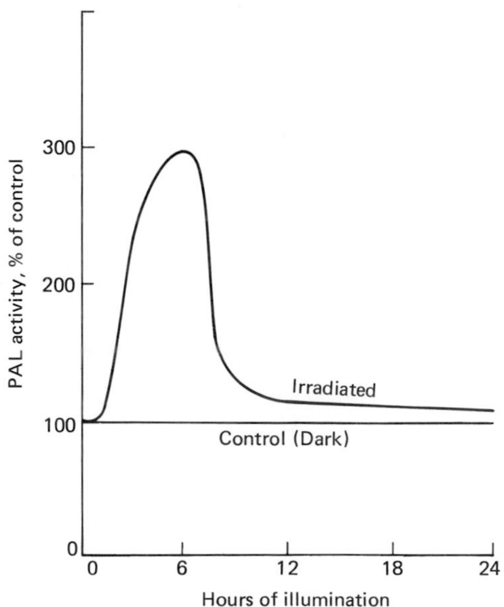

Figure 9-8. The effect of blue light on phenylalanine ammonia-lyase (PAL) in gherkin hypocotyl. [From data of H. Smith: The biochemistry of photomorphogenesis. In D. H. Northcote (ed.): *Plant Biochemistry,* Butterworths, London, 1976.]

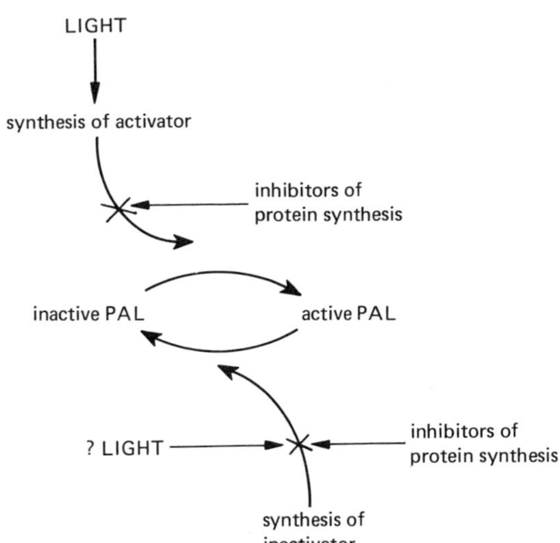

Figure 9-9. Possible mechanism of activation of phenylalanine ammonia-lyase (PAL).

It is obvious that this system is not yet fully understood. However, it forms an interesting example of a regulating enzyme that controls the relative activities of intersecting metabolic pathways. For this reason, and because it is easy to assay, it has become one of the most closely studied plant enzymes.

Lignin, after cellulose, is the most important biological substance in terms of both its total quantity in the world and its structural importance in woody tissue. Lignin is extremely difficult to study as a chemical because it cannot easily be extracted without extensive degradation. The monomers of lignin can be isolated, however, and they have been found to be mainly alcohol derivatives of simple phenolic substances, namely, coniferyl, sinapyl, and *p*-coumaryl alcohols, shown in Figure 9-7. The lignin monomers are assembled in an apparently random order by a complex system of interlocking bonds. The nature of some of them is known, but not the overall pattern of characteristic natural lignin. The important fact is that a massive amount of metabolized carbon is converted to lignins via the shikimic acid pathway. This pathway is thus of basic importance in the synthesis of many important metabolites, including such diverse compounds as protein amino acids, hormones, and the structural material of plants.

Flavones and Anthocyanins. These related compounds are derivatives of a triple ring structure shown in Figure 9-10. Ring B is derived from the shikimic acid pathway, and the rest of the structure appears to be derived by condensation of acetyl-CoA units, probably following their conversion to malonyl-CoA.

Flavones and anthocyanins are interesting because they are brightly colored. They are doubtless involved in the attraction of insects to flowers, where they chiefly occur. The anthocyanins are often similar in taxonomically related groups of plants and have been used in **chemical systematics** (the study of relationships among organisms revealed by similarities or differences in their chemical constituents or pathways). The behavior of anthocyanins is closely linked to development, and certain **leucoanthocyanins** (colorless) are suspected of acting as growth hormones in developing seeds. Environmental factors, including low nitrogen or phosphorus, cause increased anthocyanin formation in stems and leaves. The characteristic fall colors of foliage are anthocyanins, and their produc-

Figure 9-10. Anthocyanidins and anthocyanins.
Anthocyanidins are characterized by substitution on ring B:

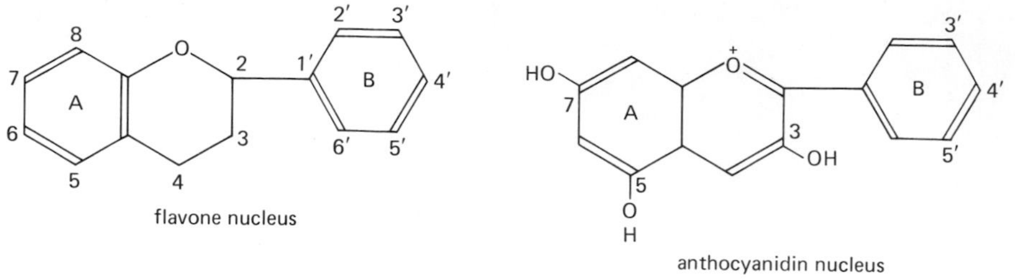

Anthocyanidin	*3′*	*4′*	*5′*	*Color*
Pelargonidin	—	OH	—	Red
Cyanidin	OH	OH	—	Red
Delphinidin	OH	OH	OH	Blue
Peonidin	OCH_3	OH	—	Red
Petunidin	OCH_3	OH	OH	Purple
Malvidin	OCH_3	OH	OCH_3	Mauve

Anthocyanins are glycoside derivatives at the 3, 5, or 7 hydroxyls on ring A and the central nucleus.

3: Several different sugars, most often glucose, often di- or trisaccharides
5: Sometimes glucose, seldom other sugars
7: Seldom glycosylated, then only by glucose

tion is much affected by temperature. Anthocyanin formation is often accelerated in senescing tissue. Their synthesis is usually promoted by light, blue light being the most effective.

Additional Reading

Annual Review of Biochemistry and Plant Physiology.

Bonner, J., and **J. E. Varner** (eds.): *Plant Biochemistry.* Academic Press, New York, 1965, Chaps. 13 and 21–28.

Pridham, J. B., and **T. S. Swain:** *Biosynthetic Pathways in Higher Plants.* Academic Press, New York, 1965.

Robinson, T.: *The Organic Constituents of Higher Plants.* Burgess Publishing Co., Minneapolis, Minn., 1967.

SECTION III

Soil, Water, and Air: The Nutrition of Plants

10 Soil and Mineral Nutrition

The Soil

The soil provides physical support and anchorage for many plants, and nutrients of various sorts for most. However, it is far more than a passive support or simply a container of water and nutrient salts. The soil is a complex medium influencing the life of the plant in many ways, because roots have not only to exist in it but to grow through it and because its chemical and physical properties can have powerful interactions with living roots. The soil-root system is a dynamic, living complex whose interrelations must be appreciated before the life of plants growing in soil can be understood.

Soil Texture and Structure. Soil **texture** refers to the size of the individual particles. Soils exist with various textures from extremely finely divided clays to coarse sand, together with varying amounts of organic matter. Soil particles are classified, according to their size, into **sand** (2–0.02 mm diameter), **silt** (0.02–0.002 mm diameter), and **clay** (less than 0.002 mm diameter). A roughly equal mixture of these three is called **loam.** Sand particles are small, often unweathered, fragments of rock and do not interact very strongly with water or minerals. Clay particles are small enough to be colloidal and exhibit the properties of colloids (see page 12). Silt particles are intermediate; their surfaces may be smooth or unweathered, but they are often coated with clay and so exhibit properties intermediate between those of sand and clay. Water percolates readily through sandy soils and evaporates from them with ease. Clay soils retain water very strongly. The water-retaining capacity is also much affected by the amount of organic matter present. The relationship between soil texture and the amount of water it holds is shown in Table 10-1.

Soil **structure** refers to the organization of the soil particles into clumps or aggregates. In sandy soils the aggregates are usually small and easily broken and often consist of single grains. However, silty or clay soils, particularly those containing much organic matter, often form into clumps or **crumbs** of 1 mm to several millimeters in diameter. Good soils have a crumbly structure. This gives them a large surface area, leading to good water-holding properties and good aeration. Soil structure can be damaged or changed by mechanical means (for example, by working clay land when it is too wet and thus forming clods or lumps). Heavy clays are particularly unstable and tend to **puddle,** losing their structure and coalescing into a solid mass. Such soils are poorly aerated and

Table 10-1. The relationship between texture and water-holding capacity of soils

Soil texture	*Water held, g/1000 cc*
Coarse sand	40–100
Sandy loam	100–175
Loam, silt loam	150–200
Clay loam	175–225
Clay	175–250
Muck and peat soils	200–300

drained and usually are not very productive. For agricultural purposes the structure of heavy clay soils has to be redeveloped by breaking up the clods and adding organic matter, which helps prevent puddling.

Organic matter (from 5 to 15 percent in most good agricultural soils) is important for the maintenance of soil structure and water-holding capacity. It is also extremely important in helping to hold nutrient minerals that might otherwise leach out of the soil. In addition, organic matter helps to provide substrates for the metabolism of soil organisms, which are unbelievably abundant in most good soils. The weight of total organisms (bacteria, molds, algae, and invertebrate animals) in the surface foot or so of soils is astonishingly large. It is usually in the range of 500–700 g/m^2 (equivalent to 4000–6000 lb/acre), but in some soils it may exceed 5 tons of living tissue per acre! These soil organisms are very important in the generation and maintenance of good soil structure, and they increase the fertility of soils by dissolving or freeing bound nutrients that might otherwise be unavailable to the plants. In addition, soil organisms are important for nitrogen fixation and other symbiotic relationships with plant roots (see Chapters 8 and 27). The soil also has a limited atmosphere that may contain 10 to 20 times more CO_2 than the air, perhaps because of the metabolic activities of the organisms it contains. This may be an important factor in the known β-carboxylating activity of roots.

The complex of events that leads to the development and formation of specific soil types and the diversity of possible soil compositions are beyond the scope of this book. It is important to recognize, however, that the origin, structure, and composition of soil have a great influence on the degree of aeration and the water relations of the soil, as well as on the spectrum, amounts, and availability of the minerals it contains. These in turn are powerful factors determining the type of plants that can grow in the soil and the physiological problems that face plants in their growth.

Soil Water. Water is present in soil in various forms and is subject to various kinds of stresses. Water is held in soils by absorptive forces or by hydrostatic pressure (that is, below the water table). It tends to leave the soil by evaporation, by gravity, or by absorption into plant roots. Since water uptake into roots takes place by osmosis, we must consider the osmotic potential of soil water as well as the various forces by which water is held in the soil.

The most effective way of examining the movement of water in soil is to consider its **water potential,** ψ (See Chapter 3). Water diffuses in a soil, just as it does between cells, from a region of high potential to a region of low potential. The components of water potential in a soil are the same as those in a cell: pressure potential (ψ_P), which includes the operation of forces of gravity; osmotic potential (ψ_π) of the soil solution; and the matric potential (ψ_M), an expression of the various chemical and physical attractions between water and the soil particles which result in the retention of water by soils. The matric potential includes capillary attraction and the intermolecular forces holding water of hydration in the soil colloids. In summary, water potential can be expressed in units of force or pressure as follows

$$\psi = \psi_P + \psi_\pi + \psi_M$$

The matric potential and osmotic potential interact quite strongly in soils due to the selective absorption of either water or solutes by colloidal particles, so their precise contribution to the water potential in the soil is rather difficult to determine.

The terms **water tension** or **water potential** have been used in the past to describe the *tendency of a soil to absorb water.* These terms are equal in value but opposite in sign to ψ, which correctly describes the potential of water rather than a characteristic of the matrix.

Water potential (ψ) of soils varies greatly. The value of ψ in a soil fully saturated with pure water at atmospheric pressure is zero. However, soil water is normally present as a solution, and in this case ψ would be below zero by an amount equal to the osmotic potential of the solution. At the other end of the scale, the water potential of a dry colloid could be as low as -3000 bars (45,000 lb/in.^{2}). More usual values range about -1 or -2 bars in normal soils. The water potential can be measured by placing soil in a container having a supported porous membrane in its bottom, and determining the applied pressure (by air or centrifugation) that is required to force water out.

When the soil has been thoroughly wetted and allowed to drain until capillary movement has essentially ceased, it is said to be at **field capacity.** Clay soils, having much larger surface area and lower matric potential, can hold far more water at field capacity than can sandy soils. The relationship between water potential and the amount of water present in typical soils is shown in Figure 10-1. Note that clay soils hold large amounts of water. They dry out slowly, but the water that remains is held with great tenacity (that is, its water potential is very low) and requires the use of large forces for its extraction by plant roots.

When the water potential falls low enough, plants are no longer able to absorb sufficient water or to absorb it rapidly enough to replace that lost by transpiration. At this point the leaves begin to wilt. If water loss is stopped by placing the leaves in a saturated atmosphere, they are able to recover and are said to be in a state of **incipient wilt.** However, a point is reached when the water content of the soil is so low that leaves cannot recover from wilting even if placed in a water-saturated atmosphere. The water content of the soil at this point is called the **permanent wilting percentage.** It is considered to be a constant of the soil, although it does vary slightly with the ability of the test plant to absorb water. Average values are about 1 percent for sand, 3–6 percent for loam, and up to 10 percent for clay, representing a water potential of about -15 bars. The water remaining in the soil at the permanent wilting percentage is unavailable to the plant, and plants held for long at the permanent wilting percentage will die.

Water may be present in soils in various forms: as water of hydration of colloids, as free water (often in capillary spaces in the soil), and as water vapor. When water is removed from the soil by roots, more water diffuses to the site of uptake to replace it. The resistance to water movement is a complex phenomenon but, in general, will be proportional to the amount of water present in the soil, the size of the spaces or **voids**

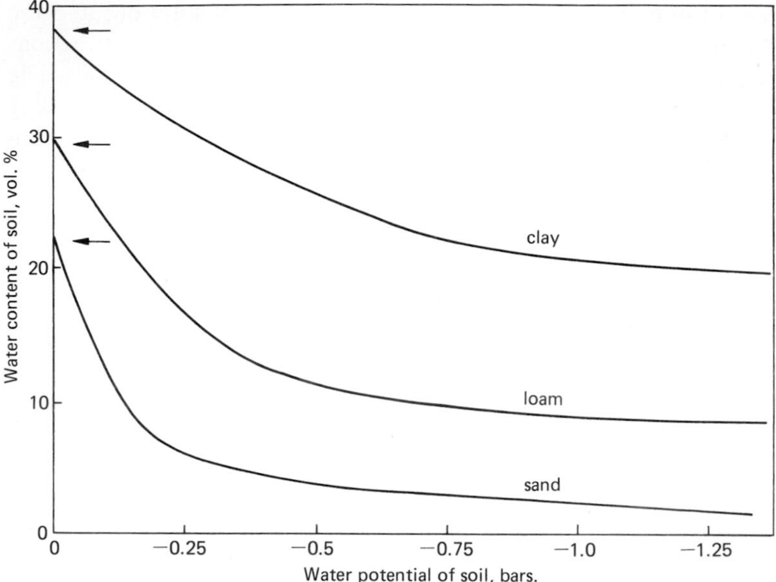

Figure 10-1. Relationship between water content and water potential, or soil moisture tension, of soils. The arrows mark field capacity of the soils.

between soil particles, and the tenacity with which water adsorbs on colloidal particles of the soil. Evidently water will move most readily in sand and most slowly in clay. When the moisture content of the soil becomes low, water may evaporate and move through the soil as vapor. Then the osmotic potential of the soil water may become an important factor in its movement because the water-air-water system introduces what amounts to a semipermeable barrier, which permits the passage of water but not dissolved solids. In general, movement of water through soils is quite slow and becomes almost imperceptible at the permanent wilting percentage. Plants absorb not only the water that moves toward them as a result of the reduced water potential around the roots but also soil water that the roots encounter as they grow through the soil. The enormous and continued growth of roots is necessary to satisfy this requirement, although of course the roots cannot grow to every water-filled or water-coated particle in the soil, because there would not be enough water to make that amount of roots!

Nutrients. Nutrients and other compounds are present in a dynamic state in the soil. They are continuously being added or removed by a variety of pathways, and the fertility of a soil is dependent on the relative rates of addition and removal of nutrient substances. In addition, elements may be held more or less tenaciously in soil, by chemical and physical bonding. Fertility thus may be affected also by the ease or difficulty with which nutrients are absorbed by roots, as well as by their tendency to be held or washed out of the soil by rain or moving ground water. Ions that are dissolved in the soil-water phase are freely available to roots. Ions that are bound to soil particles are only available as they enter solution, so the fertility of a soil depends on the concentration of nutrients in solution, not on the nutrient elements it contains.

Since soil particles are constantly being weathered and broken down, their composition and rate of decay will affect the fertility of the soil. Other factors that affect the amounts and availability of nutrients are pH, oxygen content, and the ion exchange capacity of the soil. The latter factor is dependent on the nature of the mineral and rock fragments from which the soil is made and particularly on the size of particles. A fine clay soil contains colloidal micelles having enormous surface area. Since colloidal surfaces are usually charged, they may be capable of holding large amounts of ions more or less tightly. In addition, the organic material present in the soil may have a very large ion exchange capacity. The ion exchange capacity of soils is naturally dependent on the availability and concentration of H^+ or OH^- ions, and the tendency of any given ion to be absorbed will depend very largely on the concentration of various other ions present.

The presence of soil microorganisms strongly affects the fertility of a soil. Microorganisms may be harmful by competing with plants for ions that are present in low concentrations because many of these ions become unavailable in organic form. Alternatively, the activity of microorganisms may affect the exchange of ions by changing the soil pH, so that certain elements become more available to plants. These effects may be quite striking. Microbial infestation of soil may greatly increase the growth of plants by increasing the availability of iron, boron, or molybdenum, for example.

Ions may be present dissolved in the **soil solution,** or bound by ion exchange reactions to the soil as **exchangeable nutrients** in the **exchange complex** of the soil. Alternatively they may be present in the molecular structure of the micelles of which the soil is composed, or very tightly bound to it, so that they are **nonexchangeable.** Levels of ions in the soil solution are measured by pressing out samples of soil water and analyzing this. Certain ions, such as sulfate and chloride, have only weak bonding energy and so are usually present in high levels in the soil solutions. Others, such as calcium and magnesium, may be present in larger or smaller amounts depending on the nature of the soil particles to which they are quite strongly absorbed. Levels of exchangeable nutrients in the exchange complex are measured by percolating the soil with ammonium acetate or acetic acid, which displaces ions from the exchange complex. Acetic acid usually displaces more of the micronutrients than ammonium acetate, perhaps due to their greater solubility in acid solutions. Nonexchangeable substances can be measured after suitable chemical degradation of the soil.

Both anions and cations are held in the exchange complex, indicating that soil is **amphoteric** (having both positive and negative charges). Cation exchange capacity is usually much larger and more important than anion exchange. Exchange capacity varies greatly among soils, as is shown in Table 10-2, being highest in clay soils and soils with high organic content. Different clays have different crystalline structures, which affect

Table 10-2. Cation exchange capacities of some soils

Type of soil	*Capacity, mEq/100 g*
Clays	
Kaolinite	5–15
Montmorillonite	80–120
Humus	150–300
Clay portion of loam	3–10
Organic portion of loam	10–20
Sand portion of loam	0–1

their ion-absorbing capacities as well as their behavior on hydration. **Kaolinite** particles are made of alternating layers of silica and alumina held rigidly together. Only the outside surfaces of particles are available for water absorption or ion exchange, so the exchange capacity and water retention of kaolinite are quite low and it does not swell much on hydration. **Montmorillonite** particles, on the other hand, are composed of alumina layers sandwiched between two silica layers. Each "sandwich" is only loosely bound to the next, and water or ions can be bound on all surfaces between such sandwiches within the particles. Thus montmorillonite has a high water and mineral-binding capacity and undergoes considerable swelling on hydration.

The availability to the soil solution, hence to plants, of elements in the exchange complex depends on their **bonding energy,** a measure of the tightness with which the ions are held. Aluminum, barium, and phosphorus have high bonding energies and are consequently present at low concentrations in the soil solution. Ions such as calcium, potassium, and magnesium have intermediate bonding energies, whereas sodium and most anions (chloride, sulfate, and so on) have weak bonding energies and consequently tend to be present largely in the soil solution. Phosphate is an exception, being quite tightly bound as a rule. Ions with high bonding energy will tend to displace ions of lower bonding energy. Also, adding a large amount of even a weakly bonding ion will result in

Figure 10-2. The effect of pH on the availability of nutrients to plants. The thickness of the horizontal band represents the solubility of the nutrient. The solubility is directly related to the availability of the nutrient in an ionic form that may be taken up by the plant. [From C. J. Pratt: *Plant Agriculture*. W. H. Freeman and Company, San Francisco, 1970. Used with permission.]

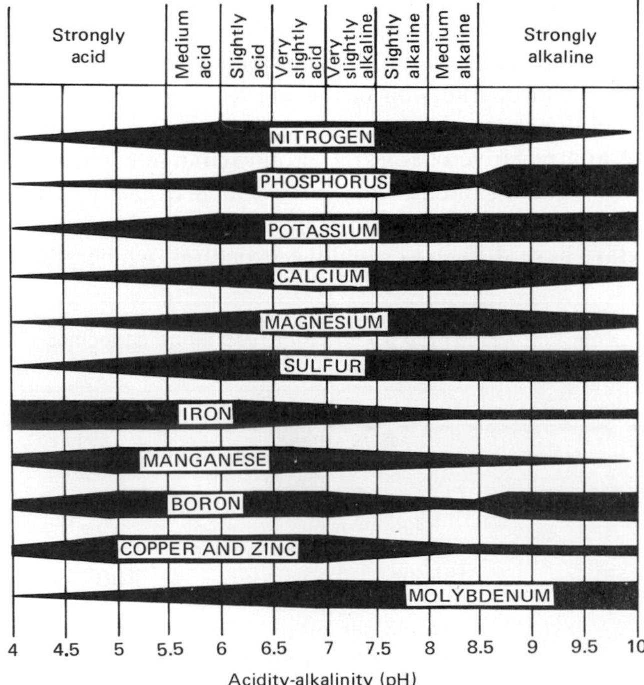

the displacement and release of a small amount of more strongly bound ions. Hydrogen and hydroxyl ions are strongly bound, thus pH is important in determining the availability of minerals that are present in soils. Figure 10-2 shows the effect of soil acidity or alkalinity on the availability of several important nutrients.

The amount of an ion present in the soil solution and exchange complex of a soil is thus the result of several factors: the total amount present (which may relate to the mineral nature of the soil particles), the exchange capacity of the soil, the soil pH, and the relative abundance of other ions. The addition of a strongly bound ion such as aluminum will release more weakly bound ions into the soil solution, thus increasing their abundance in the phase of the soil accessible to the roots. Phosphorus is often in limited supply because it is tightly bound; hydroxyl ions released by roots displace phosphate ions from the exchange complex, thus making more phosphate available. Similarly, calcium added to soil solution naturally or by fertilizing ("liming") releases magnesium and potassium from the exchange complex by displacement.

The removal of ions from the soil by roots will cause the liberation of more ions from the exchange complex. Thus, a dynamic equilibrium is maintained between the plant, the soil water, and the exchange complex, with the result that more nutrients become available to the root as they are removed, or as more water is added to the soil. A calculation by the Austrian physiologists M. Fried and H. Broeshart illustrates this point. An acre of corn producing 100 bushels of grain transpires about 10 acre-in. of water during the growing season. The crop requires 25–50 lb of phosphorus, 50–100 lb of calcium and magnesium, and 100–200 lb of potassium. Average soil solutions contain about 20 ppm phosphorus, 40 ppm calcium and magnesium, and 100 ppm potassium. Providing that these concentrations are maintained by equilibration from the exchange complex of the soil, the amount of water passing through the plant by transpiration will contain sufficient of these elements for the growth of the crop.

A summary of the important nutrients in the soil is presented in Table 10-3, together with some observations about their distribution and properties.

Mineral Nutrition

In addition to the major elements of their organic structure, carbon, hydrogen, and oxygen, plants contain a great variety of elements in various chemical forms. Evidently we cannot categorize the mineral constituents of plants according to the state in which they are present. A considerable portion of the minerals, particularly nitrogen, sulfur, and calcium, is present as part of the organic structure. Some of the metals are present as components of enzymes and catalytic molecules. Other minerals may be present simply because they are nonselectively absorbed along with soil water and tend to accumulate either as dissolved or stored ionic substances (selenium, strontium, sodium, and potassium, when present in excess) or as precipitates in the tissue (aluminum, silicon, and calcium).

Chemical Composition of Plants. The precise mineral composition of plants is of great interest when one considers the plant as a food source, because the absence or superabundance of certain elements may affect the food value of the plant. However, with certain specific exceptions, the plant's composition to a considerable extent reflects the fertility of the soil on which it is grown, so that detailed analyses of the elemental

Table 10-3. Nutrient ions in the soil

Ion	Usual levels, ppm: Soil solution	Usual levels, ppm: Exchange complex	Remarks
Ca^{2+}	50–1000	500–2000	Usually abundant in limestone soils, but may be deficient on granitic or sandy soils. Liming with CaO or $CaCO_3$ remedies this condition, as well as raising pH of acid soils and thus increasing supplies of K^+ and other cations by exchange. Excess liming may harm soils by making Fe, Mn, Zn, Cu, and B less available.
Mg^{2+}	1–100	20–150	High in limestone soils, lower in granitic.
K^+	1–50	10–50	Tends to be fixed in clay soils by absorption in the crystal lattice but remains in equilibrium with soluble K^+.
Na^+	10–500	3–50	Occasionally very high in saline soils.
Fe^{2+}, Fe^{3+}, $Fe(OH)_2^+$, $FeOH^{2+}$	0.1–25	1–500	Tends to complex with organic materials. More available in acid soils.
Mn^{2+}	0.2–2	1–4000	More available in acid soils. Oxides are precipitated, may be made available by bacterial action.
Cu^{2+}	0.01	10–1000	Very firmly bound in exchange complex.
Zn^{2+}, $ZnOH^+$, $ZnCl^-$	0.1–0.3	3–20	More available in alkaline conditions. May be firmly bound in exchange complex. Both Cu and Zn form insoluble phosphates.
PO_4^{3-}, HPO_4^{2-}, $H_2PO_4^-$	0.001–20	10–1000	Strongly held in exchange complex. Substantial amounts of P may be held in organic form. Insoluble phosphates of Ca (at high pH) or Al and Fe (at low pH) are only slowly released.
SO_4^{2-}	3–5000	low	Most of the sulfate in soils is either in the soil solution or in organic forms. Atmospheric pollution (SO_2 and SO_3) is a major source of S for plants in industrial areas.
Cl^-	10–1000	0	Largely related to salt deposition.
$H_2BO_3^-$, HBO_3^{2-}	0.1–6	very low	Liming tends to decrease the availability of B.
MoO_4^{3-}	<0.001	low	Unlike Mn, Mo becomes more available in the oxidized form in alkaline soils.

compositions of plants are not of great value except to the agricultural scientist. In addition, such data are very difficult to obtain.

The usual technique is to ash the plant at high temperatures and then analyze the ash chemically. The chemistry is difficult, and many elements, particularly nitrogen, sulfur, and chlorine, form volatile compounds, often organic, and are partially lost. A typical analysis of a corn plant is presented in Table 10-4. It may be seen that a large proportion of the mineral part of the plant is silicon, an element only required in minute amounts, if at all, and some 16 percent of the plant material was undetermined, indicating the difficulties in such an analysis. A more useful set of data is presented in Table 10-5, which gives standards for considering the nutrient status of orange trees from analyses of the leaves. Nonessential elements are not included in this table.

Table 10-4. Analysis of a corn plant

Element	*Percent of whole plant*	*Element*	*Percent of ash*
O	44.4	N	25.9
C	43.6	P	3.6
H	6.2	K	16.4
Ash	5.8	Ca	4.0
		Mg	3.2
		S	3.0
		Fe	1.5
		Si	20.8
		Al	1.9
		Cl	2.5
		Mn	0.6
		Undetermined	16.6

SOURCE: Adapted from E. C. Miller: *Plant Physiology.* McGraw-Hill Book Company, New York, 1938.

Table 10-5. Standards for classification of the nutrient status of orange trees (*Citrus sinensis*) based on concentration of mineral elements in 4- to 7-month old, spring-cycle leaves from nonfruiting terminals

*Element**	*Deficient, less than*	*Low range*	*Optimum range*	*High range*	*Excess more than*
Nitrogen, %	2.2	2.2–2.4	2.5–2.7	2.8–3.0	3.0
Phosphorus, %	0.09	0.09–0.11	0.12–0.16	0.17–0.29	0.30
Potassium, %	0.7	0.7–1.1	1.2–1.7	1.8–2.3	2.4
Calcium, %	1.5	1.5–2.9	3.0–4.5	4.6–6.0	7.0
Magnesium, %	0.20	0.20–0.29	0.30–0.49	0.50–0.70	0.80
Sulfur, %	0.14	0.14–0.19	0.20–0.39	0.40–0.60	0.60
Boron, ppm	20	20–35	36–100	101–200	260
Iron, ppm	35	35–49	50–120	130–200	250 ?
Manganese, ppm	18	18–24	25–49	50–500	1000
Zinc, ppm	18	18–24	25–49	50–200	200
Copper, ppm	3.6	3.7–4.9	5–12	13–19	20
Molybdenum, ppm	0.05	0.06–0.09	0.10–1	2–50	100 ?
Sodium, %	†	—	<0.16	0.17–0.24	0.25
Chlorine, %	?	?	<0.2	0.3–0.5	0.7
Lithium, ppm	†	—	<1	1–5	12

SOURCE: P. F. Smith: Leaf analysis of citrus. In N. F. Childers (ed.): *Fruit Nutrition.* Horticultural Publications, New Brunswick, N.J., 1966. Used with permission.

*Dry matter basis given in percent or parts per million, as shown.

† These elements are not known to be essential for normal growth of citrus.

? Indicates lack of information regarding value.

Macro- and Micronutrients. At this point we must introduce the concept of macro- and micronutrients. Evidently plants need carbon, hydrogen, and oxygen, but these are inevitably available in the normal media of plant growth—air and water—and will not be further discussed here. Certain elements—calcium, magnesium, potassium, nitrogen, phosphorus, and sulfur—are required by plants in relatively large amounts and are called **major nutrients** or **macronutrients.** Others, such as iron, manganese, boron, copper, zinc, molybdenum, and chlorine, are required in small amounts and are called **minor nutrients, micronutrients,** or **trace elements.** A few other elements are *beneficial* in the sense that they improve the growth of certain plants, but are not absolutely necessary. Certain elements can partially replace some of the essential nutrients, but usually not completely or as effectively.

Essential Nutrients. Early workers recognized only a small range of elements—the macronutrients and iron—as essential, due to the difficulties of conducting rigorous experiments. Impurities in the culture media, in the salts used, and from the vessels or pots used made it difficult to detect requirements of micronutrients, some of which are necessary only in minute quantities. In the latter part of the nineteenth century the German physiologist J. von Sachs established the technique of water culture of plants, now called **hydroponics,** that eliminated many of the difficulties with impurities. Nevertheless, it was not until the middle of the present century that sufficiently pure salts, water, and containers became available to prove the essential nature of some of the micronutrients.

In 1939 the American physiologists D. I. Arnon and P. R. Stout proposed the following **criteria of essentiality** for judging the exact status of a mineral in the nutrition of a plant.

1. The element must be essential for normal growth or reproduction, which can not proceed without it.
2. The element cannot be replaced by another element.
3. The requirement must be direct, that is, not the result of some indirect effect such as relieving toxicity caused by some other substance.

These criteria are rather strict, and some latitude is required in their application. The physiology of plants, their composition, and needs vary greatly with the circumstances. Thus an element may be absolutely essential under certain circumstances, whereas under other conditions it may be impossible to detect a requirement. It is extremely difficult to demonstrate the nonessentiality of an element. Certain micronutrients may be required at the level of a few hundred or less atoms per cell, but the most careful analysis possible could not detect the level of impurity in the culture solutions that would provide this amount. Unless a deficiency can be demonstrated, and relieved by the addition of the element, its essentiality cannot be proved. On the other hand, its nonessentiality cannot be proved merely by the inability to demonstrate a deficiency.

Culture Media. Since Sachs recognized the importance of culturing plants in a medium uninfluenced by any solid or supposedly inert supporting material, plant physiologists have been growing plants in nutrient solutions. These have the added advantage of providing exactly known and reproducible conditions for the growth of experimental plants. The basic requirement for a successful medium is that it provides

Table 10-6. A nutrient solution for plants, used in the author's laboratory

Substance	Concentration, Millimoles/liter	Substance	Concentration, mg/liter
Macronutrients		Micronutrients	
KNO_3	5	H_3BO_3	2.5
$Ca(NO_3)_2$	5	$MnCl_2 \cdot H_2O$	1.5
KH_2PO_4	1	$ZnCl_2$	0.10
$MgSO_4$	2	$CuCl_2 \cdot 2\ H_2O$	0.05
		MoO_3	0.05
Iron, $FeCl_3$ (+ chelate)	0.62		

all the nutrients in suitable amounts, that they are available to the plant, and that other conditions (for example, pH) are satisfactory. This is usually accomplished by a two- or three-part medium that is mixed as required, since some of the nutrients tend to be unstable and to precipitate on standing. Iron is often supplied as a **chelate** (from the Latin, "clawlike," a chemical, usually organic, that holds or binds the iron or other atoms as if in a claw) because it is very easily precipitated by other nutrients. Alternatively, an iron solution may be sprayed on the leaves of plants because iron is readily absorbed through foliar surfaces. The culture medium can also be used on plants growing in soil or sand or other inert supports as well as in hydroponic systems. The roots of many plants must be aerated when they are growing in water culture, since they are sensitive to low oxygen tension. A typical nutrient medium that works well for many types of plants is shown in Table 10-6. Certain plants have special requirements, and the medium may have to be modified for the growth of more fastidious species.

Macronutrients

We shall now consider the role of each nutrient substance in plants and the consequences of its deficiency. It should be emphasized that many of the symptoms of deficiency show in different ways in different plants, and the levels of substances producing deficiency may be quite different in different species. Some people who have worked in this field for many years develop the knack of recognizing mineral deficiencies from visual symptoms. It takes long experience, and even then is not always accurate. For this reason the discussion here is quite general. A photograph of plants suffering from the effects of various nutrient deficiencies is shown in Figure 10-3.

Elements can play three distinct roles in plants: electrochemical, structural, and catalytic. Electrochemical roles include balancing of ionic concentrations, stabilization of macromolecules, stabilization of colloids, charge neutralization, and so forth. Structural roles are played by elements that are incorporated into the chemical structure of biological molecules or are used in forming structural polymers (for example, calcium in pectin, phosphorus in phospholipids). Catalytic roles are played by elements involved in the active sites of enzymes.

Sometimes these categories overlap—for example, magnesium forms a part of the chlorophyll molecule and so could be said to play a structural role. However, chlorophyll is an important catalytic molecule that absolutely requires magnesium for its function; therefore, even though the magnesium ion is not directly involved in the chlorophyll

Figure 10-3. The effects of mineral deficiencies on tobacco plants. (**A**) Complete nutrient solution. The others lack (**B**) nitrogen, (**C**) phosphorus, (**D**) potassium, (**E**) boron, (**F**) calcium, (**G**) magnesium. [Courtesy U.S. Department of Agriculture.]

electron transfer reaction, it may still be said to perform a catalytic role in the plant. Some of the macronutrients perform all three roles; most of the micronutrients serve only catalytic functions.

Calcium. This element is abundant in most soils, and plants under natural conditions are seldom deficient in it. Nutrient solutions provide a lot of calcium, but it has recently been shown that plants will in fact grow well in very much lower concentrations of calcium if other adjustments are made in the composition of the nutrient medium. This may be of great importance to agriculture because the application of fertilizers is expensive and runoff from overfertilization is a major source of pollution. High concentrations of calcium, which tends to precipitate many substances, may be important in preventing toxic effects of other salts that might be present in excess.

Calcium is important in the synthesis of pectin in the middle lamella of the cell wall. It is also involved in the metabolism or formation of the nucleus and mitochondria. Calcium is thus an extremely important element to most plants, and an acute shortage causes rapid deterioration and death of the plant. Meristematic regions are affected early because a shortage of calcium prevents the formation of new cell walls, hence prevents cell division. Incomplete cell division, or mitosis without the formation of new cell walls, results in the formation of multinucleate cells, which are characteristic of calcium deficiency. Existing cell walls, particularly in supporting structures such as stems and petioles, become brittle or rigid; cell expansion is hindered. Chlorosis of the margins of younger leaves, "hooking" of leaf tips (the disease "wither tip"), and the formation of stunted, discolored roots are characteristic symptoms of calcium deficiency. Potato tubers are small and malformed, and fruits may be poorly developed or subject to such diseases as "blossom-end rot" of tomato. Since most of the calcium in the plant is immobile once deposited, calcium deficiency is most striking in young tissues; older tissue may be unaffected.

An interesting paradox has been observed in lemon trees growing in Russia. Plants growing on strongly calcareous soils nevertheless frequently show symptoms of calcium deficiency, and abnormally low calcium levels are found in the leaves. This paradox is apparently due to the fact that in highly calcareous soils iron is not available, and the plant suffers from iron deficiency. One of the consequences of iron deficiency is a reduction in the uptake of calcium. Thus, by this curious mechanism, an excess of calcium results in calcium deficiency!

Calcium serves only minor catalytic roles, being involved (though usually not exclusively) as the activator of a few enzymes such as phospholipase. It is probable that its presence in this capacity is required only in micro-amounts, so that a deficiency in this role likely never develops. It may serve an important role in detoxifying oxalic acid; calcium-oxalate crystals are often observed in the vacuoles of plant cells.

Magnesium. Magnesium is much less abundant in soils than calcium, and magnesium deficiency is not unusual in plants growing in sandy and some acid soils. It is required in quite large amounts by most plants, and the use of magnesium fertilizers (crushed dolomitic limestone is one of the best) is widespread.

Magnesium plays several important roles in the plant. It appears to be involved in the stabilization of ribosomal particles, binding together the subunits that make up the ribosome. It is involved in numerous enzymic reactions in various capacities. First, it may serve to link the enzyme and substrate together, as for example in reactions

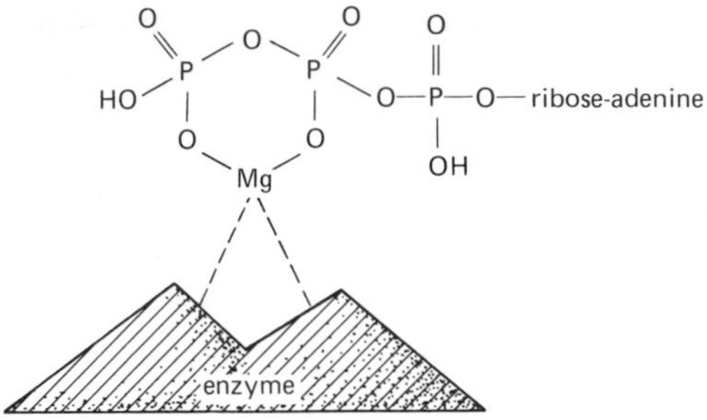

Figure 10-4. Possible role of magnesium in binding ATP to an enzyme.

involving phosphate transfer from ATP, in which magnesium serves as a link binding the enzyme to its substrate (Figure 10-4). Second, magnesium may serve to alter the equilibrium constant of a reaction by binding with a product, as for example in certain kinase reactions. Third, it may act by complexing with an enzyme inhibitor.

Magnesium is an activator, by one or more of these mechanisms, of many of the phosphate transfer reactions (except phosphorylases), of enzymes involved in the synthesis of nucleic acids, and also of many of the enzymes involving carbon dioxide transfer—carboxylation and decarboxylation reactions. As such, magnesium is crucial for the reactions of energy metabolism as well as in the synthesis of nuclear, chloroplast, and ribosomal constituents. Finally, magnesium is a component part of the chlorophyll molecule and is thus essential for photosynthesis.

Symptoms of magnesium deficiency are quite characteristic. Chlorosis develops between the veins, brilliantly colored red, orange, yellow, or purple pigments may appear, and in acute deficiency small areas or spots of necrosis appear. Since magnesium is quite soluble and readily transported around the plant, symptoms of its deficiency usually appear in older leaves first.

Potassium. Potassium is required in large amounts by plants, and a deficiency of this element may be frequent in light or sandy soils due to its solubility and the ease by which it can be leached from such soils. It is usually present in sufficient amounts in clay soils, where it is firmly bound. Potassium is the prevalent cation in plants and may be involved in the maintenance of ionic balance in cells.

Potassium appears to have no structural role in plants, but it serves a number of catalytic roles. These are mostly not clearly defined, and the exact nature of much of the large potassium requirement is unknown. Many enzymes, for example several involved in protein synthesis, do not act efficiently in the absence of potassium, although it does not seem to bind to them in the usual way. Its effect may be on protein conformation, causing exposure of active sites. However, this does not seem to account for the high specificity of potassium, which can only occasionally and inefficiently be replaced by sodium. Potassium is needed in large amounts—much more potassium is required than, for example, magnesium, for the activation of a dependent enzyme. Potassium is bound ionically to the enzyme pyruvate kinase, which is essential in respiration and carbohydrate metabolism, so potassium is very important in the overall metabolism of plants.

Potassium deficiency usually begins to show with a characteristic mottled chlorosis of older leaves that spreads to younger ones, potassium being a highly mobile element in plants. Necrotic areas develop along the margins and at the tip of leaves, which may curl in a characteristic manner, and widespread blackening or scorching of leaves may occur.

Potassium deficiency is often manifested by rosette or bushy habit of growth. Other consequences are reduction of stem growth, weakening of the stem, and lowered resistance to pathogens, so that potassium-deficient plants, especially cereals, are easily **lodged** (knocked down by weather) and attacked by diseases. Because of the reduction in protein synthesis and impairment of respiration, low molecular weight compounds such as amino acids and sugars tend to accumulate to unusually high levels in potassium-deficient plants, while proteins and polysaccharides are reduced.

Nitrogen. Nitrogen metabolism has been considered in Chapter 8, along with the process of nitrogen fixation, and need not be further considered here. Nitrogen has a special place in nutrition not only because of its high requirement by plants but because it is almost completely absent from the bedrock from which soils are made. The presence of nitrogen in the soil is almost entirely the result of biological action, artificial enrichment, or natural (resulting from lightning) fertilization. It is probable that most plants under natural conditions exist in a state of nitrogen deprivation, which, however, is not critical because of the great adaptability of plants to wide ranges of nutrition.

Nitrogen is of extreme importance in plants because it is a constituent of proteins, nucleic acids, and many other important substances. It does not, however, appear to have any specific catalytic or electrochemical roles apart from the fact that it is structurally involved in most catalytic molecules.

A deficiency of nitrogen almost invariably results in a gradual paling or chlorosis of older leaves, which may become yellow and abscise. Necrosis usually does not occur. The chlorosis spreads from older to younger leaves, which usually do not show the characteristic symptoms of deficiency until these symptoms are far advanced in older parts of the plant, indicating that nitrogen in old leaves is mobilized and transported to the younger growing parts of the plant as needed. A characteristic symptom of nitrogen deficiency is the development of anthocyanin in stems, leaf veins, and petioles, which may become red or purple. Young leaves on nitrogen-deficient plants are sometimes more erect and less spreading than normal, and branching or tillering is suppressed because of continued dormancy of lateral buds.

Plants respond in a variety of ways to high or low nitrogen supply. Overabundant nitrogen often causes a great proliferation of stems and leaves but a reduction in fruit in crop plants. Potatoes respond to overfertilization with nitrogen by producing large, dark-green, healthy looking tops but poor roots and smaller tubers. Slightly reduced nitrogen supply (but not a critical shortage), in relation to potassium and phosphorus supply, usually results in the most effective seed and fruit production of agricultural crops.

Phosphorus. The absorption of phosphorus occurs as the inorganic monovalent or divalent phosphate ion. Much of the phosphate in the plant exists in the organic form, but it is probably translocated largely in the inorganic state. Phosphate is held tightly in the soil mineral complex like potassium, and its uptake by plants may be antagonized by excess calcium. Thus phosphorus deficiency, although not often acute enough to cause gross morphological symptoms, may not be infrequent in nature. Phosphorus, like nitrogen, is extremely important as a structural part of many compounds, notably nucleic acids and phospholipids. In addition, phosphorus plays an indispensable role in energy metabolism, the high energy of hydrolysis of pyrophosphate and various organic phosphate bonds being used to drive chemical reactions.

As might be expected, phosphorus deficiency affects all aspects of plant metabolism and growth. Some of the results of low phosphorus, such as lateral bud dormancy, are in fact due to a resultant nitrogen deficiency. Symptoms of phosphorus deficiency are loss of older leaves, anthocyanin development in stems and leaf veins, and, in extreme cases, development of necrotic areas in various parts of the plant. Phosphorus-deficient plants develop slowly and are often stunted in growth. As in nitrogen deficiency, symptoms appear in older leaves first because of the great mobility of phosphorus, but, unlike nitrogen deficiency, leaves of phosphorus-deficient plants either tend to become darker green or else chlorosis spreads to the leaf veins as well as the lamella. Soluble carbohydrates may accumulate in phosphorus-deficient plants. One of the characteristics of phosphorus deficiency is a striking increase in the activity of the enzyme phosphatase; this may be related to the mobilization and reuse of available phosphate that take place under these conditions.

Sulfur. Sulfur is present as sulfate in the mineral fraction of many soils, but it is often also present in the form of elemental sulfur or sulfides of iron (FeS, FeS_2) which are unavailable to plants. A number of soil microorganisms are capable of oxidizing sulfur or the sulfides to sulfate and decomposing the organic sulfur compounds that may constitute much of the sulfur in richer soils. In industrialized areas (and areas surrounding natural phenomena such as geysers or volcanoes generating gaseous sulfur), atmospheric sulfur dioxide and sulfur trioxide (SO_2 and SO_3) may be important sources of sulfur nutrition. Indeed, it is extremely difficult to demonstrate sulfur deficiency in greenhouse plants grown in large industrial cities because of the high sulfur content of the air, since the airborne sulfur is either absorbed directly by the plant or dissolved in the nutrient medium.

Sulfur has somewhat more specialized roles than either of the other two major anionic nutrients, nitrogen and phosphorus. It forms part of the amino acids cysteine, cystine, and methionine and is an important constituent of proteins and some biologically active compounds such as glutathione, biotin, thiamine, and coenzyme A. The sulfur is often in the form of oxidizable sulfhydryl (—SH) groups, which form the active site of some redox and electron transfer agents. It is also important in forming disulfide (S—S), bridges that are involved in forming and stabilizing the tertiary structure of enzymes and other proteins. Many powerful inhibitors or poisons operate by attacking sulfhydryl groups; their action can often be reduced or relieved by the addition of an excess of some SH compound that binds the inhibitor.

Sulfur is converted into organic compounds by an adenosine derivative, 3′-phosphoadenosine-5′-phosphosulfate (PAPS) described by Franz Lipmann and coworkers in 1956. This compound is formed at the expense of ATP

$$SO_4^{2-} + \text{ATP} \xrightarrow[\text{sulfurylase}]{Mg^{2+}} \text{adenosine-5}'\text{-phosphosulfate} + \text{PPi} \quad (\text{PPi} \rightarrow 2\,\text{Pi})$$

$$\text{APS} + \text{ATP} \xrightarrow[\text{kinase}]{Mg^{2+}} \text{PAPS} + \text{ADP}$$

The sulfur moiety of PAPS is then reduced (possibly by ferredoxin but probably not by NADH or NADPH) and incorporated into organic molecules by pathways not yet clearly understood. **Sulfite reductase** is an enzyme that is in many respects similar to nitrite reductase (see page 200) and, like nitrite reductase, it may derive its electrons from the iron-containing porphyrin siroheme. The sulfur in PAPS is ultimately reduced to sulfide, and this combines with acetyl serine to make the sulfur-containing amino acid cysteine, as follows:

$$\underset{\text{acetyl-serine}}{\begin{array}{l} CH_2{-}O{-}CH_2{-}COOH \\ | \\ CHNH_2 \\ | \\ COOH \end{array}} + PAPS \longrightarrow \underset{\text{cysteine}}{\begin{array}{l} CH_2{-}SH \\ | \\ CHNH_2 \\ | \\ COOH \end{array}} + \underset{\text{acetate}}{CH_3{-}COOH} + ADP$$

The details of this reaction are not entirely clear. The substrate of reduction may be APS rather then PAPS, and unknown carrier molecules may be involved in the reduction step yielding sulfide and its transfer to acetyl serine.

Sulfur deficiency seldom occurs in nature, although the disease "tea yellows" of tea has been found to be the result of sulfur deficiency. It is characterized by a general chlorosis and yellowing of leaves, usually beginning with younger leaves, unlike nitrogen deficiency. Metabolic disturbances following sulfur deficiency may be very profound, largely because the plant is unable to make proteins as a result of a shortage of sulfur-containing amino acids. Soluble nitrogen tends to accumulate, and nitrogen-rich amino acids like glutamine and arginine reach high concentrations. Arginine breakdown may even lead to urea and ammonia production in acute sulfur deprivation, a condition not usually found under other circumstances in plants.

Micronutrients

The micronutrients usually serve in catalytic roles in plants and are only required in minute amounts. Although they are widely distributed in soils, certain micronutrients are lacking or in short supply in some areas of the world because they are absent from the base rock from which the soil is made. In addition, conditions of soil pH, the presence of other solutes, and the level of oxygen in the soil may affect their solubility or the plant's ability to absorb them, so deficiencies not infrequently occur. Deficiencies of specific micronutrient ions are responsible for many characteristic plant diseases and are of interest as such. Because of the experimental difficulties, the essential nature of many micronutrients and their relationship with specific, often well-known, diseases have not been established until recently. Indeed, the status of several substances such as sodium, selenium, silicon, chlorine, and aluminum is still not clear.

Iron. More iron is required than any other micronutrient, and it has been considered a macronutrient, or in a category by itself. However, this high requirement may be related to the strong tendency of iron to form insoluble compounds of various sorts in the soil and in the plant, which renders it unavailable or useless. Alkaline or limed calcareous soils commonly produce plants deficient in iron, even if it is abundant in the soil minerals, because of the formation of insoluble iron oxides or hydroxides. Thus, excesses of several other minerals may cause symptoms of iron deficiency by precipitating iron in unavailable forms. On the other hand, iron toxicity can also occur if soils that are high in iron become strongly acidic.

The extreme importance of iron is related to two important facts: iron is part of the catalytic site of many important oxidation-reduction enzymes, and it is essential for the formation of chlorophyll, though it is not part of the molecule. The importance of iron in the heme proteins (cytochromes and cytochrome oxidase) of the electron transport chain follows from its ability to exist in the oxidized or reduced form; that is, iron may add or lose an electron, undergoing a valency change as it does so. However, it is also present in a number of important oxidizing enzymes (for example, catalase and peroxidase) in which it does not undergo a valence change. One of the consequences of iron deficiency is a compensatory increase in noniron oxidases whose action may in part replace that of the iron-containing respiratory enzymes. Iron is a component or several nonheme enzymes, such as some flavoproteins and the extremely important electron transfer agent, ferredoxin. In addition, iron may be structurally involved in lamellar lipids in the nucleus, chloroplasts, and mitochondria and appears to be required for the synthesis of membrane proteins. It has been shown that higher levels of iron nutrition are required for cell divison than for respiration, reflecting its multiple functions.

The role of iron in chlorophyll synthesis is not clear, and it may be related to the synthesis of structural components of the chloroplast as well as to the synthesis of the chlorophyll molecule itself. It has been found that the amount of iron sufficient to maintain the growth of *Euglena* cells is lower than the amount required for maximum chlorophyll synthesis. Moreover, iron is apparently not specifically involved in those enzymic steps in the synthesis of chlorophyll that have been investigated. It may be that the synthesis of chlorophyll occurs as part of a total synthesis of the plastid structure, which would itself be limited if not enough iron was available to produce the required amount of cytochromes, ferredoxin, and other catalytic or structural iron compounds. Supporting this concept, it has been found that chlorosis resulting from iron deficiency is characterized by the simultaneous loss of chlorophyll and the disintegration of chloroplasts. This is also consistent with the fact that the degree of chlorosis in leaves is not closely correlated with the iron content, unless iron is supplied at a low, steady rate.

The symptoms of iron deficiency are easily recognized and very specific Chlorosis develops, sharply confined to the younger leaves in growing plants, without evident stunting or necrosis. The chlorosis is easily relieved in plants growing on alkaline or iron-deficient soils by spraying with an iron solution (usually iron complexed with a chelate such as ethylenediaminetetraacetic acid, EDTA). Incipient iron deficiency of fruit trees in the western United States and several Mediterranean countries used to be treated, before the discovery of the efficiency of foliar sprays, by driving iron nails into the trunks.

Manganese. Various forms of manganese may be present in the soil, but the reduced manganous ion (Mn^{2+}) is the form in which it is largely absorbed. Manganese, like iron, may become deficient in oxidizing or alkaline soils because it is converted to unavailable forms.

Manganese is widely involved in catalytic roles in plants, being the enzyme-activating metal of some respiratory enzymes and in reactions of nitrogen metabolism and photosynthesis. It is required for the operation of nitrate reductase; manganese-deficient plants usually require NH_3 for this reason. It is also required for the operation of some enzymes in the metabolism of the hormone indoleacetic acid. The most important role of manganese in photosynthesis is in the sequence of reactions by which electrons are derived from water and oxygen is liberated. Manganese may also have a structural role in chloroplasts, which become light sensitive in its absence and ultimately lose their structure and disintegrate under conditions of extreme manganese shortage. The structure of mitochondria and nuclei does not seem to be affected in the same way, indicating that the role of manganese in chloroplasts, unlike that of iron, may be quite specific.

The symptoms of manganese deficiency are the formation of small necrotic spots on leaves and the necrosis of cotyledons in leguminous seedlings. Manganese mobility is complex and depends on the species and the age of the plant, so the symptoms may appear first in either younger or older leaves. Characteristic deficiency diseases are "gray speck" of oats, "speckled yellows" of sugar beets, and "marsh spot" of peas.

Boron. Boron is present in small amounts in most soils, but its availability is often extremely low because it is tightly complexed in the structure of the soil. Boron uptake is low from calcium-rich soils, and liming tends to reduce its uptake, suggesting that calcium either causes boron to complex or precipitate in the soil or reduces the capacity of roots to absorb it.

Boron is an element whose role in plant metabolism is not clearly understood as yet, although it is demonstrably essential for plant growth. Translocation and absorption of sugars are much reduced in the absence of boron, and it has been suggested that sugar may be moved in the form of borate complexes in the plant. Radioactive sugars are found to be more readily taken up through the leaf surface if small amounts of borate are added simultaneously, and the addition of boron increases the translocation of radioactive products of photosynthesis in $^{14}CO_2$. The suggestion that the borate effect on translocation may be due to its complexing with cell membranes has also been made. The characteristic effects of boron deficiency, which usually results in the death of meristems and abortion of flowers, may result from reduced translocation of sugars to areas of high metabolism where they are most needed. There are opposing views, however. Boron may act as a natural and required inhibitor in plants, controlling the activity of enzymes that lead to the production of toxic phenolic substances. It has also been suggested that boron may be involved in many facets of cell differentiation and development.

The symptoms of boron deficiency are very characteristic. Leaves tend to thicken and darken, and meristems of shoots and roots die, giving the plant a stunted and bushy appearance as in "top sickness" of tobacco. Developmental aberrations may cause anomalous deposition of corky tissue in fruit trees ("drought spot" and "black measles" or "corky core" of apples). Metabolic disorganization leads to disintegration of cells in fleshy organs, giving rise to such disorders as "heart rot" or "water core" in beets and turnips. A common result of incipient boron deficiency often found in plants growing on acid, granitic soils (blueberries are especially susceptible) is the abortion of flowers and developing fruits.

Copper. Copper is almost universally present in small amounts in soils and is continually replenished by the weathering of copper-containing minerals. It is normally present in the exchange complex of soils where it is tightly held but available to plants, so its deficiency in nature is uncommon. However, excessive fertilization with phosphate may reduce the availability of copper by forming insoluble precipitates, and fruit orchards occasionally suffer from a deficiency.

Copper plays exclusively catalytic roles in plants, being part of a number of important enzymes such as polyphenol oxidase and ascorbic acid oxidase. It is present in chloroplasts in plastocyanin, an important member of the photosynthetic electron transport system, and it may be involved in nitrite reduction.

Deficiency of copper causes necrosis of leaf tips and produces a withered, dark appearance of leaves. Characteristic deficiency diseases are "reclamation" of crop plants, where the leaves are tightly rolled and white at the tips, and "die back" and "exanthema" of fruit trees, where the leaves wither and fall and the bark becomes rough and split, exuding gummy substances.

Zinc. Zinc is widely distributed in soils but, like many other metals, it becomes less available as pH rises. As a result, some degree of zinc deficiency is quite widespread, particularly in citrus orchards on neutral or alkaline soils. Zinc is directly involved in the synthesis of the hormone indoleacetic acid (IAA), and as such its deficiency may cause substantial changes in the form and growth habit of some species, producing shortened, stunted plants with poorly developed apical dominance. In addition, zinc is an obligatory activator of a number of important enzymes, including lactic acid, glutamic acid, alcohol, and pyrimidine nucleotide dehydrogenases. Zinc appears to be involved in protein synthesis, since its deficiency may result in a substantial increase in soluble nitrogen compounds.

Symptoms of zinc deficiency include stunting and striking reduction of leaf size, leading to "little leaf" and "rosette" of apples and peach trees, and interveinal chlorosis producing "mottled leaf" and "frenching" of citrus plants. A lack of zinc produces the disease "white bud" of maize and may lead to greatly reduced flowering and fruiting as well as stunted and poorly differentiated root growth.

Molybdenum. Molybdenum is present in minute quantities in many soils but, in spite of its low requirement by plants, incipient molybdenum deficiency appears to be widespread. It is more readily absorbed from soils of high pH (unlike most other metals) and hence tends to be in short supply in acid, granitic, and sandy soils. The most important role of molybdenum is in nitrate reduction and nitrogen fixation, and its deficiency, particularly in plants that have symbiotic nitrogen fixation, results in a reduced organic nitrogen content. This may lead to symptoms of nitrogen deficiency, which makes the detection of molybdenum deficiency rather difficult. However, molybdenum evidently has other functions because most plants require molybdenum even when grown in the presence of ammonium fertilizer.

Symptoms of molybdenum deficiency include mottling and marginal wilting of leaves, producing the disease "yellow spot" of citrus fruits. Chlorosis takes place starting with the older leaves, as in nitrogen deficiency, but, unlike nitrogen-deficient plants, the cotyledons stay healthy and green in appearance. "Whiptail" is a very characteristic molybdenum deficiency disease of cabbages and related plants, in which the young leaves are greatly distorted, having a long midrib and narrow, poorly developed, often ragged blades.

Chlorine. So far as is known, chlorine is absorbed and remains in the plant as the chloride ion. Although deficiency probably never occurs in nature, it was found in 1953 by the American physiologists T. C. Broyer, P. R. Stout, and their colleagues that it is essential for the growth of the tomato. In the absence of chlorine, plants wilt, their roots are stunted, and fruiting is reduced. Recently D. I. Arnon has demonstrated an absolute requirement for chloride ions in the photosynthesis of isolated chloroplasts. Presumably part at least of the need for chloride in plants is the result of this requirement. Chlorine requirement has now been shown to be quite widespread in plants and is probably universal. It is an interesting comment on the difficulty of establishing the absolute necessity of a micronutrient that Stout and Broyer are said to have doubted their own results (a doubt that was widely shared) until the role of chloride in photosynthesis was demonstrated!

A Key to Nutrient Deficiency Symptoms*

	Element deficient
a. Older leaves affected.	
b. Effects mostly generalized over whole plant, lower leaves dry up and die.	
c. Plants light green, lower leaves yellow, drying to brown, stalks become short and slender.	*Nitrogen*
c. Plants dark green, often red or purple colors appear, lower leaves yellow, drying to dark green, stalks become short and slender.	*Phosphorus*
b. Effects mostly localized, mottling or chlorosis, lower leaves do not dry up but become mottled or chlorotic, leaf margins cupped or tucked.	
c. Leaves mottled or chlorotic, sometimes reddened, necrotic spots, stalks slender.	*Magnesium*
c. Mottled or chlorotic leaves, necrotic spots small and between veins or near leaf tips and margins, stalks slender.	*Potassium*
c. Necrotic spots large and general, eventually involving veins, leaves thick, stalks short.	*Zinc*
a. Young leaves affected.	
b. Terminal buds die, distortion and necrosis of young leaves.	
c. Young leaves hooked, then die back at tips and margins.	*Calcium*
c. Young leaves light green at bases, die back from base, leaves twisted.	*Boron*
b. Terminal buds remain alive but chlorotic or wilted, without necrotic spots.	
c. Young leaves wilted, without chlorosis, stem tip weak.	*Copper*
c. Young leaves not wilted, chlorosis occurs.	
d. Small necrotic spots, veins remain green.	*Manganese*
d. No necrotic spots.	
e. Veins remain green.	*Iron*
e. Veins become chlorotic.	*Sulfur*

*Adapted from *Diagnostic Techniques for Soils and Crops.* American Potash Institute, Washington, D.C., 1948.

Beneficial and Toxic Elements

In addition to the essential elements that we have considered previously, there are many reports in the literature about a wide range of elements that have growth-promoting effects or that may partially replace essential elements. In addition, certain substances are present in soils that may be toxic even in small amounts (naturally any salt is toxic if present in excess).

Beneficial Elements. These include cobalt, sodium, selenium, silicon, gallium, and possibly others.

Cobalt is required by some organisms, particularly algae and other microorganisms. It is a component of vitamin B_{12} and various related compounds that are active in the metabolism of one-carbon compounds (methyl, formyl, formaldehyde, and carboxyl groups). The absolute requirement for cobalt, however, is so low that it cannot be demonstrated easily. Probably 1 part per 10^{12} is sufficient, which is beyond the limits of purification or measurement. Cobalt seems to be necessary for the bacteria involved in symbiotic nitrogen fixation, and many symbiotic nitrogen-fixing systems are unable to survive without either cobalt or added nitrogen nutrition.

Sodium has been found beneficial for the growth of many plants, particularly halophytes (salt-loving). Those plants that do respond to it tend to accumulate large amounts, whereas others, which are nonresponsive, absorb very little. The halophyte *Atriplex,* a desert plant, appears to require sodium for effective glycolysis. Some plants are faced with the problem of living in high sodium soils. Mangrove trees exemplify one solution to this problem; they do not absorb sodium. Certain *Atriplex* species, on the other hand, absorb large amounts of sodium but toxic accumulation does not occur because the sodium is actively transported out again into specialized gland cells on the leaf surfaces. Recently, it has been shown that sodium is an essential nutrient for plants that have the C_4 photosynthetic pathway and Kranz anatomy. The reason for this requirement, or for its association with C_4 photosynthesis, is not known.

Selenium has aroused much interest because it behaves in some plants as an analog of sulfur. Selenium-containing amino acids are formed, analogous to cysteine and methionine, that inhibit the synthesis or catalytic properties of proteins. On the other hand, certain plants of the genus *Astragalus* (a vetch) accumulate large amounts of selenium and appear to have a well-developed selenium metabolism that is not entirely analogous to that of sulfur. Some plants have a substantial tolerance or even a requirement for selenium, and their presence indicates a high level of selenium in the soil. It has been suggested that the beneficial effect of selenium for certain plants is, in fact, due to the reversal by selenium of phosphorus toxicity, to which these plants are susceptible.

Silicon is required for the growth of diatoms whose outer "shell" is composed of this element. It has been found that certain cereals, like rice and corn, grow better with added silicon, and this element may constitute up to 20 percent of the dry weight of these plants. Silicon is said to reduce transpiration and improve resistance to pathogens, perhaps because of its deposition in cell walls and in wounds. However, silicon does not seem to be required for the metabolism of plants. It will relieve phosphate deficiency, because silicate is more tightly absorbed than phosphate and displaces the phosphate into the soil solution. The addition of silicate also reduces iron and manganese toxicity, presumably by the precipitation of these elements.

Gallium requirement at extremely low levels has been shown for *Aspergillus niger* in careful experiments by the American physiologist R. A. Steinberg, but no one has attempted to repeat his delicate and meticulous experiments, and the requirement has not been shown for other plants.

Vanadium is required in very low concentrations (0.01 ppm) by the green alga *Scenedesmus* and possibly other microorganisms. It may be active in nitrogen fixation, partially replacing molybdenum in the nitrogenase.

Replacement. Certain metals have been shown to replace others for specific functions in the growth of plants. Often the replacement element functions less effectively or during only part of the life cycle. Barium and strontium are able to replace, usually not as effectively, the extremely low calcium requirement of some fungi and bacteria. Strontium is able to partly relieve the symptoms of calcium deficiency in corn, but it does not contribute to the formation of pectate as does calcium. A chlorosis of peach trees growing on low strontium soils has been relieved by foliar spray, and it is possible that a strontium requirement may be found for some plants. Rubidium and cesium may relieve a potassium shortage, as may sodium, but only to a limited extent. Beryllium is able to replace magnesium in some fungi and partly in tomatoes. The addition of beryllium stimulates the growth of some plants (rye grass and kale) but inhibits others (bean). Germanium may temporarily relieve boron deficiency by increasing the mobility of available boron in plants, and vanadium has been shown to replace molybdenum in the nitrogen-fixing function of *Azotobacter chroococcum,* but not of other species. The alga *Scenedesmus* exhibits improved photosynthesis in the presence of vanadium, which may thus be beneficial or required for the process.

Toxic Elements. Certain elements, for example, the heavy metals such as silver, mercury, or lead, may achieve high local concentrations in soil (for example, in mine tips) and exhibit a powerful toxic effect. Resistant genetic varieties that can tolerate very high concentrations of otherwise toxic elements sometimes arise under these circumstances. Tungsten has been found to be toxic for nitrogen-fixing organisms because it competitively inhibits molybdenum uptake. Germanium inhibits silicon metabolism and may be toxic to organisms that require or use silicon, such as diatoms, rice, and tobacco. Aluminum is also toxic and may inhibit the growth of plants under natural circumstances because it tends to precipitate in or around the roots, where it interferes with the uptake of iron and calcium. Aluminum also interferes strongly with phosphorus metabolism, causing the accumulation of large amounts of inorganic phosphate in the roots but reducing its availability for metabolism and transport.

Trace Elements in Economic Plants

Deficiency Diseases and Toxic Effects in Animals. Animals eat plants, and it follows that animals feeding on plants that are deficient in a specific mineral may themselves be deficient in that mineral. Also, minerals present in excessive amounts in feed plants, although they may not affect the plant, may prove toxic to the animals eating them. For example, certain highly organic peat soils in various parts of the world are low in copper, and this deficiency is reflected in cattle by scouring disease, a form of acute diarrhea. Most mineralized soils contain plenty of copper; in some cuperiferous soils plants may accumulate 50–60 ppm copper, and animals eating them may suffer blood disorders such as hemolysis and jaundice from copper poisoning. Man appears to be unaffected. Molybdenum excess in certain soils leads to a scouring disease (called teart disease) in cows when the molybdenum content of herbage rises to 20–100 ppm, as compared with 3 ppm or less in low molybdenum areas.

Cobalt deficiency, which occurs in certain areas particularly in Australia and New Zealand, leads to a pining or wasting disease affecting ruminants only. The cobalt is apparently required by the rumen microflora for the synthesis of vitamin B_{12}. An acute disease called phalaris staggers occurs when ruminants eat cobalt-deficient plants of *Phalaris tuberosa.* In the absence of sufficient cobalt to make vitamin B_{12} a powerful neurotoxin, apparently derived from a precursor in the *Phalaris,* is released by the rumen bacteria. The effect is roughly parallel to that of vitamin B_{12} deficiency in humans with pernicious anemia; in the absence of vitamin B_{12} neurotoxic substances are formed, in this case by the symbiotic microflora.

Humans in many parts of the world suffer from goiters because of iodine deficiency resulting from the low iodine content of plants. Seafoods, particularly seaweeds, accumulate large amounts of iodine, and people who eat these foods (for example, the Japanese) do not suffer from iodine-deficient goiters even though they live in regions where iodine-deficient soils occur. This problem has been largely overcome by the use of iodized salt. Fluorine is another element that affects man—a moderate amount of fluoride (up to about 1 ppm in drinking water) results in a great reduction in dental caries. High fluorine soils, which occur in some areas, produce plants that are high in fluorine, and animals eating these plants may suffer dental and other anomalies.

An interesting and important toxic effect is caused by the element selenium. Soils over a wide area of the Great Plains of North America from Alberta to Arizona have seleniferous soil, and the high selenium content of plants growing in these soils causes the characteristic wasting alkaline disease in farm animals. A selenium content in soils of over 0.5 ppm is potentially dangerous, and amounts up to 10 ppm are known in areas where selenium poisoning is severe. The amount of selenium in fodder plants usually ranges from 10 to 50 ppm under these conditions. However, certain plants, which appear to have a strong selenium metabolism and may even require it, may accumulate up to 10,000 ppm selenium from the same soils. If animals feed on these, they quickly develop acute selenium poisoning, called blind staggers, and die, often convulsively. Various species of the legume genus *Astragalus* (a type of vetch) and the related *Oxytropis* have been called locoweed because of their dramatic effect on animals feeding on them. Attempts to reduce selenium uptake of plants by adding excess sulfur (which might be expected to interfere with selenium metabolism) have been unsuccessful because selenium soils are usually already very high in sulfur, being derived to a large extent from gypsum. Selenium poisoning seldom affects man because selenium accumulators are not used for human food, and much of the selenium in grains and crop plants is found in the brans and other parts that are usually discarded during processing for human food.

Plants as Indicators. Certain plants grow only on soils that are high in some specific mineral. So-called **indicator plants** include *Merceya latifolia* (copper moss), which only grows on high copper soils, and *Astragalus* as well as other genera (*Stanleya, Oonopsis,* and *Xylorrhiza*), which appear to require selenium and only grow on high selenium soils. The presence or absence of some other elements sometimes causes characteristic changes in plant growth that may be diagnostic, for example, stunted shoots and thickened roots suggest high iron, white necrotic spots indicate excess cobalt, and high boron is said to cause certain shrubs to grow in a rounded shape and produce large, dark-green leaves. These and similar characteristics have been used, particularly in earlier times, by mining explorers as a first sign of the existence of minerals in the soil that might indicate commercial deposits nearby.

Additional Reading

Articles in *Annual Reviews of Plant Physiology* under the heading "Nutrition and Absorption."

Chapman, H. D. (ed.): *Diagnostic Criteria for Plants and Soils.* University of California, Davis, Division of Agricultural Science, 1966.

Epstein, E.: *Mineral Nutrition of Plants: Principles and Perspectives.* John Wiley & Sons, Inc., New York, 1972.

Fried, M., and **H. Broeshart:** *The Soil-Plant System.* Academic Press, New York, 1967.

Gauch, H. G.: *Inorganic Plant Nutrition.* Dowden, Hutchinson & Ross, Mc., Stroudsburg, Pa., 1972.

Kitchen, H. B. (ed.): *Diagnostic Techniques for Soils and Crops.* American Potash Institute, Washington, D.C., 1948.

Russel, B. W.: *Soil Conditions and Plant Growth.* Longmans, Green & Co., New York, 1961.

Steward, F. C. (ed.): *Plant Physiology: A Treatise,* Vol. III. Academic Press, New York, 1963.

11 Uptake and Movement of Water

Water Movement

The Problem of Water Loss. Most land plants require effective systems for the absorption and movement of water. This is because their primary nutrition is gaseous, and they have a highly efficient gas-exchange system. The consequence is the unavoidable loss of water by transpiration through the organs of gas exchange, the leaves (see Chapter 14, page 327). Water so lost must be continuously made up by the absorption and translocation of more water from the soil.

Much of the plant's water system may thus be looked upon as the result of a necessary evil. However, certain benefits result from the continuous flow of water through the plant. The depletion of soil water around the roots causes the influx of fresh soil solution from surrounding areas, and this undoubtedly helps the plant to tap the nutrient resources of much larger volumes of soil. Simple diffusion would eventually redistribute nutrients into areas depleted by root absorption, but the **solvent-drag** of bulk-flow of the soil solution toward roots greatly accelerates the process. A second indirect benefit is that an upward-moving stream of water is available in which solutes can move, and much of the distribution of salts and other solutes throughout the plant undoubtedly takes place in the bulk flow of the transpiration stream. A third benefit is that water loss enables some degree of control by the plant over its energy balance. Leaves must absorb energy from the sunlight falling on them. At times it may be impossible for them to utilize enough of this energy in photosynthesis, so the leaf temperature may become dangerously high. Under these circumstances, some energy may be dissipated by the evaporation of water, and the process of water loss may thus help to get rid of excess energy.

Water moves through a plant, largely entering via the roots and leaving via the leaves, in response to a potential gradient, which must thus decrease continuously from the soil to the atmosphere. Essentially, the plant acts as a link in the water system, permitting the flow of water down a potential gradient from the soil to the atmosphere. Part of the movement is by diffusion, usually by osmosis, and part of it by bulk flow.

The contributions of various processes to water movement will be discussed below in connection with the movement of water through different tissues. Not all of the processes are clearly understood. In addition to the passive diffusion of water down a potential gradient, it has often been suggested that water is caused to move from place to place, against a potential gradient, as the result of the expenditure of metabolic energy. However, no clear proof of such "active" water movement has been offered. The movement by physical or mechanical "pumping" therefore seems unlikely. The motive forces of passive water movement are, in the final analysis, those that establish potential gradients down which water may diffuse. Many of these forces are environmental and not under the control of the plant. Others, as for example the active movement of ions from cell to cell that may establish an osmotic gradient, are the result of internal metabolic activities of the plant and so may be internally controlled.

Water transport problems of very primitive and underwater plants are relatively simple and are largely related to the transport of organic and inorganic solutes. In some plants, like mosses, solutions to the problem of drought have evolved that depend mainly on their ability to survive desiccation. Higher plants, however, are mostly unable to do this, and a variety of mechanisms have evolved in them that prevent water loss and increase water uptake. Most of the following discussion will be devoted to the behavior of higher plants.

Entry of Water into Cells. The general consensus of current opinion is that water enters cells osmotically, that is, by movement down a potential gradient. The concept of **"active" water uptake,** that is, the direct transfer of water molecules across a membrane against a potential gradient or at an accelerated rate, has been invoked from time to time. Certain experiments have suggested that the expenditure of respiratory energy may be needed for the uptake of water, and this has been advanced as evidence for an active uptake process. However, it seems highly probable that the need for respiratory energy is indirect and results from the following facts: (1) water uptake requires living metabolizing tissue to maintain the organization of the cellular and subcellular structure, and (2) energy is needed to transport solutes from cell to cell to create the osmotic potential gradients that move water. Thus active water movement is better defined as the result of the energy-requiring movement of solutes that causes osmosis.

A telling argument against the active metabolic pumping or transportation of water molecules has been presented by the American physiologist J. Levitt, who showed that the permeability of membranes to water is so great that quite incredible amounts of energy would have to be expended to overcome significantly the leakage or backflow of water around such a pump.

Apparent Free Space. Water diffuses directly from the soil into the **free space** of roots. Free space is defined as that part of the root or tissue to which the solution bathing the tissue has direct, unimpeded access. In practice this space cannot be exactly measured, but a close approximation is the apparent free space (AFS) which is defined, on the basis of experimental data, as the value expressed by the fraction

$$\text{AFS} = \frac{\text{total solute in tissue}}{\text{solute concentration in surrounding solution in which tissue is bathed}}$$

when diffusion equilibrium is reached. Evidently this value cannot be obtained with absolute accuracy, but approximate measurements indicate that the AFS of roots is in the range of 6–10 percent of the total tissue volume. Direct measurement indicates that the water-filled intercellular spaces and cell walls in root tissue, excluding vacuoles, constitutes about 7–10 percent of the tissue volume. This suggests that the AFS of roots is essentially the cell walls and intercellular spaces. AFS does not include vacuoles, which are separated from the fluids surrounding the cell by the cytoplasm and the cellular membrane system of plasmalemma and tonoplast.

Entry of Water into Roots

Root Pressure. If a plant is decapitated and the roots are watered, water may exude from the cut stem. If a manometer is attached to the cut stem, it can be demonstrated that the water is exuded with a measurable pressure that may occasionally reach as high as 2–3 bars (30–45 psi). This phenomenon is called **root pressure** and has been much discussed as a component of the forces that move water from the roots to the aerial portion of the plant. Although this pressure is insufficient to move water to the top of tall trees, it may be involved in the ascent of sap in some species. However, since root pressure usually moves water rather slowly (that is, the **flux** is low), and indeed root pressure usually falls to zero when water loss is maximal, it is unlikely that it contributes significantly to the total movement of water in plants. The importance of root pressure is that it provides a mechanism for filling the xylem vessels of plants with water. This may be very important for vines, whose vessels empty of water during the winter. In addition, the continuity of water through the xylem vessels of many herbaceous plants may be broken during hot days when there is a shortage of water in the soil (see page 249), and root pressure generated during the night refills vessels with water so that the continuity of water supply is not permanently lost. In order to understand root pressure we must examine the pathways and mechanisms of the movement of water into and through roots.

Apoplast and Symplast. In 1932 the German physiologist E. Münch developed the concept of the apoplast and the symplast in the roots, illustrated in Figure 11-1. The apoplast consists of all the space in the roots that roughly equates with free space, that is, cell walls and intercellular spaces, plus the tissue in the stele that gives free access to water, mainly the xylem vessels. The important point to note is that the apoplast is discontinuous and is separated into two regions. One is the cortex and the tissues external to the endodermis. The other is the tissue of the stele, including the contents of the nonliving conducting vessels located inside the endodermis. The endodermis provides the discontinuity because of the **Casparian strip,** a heavily suberized thickening of cell walls that prevents the passage of water from outside to inside, or back, unless it passes through the cells, that is, through the cell membranes and cytoplasm. The root may thus be thought of as an osmometer, the endodermis being the osmotically active membrane. Dissolved substances or solutions may diffuse or flow unrestricted via the apoplast through the cortex to the endodermis, but to enter the stele they must pass through the differentially permeable membranes of the endodermis.

The symplast consists of the total protoplasts of the cells, that is, the portion of the cells lying within the boundary of the outer differentially permeable membrane of the cell. Vacuoles, being separated from the cytoplasm by a second differentially permeable membrane, would belong to neither system. Because protoplasts are connected from cell to cell via small strands of cytoplasm that penetrate the cell wall, called **plasmodesmata,** the symplast of the entire root can be considered as a single continuous system.

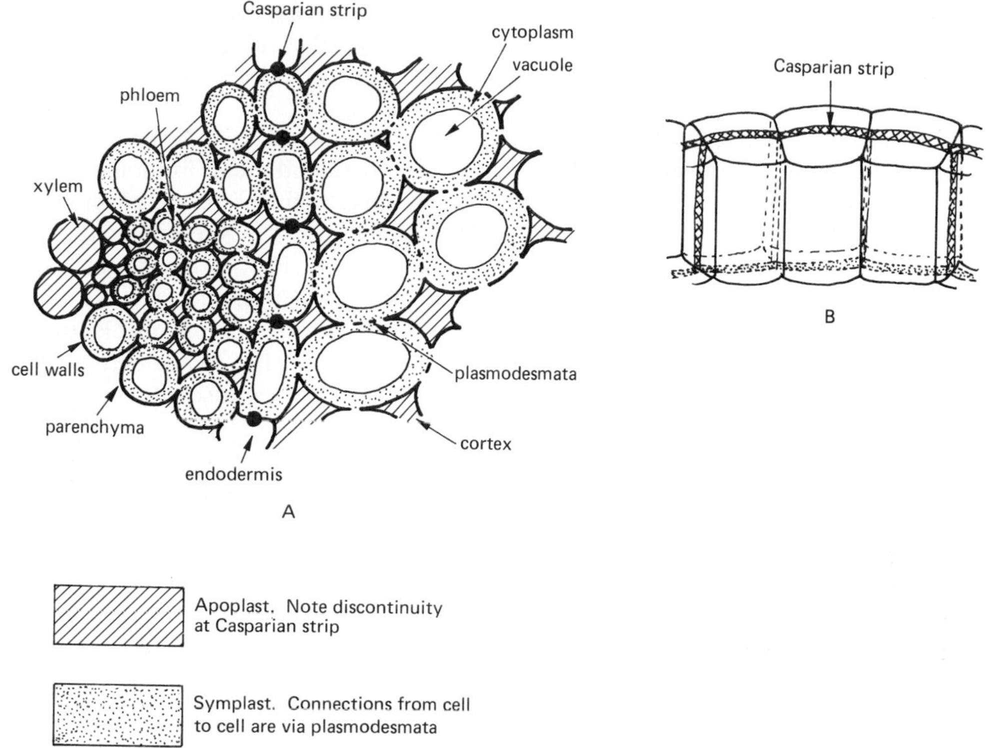

Figure 11-1. Root structure in relation to water movement.
A. Diagram of a cross section of a root to show the apoplast-symplast concept.
B. Diagram of three cells of the endodermis (viewed from inside the stele looking out), showing the Casparian strip.

Mechanism of Absorption. The apoplast-symplast concept led in 1938 to the proposal of a mechanism for water uptake and transport to the stele by the American physiologists A. S. Crafts and T. C. Broyer. Soil solution, containing dissolved ions, diffuses into the external portion of the apoplast (outside the endodermis). Ions are absorbed into the symplast from the soil solution in the apoplast by an *active* process (to be considered in the next chapter, page 293). They are actively transported through the symplast from cell to cell via intercellular connections (plasmodesmata), across the endodermis and into the stele, where they are released into the apoplast. The effect of this is to lower the concentration of ions in the outer cortical portion of the apoplast and raise their concentration in the apoplast of the stele. This sets up an osmotic gradient, since the water potential (ψ) is higher in the cortex than the stele, and water diffuses across the endodermal osmotic barrier into the stele. The resulting osmotic diffusion of water into the stele is sufficient to establish a substantial hydrostatic pressure in the stele, which causes water to flow up the xylem elements of the stem. This mechanism satisfactorily explains root pressure phenomena.

We must consider the question: Why should ions be absorbed into the symplast in the cortex, and then leaked out of the symplast into the apoplast again in the stele? The answer to this question is apparently that the oxygen concentration is sufficiently high in the cortex to permit the metabolism necessary to generate the energy required for the active absorption of solutes. This is the point at which the energy is expended that generates the water potential gradient across the endodermis necessary to move water and create a hydrostatic pressure in the stele.

It has been assumed that the oxygen concentration in the deeper tissue of the root, specifically the stele, is lower due to the long diffusion path and the utilization of oxygen in the outer parts of the root. Thus, insufficient metabolically produced energy is available to retain the ions in the symplast of the stele against a potential gradient; as a result, they leak into the apoplast. A recent report from P. J. Kramer's laboratory shows that the oxygen concentration is indeed much lower in the xylem exudate, indicating quite a sizable oxygen gradient in the root. This provides a plausible (though not proven) mechanism for the uptake of ions into the symplast in the cortex where oxygen, and hence respiratory energy, is available and for their release into the apoplast of the stele where oxygen is lower.

A second question must be considered: Is the endodermis really a true osmotic "sack" as this mechanism requires? Since the stele develops by growth (cell division and elongation) at the end, it is apparently "open ended." However, very little water absorption takes place in the region of the root where the xylem cells are differentiating. Furthermore, the young differentiating cells destined to be xylem are filled with cytoplasm at this stage and, hence, do not offer a low-resistance escape path for water, as does the mature xylem of the differentiated root. Thus the "open" end of the endodermal sac is essentially plugged. In some plants occasional thin-walled cells may be found in the endodermis. These **passage cells,** shown in Figure 11-2, are often found opposite xylem vessels and would appear to permit the free passage of solutions from the cortex to the stele. However, their radial walls are always heavily suberized, consequently they probably do not constitute pathways for the mass flow of solutes. Discontinuities do appear in the Casparian strip where side roots form their connection with the stele of the main stem (Figure 11-2). However, this may provide for the nutrition of the newly developing side root before its vascular tissue is connected to that of the main stem. In any event, provision for a certain amount of leakage, if it is not excessive, does not make the theory untenable.

We should briefly mention the effect of the osmotic potential of the soil water on water uptake and root pressure. Clearly, since water moves down a potential gradient from the soil into the xylem, the osmotic potential of soil water has a direct effect. It is possible to completely stop root-pressure exudation of water through a cut stem by placing the roots in a solution of sufficiently strong osmotic potential, at which point the potential of water in the xylem of the stem is approximately equal to the osmotic potential of the external solution. The osmotic potential of root cortical cells is not very important to this process. Regardless of whether water moves through them or around them, so long as the potential of water in the xylem is below that of soil water, water will move from the soil to the xylem. Plants can thus absorb water from solutions whose osmotic potential is greater than the osmotic potential of cortical or water-absorbing cells, provided that the water potential of the xylem is sufficiently low (that is, has a sufficiently high negative value).

Water Uptake in Transpiring Plants. When water is lost by transpiration, it must be replaced through the roots. Water loss from leaves means that the amount of water in

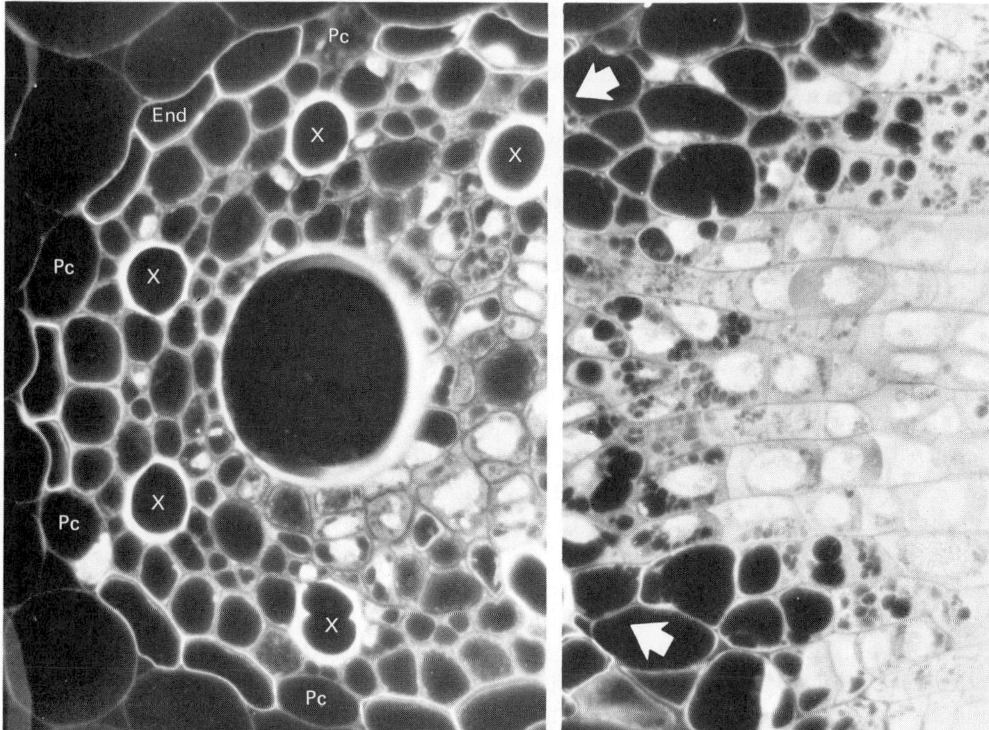

Figure 11-2. Cross section of main root of oat (*Avena sativa*) showing a branch root (right). Fluorescence stained to make cytoplasm appear grayish to white and lignified or suberized cell walls bright white. The endodermis with its heavily suberized walls (End) and the lignified xylem vessels (X) are visible. Passage cells (Pc) may be seen opposite the xylem vessels, and arrows show discontinuities in the endodermis near the branch root. [From E. B. Dumbroff and D. R. Pierson: *Can. J. Bot.*, **49:**35–38 (1971). Used with permission. Photograph kindly supplied by Dr. Pierson.]

the plant is decreased, consequently its potential is lower (becomes more strongly negative) and water diffuses into roots down the potential gradient so developed. The roots do not seem to aid actively in the process. On the contrary, they may even hinder it. If roots are removed, water uptake by the shoot is greatly increased. Killing roots by dipping them in boiling water reduces their resistance so that water can be mechanically sucked through the plant faster than when the roots are alive. Evidently the roots offer resistance to water uptake. In spite of this, roots are essential because their large absorbing surface provides the necessary contact between the aerial part of the plant and the soil water. Although roots might hinder uptake from a beaker of water, they are necessary for the uptake of the finely dispersed water that is present throughout soils.

Most plants require sufficient oxygen in order to develop a large enough root system to absorb water. Water-logged soils may inhibit root development so drastically, due to the lack of oxygen, that even though the soil is overabundantly supplied with water insufficient water is absorbed and the plant wilts. However, although they are essential, roots seem to act generally as a passive system through which water moves under the influence of the potential gradient developed by its loss from leaves.

Pathway of Water Through Tissues

We must now consider the question: Does water really move only in the apoplast or does it also move through the vacuoles of cells? It has been suggested that water moves through the cortex of the root by osmotic diffusion from cell to cell, not by free diffusion through the apoplast. Such a system would require a water potential gradient between successive layers of cells from outside to the inside of the cortex. A gradient of osmotic potential would not be necessary, so long as the combination of hydrostatic or turgor pressure and osmotic potential were such that a continuous gradient of water potential existed. However the problem of regulation and maintenance of such a gradient might be considerable.

A direct experimental approach to this problem as it applies to leaf and stem tissue has been made by the British physiologist, P. E. Weatherly, at Aberdeen. There are two possibilities: (1) water moves through free space regions of cell walls and intercellular spaces (that is, the apoplast) only; or (2) water moves through vacuoles of cells as well. The experimental system is the top of a plant under water tension. A *Pelargonium* (geranium) plant is detached above the soil and attached to a **potometer,** a sensitive device to measure water uptake ("drink" meter), with the leaves in a dry atmosphere to promote water loss, as shown in Figure 11-3. If the leaves are suddenly plunged into water or paraffin oil, water loss stops immediately and the rate of water uptake decreases. Now, if all or most of the water in the cells is involved in transport, it will all be under similar tension; thus, when the water loss is stopped, water uptake will decrease at a steady rate until the internal tensions are relieved. The experimental result, measured with the potometer, should give a curve like that shown in Figure 11-4A. However, if water is moving only in the apoplast (although necessarily in equilibrium with a substantial amount of nonmoving water in the vacuoles), a different curve will result when transpiration is stopped, as shown in Figure 11-4B. First, there will be a sudden drop in flow as the tension is released. This will occur more rapidly, since only a small volume of water will be required to take up the tension in the small volume of the

Figure 11-3. Apparatus for measuring water uptake.

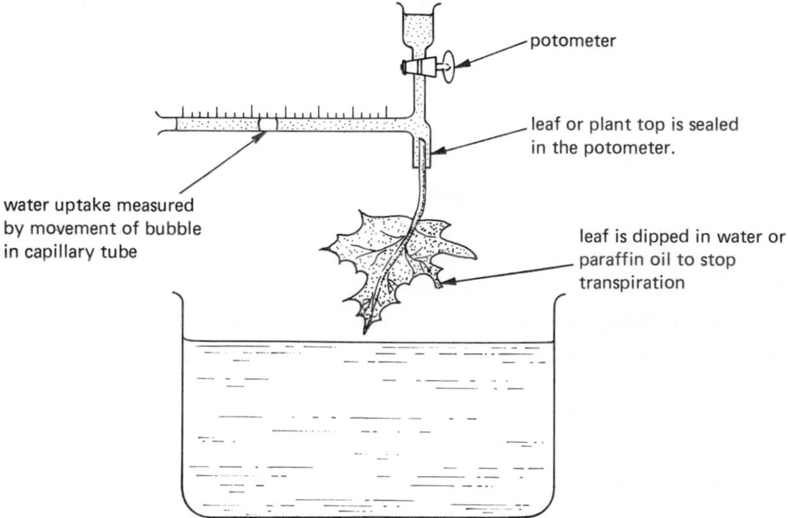

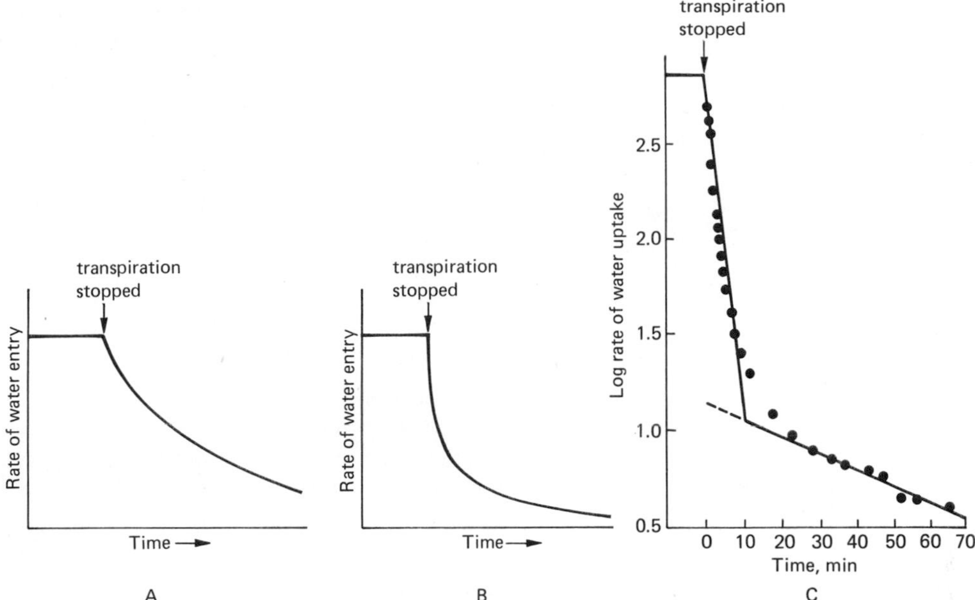

Figure 11-4. Calculated (**A** and **B**) and actual (**C**) results of an experiment on water uptake in *Pelargonium* leaf. [Data in (**C**) from results of P. E. Weatherly: In A. J. Rutter and F. H. Whitehead (eds.): *The Water Relations of Plants*. Blackwell Scientific Publications, Oxford, 1963. Used with permission.]

flowing system. Then, since the vacuolar water must ultimately be in equilibrium with the moving water in the apoplast, water will continue to enter, but much more slowly, relieving the tension in the more slowly equilibrating vacuoles. This will cause the biphasic curve shown in Figure 11-4B. Figure 11-4C shows the results of an actual experiment conducted by Weatherly. It clearly indicates that the water in motion in the stem and leaf tissue is not equivalent to the total water in the system but only to a small proportion of it. Thus, the second possibility, that water moves through the apoplast only rather than through the vacuoles as well, is most probably correct.

In recent experiments, cotton leaves were allowed to transpire with their petioles in a solution of potassium ferrocyanide. The leaves were then treated with a solution of ferric ions that precipitated the ferrocyanide as Prussian blue, visible in the light microscope and in electron micrographs. The results of this study, in which the deposition of Prussian blue crystals showed the pathways and flux of water, confirm the conclusion that water moves primarily through the cytoplasm and cell walls but not through vacuoles. A typical electron micrograph is shown in Figure 11-5.

The Ascent of Sap

The Forces Required. Water can diffuse from cell to cell down a potential gradient and may enter the xylem with sufficient force to generate pressure as high as 2–3 bars or higher. However, this root pressure is never sufficient to raise water to the top of a tall tree, and it may be very low much of the time in most plants, particularly at times when water loss is greatest. Furthermore, the flow from root pressure is not great enough to

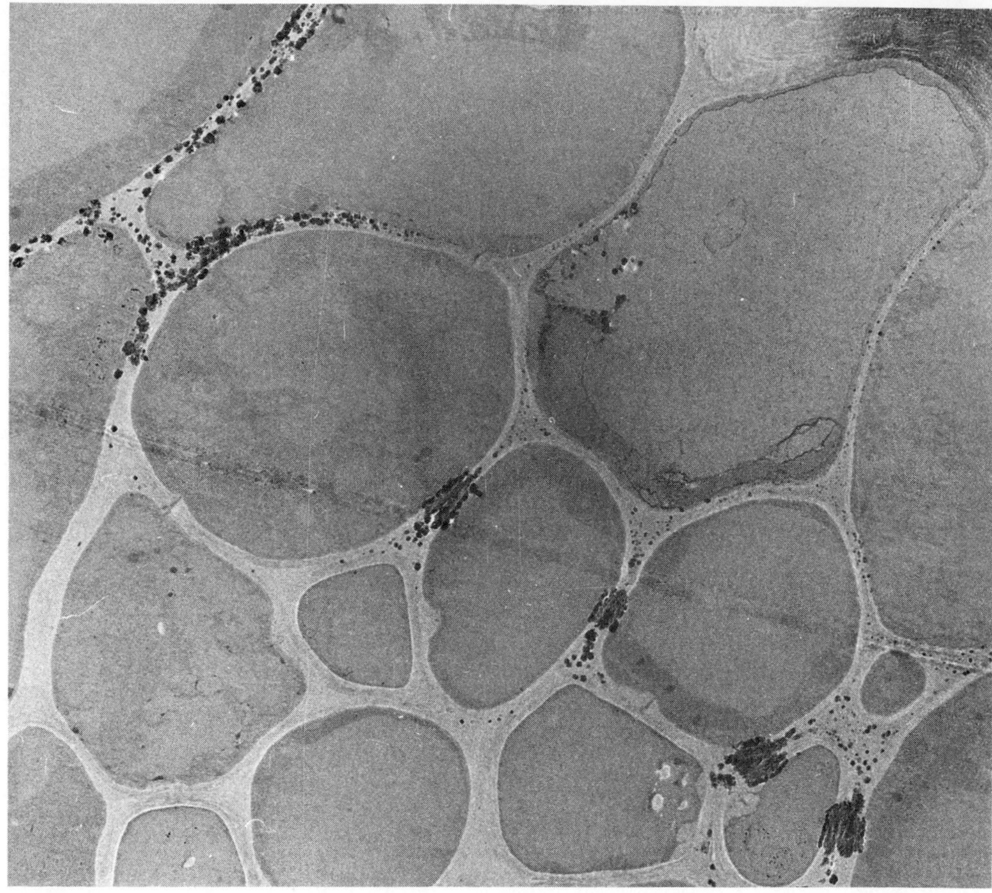

A

Figure 11-5A. Electron micrograph of phloem cells and a bundle-sheath cell (upper left) of cotton leaf (*Gossypium hirsutum*) in which Prussian blue crystals have been deposited from water moving through the leaf. Prussian blue crystals can be seen in the cell wall and cytoplasm but not in the vacuoles or (in **B,** opposite) in the cuticle. [Source: T. O. Pizzolato, J. L. Burbano, J. D. Berlin, P. R. Morey, and R. N. Pease: *J. Exp. Bot.*, **27,** 145–61 (1976). Used with permission. Photograph kindly supplied by Dr. Berlin.]

account for the volumes of water that actually move through a tree. It requires a pressure of 10 bars (150 psi) to raise water 300 ft, the height of a tall tree, and more pressure is required to overcome resistance in the trunk and to maintain an adequate flow. Clearly this pressure cannot be supplied from below since root pressures of this magnitude have never been measured. **Capillarity,** the matric potential of a system with narrow passages, may be sufficient to raise water a short distance in stems but not to the height or in the quantities required.

The solution to the problem rests in the expression of the idea that water moves down a potential gradient from soil to atmosphere via the plant. This means that a very low potential of water in the atmosphere, relative to the potential of water in the soil, supplies the force that moves water up the plant to the leaves. In other words, as water evaporates from the leaf surface, more water is "pulled" up by the tension so created.

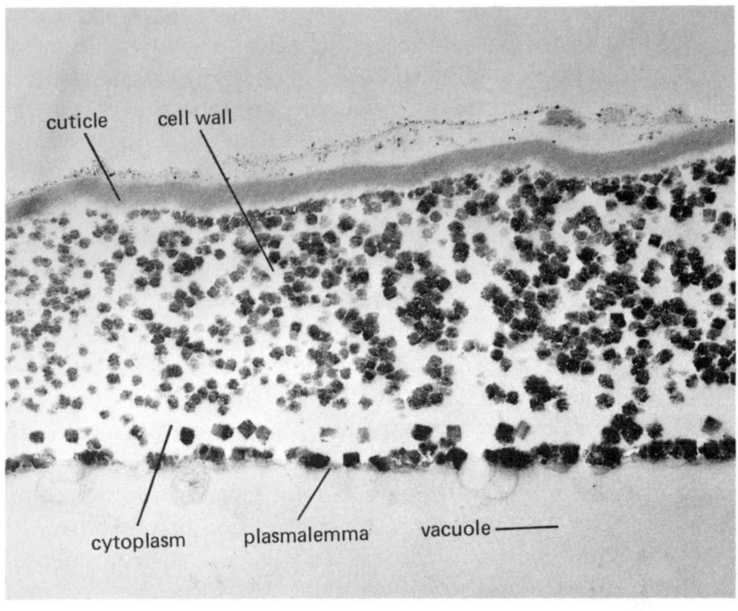

B

Figure 11-5B. Cell wall from an epidermal cell from a cotton leaf treated as in **A** (opposite) but shown at much higher magnification. [Photograph kindly supplied by Dr. J. D. Berlin, Texas Tech University.]

On first analysis the forces involved seem to be impossibly great. It seems incredible that the fragile cells of the leaf could withstand tensions of 150 psi or greater without collapsing. In fact, due to their small size, cells can withstand much greater tensions. The water potential of air is very low—at about 50 percent relative humidity (RH) it is close to -1000 bars, or 15,000 psi, and it is much greater at low RH.* The water potential difference ($\Delta\psi$) between leaf cells and the atmosphere is often very great, and water loss from leaf-cell surfaces induces tremendous tension inside the cells. This is relieved by the flow of water from internal cells, and ultimately from the xylem of leaf veins, so the tension is transmitted to the water in the xylem.

Cohesion of Water. The problem now is, how can water be pulled up a tube or pipe system, such as the xylem, for distances greater than the height of a column of water supported by 1 atmosphere (atm) of pressure (30 ft)? If one tries to pull water up a pipe by applying suction at the top, the column will break and a vacuum (or near vacuum filled with water vapor) will form when the column height reaches about 30 ft. The answer lies in the fact that water molecules have great affinity for each other, and *narrow columns* of water may withstand a tension of up to 1000 atm without breaking due to the

*Knowing the temperature in degrees absolute (T) and the RH, the water potential (ψ) of air may be calculated from the equation

$$\psi \text{ (bars)} = -10.7\, T \log(100/\text{RH}).$$

cohesive properties of the molecules. The theory that water could be pulled up very tall trees in long threadlike columns in the xylem was advanced in 1894 and 1895 by several scientists and elaborated by H. Dixon in Great Britain and O. Renner in Germany early in the twentieth century. Such columns do not normally **cavitate** (break) because the cohesive property of water molecules is sufficient to prevent their pulling apart or pulling away from the walls of the vessels.

Evidence supporting the theory that water is pulled up by the forces of evaporation from the leaves is indirect. It has been clearly shown that water does move in xylem. This can be done by injecting dyes, radioactive tracers, or small pulses of heat into a tree trunk and following the movement of the marker. The fact that the xylem water is under tension can be observed by cutting into the stem of a plant—water will snap back into the xylem, and if water is added to the cut surface it is pulled in. If a sensitive microphone is placed against the stem of a plant, the snapping of xylem water threads may be heard, particularly on dry, hot days. Extensive cavitation may cause severe wilting, because a broken column can no longer transmit to the roots the tension required to lift water. It is these broken threads that are presumed to be reunited by root pressure at night when the tension caused by evaporation of water from the leaves is low.

Perhaps the most effective evidence is derived from measurements with the **dendrometer,** a device consisting of a metal belt whose exact circumference is variable and measured by a delicate instrument, which measures the girth of a tree trunk. A piece of rubber tubing can be shown by experiment to shrink in girth under suction, that is, if its contents are under tension. The same applies to a tree trunk. Dendrometer measurements show that the girth of a tree decreases during the day when the evaporation rates are greatest, indicating that the contents are under increased tension. In contrast, it increases at night, when evaporation decreases and tension on the water in the trunk is reduced. A typical dendrometer measurement is shown in Figure 11-6.

One problem with this theory of water movement in stems is that breaks in the water column do occur, as a result of excess drought, the formation of gas bubbles from dissolved gas, and from mechanical breaks. Such breaks ought, theoretically, to inactivate the xylem strand in which they occur and reduce the capacity of the system to move water. Many such breaks do occur, but they do not appear to affect water movement seriously. Presumably when the tension is relaxed at night, the columns again rejoin. If cuts are made in a tree trunk so that no continuous vertical columns of xylem remain, water is still able to ascend in a zigzag path, though at a reduced rate. It seems probable that the lateral transport that must occur takes place as a result of diffusion in the xylem parenchyma, the living tissue of the xylem.

Figure 11-6. Dendrometer measurement of the diameter of a tree trunk. [From B. S. Meyer, D. B. Anderson, and R. H. Böhning: *Introduction to Plant Physiology*. © by Litton Educational Publishing, Inc. Reprinted by permission of Van Nostrand Reinhold Company.]

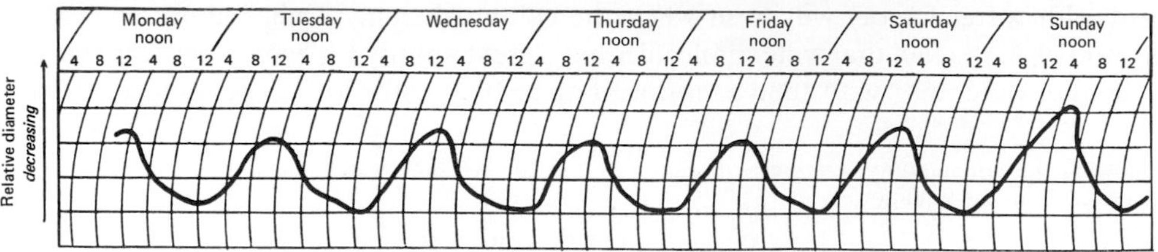

Vessel Size. Questions have been raised about the size of the xylem vessels through which water must flow. It has been shown that in small vessels the flow rate of water varies as the square of the vessel radius (from Poiseuille's law). Plants with long narrow stems, such as vines, tend to have large diameter vessels that permit high flow rates. Since flow rate varies inversely as length, this arrangement permits the efficient transfer of water through long distances in a stem of small cross section. However, such vessels are more subject to cavitation or breakage of the water column, and the high root pressure of vines may be associated with the need to refill vessels that have emptied by cavitation.

On the other hand, trees, which have a much larger cross section in relation to their length, tend to have smaller xylem-conducting elements. This means that water can be drawn to greater heights by greater forces, and the reduction in carrying capacity caused by small diameter vessels is offset by the increased diameter of the trunk and the correspondingly greater number of conducting elements.

Extremes of temperature change are likely to cause the formation of bubbles in water under tension, and freezing of the water is likely to result in the breakage of the water column because dissolved gases are frozen out of solution. This may be the reason why plants living in colder temperate or Arctic zones tend to have smaller vessel size than plants living in tropical zones. Coniferous trees, which commonly live in temperate or Arctic climates, have no vessels at all but only tracheids, which have much smaller diameter.

Alternative Theories. Several alternative suggestions have been proposed. One is that water ascends trees largely as vapor. However, most of the water in the xylem is known to be liquid, not vapor. Various suggestions of active pumping systems have been made, but no mechanical or biochemical devices that could accomplish such pumping have been found. Active water transport in living cells of trees seems unlikely. First, the resistance of living cells to flow would be very great and would add greatly to the forces required to move quantities of water. If roots are cut off a wilted plant and the stem is placed in water, the plant recovers far more quickly than when the roots of the intact plant are placed in water. This shows the much greater resistance to water transport offered by living tissue and suggests that water moves through the stem in nonliving tissue. The application of poisons or cold as metabolic inhibitors to the stem of a tree also has little or no effect on water movement, reinforcing the concept that nonliving cells are involved in water transport and that no energy input occurs in the stem itself.

Thus, the possibility that the living cells contribute substantially to the upward movement of water in plants is quite remote. The theory that water ascends in the xylem of high plants largely under the influence of the forces of evaporation from the leaf surfaces therefore seems most probably the correct explanation. A substantial amount of evidence directly supports this view, whereas alternative possibilities are unsupported or appear to contravene the known facts.

Flow of Water

Water moves under the influence of a gradient in water potential (ψ), and its movement is hindered by various resistances to flow, including the viscosity of the solution, the permeability of membranes, and the resistance to flow of narrow passages. Considering

Table 11-1. Estimated values for water potential (ψ) and water potential difference ($\Delta\psi$) in a hypothetical soil-plant-air system. The plant is a small tree, the soil is well watered, and the air is approximately 50 percent relative humidity at 22° C ($\psi = -1000$ bars)

	ψ, *bars*	$\Delta\psi$, *bars*
Soil water	−0.5	
		−1.5
Root	−2	
		−3
Stem	−5	
		−10
Leaf	−15	
		−985
Air	−1000	

these together, at steady state the flow rate (F) in any part of the system will relate to these quantities as

$$F = \frac{\Delta\psi}{\text{resistance}}$$

and for the whole system

$$F = \frac{\text{overall } \Delta\psi}{\text{sum of resistances}}$$

In a steady state system the flow rate is constant, so it is possible to relate the resistance of any part of the system to the drop in water potential across that part of the system. Some estimated measurements of water potential in various parts of the soil-water-air system of a plant are shown in Table 11-1. It can be seen that the largest value for $\Delta\psi$, by a very large factor, is at the point where water leaves the leaf and evaporates into the atmosphere. It follows that the resistance at this point is enormously greater than elsewhere in the system. Thus, this is the point at which effective control of the system must be applied. The mechanisms of control and their operation are discussed in Chapter 14 in the section on stomata (pages 331–39).

Some actual measured values of water potentials in the soil and trees under various conditions and at different times of day are presented in Table 11-2. It may be seen that the water potentials, hence the flow rates, are directly related to the conditions of irrigation (which affect the water potential of the soil) and the time of day and atmospheric conditions (which affect the water potential of the air). Thus the flow rate of water through the trees, which is directly proportional to the water potential differential between the air and the soil, varies with changing external conditions, and the water balance of plants adjusts automatically to the external conditions and the requirement for different flow rates.

Summary

The process of water movement through the plant may be summarized as follows: water enters the free space or apoplast of roots and moves by osmosis across the barrier imposed by the Casparian strip at the endodermis. The osmotic potential is created by

Table 11-2. Actual measured water potentials in soil and different parts of trees at various times

Species	*Juniperus scopulorum* (*juniper*)				*Ulmus parvifolia* (*elm*)			*Elaeagnus angustifolia* (*Russian olive*)		*Acer glabrum* (*maple*)	
Date, September	17	18	20	22	12	16	17	16	17	17	18
Time, hr	1500	0500	1200	1500	1400	1400	0400	1500	0500	1500	0500
Conditions	clear	night	rain	clear	clear	clear (+ irrigation)	night	clear	night	clear	night
Temperature, °C	18	10	12	11	27	16	7	16	7	16	7
ψ soil, bars	−5.7	−6.0	−7.1	−0.2	−4.6	−0.1	−0.1	−3.3	−3.3	−5.7	−6.0
ψ trunk, bars	−8.6	−6.6	−7.9	−5.0	−7.0	−2.6	−2.6	−7.4	−3.7	−8.9	−6.8
ψ branches, bars	−12.0	−9.4	−8.7	−8.0	−7.6	−5.3	−4.1	—	—	—	—
ψ twigs, bars	−21.8	−11.8	−12.2	−15.6	−23.0	−16.8	−5.7	−17.7	−10.7	−25.2	−8.8
ψ leaves, bars	−40.0	−25.0	—	—	−24.5	−23.9	−10.7	−31.9	−17.6	−43.0	−32.2

SOURCE: Data of H. H. Wiebe, R. W. Brown, T. W, Daniel, and E. Campbell: *Bioscience,* **20:**226 (1970). Used with permission.

the absorption of solutes from the soil solution by the protoplasts of the cortex cells and the transport of these solutes via the symplast across the endodermis, followed by their return into the apoplast inside the stele. Some positive pressure may be generated in the xylem of the lower part of the stem in this process, which may be enough to drive water slowly to a considerable height in the stem. This may be the mechanism whereby broken columns of water are repaired or empty xylem vessels are refilled. The main driving force for the upward movement of water, however, is the evaporation of water from the leaf surfaces. Water moves up the stem, drawn up by the tension created by water loss from leaves. The cohesive property of water is sufficient under normal circumstances for water columns of great height to withstand the tension of being pulled upward by the forces of evaporation.

Additional Reading

Articles in *Annual Reviews of Plant Physiology* under the heading, "Water Relations."
Slatyer, R. O.: *Plant Water Relations*. Academic Press, New York, 1967.
Steward, F. C. (ed.): *Plant Physiology: A Treatise,* Vol. II. Academic Press, New York, 1959.
Sutcliffe, J.: *Plants and Water*. St. Martin's Press, Inc., New York, 1968.

12 Uptake and Transfer of Solutes

Mechanisms for the Movement of Solutes

Solutes can move by diffusion through channels that present physical barriers, or they can be swept along by the flow of solvent (**solvent-drag** forces). However, if a physical barrier, such as a membrane or a colloidal material like cytoplasm, interferes with their free passage, a variety of mechanisms may be involved in the transfer of the solute through the barrier. If the barrier is not absolute, the solution may flow through it or the components of the solution may diffuse through it. Evidently, however, the rate of diffusion or flow will be affected by the properties of the barrier. If the barrier is a living system such as membrane or protoplasm, solutes may move across it either by passive diffusion or by active transport. (**Active transport** means the actual transfer of molecules from one place to another by some energy-requiring process.) The difference will be that if molecules move by diffusion, they can move only down a potential gradient; whereas, if they move by active transport, they appear to be moved against such a gradient.

Solutes can also move through a membrane by other processes. Material may be moved by the formation of bubbles or vesicles on one side of a membrane that discharge their contents to the other side. This process is called **pinocytosis** and is essentially the emptying of small vacuoles through a membrane. Pinocytosis is a nonselective process because solutes are moved only as part of the small bubble of solution, not independently. Active transport, on the other hand, may be highly selective.

Diffusion

MEMBRANE AND SOLUTE CHARACTERISTICS. Nonelectrolytes (uncharged particles) tend to diffuse through membranes at a rate roughly proportional to their solubility in fat or fat solvents and inversely proportional to their molecular size, as shown in Figure 12-1. The dependence on size suggests that the molecules must pass through spaces, or holes, and has led to the suggestion that cellular membranes are sievelike structures or composed of micelles or small subunits arranged in some regular pattern with spaces in between. However, the fact that the rate of diffusion of solutes varies with their fat solubility supports models of membrane structure (Chapter 3, page 51) that indicate one or more lipid layers. It is possible, however, that a composite structure, consisting of a

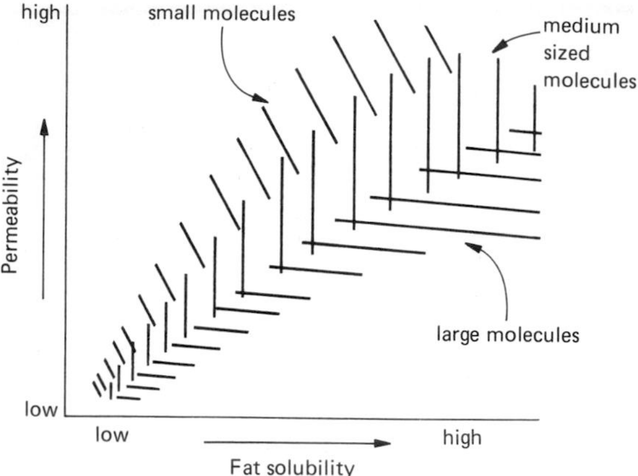

Figure 12-1. Approximate relationship between fat solubility, molecular size, and permeability, in *Chara* cells, of various nonelectrolytes. [Redrawn from data of R. Colander: *Physiol. Plant.*, **2**:302 (1943).]

discontinuous bilayer with pores of various sizes, may more correctly represent the true state of affairs. The fat-solubility effect could result from the movement of solutes either through the structure or through holes or pores in it that are lined with exposed lipophilic groups.

Cell walls appear to be freely permeable to most solutes, whereas the permeability of membranes is very much less. Hence it is the living substance of cells that affects and controls the transport of solutes into and out of cells. Cell walls and intercellular spaces that are permeable to water are also essentially freely permeable to solutes. Dissolved gases move very much more freely than other solutes; cells are nearly as permeable to gases as they are to water.

Cell walls, however, are probably not very important in solute transport. They are pierced by a large number of minute pores or holes called **plasmodesmata** (singular, **plasmodesma**) (see Figure 3-1, page 46). The outer cytoplasmic membrane (plasmalemma) of adjacent cells is in intimate contact in each plasmodesma. Since there may be as many as 5×10^8 plasmodesmata per square centimeter, this means that adjacent cells have plenty of channels through which active or passive transport of materials can take place without having to pass through the cell wall. Thus, the properties of membranes and their interaction with solutes is the most important facet of solute transfer.

The study of the diffusion of electrolytes and their diffusion or transport through membranes presents special problems. Ions usually have much lower permeability than uncharged particles because the charge makes it difficult for them to penetrate a membrane with active or charged groups that either repel or attract (and so immobilize) them. As a result of the charge, ions tend to be surrounded by a substantial layer of water molecules, and this hydration shell makes their particle size large enough to affect their penetration of pores in membranes. In addition, they are usually strongly lipophobic, hence have low rates of diffusion. Finally, the movement and distribution of ions are affected by the electrical potential of the system as well as by its chemical potential. The movement of ions is therefore considered separately, together with a discussion of the mechanisms whereby ions or solutes can be actively transferred across membranes.

Diffusion and Permeability. Diffusion of molecules is their net movement down a free energy or chemical potential gradient. The rate of diffusion varies with the chemical potential gradient or the difference in **activity** (essentially equivalent to the concentration) across the diffusion distance. Thus the flux of molecules through a membrane (J, the number of particles moving across a given area of membrane at a given time) is proportional to the driving force, which is the concentration difference on either side of the membrane (ΔC):

$$J = P\,\Delta C$$

where P is a **permeability coefficient,** measuring the ability of the substance in question to permeate the membrane. However, the permeability of a membrane is proportional to the capacity of the moving solute to diffuse through it, and inversely proportional to the thickness of the membrane. Thus the flux can be measured by

$$J = \frac{D}{\chi}\Delta C$$

where D is the **diffusion coefficient** of the particular substance through the membrane, and χ is the thickness of the membrane. From this it is clear that the membrane thickness is extremely important. Since the diffusion coefficients of most substances are very small, the thinnest possible membranes are required for efficient transport. In fact, the values for D are so small for many solutes that they could not penetrate at the required rate unless some active transport mechanism were in operation.

Accumulation by Diffusion. The term **accumulation** means the acquisition of a higher concentration of the substance on one side of the membrane (usually inside the cell) than on the other. Evidently, since the motive force of diffusion is a concentration difference, it is impossible to achieve accumulation by simple diffusion. No concentration difference is possible at equilibrium. However, if conditions are such inside the membrane that the physical or chemical state of the substance is changed on entry, substantial accumulation may occur.

For example, if the vital stain neutral red is supplied to cells, the dye accumulates inside them to a concentration that may be over 30 times the external concentration. This is because the dye is supplied at about pH 8, at which the molecules are undissociated. On entering the cell, whose sap has a pH of about 5.6, the dye molecules dissociate. Since the cell walls are impermeable to the ionic form of the dye, it cannot diffuse out again, and accumulation occurs. The process does not, however, take place without the expenditure of energy. The motive force is supplied by the cellular system, whatever it may be, that maintains the internal pH sufficiently low to keep the dye molecules in the ionized form. Similarly, if a molecule is precipitated, adsorbed, or chemically changed on entering a cell, it may be concentrated without any active process of transport being required. Again, the motive force is supplied by the energy required to maintain whatever conditions inside the cell may be necessary to inactivate the entering molecules.

Movement of Ions

Special Problems. The movement of ions and their transport across membranes present special problems. Ions usually have very low permeability through membranes because of the large size of their hydration shell and their low lipid solubility. More important than this, the forces acting on ions include electric potential gradients, or charge potential gradients, so that their movement is influenced by charge distribution as well as by concentration. Further, the movement of one ion automatically influences the charge pattern of the system, so that the movement of other ions, regardless of the sign of their charge, is affected. Thus the active movement of ions may be brought about by the creation of electrical gradients and vice versa.

Antagonism. This is an important phenomenon which may protect plants from toxic effects of certain ions. A plant placed in a dilute solution of potassium chloride will rapidly accumulate potassium ions until toxic levels are reached, and the plant may die. However, if trace amounts of calcium are present in the solution, the absorption of potassium is greatly reduced and no toxicity occurs. The calcium is said to **antagonize** the uptake of potassium. Similarly, calcium will antagonize sodium, and either sodium or potassium, added in small quantities, will antagonize calcium uptake. Apparently ions must be unrelated (that is, not be in the same group in the periodic table) for effective antagonism. Sodium will not antagonize potassium uptake, and barium will not antagonize calcium, but either sodium or potassium will antagonize either barium or calcium. It is thought that calcium is necessary for the structural integrity of membranes. In its absence, selective transport mechanisms break down and the indiscriminate permeability of the membrane increases. This may be the basis for the calcium antagonism effect.

Only small concentrations of the antagonizing ion are required, and antagonism is freely reversible. Thus the antagonism is unlikely to act at the level of specific ion carriers. It has been suggested that antagonists may affect the colloidal structure of the absorbing surface, thus exerting an influence, but the quantities that would be required to effectively change the permeability of membranes appear to be too large. No satisfactory explanation for antagonism has been put forward. The process is undoubtedly of value in the field. Many soils have an excess of certain elements, particularly potassium or calcium, and toxic effects would certainly occur if some regulating mechanisms such as antagonism had not evolved. However, there is a negative side, too. Excesses of certain ions may prevent uptake of other required ions and so induce deficiency symptoms, even though the required ion is present in sufficient amounts in the soil. Thus, excess sodium in the soil might induce calcium deficiency through antagonism.

Electrochemical Potential. Just as nonionic particles diffuse down a chemical potential gradient, so ions diffuse down an electrochemical potential gradient. Such a gradient has a chemical component and an electrical component. A chemical potential gradient exists if the concentration of an ion on one side of a membrane is greater than on the other. An electrical potential gradient may result from the presence of charged particles or ions, but may also result from the charge on one side of the membrane with respect to the other (that is, the charges may be associated with the membrane surface or with some fixed or nondiffusable component on one side of the membrane or the other).

Thus a situation may exist where, for example, a cation is more concentrated inside the cell than outside, but the inside of the cell is negatively charged with respect to the outside. The ion will thus tend to diffuse *out* down the chemical potential gradient, but it will tend to diffuse *in* down the electrical potential gradient. The final direction of movement will be determined by the component of the gradient (electrical or chemical) that is steepest.

The relationship between electropotential and chemical potential is defined by the **Nernst equation,** which has been derived from basic physical chemical laws

$$\Delta\varepsilon = -\frac{2.3RT}{z\mathfrak{F}}\log\frac{a_i}{a_o}$$

where $\Delta\varepsilon$ is the electropotential difference across a membrane and a_i/a_o is the chemical potential difference, being the ratio of activities inside and outside (essentially, the activity is equivalent to the molar concentration). R is the gas constant, $\mathfrak{F}$ the Faraday constant, and z is the charge per ion, or valency. Assuming a constant temperature, it may be seen that for any ion

$$\Delta\varepsilon = -K\log\frac{\text{concentration inside}}{\text{concentration outside}}$$

or, in other words, the electrochemical potential across a membrane varies as the log of the ratio of the ion concentration on either side of the membrane. This relationship is extremely valuable in the study of active transport (page 295).

DONNAN EQUILIBRIUM. The Donnan equilibrium is an ion accumulation phenomenon named for its discoverer, F. G. Donnan. If there is a negative, nondiffusing charge on one side of a membrane (for simplicity's sake, let us say inside a cell), this will create a potential gradient across the membrane down which ions will diffuse. The result will be that at electrochemical equilibrium, the *concentration* (chemical potential) of ions will not necessarily be the same inside as outside. Thus, as the result of an electrical disequilibrium maintained because of nondiffusing charges, a concentration disequilibrium is established.

The equation describing a Donnan equilibrium states that the ratio of positively charged ions, inside to outside, must equal the ratio of negatively charged ions, outside to inside (square brackets indicate concentration).

$$\frac{[\text{positive ions inside}]}{[\text{positive ions outside}]} = \frac{[\text{negative ions outside}]}{[\text{negative ions inside}]}$$

A typical situation is shown in Figure 12-2. Figure 12-2A represents a cell in which the internal fixed charges are balanced by potassium ions. The cell is placed in a solution of potassium chloride (Figure 12-2B), and it is initially out of equilibrium with its surrounding solution. After equilibration takes place, as shown in Figure 12-2C, potassium has concentrated inside the cell and chloride is excluded as a result of the influence of the fixed negative changes inside. Such a Donnan type equilibrium may concentrate substances in a cell up to 30-fold over the concentration on the outside or in the environment. The accumulation of zinc by roots, for example, is substantially due to the formation of a Donnan equilibrium as well as to the formation of stable or nonionized zinc derivatives inside the cell.

MEMBRANE POTENTIAL. Most biological membranes are found to have a potential or charge difference from one side to the other; usually the inside of cells is negative with respect to the outside. Such membrane potentials influence the flow of ions, but they may in fact be established by unequal diffusion of ions. A potential will be formed, for example, if fixed charges exist on one side of the membrane. Similarly, unequal rates of diffusion of the ionic components of a salt across the membrane, unequal rates of

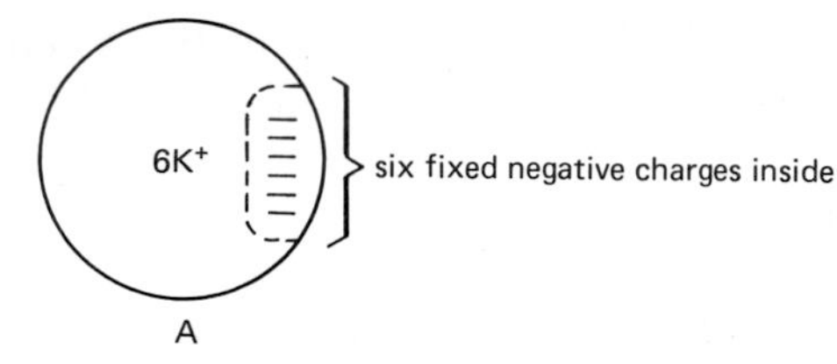

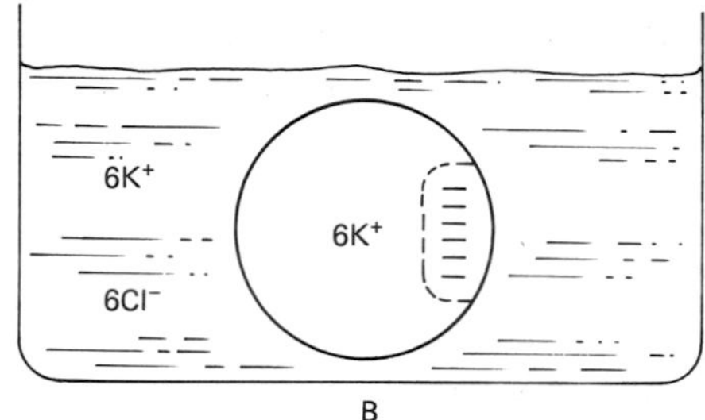

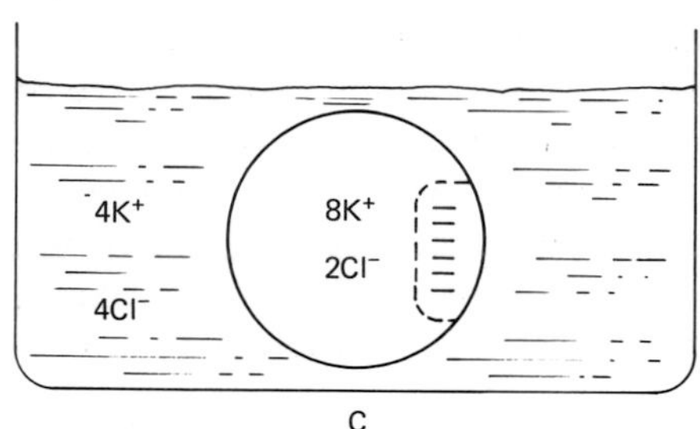

Figure 12-2. Donnan equilibrium.
A. Cell with fixed charges neutralized by K^+ ions.
B. Cell is placed in a solution of KCl.
C. Donnan equilibrium is established. K^+ has been accumulated inside the cell, Cl^- has been excluded.

$$\frac{[K_i^+]}{[K_o^+]} = \frac{[Cl_o^-]}{[Cl_i^-]} : : \frac{8}{4} = \frac{4}{2}$$

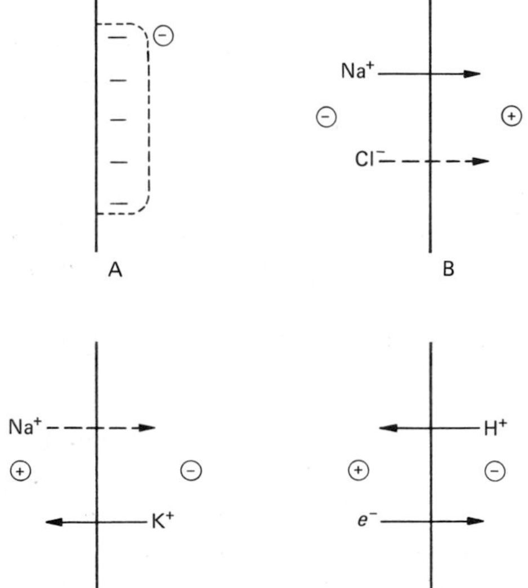

Figure 12-3. Membrane potential.
A. Fixed charges result in a negative charge on the right side of the membrane.
B. Unequal rates of diffusion of anions and cations result in a positive charge on the right side of the membrane.
C. Unequal rates of diffusion of ions of the same charge result in a negative charge on the right side of the membrane.
D. Active transport of charges or ions: Here two processes are contributing to create a negative charge on the right side of the membrane.

diffusion of different ions on opposite sides of the membrane, or active transport of a charged particle, ion, or electron may result in a potential, as shown in Figure 12-3. Regardless of the source of the potential, it affects the electrochemical potential gradient across the membrane, and, hence, the diffusion of ions. It is possible, also, that electric potentials or potential gradients in plants may be important in establishing patterns of development (see Chapters 19 and 20, pages 455, 487). To date the physiology of potentials and their generation has not been studied as intensively in plants as it has been in animals.

Active Transport

Definition. Active transport is the transfer of ions or molecules at rates or in quantities that appear to defy the laws of diffusion and electrochemical equilibrium. This can only be achieved by an input of energy. Thus, active transport can be defined as the movement of ions against an electrochemical potential by the use of energy derived from metabolism. It is the coupling of metabolism to drive transport.

One of the problems inherent in the concept of active transport is that the input of energy may be quite remote from the actual process of transport, and the degree of remoteness must determine whether the process is to be considered active or passive. For example, even to establish a Donnan equilibrium, a passive chemical concentration of ions down an electrochemical potential gradient, ultimately requires the expenditure of metabolic energy to establish and maintain the system and the fixed charges that make the Donnan concentration possible. Similarly, solutes could be actively transported across a membrane, and as a result water would diffuse across the membrane by osmosis. However, the mechanism by which water moves is essentially a passive one, down a chemical potential gradient. The term "active" might thus be applied to the transfer of solutes in this case, but not to the transfer of water.

The American plant physiologist J. Levitt has established four criteria for distinguishing active transport:

1. The rate of transport exceeds that predicted from the permeability and electrochemical gradient.
2. The final steady state electrochemical potential is not in equilibrium across the region of transport.
3. A quantitative relationship exists between the amount of transport and the amount of metabolic energy expended.
4. The mechanism of transport depends upon cell activity.

Confusion has often arisen because experimenters have not critically examined the system to determine whether the accumulation might not have resulted from some nonactive process, such as a Donnan type equilibrium, or from the formation of some electrochemically inactive derivative (that is, a precipitate, ions from an uncharged particle, a chemical derivative, and so on). Many so-called active processes may finally be recognized as passive. Obviously it is difficult or impossible to study ions that are readily adsorbed to, or form derivatives with, cellular components, such as phosphate or iron. As a result, the study of active transport is quite difficult.

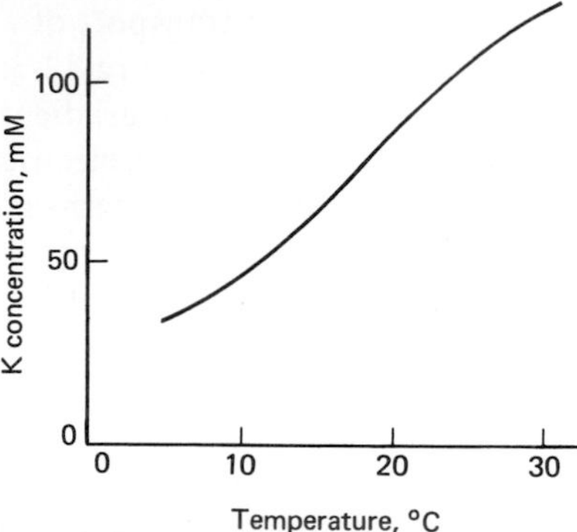

Figure 12-4. The effect of temperature on K^+ uptake by barley roots. [From data of D. R. Hoagland and T. C. Broyer: *Plant Physiol.*, **11**:471–507, 1936.]

Experimental Support for Active Transport. One of the most important recent techniques that has enabled physiologists to tackle these problems has been the use of radioactive isotopes of unnatural ions such as barium or rubidium. The isotopes are easily traced and measured in minute quantities, and the situation is uncomplicated by the presence of large cellular pools of the ions in question. Of course, the situation is complicated because many of the transport systems are quite ion-specific, but they are usually capable of transporting closely related ions. Thus, a potassium-transporting system might well be expected to transport rubidium, and radioactive rubidium has been much used to study the potassium-transport system.

Earlier experiments that implied active transport included studies of the effects of various factors that are important in metabolism. The accumulation of potassium was early found to be strongly affected by temperature (Figure 12-4) and O_2 (Figure 12-5), both of which affect metabolism. Adding the substrate of respiration, sugar, also stimulates ion uptake by roots (Figure 12-6). Finally, the uptake of ions is proportional to the rate of respiration itself, as measured by O_2 uptake or CO_2 evolution (Figure 12-7). All these data and the results of many similar experiments strongly suggest that the transport of the ions in question is mediated by metabolic energy released in the process of respiration.

The addition of salts or ions to roots and other tissues usually causes a decided increase in respiration. The increase over ground-level respiration is known as **salt respiration.** The inference is that salt respiration represents the increased metabolism needed to generate energy for the active transport of ions. Unfortunately the relationship is not always linear, and salt respiration may persist after the salts are removed. Consequently, salt respiration does not offer many useful clues to the nature of the coupling of respiration and ion transport.

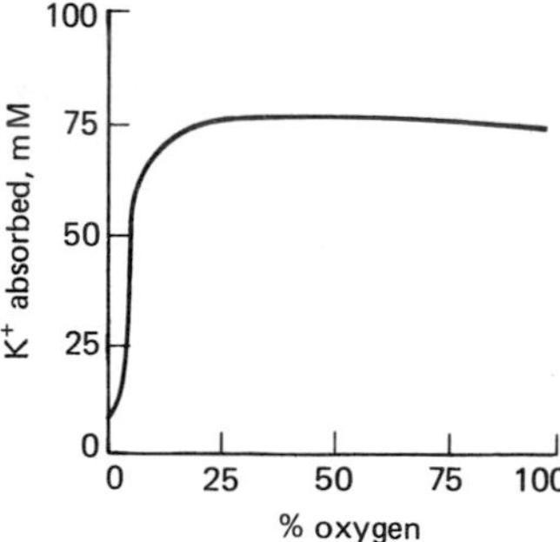

Figure 12-5. The effect of oxygen on K^+ uptake by barley roots. [From data of D. R. Hoagland and T. C. Broyer: *Plant Physiol.*, **11**:471–507, 1936.]

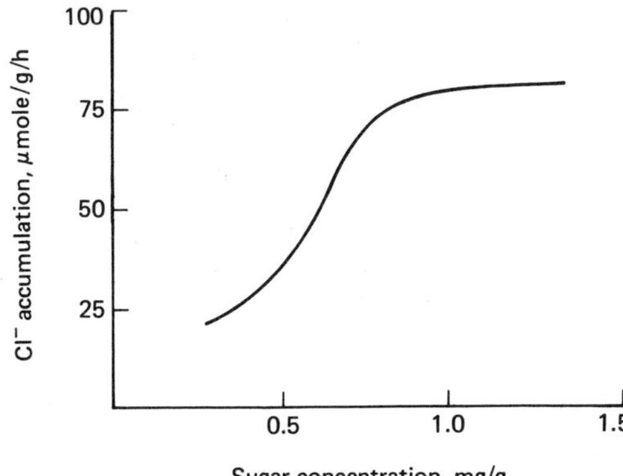

Figure 12-6. The effect of tissue levels of hexose on the uptake of Cl^- by barley roots. [From data of M. G. Pitman, J. Mowat, and H. Nair: *Aust. J. Biol. Sci.*, **24**:619–31, 1971.]

Demonstration and Proof of Active Transport. Although the data presented previously are strongly suggestive, we need a more rigorous quantitative proof of the existence of active transport and a way of determining unequivocally whether it is taking place in any given situation. Earlier we introduced the Nernst equation, which relates electropotential difference across a membrane to the chemical activity of the substance

$$\Delta\varepsilon = -\frac{2.3RT}{z\mathfrak{F}}\log\frac{a_i}{a_o}$$

We can simplify this, inserting numerical values for the constants and assuming a temperature of approximately 20° C

$$E(\text{mv}) = -\frac{58}{z}\cdot\log\frac{\text{conc. inside}}{\text{conc. outside}}$$

or, rearranging

$$\log\frac{\text{conc. inside}}{\text{conc. outside}} = -\frac{E\,(\text{mv})z}{58}$$

The numerical constant 58 would increase to 59 at a temperature of 25° C. z represents the valency charge of the particle in question, negative for an anion and positive for a

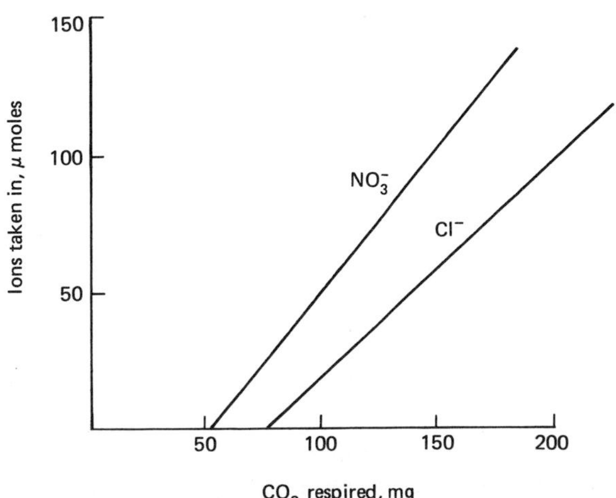

Figure 12-7. The relationship between tissue respiration and the uptake of NO_3^- and Cl^- ions by wheat roots. [From data of H. Lundegårdh and H. Burström: *Biochem. Z.*, **277**:223–49, 1935.]

cation. The value of z for sodium or potassium would be 1, for calcium or barium it would be 2, for chloride it would be -1, and for sulfate it would be -2.

Providing that a genuine equilibrium can be established, and that the ions are able to move both ways across a membrane, this expression can be used to determine explicitly whether active accumulation or expulsion of an ion has taken place. First, one must establish that the ions in question really are free to move in either direction across the membrane—radioactive isotopes have proved very useful in doing this. Then, it is necessary to measure the charge difference across the membrane, using a sensitive voltmeter and microelectrodes (by no means an easy technique). Finally, one must measure the concentration of the ions in question inside and outside the cell, which requires sophisticated microchemistry. If these values do not fit the Nernst equation, then active transport must have occurred.

Two characteristic sets of data are shown in Tables 12-1 and 12-2. In Table 12-1, the electropotential values of roots compared with that of the solution were measured, and from this the values for the internal concentrations of a number of ions (the external concentrations being known) were calculated. The calculated values are compared in Table 12-1 with the actual values. It may be seen that in the pea roots, K^+ was not actively transported, Na^+, Mg^{2+}, and Ca^{2+} were actively excluded, whereas the anions were all actively absorbed. The results for oat roots were similar, except that a small accumulation of K^+ occurred. A different experimental situation is shown in Table 12-2, where the potentials across the membranes of *Nitella* cells were calculated from observed concentrations of ions inside and outside and then compared with the actual

Table 12-1. Determining active or passive transport of ions using the Nernst equation

	Pea root ($E = -110$ mv), μmoles/g tissue		*Oat root ($E = -84$ mv), μmoles/g tissue*	
	Predicted	*Measured*	*Predicted*	*Measured*
K^+	73	75	27	66
Na^+	73	8	27	3
Mg^{2+}	1350	1.5	175	8
Ca^{2+}	5400	1	700	1.5
NO_3^-	0.0272	28	0.0756	56
Cl^-	0.0136	7	0.0378	3
$H_2PO_4^-$	0.0136	21	0.0378	17
SO_4^{2-}	0.000047	9.5	0.00035	2

SOURCE: Data from N. Higinbotham, B. Etherton, and R. J. Foster: *Plant Physiol.*, **42**:37 (1967). Used with permission.

Roots were allowed to equilibrate in solutions of known composition for 24 hours at 25° C, then analyzed. Predicted values were obtained from the Nernst equation

$$\log \frac{\text{conc. inside}}{\text{conc. outside}} = -\frac{Ez}{59}$$

Calculation for K^+ in pea roots (conc. outside = 1 μmole/ml):

$$\log \frac{C_i}{C_o} = -\frac{110 \times 1}{59} = 1.865$$

antilog 1.865 = 73, predicted concentration inside = 73 μmoles/g.

Table 12-2. Active and passive transport of ions in *Nitella* cells

	Calculated ΔE (mv) for observed concentrations of ions		*Measured ΔE (mv)*	
Ion	*Plasmalemma*	*Tonoplast*	*Plasmalemma*	*Tonoplast*
Na^+	−66	+39		
K^+	−178	−12	−138	−18
Cl^-	+99	−23		

SOURCE: Data of R. M. Spanswick and E. J. Williams: *J. Exp. Bot.,* **15:**193–200 (1964). Used with permission.

INTERPRETATION. It may be concluded that Na^+ is excreted through the plasmalemma, because the measured ΔE is more positive than the calculated value. Na^+ is secreted into the vacuole through the tonoplast because the ΔE is more negative than calculated. K^+ and Cl^- are actively absorbed through the plasmalemma because the measured ΔE for K^+ is more positive than calculated, and the measured ΔE for Cl^- is more negative than calculated. K^+ and Cl^- are probably not actively moved across the tonoplast to any great extent.

measured potentials. It may be seen that Na^+ ions were actively excluded from the cell and actively transported across the tonoplast into the vacuole. Both K^+ and Cl^- were actively transported into the cell, but neither was transported actively into the vacuole in significant amounts. A diagram interpreting the data of Table 12-2 is shown in Figure 12-8.

Active transport systems that expel sodium are commonly found in plants. Many roots appear to absorb anions actively; cations diffuse along with them down the electropotential gradient thus created. Active absorption of K^+ seems to be quite common as is the expulsion of Ca^{2+} and Mg^{2+}. Active transport of many substances other than ions occurs. Sugars and other organic compounds to which membranes are relatively impermeable (to the extent that these substances can be used to plasmolyze cells) are nevertheless rapidly absorbed if the cells are aerated and metabolizing. Active transport occurs across many intracellular membranes, such as into and out of mitochondria, chloroplasts, and other organelles, as well as across endoplasmic reticulum and other cellular membranes.

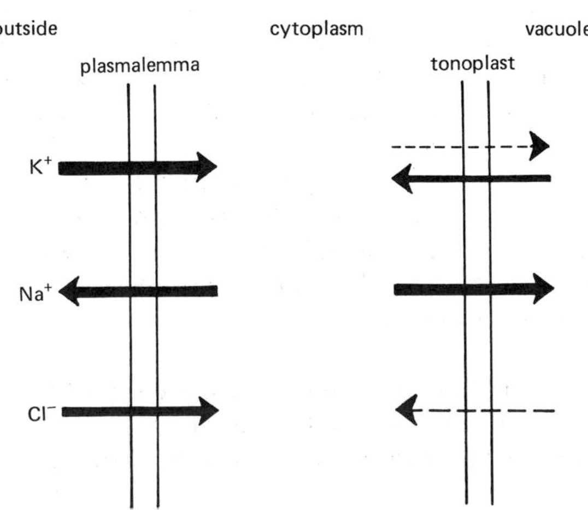

Figure 12-8. Active transport of ions across the cellular membrane of *Nitella* cells, as shown by the data in Table 12-2. Active transport shown by solid arrows, the thickness indicating the intensity of transport; passive transport shown by dashed arrows.

Charge Balance. It is evident that, since excessive potential differences do not occur, and approximate electrical neutrality is maintained across a membrane, the movement of anions and cations must be approximately equivalent. Alternatively, for each anion or cation that moves in, an ion of the same charge must move out. It has been suggested that H^+ ions are used as **counterions,** being actively transported and so establishing an electropotential gradient down which other ions diffuse. Malate appears to be an effective counterion, since it is often metabolically produced at the same time that cation secretion into the vacuole is taking place, thus maintaining charge balance. Evidently pH as well as charge balance must be maintained within the cell by this mechanism.

Mechanisms of Active Transport

Source of Energy. The decrease in free energy ($\Delta G°$) required to transfer 1 mole of solute against a tenfold concentration barrier ($C_2/C_1 = 10$) can be calculated from the relationship

$$\Delta G° = 2.303RT \log \frac{C_2}{C_1}$$

where R is the gas constant, 1.987 cal/(mole)(degree), and T is the temperature, 293° A. Thus

$$\Delta G° = 2.303 \times 1.987 \times 293 \times 1 = 1340 \text{ cal/mole}$$

Now the hydrolysis of 1 mole of ATP will yield over 7 kcal/mole. Thus there is ample energy from the hydrolysis of one molecule of ATP to transport one or more ions or particles even against a substantial concentration gradient. It is probable that, in fact, the relationship is one ATP per ion transported, since mechanisms whereby the energy of one ATP could cause the transfer of more than one charge or particle are hard to envision.

Recently it has become clear that ions may be transferred across membranes by the systems that derive energy directly from the membrane electron transport system. This idea was first expressed many years ago by the Swedish physiologist, H. Lundegårdh, who suggested that the cytochrome system could be used to transfer ions across membranes, energy being supplied directly by the oxidation of respiratory intermediates. This idea, although probably not correct as expressed, nevertheless foresaw the development of the Mitchell hypothesis of ATP synthesis (see Chapter 5, page 99, and Chapter 7, page 162). The ideas of Mitchell elegantly accommodate the transport of ions, as we shall see in the following section.

Alternative uses of the energy of ATP by hydrolysis may be through an ATPase operating in such a way as to generate a pH gradient across a membrane, down which charged or ionized particles would diffuse. It is also thought, mainly as the result of studies on animal membranes, that the ATPase may itself catalyze a direct transport of K^+, or a Na^+-K^+ exchange pump. We shall now look at details of some of the recent models of ion and solute transfer.

Possible Mechanisms. There are probably several mechanisms that transfer solutes across membranes, and different systems may work in different places. Not all membranes contain the enzymes of electron transport. These are well known in mitochondria and chloroplasts but have not been found in other cellular membranes such as the tonoplast or plasmalemma. Just how these membranes accomplish active transport is not clear. Perhaps the respiratory metabolism is done by adjacent mitochondria, and the energy so released in the form of ATP may be used by ATPase-mediated carrier systems located in the cell membranes.

The major hypotheses that seem relevant today are

1. Transport by a carrier protein, possibly ATPase.
2. Transport down an electrochemical gradient generated by electron transport.
3. Transport down a pH gradient generated by the electron transport system or ATPase.

The first, transport by a carrier molecule, is shown diagramatically in Figure 12-9A.

Figure 12-9. Models of possible ion transport systems. **A.** An ion carrier driven by respiratory ATP. **B.** A (NA^+, K^+)-ATPase transport system.

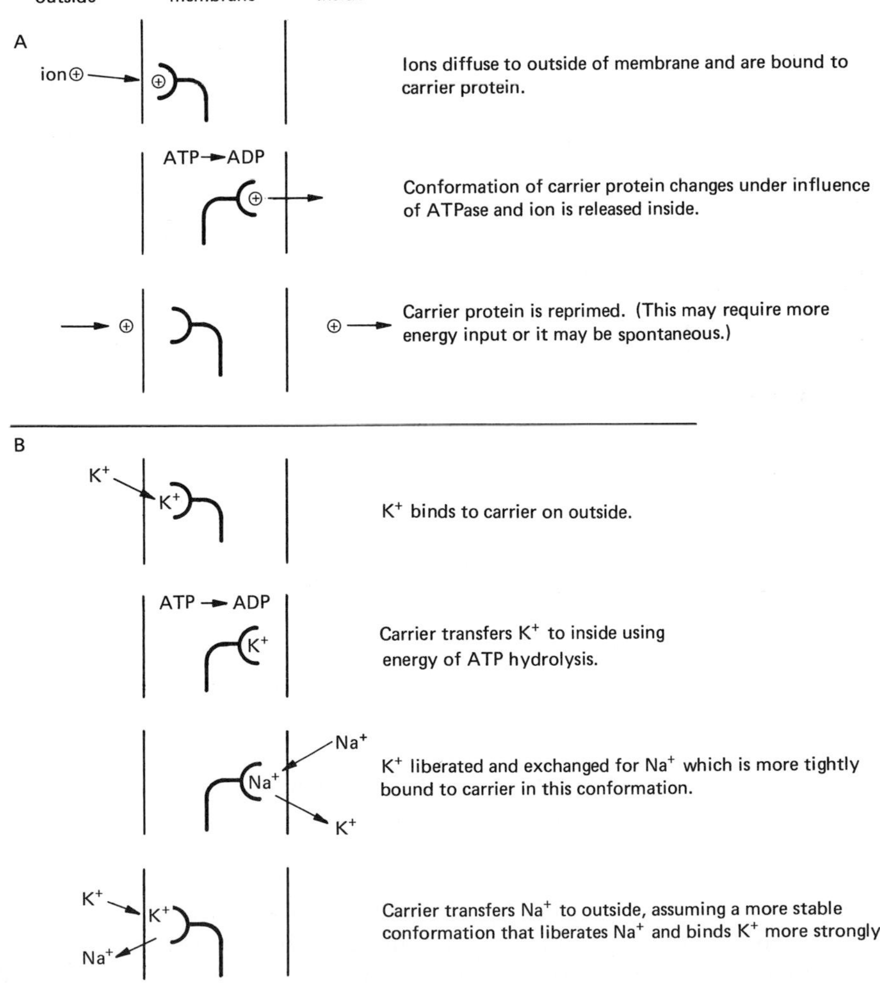

The energy of hydrolysis of ATP is used to change the conformation of the carrier protein (which may be the ATPase itself) so that the ion is picked up on one side of the membrane and discharged on the other. The alternating pickup and discharge may relate to the strength of the carrier-ion binding, which could be different in one conformation than in the other. Alternatively, it might be effected by changes in the sites of the membrane surface through which the ion must pass. The ATPase transport system in animal membranes has been shown to exchange Na^+ for K^+, and evidence for a similar (Na^+, K^+)-ATPase membrane transport system in plants has been presented, as shown in Figure 12-9B. Since there is good evidence that the plasmalemma and possibly the tonoplast contain ATPases that could mediate transport of this sort, it seems more likely that this is the mechanism whereby ions are actively transported into and out of cells.

It should be noted that the active transport of cations automatically sets up a charge gradient down which anions will diffuse. Differential permeability of the membrane or of the site of anion penetration would allow for some degree of selectivity of anion transport. Selectivity of cation transport would be due to the selective binding by the ATPase or selective permeation of the ion from the outside to the binding site of the carrier.

The second and third mechanisms depend on the Mitchell chemiosmotic hypothesis described in Chapters 5 and 7, illustrated in Figures 5-8 and 7-10, pages 101 and 164. Figure 12-10 shows how this scheme can be linked to ion transport. The electron-transport system could be used to generate a proton gradient which would drive anion or cation transport as shown in Figure 12-10A. This system might operate in mitochondria where the electron transport system is known to be located in the membranes surrounding the organelle. However, it does not seem likely to occur in other cellular membranes that lack electron transport enzymes.

The system outlined in Figure 12-10B shows how ATPase could generate a proton gradient down which ions could move. This system could function in chloroplasts and mitochondria, as well as in other membranes having the necessary spatial organization of the ATPase. A further possibility is that the active transport of K^+ ions by an ATPase carrier system could be used by an exchange system to generate a proton gradient that would then permit transport of other ions using the motive force generated by the K^+-transporting ATPase.

It should be noted that when protons (H^+) or hydroxyl ions (OH^-) are transported across the membrane, they immediately form water. Consequently, when cations are transported across as counter ions for H^+, anions must also diffuse across passively to satisfy the charge imbalance. Similarly, active transport of anions by exchange for OH^- requires the simultaneous movement of cations. Thus, by coupling the proton gradient to either cation or anion transport directly, a net movement of both cations and anions either out of or into the cell or organelle can be established. This is illustrated in Figure 12-10C.

The fact that active transport systems for certain ions can be saturated independently suggests that there are specific carriers or specific binding sites for certain ions. However, many ions do interact and appear to compete for the same binding site. Thus the American physiologist, E. Epstein, using corn roots, has shown that K^+, Cs^+ (cesium), and Rb^+ (rubidium) all compete, but Na^+ and Li^+ (lithium) have different sites, and that the divalent cations Ca^{2+} and Mg^{2+} are also absorbed at separate sites. It should be noted that competition of ions for active transport is a different phenomenon from

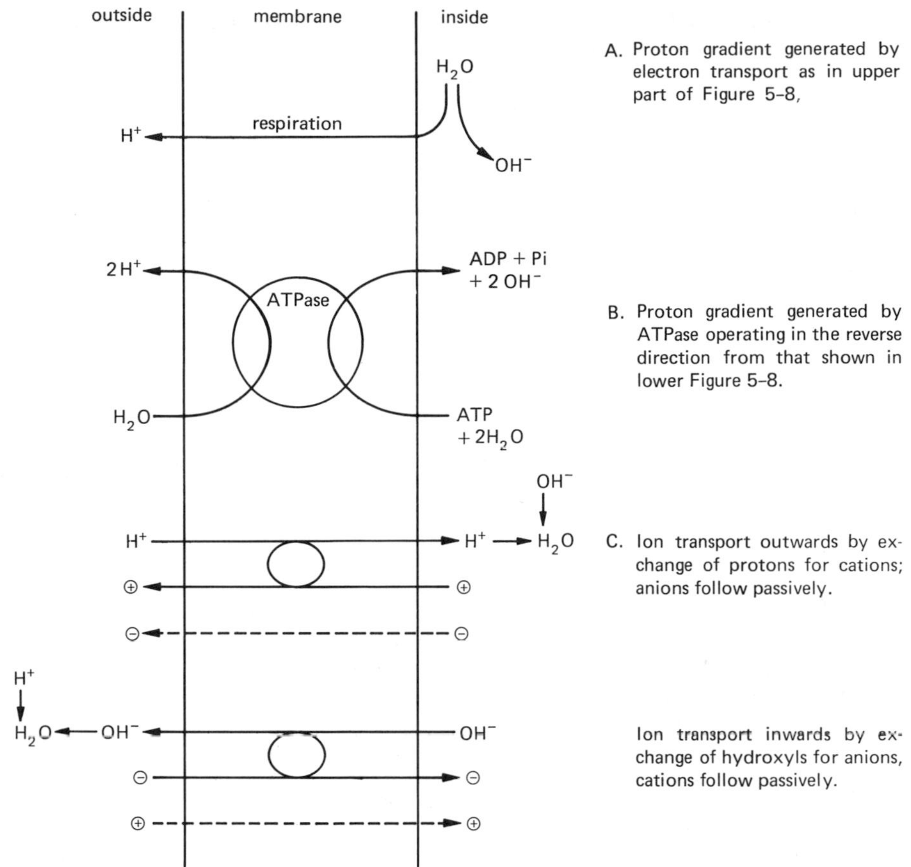

Figure 12-10. Schemes for ion transport (**C**) across membranes, coupled to (**A**) electron transport or (**B**) ATPase.

antagonism, since large concentrations of usually similar ions are required for competition, whereas extremely low concentrations of an entirely dissimilar ion are required to cause antagonism.

There is evidence that two mechanisms may exist for the movement of certain ions. One, called System I, is apparently located in the plasmalemma and absorbs ions from very dilute solutions. The location of System II is not certain, but it is also probably the plasmalemma. System II moves ions much faster from more concentrated solutions. Some physiologists have suggested that there is in reality only one system which is affected by phase changes in the membranes caused by the different ion concentrations. This matter has not been resolved as yet.

Importance of Active Transport. Active transport is necessary for living cells because certain substances must be concentrated and others must be excluded. Further, the concentrations needed for metabolism, growth, and perhaps also for the activation of enzyme systems could not be maintained unless ions were being actively pumped in to replace those that continuously leak out by diffusion. A sodium pump is necessary in most organisms because the sodium levels in the environment are far too high. Sodium cannot be kept out by permeability barriers alone—eventually, if any diffusion were possible at all, an equilibrium would be reached and the internal concentration would equal the external, unless a pump was in operation.

Active transport may also be important as an electrogenic device. So far, little evidence has been found that plants contain complex electronic systems like animal nerves, whose pulses are generated by ion pumps. However, fluids such as water tend to flow down narrow tubes whose walls and ends have a charge differential (the process of **electroosmosis**), and it is possible that ion pumps may be involved in the process of water (and solute) movement (see Chapter 13, page 315). It is possible, also, that electrical potential gradients are involved in developmental phenomena in some plants, and, if so, their establishment and maintenance may be of importance in development (see Chapters 19 and 23).

Additional Reading

Articles in *Annual Reviews of Plant Physiology* under the heading "Nutrition and Absorption," and in *Current Topics in Membranes and Transport,* Academic Press, New York, 1970 and 1971.

Baker, D. A., and **J. L. Hall** (eds.): *Ion Transport in Plant Cells and Tissues.* North-Holland/American Elsevier Publishing Companies, Amsterdam and New York, 1975.

Bowling, D. J. F.: *Uptake of Ions by Plant Roots.* Chapman and Hall, London, 1976.

Briggs, G. E., A. B. Hope, and **R. N. Robertson:** *Electrolytes and Plant Cells.* Blackwell Scientific Publications, Oxford, 1961.

Schutte, K.: *The Biology of the Trace Elements.* J. B. Lippincott Co., Philadelphia, 1964.

Steward, F. C. (ed.): *Plant Physiology: A Treatise,* Vol. II. Academic Press, New York, 1962.

Wardlaw, I. F., and **J. B. Passioura** (eds.): *Transport and Transfer Processes in Plants.* Academic Press, New York, 1976.

Zimmermann, U., and **J. Dainty** (eds.): *Membrane Transport in Plants.* Springer-Verlag, New York, Heidelberg, Berlin, 1974.

13 Translocation

The Problems of Translocation

The difficulty in studying translocation experimentally is primarily the difficulty of analyzing a kinetic system. It is easy to determine, either by direct measurement or by inference, that a given amount of material has been moved from one place to another in a given time. Finding out *what* is moving, where, or in what tissue it moves, how fast it goes, and by what agency is another matter. Flow of material may be continuous or discontinuous in time and space. Rate of flow is a product of the velocity of movement of individual molecules and the amount of material moving at any time. These parameters in turn are affected by, or related to, the size (particularly cross-sectional area) of the translocating tissue, a value that is extremely difficult or impossible to measure.

It is often not even clear what is being translocated (for example in ^{14}C-tracer experiments, where the chemical nature of small amounts of moving radioactivity may be hard to determine) or in what tissue movement is taking place. The manner in which the translocation system is loaded and unloaded is not clear, nor is the nature of the signal system that controls the traffic patterns of substances moving in plants. Many conflicting data have been produced, and several theories describing translocation processes are presently under consideration. It is not presently possible to unravel this tangled skein of contradictions. We shall briefly consider the basic phenomena and some of the explanations that have been offered.

The rates and velocity of translocation may be very great. The American physiologists A. S. Crafts and O. Lorentz measured the total amount of organic material in a mature pumpkin. Knowing the approximate cross-sectional area of the translocating tissue of the stem and the concentration of the solute, they calculated that an average translocation velocity of 110 cm/hr would have to be maintained during the growth of the pumpkin. Evidently peak velocities would be much higher. Tracers injected into the translocatory stream normally move with velocities of up to 100–200 cm/hr; velocities approximately ten times this value have been measured for small amounts of photoassimilated carbon in soybeans by the Canadian physiologist C. D. Nelson.

Small amounts of substances can move more or less slowly from cell to cell in most tissues by diffusion or active transport. We shall not consider this process in the present discussion; but confine our attention to the long-distance movement of substances in well-defined translocation tissues.

Tissues of Translocation

Ringing Experiments. As long ago as 1671 the English scientists Thomas Brotherton and Robert Hooke (discoverer of cells) conducted ringing experiments, which showed that the nourishment plants received from the air was transported downward in the bark while the nourishment derived from the soil moved upward in the wood. In essence, completely ringed trees continued for some time to grow in diameter and the branches continued to flourish in the region above the girdle, whereas the lower part of the tree did not grow. Eventually the trees died from lack of nourishment from the leaves to the roots. These early experiments have been repeated and refined by using steam or narcotizing chemicals to kill or inhibit the phloem. The principal conclusion still stands: The sap or water of transpiration containing inorganic salts is moved upward in the nonliving wood or xylem tissue, and the translocation of organic materials downward from the leaves to the roots takes place in the living, metabolizing bark or phloem. Some modification of these generalizations is necessary, and there are exceptions, but the basic concept has remained essentially unaltered for 300 years.

Analysis of Tissues. It is possible to extract xylem vessel or tracheid sap by centrifugation, since the xylem cells have no cross walls, and this sap has been subjected to chemical analysis. Phloem sieve-tube sap is much harder to obtain pure, because of the anatomical limits of the phloem tissue and because the cells need to be emptied individually. In addition, it is difficult to extract cell sap uncontaminated by cytoplasm, and the presence of a relatively large proportion of nonconducting parenchyma cells in phloem makes it impossible to obtain pure sap by maceration.

However, the discovery that aphids have the ability to insert their proboscis specifically into the lumen of sieve tubes has provided a valuable tool. Because sieve-tube contents are under pressure, these contents flow into the aphid's body without any effort on the part of the aphid. If the proboscis is cut, the phloem sap will continue to flow through it for long periods, up to several days, and considerable quantities can be collected essentially pure. This sap can be analyzed for kinds and quantities of chemical constituents by means of chemical and chromatographic techniques. Analyses of xylem vessel sap and phloem sap are shown in Table 13-1.

Although these analyses are not comparable, having been done on different species at different times, they emphasize the facts that xylem sap has a low concentration of mainly inorganic salts whereas phloem sap has much higher concentrations of total solids, of which organic compounds constitute the bulk. It is important to note that the concentration of inorganic salts in the phloem sap was about ten times greater than that in xylem sap. Concentration does not necessarily reflect the amount translocated, since velocity of translocation in the two tissues may be quite different.

Clearly, the early conclusion that inorganic substances move only in the xylem is not strictly correct. Also, the xylem may at times (particularly in the spring, in trees) contain considerable amounts of sugars and other organic compounds. The spring flow of sap—the best-known example is in maple trees—is influenced and may be largely caused by the addition of large amounts of sugar to the xylem sap. The sugar greatly lowers the osmotic potential of the sap and causes the absorption of water. The consequence is the movement of the solution to all parts of the tree by the hydrostatic pressure so generated. In addition, organic nitrogen compounds appear to move upward in the

Table 13-1. Analyses of vessel sap of a pear tree and phloem exudate of *Robinia pseudo-acacia*

	Concentration in sap, mg/liter	
Substance	*Xylem*	*Phloem*
Ca	85	720
Mg	24	380
K	60	950
SO_4^{2-}	32	—
PO_4^{3-}	25	—
Sugars*	—	200,000
N, organic	—	425
N, inorganic	—	135

SOURCE: Data recalculated from F. G. Andersen: *Plant Physiol.*, **4**:459–76 (1929), and C. A. Moose: *Plant Physiol.*, **13**:365–80 (1938).

*Mostly sucrose.

xylem of trees, although they are more generally transported in the phloem of herbaceous plants.

The nature of the organic substances in the phloem is of interest. Nitrogen compounds often include amino acids and the amides, glutamine and asparagine. Occasionally other nitrogenous compounds are translocated in large amounts; the place of the amides appears to be taken by the nonprotein amino acid homoserine in pea seedlings. The commonest sugar of translocation appears to be sucrose; reducing sugars (glucose and fructose) are seldom found. However, sucrose-based oligosaccharides are frequently found as major translocation sugars. Raffinose (a trisaccharide), stachyose (a tetrasaccharide), and verbascose (a pentasaccharide) appear to be species specific in a number of plants. These oligosaccharides are essentially sucrose with additional galactose units attached by glycosidic linkage. It appears that sucrose is used for translocation rather than the monosaccharides because the nonreducing sugar is much more easily transported and less prone to being "hung up" by adsorption. The reason why larger oligosaccharides are used in some plants is not known.

Tracer Experiments. The discovery of radioactive tracers of many common elements has provided an extremely powerful tool for measuring velocities, rates, and paths of translocation. Extremely sophisticated versions of the early girdling experiments have been conducted by several research groups.

The most informative type of experiment has been to separate the phloem from the xylem in the stem (the tissue breaks easily, and the bark separates along the natural shear line of the cambium). This can be done by making an incision in the bark, then inserting a sheet of wax paper or some similar material so as to form a cylindrical barrier around the stem between the xylem and the phloem, as is shown in Figure 13-1. This prevents any lateral translocation between the two tissues. Then radioactive tracers can be applied either above or below the operated area, and either xylem or phloem may be interrupted independently if necessary.

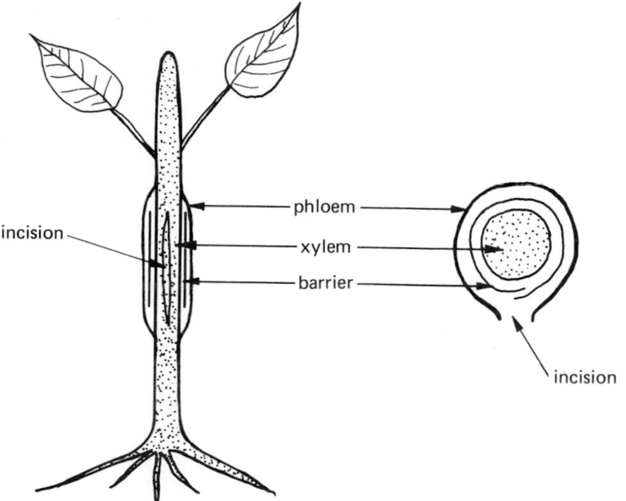

Figure 13-1. Preparation for a translocation experiment.

Various substances can be supplied to leaves, either as solutions of radioactive ions or ^{14}C-labeled organic compounds or else as $^{14}CO_2$, which is converted by photosynthesis into organic substances normally translocated, as shown in Figure 13-2. Alternatively, labeled substrates can be supplied to roots. After a suitable period of time the plant can be cut into small segments and the distribution of labeled translocate determined in the various tissues of the plant at various levels above or below the operated zone. A

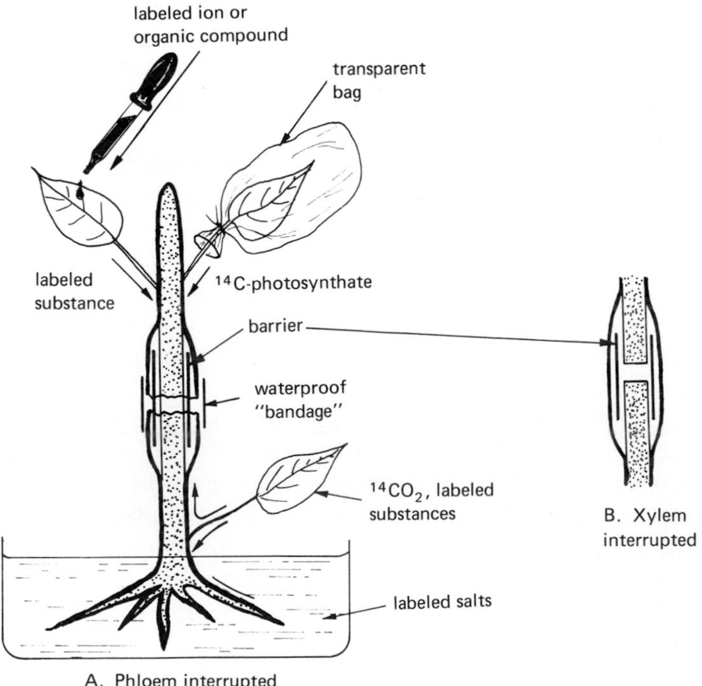

Figure 13-2. Translocation experiments.

comparison of the results for intact plants (control), plants with phloem cut, and plants with xylem cut will provide information on the pathway or tissue of up and downward movement of various substances. In addition, lateral transport from one tissue to another can be studied.

Two classical experiments are worth examining in detail. In the first, P. R. Stout and D. R. Hoagland separated the bark from the wood over a short zone in a willow shoot with waxed paper ("stripped") and supplied its roots with radioactive potassium (^{42}K), as shown in Figure 13-3. The results, shown in Table 13-2, indicate that the ^{42}K ascended in the xylem. When xylem and phloem were in normal contact, as above or below the stripped zone or in the control (unstripped) plant, lateral transport from xylem to phloem occurred, and the phloem contained as much ^{42}K as the xylem. However when they were separated, the amount of radioactivity in the phloem was extremely small, indicating little upward or downward movement of potassium in the phloem tissue.

In the second experiment, O. Biddulph and J. Markle supplied radioactive phosphorus (^{32}P) as phosphate ($^{32}PO_4$) to a leaf and followed its downward movement using the same technique. The results, shown in Figure 13-4 and Table 13-3, indicate that

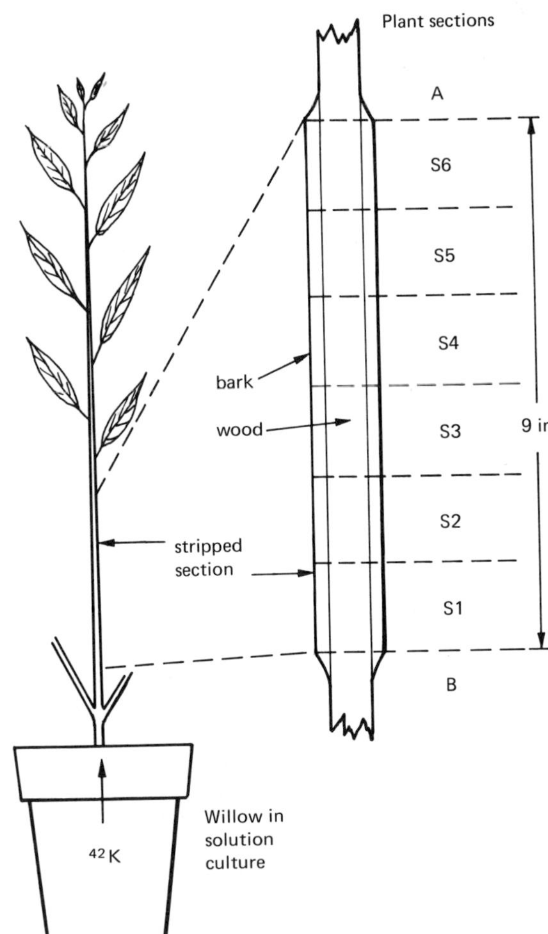

Figure 13-3. An experiment in upward translocation of radioactive potassium (^{42}K). [From P. R. Stout and D. R. Hoagland: *Am. J. Bot.*, **26**:320 (1934). Used with permission.]

Table 13-2. Distribution of ^{42}K in sections of willow (*Salix lasiandra*) after 5 hr absorption

	^{42}K *in each section, ppm*			
	Stripped plant		*Control plant*	
Section	*Bark*	*Wood*	*Bark*	*Wood*
A	53	47	64	56
S6	11.6	119		
S5	0.9	122		
S4	0.7	112	87	69
S3	0.3	98		
S2	0.3	108		
S1	20	113		
B	84	58	74	67

SOURCE: Data of Stout and Hoagland, see Figure 13-3.

Figure 13-4. An experiment in downward translocation of radioactive phosphorus (^{32}P). [From O. Biddulph and J. Markle: *Am. J. Bot.*, **31**:65 (1944). Used with permission.]

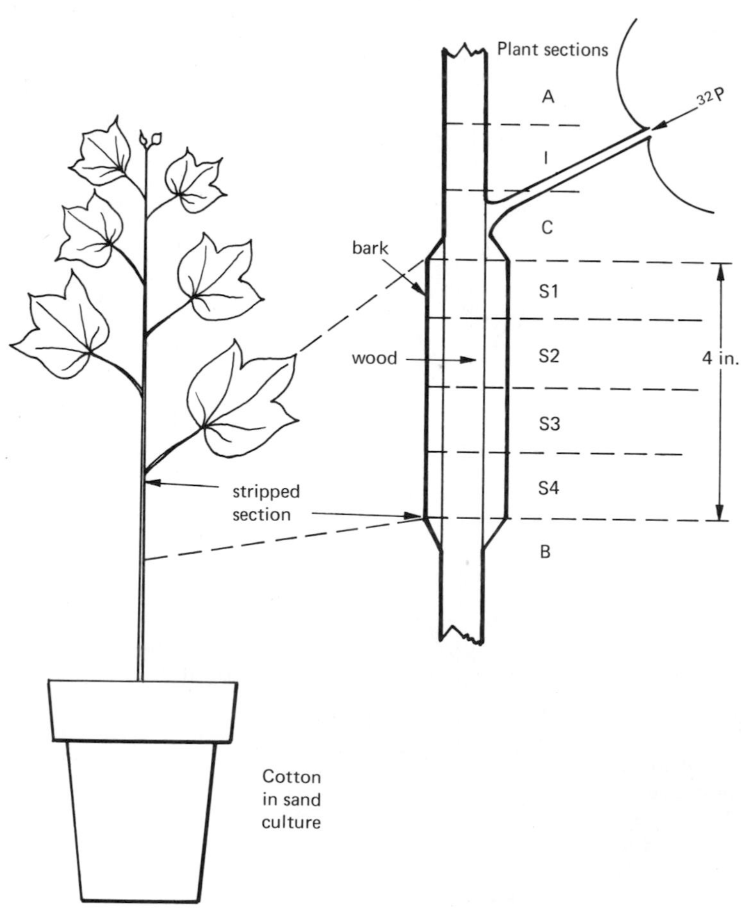

Table 13-3. Distribution of ^{32}P in sections of a cotton stem 1 hr after application of $^{32}PO_4$ to a leaf

	$^{32}PO_4$ in each section, μg			
	Stripped plant		*Control plant*	
Section	*Bark*	*Wood*	*Bark*	*Wood*
A	1.11			
1	0.458	0.100	0.444	
C	0.610			
S1	0.544	0.064	0.160	0.055
S2	0.332	0.004	0.103	0.063
S3	0.592	0.000	0.055	0.018
S4	0.228	0.004	0.026	0.007
B	0.653		0.152	

SOURCE: Data of Biddulph and Markle, see Figure 13-4.

phosphate moves downward in the phloem, not in the xylem, and is also able to move laterally from the phloem to xylem.

The technique of microradioautography has proved useful in studying translocation. Tracers are supplied to the plant, then thin sections of the stem are cut and immediately fixed by being frozen or treated chemically. A thin layer of photographic emulsion is then spread over the tissue slice on a glass slide, and the resulting "sandwich" is left in darkness until an image is formed on the film from the radioactive tracers present. The film is developed while still adhering to the specimen on the slide, and microscopic examination permits location of the radioactivity in the tissues or even in the cells of the specimen. A typical radioautograph of the petiole of a sugar beet leaf after the supply of $^{14}CO_2$ to the leaf is shown in Figure 13-5. It can clearly be seen that the darkening of the film resulting from the presence of ^{14}C-labeled photosynthetate being translocated from the fed leaf is associated with phloem tissue. In other words, the organic products of photosynthesis were being translocated in the phloem.

SUMMARY. The general conclusions that can be reached about the pathways and tissues of translocation are as follows:

1. Salts and inorganic substances move upward in the xylem.
2. Salts and inorganic substances move downward in the phloem.
3. Organic substances move up and down in the phloem.
4. Organic nitrogen may move up in the xylem (for example, in trees) or phloem (herbaceous plants).
5. Organic compounds like sugar may be present in the xylem sap in large concentrations during the spring when sap rises in trees before the leaves emerge.
6. Lateral translocation of solutes from one tissue to another occurs, presumably by normal mechanisms of transfer (diffusion, active transport, and so on).
7. Exceptions to these generalizations are known to occur.

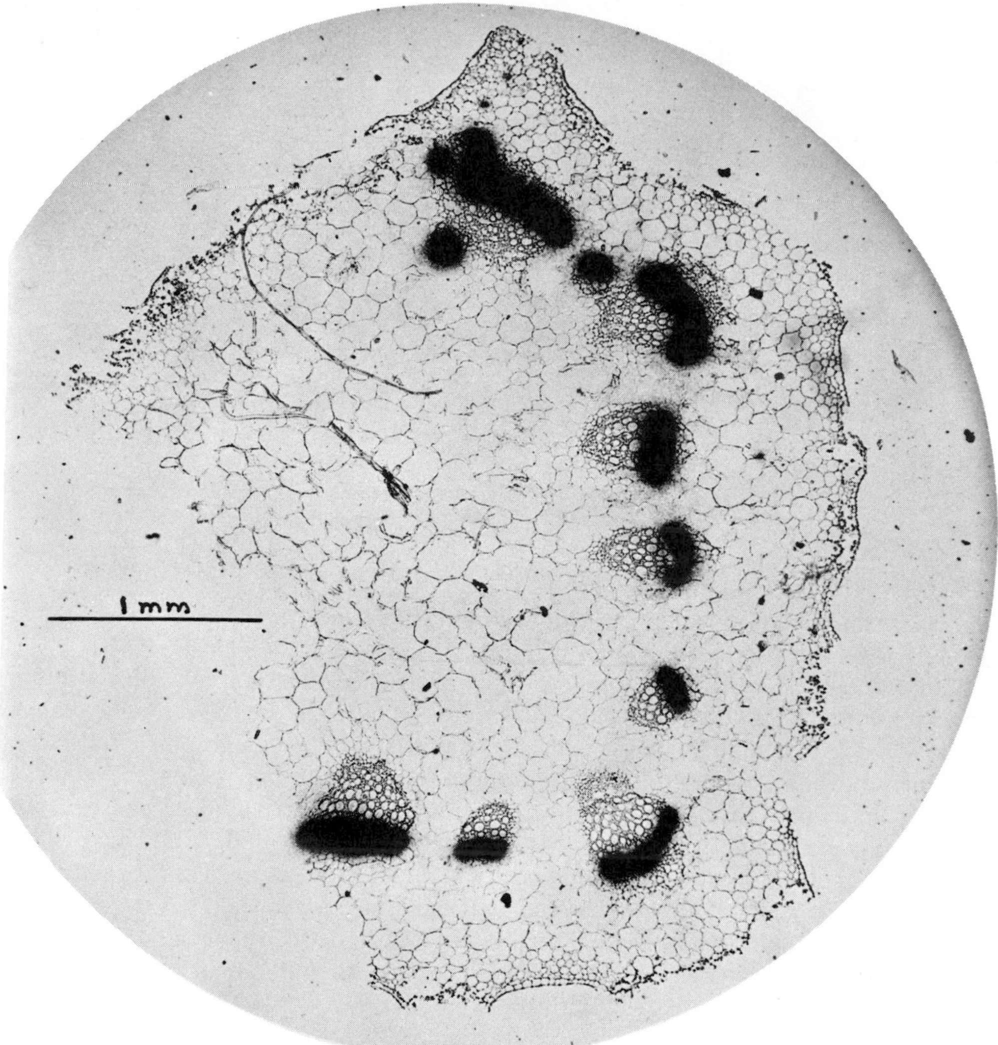

Figure 13-5. Radioautograph of cross section of stem after photosynthesis. [Reproduced by permission of the National Research Council of Canada from *Can. J. Bot.*, **43:**269–80 (1965). Photograph courtesy Dr. D. C. Mortimer, N.R.C., Ottawa, Canada.]

Translocation in the Xylem

STRUCTURE OF XYLEM. Xylem conducting tissue consists mainly of vessels and tracheids, or entirely of tracheids as in the conifers. These cells, when functional, are nonliving and usually form part of a continuous open tube system through the stem. Vessels lose their end walls and tracheids develop large areas of perforations in the walls between cells that are not blocked by living membranes. This fact has often confounded shipbuilders—red oak is most unsatisfactory for ships, for example, because one can suck water through the wide xylem vessels in this wood as through a straw! Luckily for shipbuilders, deposits that effectively plug the old xylem vessels normally form in many other woods. Some of the xylem vessels or tracheids, although they are themselves not living, are in direct contact with living cells in the xylem and parenchyma wood rays. Thus, either metabolic or physical-chemical mechanisms may function to control the contents of the xylem translocation streams.

Sheets or blocks of parenchyma tissue are interspersed among the tracheids and vessels of most stems. These appear to function largely for transverse or horizontal movement of solutes and for the removal or addition of solutes to the xylem translocation stream. They do not appear to take part in a large way in long-distance translocation.

Xylem Transport. Since the xylem represents an open-ended water transport system with essentially one-way movement, it seems likely that solutes move passively in the xylem by solvent-drag. They need not move at the same rate as the water, since their movements may be influenced by adsorption to the walls of the vessels or by diffusion down a potential gradient within the flowing system. Provided that some active transport system is available to transfer solutes to the xylem apoplast, in other words to load the translocation stream at the bottom, no motive power other than the flow of water is required to move them to the top of the open xylem system. At this point solutes could be removed by active transport or by diffusion, depending on the concentration of solutes in the leaf cells and their requirements.

The work of the American physiologist W. Lopushinsky supports this concept. He applied pressure to the solution surrounding roots of detopped tomato plants and found that increased pressure increased the transport of radioactive ions in the solution through the xylem. However, the increase in solute transferred was not equal to the increase in solvent (water) flow, suggesting a selective process in the absorption or root-transport part of the process.

Experiments with whole plants support the view that increased transpiration usually results in increased salt uptake and transport. It is not clear whether this results solely from an increased solvent-drag or because steeper diffusion gradients are established between soil and stele solutions, which would increase diffusion rates of solutes. However, there seems no good reason for rejecting the idea that solutes move passively in the xylem itself, carried along with and diffusing through the transpiration stream.

Translocation in the Phloem

Structure of Phloem. Phloem is a much more complex tissue than xylem. Its exact structure is still a matter of conjecture, because its fine structure seems to be unusually delicate and is changed or destroyed during preparation for microscopic examination. In fact, it is so delicate that it may suffer serious changes in structure as the result of hydraulic surges of the phloem contents when stems or petioles are cut. As a result, studies have been done with phloem from callus or from starved plants, where no translocation is taking place and surges are reduced to a minimum. However, then the phloem may not be in a normal condition. Some success has been had with ultrarapid freezing of sections, and certain observations have been possible with the microscope using living tissues.

The main components of phloem are **sieve tubes,** longitudinally arranged rows of individual cells called **sieve elements,** separated by perforated end walls called **sieve**

plates. In addition, there are small parenchymalike cells that are oriented longitudinally among the sieve tubes called **companion cells.** Companion cells are clearly living and contain the usual complement of cellular inclusions and organelles. The state of the sieve elements is not so clear. Unlike xylem vessels, mature sieve elements appear to contain living protoplasm, although they do not have a nucleus. Sieve elements can be plasmolyzed, indicating the presence of differentially permeable membranes, but they do not have a tonoplast. Much of the protoplasm is present in the form of **P-protein,** a fibrillar protein whose exact structure and function are presently the subject of intense debate among plant physiologists.

Microscopic preparations of phloem usually show the sieve plates plugged with proteinaceous masses that were earlier called **slime bodies.** More recent studies suggest that in living tissues either the sieve plates are open and unplugged, or they are partially plugged with strands or fibrils of P-protein that are intimately concerned with the process of translocation. Figure 13-6 shows an electron micrograph of a sieve plate;

Figure 13-6. Electron micrograph (×25,000) of a functional sieve plate in a soybean petiole (*Glycine max*). The tissue was quick frozen to −170° C prior to fixing and staining with uranyl acetate and lead citrate. Note the strands of fibrillar protein suspended from the sieve plate and the layer of cytoplasm around the walls of the sieve tube. [From D. B. Fisher: *Plant Physiol.*, **56**:555–69 (1975). Used with permission. Photograph kindly provided by Dr. Fisher.]

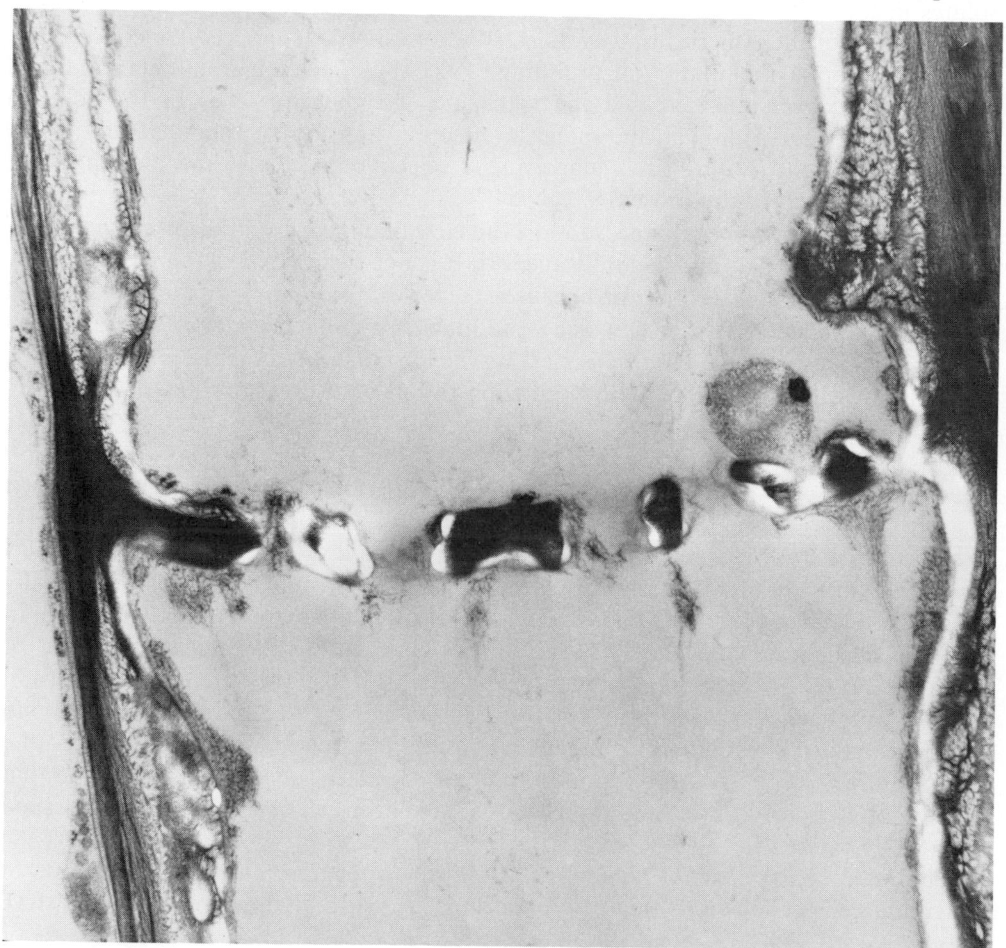

some endoplasmic reticulum and microfibrils are clearly visible, and the pores appear to be largely unplugged. This question is not finally settled; other micrographs show sieve plates essentially blocked either by carbohydrate deposits **(callose)** or by proteins. However, the weight of recent evidence does suggest that some, at least, of the sieve plates in a translocating stem are unplugged.

Mechanisms of Phloem Transport. There are several possible mechanisms whereby phloem transport might take place. Some experiments favor one and some another. Some evidence appears on the surface to exclude one system or another, but the full implication or meaning of experimental evidence in this difficult and highly complicated field is often unclear. The problem is not resolved as yet; it is possible that phloem transport may have a number of mechanisms or that different mechanisms may operate at different times or in different tissues in the same plant.

The five main theories for phloem transport are

1. Bulk flow.
2. Activated diffusion and pumping.
3. Cytoplasmic streaming.
4. Interface diffusion.
5. Electroosmosis.

We shall consider these in turn. It must be recognized that, although the basic concepts of some of these ideas appear to be contradictory or mutually exclusive, it is possible—even probable—that there is no one "right" solution to the problem of phloem transport.

Bulk Flow. In 1931, E. Münch in Germany suggested a flow system involving the circulation of water to account for phloem translocation of solutes from leaves to roots. Models of the system, in terms of osmometers and an idealized plant, are shown in Figures 13-7 and 13-8. Photosynthesis causes the production of a large amount of sugars that are actively loaded into the phloem cells of the leaf veins. The osmotic potential in the phloem becomes very low (negative) as a result, and water enters from the apoplast or xylem tissue by osmosis. The hydrostatic pressure so produced in the phloem forces the phloem sap down to the roots through the sieve tubes. Some water or solutes may leak out en route, but the bulk of the solution is carried to the roots. In the roots, the solutes are unloaded from the phloem translocation stream by active transport, with the result that the osmotic potential of the remaining phloem solution rises. In response, water diffuses out of the phloem and moves back up to the leaves along the water potential gradient of the xylem. The net effect is a circulation of water; the motive forces for this circulation are supplied by the active addition of solutes in the leaf by photosynthesis and their active removal in the roots. In fact, circulation is not always necessary and may not occur if sufficient water is removed from the system as the result of growth or transpiration.

The main point to note about this theory is that it requires a positive hydrostatic pressure in the phloem and, therefore, a continuous supply of sugars in the leaves that can be used to generate the pressure. Phloem sap certainly does have a positive pressure. If the stylet of a feeding aphid is cut off, phloem sap continues to run out of the stylet for some time as the result of internal pressure. Analysis of phloem sap has shown that it does indeed have a very high concentration of sugar, as the theory demands. A very strong argument in favor of the bulk flow hypothesis is that growth substances or virus

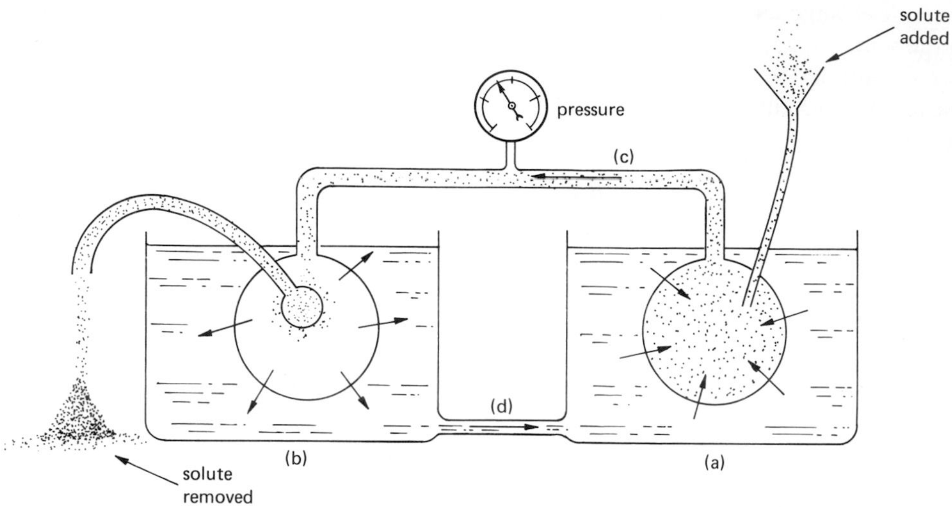

Figure 13-7. A double osmometer system illustrating the bulk flow hypothesis. Solutes (for example, sugars produced in photosynthesis) are added to osmometer (a), so water diffuses into bulb (a) by osmosis. The hydrostatic pressure so generated forces water out of bulb (b), and the solution in bulb (a) flows along path (c) to bulb (b). Solutes are removed from bulb (b), so that its osmotic potential is always higher than that of bulb (a). As a result, water continuously moves into bulb (a), through path (c) to bulb (b), and out of bulb (b), returning to (a) via path (d). The solutes added to bulb (a) are translocated with the water along path (c) to bulb (b), where they are removed.

particles applied to leaves are translocated rapidly if the leaf is illuminated but not if it is in darkness. However, if the darkened leaf is supplied with sugar, translocation of the added material takes place. This suggests that sugars, either produced in photosynthesis or supplied in darkness, are used to generate the pressure gradients required for the operation of bulk flow.

The second important point about this theory is that it requires low diffusion resistance along the translocation path—that is, the sieve plates must not be plugged. The proponents of this theory cite evidence like the electron micrograph in Figure 13-6 to support mass flow. However, many studies show a high degree of organization of the sieve element contents and strong evidence of strands of protein or other structures in and across the sieve plates. These are clearly not consistent with the mass flow hypothesis.

Another problem with bulk flow is that movement could take place in only one

Figure 13-8. Diagram of the bulk flow system in a plant: (a) leaf, (b) root, (c) phloem, and (d) xylem are as in Figure 13-7.

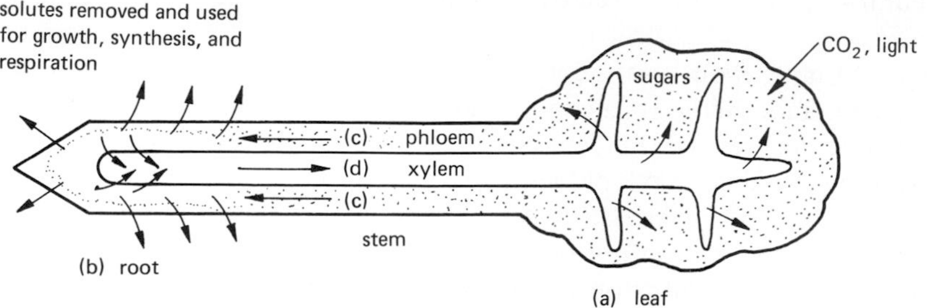

direction at a time in the phloem. There are many experiments indicating clearly that different substances can move in opposite directions at the same time in the phloem tissue, and that different substances may move simultaneously in the same direction but with widely differing velocities. These facts are incompatible with the bulk flow hypothesis as outlined. However, it is possible that different vascular bundles or even different sieve elements might be simultaneously engaged in pressure-flow translocation in opposite directions. This would only require controlled loading and unloading of the appropriate solutes from opposite ends of the same sieve tube. The theory does not require that all sieve tubes translocate at the same speed and in the same direction at the same time.

Activated Diffusion and Pumping. It has been observed that the distribution of a solute along the stem a short time after its supply at the source (usually measured by radioactivity after supplying a labeled substance) is a logarithmic function of the distance from the source, as shown in Figure 13-9. This is the expected pattern resulting from diffusion, but not from pumping or flow mechanisms. The British physiologists T. G. Mason and D. J. Maskell observed in the 1930s that the translocation rate varied as the gradient of sucrose concentration in the phloem. These facts are not consistent with mass flow, and led Mason and Maskell to suggest that some undefined property of the protoplasm reduces the resistance to diffusion. They suggested that the movement of solutes is thus essentially similar to normal diffusion but is activated or increased in rate as the result of the reduced resistance.

Various pumping or peristaltic mechanisms have been suggested recently as more information (and, one sometimes suspects, misinformation) about sieve elements is amassed. The main point about these mechanisms is that pumping "stations" are thought to occur at intervals along the sieve tubes, with power input from the metabolism of companion cells or of the sieve tubes themselves. None of these hypotheses has been generally accepted, but this may be because of their newness as much as the difficulty of providing proof.

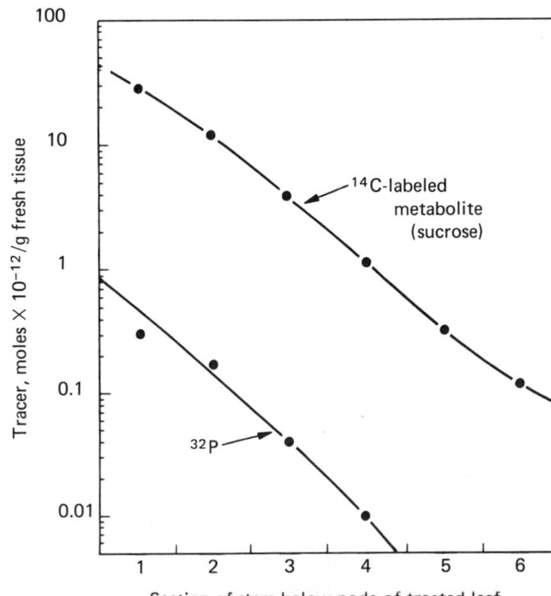

Figure 13-9. Movement of tracers (^{32}P and ^{14}C-sucrose) in a bean plant through the stem 15 min after the application of $H_2{}^{32}PO_4^-$ and $^{14}CO_2$ to the leaf. The ^{14}C-sucrose was made from $^{14}CO_2$ in the leaf by photosynthesis. [Adapted from data of O. Biddulph: In F. C. Steward (ed.): *Plant Physiology: A Treatise*, Vol. II. Academic Press, New York, 1959, p. 591. Used with permission.]

Recent data suggest that the P-protein in sieve elements has many of the characteristics of **actin,** a contractile protein involved in a variety of animal cytoplasmic systems including muscle, ameboid movement, and cytoplasmic streaming. It has been suggested that microtubules or strands of protein extend through the pores of sieve plates for distances of up to several sieve elements. These may be involved in the transfer of solutes in various ways. Three British physiologists, R. Thaine, M. J. Canny, and E. A. C. MacRobbie, have elaborated models along these lines that differ in the way solutes are moved. In Thaine's hypothesis, solutes are pumped through strands or tubules. In Canny's model the strands themselves move in opposite directions over distances of several cells, carrying materials up or down. Solutes diffuse into or out of the strands depending on the relative concentration in the vacuole (stationary phase) and the moving strands. Thus, a strand moving up through a region of low sugar concentration to one of high concentration will lose sugar initially to the sap. A downward-moving strand will gain it initially and move the sugar down to the area of low concentration. The result will be a greatly increased rate of diffusion of the sugar from the area of high concentration to the area of lower concentration. In MacRobbie's model the tubules of protein are, in fact, contractile and move solutes by peristaltic contractions. The Canadian physiologist D. S. Fensom has elaborated this concept on the basis of microscopic examination of living phloem strands in the stem of *Heracleum* (cow parsnip) and by biochemical experiments that suggest the participation of actinlike proteins.

Cytoplasmic Streaming. The Dutch botanist H. de Vries suggested in 1885 that cyctoplasmic streaming or **cyclosis** might be a mechanism of active translocation, and the idea has been developed by the American physiologist O. F. Curtis. This concept has a number of attractive features, since substances would move essentially down diffusion gradients but at rates increased by the flow of cytoplasm; thus some substances could move at different speeds and even in opposite directions in the same sieve element, as may be the case (see Figure 13-10).

It has been observed that rates of cyclosis are affected by temperature change and other factors affecting metabolism in the same way as the rates of translocation, and very high rates of cyclosis have been observed in some phloem cells. However, experimental evidence shows that as sieve elements mature the rate of cytoplasmic streaming in them decreases and finally stops altogether. Further, it has been observed that, in parenchyma cells, cyclosis does not increase the rate of Cl^- ion translocation, which is in fact independent of streaming rates. The cytoplasmic streaming concept is in direct conflict with activated diffusion or metabolic pump hypotheses, because these require structural organization of phloem cytoplasm, whereas theories involving cyclosis require complete mobility of the cytoplasm. With the exception of Canny's specialized model (see preceding section) the cyclosis hypothesis has been generally discarded on these and energetic grounds.

Interface Diffusion. Substances that lower the surface tension at a boundary between two immiscible liquids, or between a liquid and a gas, are rapidly distributed across the whole area of the boundary. The substance in fact diffuses down a potential gradient, from a region of low surface tension to a region of high surface tension. An example is the rapid spreading of a thin layer of gasoline across a water surface, the boundary in this case being water-air. Diffusion at the boundary of such a system may be more than 50,000 times faster than diffusion through either component of the boundary system. The possibility that the cytoplasm of phloem elements contains such boundaries, or that the cytoplasm vacuole offers an effective boundary system, has been

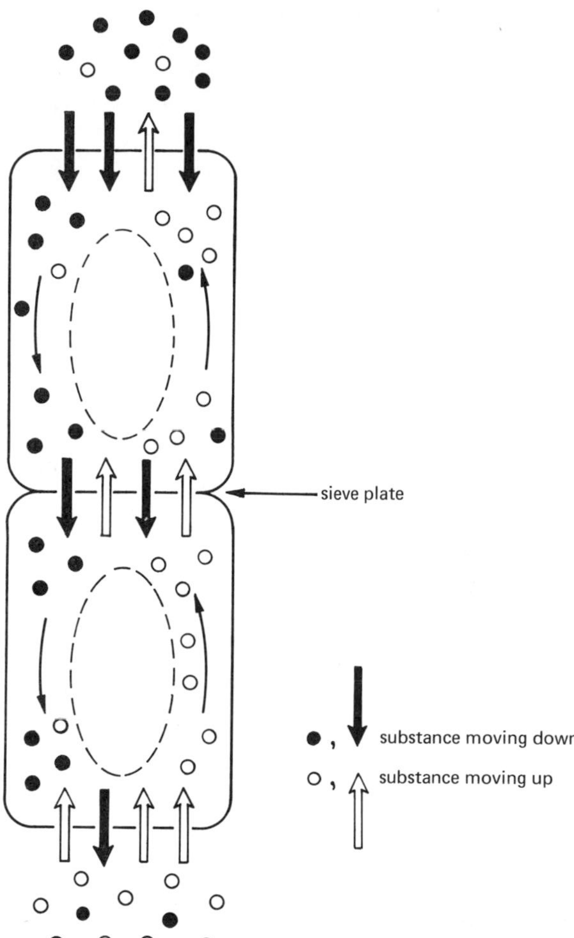

Figure 13-10. Diagram to illustrate how cyclosis (cytoplasmic streaming) might result in simultaneous two-way transport along diffusion gradients in phloem sieve elements.

suggested by the Dutch physiologist J. H. van den Honert. Such a mechanism might account for the extremely rapid translocation of small amounts of photosynthate observed by Nelson. Diffusion gradients would be maintained by the active addition and removal of solutes from the boundary layers by active transport. However, the possibility has been questioned that a sufficient area of boundary layer could exist in phloem cells for interface diffusion to be involved in the transport of large amounts of material. The recent discovery that sieve elements contain a substantial number of longitudinally oriented proteinaceous strands has once again revived the possibility that this theory is correct, but no clear proof of this type of mechanism exists at present.

Electroosmosis. If a membrane bearing fixed charges is flooded with a salt solution on both sides and an electric potential applied across the membrane, salts and water will diffuse through the membrane as the result of **electroosmosis.** This process requires quite specialized conditions, as shown in Figure 13-11A. The pores must be small and lined with fixed charges (negative charges are shown in Figure 13-11A). Since there are fixed negative charges in the pores, cations will move into the spaces within the pores. If the pores are small enough, anions will be repelled and will not enter. Now, when a potential difference is applied across the membrane, the cations will move through the membrane under its influence, carrying water or sugar molecules with them

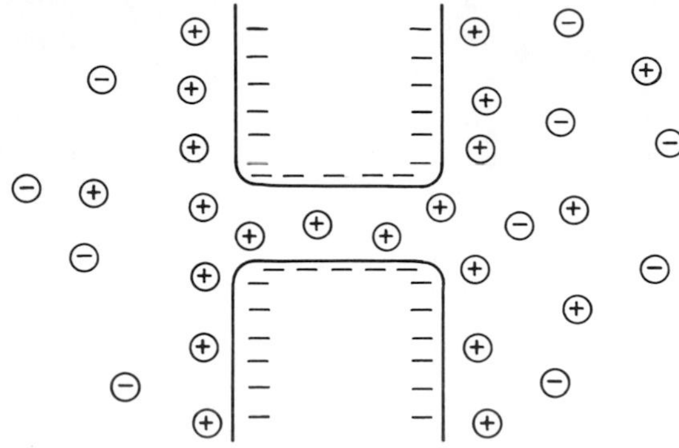

A. The distribution of ionized particles in and around a small pore in a membrane bearing fixed negative charges.

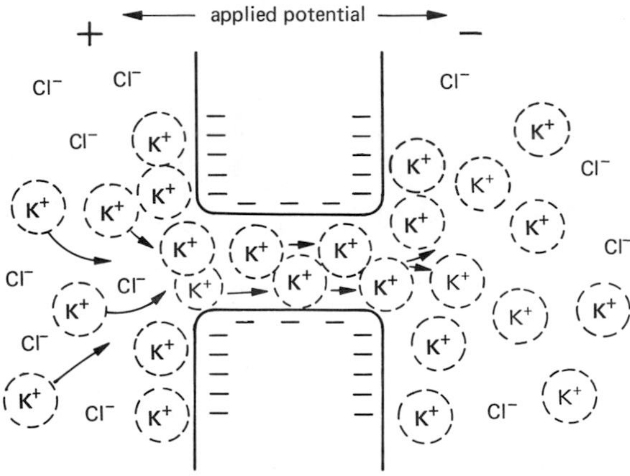

B. An electric potential is applied to a negatively charged membrane bathed in a KCl solution. K^+ (but not Cl^-) move through the pore causing electroosmosis. The hydration shells of K^+ ions are shown by dotted lines.

Figure 13-11. Diagram to illustrate electroosmosis.

in hydration shells and by solvent drag. The movement will be one way because anions, being repelled by the charges in the pores, will not pass through; only cations will move. A substantial flow of solute can be maintained by such a system.

The British physiologist D. C. Spanner has proposed a model based on electro-osmosis in which the electromotive force is generated by the active transport of K^+ back around the sieve plates (perhaps via companion cells). The return of K^+ ions through the sieve plate pores sets up an electroosmotic flow that could drive translocation, as shown in Figure 13-11B. The major problem with this hypothesis is the large energy requirement to move K^+ and the fact that it does not permit bidirectional movement or the selective movement of different substances at different rates.

Summary. The evidence available today tends to support the mass or pressure flow model of translocation more strongly than other models. However, there is little doubt that certain components of phloem translocation may be activated or driven by other mechanisms or pumps. It is possible that while some compounds are being translocated by a mass-flow system in certain parts of the phloem others are being actively pumped, or moved by activated diffusion, electroosmosis, or cyclosis, in other regions of the phloem.

The main questions that need to be answered before translocation is clearly understood concern the resistance (or plugging) of sieve plates, the nature and role of P-proteins, and the sites of energy input. Mass flow requires energy input only at the loading and (or) unloading sites, and requires unplugged sieve plates. This hypothesis provides no role for the P-proteins or companion cells. Activated diffusion or pumping models as well as mass flow by electroosmosis require energy input along the sieve tubes, a process in which companion cells may be active. They also require varying degrees of organization of sieve element contents and usually of the sieve plates.

Fensom has suggested that translocation proceeds simultaneously by two, three, or even four modes. In his postulation, the main driving force is contractile microtubules, but a varying amount of mass flow takes place in the sieve element contents outside the microtubules. Fensom also envisages the possibility of a minor electroosmotic component and suggests that interface diffusion of certain compounds other than sugars may also take place. This hypothesis has certain attractive points, particularly that it resolves the difficulties of apparently mutually exclusive data supporting one model or another. It explains the experimental observation that different substances (for example, ^{14}C-sugars and $^{42}K^+$) move in different directions and at different velocities when simultaneously injected into or applied to sieve tubes. Perhaps the most important conclusion yet to be reached is the relative importance or prevalence of the various modes of translocation under specified conditions in plants.

Control of Translocation

It is important to recognize that the patterns of translocation are not haphazard but are highly directional. There are clear-cut relationships between **sources** and **sinks** of translocation, and the direction of translocation appears to be under precise control. The whole subject of the traffic patterns of translocation will be considered in Chapter 21, but a few observations are relevant here.

Any mass flow or diffusion mechanism of translocation requires a gradient of solute concentration from source to sink. It may thus be controlled by sink demand, which creates a gradient by unloading, or by source supply, which creates a gradient by loading. Both these factors may be operative. There is considerable controversy at the moment about whether loading (or photosynthesis) is controlled via translocation by sink demand. Alternatives included some sort of sink-signal in the form of hormones or electrical phenomena. Since ultimately growth and sink requirements are met by photosynthesis, the rate of photosynthesis may control translocation without the intervention of specific mechanisms.

There is, however, much evidence (to be discussed in more detail in Chapter 21) that the distribution of material throughout the plant does not take place merely on the basis of the avidity of sinks, but is part of the overall system that correlates and coordinates the growth of plant parts. The nature of the correlative growth mechanism is complex and concerned with hormone action (see Chapter 19), but it is quite clear that nutritional traffic is mediated to some extent by the growth requirements of various parts of the

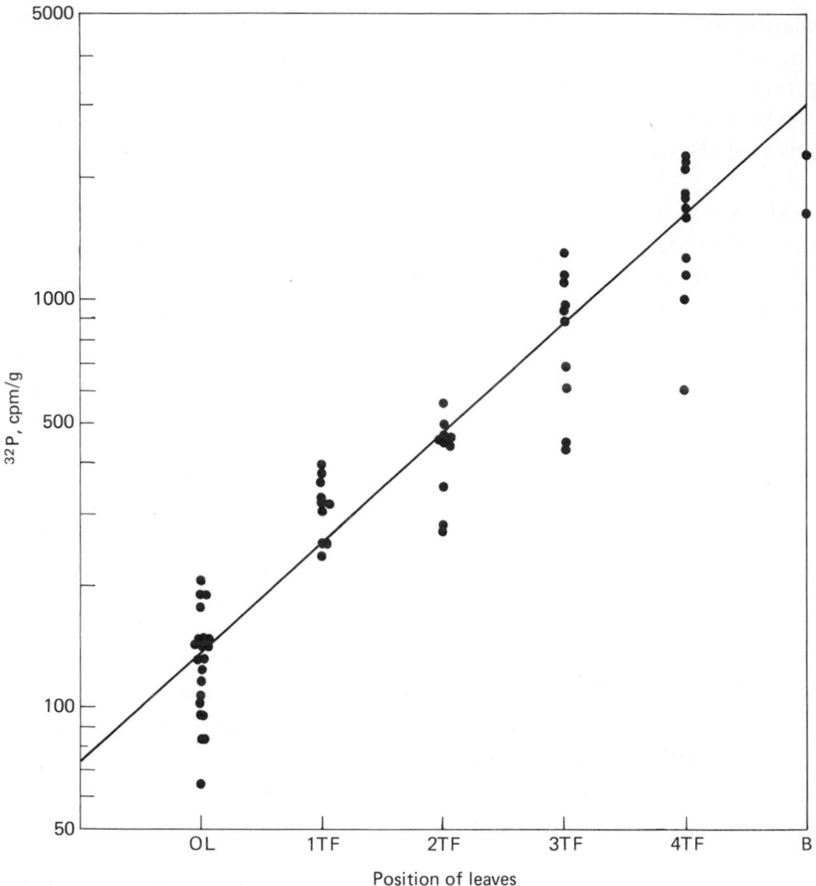

Figure 13-12. The amount of ^{32}P moving into leaves at various positions on the stem of the red kidney bean (*Phaseolus vulgaris*) plant. The ^{32}P was absorbed as phosphate from the nutrient solution during a 4-day interval. The data are from the leaves of ten plants, all grown in the same tank. The oldest (opposite) leaves are to the left (OL); progression is toward younger leaves to the right, through the first to the fourth trifoliate leaf (1TF-4TF). Two very small leaves are shown at B. [From O. Biddulph: In F. C. Steward (ed.): *Plant Physiology*: *A Treatise*, Vol. II. Academic Press, New York, 1959. Used with permission.]

Figure 13-13 (opposite). A sequence of six autoradiograms showing the fate of an aliquot of ^{35}S absorbed as $^{35}SO_4$ during a 1-hour absorption period. The plants, after the hour in the nutrient solution containing the tracer, were removed to a normal (nonradioactive) solution where they remained for the following periods: **A,** 0 hr; **B,** 6 hr; **C,** 12 hr; **D,** 24 hr; **E,** 48 hr; and **F,** 96 hr. Most of the ^{35}S, which moved directly into the mature leaves, was withdrawn within 12–24 hr. It moved predominantly into younger leaves near the stem apex, where it remained. [From O. Biddulph: *Plant Physiol.*, **33:**295 (1958). Used with permission. Photographs courtesy Dr. Biddulph.]

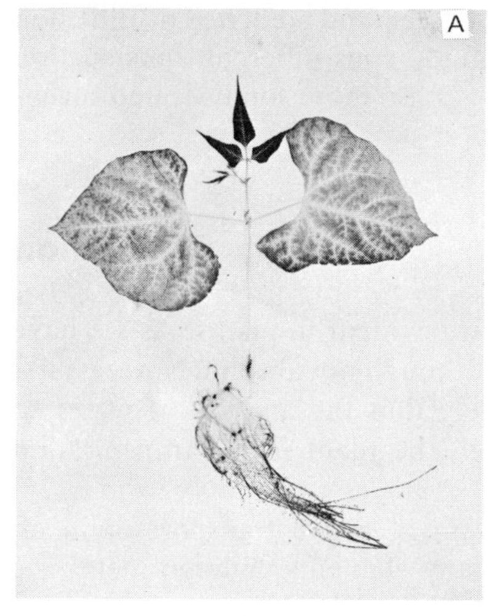
A

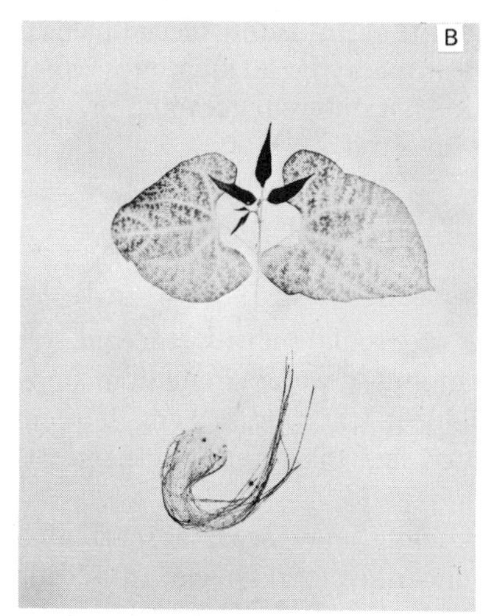
B

C

D

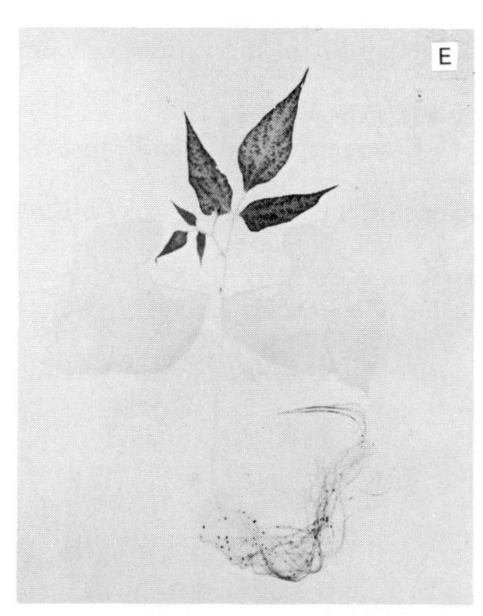
E

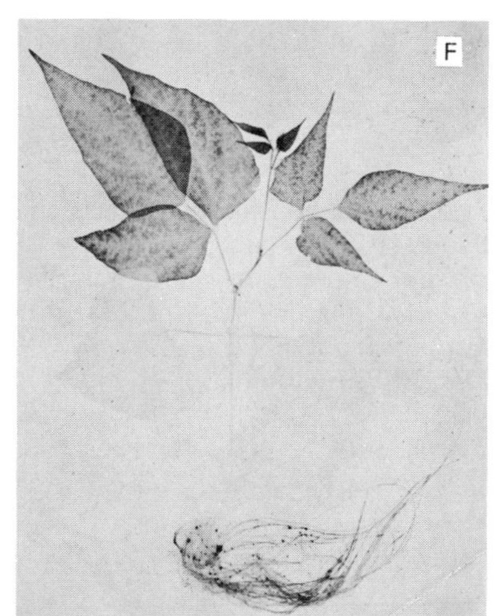
F

plant. Simple loading or unloading mechanisms based on demand mediated by diffusion gradients would probably be adequate to explain simultaneous different translocation rates and directions of different compounds. However, other more sophisticated mechanisms may be involved.

Circulation

Certain patterns of circulation of water and solutes are evident in plants. As we have seen, a certain amount of water circulation is possible, though not absolutely necessary, in a pressure-phloem transport system. It seems likely that the amount of water so circulated is much less than normally passes through the plant in the transpiration stream.

Circulation of ions may occur as part of a system of active transport or in an electrogenic or electroosmotic system. In addition, many elements undergo metabolic circulation through the plant. This results from their initial translocation to young, actively growing tissue, as shown in Figure 13-12. As the tissues mature, the elements are withdrawn from the tissue where they were originally located and moved to newly developing young tissue. An example of this may be seen in Figure 13-13, showing how radioactive sulfate applied to roots moves first to the primary leaves and the first trifoliate leaf of a bean plant, then to the second trifoliate leaf, and finally to the third trifoliate leaf, leaving none at all in the primary leaves.

Circulation of carbon also occurs. Organic compounds formed in photosynthesis move to the roots, and a certain proportion is chemically combined with ammonia and returned to the leaves or growing tips as organic nitrogen compounds. This and other aspects of the nutritional traffic of the whole plant will be discussed in greater detail in Chapter 21.

Additional Reading

Articles in *Annual Reviews of Plant Physiology* under the heading, "Translocation."

Aronoff, S., J. Dainty, P. R. Gorham, L. M. Srivastava, and **C. A. Swanson:** *Phloem Transport.* Plenum Press, New York, 1975. (This is a particularly interesting and useful source, being a series of free-wheeling discussions by protagonists and antagonists of the various phloem translocation theories.)

Canny, M. J.: *Phloem Translocation.* Cambridge University Press, New York, 1973.

Crafts, A. S., and **C. E. Crisp:** *Phloem Transport in Plants.* W. H. Freeman and Co., San Francisco, 1971.

MacRobbie, E. A. C.: Phloem translocation—facts and mechanisms: a comparative survey. *Biol. Rev.,* **46:**429–81 (1971).

Peel, A. J.: *Transport of Nutrients in Plants.* John Wiley & Sons, New York, 1974.

Richardson, M.: *Translocation in Plants.* St. Martin's Press, Inc., New York, 1968.

Steward, F. C. (ed.): *Plant Physiology: A Treatise,* Vol. II. Academic Press, New York, 1959.

Wardlaw, I. F., and **J. B. Passioura** (eds.): *Transport and Transfer Processes in Plants.* Academic Press, New York, 1976.

14 Leaves and the Atmosphere

Having got water into the root and moved it up the stem, we must now get rid of it. As we have seen, the motive force for upward movement of water comes from its evaporation from leaves to the atmosphere. In fact, the problem is not so much to get rid of water, but to prevent its loss at rates and in amounts that would be harmful to the plant.

Leaves

The appearance and anatomy of typical leaves (see Chapter 4) reflect their capacity for gas exchange and radiation absorption. To be maximally efficient, a radiation absorber needs to be broad and thin and oriented at right angles to the source of radiation. Similarly, for efficient gas exchange, a thin sheet giving maximum area per unit weight is required. Only the requirement for some storage tissue, some support tissue and transport systems limits the thinness of the leaf. However an efficient gas exchanger is also an efficient evaporator, and a leaf-shaped structure with no protective covering would rapidly dry out. The epidermis with its cuticle protects the leaf from drying out, but it also cuts down gas exchange to a very low level. The system of small holes or stomata, through which the gases diffuse, and air passages inside the leaf, are surprisingly effective for carbon dioxide exchange while reducing evaporation, as we shall see.

Leaves in many plants are also modified in ways that permit them to store food and water or perform other unrelated functions (for example, tendrils, thorns, and so on) with which we shall not be concerned. The primary function of the normal sort of leaf is the manufacture and export of food. Since food is made in parenchyma cells, it must be collected and transported. Leaves are highly vascularized, the major veins breaking up into many tiny veinlets that ramify in a variety of patterns through the leaf tissue, so that no parenchyma cell is more than a few cells' breadth from a vein. The bundles are sheathed in parenchyma cells; this **bundle-sheath** appears to be involved in the transport of food from leaf cells to the bundle and, in some species, to be involved in the final stages of photosynthesis (see page 358).

Leaves are adapted in various ways to particular environmental conditions. In extreme cases, as in some desert cacti, leaves disappear altogether as photosynthetic organs and the stem performs this function. However, in plants having more ordinary leaves, wide variations are possible. Species modifications include a variety of mechanisms that reduce water loss in xerophytic plants—surface hairs, sunken stomata, leaf rolling, and the like (see Figure 14-1). Even in a single species, or on a single plant,

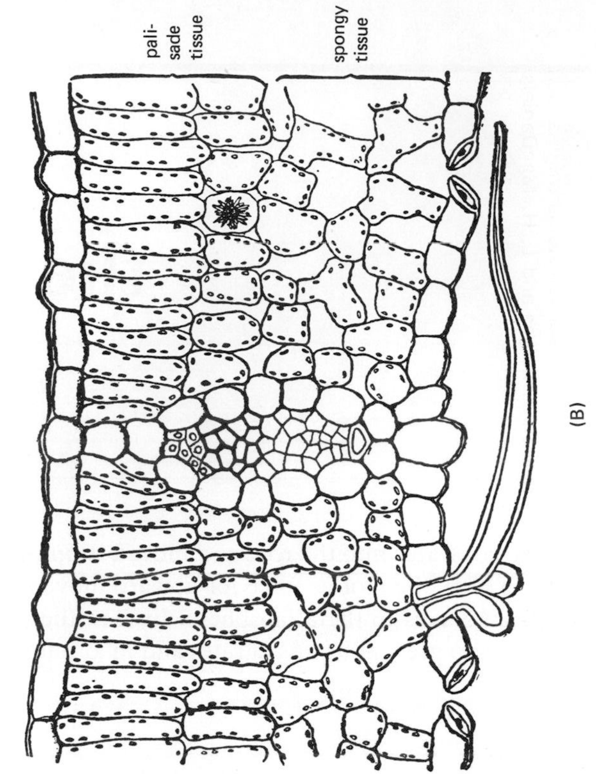

(B)

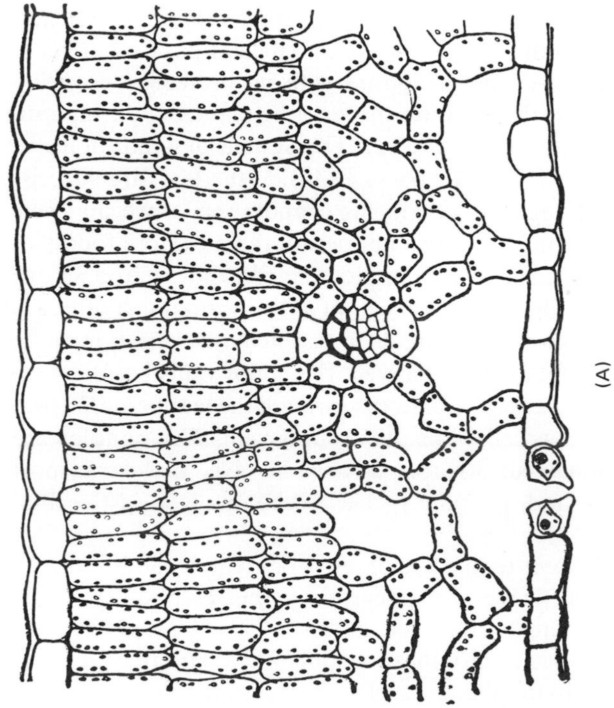
(A)

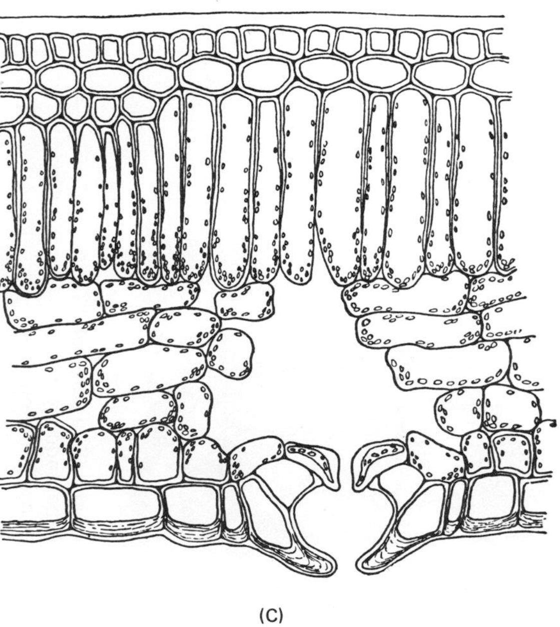
(C)

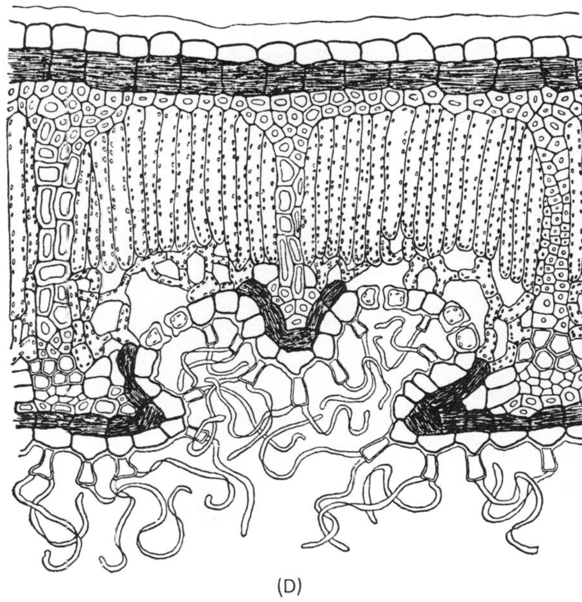
(D)

Figure 14-1. Diagrams of stomata. (**A**) A normal leaf, apple; (**B**) hairs, oak; (**C**) a sunken stoma, *Cycas* sp.; and (**D**) a sunken stoma with hairs in the stomatal pit, *Banksia* sp. [From A. J. Eames and L. H. MacDaniels: *Introduction to Plant Anatomy*. McGraw-Hill Book Co., New York, 1925. Used with permission.]

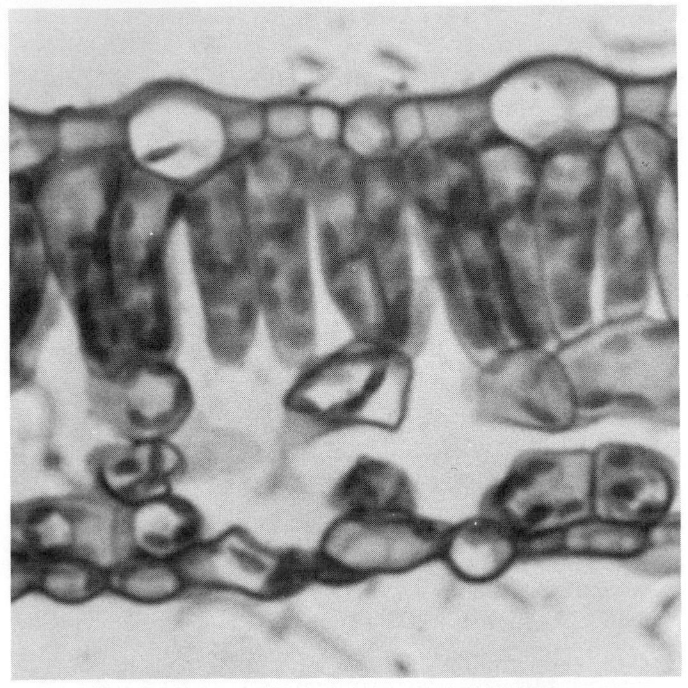
A

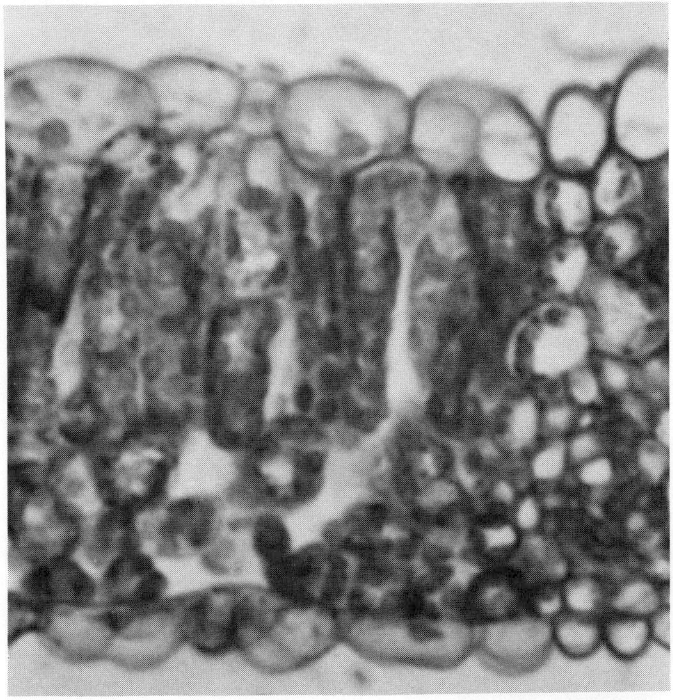
B

Figure 14-2. Cross section of (**A**) a shade leaf and (**B**) a sun leaf of maple (*Acer*). Note the enlargement of the palisade layer, the chief photosynthetic tissue, and the great increase in the number of chloroplasts in sun leaves.

considerable variation may occur in response to different environmental and physiological situations. Leaves grown in the shade tend to be thinner with a smaller palisade layer (the parenchyma cells mainly responsible for photosynthesis) and large, well-developed air spaces. Sun leaves, on the other hand, tend to have a much thicker and more elaborate palisade layer, which improves their capacity for light trapping, but they have less well-developed air spaces (Figure 14-2). Other less obvious modifications occur, but these are more effectively treated in the subsequent discussion.

Gas Exchange

Diffusion Through Pores. Early physiologists wondered that gases seemed able to diffuse freely into and out of leaves in spite of the fact that their surfaces were covered by an impermeable cuticle perforated only by very small holes. Because the holes, **stomata** or **stomates** (singular **stoma** or **stomate**), amount to no more than about 0.1 percent of the leaf surface area, diffusion would be expected to be extremely slow. However, experiments showed that, on the contrary, gases could enter and leave very quickly. Moreover, it was soon apparent that gas does indeed pass through the stomata, since little gas exchange occurs through leaf surfaces lacking stomata. Most, though not all, cuticle appears to be relatively impermeable to oxygen, carbon dioxide, and water, the main gases under consideration.

Since no pumps are apparent in leaves, gases must enter and leave by diffusion. Some pumping effect may result when leaves are bent and folded, as in a high wind, but such conditions are not necessary to attain high rates of gas exchange. Clearly the important process is diffusion, so the motive force must be a chemical potential gradient. Here we encounter a dilemma that must be borne in mind during the discussion. The concentration gradient down which CO_2 diffuses into a leaf is a rather shallow one. The external concentration of CO_2 in air is 0.03 percent, the internal concentration (at the surface of the leaf cell) cannot be less than 0. The resulting gradient therefore cannot be greater than 0.03 percent/d, where d represents the diffusion path length. The diffusion gradient for water vapor, however, is an extremely steep one. Air at 21° C and 50 percent relative humidity contains about 10 g/m^3 H_2O or about 1.25 percent H_2O. Saturated air, which would exist at the cell surface, would contain 2.5 percent H_2O, and a gradient of 1.25 percent/d would be established. This is a 40 times steeper gradient. Further, the water concentration within the cells is close to 1000 g/liter (ψ close to 0 bars), whereas that of air at 50 percent relative humidity is 0.125 g/liter (ψ roughly $-$ 1000 bars), giving a potential gradient of about $-$ 1000 bars/d. The potential gradient of CO_2 from 0 concentration to atmospheric is 0.0003 bars/d. Thus it can be seen that the forces motivating the outward diffusion of water vapor are enormously greater than those moving CO_2 into the leaf. Yet the plant absorbs CO_2 at a maximum rate while at the same time the rate of water loss is reduced to a minimum.

Oxygen diffusion gradients are more difficult to measure, but, due to its low solubility in water, oxygen probably diffuses out of photosynthesizing cells rapidly. Small increases in oxygen concentration do not greatly affect metabolic processes, so the exact diffusion rate of oxygen is probably not important to tissues like leaves. It may be an important limiting factor in the respiration of some bulky tissues, but we need not consider it further here.

Gas Exchange Through Stomata. It was observed in about 1900 by the English physiologist F. F. Blackman, and independently by H. T. Brown and F. Escombe, that CO_2 diffusion through leaves was closely correlated with the presence and number of stomata. Some figures are given in Table 14-1, which show that CO_2 diffusion usually takes place only through leaf surfaces that have stomata and roughly in proportion to the number of stomata present. The conclusion seems inescapable—CO_2 diffuses mainly through stomata. Brown and Escombe calculated that, allowing for the area of all its stomata when wide open, a photosynthesizing leaf absorbs about 70 times more CO_2 per unit area of stomatal holes than an open dish of *N* NaOH, one of the strongest CO_2 absorbers known. They reasoned that this extraordinarily high efficiency must be related to the size of the pores and tested this as follows.

A series of cups containing a CO_2-absorbing agent were covered with a thin membrane having in it a perforation or aperture whose size was carefully measured. The rate of CO_2 diffusion through the aperture was measured, with the results shown in Table 14-2. As the aperture decreased in size, its efficiency in terms of diffusion per unit area increased. Diffusion, in fact, varied approximately as the diameter of the pore, not as its area. They also measured the efficiency of a number of small pores as compared with only one, with the results shown in Table 14-3. They found that so long as the pores were at least 10 diameters apart, high efficiency was maintained. If the pores were placed closer together, the efficiency was lower.

Brown and Escombe developed the idea that gas molecules diffuse through small pores in a membrane following a pattern similar to that diagrammed in Figure 14-3A. Gas molecules diffuse through the hole forming a **diffusion shell:** the molecules momentarily achieve a high concentration in the opening but disperse again in a diffusion shell on the other side (Figure 14-3A, D). Thus widely spaced small pores are extremely efficient in relation to their area. In an open vessel (Figure 14-3C) or one with a septum

Table 14-1. Relationship between stomatal distribution and CO_2 transferred

Plant	*Ratio of stomata upper/lower surface*	*CO_2 transferred upper/lower surface*
Catalpa bignonioides	0/100	0/100
Nerium oleander	0/100	3/100
Hedera helix	0/100	4/100
Prunus laurocerasus	0/100	0/100
Polygonum sacchalinense	0/100	6/100
Ampelopsis hederacea	0/100	3/100
Phaseolus vulgaris	0/100	8/100
Nuphar advenum	100/0	100/0
Alisma plantago	100/74	100/74
Iris germanica	100/100	100/95
Colchicum speciosum	100/119	100/75
Senecio macrophyllus	100/126	100/92
Rumea alpinus	100/144	100/269
Tropaeolum majus	100/200	100/265
Helianthus tuberosus	100/240	100/273
Populus nigra	100/575	100/375

Calculated from data of F. F. Blackman, H. T. Brown, and F. Escombe, courtesy the late Prof. G. Krotkov, Queen's University, Kingston; also data of R. G. S. Bidwell and W. Levin.

Table 14-2. The diffusion of CO_2 through a small aperture

Diameter of aperture, mm	CO_2 diffused: μg/hr	CO_2 diffused: μg/(hr)(cm²) of aperture	Relative CO_2 diffusion	Relative area of aperture	Relative diameter of aperture	Efficiency: diffusion per unit of area
22.7	238	58.8	1.00	1.00	1.00	1.00
12.1	101	89.1	0.42	0.28	0.53	1.50
6.03	62.5	219	0.26	0.07	0.26	3.70
3.23	39.9	486	0.16	0.023	0.14	7.0
2.12	26.1	825	0.10	0.008	0.093	12.5
2.00	24.0	763	0.10	0.007	0.088	14.2

Calculated from data of H. T. Brown and F. Escombe, courtesy of the late Prof. G. Krotkov, Queen's University, Kingston.

having its perforations close together (Figure 14-3B), the diffusion shells overlap and so cannot form. Thus the system is less efficient on an area basis.

The pores tend to permit the passage of gas roughly in proportion with their *diameter* instead of their *area* (Table 14-3) because the important factor is actually the *circumference* of the pores (which is linearly related to the diameter), not the area (which has an exponential relation to the diameter). This is because molecules can diffuse straight through (at right angles to) the pore, in which case diffusion is proportional to pore area, and they can also "spill over" the edges of the pore, in which case diffusion is proportional to the amount of edge, or the circumference, of the pore. In large pores the area factor is of major importance, but in small pores like stomata the circumference is relatively much greater and diffusion is more nearly proportional to circumference than to area. Thus the smaller the pore, the more efficient it is for diffusion, per unit of area. Because the shape of stomatal pores is not round but oval, the relationship to the circumference is not exact. However, it is a good approximation, and extensive mathematical analysis is required to get a better one.

The consequences of this arrangement of small, widely spaced holes in the surface of leaves are very important. Because carbon dioxide is present only in extremely small quantities, it forms diffusion shells on both sides of the membrane (epidermis) covering the leaf so that its diffusion is maximally efficient. In other words, the movement of carbon dioxide is only minimally hindered by the diffusion barrier of the epidermis. Water vapor, on the other hand, is normally at or near saturation inside the leaf so that diffusion shells do not form inside. Moreover, the effect of wind is enormously reduced

Table 14-3. The diffusion of CO_2 through a multiperforate septum

Separation of apertures, diameters	$\frac{\text{Area of holes}}{\text{Area of open tube}} \times 100$	$\frac{\text{Septum diffusion}}{\text{Open tube diffusion}} \times 100$	Efficiency: diffusion per unit of area
2.63	11.3	56.1	5
5.26	2.82	51.7	19
7.8	1.25	40.6	31
10.5	0.70	31.4	45
13.1	0.45	20.9	46
15.7	0.31	14.0	45

Calculated from data of H. T. Brown and R. Escombe, courtesy of the late Prof. G. Krotkov, Queen's University, Kingston.

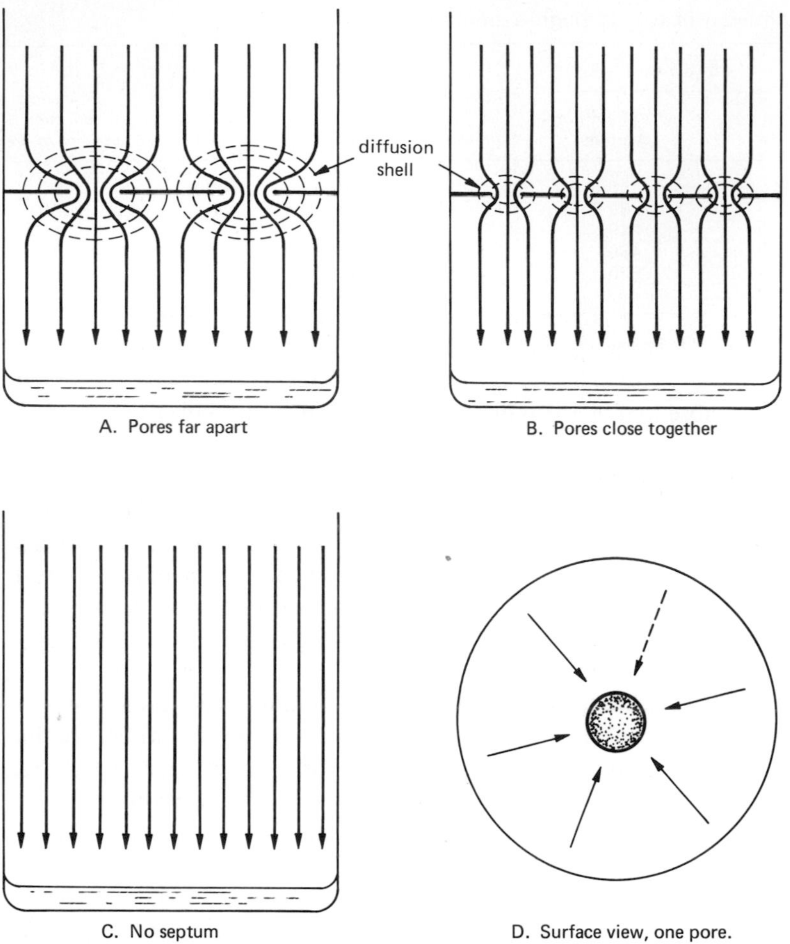

Figure 14-3. Diffusion of gas through perforated septa or membranes into an absorbing fluid.

because air currents are prevented from reaching the evaporating surface, removing water, and so steepening the diffusion gradient. Thus the system is remarkably efficient: it offers very substantial resistance to the evaporation of water while still permitting high rates of carbon dioxide absorption.

Stomatal Movement. A typical stomate is diagrammed in Figure 14-4. The guard cells, unlike other epidermal cells, contain chloroplasts and they have a curious thickening on their adjacent surfaces. When the turgor pressure inside the guard cell increases and the cells become turgid, they assume a banana shape, the thickened walls separating to form a pore or opening. This is because as the cells become turgid they tend to expand in all directions; therefore, as they elongate they are forced to assume the banana shape because the thickened walls cannot stretch. When turgor pressure decreases, the guard cells become flaccid, the thickened walls come together, and the pores close. Guard cells also have thickened bands that form a network radiating outward from the pore around their circumference. These prevent the guard cell from enlarging

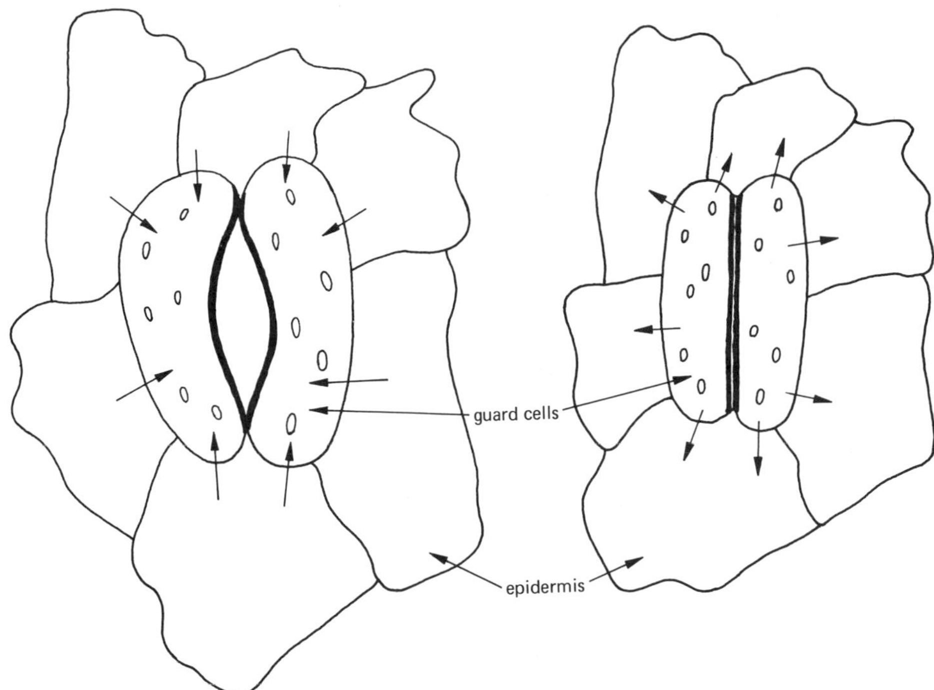

Figure 14-4. Diagram of a stomate. Arrows show direction of water movement.

crosswise, so that they must elongate and bend when their turgor pressure rises. This causes the pore to open when their turgor pressure increases, and to close when it decreases.

Stomata are found in various shapes and forms, as shown in Figure 14-5, but the essentials of their movements are the same. Stomata open when water diffuses by osmosis into guard cells from the surrounding epidermal cells, which may be undifferentiated (Figure 14-5B, C, D) or specialized **subsidiary cells** (Figure 14-5A, E). The increased turgor pressure in the guard cells causes them to swell and the stomata open. The osmotic potential may be created by several different agents or mechanisms: the active pumping of potassium ions (accompanied by chloride or organic acid counterions), the synthesis of sugars or organic acids in the guard cells, or the hydrolysis of starch to sugars. We shall consider these mechanisms and attempt to reach an understanding of how stomata are controlled after examining the effects of various environmental parameters.

Factors Affecting Stomatal Action. Stomata evidently constitute a homeostatic mechanism for regulating the conflicting demands of increased CO_2 uptake and decreased water loss. We may therefore expect to find feedback-type controls through which water and CO_2 cause opening and closing of stomata, respectively. We might expect that when CO_2 is very high, stomata will tend to close, thus minimizing water loss without jeopardizing photosynthesis. On the other hand, low CO_2 would be expected to cause stomatal opening to increase the CO_2 supply for photosynthesis. Since excessive loss of water could cause much more severe damage than temporary reduction in photosynthesis, we might expect to find an overriding control that would shut stomata

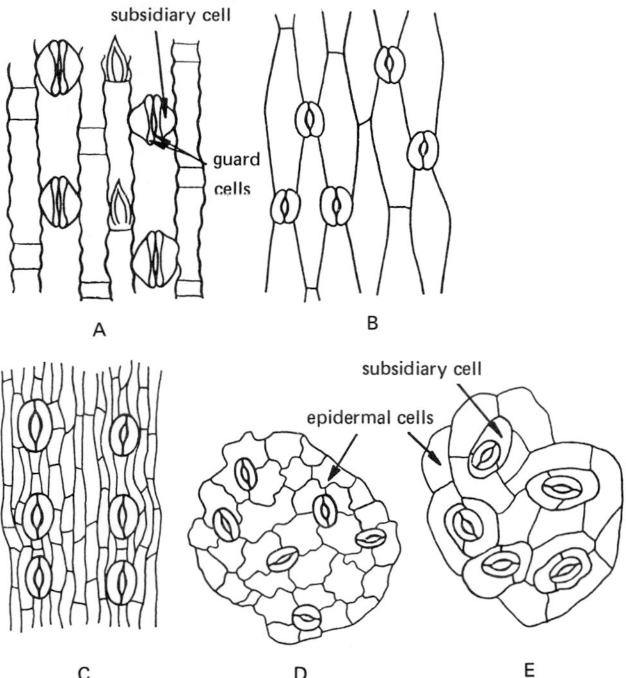

Figure 14-5. Stomatal arrangement in: (**A**) *Graminae* (maize), (**B**) *Liliaceae* (onion), (**C**) *Gymnospermae* (pine), (**D**) plants with elliptical stomate (*Vicia faba*), (**E**) most succulent plants (*Sedum spectabillis*). [From H. Meidner and T. A. Mansfield: *Physiology of Stomata*. McGraw-Hill Co., New York, 1968. Used with permission.]

when water supplies are low. Furthermore, since photosynthesis goes on only in light, we should expect stomata to open in light and close in darkness. It is a triumph of the evolutionary logic of plants rather than the logic of our thinking processes that these predictions we have made are all correct.

Water. There appear to be two main types of stomatal control by water. The first, called **hydropassive control,** results from the effect of the overall water potential of the plant on the stomata. Its effect is usually rapid and complete. When the critical leaf water potential is reached (this varies among plants; it is -8 to -11 bars in beans and may be below the wilting point) the stomata close quite tightly and completely. They usually do not open until the water potential of the plant has been restored to a normal operating level. This mechanism prevents plants from damage due to extreme water shortage.

Hydroactive control involves the measurement of water potential by the plant, the detection of a shortage of water, and the operation of a specific mechanism or action that closes stomata. One such hydroactive control mechanism is mediated by the hormone **abscisic acid (ABA),** whose mode of action will be considered in the next section. The sequence of ABA hydroactive control is as follows: When there is plenty of water no ABA is formed and stomata are open. When a slight water shortage develops (much too slight to cause hydropassive stomatal closing), a small amount of ABA is formed and stomata close slightly. At the same time the action of ABA makes the stomata much more sensitive to CO_2 requirements, so that photosynthesis is not unnecessarily hindered. On the development of severe water shortage (though still less than that required for hydropassive closing), larger amounts of ABA are formed and the stomata close.

The mechanism whereby decreasing water potential causes synthesis of ABA is not known. However, the response is very fast, and ABA may be formed within as little as 7 min in wilting leaves. When water is supplied to wilting leaves, synthesis of ABA stops at once. However, although all the ABA does not immediately disappear, stomata usually reopen. Again, the mechanism is unknown; it is possible that ABA may be sequestered or compartmented in these conditions to a place in the cell where it is no longer active. Recent data of P. E. Kriedemann and his associates in Australia suggest that **phaseic acid,** a compound closely related to ABA, may be responsible. It causes stomata to close and inhibits photosynthesis. It is formed in leaves that have suffered drought and may linger even after ABA has disappeared.

Carbon Dioxide. CO_2 has a pronounced effect on stomata. Low CO_2 concentrations promote stomatal opening, and high concentrations cause rapid closing in light or in darkness. As might be expected, if stomata are forced to close by high CO_2 treatment of leaves, they cannot be forced to reopen simply by flushing the leaf with CO_2-free air because of the high CO_2 concentration trapped behind the closed stomata inside the leaf. However, exposure to light under these conditions will soon cause opening, as the CO_2 inside the leaf is used up in photosynthesis. Thus, control mechanisms have evolved that effectively prevent stomata from unduly inhibiting the rate of photosynthesis, so long as water is not limiting, but shut them to prevent unnecessary water loss when no photosynthesis can take place because of the absence of light. The presence of cuticle, which is relatively impermeable to CO_2, on the outside of guard and epidermal cells ensures that the stomata respond to the CO_2 concentration inside the leaves, where it matters, rather than outside.

Light. A strong controlling factor is light. Stomata normally open in light and close in darkness. When a darkened leaf is illuminated, photosynthesis does not normally begin or reach maximal rate for some minutes. This may be at least partly due to the lag time in stomatal opening. The amount of light required to cause stomatal opening varies among species. Some, like tobacco, require only low light intensities, as little as 2.5 percent of full daylight. Others may require almost full sunlight for full opening. Stomata usually close in light intensities below compensation point. Certain exceptions are known—the stomata of plants exhibiting CAM open at night and close in the daytime. This fits with their tendency to absorb carbon dioxide and store it in the form of organic acids at night, then to reduce it photosynthetically in the daytime (see page 184, Chapter 7).

Light appears to have a dual function. The action spectrum of the light effect on stomata offers some clues: it appears to be essentially that of photosynthesis with an added blue light sensitivity. A few plants lack the photosynthetic spectrum and are only sensitive to blue light. The photosynthetic component may well be due to photosynthesis in the guard cells (which, unlike other epidermal cells, have chloroplasts). This might affect stomatal opening in three ways. First, photosynthesis reduces the concentration of CO_2, which is a powerful stimulus for stomatal opening. Second, osmotically active substances such as sugars are made in photosynthesis, which can contribute to the lowered water potential of guard cells. Third, photosynthetic phosphorylation could provide ATP required to drive the ion-transport pumps that move K^+ or other substances into guard cells. The blue light component may relate to a different photoactive control, possibly through the pigment **phytochrome,** which is known to mediate other plant movements caused by K^+-driven osmotic changes (see Chapter 20, page 490). Phytochrome is known to absorb blue light. The stomata of some plants also react to alternating red and far-red light, another characteristic of mechanisms mediated by phytochrome.

Temperature. Temperature appears to influence stomatal opening, but the effect is not so clear as that of light. Generally, increasing the temperature increases stomatal opening, so long as water does not become limiting. This appears to be a protective mechanism against heating, because the evaporation of transpired water has a strong cooling effect. In line with this idea, stomata of some plants (particularly desert plants) become insensitive to CO_2 at high temperature. Thus the plant is protected against overheating regardless of the photosynthetic activity. If this were not so, stomata might close on overheating due to the rise in CO_2 content resulting from excessive respiration and heat-impaired photosynthesis. At the other end of the scale, stomata of some plants do not open at very low temperatures, even in strong light.

Other factors affecting stomata, such as wind, are usually related to a combination of the factors previously mentioned.

Measurement of Stomata. The interaction of factors and the problems in effectively measuring stomatal opening have created difficulties in obtaining a clear understanding of stomatal mechanisms. Stomata are frequently measured by direct microscopic observation, but experimental conditions are hard to maintain. Other techniques include dropping liquids, such as dye solutions or oils, on leaves and measuring their rate of penetration, or applying a film of a quick-setting substance like collodion or silicone rubber to make a replica of the leaf surface from which subsequent direct measurements can be made. Both of these techniques are subject to the criticism that added penetrating fluid or film may cause changes in the stomatal environment. However, they have both given useful results.

The most frequent method has been to use a **porometer.** A small cup is glued to a leaf surface and a weak suction is applied. The rate of flow of air through the leaf surface into the cup can be measured accurately and is proportional to the degree of stomatal opening. However the relationship between air flow and stomatal opening is a complex one and almost certainly varies with different degrees of stomatal opening. Further, conditions inside the porometer cup, such as CO_2 concentration, relative humidity, and temperature, may change rapidly and affect stomatal opening. Thus, even this simple technique has its limitations.

Mechanism of Stomatal Action. As was mentioned earlier, stomatal movement results from changing turgor pressure in the guard cells. This is caused by changing water potential in guard cells relative to the cells that surround them. The problem then is to discover the mechanism that enables the plant to detect environmental situations that require the opening or closing of stomata and the mechanisms whereby the required water-potential changes are brought about. A model of the ways in which this system might by controlled is shown in Figure 14-6. A number of observations related to the facts outlined in the previous section, and which are included in the model in Figure 14-6, must be examined.

Photosynthesis appears to be necessary for stomatal opening. In etiolated leaves, when guard cells are devoid of chlorophyll, no stomatal action takes place under the

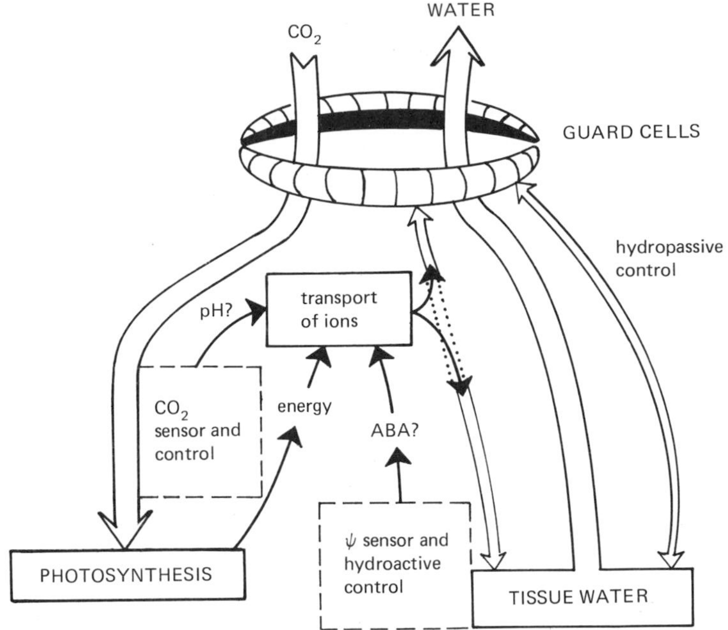

Figure 14-6. Model of possible stomatal control systems. [Modified from K. Raschke: *Ann. Rev. Plant Physiol.*, **26**:309–40, 1975.]

influence of light. Moreover, the action spectrum of stomata is often that of photosynthesis. Poisons such as DCMU, which specifically inhibit photosynthesis, also inhibit stomatal movement. Even in Crassulacean plants, in which stomata open in darkness, stomatal movement is proportional to photosynthesis in the previous light period. Stomatal opening, though dependent on water movement, is often related to carbon dioxide concentration, and mechanisms in which stomatal control by carbon dioxide and light are related to photosynthesis have been proposed.

Nest, we must consider factors affecting osmotic regulators. Substantial changes in pH of guard cells between light and darkness have been observed, and floating leaves on solutions of high or low pH has been found to cause stomatal opening or closing, respectively. Many years ago it was observed that pH affects the starch phosphorylase reaction

$$\underset{(\text{high } \psi_\pi)}{\text{starch} + \text{Pi}} \underset{\text{low pH}}{\overset{\substack{\text{phosphorylase} \\ \text{high pH}}}{\rightleftharpoons}} \text{glucose-1-P} \rightleftharpoons \underset{(\text{low } \psi_\pi)}{\text{hexose} + \text{Pi}}$$

The change toward more basic conditions promotes hydrolysis of starch; more acid conditions promote its synthesis. This would result in a decrease or increase in osmotic potential (ψ_π), resulting from high or low pH, respectively. It has frequently been observed and is very well documented that starch disappears when leaves have a low water potential and stomata are closed, and much starch is usually present in the guard cells of open stomata.

These data gave rise to the original "classical theory" that depends on the pH effect on starch phosphorylase. Low CO_2 (the result of photosynthesis in light) causes high pH (since CO_2 is in equilibrium with carbonic acid, H_2CO_3), which in turn causes starch hydrolysis, glucose production, and a lower or more negative osmotic potential in the guard cells. Water moves into the guard cells by osmosis, and the stomata open. Darkness reverses this situation; photosynthesis stops, respiration causes a build-up of CO_2 and H_2CO_3, pH decreases, sugar is converted to starch, and the osmotic potential rises. This results in stomatal closing. The American physiologist J. Levitt has suggested that dark acidification may result from the formation of organic acids by dark fixation of CO_2, since the change in acidity that might be caused by changes in CO_2 partial pressure over the normal range is very low. Osmotic potential might also be decreased in light simply as the result of the photosynthetic production of sugars in the guard cells. These ideas are shown on the left-hand side of the diagram in Figure 14-7.

Figure 14-7. Various hypothetical mechanisms for stomatal opening. ATP synthesis might be by normal photosynthetic phosphorylation (a) enhanced by low CO_2 (b) or by increased Δ pH between chloroplasts and cytoplasm (c). Another possible mechanism at (b) would follow from the increased formation of glycolic acid (caused by low CO_2) which might then be oxidized by a NAD-coupled reaction. Reoxidation of NADH would produce ATP. The reduction of pH would follow CO_2 depletion (d) or H^+ transport resulting from increased photosynthetic phosphorylation (e). [See J. Levitt: *Planta,* **74:**101–18 (1967) and *Protoplasma,* **82:**1–17, (1974); I. Zelitch: *Ann. Rev. Plant Physiol.,* **20:**329–35 (1969); K. Rashke: *Ann. Rev. Plant Physiol.,* **26:**309–40 (1975).]

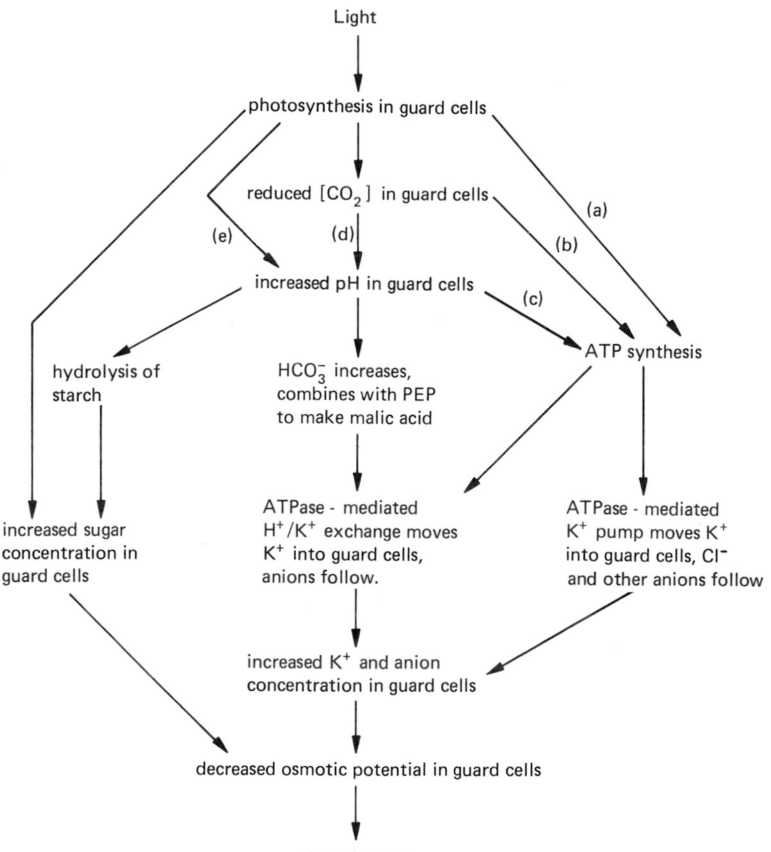

An alternative possibility, now much more widely accepted than the classical theory, is that osmotic conditions could be regulated by active ion pumps. A number of observations support this concept. The Japanese physiologist M. Fujino observed that the guard cells of open stomata in light contain much higher concentrations of K^+ than do those of closed stomata in darkness. This situation may be visualized with K^+-specific stains, as shown in Figure 14-8. That K^+ ions are indeed pumped (that is, they do not move passively) is supported by the fact that the addition of ATP to epidermal strips floating in a KCl solution greatly increased the rate of stomatal opening in light. This suggests that the ion pump is operated by ATP, which could be generated by photosynthesis. Recent analyses with the very sensitive tool, the **electron beam microprobe,*** have shown that K^+ fluxes do occur in opening or closing stomata, and, together with their counterions, they are quite large enough to account for the osmotic potentials required to open and close stomata. This scheme is represented by the various hypothetical mechanisms that are shown on the right-hand side of the diagram in Figure 14-7.

The very strong relationship between pH and the opening and closing of stomata has suggested still other hypotheses that interrelate the light and CO_2 effects through pH. The pH is usually high, or basic, in open stomata, and low in closed stomata. It now seems unlikely that the CO_2/bicarbonate/carbonic acid system markedly affects pH as CO_2 concentration rises or falls. However, as CO_2 is used up, the pH does rise somewhat, and the result is the rapid conversion of much of the remaining CO_2 into bicarbonate. Bicarbonate is the substrate for phosphoenolpyruvate carboxylase (PEP carboxylase), which produces malic acid. The formation of malic acid would produce protons that could operate in an ATP-driven proton-K^+ exchange pump, moving protons into the subsidiary or epidermal cells and K^+ exchange pump, moving protons into the subsidiary or epidermal cells and K^+ into the guard cells. The increased pH of the guard cells could also be brought about by increased photosynthetic proton transport in the chloroplasts of the guard cells as CO_2 concentration decreases and photosynthetic phosphorylation increases (see Chapter 5, page 100, and Chapter 7, page 164). These hypothetical schemes are also incorporated in Figure 14-7.

It seems clear, however it is driven, that a K^+ pump is responsible for the transport of K^+ ions into and out of the guard cells, and this in turn brings about the change in osmotic potential that opens and closes the stomata. The classical hypothesis still has proponents, and it seems probable that this mechanism does operate under certain circumstances. However, the ready demonstration of K^+ fluxes, the fact that some guard cells have no starch, and that even when starch is hydrolyzed it is difficult to demonstrate the presence of sugars, all mediate against the widespread adoption of the classical hypothesis.

As yet there is no clear explanation for the way in which the control systems are driven by the sensing devices that detect environmental parameters requiring stomatal opening or closing. It seems likely that the light effect operates primarily through the influence of light on CO_2 concentration as the result of photosynthesis, although we cannot rule out the possibility of a phytochrome-sensitized K^+-transfer pump. CO_2 seems to be one of the two primary controlling factors, and it (together with light)

*An electron microprobe is an instrument that irradiates the tissue with an extremely narrow (5 μ or less in diameter) beam of electrons. Various ions or elements in the tissue emit characteristic secondary radiation as a result, and the presence and amount of the element or ion can be determined by measuring the amount of such secondary radiation. By the use of an electron microprobe, it is possible to measure, for example, the distribution and amount of potassium in a single cell or even in a part of the cell.

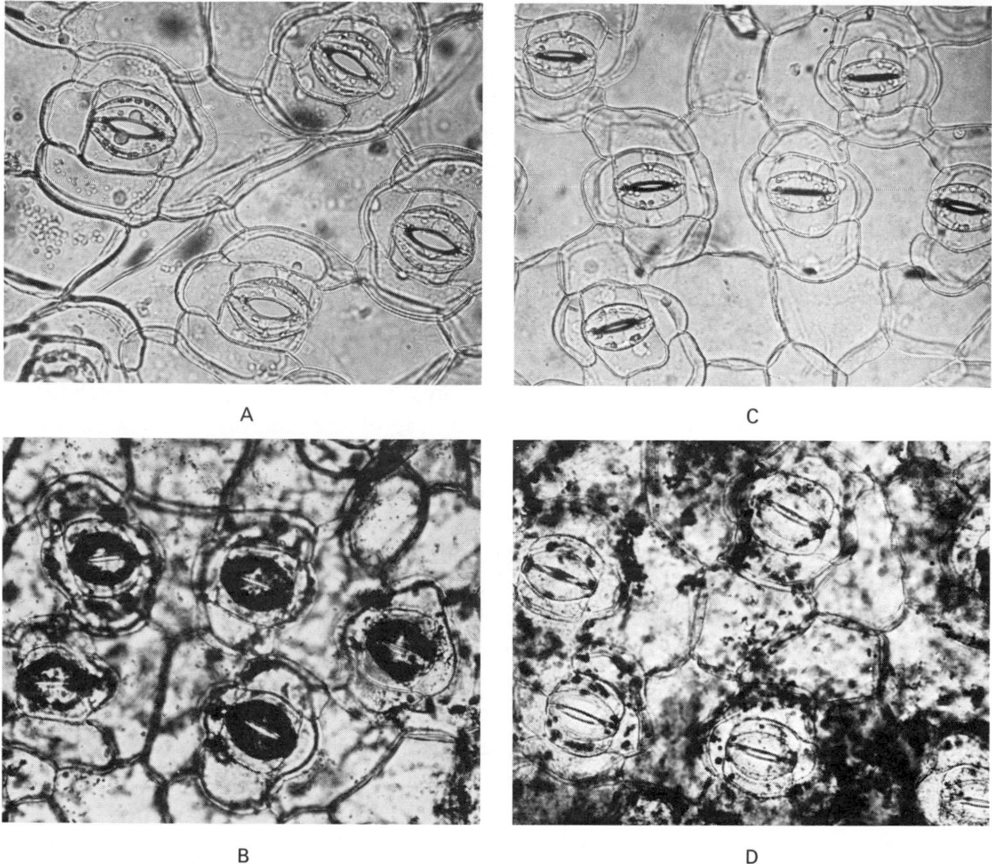

Figure 14-8. Photographs of stomata of a *Commelina communis* leaf (**A**) and (**B**) in light, (**C**) and (**D**) after treatment with abscisic acid. (**A**) and (**C**) are unstained, (**B**) and (**D**) are stained to reveal potassium. Stomata are open (**A**) and guard cells full of K^+ (**B**) in light. ABA causes stomata to close (**C**) and K^+ to move into the surrounding epidermal cells (**D**). [Source: T. A. Mansfield and R. J. Jones: *Planta*, **101:**147–58, (1971), in O. V. S. Heath: *Stomata*. Oxford University Press, London, 1975. Used with permission. Photographs kindly supplied by Professor Mansfield.]

presumably operates in some way through a pH-sensitive mechanism. The hydroactive mechanism operates in ways that are not yet understood through the synthesis and destruction or compartmentation of the hormone abscisic acid.

Control of Stomata. The mechanisms of stomatal control are adapted for the maintenance of the integrity of the plant in an essentially hostile environment. Thus, the primary response is to water, since the control of water loss is of first importance for survival. A secondary response based on CO_2 concentration satisfies the synthetic requirements of the leaf. Since the need for water conservation is more important, and opposite to, the requirement for photosynthesis, it is necessary that the control for water loss overrides the control for photosynthesis. A multiinput feedback control circuit with overriding inputs essentially equivalent to a modern electronic environmental control system has developed in plants. Such a system is an absolute prerequisite to successful invasion of terrestrial habitats by plants.

Stomata also appear to be under some sort of intrinsic control by the plant. In many plants the diurnal rhythm of stomatal opening and closing will continue for some days under constant conditions. This is presumably related to the fact that many aspects of plant behavior exhibit specific temporal patterns (rhythms of activity); these are discussed in greater detail in Chapter 20. Many of these phenomena, like stomata, are rapid movements of the plant or its parts brought about by the transport of K^+ ions from one cell or group of cells to another. Most of these mechanisms are under the control of phytochrome, and the overall control of these rhythmic movements of plants will be dealt with in Chapter 20.

Water Loss

Water loss, as we have seen, is unavoidable in photosynthesizing organs, but its regulation is essential for the well-being of the whole plant. We shall therefore examine the process of water loss in some detail.

Guttation. Loss of liquid water through the leaf surface (often through specialized structures called **hydathodes**) is called guttation. Guttation usually takes place at night, particularly in humid weather, when transpiration is reduced. Guttation is usually the result of root pressure, although water appears to be exuded from certain glandlike structures by osmotically generated hydrostatic pressures that may arise within the leaf cells of the glands themselves.

The quantity of water lost in guttation is not great, usually only a few drops on the blade of a leaf. However, large amounts of water may be lost by tropical plants. Guttation fluid often contains both organic compounds (nitrogenous compounds like glutamine and sugar) and inorganic salts (calcium, potassium, and magnesium salts, often as nitrates but also as sulfates or chlorides). Occasionally, sensitive young leaves are damaged by the drying of guttation fluid, which subjects the blade to highly concentrated solutions. Otherwise, guttation seems to have little significance in the water regulation of plants.

Transpiration. Most of the water lost by plants evaporates from leaf surfaces by the process of transpiration. Transpiration is essentially the evaporation of water from cell surfaces and its loss through the anatomical structures of the plant (stomata, lenticels, cuticle). Total water loss by transpiration may be very great. The daily water loss of a large, well-watered, tropical plant such as the palm may run as high as 500 liters. A corn plant may loose 3–4 liters/day, whereas a tree-sized desert cactus loses less than 25 ml/day. It has been calculated that over 99 percent of the water absorbed by a corn plant during its growth is lost in transpiration. Water lost by a growing field of corn would amount to 8–11 in. of water per acre during the growing season, and the loss from a hardwood forest may be twice as great.

Loss of water through the epidermis of the plant, which is usually covered with a cuticle, is called **cuticular transpiration.** In some plants, quite considerable amounts of water may be lost by this pathway. About 5–10 percent of water loss in temperate zone plants takes place through the cuticle; much lower values are found in xerophytic or desert plants, whereas tropical plants normally growing in damp climate tend to transpire more vigorously through the epidermis. The cuticle is apparently not a smooth or amorphous layer but has a complex ultrastructure with pores or passages that permit the transfer of gas. Water content apparently affects the ultrastructure of the cuticle. When cells are dehydrated, the overlaying cuticle become less permeable to water.

Small amounts of water are lost through the bark of trees, and particularly through the lenticels. Although **lenticular transpiration** is relatively unimportant normally, damaging water loss by this pathway can occur in evergreen trees during the winter. If soil moisture is low, water stress and injury from desiccation may thus result.

Most of the water loss that occurs in plants takes place through the stomata of the leaves. This process is under control of the plant, as dictated by environmental conditions, and represents one of the major points of interaction between the plant and its environment. Because of the plant's capacity to control stomatal transpiration, the rates of water loss by plants are frequently very different from comparable rates of evaporation from an open dish or special devices for measuring the rate of evaporation called **atmometers.**

Transpiration

Factors That Affect Transpiration. Since most transpiration occurs via stomata, the degree of stomatal opening is a major factor in its control. The data plotted in Figure 14-9 show that transpiration is affected by stomatal opening under widely differing conditions. Thus conditions which influence stomatal opening (page 331) also affect transpiration, particularly when stomata are nearly closed (that is, less than 2 μ in Figure 14-9).

Figure 14-9. Rate of transpiration as affected by stomatal opening. The fact that almost identical curves (not shown) were found under conditions causing evaporation rates from blotting paper ranging from 4 to 24 mg/(hr)(cm²) shows that stomata affect transpiration equally under various conditions of humidity. [Data recalculated from M. G. Stalfelt: *Planta,* **17:**22 (1932).]

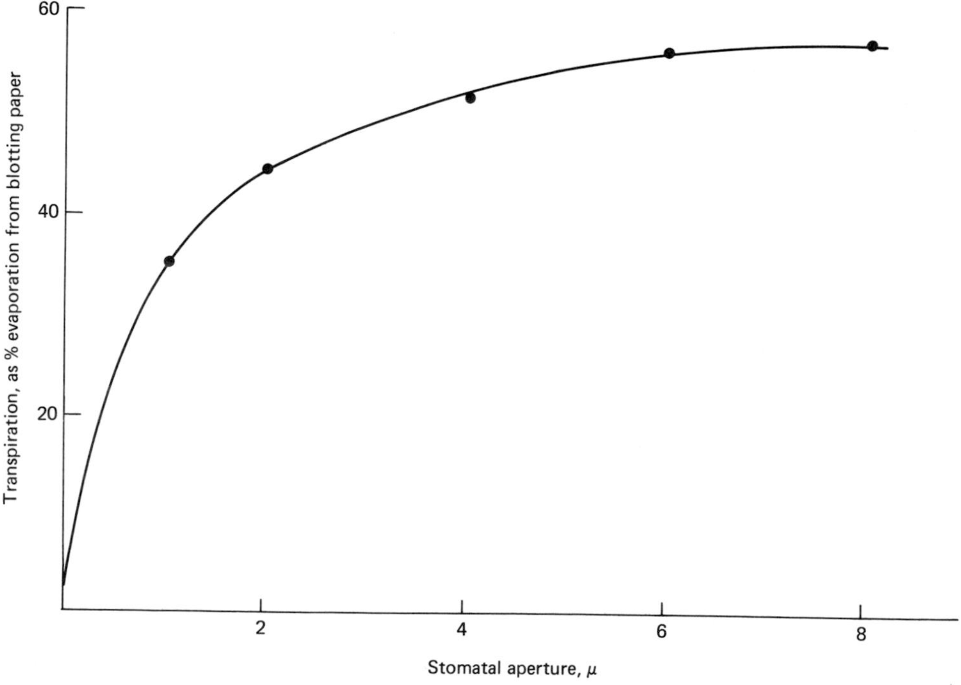

Table 14-4. Relationship between relative humidity (RH) and vapor pressure of water in air at different temperatures

	Vapor pressure, mm Hg, at			
Temperature, °C	*10% RH*	*50% RH*	*70% RH*	*Saturation (100% RH)*
10	0.92	4.60	6.45	9.21
20	1.75	8.77	12.28	17.54
30	3.18	15.91	22.27	31.82
40	5.53	27.66	38.72	55.32

The water content of the plant may affect transpiration in two ways: indirectly by affecting stomatal opening, and directly by affecting the gradient of water vapor concentration from the cell surfaces in the leaf to the air. It is not clear whether small changes in the water potential of leaves influence transpiration directly or only by the effect on stomatal opening. It appears likely that stomata respond to small changes of water potential that do not materially affect the vapor pressure of water in the cell walls. However, severe dehydration undoubtedly reduces evaporation from cell walls into the intercellular spaces.

Water content or humidity of the air has a marked effect on transpiration because it modifies the gradient down which water vapor diffuses. Temperature greatly affects the vapor pressure of water required to saturate air, as shown in Table 14-4. The intercellular spaces in plants not under water stress are probably close to saturation much of the time, whereas the humidity of the surrounding air varies around a much lower value, usually between 30 and 80 percent relative humidity (RH). Thus a change in temperature will greatly change the vapor pressure gradient from inside to outside of a leaf, as shown in Table 14-5. Even if the air vapor pressure comes to a new equilibrium at a constant relative humidity, a substantial change occurs in the vapor pressure gradient. A much greater change occurs if the water content of the air remains constant.

Wind speed has a marked effect on transpiration because it influences the water vapor gradient near the leaf surface. Normally a **boundary layer** exists at the surface of the leaf—a layer of undisturbed air through which water must diffuse from the leaf to the external atmosphere. The thinner the boundary layer, the steeper is the vapor

Table 14-5. Effect of changing temperature on the vapor pressure gradient (Δ VP) between leaf and air

	Temp °C	*Sat'n VP (inside leaf), mm Hg*	*Air VP at 50% RH, mm Hg*	*Δ VP from leaf to air, mm Hg*
A*	10	9.21	4.60	4.61
	20	17.54	8.77	8.77
	30	31.82	15.91	15.91
B†	10	9.21	8.77	0.44
	20	17.54	8.77	8.77
	30	31.82	8.77	23.05

*Air vapor pressure at constant RH; that is, a different water content at each temperature.

†Air water vapor content constant at the value for 50 percent RH at 20° C; that is, water content remains constant regardless of temperature.

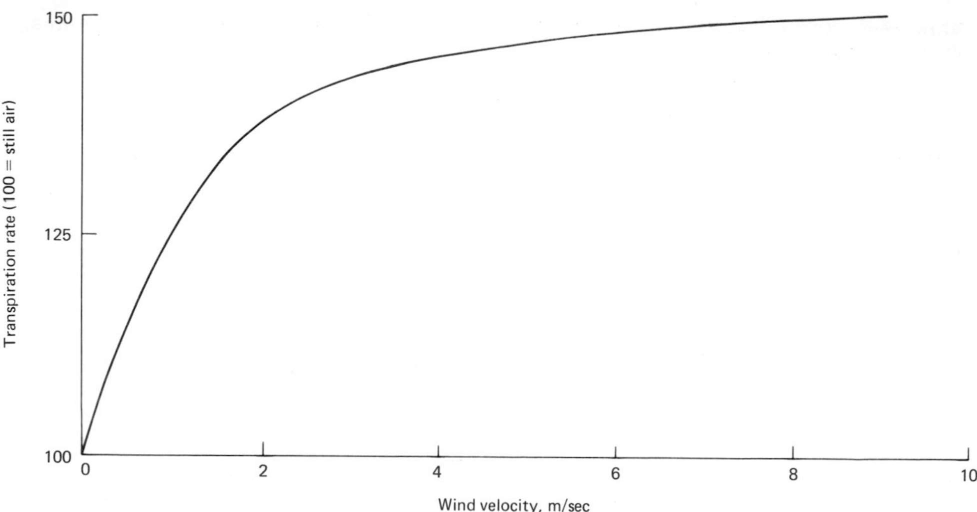

Figure 14-10. The effect of wind velocity on water loss by transpiration from a geranium leaf. [Data from an undergraduate plant physiology exercise, Botany Dept., University of Toronto, 1961.]

pressure gradient, and, hence, the faster is transpiration. Wind, by disturbing the boundary layer, increases transpiration. However this is usually a secondary effect. As the tissues dry out the stomata close, thus limiting transpiration. The wind effect seems to be maximal at wind velocity below 2 m/sec (5 m/p/h) as is shown in Figure 14-10. This is presumably because the lower wind speeds disturb the boundary layer without closing stomata; higher wind speeds are sufficiently desiccating to close the stomata.

Control of Transpiration. Transpiration control is largely achieved through stomatal control. However, a number of striking anatomical and behavioral modifications of leaves have developed that limit transpiration. The most obvious modifications include reduction in leaf size, decrease of surface area per unit mass, and a variety of surface modifications found in plants growing in xerophytic conditions. The latter include sunken stomata, reduction of stomatal size and number, and the presence of epidermal hairs (Figure 14-1). The latter are effective, as are sunken stomata, when the plant is often subject to high winds, since they prevent the disturbance of the boundary layer and the consequent shortening of the water vapor diffusion path. Mechanisms that reduce transpiration under conditions of water stress include dropping, rolling, curling or folding of leaves. It must be emphasized that these responses do not relieve severe water stress unless water is still available in the soil. Indeed, by the time such reactions occur, the plant may already be suffering damage from drought. However, they serve to prevent further and more severe damage due to extreme desiccation.

Necessity for Transpiration. As we have mentioned earlier, transpiration must occur in organisms that depend on gas exchange and incident energy for their major nutrition. However the process has some useful side effects. The flow of water through the plant induced by transpiration provides a transport system for minerals from the soil (Chapter 13). Further, the constant removal of water from the soil has the effect of mobilizing soil nutrients and carrying them to the roots, thus enabling the plant to tap a large volume of soil without the necessity of root growth completely throughout it.

Another possible beneficial effect of transpiration is that it effectively cools the leaf. The heat of evaporation of water is close to 600 cal/g; this amount of heat loss may help maintain physiologically effective temperatures in bright sunlight. However the actual temperature reduction from transpiration is normally about 2–3° C. Temperature loss by radiation and convection appears to be more effective in keeping leaves cool, except under special conditions (see page 344).

It has been suggested that transpiration is necessary for normal growth of plants. Some plants appear to develop more slowly at 100 percent relative humidity, whereas others survive normally under these conditions. However it should be noted that transpiration does occur even in saturated air, because the leaf temperature in sunlight is usually somewhat higher than the temperature of the surrounding air. Thus, the interior of the leaf will normally have a higher vapor pressure than the air around it, even at 100 percent relative humidity.

MEASUREMENT OF TRANSPIRATION. Transpiration can be measured by determining the water lost from a plant in a monitored stream of air or by measuring the wieght loss of an enclosed plant-soil system. The uptake of water by a transpiring leaf or plant top can be measured by a **potometer,** which measures the rate of water removal from a reservoir (see Figure 11-3, page 278).

Water content of an air stream can be measured by several devices such as **psychrometers** (wet-dry bulb thermometers), **hygrometers** (a fiber, often a hair, that expands or contracts on humidity change), **infrared analyzers** (which measure water vapor directly by its characteristic absorption of infrared radiation), or by the use of water-absorbing desiccants that can be weighed. These methods all require the use of a closed container, which may range from a laboratory cuvette to a large plastic tentlike structure used in the field. It is, of course, extremely important that the environmental conditions inside the cuvette should be precisely controlled and, if field measurements are being made, as nearly as possible identical with natural conditions.

Accurate measurements of a plant's transpiration under natural (that is, nonenclosed) conditions can be made if the roots and soil are enclosed in a waterproof pot or bag. Thc whole plant plus medium can then be weighed, and water loss can be measured directly as weight loss. This method has been extended to field operations by the use of very large balances called **lysimeters.** The weighing method can be used to measure the response of a single leaf with great sensitivity; however, doubt exists about the validity of extending the results obtained with a detached leaf to whole plants.

The potometer method of measuring transpiration is simple and direct. Usually the severed plant part is sealed into a small water-filled reservoir having a calibrated capillary inlet tube. A bubble of air is introduced into the capillary, and the rate of its movement measures the rate of water uptake. If the plant is in steady state conditions, that is, losing water at the same rate that it is absorbing it, the rate of transpiration is equivalent to the rate of water uptake. Unfortunately, this method cannot be used for intact plants unless they were grown in water culture, and the condition that the root is totally immersed is experimentally acceptable.

Heat Exchange

Heat is gained or lost by a leaf by three main pathways: **radiation** (direct transfer of heat to or from surrounding objects), **convection** (heating or cooling the ambient air), and

latent heat exchange (the energy used to evaporate or condense water). Minor amounts of heat may be produced by metabolic activity, but these are not normally large enough to be important. This factor cannot always be ignored, however. The heat of germinating seeds and the very high temperatures achieved by the *Arum* spadices are due to respiratory activity. This factor may be important in the survival of fleshy-leafed desert plants, which are subject to freezing temperatures at night.

Leaves are subject to radiation over a wide spectrum, but they do not absorb all of the radiation that falls on them. Their green color results from the fact that they reflect or transmit green light (λ = 500–600 nm) in the visible range. In fact, only about one half of the incident visible light is absorbed. Plants do not absorb much short-wave infrared light (that is, infrared radiation whose wavelength is only slightly greater than visible light, in the range from 700 to 2000 nm). However all objects radiate very long-wave infrared radiations (wavelength greater than 2000 nm), or heat, and the plant may absorb a very large amount of heat from its surroundings.

Of course, a plant is also radiating energy itself; when the amount of radiant energy leaving the leaf is greater than the amount entering it, its temperature will fall. It is quite possible, during a clear night, for a leaf to radiate sufficient energy—essentially to outer space—that its temperature falls below that of surrounding air, which absorbs or emits very little energy by radiation. When this occurs, water condenses from the air onto the leaf, resulting in the familiar clear-night phenomenon of dew or frost formation.

Convection and conduction of heat to and from leaves constitute a complex phenomenon. The amount and direction of heat transferred depend on the relative temperature of leaf and air. However, the efficiency of heat transfer depends on the thickness of the boundary layer as well, and this is determined by the size, shape, and orientation of the leaf as well as by the wind velocity. Thus the efficiency of heat transfer will be greatest, and the leaf temperature will most rapidly and nearly approach air temperature, under conditions that cause a thin boundary layer. Small size in leaves, particularly bulky ones like conifer needles, and high wind velocities result in a thinner boundary layer, hence more rapid convection heat exchange.

Heat loss by transpiration can be quite large, up to 50 percent of the total heat loss to the environment. It is true that if transpiration stops and the leaf temperature rises, then more heat will be lost by the increased convection and radiation that result. However, it is likely that transpiration may mean the difference between survival and heat damage or death to some leaves. For example, it has been reported that the temperatures of leaves of *Citrullus colocynthis* growing in a North African oasis were as much as 15° C below air temperature (50° C) because of transpiration. This effect of transpiration fits with the observation that at high temperatures stomata tend to open and to stay open in spite of the increased CO_2 concentration that might result from heat-damaged or reduced photosynthesis and greatly accelerated respiration.

Plants and the Weather

Plants interact with the weather. The ability of plants to grow is dependent on the climate and the extremes of various environmental conditions (Chapter 28). Plant growth is affected directly by conditions of temperature, light (cloud cover), wind, humidity, and precipitation. Not only the absolute values but their periodicity is important in determining the ability of plants to survive or thrive. Various aspects of this subject will be considered in greater depth in Chapters 16, 22, and 28.

Plants also affect the weather in many ways. A dense stand of plants greatly increases the depth of the boundary layer between the soil and moving air masses. The importance of the plant as a constituent of the water-transport pathway from soil to air is thus increased. Plants, as we have seen, control their rate of water loss to a very considerable degree. As a result, the rate of transfer of water from soil to air masses is substantially affected by transpiration. Air masses contributing to local weather may travel vast distances over regions of forest or plain that are heavily covered by plants. It follows that the weather conditions in any given locality may depend to a surprising extent on the nature and behavior of plants in the regions over which the air has travelled. Thus, preceding or distant weather phenomena, such as rainfall, high temperature, and so on, which do not contribute directly to the current weather pattern, may nevertheless greatly affect the weather indirectly via their influence on the behavior of plants.

A number of local conditions may also depend to a great extent on plants. Mist, fog, or even precipitation may occur in dense forests under the right conditions—particularly when the air is at or near saturation and usually relatively cool. Leaves absorb heat from their environment—from the soil, tree trunks, or from radiant energy penetrating from the sun. This heats up the leaves, their transpiration increases, and a boundary layer of warmer, saturated air is formed. As this boundary layer air diffuses or blows away from the leaf surface, it cools and water condenses, causing precipitation.

The cooling effect of vegetation on a warm wind due to the absorption of energy in evapotranspiration, and the effect of vegetation in raising relative humidity of the air (most noticeable in tropical regions or in hot, still weather) are well known. Plants may also affect weather conditions by the evolution of volatile substances. A considerable amount of air "pollution" is caused by clouds of terpenes and other volatile organic substances liberated by trees, particularly at high temperatures. These substances form haze and provide the nuclei for water droplet and, hence, cloud formation. It is very probable that extensive weather modification takes place through the effects of volatile plant products entering the atmosphere.

Additional Reading

Articles in *Annual Reviews of Plant Physiology*.

Heath, O. V. S.: *Stomata*. Oxford Biology Readers No. 37, Oxford University Press, London, 1975.

Kramer, P. J.: Transpiration and the water economy of plants, In F. C. Steward (ed.): *Plant Physiology: A Treatise,* Vol II. Academic Press, New York, 1959.

Lemon, E.: Micrometeorology and the physiology of plants in their natural environment. In F. C. Steward (ed.): *Plant Physiology: A Treatise,* Vol. IVA. Academic Press, New York, 1965.

Lemon, E., D. W. Stewart, and **R. W. Shawcroft:** The sun's work in a cornfield. *Science,* **174:**371–78 (1971).

Levitt, J.: Physiological basis of stomatal response. In O. L. Lange, L. Kappen, E.-D. Schulze (eds.): *Ecological Studies, Analysis and Synthesis,* Vol. 19 *Water and Plant Life*. Springer-Verlag, Berlin, 1976.

Lowry, W. P.: *Weather and Life*. O.S.U. Book Stores, Inc., Corvallis, Oregon, 1968.

Munn, R. E.: *Descriptive Micrometeorology*. Academic Press, New York, 1966.

Shaw, R. H. (ed.): *Ground Level Climatology*. American Association for the Advancement of Science, Washington, D.C., 1967.

15 Carbon Nutrition—A Synthesis

Introduction

Photosynthesis appears in many ways to be backward respiration. The enzymes of the Calvin cycle are similar to many of those in glycolysis and the pentose phosphate shunt. Most of the steps in photorespiration and the C_4 photosynthetic cycle are common with steps in the dark metabolism of carbon or nitrogen. The possibilities for metabolic mix-up are greatly reduced by the fact that the enzymes of different pathways, although they are often similar, are seldom identical, so they can be independently controlled. Nevertheless, the overall metabolism of a respiring photosynthetic cell would seem to offer considerable potential for biochemical interference.

At one time it was thought that respiration and photosynthesis were entirely different processes. Then it was realized that many of the metabolic sequences were the same, but it was recognized that they were spatially separated. The basic arrangement seemed simple—respiration took place in the cytoplasm and mitochondria and photosynthesis was confined to the chloroplasts. All was neatly separated, even though the two processes involved the same sort of reactions and the same reactants. Thus there was no need for regulatory mechanisms to keep the processes separate.

Now, however, we have begun to realize that the whole cell is involved in both respiration and photosynthesis and that several organelles and even several cells may be involved in the totality of each process. Photosynthesis, once thought of as a simple pathway for carbon reduction, is now known to be far more complex. It involves several different possible pathways and metabolic sequences, and different carboxylating enzymes. The integration and regulation of all the metabolic activities of cells comprise a vastly complex subject beyond the scope of this book. Indeed, when it is fully understood, we shall fully understand the whole of metabolism. However, some overview of the subject is essential for the study of plant physiology.

The objective of this chapter is to put into perspective the main lines of carbon and nitrogen metabolism that together constitute the integrated totality of photosynthesis, photorespiration, and respiration. We shall also examine the processes of gas (particularly CO_2) exchange in leaves, techniques of measuring and studying them, and something of the methodology that has led to our present understanding of leaf metabolism.

One thing must be emphasized: we do not yet know all about it. There is real likelihood of new and major discoveries, and many of our present ideas may be proved

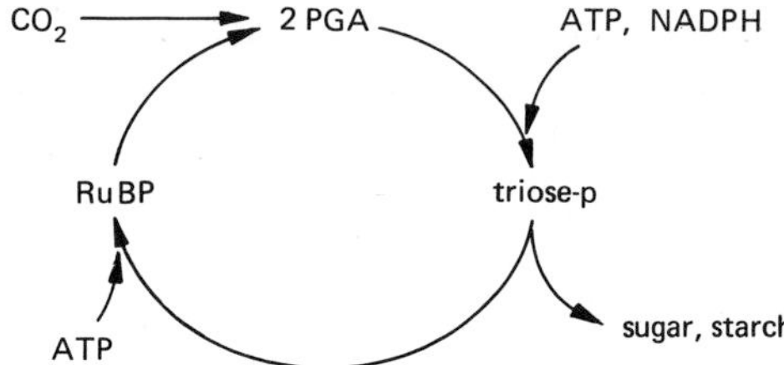

Figure 15-1. Outline of C_3 cycle.

wrong. Much of the metabolism that goes on in leaves (such as photorespiration) appears to be pointless or even deleterious. Scientists tend to ask questions like "Why does the leaf photorespire so wastefully? Why, if this is an unhealthy or even harmful process, has it not been lost in evolution?" Then they invent ingenious reasons why leaves behave as they do. The fact that scientists are still arguing why leaves do this or that indicates clearly that we do not yet really understand either the metabolism or the behavior of the leaf. So this account may contain unrecognized errors and omissions that can only be corrected by continuing research and by the development of understanding through experiment.

The process of photosynthesis was considered in detail in Chapter 7, and respiration was examined in Chapter 6. The overall integration of photosynthesis and respiration into the patterns of plant development will be considered later in Chapter 21. Here we shall review briefly the main points of photosynthetic metabolism and their integration with photorespiration and with dark respiration.

The C_3 Photosynthetic Cycle

Outline of Reactions. An outline of the Calvin cycle, convenient for the following discussion, is shown in Figure 15-1. Details of the reactions are shown in Figure 7-13, page 170. CO_2 diffuses from outside the leaf through the stomata into the intercellular spaces and is absorbed through the cell surfaces. It then diffuses (either as CO_2 or as bicarbonate ion, HCO_3^-) through the mesophyll cells until it reaches the chloroplasts, mostly in the palisade layer. At this point, since the substrate of the carboxylase is CO_2, any bicarbonate must be converted back to CO_2, a reaction that may be mediated by **carbonic anhydrase** as follows:

$$CO_2 + H_2O \xrightleftharpoons{\text{carbonic anhydrase}} HCO_3^- + H^+$$

This reaction occurs spontaneously, but it is greatly accelerated by carbonic anhydrase.

The carboxylating enzyme of the Calvin cycle is ribulose bisphosphate carboxylase (RuBPcase). The products of the carboxylation are reduced and directed to the reformation of the substrates of carboxylation, RuBP, and to sugars or starch. The stoichiometry is such that for each CO_2 fixed two C_3 molecules are formed. The carbons are rearranged so that for every six C_3 molecules formed, one is converted to the carbohydrate end product and five are used to regenerate the CO_2 acceptor. The cycle may thus be represented (starting with six molecules of CO_2) to make one molecule of hexose end product as follows:

$$6\,C_1 + 6\,C_5 \longrightarrow 12\,C_3 \longrightarrow 6\,C_5 + C_6$$

(with an arrow returning $6\,C_5$ to the $6\,C_5$ on the left)

This constitutes an abbreviated version of the cycle shown in Figure 7-13. The identity of intermediates will be found in this figure.

Autocatalysis. The cycle as drawn in the preceding scheme works well so long as there is sufficient RuBP available. If there is not (for example, if RuBP leaks out of the chloroplasts or was metabolized away during a prolonged period of darkness), then the plant is in difficulty. This is because the reaction rates depend on the concentrations of the reactants. If there is not enough RuBP, the carboxylation reaction will proceed very slowly. Unless more RuBP can be formed, the reaction cannot be made to go any faster.

Examination of the reactions of the Calvin cycle shows that it is possible to rearrange them so that the product molecules are also converted to RuBP. Using our simplified outline, the cycle can then be altered as follows:

$$5\,C_1 + 5\,C_5 \longrightarrow 10\,C_3 \longrightarrow 5\,C_5 + C_5$$

In other words, the cycle has been modified to produce an extra molecule of RuBP as an end product instead of a molecule of hexose. Thus the cycle can be described as **autocatalytic;** that is to say, in this arrangement it will continuously build up the concentration of its own intermediates, and, hence, the rate of its reaction.

The autocatalytic nature of the Calvin cycle is most important because it permits simple and rapid regulation of photosynthetic rates. Because of this, there is no need for elaborate mechanisms to protect the required supply of cycle intermediates during periods of nonoperation, which might occur for lack of either CO_2 (as when stomata close during water stress) or light. In fact, the cycle could be allowed to "run down" by reversal of the autocatalytic process, converting its intermediates into end products (during periods of CO_2 deprivation, for example) without danger. Autocatalysis is a most important characteristic of a reaction sequence such as photosynthesis that is intermittent but is required to get quickly to high speed when conditions are right.

Regulation. Since CO_2 at its normal concentration in air (0.03 percent) appears to limit photosynthesis under normal conditions, regulation of the cycle would seem to be unnecessary. However, a supply of CO_2 acceptor would need to be maintained in darkness to enable photosynthesis to start up quickly on illumination. Since the carboxylation of RuBP does not require light energy and CO_2 concentration is normally high in darkness because of respiration, a regulating mechanism is necessary to prevent the carboxylation continuing in darkness and all the RuBP being used up. This requirement is met by the need for light to activate the carboxylase, which rapidly becomes inactive on darkening. Light activation may be related to the need for Mg^{2+} ions to maintain carboxylase activity. On illumination Mg^{2+} moves from the thylakoids into the stroma of the chloroplasts (where the carboxylase is located) in exchange for protons that enter the thylakoids.

Some data suggest that certain intermediates of the Calvin cycle can regulate RuBPcase activity, but the evidence is by no means clear. The phosphatases that attack fructose diphosphate and sedoheptulose diphosphate are likely candidates for regulation since they catalyze strongly exergonic reactions, and it has been found that Mg^{2+}, a reducing compound, and the substrate of the reaction will all activate these enzymes.

Other enzymes of the cycle may be activated by light, by the energy charge (relative concentrations of ATP, ADP, and AMP, described in Chapter 5, page 103), or by various small metabolites.

Clearly, since the cycle can be adjusted to produce various end products (including hexose phosphate, triose phosphate, phosphoglycolate, and the CO_2 acceptor RuBP), its activities must be internally regulated to balance the end products against the needs of the cell. Exactly how this is accomplished is not yet clear, and this is one of the important areas of study open to plant physiologists today.

Location of Activities. All the reactions of the Calvin cycle are carried out in the chloroplast. Carbon dioxide enters the chloroplast and the end products (mainly triose or hexose) must leave it. By and large, the intermediates of the Calvin cycle do not permeate the chloroplast envelope readily. The site of sucrose synthesis is still something of an enigma. Isolated chloroplasts cannot make sucrose, but its synthesis is associated with chloroplasts. It has recently been suggested that sucrose synthesis takes place in the cytoplasm at or close to the outside of the chloroplast envelope, from photosynthetic carbon that diffuses or is transported out of the chloroplast. However, the reactions of the cycle are all confined to the chloroplast.

The C_2 Cycle—Photorespiration

Measurement of CO_2 Exchange. Some years ago the American physiologist J. P. Decker observed a brief burst of strong respiration on darkening leaves, a phenomenon that was oxygen dependent and directly related to the intensity of previous photosynthesis. He suggested that he was measuring the dying end of a respiratory process that occurred in light and was different from dark respiration. He called it **photorespiration,** but was unable to publish his findings in a well-known scientific journal because the editor did not believe that such a phenomenon could exist! It was not until some years later that the term and the idea of photorespiration came to be accepted.

At first photorespiration was observed by measuring CO_2 release in light into a stream of CO_2-free air. However, photorespiration appears to be closely connected with photosynthesis and may be affected by the absence of photosynthesis in CO_2-free air. Measurements can now be made at normal (or any) CO_2 concentration by using two isotopes of carbon. The double isotope technique for measuring CO_2 exchange was developed by the joint efforts of scientists from Canada and Singapore (C. S. Hew) working in the laboratories of G. Krotkov and R. G. S. Bidwell.

In essence, an illuminated leaf (grown in normal air and made entirely of ^{12}C-containing compounds) is exposed to a mixture of $^{12}CO_2$ and $^{14}CO_2$ in a flowing gas stream. The leaf does not discriminate between the two isotopes (except in a very small way due to the slightly higher mass of $^{14}CO_2$) and absorbs them both in proportion to their relative abundance in the gas stream. However, since the leaf is made (initially) of ^{12}C only, it gives off $^{12}CO_2$ in respiration. Thus the uptake of $^{14}CO_2$ from the gas stream will reflect the full rate of photosynthesis (variously called **gross photosynthesis** or **total photosynthesis**). On the other hand, the uptake of $^{12}CO_2$ from the gas stream will reflect the rate of photosynthesis less the rate of respiration (usually called **net photosynthesis** or **apparent photosynthesis**). The difference between gross and net photosynthesis represents the production of CO_2 by photorespiration and other respiratory processes

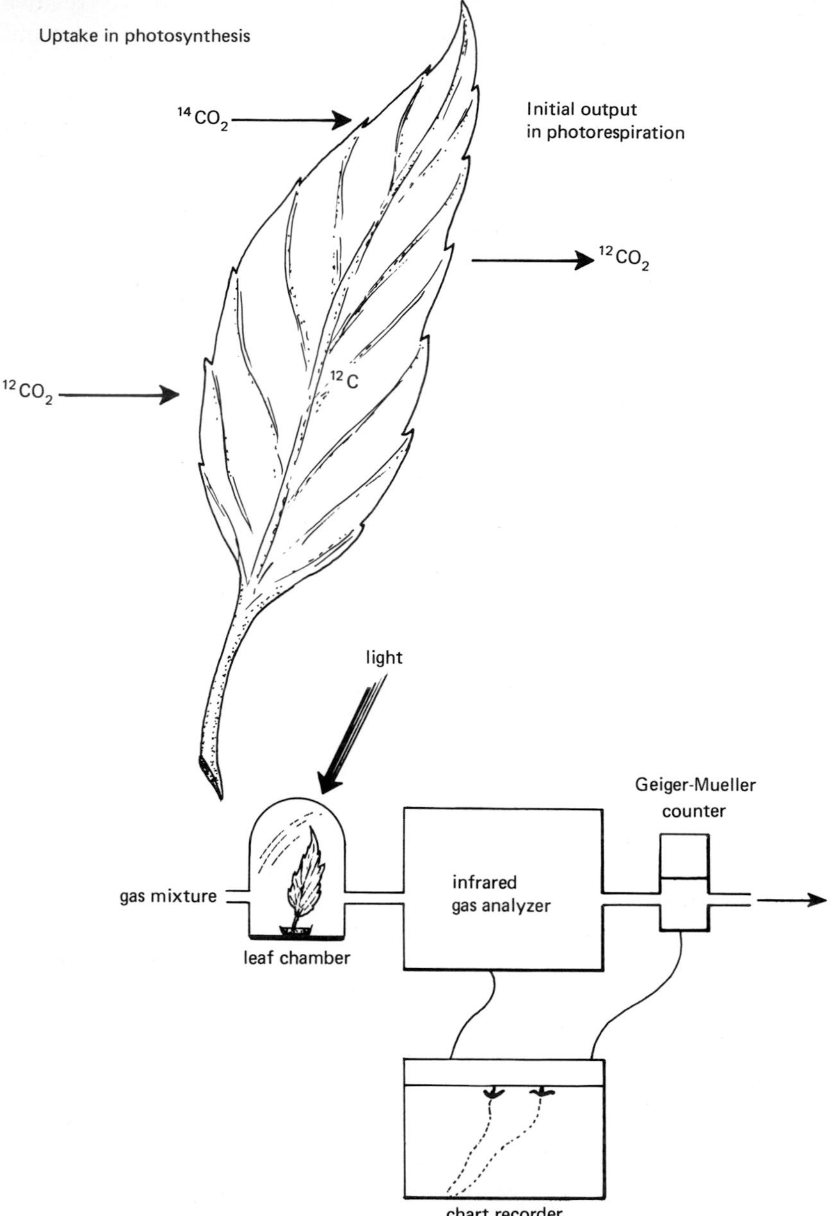

Figure 15-2. Double isotope method of measuring gas exchange. Initial uptake of $^{14}CO_2$ exceeds that of $^{12}CO_2$ by the rate of $^{12}CO_2$ output in photorespiration. $^{14}CO_2$ measures gross photosynthesis; $^{12}CO_2$ measures net photosynthesis; the difference between the two measures photorespiration.

that may be going on. This technique is illustrated diagrammatically in Figure 15-2. The abundance of $^{12}CO_2$ and $^{14}CO_2$ in the flowing gas stream is measured independently—$^{12}CO_2$ by an infrared gas analyzer (which measures CO_2 concentration by infrared absorption) and $^{14}CO_2$ by Geiger-Müller counters or an ionization chamber.

Similar techniques have been developed using a mass spectrometer to differentiate $^{16}O_2$ and $^{18}O_2$ exchange in photosynthetic and respiratory oxygen exchange, also to measure the nonradioactive isotope of carbon, $^{13}CO_2$. It must be noted that all of these techniques suffer from the problem that some CO_2 or O_2 may be reused in the leaf before it escapes to the outer atmosphere, and the degree of this **recycling,** as it is called, may depend on the internal structure of the leaf, the resistances offered to the passage of CO_2, the degree of stomatal opening, and the activity of the carboxylase at the time. For these reasons it is still difficult to estimate a true rate of photorespiration. However, much research has now put the phenomenon of photorespiration on a sound footing, and we are beginning to understand the apparent contradiction of leaves actually giving off CO_2 at the same time as they are absorbing it in photosynthesis.

Characteristics of Photorespiration. Photorespiration is oxygen sensitive in quite a different way from dark respiration, as is shown in Figure 15-3. Photorespiration has a much lower affinity for O_2 and apparently saturates at a very high oxygen concentration, whereas dark respiration saturates at a much lower level of O_2. The rate of photorespiration is usually higher than that of dark respiration, but lower rates have been recorded. Characteristically, the substrate of photorespiration is different from that of dark respiration and appears to be closely derived from recent products of photosynthesis. Thus, if a leaf is supplied with $^{14}CO_2$ and then placed in CO_2-free air, the measured specific radioactivity (relative proportion of ^{14}C) in respired CO_2 is high and close to that of the supplied $^{14}CO_2$. If the lights are turned off, CO_2 of a low specific activity (that is, derived from older, stored substrates formed prior to the $^{14}CO_2$ feeding) is released. Results of a typical experiment that demonstrate this fact are shown in Figure 15-4.

Photorespiration thus differs from normal dark respiration (which may also operate in light) by being oxygen sensitive and by having different substrates drawn from recent photosynthate. Rates of photorespiration are normally approximately one fifth to one quarter the rate of CO_2 fixation. This process is thus of great importance in the carbon economy of the plant and has received much study in the past decade.

One interesting point about photorespiration that will be dealt with in detail later (page 363) is that plants having the C_4 cycle apparently do not photorespire. They do not release or exchange CO_2 in light as do C_3 plants, and their O_2 exchange is much

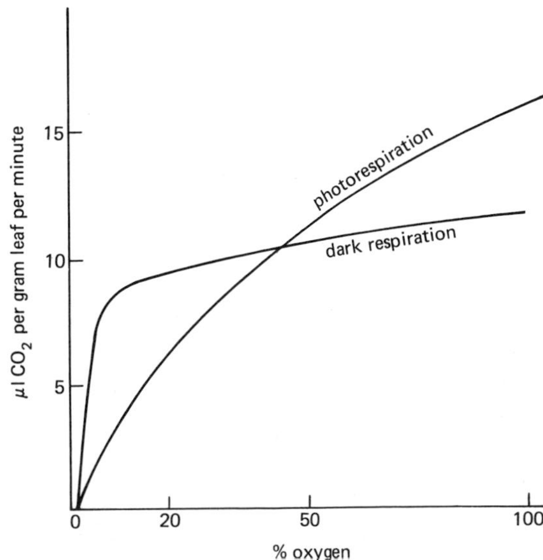

Figure 15-3. Oxygen sensitivity of dark respiration and photorespiration.

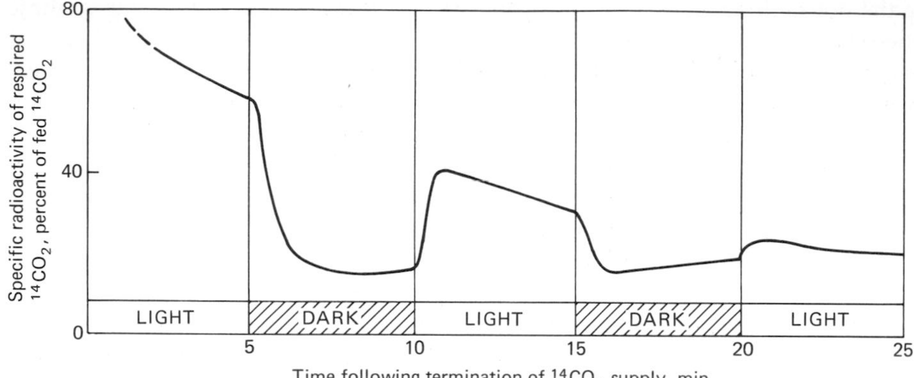

Figure 15-4. Specific radioactivity of respired CO_2 from a bean leaf after 15 min photosynthesis in $^{14}CO_2$. Photorespired CO_2 (in light) had high but declining specific radioactivity, whereas dark-respired CO_2 had low but increasing specific radioactivity. The oscillation of specific radioactivity from high to low in light or darkness indicates that the substrates of light respiration (high specific radioactivity) were separate and distinct from the substrates of dark respiration (low specific radioactivity). [Data of W. B. Levin and R. G. S. Bidwell.]

reduced. The question whether they truly lack photorespiration (that is, lack the reactions of photorespiration) or their photosynthetic system prevents the expression of photorespiration (that is, the release of CO_2 in light) will be discussed later.

Reactions of the C_2 Cycle. Experiments by the American physiologist I. Zelitch showed that the substrate of photorespiration is probably glycolic acid. This two-carbon acid is most probably produced by the oxygenase function of RuBPcase. Its subsequent metabolism, described in detail in Figure 7-15, page 175, involves peroxisomal and mitochondrial enzymes to achieve the net reaction

$$2\,C_2 \longrightarrow C_3 + CO_2$$

Figure 15-5. Integration of the C_3 and C_2 cycles. The C_2 cycle is so-called because the product of RuBP oxygenase is a C_2 compound, as are glyoxylate and glycine. [Adapted from G. H. Lorimer, K. C. Woo, J. A. Berry, and C. B. Osmond: *Photosynthesis 77: Proceedings of the IV International Congress of Photosynthesis* (D. O. Hall, J. Coombs, and T. W. Goodwin, eds.), The Biochemical Society, London, 1978.]

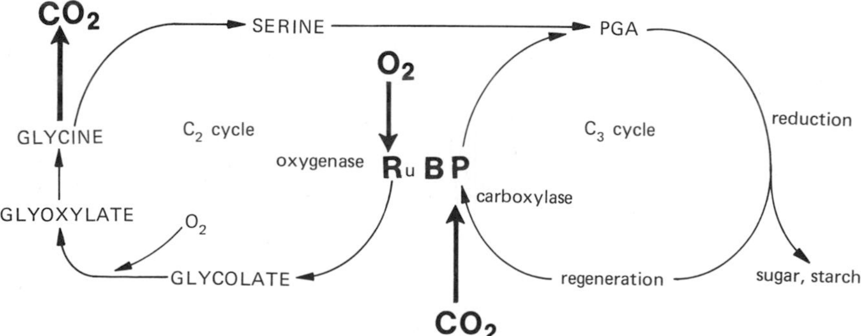

No ATP, NADPH, or NADH is synthesized in the reaction sequence. Its main function seems to be the reclamation in the C_3 cycle of carbon lost by the oxygenase reaction. The integration of the C_2 and C_3 cycles, as well as the reason for calling the photorespiratory metabolic pathway the C_2 cycle, is shown in Figure 15-5.

Various alternative sequences have been suggested. These include the total oxidation of glyoxylate to CO_2 and the conversion of the glyceric acid formed in the C_2 cycle to sugars in the cytoplasm, instead of its reentry into the C_3 cycle. However, much evidence, including tracer studies with $^{18}O_2$ and $^{14}CO_2$, the study of individual enzymes, and the localization of activities in specific organelles, now favor the general outline of the C_2 cycle as shown in Figure 15-5.

Location of Activities. It is now quite clear that the first oxidative step of the C_2 cycle, leading to the formation of glycolate, takes place in the chloroplast. The second oxidative step, the oxidation of glycolate, takes place in peroxisomes, and the decarboxylation of glycine and the synthesis of serine take place in the mitochondria. Thus carbon passing round to the C_2 cycle travels from chloroplast to peroxisome to mitochondria and back. Recent evidence suggests that the carbon does not meander freely about the cell, but moves through organelles in close proximity to each other, that may, perhaps, be held together in some sort of loose association in the cytoplasm.

Integration of C_2 and C_3 Cycles—Oxygen and Photorespiration. The diagram in Figure 15-5 shows how the two cycles are connected. It does not show, however, the ratio of activities of the two cycles. It has been suggested that photorespiration and photosynthesis are linked with a fixed stoichiometry, but in fact the rates of photosynthesis and photorespiration have been shown to vary independently during the day and during the ontogeny of the plant, or to be independently affected by its physiological status. In addition, relative rates of photosynthesis and photorespiration are strongly affected by relative concentrations of O_2 and CO_2 in the chloroplast.

This follows because the carboxylating enzyme of the C_3 cycle is also the oxygenase of the C_2 cycle. In other words, CO_2 and O_2 compete as substrates for the enzyme RuBPcase. From the point of view of the photosynthesizing plant, oxygen is a competitive inhibitor of CO_2 fixation. As the oxygen concentration decreases, activity of the oxygenase and the C_2 cycle decreases; and it ceases altogether at O_2 levels below 2–5 percent. Conversely, as CO_2 concentration increases, the proportional activity of the carboxylase and the C_3 cycle increases. This effect is illustrated in Figure 15-6. At very high levels of O_2, oxidative damage to the photosystems causes irreversible loss of photosynthetic activity, but at levels below about 70 percent the O_2 effect is a reversible inhibition of CO_2 fixation.

The O_2 effect on photorespiration has been carefully studied. It is clear, since O_2 causes a loss of photorespired CO_2 which reduces photosynthesis, that in the absence of O_2 the productivity of plants should be greatly increased. It has indeed been shown that plants will grow much faster at low O_2, but unfortunately O_2 is required for normal plant development and seed production, so productivity (in terms of seed or fruit) is greatly reduced even though the plant grows bigger.

Nitrogen Metabolism in the C_2 Cycle. Since glycine and serine are interconverted in photorespiration, a transaminase is essential for the operation of the C_2 cycle; it has been found in peroxisomes as expected. However, the decarboxylation of glycine in

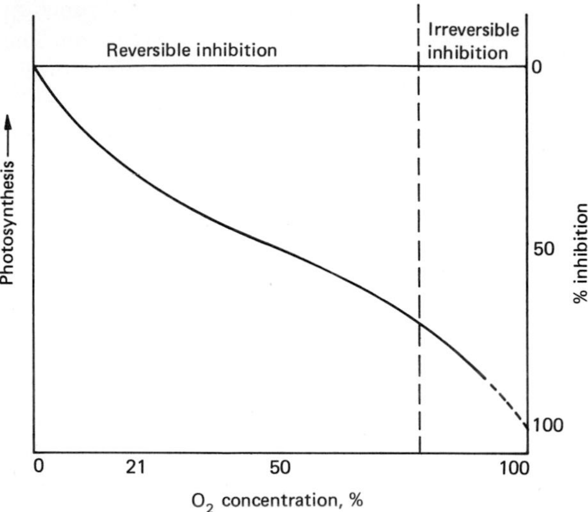

Figure 15-6. Effect of O_2 concentration on the rate of net photosynthesis of a bean leaf.

mitochondria releases NH_3, and additional amino groups are required for glycine synthesis in peroxisomes because two molecules of glycine are needed for each molecule of serine produced. Thus, some means is needed to transport NH_3 from mitochondria and deliver it, in the form of amino nitrogen, to the peroxisomes.

Recent work of the group associated with C. B. Osmond in Australia has shown that the amino groups are generated in the chloroplasts, using light-produced ATP, through the operation of the glutamine synthetase–glutamic acid synthetase system described in Chapter 8 (page 204). 2-Oxoglutaric acid* is converted to glutamic acid in chloroplasts, which is then transferred to mitochondria. There it undergoes transamination with glyoxylate to make glycine, and the resulting 2-oxoglutaric acid is returned to the chloroplast. The enzyme systems required for these reactions have been demonstrated in chloroplasts or peroxisomes as appropriate.

These two cycles of nitrogen metabolism are shown in Figure 15-7. It is immediately apparent that the traffic of metabolites among organelles must be even more intense than we had formerly imagined. Also, it is clear that photorespiratory metabolism makes further demands on the energy economy of the plant, since photosynthetic ATP is needed to drive the glutamine synthetase reaction. Thus an additional ATP is required per CO_2 released, in addition to the reducing power and ATP needed to operate the C_2 cycle. This means that photorespiration would seem to be even more wasteful than was thought.

It should be noted that while the serine-glycine cycle of nitrogen is probably quite tight (that is, not many molecules of intermediates leak in or out of it), the glutamic acid cycle is not. Quite probably much of the NH_3 produced in glycine decarboxylation is used for the nitrogen demands of the cell. This means that new NH_3 must be produced by nitrate-nitrite reduction, a further demand on photosynthetic energy production. It is also evident that metabolic controls are necessary to regulate the division of available products of the photosynthetic light reaction among the requirements for carbon reduction, nitrate reduction, and glutamine synthesis. This further emphasizes the close

*2-Oxoglutaric acid is a new (and chemically more precise) name for α-ketoglutaric acid.

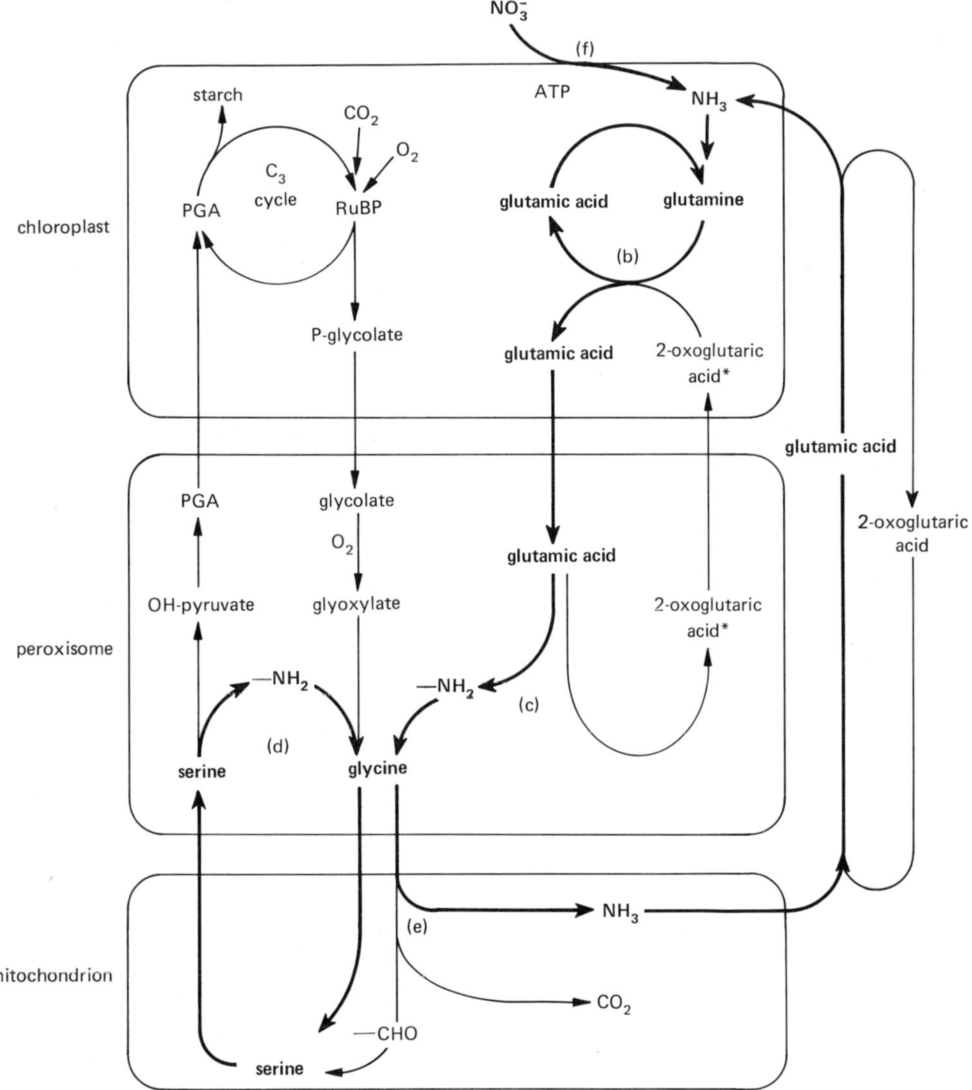

Figure 15-7. The nitrogen cycles associated with the C_2 photorespiratory cycle.
(a) glutamine synthetase
(b) glutamic acid synthetase
(c) glutamic acid–glycine transaminase
(d) serine–glycine transaminase
(e) glycine decarboxylase
(f) nitrate/nitrite reductase

cooperation required between all phases of cellular metabolism in the various processes that we now know to be associated with photosynthesis.

Control of Photorespiration. Much research effort has been spent to find ways of eliminating the "wasteful" reactions of photorespiration from crop plants. Breeding programs have not been successful, in that varieties with consistently low photorespiration and high photosynthesis have been difficult to find or disappointingly low in fruit/seed productivity. Certain chemicals have been found that will reduce photorespiration by inhibiting glycolic acid oxidase. However, these poisons are expensive to apply and tend also to inhibit photosynthesis. Projects are presently underway in various laboratories to select high-photosynthesis, low-photorespiration plants by breeding, by cell fusion, by selection from tissue cultures, and by chemical control. Whether these programs will be successful remains to be seen.

Possible Roles of Photorespiration. Until very recently it was thought that since photorespiration is an apparently wasteful loss of photosynthetic carbon, it must be a useless and unavoidable process caused by the poisoning effects of O_2. Recently some physiologists have suggested that this highly teleological view may not be correct. However, the arguments for a useful role for photorespiration are not conclusive and tend also to be rather teleological.

It has been argued that if photorespiration were entirely useless or detrimental it would have been lost during the eons of evolutionary time. On the other hand, the oxygenase characteristic of RuBPcase may be inescapably inherent in the nature of the carboxylase. It has also been claimed that photorespiration is unnecessary because C_4 plants do not photorespire. However, this is still a matter of debate, as we shall see later.

Possible beneficial roles for photorespiration have been suggested. Algae do not have glycolic acid oxidase as do higher plants, but have an NAD-linked glycolic acid dehydrogenase instead. Thus, glycolate metabolism in algae could lead to ATP formation as in dark respiration. However, it now appears, after a considerable controversy on the matter, that algae do not exhibit normal photorespiration.

Photorespiration appears to increase during the rapid translocation of photoassimilates, for example, during the early development of a new leaf or flower bud or during fruit set. Bidwell has suggested that photorespiration is in some way associated with the transfer or formation of sugars at the loading site for translocation. Another view stems from the fact that photorespiration maintains the concentration of CO_2 in a leaf when the stomata are shut because of water stress. This might have two different kinds of beneficial effect. First, RuBPcase requires CO_2 for its activation; in the absence of CO_2 it becomes inactive, and photorespiration might provide enough CO_2 to maintain it in the active state so that photosynthesis would resume immediately on stomatal opening. Alternatively, the CO_2 produced by photorespiration could equally serve to keep the C_3 cycle turning over and maintain the levels of intermediates. This would also serve to keep the cycle in a state of readiness so that photosynthesis could proceed at once when stomata opened. Finally, photorespiration could be beneficial simply because it is wasteful. That is, it might serve to dissipate unwanted energy at times when the light intensity is too high—for example, when stomata are closed and CO_2 is in short supply.

These are merely suggestions. At the moment we cannot say why plants photorespire or why (if it is a deleterious process) photorespiration has not been lost during evolution. There are many questions about photorespiration still to be answered: the metabolic pathways are not known with certainty, there are still questions about the nature and source of the substrates, its true rate or intensity is difficult to measure, and at best we are guessing about its role.

The C_4 Photosynthetic Cycle

Outline of Reactions. The reactions of the various possible alternatives of the C_4 cycle are presented in detail in Figure 7-19, page 180. Here we shall consider the

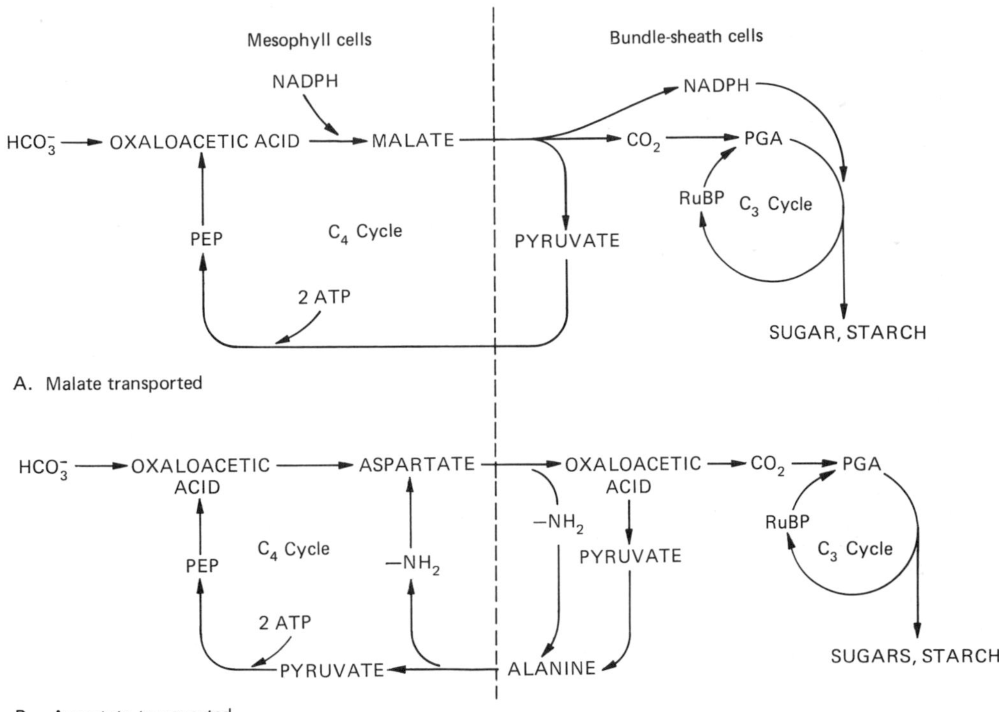

Figure 15-8. Outline of schemes for operation of the C_4 cycle of photosynthesis.

physiological implications of the cycle. A simplified scheme is shown in Figure 15-8 as a background for this discussion.

As we now know, different parts of the C_4 cycle take place in different parts of the leaf. The main accomplishment of the cycle appears to be to trap CO_2 in the mesophyll cells that are close to the stomata and then to pass it in the form of the β-carboxyl of a C_4 acid to the bundle-sheath cells inside the leaf. There it can be released again as CO_2, to be fixed and reduced by the C_3 cycle. The separation of the two CO_2-fixing mechanisms and the special nature of the C_4 carboxylating enzyme are key points in the C_4 cycle.

The important reaction is the carboxylating enzyme, phosphoenolpyruvate carboxylase (PEP carboxylase). Its cellular location is not known with certainty, but it is thought to be in the cytoplasm. The key point about this enzyme, and the major point at which it differs from RuBPcase, is that it uses bicarbonate ions (HCO_3^-) instead of CO_2 as a substrate. It has a somewhat higher affinity for bicarbonate than RuBPcase has for CO_2, so that it is able to maintain higher reaction rates at low CO_2 concentrations. Much more important, however, is that a carboxylase that uses bicarbonate is not sensitive to O_2. The use of this enzyme thus frees the photosynthetic carboxylation reaction from the poisoning effect of O_2. Furthermore, because the O_2 effect on RuBPcase is competitive, that carboxylase becomes less and less effective as the CO_2 concentration declines because of the increased O_2/CO_2 ratio. PEP carboxylase does not suffer from this effect. As a result, plants having the C_4 cycle are able to absorb CO_2 much more efficiently at low CO_2 levels than are C_3 plants.

The product of the carboxylase is oxaloacetic acid. This unstable C_4 acid is rapidly converted by reduction or transamination to malate or aspartate, which are then transported to the bundle-sheath chloroplasts. There the C_4 acid is decarboxylated by one of several mechanisms (see Figure 7-19) and the CO_2 so released is fixed by RuBPcase in the Calvin cycle in the usual way. The C_3 acid that remains after the decarboxylation, either pyruvate or alanine, is returned to the mesophyll cells and converted back to PEP by pyruvate, phosphate dikinase, a reaction that requires two molecules of ATP.

The operation of the C_4 cycle has an additional energy requirement of two ATP per CO_2 fixed above the requirement of the Calvin cycle. The reducing power required to convert oxaloacetic acid to malate is regenerated during the decarboxylation of malate, so it is not accountable as an extra requirement of the C_4 cycle. It is generally considered that the advantages of the cycle outweigh the disadvantage of the requirement for extra energy input.

Special characteristics that should be noted are (1) the separation of C_4 and C_3 carboxylations, (2) the participation of nitrogenous compounds, and (3) the wide variety of reaction mechanisms used by different plants to accomplish the same basic result.

Location of Activities—Kranz Anatomy. It was early noted that one could distinguish C_4 plants because they have dark green veins in their leaves. The specialized arrangement of tissues that almost invariably accompanies C_4 activity (only one or two exceptions have been reported, and their significance is not clear as yet) is called the **Kranz anatomy.** It is characterized by small intercellular spaces, frequent veins, and a pronounced ring of bundle-sheath cells around each bundle that are liberally endowed with chloroplasts (Figure 15-9).

In some plants (notably in corn, *Zea mays*) the bundle-sheath chloroplasts appear poorly developed and lack grana, as seen in Figure 15-10. Granal thylakoids appear to be required for the efficient cooperation of the two photosystems, and agranal chloroplasts are often deficient in photosystem II. This means that in agranal chloroplasts the production of reducing power and the evolution of oxygen are much reduced, although ATP production by cyclic phosphorylation is usually unaffected.

The presence of agranal bundle-sheath chloroplasts correlates with the presence of a NADP-malic enzyme type of C_4 cycle (reaction d*i*, in Figure 7-19, page 180) in which the decarboxylation of malate leads to NADPH formation in the bundle-sheath chloroplasts. The Calvin cycle requires two NADPH for each CO_2 fixed. Thus the bundle-sheath chloroplasts are required to produce only half the normally required amount of reducing power, the other half being provided by the C_4 cycle.

The NADP-malic enzyme type of C_4 cycle combined with chloroplast dimorphism represents the highest level of evolutionary development in photosynthesis. Not only are the cells and organelles specialized to perform the specific parts of the reactions of photosynthesis, but the reduction of noncyclic electron transport associated with the lower NADPH requirement lowers the O_2 production in agranal bundle-sheath chloroplasts. Since this is the location of the RuBPcase, which is poisoned by O_2, a further increase in operating efficiency is gained.

Many surveys have shown that the C_4 cycle is virtually always associated with Kranz anatomy. There are both C_3 and C_4 species in the genus *Atriplex,* and crosses of these produce plants that are intermediate. Some hybrids appear to be normal C_3 plants, some are intermediate in their anatomy, and some have what appears to be normal Kranz

A

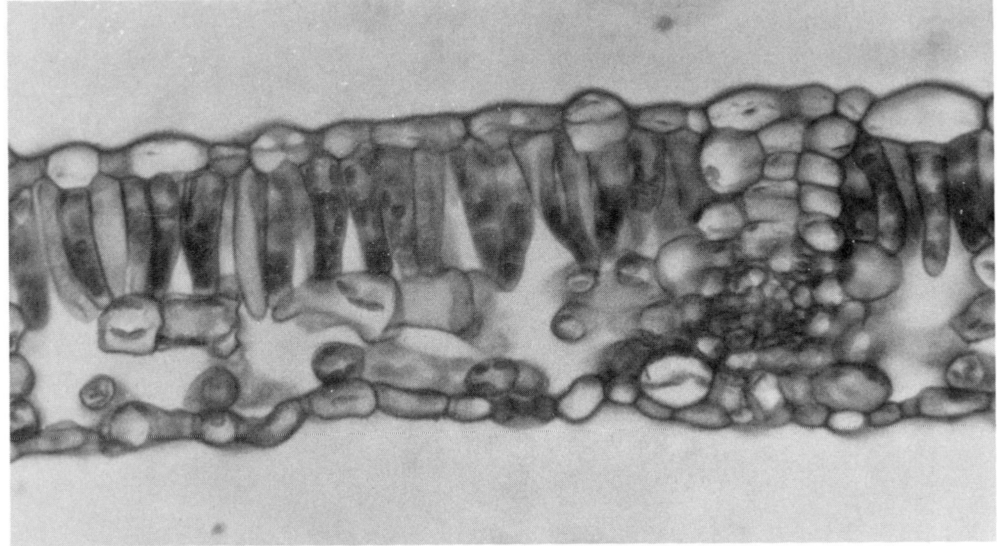

B

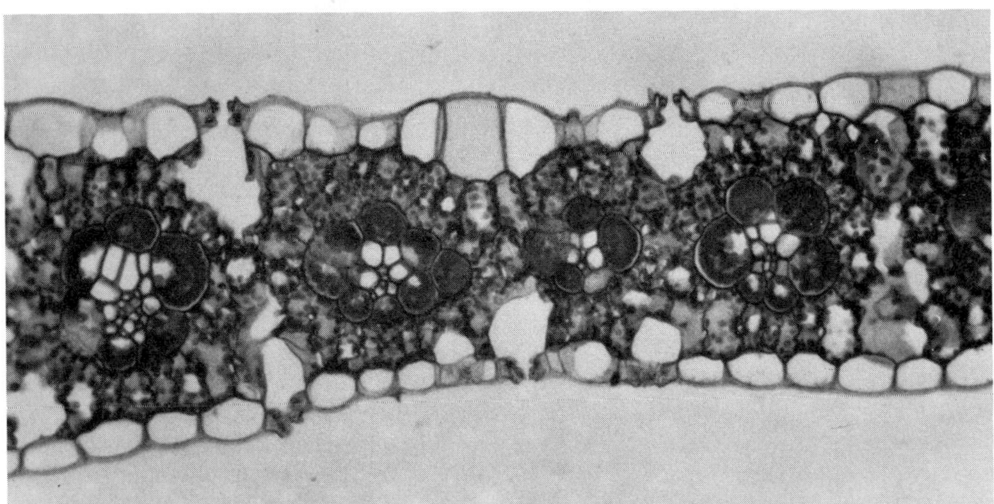

Figure 15-9. Cross section of (**A**) a C_3 leaf (*Acer,* maple) and (**B**) a C_4 leaf (*Zea mays,* corn). Note the loosely structured spongy parenchyma and the chlorophyllous palisade layers of the C_3 leaf, compared with the dense mesophyll, small air spaces, and pronounced chlorophyllous bundle sheath of the C_4 leaf.

anatomy, but they do not have a C_4 cycle. Evidently the organizational requirement for C_4 photosynthesis is very precise. Most plants have PEP carboxylase, and most plants fix some CO_2 by this reaction. However, they lack the cooperative organization needed for the effective operation of a C_4 cycle. The point to be stressed is that the evolution of C_4 photosynthesis required not the evolution of new enzymes or new metabolic pathways but the proper coordination of metabolism in various organelles in different cells to produce a whole-plant, coordinated metabolic system that transcends, under appropriate conditions, the best possible efforts of simpler systems.

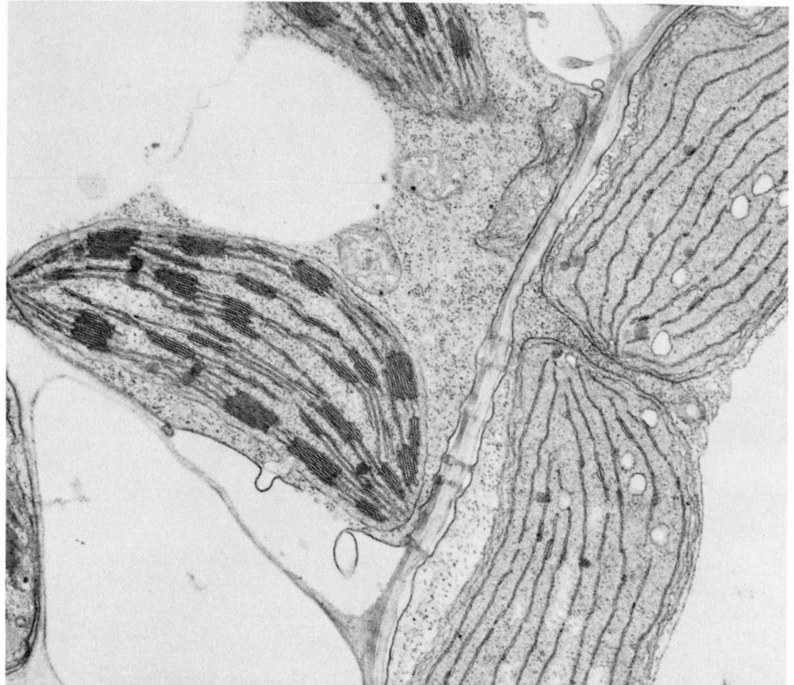

Figure 15-10. Electron micrograph of chloroplasts from corn (*Zea mays*) leaf showing portions of a mesophyll cell (left) and a bundle-sheath cell (right). The mesophyll chloroplast has many grana whereas the bundle-sheath chloroplasts are agranal. Note the plasmodesmata between the cells. [Photograph kindly supplied by Dr. C. R. Stocking, University of California, Davis, from a preparation by S. Larson.]

Nitrogen Metabolism in the C_4 Cycle. The nitrogen metabolism of the C_4 cycle is less extensive than that of the C_2 cycle, but it is nevertheless important. The required transaminases have been demonstrated in mesophyll or bundle-sheath chloroplasts as required in plants that transport aspartate, and the enzymes appear to be adequate for the heavy metabolic traffic they mediate.

The reason why nitrogen compounds should be involved in C_4-cycle metabolism, or in C_2 metabolism, has been questioned. It may be that the compounds are appropriate in terms of the free-energy changes associated with their required metabolism. For example, no reaction of hydroxypyruvate parallel to the serine hydroxymethyl transferase of the C_2 cycle is known, so that glyoxylate could not be converted directly to hydroxypyruvate. It must go via glycine and serine. The same argument cannot hold for the C_4 cycle, since a nitrogen-free version does occur. Instead, it has been suggested that either some plants have not made the adjustment necessary to deal with the NADPH synthesis and use associated with malate transfer, or else the amino compounds are more readily transportable because they are less reactive. There may be no good reason. Perhaps some plants do it one way and some another because of accidents of evolution.

Integration and Regulation of the C_4 Cycle. As might be expected, a metabolic system as complex and highly ordered as C_4 photosynthesis has several regulated steps.

The C_4 acids aspartate and malate act as feedback inhibitors on PEP carboxylase. The carboxylase itself is closely regulated by light, its activity depending on the intensity of illumination in such a way that the rate of β-carboxylation is closely proportional to the demand for CO_2 by RuBPcase.

It will be remembered that the C_3 cycle is autocatalytic; that is, it can serve to build up the concentration of its own intermediates. The C_4 cycle lacks this property. Yet, C_3 compounds such as PEP and pyruvate are mobile and in demand in metabolizing cells. As a consequence, if the C_4 cycle stopped because of darkness or lack of CO_2, there is danger that intermediates would drain away and render the cycle slow or ineffective when carboxylating conditions again prevailed. It appears that this difficulty is overcome by the capacity of the C_3 cycle to leak C_3 intermediates (perhaps produced by autocatalysis), which can then be shunted to the bundle-sheath cells and serve to prime the C_4 cycle.

Data illustrating this point are shown in Figure 15-11 taken from a student exercise in the author's laboratory. Illuminated corn (*Zea mays*) leaves were fed $^{14}CO_2$ for a short time and then switched to $^{12}CO_2$ (a **pulse-chase** experiment, of the type that has provided important data about the kinetics and operation of the C_4 cycle). Radioactivities of the relevant intermediates of the C_3 and C_4 cycles were measured at intervals during the pulse and the chase. The rapid appearance of radioactivity in the C_4 acids

Figure 15-11. Pulse-chase experiment with corn leaves. Data from experiments in the author's laboratory. Compounds associated with the C_3 cycle are solid lines, and compounds of the C_4 cycle are dashed lines.

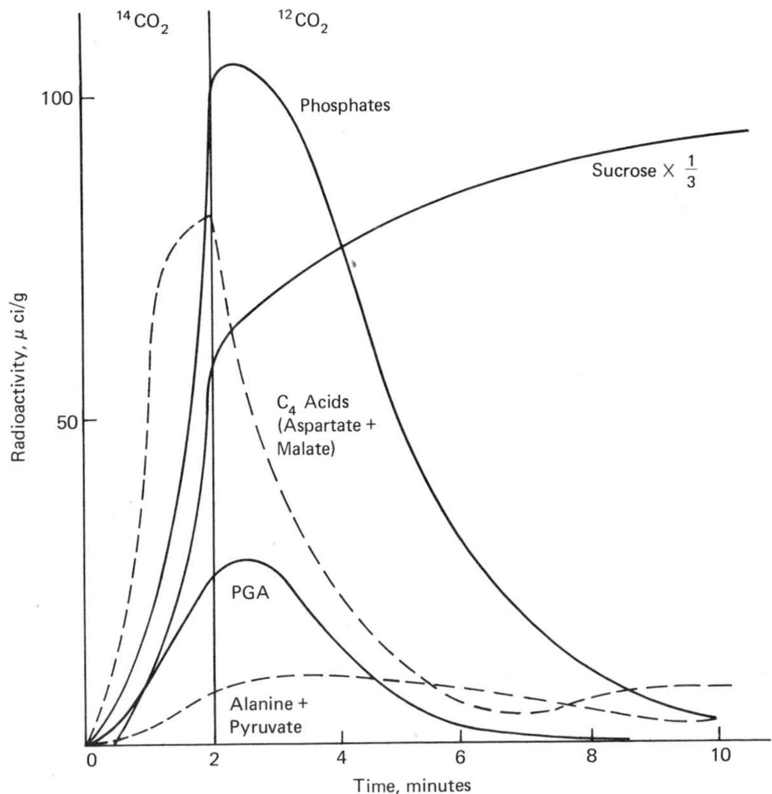

and its transfer to PGA and the Calvin cycle intermediates during the $^{14}CO_2$ pulse can clearly be seen. Note also the way the $^{12}CO_2$ chases the $^{14}CO_2$ out of the C_4 and then the C_3 cycle intermediates. Note also the behavior of the C_3 compounds of the C_4 cycle, pyruvate and alanine. If $^{14}CO_2$ was transferred only by the β-carboxylation reaction and decarboxylation and the cycle was tight (that is, no compounds leaked into or out of it), then these C_3 compounds should never acquire radioactivity. The fact that they did so indicates that new carbon was entering the C_4 cycle, presumably by leakage from the C_3 cycle.

Recent data, to be discussed in the section on Dark Respiration later in this chapter, show that C_4 acids synthesized by the operation of the Krebs cycle may also be used to prime the C_4 photosynthetic cycle. Thus it seems that the absence of autocatalysis is not a serious drawback to the operation of the C_4 photosynthetic cycle.

Regulation of the integration of C_3 intermediates must take place, but details have not yet been worked out. However, it is known that the C_3 cycle can exert a regulatory effect on the C_4 cycle through the stimulation of PEP carboxylase by glucose-6-phosphate. Glucose-6-phosphate is a product of the C_3 cycle that leads to starch formation, and its activation of PEP carboxylase would tend to prevent the loss of PEP to other reactions, including starch synthesis.

Regulation of the C_4 cycle is also accomplished by the energy charge level of the cell. This may be done by allosteric effects of adenylates or through the direct effect of ATP, ADP, and AMP concentrations on the pyruvate, phosphate dikinase reaction that regenerates PEP. There are almost certainly other regulated steps in C_4 photosynthesis—the integration of this complex sequence of metabolism demands them. A major area of research for the future is the discovery and clarification of these control mechanisms.

Productivity and Ecological Significance of C_4 Plants

Advantages of the C_4 Cycle. The C_4 cycle provides two distinct advantages: a mechanism for gathering CO_2 more efficiently, and a mechanism for transporting it to the site of the reductive photosynthetic cycle. We shall see that these advantages outweigh the increased energy cost of the C_4 cycle under certain circumstances. However, C_4 metabolism is not always advantageous, and many C_3 plants have as high rates of productivity as C_4 plants under appropriate circumstances.

Gathering CO_2 and Conserving Water. PEP carboxylase has a somewhat higher affinity for CO_2 than RuBPcase. However, it has been calculated that RuBPcase is adequate under normal conditions, in its amount and in its affinity for CO_2, for the highest rates of photosynthesis that have been observed. Thus the C_4 cycle offers an advantage specifically when the ambient CO_2 concentration is very low. This occurs when stomata are nearly shut as a result of water stress. Then the C_4 cycle can maintain high rates of photosynthesis even though the CO_2 concentration inside the leaf may fall to levels low enough that C_3 photosynthesis would be severely reduced. The C_4 syndrome has evolved primarily in tropical plants that occupy dry habitats and thus need to conserve water. High growth rates and productivity under these conditions confer a decided advantage. Furthermore, C_4 plants attain rapid rates of photosynthesis and growth under the very high light intensities found in the tropics, which would more than saturate C_3 plants. They can efficiently use light at intensities that would be wasted on C_3 plants.

Concentration of CO_2. In C_3 leaves, CO_2 must diffuse down a shallow concentration gradient from outside the leaf to the site of the carboxylase in the chloroplasts of photosynthetic cells. In C_4 plants, the Kranz anatomy provides for a short CO_2 diffusion pathway because the substomatal spaces are small and CO_2 need only diffuse to the cytoplasm of mesophyll cells. The C_4 acids that carry CO_2 to the bundle-sheath cells diffuse down steeper gradients maintained by the concentration differentials at their sites of synthesis and decarboxylation. The net result is that the CO_2 concentration at the bundle-sheath chloroplasts in a C_4 plant may be as high as an estimated 200–500 ppm. In a C_3 plant the CO_2 concentration inside the leaf spaces is usually at the compensation point (about 50 ppm), and its concentration at the chloroplast surface is thought to be much lower. Thus the C_4 cycle serves to maintain a sufficiently high level of CO_2 inside the bundle-sheath cells for the RuBPcase to operate at full speed. The high CO_2 concentration also produces an additional advantage: it greatly increases the CO_2/O_2 ratio at the site of the carboxylase, thus reducing the oxygen effect on photosynthesis and greatly decreasing photorespiration. These points are illustrated in Figure 15-12, which shows photosynthesis rates plotted against CO_2 concentrations for two C_4 plants having different rates of photosynthesis and a C_3 plant closely related to one of the C_4 plants.

Photorespiration in C_4 Plants. The RuBPcase in C_4 plants is not different from that in C_3 plants, and it is sensitive to O_2 in the same way. Yet C_4 plants do not exhibit detectable photorespiration. C_4 leaves do contain peroxisomes and the enzymes of the glycolate pathway, though sometimes in reduced amounts. Tracer studies show that C_4 plants can metabolize glycolate, but much less ^{14}C normally passes through glycine and serine during photosynthesis with $^{14}CO_2$ in C_4 plants than in C_3 plants. It is now generally accepted (though not finally proved) that C_4 plants may have the reactions of photorespiration, but they proceed only slowly because of the high CO_2/O_2 ratio in the bundle-sheath cells. When C_3 plants are given a CO_2/O_2 mixture equivalent to that calculated for the bundle-sheath cells of C_4 plants, they also show high rates of photosynthesis and very low rates of photorespiration. In addition, any CO_2 that may be produced by photorespiration in C_4 plants is trapped and reused by the C_4 carboxylation system and so does not escape outside the leaf.

Temperature Effects. Many C_4 plants are tropical, and many have quite high optimum temperature ranges for photosynthesis and growth. However, some C_3 plants have high temperature adaptation also, and a few C_4 plants lack it. So it may be concluded that high temperature adaptation does not automatically accompany the C_4 cycle, although the two are often linked. This point is illustrated in Figure 15-13, showing photosynthesis of the same three plants that were illustrated in Figure 15-12, but as affected by temperature. (Both Figures 15-12 and 15-13 show data from the laboratory of O. Björkman at the Carnegie Institution of Washington.) The curve showed by *T. oblongifolia* is characteristic of high-temperature C_4 plants. The two

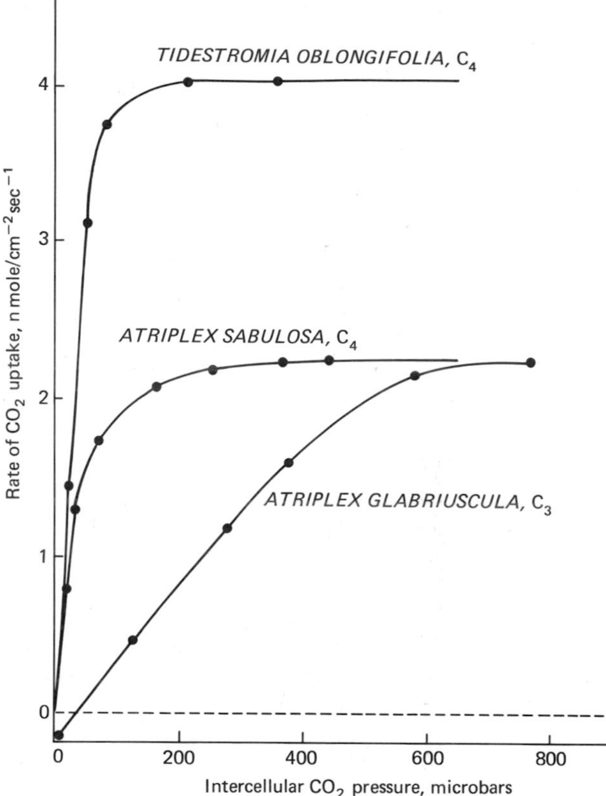

Figure 15-12. Photosynthesis as a function of the CO_2 concentration in the intercellular spaces in C_3 and C_4 species, grown under a temperature regime of 40°C day/30°C night. Measurements were made at a leaf temperature of 40°C, a light intensity of 160 nanoeinstein cm^{-2} sec^{-1}, and an O_2 concentration of 21 percent. [Microbars (partial pressure) CO_2 are roughly interconvertible with parts per million (ppm) of CO_2.] Note that the C_4 plants reach maximum rates of photosynthesis at much lower CO_2 concentrations than the C_3 plant. The compensation point of the C_3 plant is about 40 ppm CO_2, whereas that of the C_4 plants is 0, showing that they do not lose CO_2 by photorespiration. [From O. Bjorkman, H. Mooney, and J. Ehleringer: *Carnegie Inst. Wash. Year Book,* **74**:760–61 (1975). Used with permission.]

Atriplex species are characteristic of a lower temperature regime. Their photosynthetic responses to temperature are very similar, even though one is a C_4 plant and the other is a C_3 plant.

Some high-temperature adaptation is conferred by the C_4 cycle because photorespiration is strongly temperature dependent, and its rate increases proportionately more at high temperature than does the rate of photosynthesis. Thus C_3 plants face an increasingly higher percentage loss of fixed carbon by photorespiration with increasing

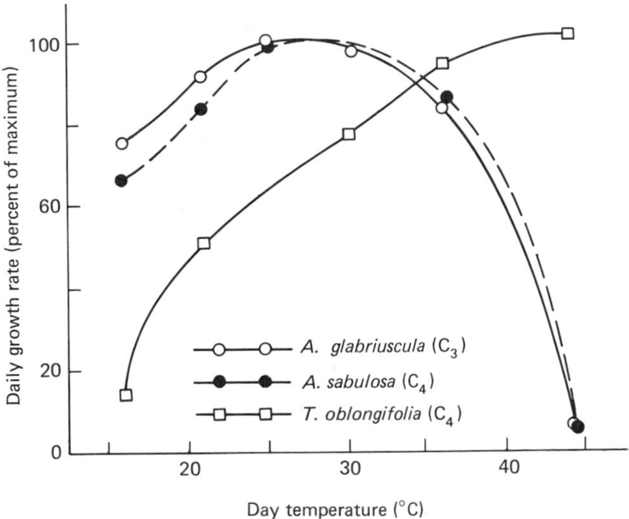

Figure 15-13. Daily relative growth as a function of day temperature in *Atriplex glabriuscula*, *A. sabulosa*, and *Tidestromia oblongifolia*. Growth rates were measured as daily gain in grams per gram of total plant dry weight. Note that one C_4 plant is high-temperature adapted but the other behaves exactly like a C_3 plant with respect to temperature. [From O. Bjorkman, B. Marshall, M. Nobs, W. Ward, F. Nicholson, and H. Mooney: *Carnegie Inst. Wash. Year Book,* **73:**757–67 (1974). Used with permission.]

temperature. C_4 plants, which do not lose CO_2 by photorespiration, do not face this problem.

Ecological Adaptation. C_4 plants can maintain high photosynthetic rates under conditions of water shortage that would stop photosynthesis in C_3 plants. This seems to be the main advantage of the C_4 syndrome, and it permits more efficient growth in a situation where high light and low humidity are common. At low light intensities the C_4 cycle provides no advantage because of its need for extra light energy. Similarly, when stomata are wide open, many C_3 plants show as high rates of photosynthesis and growth as do C_4 plants. Ultimately, the rate of photosynthesis in both C_3 and C_4 plants is limited by the amount and performance of RuBPcase, which controls the rate of fixation and reduction of carbon.

Productivity. Because some C_4 crops are among the most productive in the world (for example, corn, sugarcane), it is often stated that C_4 photosynthesis confers a high level of productivity. This is not strictly true; many C_3 plants are as productive as the better C_4 plants, and many C_4 plants are considerably less so. In fact, if the major crops of the world are listed in order of productivity, little relationship with their photosynthetic mechanism will be found, as shown in Table 15-1.

This does not necessarily mean that the C_4 cycle is not advantageous. Crops are usually grown under conditions that are at least favorable and usually approach

Table 15-1. Maximum rates of productivity attained by some crop plants under optimum conditions.

	Production *($g/m^2/day$)*	*Photosynthesis type*
Sunflower	68	C_3
Elephant grass	60	C_4
Wheat	57	C_3
Cattail (*Typha*)	53	C_3
Corn	52	C_4
Sugarcane	38	C_4
Potato	37	C_3
Beet	31	C_3
Soybean	17	C_3

optimum. The important point is that under optimum conditions both C_3 and C_4 plants are limited in photosynthesis by the operation of their RuBPcase, and under optimum conditions this enzyme is capable of handling carbon at rates consistent with the maximum rates of photosynthesis attainable. Thus the factor limiting crop productivity is not likely to be the presence or absence of C_4 photosynthesis, but some other intrinsic factor like the amount of RuBPcase, the stomatal resistance, or the efficiency of conversion of fixed carbon into production.

However, one should not conclude that C_4 photosynthesis is not advantageous to crop plants that possess it. On the contrary, plants having the C_4 syndrome can be grown in locations where their productivity would be low if they were not C_4. There would be little advantage from the C_4 cycle to plants normally growing in moist earth or reduced light, for example. But if they had C_4 photosynthesis, they might then be grown in dry ground and in more tropical areas where sunlight is more intense and the growing season longer. These factors would combine to produce much higher productivity. It is for this reason that scientists in many agricultural and plant physiological research laboratories in the world today are studying the possibility, by traditional genetics, genetic engineering, or cell biology, of transferring C_4 photosynthesis to valuable crop plants that do not now possess it.

Crassulacean Acid Metabolism

Outline of Reactions. **Crassulacean acid metabolism (CAM)** is characterized by CO_2 fixation at night by β-carboxylation of PEP to make malate, and its decarboxylation by day to yield CO_2 which is fixed by the C_3 cycle. The source of PEP at night is starch; starch is made in the daytime so starch content tends to be a reciprocal function of acidity in CAM plants. The basic reaction is schematically outlined in Figure 15-14. Details of the reaction were considered in Chapter 7 (Figure 7-22, page 184). CAM thus permits plants to operate under extremely dry conditions when the stomata must be kept closed all day in order to conserve water.

Two facts emerge. First, the stomatal mechanism must be closely linked to photosynthetic metabolism, since CAM requires that stomata be open at night and (usually) closed in the daytime. The second is that in CAM there is close integration between

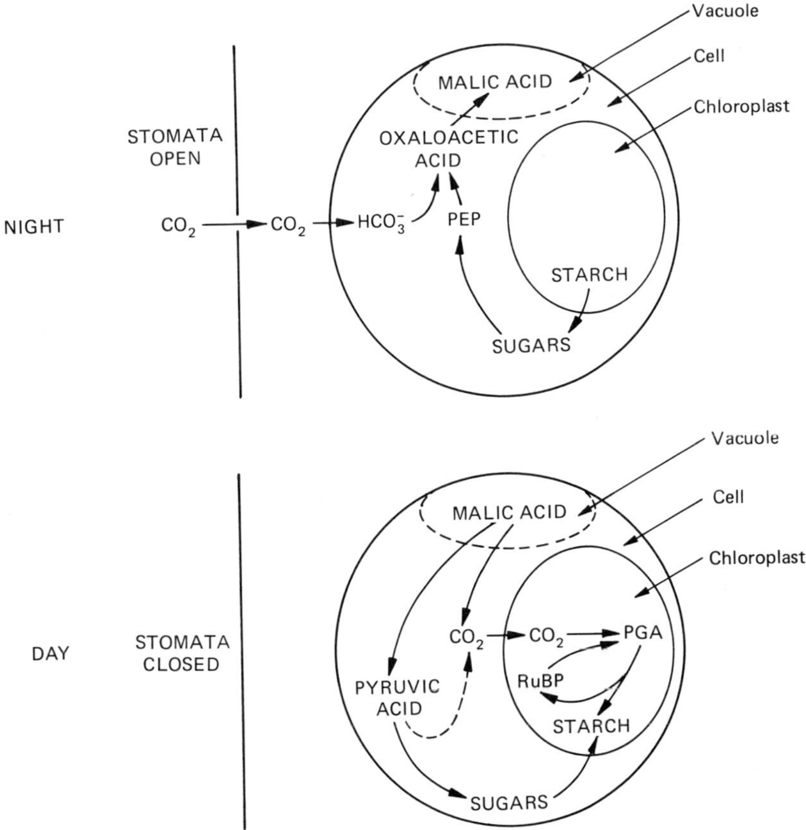

Figure 15-14. Simplified diagram of CAM (crassulacean acid metabolism).

aspects of cellular metabolism (interconversion of starch and acids, gluconeogenesis) and photosynthesis. These facts imply precise regulation of metabolism.

However, it is not sufficient merely to maintain constant conditions. The alternations of the metabolic patterns of day and night require very different levels of enzyme activities and metabolites at different times of day. To make matters even more complicated, CAM is not simply an alternation of day and night patterns. There are several phases of metabolism, each characterized by different metabolic activities, and there are also long-term changes in CAM patterns as the seasons advance. Furthermore, some plants exhibit CAM only at certain times in their life cycle, or at certain times of the year, or under the influence of certain environmental conditions such as salinity or water stress.

Many aspects of CAM are not yet clear, and, as with C_4 photosynthesis, different patterns are found in different plants. The Australian physiologist C. B. Osmond has made a detailed study of the typical CAM plant, *Kalanchoe daigremontiana,* and we shall briefly examine the metabolic patterns and controls presented by this plant.

CAM Patterns. Figure 15-15 shows the CO_2 uptake, malic acid content, and stomatal behavior of *K. daigremontiana*. The pattern is divided into four phases. In phase 1, at

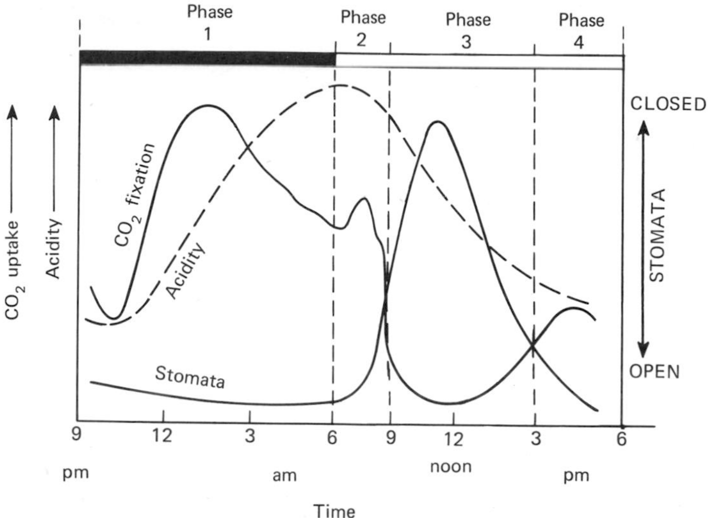

Figure 15-15. CAM patterns. [Modified from data of C. B. Osmond in R. H. Burris and C. C. Black (eds.): *CO_2 Metabolism and Plant Productivity.* University Park Press, Baltimore, 1976.]

night, CO_2 is actively absorbed and fixed by PEP carboxylase to make malic acid, which accumulates. The stomata are wide open. When the light is turned on a brief gulp of CO_2 occurs, which constitutes phase 2. Then phase 3 begins with the closing of stomata, the cessation of CO_2 assimilation, and the beginning of deacidification. This phase may continue well into the day, when (the exact time depending on external conditions) stomata may open and CO_2 fixation again begin, but now depending on RuBPcase and the C_3 cycle of photosynthesis. This constitutes phase 4, which may last until darkness and the reinitiation of phase 1.

CO_2 fixation in phases 1 and 2 is insensitive to O_2 concentration, showing that PEP carboxylase is the main pathway of CO_2 fixation at these times. In studies with leaves from which the epidermis (and the closed stomata) had been stripped, it was found that CO_2 fixation in phases 3 and 4 is competitively inhibited by O_2 and has a CO_2 compensation point of about 50 ppm CO_2. This shows that fixation in these phases is by RuBPcase. Kinetic studies with $^{14}CO_2$ show the transfer of carbon to malate in the dark and its transfer to the C_3 cycle in daylight. These and many similar experiments have confirmed the basic outlines of the reactions shown in Figure 15-14.

The important requirements for CAM are thus mechanisms that permit regulation of the carboxylases and decarboxylases and of the metabolism associated with the production and utilization of the C_3 components. In addition, the stomatal mechanism needs to be closely regulated.

Control. A number of hypotheses for the control of CAM have been presented, based on various factors. PEP carboxylase and malate dehydrogenase have temperature coefficients such that high temperature favors malate decarboxylation and low temperature favors its synthesis. Although it is true that many CAM plants live in desert environments where it is indeed warm in the day and cool at night, it has been shown that CAM works equally well under constant temperature. Thus, although temperature may affect CAM, it is unlikely to be the final controlling factor.

Alternative models suggest that CAM is regulated by competition, either for CO_2 or for C_3 skeletons. The known differences in activity and affinity for CO_2 of PEP carboxylase and RuBPcase, however, make competition between them unlikely as a regulating mechanism. Rather, mechanisms are needed to prevent one-sided competition between these two carboxylases. The C_3 competition model depends on the fact that PEP in the C_4 cycle and PGA in the C_3 cycle are interconvertible. Thus C_3 skeletons could be diverted to the C_3 cycle in light, so controlling PEP supply. However, CAM is strongly periodic and will continue to pass through its various phases even if the plant is placed in constant light. That is to say, CAM has a strong endogenous rhythm (for further discussion of rhythms, see Chapter 20). So, even in continuous light, where the C_3 cycle would be continuously active, the alternation of acidification and deacidification (of β-carboxylation and C_3 carboxylation) continues.

The possibility exists that some unknown rhythmic control mechanism (that is, a biological clock) regulates CAM. Unfortunately, such a model does not help our understanding of CAM until the intricacies of biological clocks are worked out. This has not happened yet.

Yet another model suggests that PEP carboxylase, the more active and effective carboxylating enzyme, is regulated through feedback control by malate, its product. In order to allow for high build-up of malate in phases 1 and 2 followed by sudden control of PEP carboxylase in phase 3, it is necessary to postulate that malate, which is largely stored in the vacuoles, is suddenly allowed to flood into the cytoplasm at the start of phase 3. Parallel situations exist in which small changes in turgor pressure (affected by malate build-up) at some critical level allow or cause sudden changes in membrane permeability. However, this model, in common with the others mentioned, has not yet been proved.

Respiration and Photorespiration in CAM. CAM plants respire normally, and it has been suggested that much of the C_3 compound remaining after malate decarboxylation in phase 3 is respired away by the Krebs cycle, thus producing more CO_2 for C_3 fixation. However, many physiologists are of the opinion that this is not an important reaction, partly because in a number of plants the rates of dark respiration are substantially decreased in light (see page 373). More likely (but still not certain) the pyruvate formed from malate decarboxylation may be used in gluconeogenic reactions leading to the formation of starch.

Photorespiration most likely goes on in CAM plants in light, as shown by the fact that when the epidermis is stripped off (so that stomatal resistance is removed) they show a compensation point of about 50 ppm, characteristic of normal C_3 plants. ^{14}C passes through glycolate, glycine, and serine during light supply of $^{14}CO_2$, and phase 3 or 4 CO_2 fixation is inhibited by O_2. Of course, since stomata are normally closed in light no CO_2 is evolved. Instead, any CO_2 produced in photorespiration would be refixed.

Ecological Significance of CAM. Unlike C_4 photosynthesis, CAM does not confer high rates of photosynthesis on plants that possess it. However, like the C_4 cycle, it does confer decided advantages under specialized conditions. Although the C_4 cycle permits high rates of photosynthesis under drought conditions, CAM permits photosynthesis and survival under conditions of extreme desiccation. Many cacti and other desert plants having CAM can survive long periods under such extreme conditions that no net CO_2 fixation takes place at all. Other plants would lose CO_2 by respiration, but CAM plants refix all that they lose. Thus, although they may not grow, they can survive when other plants would starve. Consequently, although the interest is not so great as with C_4 photosynthesis, some effort is being devoted to conferring on crop plants thc CAM ability to conduct photosynthesis behind closed stomata. Whether this can be done remains to be seen.

Dark Respiration

Role of Dark Respiration. Recently the attention of plant biologists, and of agricultural scientists in particular, has been turned to the question of why plants respire, and whether some respiratory metabolism might not be "bred out" of plants as wasteful and unnecessary. This viewpoint forces us to examine the question, what is respiration for? In the discussion in Chapter 6 we were concerned primarily with the role of respiration in providing energy and reducing power (ATP, NADH, NADPH), on the one hand, and intermediates for synthetic metabolism on the other.

Another way of looking at this is to consider that respiration provides for two different sorts of processes in plants: maintenance and growth. Maintenance is primarily concerned with repair and restoration, and the operation of all the metabolic systems necessary for the normal functioning of the plant. These include turnover, transport, maintenance of gradients of all sorts, and the requirements for operating controls, signal systems, and the like. Growth is mainly the synthesis and accretion of all the material and operational systems that constitute the plant.

Growth and maintenance are two quite different processes, and the American agriculturalist K. J. McCree has shown how they can be quantified independently. Maintenance respiration is clearly a function of plant size (at least in herbaceous plants that do not possess large masses of metabolically inert tissue), so it can be represented as cW, where c is a constant and W the dry weight of the plant. Growth respiration, however, depends only on the actual new growth being made by the plant, which is best measured as the net rate of photosynthesis. So growth respiration may be represented as kP, where k is another constant and P is the rate of photosynthesis. The equation for respiration is then:

$$R = kP + cW$$

It is possible to get experimental values for k and c. For example, plants may be starved for 48 hr, after which growth essentially ceases and only maintenance respiration (cW) is left, from which c can be calculated. Alternatively, the plant may be supplied briefly with $^{14}CO_2$ in light, and the amount of ^{14}C remaining in the plant after 24 hr then gives the value for k, the proportion of carbon being used in growth. Calculations for k based on energy requirements to make a known amount of plant body agree very closely with experimentally obtained values of k and indicated that 25–30 percent of absorbed carbon is used for growth respiration. Maintenance respiration is much lower and is generally in the range of 1–4 percent of the plant's dry weight. These data give values of 0.25–0.30 for k, and 0.01–0.04 for c in the preceding equation.

Clearly the respiration necessary to support growth is much the larger component of total respiration; it is here that any large improvement in crop growth efficiency must be sought. Crop physiologists have speculated that some of the growth respiration may be used to make metabolic or regulatory mechanisms or plant parts that are unnecessary to agriculturally maintained crops. This approach requires the application of genetic engineering principles rather than, or in conjunction with, those of plant physiology. Since calculated overall conversion efficiencies of crop plants (grams of carbon built into the plant per gram of carbon fixed in photosynthesis) are quite high, in the range 70–75 percent, it is unlikely that much improvement can be made. In fact, insofar as respiration is a measure of metabolic activity, high respiration is a characteristic of a healthy plant! It should be noted that as the plant passes maturity its growth respiration becomes smaller and its maintenance respiration then constitutes a larger proportion of the total. At that point, maintenance respiration becomes a more worthwhile target for studies on the efficiency of carbon utilization in crop plants.

Integration of Respiration and Photosynthesis. The British physiologist J. A. Raven has analyzed data from his own and other laboratories to consider the question of how much of the cell's energy requirements for growth and maintenance can be supplied directly from photosynthetic ATP and reducing power in light, and how much is supplied by dark-type respiration continuing in light. Possible points of interaction between photosynthesis and respiration in this way are shown in Figure 15-16. It should be noted that photorespiration cannot be considered as a respiratory process in this context since it does not generate usable energy. It is important not to confuse the manifestations of respiration (CO_2 production, O_2 uptake) with the process of respiration (the oxidation of substrates to produce usable energy).

Many algae can be grown in darkness on a carbon source—that is, heterotrophically—as well as photosynthetically. Raven examined the efficiency of conversion for respiration (which can be expressed as C/CO_2, the ratio of carbon incorporated into plant material per CO_2 produced in respiration) for a number of algae. He found that C/CO_2 was always much higher when the organisms were grown photosynthetically than when they were grown heterotrophically, as shown in Table 15-2. Values for phototrophic growth averaged more than twice as high as corresponding values for heterotrophic

Figure 15-16. Alternative pathways for the supply of ATP and reductant in plants. Solid arrows represent conventional dark or light pathways; dashed arrows indicate direct use of photosynthetic products in cellular reactions. [Modified and adapted from J. A. Raven: *Ann. Bot.,* **40:**587–602, 1976.]

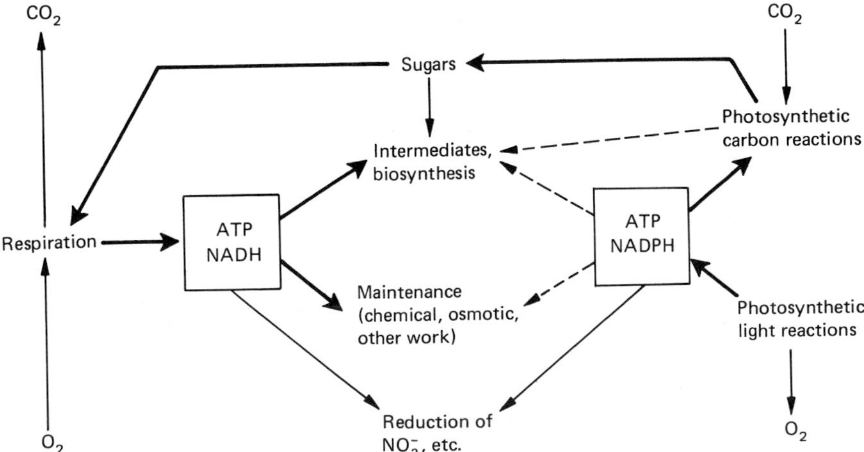

Table 15-2. C/CO_2 ratios for heterotrophic (dark, glucose) or phototrophic (light, CO_2) growth of algae. The ratios indicate the amount of growth (as carbon) achieved for a given amount of respiration (as CO_2). The phototrophically grown algae achieve higher values because photosynthetic ATP or reducing power contributes to their growth, so less respiration is required.

	C/CO_2 ratio	
Organism	*Heterotrophic*	*Phototrophic*
Chlorella spp.	0.8–1.8	4.0–6.7
Euglena gracilis baccilaris	2.5	3.3–5.0
Euglena gracilis strain Z	2.0	2.9–3.3
Average of several algae	1.3	3.6

SOURCE: Adapted and modified from J. A. Raven: *Ann. Bot.*, **40**:586–607 (1976).

growth. This shows that a substantial part of the energy required for growth can be derived directly from light; in dark-grown algae all the energy would have to be derived from respiration. Furthermore, the C/CO_2 values found in light were often higher than values calculated for the known growth and respiratory efficiency of the organisms, showing that some of the energy must have come from a source other than respiration. Similar analyses for higher plants are harder to make, and the results more difficult to interpret. However, it seems probable that some of the growth and maintenance requirements of higher plants may also be met by the photosynthetic production of ATP and reductant.

These ideas suggest that, under conditions when light is not limiting, excess light energy can be used directly in the form of ATP or reductant for growth and maintenance of the plant body. Clearly this requires that light is in excess, and that there are not other direct requirements such as nitrate reduction or nitrogen fixation (also indicated in Figure 15-16) that would compete for photosynthetic energy. It also suggests that regulation of this aspect of photosynthetic metabolism should be important. There are some indications that such controls work rather through the requirement for ATP than through its supply (that is, its concentration). This suggests that controls may be exercised by some alternative signal system, but little is known of this at present.

It is, of course, necessary that ATP and reducing power from photosynthesis (that is, formed in the chloroplasts) should be available to the cytoplasm and even to adjacent cells. ATP (and ADP) can penetrate the chloroplast membrane, so this appears to present no problem. NADPH, on the other hand, cannot pass the chloroplast boundary, and, in any case, the required reductant for many cellular syntheses is NADH. However, several reducing shuttles have been proposed that permit the export of reducing equivalents from the chloroplast to the cytoplasm. Some of these are illustrated in Figure 15-17, showing how NADPH in the chloroplast can be used, via the shuttle systems, to generate NADH in the cytoplasm. It is apparent that the export of photosynthetic energy in the form of chemical potential to distant parts of the cell, or to adjacent cells, may go on as needed.

Light Control of Respiration. Early workers were concerned by the problem of whether plants respire in light, a question that could not be unequivocally answered

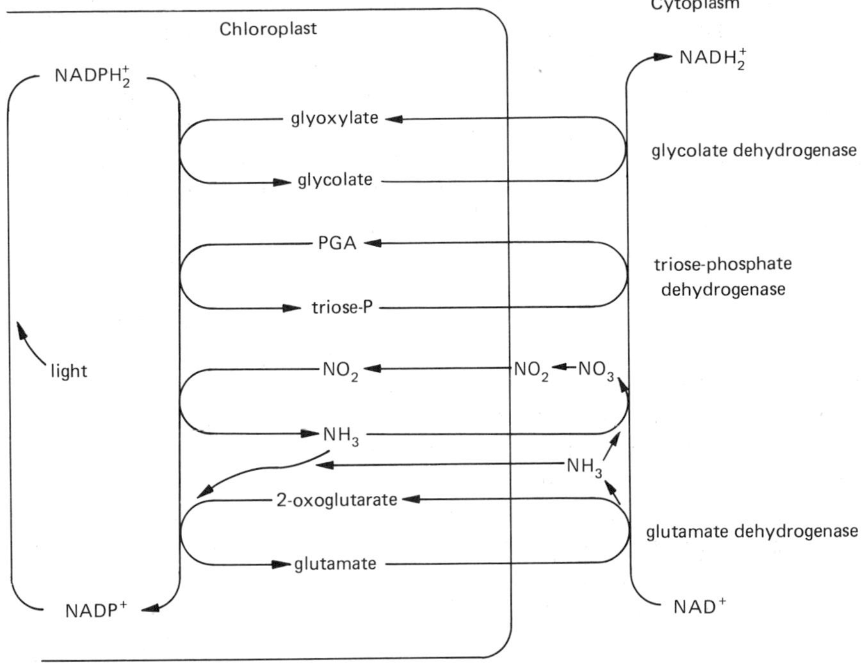

Figure 15-17. Metabolic shuttles that might accomplish the transfer of photosynthetic reducing power to the cytoplasm.

until the use of isotopes. Now we know that plants do indeed respire in light, and photorespiration is an accepted and verified phenomenon. However, controversy still exists about whether *dark respiration* continues in light in addition to photorespiration, and if so, at what level.

The preceding discussion suggests reasons why dark respiration should go on in light. There may be a continued need for ATP and reducing power for syntheses that is not entirely met by photosynthesis, particularly if light is low or limiting. Many of the carbon skeletons used to synthesize the substance of the growing plant body are derived from respiratory pathways. These include intermediates for the synthesis of lipids, isoprenoids, proteins, lignin, and various polysaccharides whose monomers are sugars other than hexose. Finally, there may also be nonphotosynthetic cells in the plant body that have no (or insufficient) access to the immediate products of photosynthesis for their maintenance and growth.

In spite of this, data for rates of dark respiration in photosynthetic tissues frequently show a marked decrease in respiration in light. Data of the type shown in Figure 15-4 (page 352) suggest that dark respiration is considerably reduced in light, and it has been calculated that in bean leaves dark respiration may be reduced on illumination to 25 or 30 percent of its dark rate. Data on sunflower leaves, on the other hand, indicate that dark respiration is little affected by light, and photorespiration is an "add on" component of respiration in light. The data of Raven (Table 15-2) suggest that requirements for respiratory energy are to a large extent met by photosynthesis, but other results indicate that algae may, under certain conditions of growth, show a high rate of dark respiration in light.

The nature of the light control that reduces or switches off dark respiration, when this does occur, is not very clearly understood. Respiration is undoubtedly controlled by the energy charge of the cell (relative amounts of ATP, ADP, and AMP; see page 104), and also by the NADH/NAD ratio. Thus it is likely that light-generated ATP or reducing power spilling over into the cytoplasm would have a controlling effect on respiration. Various investigations have implicated both photosynthetic ATP and photosynthetic reducing power as controlling agents. In truth, we do not yet clearly understand the conditions under which light control of respiration takes place or the nature of the controlling mechanism.

Several investigations have been made over the past two decades into the question of the activity of the Krebs cycle in light. It was initially thought that this cycle was largely shut down by the generation of reducing power that would hold the pyridine nucleotide cofactors of the cycle in the reduced form. However, recent investigations with specific inhibitors of mitochondrial Krebs cycle metabolism, and the use of ^{14}C-labeled Krebs cycle intermediates as substrates, have shown that the cycle is active in light. There appears to be a short period on illumination when the cycle activity is reduced, then it resumes operation at full speed.

The reason for Krebs cycle operation in light appears to be a requirement for biosynthesic intermediates. Either the cofactor controls are overridden by the need for intermediates, or the cycle is uncoupled from oxidative phosphorylation so that its activity can continue even in light. In C_4 plants it has been clearly shown that malate and aspartate generated in the Krebs cycle may be shunted into the C_4 cycle. In this way, perhaps, the C_4 cycle can compensate for its lack of autocatalysis.

Summary

The key points discussed in this chapter are integration of metabolism among different organelles, cells, and parts of plants and between different organelles, cells, and parts of plants and between different processes of carbon and nitrogen metabolism. This integration of metabolic processes has permitted the evolution of larger, more complex, and far more effective strategies of metabolism than were possible even with the relatively complex sequences of the Calvin cycle, the Krebs cycle, and similar systems. This in turn has permitted the integration of requirements for the control of water loss concomitant with the need to admit maximum amounts of carbon dioxide, as in CAM and the C_4 photosynthetic syndrome. In these ways plants have been able to optimize their use of available, free resources of material (CO_2, H_2O, minerals) and energy (light) to produce maximum results consistent with operating conditions. Many anxious hours are spent in big business with production charts, statistics, and computers to achieve this goal that in plants has been achieved by natural selection.

What is now becoming apparent is that the strategies of growth, development, production, and reproduction that best suit the plant in the wild, and that have become genetically fixed by eons of evolution, may not be the best ones for plants grown in agriculture. Indeed, the strategy of plants in the wild is generally to grow fast and large in spring and early summer in order to lay claim to the maximum territory (minerals and water) and space (light). Then they must store an excess of minerals against the time of need during the reproductive period. Finally, they must produce the right amount (not necessarily the most) of seed late enough in the season to ensure survival until next year. Agricultural plants have no such need. They are spaced, fertilized, and under no compulsion for their seeds to survive. The earlier they set seed, the better. As a result, the attention of agricultural scientists is turning more and more to the problem of genetically engineering plants to develop a strategy for growth and reproduction that is better suited to agricultural practice than to survival in the wild. It is likely that the future results of this kind of research will have far more impact on agriculture than merely improving photosynthesis rates or attenuating unwanted respiration.

Additional Reading

Burris, R. H., and **C. C. Black:** *CO_2 Metabolism and Plant Productivity.* University Park Press, Baltimore, Md., 1976.

Cooper, J. P.: *Photosynthesis and Productivity in Different Environments.* Cambridge University Press, Cambridge, England, 1975.

Hatch, M. D., C. B. Osmond, and **R. O. Slatyer:** *Photosynthesis and Photorespiration.* John Wiley & Sons, Inc., New York, 1971.

Sestak, Z., J. Catsky, and **P. G. Jarvis:** *Plant Photosynthetic Production.* Dr. W. Junk N. V., The Hague, 1971.

Zelitch, I.: *Photosynthesis, Photorespiration and Plant Productivity.* Academic Press, New York, 1971.

SECTION IV

The Developing Plant—Plant Behavior

16 Interpretation of Growth and Development

Introduction

Growth and development are a wonderful coordination of many events at many different levels, from biophysical and biochemical to organismal, which result in the production of a whole organism. The subject is very complex, and there are many different ways of approaching it. Several recent texts and monographs deal with the subject from the point of view of the mechanisms and agencies of development. We shall emphasize the plant and its behavior, dealing with the mechanisms of development as we come to them. In this chapter we shall examine the concepts of growth and development and their control in a general way. This will be followed in Chapters 17 to 22 by an examination in detail of various events in the growth and development of plants. Then, in Chapter 23, we shall summarize Section IV from the mechanistic viewpoint, reviewing the major classes of plant growth substances and their mode of action and the types of process they control.

It is not possible to cover all aspects of development and its control—the primary literature on the subject is immense, and there is much yet to be discovered. Our understanding of development is growing rapidly, but many areas are still the subject of debate or are frankly unknown. For this reason we shall have to leave many questions unanswered. Indeed, often it is not even certain what questions we should be asking.

Growth and Its Measurement

Parameters of Growth. Growth is defined in a dictionary as "the advancement toward or attainment of full size or maturity; development; a gradual increase in size." We shall arbitrarily separate the concepts of growth and development, reserving the term *growth* to denote increase in size, leaving out any qualitative concepts such as "full size" or "maturity," which are clearly unrelated to the process of increase. However, even the simple concept of size increase has difficulties because there are several possible ways of measuring it. Growth can be measured as increase in length, width, or area; often it is measured as increase in volume, mass, or weight (either fresh or dry weight). Each of these parameters describes something different, and there is seldom a simple relationship among them in a growing organism. This is because growth often occurs in different directions at different and possibly unrelated rates, so that simple linear-area-volume ratios do not persist in time.

This problem, the difficulty of defining growth and size, is further emphasized by the fact that it is quite likely, during certain kinds of growth, that one of the parameters may increase while another decreases. For example, during germination of a seed there is an initial uptake of water that is not accompanied by any significant growth as we normally consider it. There is an increase in volume and fresh weight, but not in dry weight. Subsequently the seedling increases greatly in length (it grows), but there is a net decrease in dry weight! Nevertheless, growth by any reasonable definition has occurred.

These two examples are extremes; however, intermediate situations occur in which it is very difficult to determine intuitively whether or not growth has occurred. Examples are an increase in size due to water absorption, which may be permanent or temporary, or cell division resulting in a large increase in cell numbers without any substantial change in shape or form. The first instance may be a physiological manifestation other than growth; the latter would probably be described as development (see next section). There seems to be no acceptable way out of this dilemma at the moment because growth, development, and simple changes in size (resulting from the absorption of water or other inert substances) frequently overlap. These processes all graduate into each other and can occur separately or together in any combination.

As a consequence we must define the term *growth* when we use it. This is commonly done unconciously: we speak of a child "growing *up*" (that is, he increases in length). When a boy becomes a young man, he "stops growing" (frequently a delusion—he may continue to grow in girth and weight, and hopefully also in wisdom). Much later we say he is "growing *old*." Each use of the word *grow* involves different concepts and is explicitly defined. So in plant physiology we must define our term as we use it, as increase in weight, or length, or some other measurable parameter. Alternatively, we may dodge the issue by using such terms as "increase" or "elongation" which do not carry the physiological implications of growth. These terms may also be used with validity when we feel intuitively that growth, as usually understood, has not occurred; an example would be the increase in weight by cells absorbing water but not dividing or otherwise changing.

Growth Versus Development. Development may be defined as ordered change or progress, often (but not always) toward a higher, more ordered, or more complex state. Thus development may take place without growth and growth without development, but the two are often combined in a single process. This is particularly true when some degree of rigidity prohibits the development of one form into another without the addition of new material to effect the change. Here development cannot take place without concomitant growth, and the two are part of the same process. This underlines the problem of isolating and understanding the causes of growth and of development—they may be separate and distinguishable, or they may result from only one stimulus. This, in turn, raises further problems in the mathematical analysis of growth, in that the ordered progress of growth may be perturbed in unpredictable ways by developmental events. This consideration will be examined in the following section.

Development implies change, and changes may be gradual or very abrupt. Certain important developmental events such as germination, flowering, or senescence result quite suddenly in a major change in the life or growth pattern of a plant. Other processes of development continue more or less slowly or gradually during part or all of the plant's life. The greater part of Secion IV will deal with development in terms of *where* in the plant growth occurs, *what sort* of growth takes place (that is, the changes in form and composition resulting from development), and *how much*.

Kinetics of Growth. It has long been considered that if the growth of an organ or organism could be exactly described by a mathematical formula or model, one would then have an explanation of the pattern of growth. Such a model (which need not describe the whole growth of an organism), if incomplete, could be used to test hypotheses about unknown or suspected factors, and, if complete, could be used to test its own validity by experiments in which the effects of specific perturbations to the model and the living system were compared. Unfortunately the processes of growth and development are so complex that a satisfactory formulation having these capabilities is probably still far in the future.

A number of attempts have been made to describe growth in mathematical terms. In some respects many of these have not been very successful, since they have accurately described growth for a short time only, usually when no major developmental change is taking place. Such models do not add to our knowledge of the causes of development. However, recently a number of mathematical models for the growth of important crop plants have been elaborated. These apply environmental parameters (light, temperature, water, and so on) to a simple growth model for individual parts of the plant (roots, leaves, stem). The contribution of each part to all others (or the interdependence of parts) is also applied. Some very interesting models of plant growth have resulted that relate actual growth to the physiological or biochemical capabilities (such as photosynthesis, respiration, translocation) of the developing plant parts. These models have been of great use in developing breeding programs to design plants best suited to particular environmental parameters and to program the most efficient application of fertilizer and water. To the present they have not contributed in a major way to our understanding of development and its control, but perhaps this will come as models are improved and refined.

Nevertheless it is valuable to make a brief mathematical analysis of the simple aspects of growth, because doing this clearly reveals the nature of some of the factors that govern growth (but not differentiation). A typical growth pattern of an annual plant is represented in Figure 16-1. This can be divided into three phases: (1) a logarithmic or exponential phase, (2) a linear phase, and (3) a phase of declining growth rate called senescence. The rate of growth is shown in Figure 16-2. It increases continuously during the exponential phase, is constant during the linear phase, and declines to zero during senescence. The curves in Figures 16-1 and 16-2 are idealized; the perturbations caused by environmental variation and developmental events have been smoothed out. Also, many plants exhibit quite a different form of growth curve. One or another of the phases may be emphasized or even left out, and growth rates may fluctuate substantially from time to time, introducing bends, plateaus, and sections of very steep slope into the curve. However, such variations are usually due to developmental events and are very difficult to treat in simple mathematical terms.

The analysis of the log phase is relatively simple. If each cell (or a fixed proportion of them) in a colony or an organism divides into two at regular intervals and the daughter cells grow to the same size as the original cell, then the colony or organism increases by the compound interest law:

$$Mt_2 = Mt_1 e^{r(t_2 - t_1)}$$

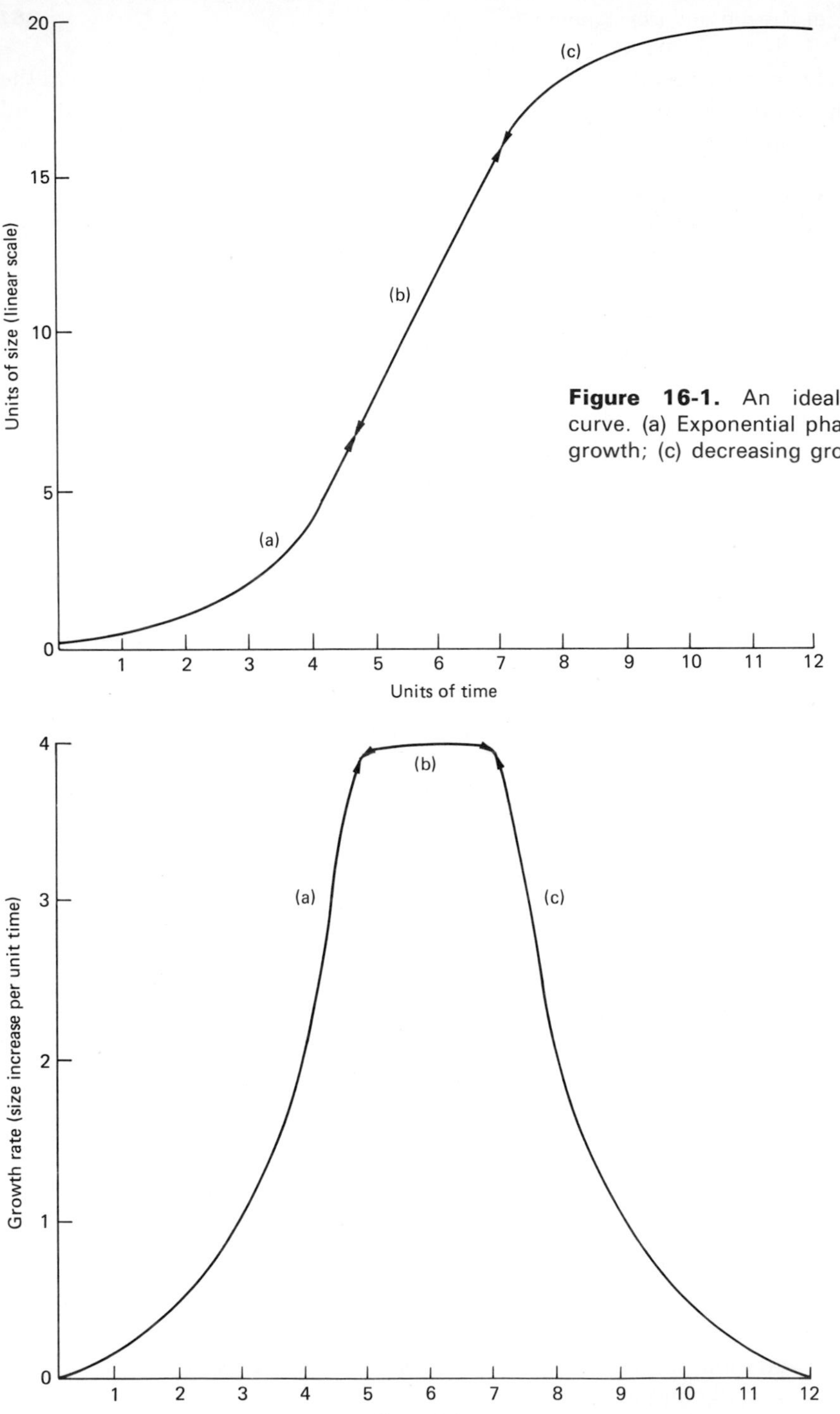

Figure 16-1. An idealized growth curve. (a) Exponential phase; (b) linear growth; (c) decreasing growth rate.

Figure 16-2. Growth-rate curve derived from Figure 16-1. (a) Exponential growth phase; (b) linear growth; (c) decreasing growth rate.

Note: This is an idealized curve. In many plants the linear phase may be very short, it need not occur at the peak of the growth rate, and some plants may exhibit distorted curves with several peaks and linear phases.

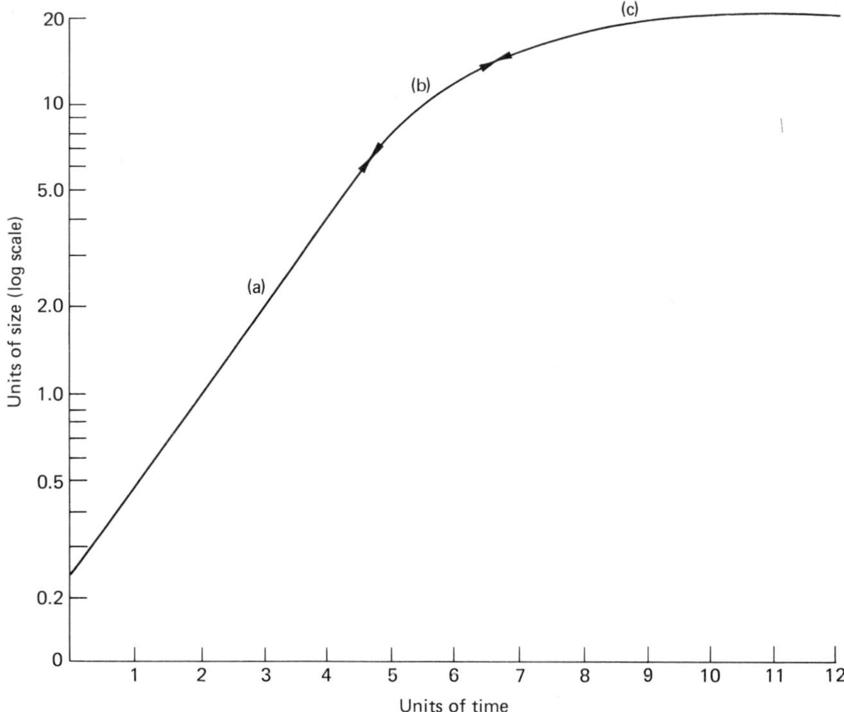

Figure 16-3. Logarithmic plot of the growth curve in Figure 16-1. (a) Exponential phase; (b) linear phase; (c) decreasing growth rate.

where M = mass, t_1 = the initial time, t_2 = the final time, e = base of natural logarithms, and r = growth rate. This equation can be rewritten as

$$2.303 \log (\text{size increase}) = r \times \text{time interval}$$

and a log plot of growth gives a straight line, as shown in Figure 16-3a.

This situation is realized in a culture of growing single-celled organisms, such as bacteria, or in the growth of a mass cells increasing in every direction. Evidently it must end in the bacterial culture either when some nutrient or nutrients run out or when a toxic waste product builds up. Then the culture goes rapidly into the phase of senescence and does not exhibit a linear phase at all. In a mass of tissue or a plant the situation is more complex. As the cell mass increases, nutrients in the medium become less accessible to cells in the center and the presence of outer cells hinders the division of the cells in the middle. Then the rate of increase falls, probably becoming proportional to some geometric function (for example, the area or the mass). Again, the growth will finally become limited by shortage of substrates or build-up of end products, and the senescence phase will begin.

Most higher plants, however, do not follow this pattern of growth for very long. They quickly develop a meristematic growth plan. Growth takes place only at discrete places and is essentially one-dimensional, that is, an increase in length. It is apparent that growth occurring in one or more meristems of constant size results in a linear function. That is, the rate is constant and not related to the size of the organism. This is the linear phase of growth, described by the nineteenth century German physiologist, J. Sachs, as the *grand phase* of growth (b in Figures 16-1 to 16-3). The expression for this type of growth

$$Mt_2 = Mt_1 + r(t_2 - t_1)$$

contains no terms for initial or final size, so no predictions can be made from it.

Some growth curves resemble the kinetic expression for a closed autocatalytic monomolecular reaction—a reaction in which one of the products accelerates or catalyzes the reaction. The expression for this type of reaction is

$$\log \frac{M_t}{M - M_t} = K(t - t_{\frac{1}{2}})$$

where M_t = mass at time t, M = final mass, and $t_{\frac{1}{2}}$ is the time required to reach one half final size. The curve produced by this relation approximates a growth curve (Figure 16-4). Because it relates to the present size and the final size of the organism, the curve can be used to predict or interpret growth. The expression can be restated

$$\frac{dm}{dt} = Km(M - m)$$

where m represents present size and M final size. This expression predicts a high growth rate (dm/dt) when m is small that increases rapidly as m increases. However, as the organism approaches final size ($m = M$), growth rate falls to zero.

Unfortunately the similarity between this expression and plant growth curves does not throw very much light on the process of growth. We know that growth is limited by

Figure 16-4. A plot of a closed autocatalytic monomolecular reaction. (a) Phase of exponential increase in rate; (c) phase of exponential decrease in rate. There is no linear phase.

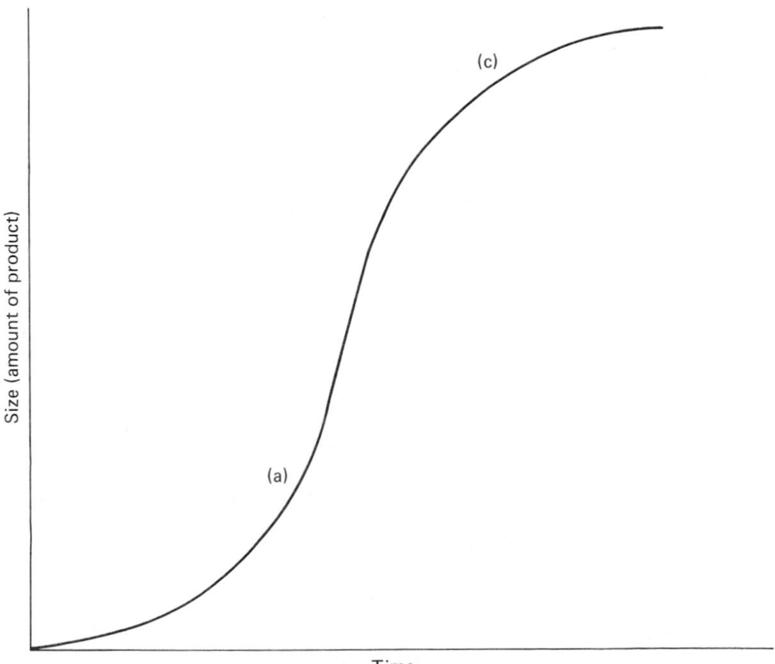

many reactions, not by one. Further, no matter how carefully one continues to add all possible reactants (that is, by improving growing conditions and maintaining them as nearly perfect as possible), there are inherent size limitations in organisms that prevent them from achieving the final size that the equation would predict. One cannot, by overfeeding it, make a mouse as big as an elephant or a daisy as tall as an oak! Moreover the equation does not predict linear growth.

A wide variety of mathematical expressions have been developed that simulate parts of the growth curve either of whole plants or of some specific organs. As yet, however, they have not contributed very much to our understanding of the processes governing development. On the other hand, they have been most useful in helping to clarify the roles of growth factors and nutrients in growing plants. Intuitively, when an essential factor is limiting, one would expect growth to be limited by its supply. Similarly, if it is presented in excess, one would expect growth to approach some maximum rate. But the supply of an essential factor often varies over only a narrow range between these limits, or two or several factors may vary in different ways at the same time. Then only precise mathematical analysis of growth in terms of a mathematical model, which must often be developed for the situation, can safely determine the true role of the factor or factors in question. This kind of treatment is often very difficult and sometimes unprofitable. However, plant physiologists should certainly not give up trying. Great advances are currently being made, and ultimately, if the understanding of growth and development is to be complete, it must be possible to describe them in mathematical terms.

Measurement of Development. Development is not easily measured quantitatively, because it tends to take place by a series of more or less discrete events. This makes an evaluation of *how much* difficult. For example, a plant may have flowered, or it may not. If (like a tulip) it produces only one flower, the question of how much is unanswerable. Similarly, the question *when* is often hard to answer, because developmental events may have been initiated long before their external manifestations become visible. Finally, the question of *what has transpired* is essentially one of quality, which is notoriously hard to evaluate in quantitative terms.

As a result, natural and experimental observations in development must frequently be expressed in terms of average responses or degree of response. This means that large numbers of individual responses or instances must be analyzed statistically so that the significance of observed and experimentally induced differences can be determined. For this reason, experimental design and experimental plan become more important in developmental studies than in almost any other aspect of plant physiology.

Kinds of Developmental Control

Genetic Controls. All the information that is ultimately responsible for the activities, growth, and development of cells is stored in the genetic complement of the cell. The majority of such information is stored in the nucleus, but some characteristics are inherited cytoplasmically, indicating that some genetic material is extranuclear. In particular, certain organelles, such as chloroplasts and mitochondria, have a degree of genetic independence, and it is known that some of their genetic information is stored in them. Each cell receives a full complement of the original genetic information at division, so the problem becomes ones of selection of information. At any given moment

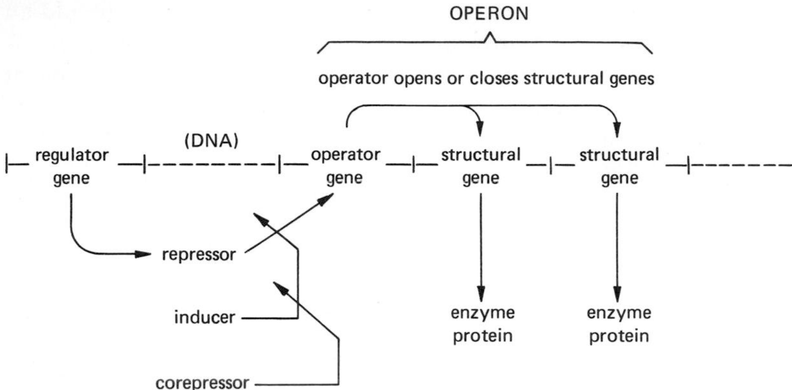

Figure 16-5. Model of the mechanism proposed by Jacob and Monod for the control of protein synthesis.

in the life cycle of a developing organism the proper information must be selected in all the cells that are undergoing division, growth, or development so that each organ in each part of the plant develops in the proper way. All irrelevant or unnecessary information must be ignored (that is, suppressed or stored so that it is inaccessible).

There are two points to be considered: the revelation of the appropriate gene or information package, and the capacity of the cell to obey. Revelation of information implies the activation of genes to make mRNA, which will program the synthesis of specific required enzymes (see Chapter 3). Evidently development requires that genes must be activated in an appropriate sequence; that is, they must be programmed so that each step in development activates the next.

A general mechanism whereby this can be accomplished has been proposed by the French scientists F. Jacob and J. Monod, as shown in Figure 16-5. Briefly, they proposed that **structural genes** (which program mRNA for specific enzymes) exist in groups or singly, each in combination with an **operator gene,** which functions to maintain the structural gene either in the active or **open** state, or in the inactive or **closed** state. The combination of structural gene and operator gene in cells is called an **operon.** A separate **regulator gene** (not part of the operon) forms a regulating molecule (a protein) called the **repressor** that maintains the operator gene in the closed state, thus holding the operon inactive. The presence or addition of a molecule called the **inducer,** which combines with or inactivates the repressor, allows the operator gene to go into the open state, activating the operon. Some other molecule called a **corepressor** may act to shut down the gene by activating the repressor so that the operon becomes closed and the operon inactivated. Inducer and corepressor molecules may be simple metabolites involved in specific reactions or metabolic sequences.

It is not hard to imagine that some metabolic activity of the cell associated with growth (for example, cell wall synthesis) produces molecules that, besides being intermediates in cell wall synthesis, can act as inducers of the operons that program the formation of mRNA synthesizing cytoplasmic enzymes. These, in turn, might produce intermediates inducing the synthesis of structural components, and so on. At some stage, some intermediate or product of metabolic activity might also act as a corepressor of an earlier operon concerned with the sequence, and so terminate the process of growth. Thus, as a result of the programming of sequences of operons by simple molecules or intermediates of the various processes of growth, an orderly pattern of activation and repression could ensure the orderly process of growth and differentiation.

It must be emphasized that no such complete sequence has ever been elucidated, and the above description only represents a possible mechanism. Specific examples of enzyme induction and repression are known, particularly in the control of microbial metabolism, but the exact mechanisms by which plant growth and development are programmed are not known. It has been suggested that certain hormones or hormone-like compounds may act to stimulate enzymes, also that members of a class of proteins called histones may act in a specific capacity to mask or inactivate genes or operons. However, details of how these might work are not clear.

Some sequence of order or regulation may be imposed by the capacity of the cell to react to genetic information. Thus, activation of a specific operon or group of operons at one stage in a cell's development could lead to one pattern of development; at another stage in the cell's life, activation of that same operon might lead to a quite different developmental reaction. It is possible to imagine changes in the pattern of reaction to the commands of a gene or group of genes as the cell or organism develops, as in the example shown in Figure 16-6. Thus it may not be necessary to have either a complete

Figure 16-6. A hypothetical model showing how the same cell might react differently to the activation of a single gene (gene A) in a juvenile (growing) organism or in a full-sized (developing) organism.

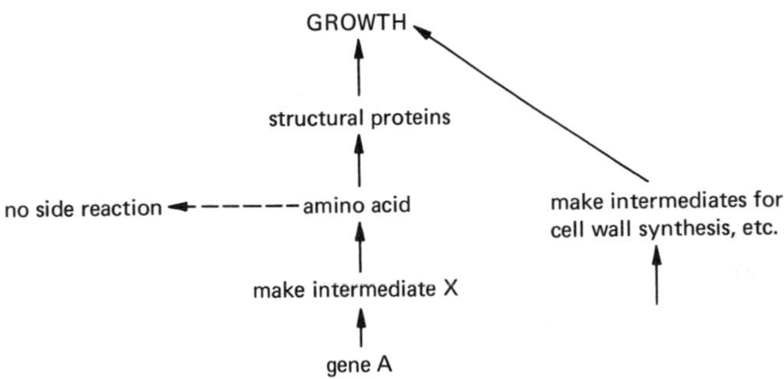

Juvenile (Growing) Organism

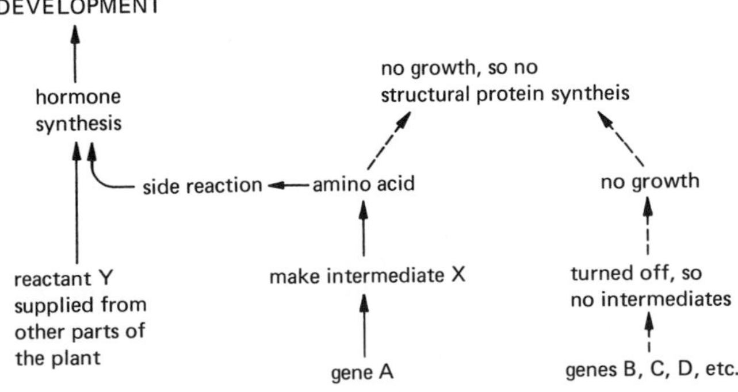

Full-sized (Developing) Organism

program for every step in the development of the organism or controls for every bit of genetic information carried by the cell. It is only necessary that the genes to which the cell is capable of reacting at a given time be under control at that time. It must be recognized that the foregoing discussion is entirely hypothetical; the nature of developmental programs at the genetic level and how they work are not yet known.

Organismal Controls. Much of the development of plants is mediated by internal stimuli generated within the organ or resulting from the organization it has already achieved. For example, a cell isolated from a developing plant organ and cultured in vitro will usually divide and grow as it would in vivo. However, in the culture this usually leads to tumorous growth, resulting in a shapeless mass of cells, whereas in the intact tissue it would lead to the production of leaf, root, or stem as dictated by the cell's position in the organism. Thus, this kind of control is a result as well as a cause of organized growth. The proximity of cells or groups of cells—tissues or organs—permits transfer of metabolites and other compounds so that metabolic reactions may be influenced by gradients of metabolites superimposed on the cell's metabolism by its position in the organism.

In addition, development may be affected or controlled by **hormones**—chemicals synthesized at one location in the organism and translocated to another, where they act in specific ways and at very low concentration to regulate growth, development, or metabolism. The effect of hormones is usually indirect, and they are active in very small amounts. Thus, the effect of metabolites, nutrients, and such compounds, either directly on metabolism or indirectly as inducers or corepressors, does not constitute hormone action. In actual fact it is very difficult to define the term **plant hormone** precisely. Many hormonelike substances may act at their site of synthesis, may act in what appears to be nonspecific ways, or may act (as do some metabolites) at the genetic level as inducers or repressors. The term **plant growth substances** is often preferred and refers to compounds, both natural and synthetic, that elicit growth, developmental, or metabolic responses. These substances are usually not metabolites in the sense that they are not intermediates or products of the pathways they control, and they are active at very low concentrations.

Several classes of hormones are known, some of which are growth or development *promoting* substances, and others are *inhibitors*. In addition, the existence of a number of hormones is postulated on the basis of experiments whose results apparently cannot be interpreted without the implication of, as yet unknown, translocated stimuli. We shall briefly describe the main groups of hormones without, at this time, going into details about their mode of action.

Auxins. The presence of a growth substance affecting the extension of oat coleoptiles was first suspected by Charles Darwin in the latter part of the nineteenth century. Definitive experiments proving the existence of a diffusible substance that stimulates cells' enlargement were done by Fritz Went in Holland in the 1920s, and by the 1930s the structure and identity of **auxin**—indole-3-acetic acid (abbreviated **IAA**)—were known. A pictorial representation of typical experiments that led to the discovery of IAA, together with the structures of important natural and synthetic auxins, is shown in Figure 16-7.

Auxin is characteristically synthesized in the stem tip (at or near the terminal meristem) and in young tissues (for example, young leaves) and moves mainly down the stem. It thus tends to form a gradient from the shoot tip to the root. Its activities include

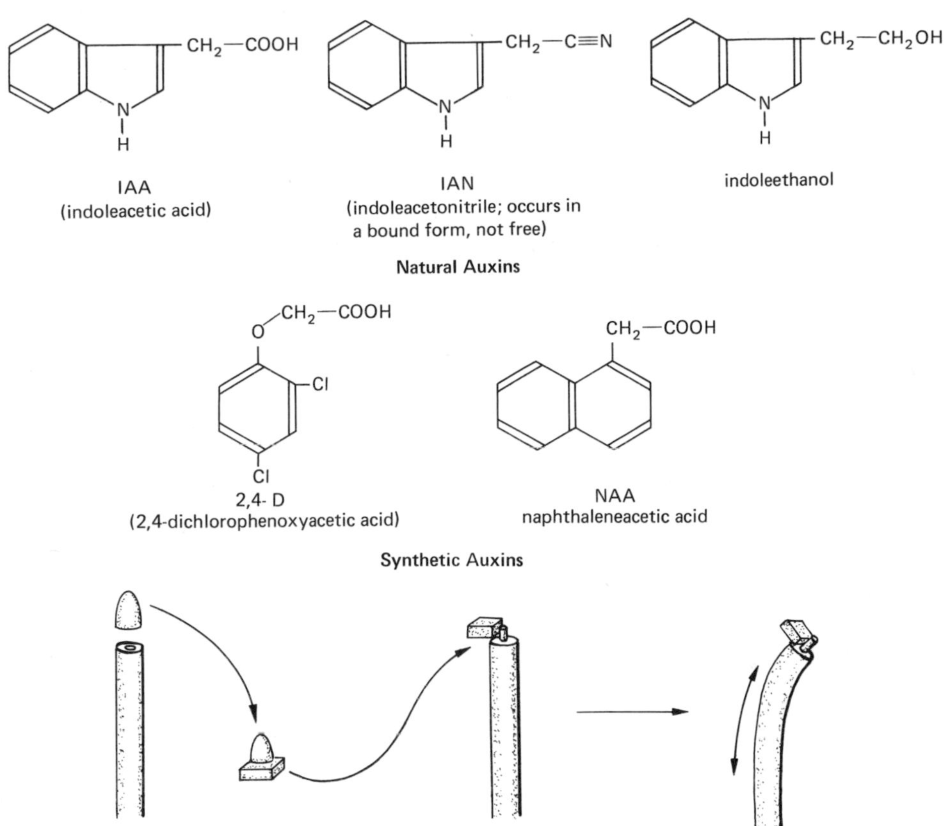

Figure 16-7. Some natural and synthetic auxins, and a pictorial representation of an experiment by Went demonstrating the presence of a substance (the hormone, IAA) that could diffuse from the coleoptile tip into an agar block and cause extension growth of the coleoptile.

both stimulation (principally cell elongation) and inhibition of growth, and the same cell or structure may exhibit opposite responses, depending on the concentration of IAA. Furthermore, different tissues respond to very large differences in concentration—roots are stimulated by concentrations several orders of magnitude below those that stimulate shoots. A generalization of these facts, as they are often presented, is shown in Figure 16-8.

As a result of these patterns of activity, the gradient of IAA found in plants may exert a variety of developmental effects, from suppression of lateral buds or shoots to stimulation of shoot or root elongation, in different parts of the plant. In addition, auxin acting alone or in concert with other hormones stimulates or inhibits a variety of other events, from individual enzyme reactions to cell division and organ formation. Thus, its effects are many and diverse, and one of the greatest problems in plant physiology is to develop an understanding of how one small, relatively simple molecule like IAA can have so many different kinds of effects, and how this apparent jumble of miscellaneous effects is coordinated into ordered control of growth and development.

One of the great problems with hormones is their assay. They are usually present in minute amounts and are very difficult to detect or characterize chemically. Many biological assays have been developed for auxins. In one, the degree of curvature in oat

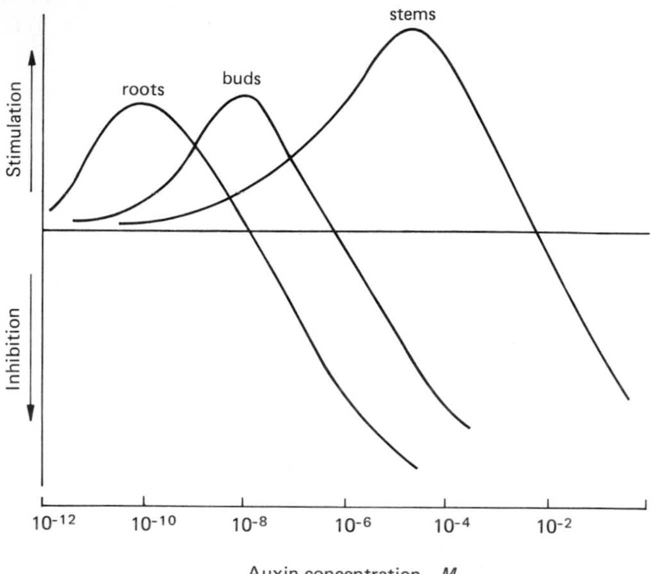

Figure 16-8. Differential action of IAA on roots, buds, and stems as envisioned by K. V. Thimann.

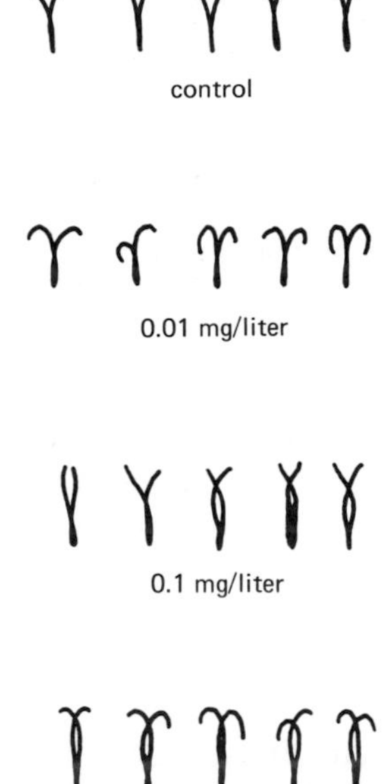

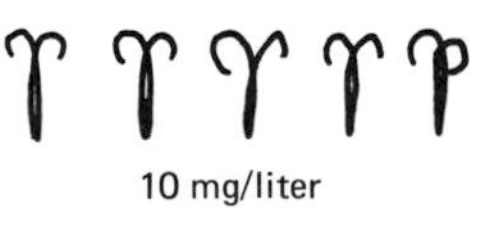

Figure 16-9. Split-pea stem test for IAA.
Short sections of stem from the third internode of pea plants are split with a razor blade and incubated in a dish containing the hormone solution. The split ends curl, the amount of curvature being related to the IAA concentration as shown.

coleoptiles is measured following asymmetric application of agar blocks containing extracts or diffusates of plants. This makes use of the phenomenon observed by Darwin and employed by Went in his experiments. Other assays use the stimulating effect of auxin on the elongation of pieces of coleoptile or of stem sections (usually from the epicotyls of etiolated pea seedlings) or the curling of the ends of split sections of pea stems (Figure 16-9). Auxins can be separated by chromatography and detected either by bioassays of extracts from pieces of the chromatogram or by chemical reactions. The chromatographic procedures are much favored because bioassays are frequently non-specific and do not distinguish between IAA and other substances that have auxin or auxinlike activity.

Gibberellins. These compounds were discovered when it was shown in Japan that extracts of a disease-producing fungus (*Gibberella fujikuroi*), which attacks rice plants, would duplicate thc symptoms of the disease. The characteristic of the disease is excessive elongation of the internodes leading to lodging or falling over of the shoots, and the main action of gibberellins is to promote elongation. Many dwarf or bushy plants (for example, dwarf corn mutants, "bushy" or dwarf garden peas and beans) grow tall when supplied with minute amounts of gibberellins, as shown in Figure 16-10. Gibberellins are also involved in flowering and the **bolting** that precedes it in rosette-type plants, in certain phases of seed germination, in the breaking of dormancy, and in several formative effects. They also interact in their effects with other hormones. Unlike auxins, gibberellins seem to move freely about the plant, and their transport and distribution patterns are not polar like those of auxins.

There are now many gibberellins known; all have the same basic structure as gibberellic acid (GA_3, Figure 16-11) but differ in the nature of various side chains or substitutions. Different gibberellins are found in different plants and, although many of them produce similar results, a number of species- or compound-specific effects are known.

Cytokinins. For many years it has been known that soluble substances from various plant sources are necessary for cell division in cultures of isolated cells or plant parts. In the 1950s work in the laboratory of F. Skoog led to the isolation of a nonnatural degradation product of animal DNA that has this effect. The compound, **6-furfuryl-aminopurine,** was named **kinetin.** Other synthetic compounds, such as 6-benzylamino-purine, have been found to stimulate cell division, and such compounds have been called **kinins.** However, this term has other connotations in animal physiology, and the accepted term for this class of hormones is now **cytokinin** (a hormone stimulating cytokinesis). It appears likely that many naturally occurring cytokinins exist; only a few have been isolated and identified. One of these is **zeatin** (extracted from endosperm of corn, *Zea mays*), which appears to be widely distributed among plants. Its structure, together with those of some synthetic cytokinins, is shown in Figure 16-12.

Cytokinins do not appear to be as mobile in plants as auxins and gibberellins. They mediate a wide range of responses besides stimulating cell division. In the presence of auxin, various concentrations of kinetin cause either root, shoot, or completely disorganized callus growth in cultured tobacco pith tissue (Figure 16-13). Perhaps related to their effect on cell division, cytokinins also prevent the onset of senescence in aging tissues. Many of the bioassays that have been developed make use of this aspect of their activity; when applied to detached leaves, they prevent chlorophyll loss, and this can

A

B

Figure 16-10 (opposite). The effect of gibberellic acid on the growth of (**A**) dwarf peas and (**B**) beans.
In each photograph, the plants on the right were treated with a drop of gibberellic acid solution soon after they sprouted. The gibberellic acid-treated plants have the same number of internodes and leaves, but their stems are much elongated.

Figure 16-11. Gibberellic acid. Other gibberellins have various side chains on the same nucleus.

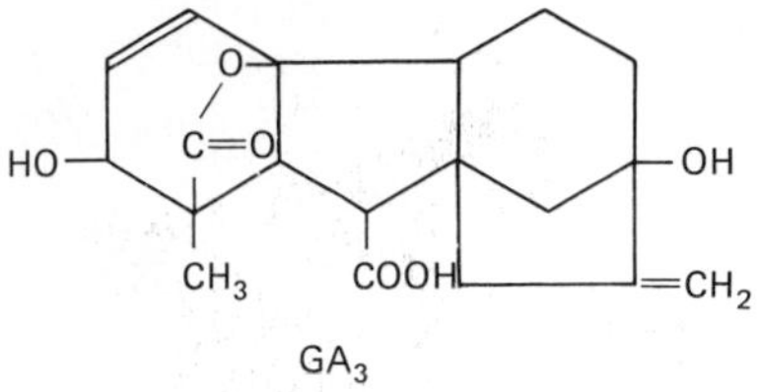

GA$_3$

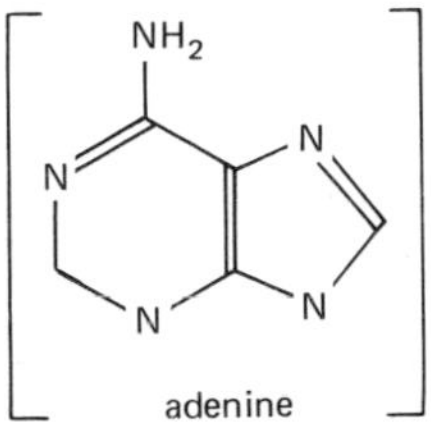

adenine

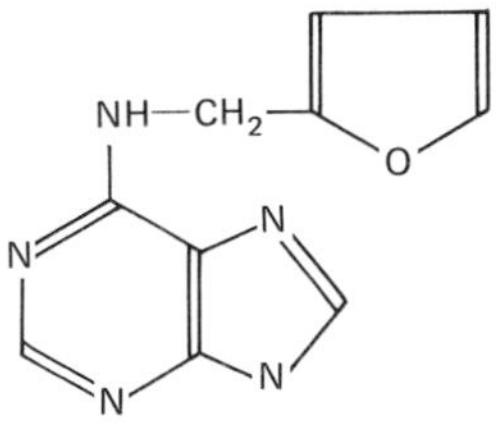

kinetin
(6 furfurylamino purine)

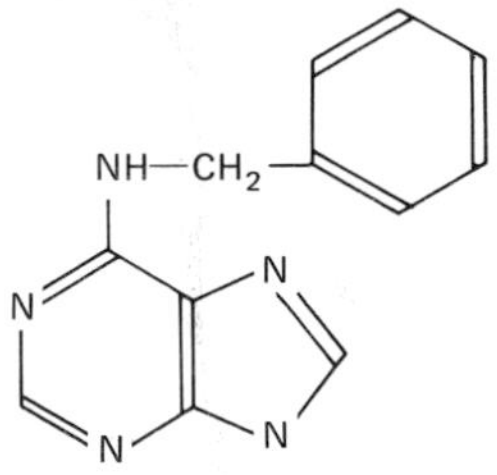

6-benzylamino purine

Synthetic Cytokinins

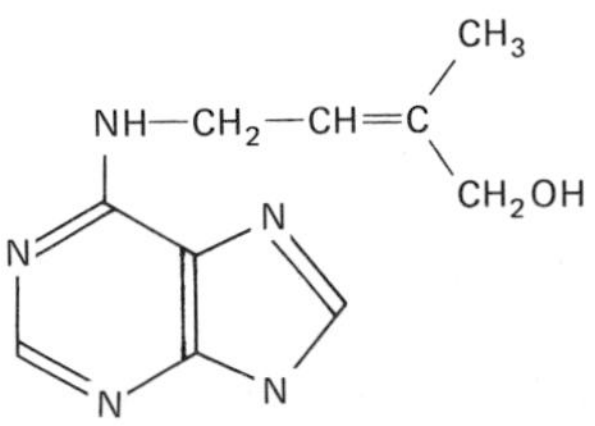

zeatin

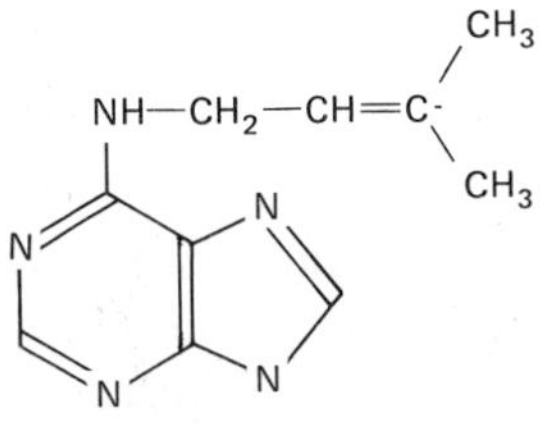

isopentenyl adenine

Natural Cytokinins

Figure 16-12. Some cytokinins. They are all adenine derivatives.

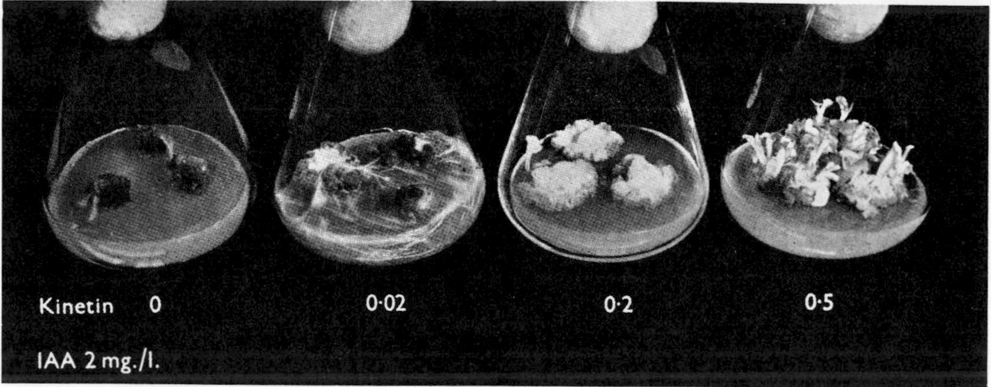

Figure 16-13. The effect of different concentrations of kinetin on growth and development of callus tissue from tobacco pith in the presence of 2 mg/liter IAA. With no kinetin, little growth occurs. At low levels, roots develop. At an intermediate level, unorganized growth continues. At a higher level buds and shoots develop. [From F. Skoog and C. O. Miller: In *Symp. Soc. Exp. Biol.*, **11**:118–31 (1957). Used with permission.]

easily be detected and measured. Other bioassays depend on their growth-stimulating effect on cultured pieces of tissue or callus from various plant sources.

Ethylene. Ethylene is a simple gaseous compound that elicits a wide range of responses in plants (Figure 16-14). It is produced in leaves, where it acts powerfully in inducing or promoting senescence, and in fruits, where it greatly affects the ripening process. Ethylene also causes or mimics many of the formative effects of auxin. Its synthesis is strongly stimulated by auxin, and it has been suggested that many of the formative effects of auxin, particularly in roots, are really due to ethylene production resulting from auxin stimulus. It may also affect or interfere with normal responses to auxin, so there is a possibility for complex interactions between these two growth substances.

Abscisic Acid. This compound, in spite of its name, seems to be more involved in the maintenance of dormancy than in the abscission of leaves. It was independently discovered by the British physiologist P. F. Wareing and his group and by an American group under F. T. Addicott, who called it **dormin** and **abscisin II,** respectively. Abscisic acid **(ABA),** as it is now called, appears to induce dormancy in perennial plants and trees, and it causes or maintains dormancy in many seeds. ABA appears to counteract the effect of gibberellin in some plants, and it is somewhat similar to gibberellic acid in structure (Figure 16-15). It also induces stomatal closure.

Hypothetical Growth Substances. Many different kinds of hormones and inhibitors have been proposed or appear to be required in various aspects of plant development, without having been isolated or their existence formally proved. Many experiments indicate the existence of a **flowering hormone,** or **florigen,** but none has yet been isolated. Analysis of developing and dormant buds of trees usually shows a range of substances that can be separated by chromatography and act either as stimulators or inhibitors in various bioassays. A wide variety of naturally occurring chemicals act as growth inhibitors *in situ* or on other plants after extraction. Whether these chemicals can

Figure 16-14. Ethylene, and an illustration of the epinasty resulting from gassing the center part of a tomato plant with 100 ppm ethylene.

(or should) all be classified as hormones is an unprofitable question. It is even difficult to determine whether all or any of these play a real part in normal development. However, it seems very likely that such complex and precise events as flowering do indeed require or involve the presence and activity of growth-promoting and -inhibiting substances. It therefore seems possible that such concepts as flowering hormones or root-growth hormones may refer either to single substances or groups of interacting substances. Much basic work still needs to be done in this area of plant physiology.

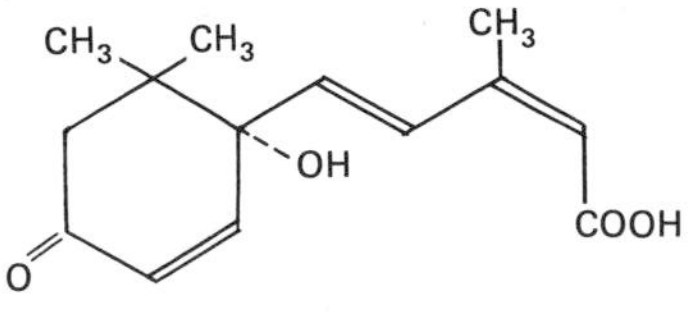

ABA
(abscisic acid)

Figure 16-15. Abscisic acid, ABA.

Environmental Controls. Many environmental or external stimuli affect plant development. Chemical substances produced by other organisms may be involved, but the range of factors usually considered is mainly physical: light, temperature, nutrients, and so on. These are superimposed upon and often override the genetic and organismal controls of the organism. Environmental stimuli often initiate events, as might be expected, since successful growth and reproduction require effective coordination with the seasons.

The function of an environmental control mechanism requires three steps: (1) It must be perceived or measured by the plant. (2) There must be a mechanism whereby the plant reacts to the stimulus. (3) Some degree of permanence must be achieved during which reaction to the stimulus can take place. Our understanding of how plants react to environmental stimuli is far from complete. For example, we have some idea how plants perceive and measure light periods, but we do not yet understand their mechanism of reaction. Many stimuli have permanent or semipermanent effects; that is, the plant continues to react long after the stimulus has ceased to act. Just how permanence is achieved is not yet known. Nor do we understand how plants perceive or react to temperature fluctuations, which have powerful physiological effects in certain species.

The main environmental stimuli that affect plant development are as follows:

1. Light—intensity, quality (color), duration, periodicity.
2. Temperature—absolute and periodicity.
3. Gravity.
4. Sound.
5. Magnetic field.
6. Electromagnetic radiations.
7. Humidity.
8. Nutrients.
9. Mechanical (for example, wind).

The first three are the factors of major importance. Experiments that appear to demonstrate that sound waves affect plant development have been reported from time to time, but they are not always convincing. Some phenomena appear to indicate that plants can respond in their growth and development to various electromagnetic fields, including radar and the earth's magnetic field. However, data are scant and their interpretation is equivocal. Humidity and nutrients are included because not only are all other responses dependent on the nutritional state of the plant, but also some developmental effects appear to be initiated by changes in the nutritional or water status of the plant.

Level of Action of Controls

The Genetic Level. The idea that all cells are **totipotent** (that is, that each cell carries all the genetic information for the whole plant) was put forward many years ago by the German physiologist G. Haberlandt. Many recent experiments bear out his foresight. For example, F. C. Steward, at Cornell University, has shown that single cells of carrot phloem, properly cultured, can grow into whole new carrot plants, complete with flowers and roots. The secret appears to be in the supply of the proper nutrients and growth

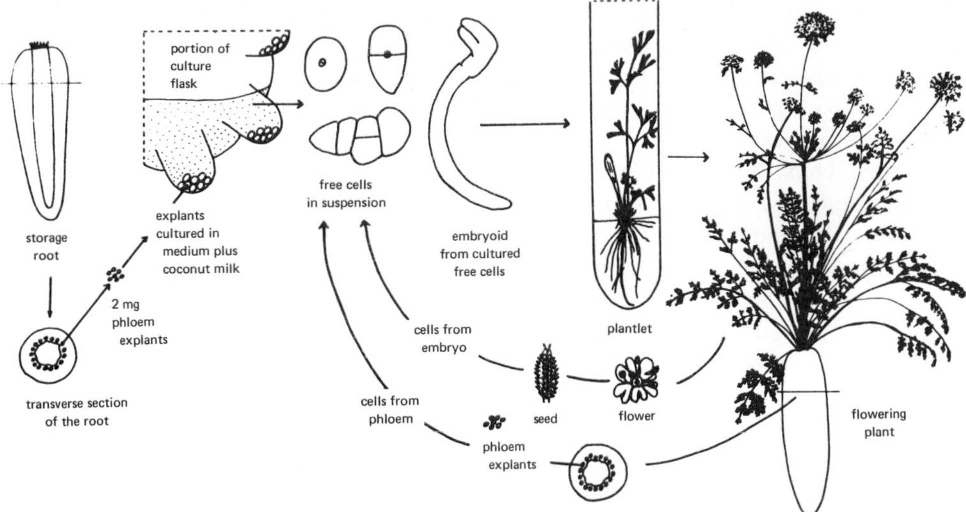

Figure 16-16. Diagrammatic representation of the cycle of growth of the carrot plant; successive cycles of growth are linked through free cultured cells derived from phloem or from the embryo. [From F. C. Steward, M. O. Mapes, A. E. Kent, and R. D. Holsten: *Science*, **143:**20–27 (1964). Copyright 1964 by the American Association for the Advancement of Science. Used with permission. Photograph courtesy F. C. Steward.]

substances to stimulate cell division and growth and in certain external stimuli (that is, a solid support medium properly oriented in a gravitational field) that permit the establishment of polarity, hence, differentiation. The sequence of events in Steward's experiments is shown in Figure 16-16. This sort of experiment shows that the information for all the developmental events in the life of the whole plant is present in every cell, and it emphasizes the importance of genetic mechanisms for uncovering or selecting the right information at the right time.

We have already commented on the operon theory of Jacob and Monod, and how a variety of small, possibly metabolic, chemical species may act as activators or corepressors. Another type of control mechanism has recently been suggested. As was mentioned earlier (Figure 16-12) cytokinins are derivatives of the purine adenine, and they may form ribosides structurally similar to the ribosides of RNA. Cytokinin ribosides have been found in very small amounts in tRNA. At first it was thought that this provided the key for cytokinin activity in cell division. More recent experiments have shown that this is not correct and that the activity of cytokinins is not directly related to their presence in RNA. However, there are other possible ways in which the presence of cytokinins in RNA may be involved indirectly in the control of enzyme synthesis. These will be explored in Chapter 23.

One of the interesting problems of the genetic level of developmental control is the acquisition of permanence or the "fixation" of information. An example of this is **plagiotropism,** the fixed angle of growth of the branches of many trees (for example, pine or spruce). Plagiotropism may be temporary and may be released when the terminal bud is cut off. This is commonly observed in spruce trees, where adjacent branches rapidly assume an upright position when the terminal bud dies or is cut off. However, in certain plants plagiotropism is permanent and is retained even if a twig is removed from a plant, rooted, and subsequently planted as an independent individual. In such situations as this, evidently certain information, once "learned" (that is, programmed or released from repression), cannot be "forgotten" or rerepresseds Differences in habit of growth among species may be due to degrees of permanency in the retention of specific programs. Thus, the rigid plagiotropism of fir trees may be due not to a continued supply of information from the stem tip (the tips of lower branches, even though far away from the leader, are still under complete control) but to permanence of the original information or program. Other trees may show strong apical dominance initially but, as the tree grows taller and the leader is further away, dominance is lost. This allows the bushy, rounded crown typical of many deciduous trees. Here the imprint is not permanent, and as a result lower limbs are released from their initial program.

The mechanisms of permanence are not understood, but are obviously important. Some types of environmental stimuli appear to result in modifications that are inherited for several generations. The American physiologist H. Highkin showed that peas grown under low temperature for several generations produce successively smaller plants. When released from low temperature stress, the plants gradually revert to their former size over several generations. Genetic explanations for this degree of semipermanence are lacking.

Biochemical Level. The idea that hormones might affect growth by directly affecting specific enzyme activities or biochemical pathways has attracted physiologists and biochemists for a long time. However, few clear examples have developed, in spite of much intensive research. The American physiologist I. Sarkissian has presented evidence to show that IAA acts directly, perhaps allosterically, to activate the citrate-condensing enzyme of the Krebs cycle. Since this is a key enzyme in energy metabolism, it could be an important regulating mechanism. Unfortunately, further evidence of how this might work is lacking.

IAA has also been found to affect certain other enzymes of metabolism as well as photosynthetic rate, but again we have no knowledge of its mechanism of action. One well-authenticated mechanism is the stimulation of α-amylase synthesis in germinating cereal seeds by gibberellin, first noted independently by H. Yomo in Japan and L. G. Paleg in Australia. The American physiologist J. E. Varner has shown, using radioactive amino acids and also inhibitors of RNA synthesis, that gibberellins act to release a formerly repressed operon so that it becomes active, and the enzyme α-amylase is synthesized. This enzyme is involved in the breakdown of starch reserves during the germination of seeds.

Other biochemical actions of specific hormones or inhibitors will doubtless be found. However, it is difficult to see how all hormone action could be understood by action only at the genetic and biochemical level.

Cellular Level. A bewildering number of control mechanisms operate at the cellular level. Although it seems likely that many of these are the direct consequence of control of some biochemical system, most of them now cannot be understood in these terms. We shall consider a few examples here, in order to provide an overview of the variety of possible effects.

Cell Division. Cell division appears to be very much under the control of hormones. In the absence of kinetin, auxin causes cell enlargement in cultured tissue. If kinetin is present, then cell division takes place. However, even in the presence of cytokinin, excess auxin suppresses cell division and growth. Thus, the balance of the hormones is important in regulating growth by cell enlargement or by cell division. However, other agencies can modify this response. Calcium ions prevent the auxin-induced expansion of cells, presumably by combining with pectate in the walls and rendering them less plastic. The addition of calcium to a medium that favors cell enlargement causes a switch to cell division in cultured tissues. The calcium thus modifies the response of the cell to the hormone. Growth still continues, but the pattern of growth is changed.

CELL ENLARGEMENT. The enlargement of cells, as we have seen, is under hormonal control. The original observations on grass coleoptiles that led to the discovery of auxin are all elongation responses to auxin, which is produced in the tip of the coleoptile. Cell enlargement requires an increase in the cell's contents, and it was once thought that IAA activated a metabolic water pump that literally forced the cell to expand from the internal pressure generated by water uptake. However, it is now known that auxin acts by causing a relaxation of the cell wall structure so that plastic (that is, irreversible or nonelastic) growth is possible. Cell enlargement may be a fundamental process: plants or tissues in which cell division is inhibited (for example, by the exclusion of calcium, or by intense gamma irradiation) nevertheless continue to grow by cell enlargement. Thus the stimulus to grow provided by the hormone is obeyed even though one of the primary processes involved is inoperative.

POLARIZATION. Polarization in an organism results from inequality and first becomes apparent at the subcellular level. Many different stimuli may impose polarization upon a cell. Egg cells in the ovary are highly polarized because of their position in a polarized structure. That is, the chemical (and perhaps mechanical) stimuli of the ovary are applied preferentially at one end or the other of the egg, which thus lies within a gradient of substances originating in the ovary.

Other environmental stimuli affect free-floating cells. The eggs of the brown marine alga *Fucus* are free-floating in the sea, and after fertilization an essentially spherical and unpolarized zygote is formed. This begins its development by dividing into two unequal cells; the larger will eventually form the thallus and the smaller the rhizoid (see Figure 16-17). Unequal cell division of this sort inevitably results in polarization; however, it also requires polarization before it can take place. In *Fucus* zygotes, polarization is apparently determined entirely by environmental stimuli, which include tactile response, pH, light (both its presence and its plane of polarization), temperature, oxygen content, auxins, and the presence of other zygotes. The zygote does not necessarily react equally, or always, to all of these stimuli, but all of the stimuli are capable of establishing polarity.

Similarily, growing cultured free cells of higher plants do not differentiate readily if kept in liquid culture where they are unpolarized. However, if the free cells are placed on a suitable nutrient agar surface, they are individually polarized with respect to the solid surface and the nutrients it contains (which thus form gradients) and differentiation begins. Evidently many or all of the cells in an organized tissue are polarized with respect to their position in the organism. Various gradients of nutrients or growth substances are thus possible which contribute to the polarity and to the continued organization of the tissues.

CELL MATURATION. Much evidence suggests that this process is affected by hormones as well as other factors. Exactly what gradients of growth substances and

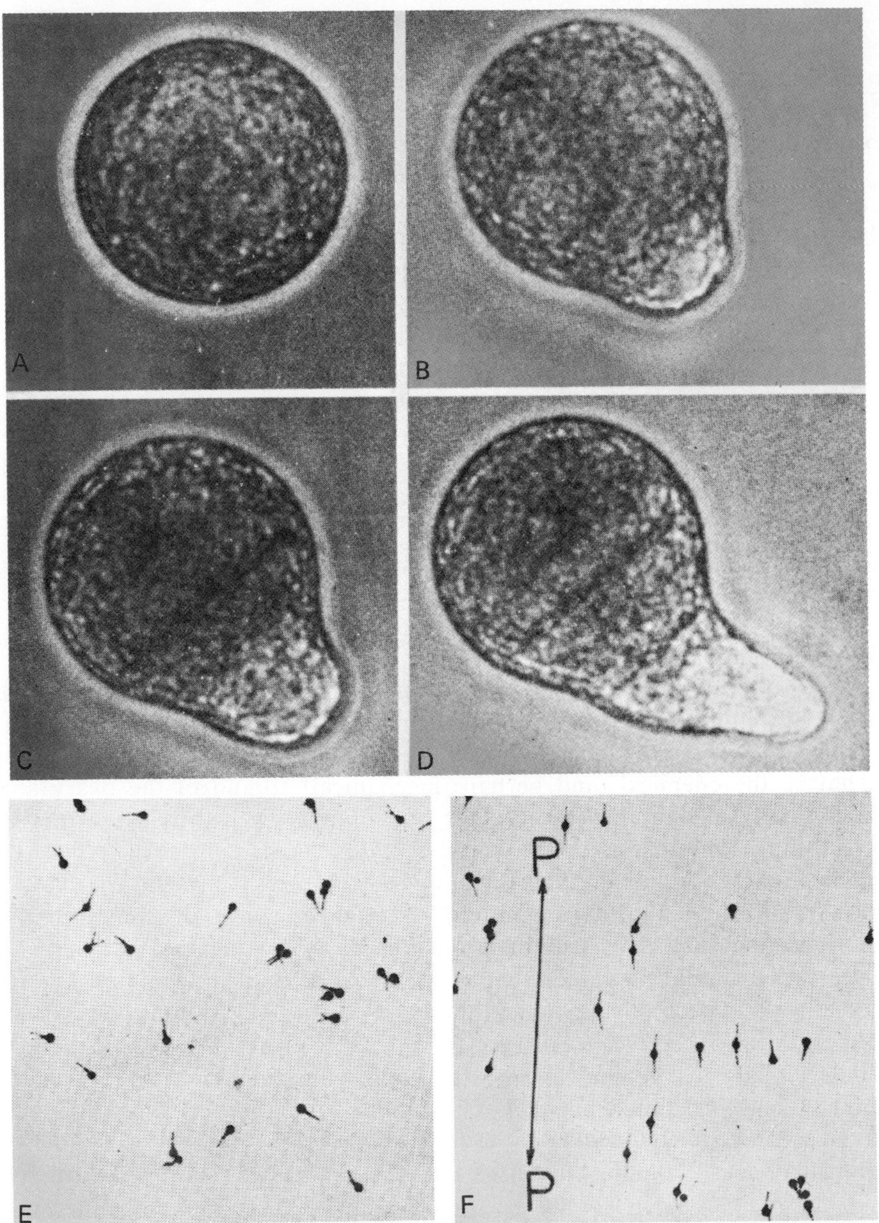

Figure 16-17. Polarity in *Fucus* zygotes.
A to **D,** development of a *F. vesiculosus* zygote at 4, 16, 18, and 26 hr from fertilization. Unequal cell division into thallus and rhizoid cells has taken place in **C** (18 hr). **E** and **F,** 4-day-old *F. furcatus* embryos in nonpolarized (E) and plane polarized (F) light. The arrow P ⟷ P shows the plane of polarization. [A to D from L. Jaffe: *Adv. Morphol.,* **7:**295–328 (1968). Photos by Dr. B. Bouck. E and F from L. Jaffe: *Science,* **123:**1081–82 (1956). Used with permission.]

nutrients are involved in the maturation of cells as they age in various tissues of the root or stem tip is not known. The control mechanisms must be precise, since adjacent cells mature into phloem, cambium, xylem, pith, cortex, and so on, in a very precise pattern.

Some indications of how this might occur can be found in the work of the American physiologist R. H. Wetmore and his collaborators. He cut approximately cubic pieces of callus tissue grown from lilac phloem, and implanted an excised bud into the callus. After a time, examination revealed that patches of conducting tissue had differentiated in the callus block beneath the bud, evidently induced by the bud. If the callus was placed on nutrient agar containing sucrose and a drop of IAA solution was placed on the callus (instead of the excised bud), differentiation of conducting elements occurred in the callus. The location of different tissues was affected by the concentration of IAA in the drop, but the nature of differentiation was controlled by the concentration of sugar in the agar. Low sugar caused the production of xylem, high sugar induced phloem, and intermediate concentrations induced both xylem and phloem with a cambial layer between. Many similar systems, some of which will be discussed later, suggest that the interaction of nutrient and hormone gradients is important in establishing patterns of cell differentiation.

Organizational Level. Organization is the result of polarized cell division and cell specialization. However the organization of tissues into **organs** transcends these, because differential patterns of growth must be superimposed on individual cell reactions to achieve control of shape, size, and form. Again, as in cell and tissue organization, gradients of growth substances are probably involved. However, their influences are felt over much larger distances, and translocation of growth substances becomes important.

The translocation of auxin has been extensively studied. It was early recognized that auxin is formed at the tip of the grass coleoptiles and moves downward. If it is introduced at the bottom, it does not move upward. This phenomenon has been found to be quite general in plants—auxin usually moves from the tip to the base of a plant (**basipetal** translocation), but not from the base to the tip (**acropetal** movement). This polarity of transport is maintained even if the plant tissue is excised or placed upside down. One consequence is illustrated in Figure 16-18. A piece of willow stem grows roots

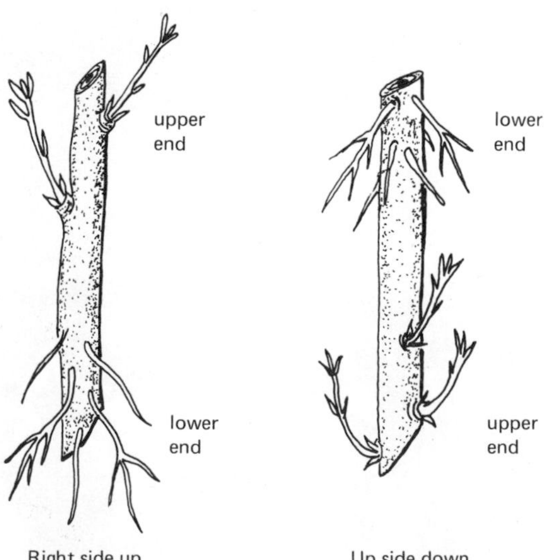

Figure 16-18. Polarity of root and shoot formation in a cutting of willow. This polarity is maintained because of polar (basipetal) auxin transport, regardless of the direction of operation of the force of gravity.

at the bottom end and shoot buds at the upper end as a result of the auxin gradient that develops in the shoot. But the same gradient is formed even when the shoot is held upside down. Roots are formed on the *physiological* lower end and shoots on the *physiological* upper end, regardless of the orientation of the stem.

Auxin translocation is not absolutely basipetal; instances of acropetal transport are known. Also, even basipetal translocation does not require the exclusive downward movement of every molecule. Only a small preferential downward movement in each cell of a tissue composed of columns of cells will result in an almost complete downward translocation in the whole tissue. Each cell in the column would act in a small way to cause polarity, and their activities would be cumulative.

Other plant growth substances do not undergo polar translocation. Gibberellins are quite rapidly moved about the plant, apparently without restriction. Cytokinins seem to move relatively slowly. The movement of hormones requires the expenditure of metabolic energy derived utlimately from respiration. It appears to be independent of the simultaneous movement of other substances, because these may move at different velocities or in different directions at the same time in the same tissue. Factors governing the translocation of hormones are not well known.

Whenever morphological events take place, hormones seem to be involved. Conversely, if hormones are applied to plants, various morphological or developmental events ensue (see Figure 16-19). The effects of auxins in causing root initiation are well known and are used commercially for rooting cuttings (see Figure 16-20). Other effects

Figure 16-19. Auxin effects on a tomato plant.
Note the adventitious roots and epinasty caused by the application of 2% α-naphthalenacetic acid in lanolin paste. [From P. W. Zimmerman and F. Wilcoxon: *Contrib. Boyce Thompson Inst.*, **7**:209–29 (1935). Used with permission.]

Figure 16-20. Effect of auxins in root initiation.

A. Influence of auxin on root initiation in bean (*Phaseolus vulgaris*) seedlings. The seedlings were cut off at the soil surface, placed in water (control) or IAA solution for 48 hr, then in nutrient solution for 1 week. The short stem of the control is accidental and not related to the auxin treatment. [Courtesy Miss Gail Bebee.]

B. Photographs of a historic experiment conducted by W. C. Cooper in April, 1935, the first attempt to root cuttings of horticulturally important plants (lemon) using auxin (0.05% in lanolin paste). [Photographs courtesy Dr. Cooper, U.S.D.A., Orlando, Florida.]

of auxins include **epinasty** (deformation by bending downward) and various other growth anomalies, abnormalities of fruit formation, and many others. The phenomenon of apical dominance and various responses of plants to environmental factors such as light **(phototropism)**, gravity **(geotropism)**, or touch **(thigmotropism)** all invoice auxin-mediated responses. In summary, almost every morphogenetic and organizational pattern in plants is affected by, or responds to, plant growth regulators. How these are coordinated is one of the great questions in plant physiology still to be answered.

Distribution, Formation, Breakdown, and Compartmentation of Regulators. Hormones have different effects at different concentrations, and concentration gradients of hormones are important in establishing polarity and organization in the development of a plant. The absolute quantities of hormones in a tissue are thus very important. But it must be recognized that hormone levels may be established in several ways: by synthesis, by transport to or from a site, or by destruction. Furthermore, hormones and other regulators are only active if they are accessible to their site of action. High concentrations of hormones may be kept inactive by being located or stored in compartments (for example, the vacuole) apart from their active site. They may also be chemically inactivated without being destroyed by being complexed with some other molecule. Thus, hormone effects may be mediated by the metabolism that makes or destroys hormones, by their transport about the plant or from place to place within the cell, or by metabolism that chemically masks or unmasks their activity.

The understanding of hormone control of growth and development thus requires much more than the knowledge that hormones are being elaborated or transported. Even the knowledge of the amounts of hormone present in a tissue may be insufficient if it is compartmented or otherwise held in an inactive form or site. These problems are only now being clearly recognized. Doubtless much of the present confusion and contradiction in the literature on growth and development will be resolved as we achieve a better understanding of these mechanisms of hormone control.

Initiation of Events. Development in plants is not a smooth, continuous process but is marked by a series of events or crises. Some of these are not of great consequence and may be often repeated, such as branch, bud, lateral root, or leaf formation. Others are of great importance, perhaps once-only events, such as germination or flowering. The initiation of these events is under controls of various kinds, some of which have been worked out in detail and some of which are still quite obscure.

Perhaps the most-studied and best-understood event in the development of plants is flowering. The timing of flowering is critical because it must be early enough in the season for fruit and seed development to take place yet late enough for adequate growth and maturation of the plant. Biennial plants must flower in their second year, not their first. Experiment has shown that flowering is frequently controlled by day length (or more strictly, night length). Plants measure the length of night, and flowering takes place at the proper season of the year as determined by the night length characteristic of that season. In addition, many plants must undergo a cold treatment (either in the seed or seedling stage or in the form of some storage organ as in biennials) before flowering can take place. This prevents the plant from flowering prematurely, that is, in the period of long nights in the fall when it should flower in the spring. The mechanism by which the plant perceives, measures, and reacts to light and to temperature will be discussed in Chapters 19 and 20.

Figure 16-21. Morning glory flowers (**A**) open in the morning (photo taken in the author's garden at 9 AM) and close again (**B**) in late afternoon (photo taken at 6 PM the same day).

Other events whose initiation is mediated by external, environmental factors include germination and leaf abscission. Germination starts with the addition or absorption of water, but other factors such as low-temperature treatment may be necessary to put the seed into the proper physiological state to react to this stimulus. Leaf abscission, like dormancy, is largely initiated by external or environmental factors, but it can be brought about or influenced by physiological factors as well.

The initiation of bud or lateral root formation is largely an internally controlled process, apparently determined by levels or gradients of growth substances in the meristem or root. External conditions do not appear to have much effect on the initiation of these structures, although they have a strong influence on their subsequent growth patterns.

Rhythmic Behavior. One of the most fascinating aspects of plant behavior is that it is often periodic or rhythmic. For example, some flowers open in the morning and close at night (morning glory, see Figure 16-21), and many plants lower their leaves at night (sleep movements or **nyctinasty**). Rhythmic changes in many physiological functions, ranging from metabolic processes like photosynthesis or respiration to physiological states like the ability to flower, are also found. Many of these rhythms are related to rhythmic environmental stimuli, such as the alternation of day and night, but some develop or continue if the plant is put in an absolutely constant environment. It appears at present that organisms have an innate clock that measures time and controls rhythmic behavior. However, the nature of the clock and how it controls plant behavior are still great mysteries.

All the processes mentioned in this chapter are of sufficient importance to warrant further careful study, and they will be considered separately in the chapters of Section IV that follow. This section will form a "life history" of plants, from germination to death, which will be followed in Chapter 23 by a brief summary of major plant growth substances and their mode of action.

Additional Reading

Evans, G. C.: *The Quantitative Analysis of Plant Growth.* University of California Press, Berkeley, Calif., 1972.

Halperin, W.: Morphogenesis in cell cultures. *Ann. Rev. Plant Physiol.,* **20:**395–418 (1969).

Lockhart, J. A.: The analysis of interactions of physical and chemical factors on plant growth. *Ann. Rev. Plant Physiol.,* **16:**37–52 (1965).

Richards, F. J.: The quantitative analysis of growth. In F. C. Steward (ed.): *Plant Physiology: A Treatise,* Vol. VA. Academic Press, New York, 1969, pp. 3–76.

Stange, L.: Plant cell differentiation. *Ann. Rev. Plant Physiol.,* **16:**119–40 (1965).

Stern, H.: The regulation of cell division. *Ann. Rev. Plant Physiol.,* **17:**345–78 (1966).

Thimann, K. V.: Plant growth substances; past, present, and future. *Ann. Rev. Plant Physiol.,* **14:**1–8 (1963).

Thornley, J. H. M.: *Mathematical Models in Plant Physiology.* Academic Press, New York, 1976.

General References for Section IV

Bieleski, R. L., A. R. Ferguson, and **M. M. Cresswell:** *Mechanisms of Regulation of Plant Growth.* Royal Society of New Zealand, Bulletin 12, Wellington, New Zealand, 1974.

Galston, A. W. and **P. J. Davies:** *Control Mechanisms in Plant Development.* Prentice-Hall, Englewood Cliffs, N.J., 1970.

Graham, C. F., and **P. F. Wareing:** *The Developmental Biology of Plants and Animals.* W. B. Saunders Co., Philadedphia, 1976.

Laetsch, W. M., and **R. E. Cleland** (eds.): *Papers on Plant Growth and Development.* Little, Brown and Co., Boston, 1967.

O'Brien, T. P., and **M. E. McCully:** *Plant Structure and Development.* Macmillan Publishing Co., Inc., New York, 1969.

Steeves, T. A., and **I. M. Sussex:** *Patterns in Plant Development.* Prentice-Hall, Englewood Cliffs, N.J., 1972.

Steward, F. C. (ed.): *Plant Physiology: A Treatise,* Vols. VA and B, VIA, B, and C. Academic Press, New York, 1969–73.

Steward, F. C.: *Growth and Organization in Plants.* Addison Wesley, Reading, Mass., 1968.

Steward, F. C., and **A. D. Krikorian:** *Plants, Chemicals and Growth.* Academic Press, New York, 1971.

Torrey, J. G.: *Development in Flowering Plants.* Macmillan Publishing Co., Inc., New York, 1967.

Wareing, P. F., and **I. J. D. Phillips:** *The Control of Growth and Differentiation in Plants.* Pergamon Press, Oxford, England, 1970.

Wilkins, M. B. (ed.): *Physiology of Plant Growth and Development.* McGraw-Hill, New York, 1969.

The books listed above serve as general reference material for all of Section IV, and this list will not be repeated after each chapter.

17 Sexual Reproduction in Higher Plants

The Gametophyte Generation

We shall begin the description of the reproductive process with the formation and development of the male and female gametophyte generations. The subjects of flower initiation and development are separate stories and will be told in the following two chapters. Patterns of flowering and reproduction differ very widely in higher plants; because very little is known of the underlying causes or mechanisms of the processes involved, this treatment must of necessity be rather general.

Carpel and Egg. The carpels, like other flower parts, appear to be derived from modified leaflike structures. However, although they are derived from the same meristem, carpels differ greatly from leaves. Evidently new and different forces are in action that modify the expression of genes already active or bring new genes into play.

Ovule development takes place within the carpel. One or many ovules form within their layered nucellus and integuments (see Figure 17-1). Within the meristematic region of the ovary an enlarged central cell, in some way different from those around it, undergoes meiosis, forming four haploid daughter cells, the beginning of the gametophyte generation. Only one of these cells develops; the other three abort and ultimately disintegrate. The nucleus in the remaining haploid cell undergoes mitosis successively until the cell (now called the **embryo sac**) holds eight nuclei, including an egg-cell, two synergids, two polar nuclei, and three antipodal cells (Figure 17-1). This eight-nucleate structure is the full extent of the female gametophyte.

Factors that initiate and control the events leading to gametophyte formation are unclear. Just what stimulus is required to cause one cell to undergo reduction division—essentially, what tells the cell to do this—is unknown. Why the gametophyte should develop from one cell only of the four haploid daughter cells and why it should develop only to the eight-nucleate stage are also mysteries. Undoubtedly regulatory substances from the ovule are involved because the female gametophyte is highly polarized in its relation to the ovary, the egg cell being nearest to the micropyle and the antipodal cells opposite. This polarization is important because it establishes the polarity of the embryo that will develop in the embryo sac.

Anther and Pollen. The stamens develop in the apical meristem in many different patterns, groupings, and shapes in different plants. However, the general pattern of

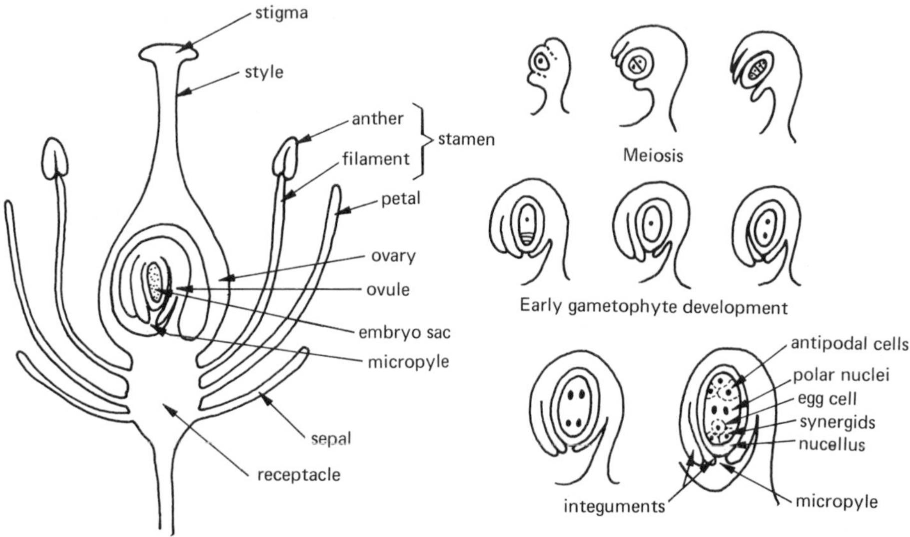

Figure 17-1. Diagram of a flower, and stages in ovule development. [Adapted from J. G. Torrey: *Development in Flowering Plants*. Macmillan Publishing Co., Inc., New York, 1967.]

pollen development is quite similar. The anther is often four-lobed, each lobe containing a central core of sporogenous tissue—tissue that will give rise to a number of microspore mother cells. These cells are diploid, and each package or group of them in the anther is surrounded by a layer of specialized cells called the **tapetum.** This layer of cells may be closely associated or concerned with the events leading to meiosis in the microspore mother cells and to the formation of pollen grains. Cells of the tapetum undergo a substantial increase in RNA and sugar content, then they break down shortly before meiosis takes place. It has been suggested that both DNA and carbohydrate material, or breakdown products of these polymers, are transferred to the developing spore mother cells. However, it seems unlikely that these compounds provide anything but nutrition. The stimulus that leads to meiosis most likely comes from hormones or growth factors released simultaneously during tapetal breakdown.

Experiments on the excision and culturing of anthers and sporogenous tissue have revealed that pollen development in vitro requires a very complex nutrient medium containing many organic nutrients, vitamins, nucleotides, and hormones. Even so, the factors leading to meiosis and to the final stages of pollen development have not been identified, and these events do not take place in cultured tissue. As in egg production, nothing is presently known about the factors that induce meiosis or reduction division. This is one of the major gaps in our knowledge about the controlling mechanisms of plant development.

At meiosis each of the microspore mother cells produces four haploid microspores that develop into pollen grains, which form a heavy, often conspicuously patterned cell wall. The single haploid nucleus divides by mitosis producing two nuclei; one of these will be the **tube nucleus,** and the other, the **generative nucleus,** will further divide to produce two **sperm nuclei** (Figure 17-2). This pollen grain with three nuclei is the complete male gametophyte—it never exceeds this number of nuclei, although it may increase substantially in size during the growth of the pollen tube.

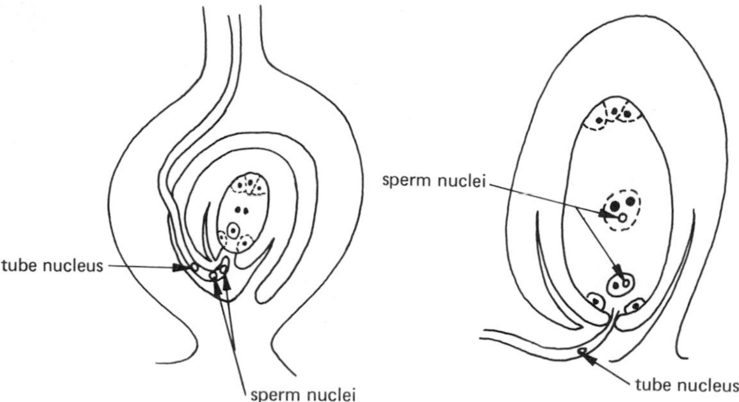

Figure 17-2. Growth of the pollen tube to the ovule and double fertilization.

Sex Determination. Many plants form both anthers and carpels on each floral meristem, producing so-called **perfect** flowers. Some plants produce two types of flowers containing either anthers (staminate flowers) or carpels (pistillate flowers). **Monoecious** plants (including squash, corn, and many trees) produce both kinds on one plant; **dioecious** plants such as maple, willow, ginkgo, or spinach produce only one kind of flower on one plant. The question of sex determination in plants has received considerable attention.

The fact that stamens always develop further from the center of the meristem than carpels suggests that gradients of growth substances or hormones, derived from or directed toward the center of the meristem, are important in controlling sex expression in flowers. This has been confirmed in experiments with monoecious plants such as cucumber and squash. In these plants, primordia for both stamens and carpels are present at early stages of flower development, but only one develops. Usually the first-formed flowers (lower in the spike) are male, and later female flowers are formed.

The application of auxin sprays (either natural auxin, IAA, or synthetic compounds like naphthaleneacetic acid, NAA) greatly increases the proportion of female flowers. A. Lang, now working in the United States, cultured isolated flower primordia and showed that auxins caused female flowers to develop even in primordia that are genetically disposed to be male and that gibberellic acid has the opposite effect. Once ovaries have begun to develop, however, further stamen development is inhibited.

Even in dioecious plants, sex reversal is possible. The British physiologist J. Heslop-Harrison has shown that the application of auxins to lower primordia of the male plant will induce female flower development in hemp (*Cannabis*). Thus, it appears likely that the determination of sex in flowers is influenced and perhaps determined by hormones. It is not clear, however, whether primary sexual determination of a plant or flower primordium is under direct genetic control or the result of a disposition (itself genetically controlled) to the appropriate hormone balance for maleness or femaleness.

Pollination and Fertilization

Pollen Tube Growth. For many years plant scientists have been interested in the rapidity and accuracy with which pollen tubes grow through the great distances from the stigma to the egg (see Figure 17-2). Chemical direction appeared probable, and early experiments confirmed that, when pollen grains were germinated on an artificial medium, the tubes would grow toward pieces of ovule or ovary placed nearby.

Experiments by the American physiologist L. Machlis have shown that pollen tubes of snapdragon and some other plants are strongly responsive to calcium ions (Ca^{2+}) and grow in a gradient of Ca^{2+} toward the region of highest concentration. It has been found in at least one species (snapdragon) that the concentration of Ca^{2+} is lowest in the stigma, higher in the style, still higher in the ovary, and highest in the ovule. However, it is quite clear that Ca^{2+} is not the only chemotrophic agent directing pollen tube growth. The American physiologist W. Rosen found that lily pollen tubes are insensitive to Ca^{2+} but are attracted by some other, apparently organic, stimulus. Certain amino acids have been found to act as chemotrophic agents, but extracts of pistils appear to be highly specific. Rosen's group has not observed chemotropism when pollen of one genus is tested against the pistil parts from outside the same genus. It is thus apparent that there is no single, universal chemotrophic agent.

It is a curious fact that the pollen of many species will germinate and grow readily on a chemically defined medium that does not contain extract of floral parts or other special components, yet the specificity of pollen activity is frequently very precise. Some pollen will germinate on the style of the wrong sort of plant, but most will not. Moreover, many species of plants are **self-incompatible**—that is, the pollen of a given flower will not germinate or grow on the stigma of the same flower. This prevents self-pollination and ensures cross-pollination, with the resultant advantages of more effective reassortment of genetic material.

Some workers have sought for inhibitory substances that might prevent the wrong pollen from growing, and a number of such substances have been postulated. However, no identifiable substance has been isolated, and the number of such compounds (as well as the number of specific reactions or sensing mechanisms) would have to be stupendous if they alone were responsible for self-incompatibility or species specificity. It seems more likely that nutritional balances are involved as well. Self-incompatibility may result from timing differences in pollen and stigma development, such that conditions (nutritional or hormonal) for pollen germination are not correct in the stigma of a flower at the time when pollen is being shed in that flower. Still another possibility exists in certain members of the *Cruciferae,* where pollen is irreversibly *activated* only when placed on a compatible stigma rather than being *inactivated* on an incompatible one!

Fertilization. The growth of a pollen tube continues until it reaches the embryo sac (usually it enters the ovule through the micropyle, as shown in Figure 17-2). The sperm nuclei, together with most of the cytoplasm in the pollen tube, stay at the tip of the tube. As soon as the pollen tube enters the embryo sac its tip breaks, perhaps by the agency of enzymes secreted there or, alternatively, as a result of a stimulus for autolysis or self-disruption received from the embryo sac. Sperm nuclei enter the embryo sac and double fertilization takes place. One sperm nucleus unites with the egg nucleus to form the zygote, and the other unites with two (occasionally four) polar nuclei to form a triploid or pentaploid endosperm nucleus.

The consequences of double fertilization are very important. First, there are two (or three, counting the sporophyte tissue of the ovule and ovary) different cell lines developing simultaneously, and they differ in their genetic constitution. It is not now known, but it seems likely, that these differences make possible the very complex sets of signals needed to program correctly the sequence of events in the growth and development of the embryo. The endosperm and probably the ovary as well are very powerful sources of all the known classes of hormones that regulate and control the embryo's development before it becomes independent of outside influences.

A second consequence of double fertilization, also related to the genetic differences of the embryo and the endosperm, is that many hybrid crosses produce no viable seeds. Although this may be due to incompatibility that prevents fertilization, it is often due to a failure of the endosperm to develop or to an incompatibility of endosperm and embryo that prevents the proper development of the embryo. In the latter event, it is sometimes possible to excise and culture embryos artificially, so that the results of a cross need not be lost even though normal seed development cannot take place.

In some plants (including certain species of *Rosaceae* and some composites like dandelion and hawk weed) although pollination takes place, fertilization does not. In these plants reduction division of the megaspore mother cell does not take place, and the embryo develops, without meiosis and fertilization, directly from tissue of the previous sporophyte generation. This is an example of **apomixis,** reproduction without the union of gametes. In some apomictic plants the endosperm results normally from the fertilization of polar nuclei, and in this case the requirement for pollination is obvious. In other apomicts, however, the endosperm, as well as the embryo, develops directly without the requirement of fertilization. Nevertheless, pollination is frequently necessary. Evidently the act of pollination in some way releases a stimulus that causes the development of embryo and endosperm to start. The consequences of apomixis are very important in genetics, because all the offspring of an apomictic plant are genetically identical with the female parent. It has recently been suggested that apomixis is much more common among plants than has hitherto been suspected. It is very difficult to detect in plants that require pollination.

Embryo Development

Capacity to Grow. The zygote, the first cell of the new sporophyte generation, has a maximum potential for growth because it is capable of giving rise to an entire new organism. After one division, however, each daughter cell has a much reduced "morphogenetic propensity," as it has been termed by Steward. That is, each daughter cell can only produce a limited part of the organism. However, if the two daughter cells were to be separated, each would then recover the capacity to grow a complete organism. This capacity to grow, or morphogenetic propensity, is not just a characteristic of isolated cells. The gametes, for example, cannot grow unless fertilization occurs. Nor is morphogenetic propensity associated specifically with the $2\,N$ number of chromosomes—haploid cells grow very well, particularly in more primitive organisms, and the triploid or pentaploid endosperm is much restricted in its growth when compared with the embryo. The source or driving force of this mysterious propensity to grow and undergo morphogenesis must be sought in some other direction.

It has been frequently observed that "living systems feed on negative entropy"; that is, they use external sources of energy to decrease their own entropy, to build up their own complexity. In earlier times it was believed that a mysterious vital driving force, which was peculiarly biological, drove cells to divide and organisms to grow. Now we recognize the cell as a machine that is capable of harnessing external energy sources to grow. All that need be added is the idea of a program that produces and regulates an irreversible sequence of events in the cell, and the capacity of each cell to react to external stimuli, which can modify the program. Then this difficult and mysterious concept of "the capacity to grow" becomes recognizable as the natural result of adding fuel to a machine in good running order and turning the starter switch—it goes. The reduced capacity of the two daughter cells of the original zygote follows because each cell is now no longer isolated but operates within the influence of another adjacent cell, which is joined to it by protoplasmal connections (the plasmodesmata). As the organism increases in complexity, the capacity for subsequent growth of each cell in it is reduced accordingly by the influence of all the cells and tissues that surround it.

EMBRYO GROWTH. Shortly after it is formed the zygote begins to grow more or less rapidly and develop into an embryo, the form in which it will rest until germination of the seed. Endosperm development precedes embryo growth; the endosperm is the tissue that nourishes the embryo during development. Nuclear divisions form an amorphous, often fluid, endosperm (an extreme case, with respect at least to the volume of fluid, is the coconut). The "milk" stage of corn and many other cereal grains is well known. Later, cell walls are usually formed and the endosperm becomes solid. After this, embryo development normally begins.

Patterns of embryo growth differ from plant to plant, but some generalizations may help. The initial cell division of the zygote results in two cells, one of which will develop into the embryo and the other into the **suspensor.** The suspensor serves to maintain the anchorage or orientation of the embryo and to thrust it into the mass of the endosperm from which it derives its nutrition (Figure 17-3). The embryo develops through the **globe, heart,** and **torpedo** stages, so-called because of their appearance. The two lobes of the heart stage develop into the cotyledons, and in the cotyledonary stage the embryo has developed a radicle or root meristem and a shoot meristem (Figure 17-3).

As the embryo grows the endosperm is digested away; its substance is used for the nutrition of the embryo. This process may continue without pause until no endosperm is

Figure 17-3. Stages in the development of the dicot embryo.

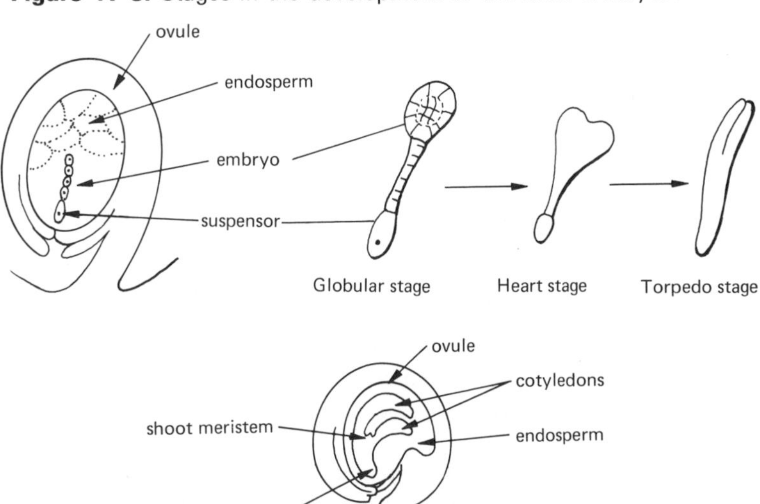

left, and all the residual storage material is transferred to the cotyledons, as in the bean seed (see Chapter 4, page 73). Alternatively, as in corn, the endosperm may remain in the seed until its germination, the cotyledon functioning mainly as an absorption rather than as a storage organ.

During early growth, the level of cellular organization of the embryo changes from essentially zero in the zygote to a very high level in the mature embryo. In this period the embryo develops from a fully heterotrophic single cell into an autotrophic unit capable of further development on its own. Evidently the embryo relies heavily on direction from outside (supplied by the endosperm and ovary tissue) in its early stages, but this reliance gradually decreases to zero. For this reason, the embryo has been a favorite subject of study by plant physiologists who wished to learn and understand the nature and extent of the factors that control growth and development. Two very fruitful approaches have been the study of excised embryos in culture and the induction of embryogenesis in cultures of cells.

Embryo Growth in Vitro. The rearing of isolated embryos began very soon after the turn of the present century, and by the 1930s the American physiologist P. White had developed media in which embryos as small as the early heart stage could be successfully cultured. However, whereas later embryos grow on a relatively simple medium of inorganic nutrients and sucrose, earlier stages require the addition of many other factors and such undefined (and at the time, undefinable) supplements as yeast extract. Later it was discovered that the best possible source of nutrient for embryo growth was endosperm, and the addition of coconut milk to culture media made possible the culture of embryos excised in even earlier stages of development.

Although the addition of coconut milk solves one problem, it raises another because its composition is very complex and largely unknown. It is rich in organic nitrogen, sugars, a variety of growth substances and hormones of every known class, and many other compounds with greater or lesser degree of physiological activity. Thus, although it permits in vitro culture of embryos (as well as many other cells and tissues), the use of coconut milk really represents supplanting one unknown complex medium by another! Determined efforts to isolate all the growth-regulating substances from coconut milk have met considerable success in the laboratory of F. C. Steward at Cornell University, but much remains yet to be learned.

Some success has been achieved in the culture of early stages of embryos in complex (but defined) synthetic media, provided that various additional conditions are right. Thus very early embryos require an unusually low osmotic potential (perhaps the degenerating endosperm provides high concentrations of soluble compounds that lower the osmotic potential around the embryo in its natural surroundings), but this requirement lessens as the embryo matures. Recent experiments indicate that if the correct balance of hormones of the various types is maintained, much of the complexity of the nutrient medium becomes unnecessary. Thus we come again to realize that the major factors influencing development are the plant-growth substances or hormones, and these act together in balanced concert, not singly or alone.

Embryogenesis in Cell and Tissue Culture. In the late 1950s, Steward's laboratory made great progress in the culture of explanted tissue and cells of several plants, particularly carrots, using White's basal medium fortified with coconut milk. It was found that free-floating cells, which are sloughed off by the masses of growing tissue, can be planted on agar or suspended in appropriate media. When treated in this manner, the cells undergo essentially the normal embryological development of carrot plants; they produce roots and shoots and finally mature plants complete with flowers. Illustrations of this process are shown in Figures 17-4 and 17-5.

It later became apparent that cells from various sources, if properly treated, could revert to the unrestricted behavior characteristic of a zygote and undergo embryogenesis, or something very like it, to produce whole plants. Cells that can do this most readily, requiring only natural endosperm or coconut milk in addition to the basal medium of salts and sugars, are the zygote itself and cells derived from the culture of excised embryos. Cells derived from more mature or differentiated parts of a plant (stem, root, and so on) are restricted in their response and require additional treatment. They can be restored to the freely dividing state by coconut milk, but require the addition of auxin

Figure 17-4. From carrot cell to carrot plant.
Sequence in the development from (**A**) freely suspended carrot phloem cells to (**B**) cell aggregates, (**C**) colonies with organized growth centers or nodules, (**D**) roots. (**E**) Plantlets grown first in tubes on agar, later in pots in vermiculite. (**F**) Plants with massive storage roots with normal carotene content. [From F. C. Steward, et al.: In D. Rudnick (ed.): *Synthesis of Molecular and Cellular Structure*. Copyright 1961, The Ronald Press, New York. Used with permission. Original photographs courtesy Prof. Steward.]

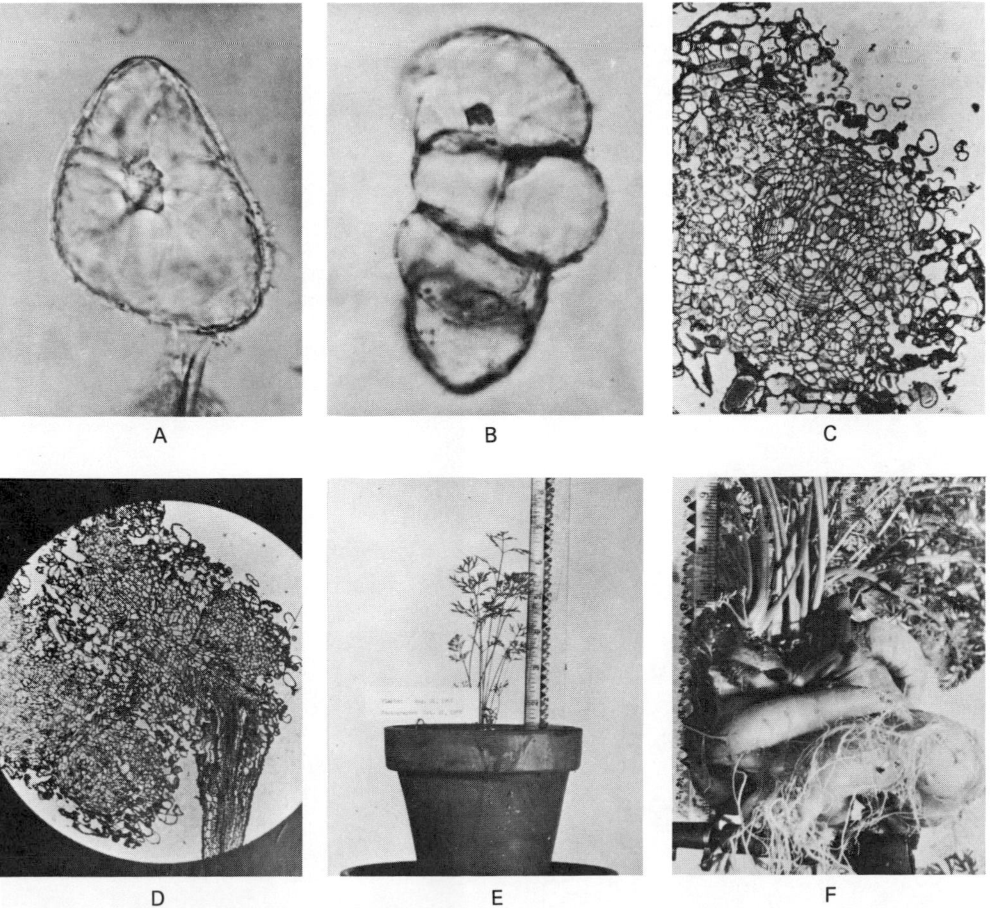

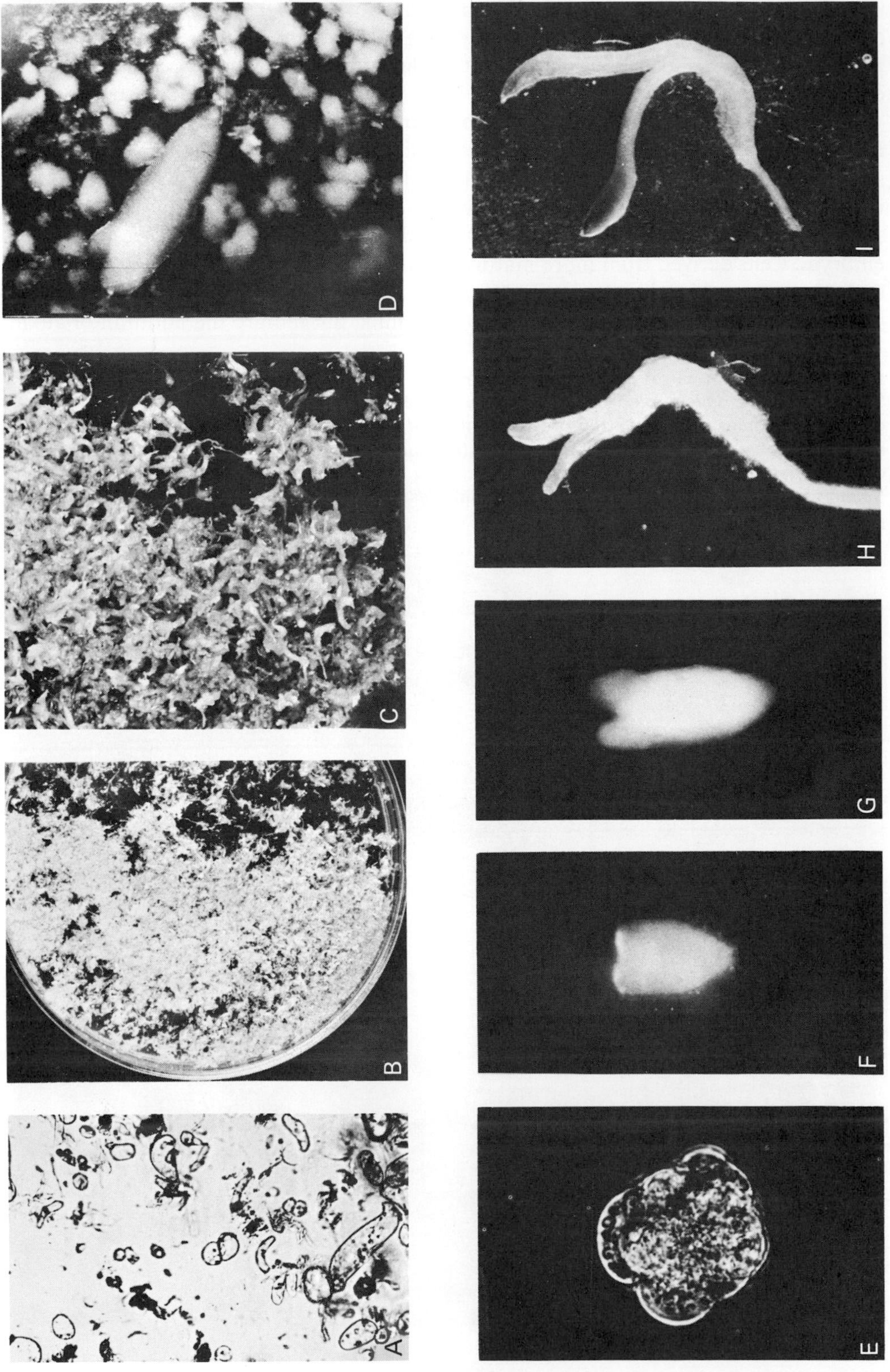
A
B
C
D
E
F
G
H
I

Figure 17-5. Development of carrot plants from freely suspended cells of embryo origin.
A. Cultured carrot cells in liquid medium; the suspension is filtered through bolting silk. **B.** Growth of a large number of embryoids on a Petri dish on which a suspension of carrot cells was dispersed. An estimated 100,000 embryoids were on this plate, and all of them were derived from part of the cells from one embryo of the wild carrot. **C.** Higher magnification of the plate shown in B. **D.** Growth in a liquid medium of many units from a cell suspension similar to that shown in A. Many globular masses in various stages of development and one clear torpedo stage may be seen. In the second row of photographs are shown selected stages of carrot embryogeny developed from free cells: **E.** Globular stage. **F.** Heart-shaped stage. **G.** Torpedo stage. **H,I.** Cotyledonary stages. In the bottom row are shown stages in development of plants reared from embryoids grown, in turn, from cells. **J.** Plant on agar. **K.** Plant bearing inflorescences after six months' growth. **L.** Detail of inflorescence on a plant of Queen Anne's Lace (*Daucus carota* L.) reared from cells of embryo origin (see A). Note the few typical red flowers at the center of the inflorescense. [From F. C. Steward, M. O. Mapes, A. E. Kent, and R. D. Holsten: *Science*, **143**:20–27 (1964). Copyright 1964 by the American Association for the Advancement of Science. Used with permission. Photograph courtesy Prof. Steward.]

(IAA) or the synthetic auxins naphthaleneacetic acid (NAA) or 2,4-dichlorophenoxy-acetic acid (2,4-D) before they can develop into new plants, as shown in Figure 17-4.

Evidently the capacity to grow exhibited by the zygote is not lost by cells that result from its division. The morphogenic propensity becomes masked as the organism develops, but it can be released again if the cells are freed from their tissue and provided with the right sort of medium. Once again, the importance of the appropriate balance of hormones and growth substances as well as nutrition (in this case provided by the use of coconut milk with the further addition of synthetic hormones) is emphasized in the induction and maintenance of orderly development.

Totipotency of Plant Cells. In the preceding section we added a new dimension to the discussion of development. This is the fact that somatic cells from various parts of the plant can, given appropriate conditions, develop into a whole new plant. That is to say, these cells are **totipotent;** they have the full potential for development exhibited by the zygote. It is apparent that all plant cells cannot undergo "rejuvenation" in this way because some have been locked into a specific pattern by an irreversible event, such as the deposition of a massive and insoluble cell wall. However it seems probable that all plant tissues contain some cells that are totipotent, given the right conditions. There are many examples in the normal development of plants where cells are rejuvenated and begin to divide again. Lateral root formation, the formation of interfascicular cambium in dicots, and the formation of phellogen in phloem tissues of trees are good examples (see Chapter 4). The reversion of single cells to embryogenesis in laboratory culture is thus an experimentally induced example of a normal, natural type of behavior.

This implies that every living cell in the plant carries within itself the totality of the genetic information necessary to make that plant. What the cell loses during development is not the information but the ability to use that information. Thus every cell has innately the full capacity to grow, or the morphogenetic potential, of the zygote. This is a natural consequence of the fact that all the genetic material is duplicated and apportioned equally between daughter cells in mitosis. However, the capacity to grow must be controlled for orderly development to take place. Each cell in a tissue restricts, to a greater or lesser degree, the capabilities of every other cell, so that orderly cooperation in development and function becomes possible. This restriction is accomplished by the complex interaction of cells with each other through the creation of fields or gradients of nturients, metabolic intermediates, and hormones.

One-way Streets in Development. This problem can be viewed from another angle. Development and growth are essentially one-way processes with only occasional partial reversions (returns to an earlier state) being permitted. It is very difficult to conceive of a mechanism within the genetic information bank for programming the totality of information stored there. Such programming would take a very complex system of switches. Development has been likened to a computer program, in which each stage must be successfully completed before the program readout can proceed to the next step. But this implies a very much more complex system of programming than is normally considered to be present in the nuclei of cells. It is much simpler to envisage development as a one-way program, because when a cell divides into two, there is no pathway of return. An irreversible step has been taken, and a new set of conditions has been applied. Thus the *organism itself* can be looked upon as the program. The genetic information is what determines the differences between organisms or species. But just what fraction of the total genetic information is being used at any moment by a cell depends upon its position in the organism and the developmental stage of the organism at the time.

Again, this emphasizes the interaction of cells—perhaps through the plasmodesmata that connect them—in the development of the organism. It also follows that, if cells are removed from their program and thus freed from this restriction, they may revert to the beginning of the program, as befits single cells, and begin to develop again as if they were zygotes. All that is required is precisely the right environment. It is the difficulty of providing this environment, not the recalcitrance of cells, that has made artificial embryogenesis from vegetative cells so complicated. The German physiologist G. Haberlandt predicted in 1902 that embryogenesis would be possible; it was not until 60 years later that W. Halperin, F. C. Steward, and others in America proved him to be right.

Fruit and Seed Formation

Fruit Set. Following pollination the development of the fruit and seed begins. If pollination does not take place, the flower senesces quickly and dies. In apomictic plants, the stimulus of pollination alone is sufficient to start embryo development; presumably the pollen provides growth substances or hormones that stimulate embryo development. Pollen-produced growth hormones are also involved in fruit set—the prevention of abscission. Auxin (normally produced by pollen) can be sprayed on unpollinated flowers, with the result that **parthenocarpic** (seedless) fruit develop. In some plants, particularly stony-fruit species like peach, plum, cherry, and grape, gibberellic acid acts instead of auxin. In at least some of these species the pollen is known to produce a gibberellin rather than an auxin.

Fruit and Seed Development. The first stage in fruit and seed development is rapid cell division without much enlargement. The principal factor involved appears to be cytokinin, much of which may be produced by the triploid (or pentaploid) endosperm, which is growing at this stage. Various tissues of the parent plant—the ovary, receptacle, and sometimes parts of the floral tube—may be involved in the formation of fruit. It is beyond the scope of this book to describe the different kinds of development that result in the bewildering diversity of fruit. The main processes are somewhat similar in most plants.

Following the cell division stage comes a phase of growth mainly by cell enlargement. The evidence of a number of experiments suggests that this is caused by auxins produced in the seeds. If the seeds are removed from a developing fruit, development stops; however it can be restarted again by the application of auxins. The late French physiologist J. P. Nitsch showed that fruit development in the strawberry and cucumber is dependent on auxins which originate from the ovule. Strange and misshapen strawberries can be made by removing all but one or a few of the seeds, and nearly normal growth is restored by the addition of IAA in a lanolin paste (Figure 17-6).

As with fruit set, some fruits respond rather to gibberellins than to auxin treatment. This may signify a difference in response mechanisms. However, a more likely explanation is that the response is most fruits depends on both auxins and gibberellins, and

Figure 17-6. Growth of strawberry fruits.
In the top photographs only one (**A**) or three (**B**) achenes were fertilized. In (**C**) and (**D**), only a few achenes in vertical or horizontal rows were fertilized. (Normal strawberries outside, treated ones in the center.) In (**E**) the strawberry at the left is normal; the other two had all their achenes removed and were treated with lanolin paste. The one in the center received only lanolin; the one on the right received lanolin paste containing 100 ppm β-naphthoxyacetic acid. [From J. P. Nitsch: *Am. J. Bot.*, **37:**211–15 (1959). Used with permission.]

what distinguishes between the different responses is the natural concentration of one or another hormone that is present. Alternatively, there may be differences in the required balance between the hormones. At this stage in the development of fruits the concentration of organic acids and sugars begins to increase, and the resulting decrease in osmotic potential is probably related to the increasing absorption of water and growth by enlargement of the cells.

Fruit Ripening. The ripening process of fruits involves many complex chemical and physiological changes. Fruit-bearing plants have been cultivated and subjected to intense selection processes for specific characteristics in the fruit (taste, size, color, texture) so that many familiar fruits probably exhibit developmental patterns that are far from the "natural" state. The ripening process involves the conversion of acids and starch to free sugars, the elaboration of pectinases which soften and ultimately break down the cell walls, and frequently the elaboration of various pigments, usually anthocyanins, and the loss of chlorophyll. Many of the changes are induced or caused by ethylene, which is produced by the fruit itself.

The production of ethylene by fruit has a profound consequence in storage. The ethylene produced by each fruit has a cumulative effect and stimulates other fruit to ripen faster and, coincidently, to produce more ethylene. Thus the ripening process becomes "autocatalytic" in a stored mass of fruit, with potentially catastrophic results! Cold storage and the technique of flushing the fruit with an inert gas (nitrogen or CO_2), which retards ripening by removing ethylene as it is produced, are important methods of prolonging the storage life of fruit.

A major change in physiology that occurs during ripening is the respiratory climacteric (see Chapter 6, page 132). Many fruits undergo a climacteric, and it can be induced by the addition of ethylene in some of those that normally do not. The climacteric is usually accompanied by a short-lived but often dramatic increase in ethylene production. Ethylene production may rise again later (Figure 17-7).

Ethylene is thought to affect the process of ripening, including the respiratory climacteric, in two ways. It has a strong effect on the permeability of cell membranes, and cell permeability certainly increases greatly during ripening. This permits softening of the fruit and the intermingling of metabolites and enzymes normally kept separate so that respiratory metabolism is greatly speeded up. In addition, it has been shown that there is an increase in protein content during the climacteric of some fruits, and it has been suggested that ethylene stimulates protein synthesis at this time. Presumably the proteins so formed are involved in the processes of ripening, and the climacteric may be in part a reflection of an increase in respiratory enzymes. In any event, the climacteric is an aerobic process and can be prevented or postponed by storing the fruit at reduced oxygen tension.

Many of the important physiological studies on ripening have been done on fruit like apples and bananas which must be stored for prolonged periods. This work has resulted in practical discoveries that have enormous commercial value, as well as being of fundamental importance to plant physiology. Techniques developed for the controlled ripening of fruit in storage include the use of hormones and growth factors, and of high nitrogen, high carbon dioxide, and low oxygen gas mixtures as well as carefully controlled temperatures. The recently popular use of polyethylene bags to package fruit has advantages (carbon dioxide builds up and oxygen decreases inside); but if the fruit ripens to the point where ethylene production increases to a high level, "autocatalytic"

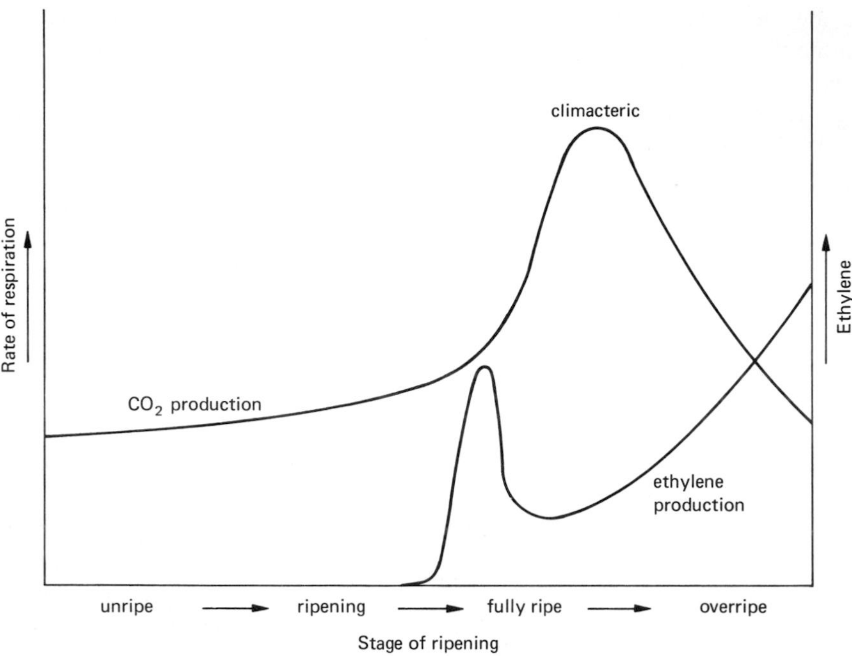

Figure 17-7. Relationship between ethylene production and the respiratory climacteric in ripening fruits.

ripening takes place unless the bags are punctured to let the ethylene escape. This area of plant physiology is a fine example of the effective interchange of ideas between scientists engaged in pure and in applied research. The important discoveries, the technological advances, and the progress of knowledge have come from a synthesis of the work and ideas of both kinds of science.

Germination

The seed matures within the fruit. After the ripening and shedding of the fruit, the seed usually becomes dormant for a shorter or longer time. This means that, even if it is moistened and supplied with normal conditions that favor germination, it will not germinate. Dormancy is imposed by the formation in the seed of chemical inhibitors, by the lack of necessary stimulating substances (which will later be provided by the embryo), or by mechanical resistance of the seed coat to the entry of water and oxygen. Dormancy is broken following the subjection of the seed to various environmental conditions, which may include frost, prolonged periods of deep cold, prolonged exposure to cool, moist conditions in the presence of oxygen **(stratification),** intense heat (as in a fire), passage through the intestine of birds or mammals, physical abrasion, or fungal attack. All these requirements ensure that the seed will survive through ensuing periods of conditions in which the seedling could not grow and ensure that it does not germinate until conditions are once again right for growth. They also help to prevent germination, which might have disastrous results, during unseasonably clement weather that might occur in winter time. The mechanisms of dormancy will be discussed in detail in Chapter 22.

When the conditions required to break dormancy occur, the embryo begins the formation of gibberellins and cytokinins that are required to overcome the action of growth inhibitors and start growth. At this point, with the addition of water, the seed will germinate.

Conditions for Germination. Germination will not proceed unless conditions are correct. The major factors are water, oxygen, temperature, and light.

Water is of prime importance because seeds are extremely dehydrated. They normally contain only 5–20 percent of their total weight as water and must absorb a substantial amount before germination can begin. The first stage of germination is rapid water uptake called **imbibition.** There are indications that until a certain, critical level of water is reached (it would be different for different kinds of seeds) no growth can take place. If water is withdrawn before this critical point is reached the seeds may be undamaged, but after this point has been passed and metabolism has started, the germinating seed will die if it is dried again. After imbibition, water uptake slows, germination proceeds, and the irreversible processes leading to the growth and development of the seedling begin.

Oxygen is necessary for seedling germination. Metabolism during the initial stages of germination may be anaerobic, but it quickly switches to aerobic as soon as the seed coat is ruptured and oxygen can diffuse in. The importance of oxygen is exemplified by the experiments of Yemm on pea seeds shown in Chapter 6, Figure 6-18 (page 138). Seeds with the testa intact require much higher oxygen for maximum respiration than those from which the seed cover has been artificially removed.

Correct temperature is important for germination; seeds usually will not germinate below a certain temperature specific for the species. Light is also important in germination of some seeds. Very small seeds have only minimal amounts of stored food for early embryo growth, and it is therefore necessary for them to become autotrophic as quickly as possible. If they germinate too deep in the soil, they might exhaust their stored reserves before penetrating to the surface. The light requirement prevents this from happening and ensures that germination will only take place at or near the surface. The light-sensitive pigment is **phytochrome,** which will be discussed in detail in Chapter 20. Only certain seeds are responsive to light; Grand Rapids lettuce has been much studied because of its strong response. The germination of some other seeds is inhibited by light. This may also be a protective mechanism that prevents slower-growing seeds from germinating on the surface during a brief shower of rain, when they might dry out before their roots reached a steady or sufficient supply of water.

The age of seeds is an important factor in germination. Contrary to popular belief, few seeds are capable of surviving for a very long time in storage. There are some authentic records of seeds surviving storage for periods of up to 100 years, but the majority of seeds can last at best a few years. Some are able to survive only a few days or weeks. Seeds kept in deep freeze or under anaerobic conditions appear to last longer. There have been many attempts to study the metabolism of seeds that are dormant or **quiescent** (that is, seeds which do not germinate because conditions are not right—dormancy implies inability to germinate even under ideal conditions). However, it seems that the very low oxygen uptake of such seeds is probably the result of nonmetabolic and destructive processes of slow autooxidation. It would seem most probable that the slower this process goes, the longer will be the life of the seed in storage.

Mobilization of Reserves. In some seeds (primarily monocots) the endosperm is retained until germination, whereas in others (mainly dicots) the cotyledonary leaves of the embryo grow and absorb all the nutrient material contained in the endosperm during the ripening of the fruit prior to the shedding of the seed. In either event, on germination a large amount of reserve material in the form of protein, fats, starch, or other carbohydrates has to be digested and mobilized for the nutrition of the growing seedling. This means that digestive enzymes must be activated or synthesized immediately after germination begins.

It appears that gibberellins are very important in this process. The seeds of a number of cereals (monocots) have a structure that makes them amenable for studies on germination. The endosperm consists of a starchy tissue that is surrounded by a proteinaceous cellular tissue called the **aleurone** layer. It is in the aleurone layer that many of the digestive enzymes are elaborated or secreted.

Of the two enzymes required for the digestion of starch (see Chapter 6, page 120) β-amylase is present in the seed prior to germination, but α-amylase and proteases appear immediately after the start of germination. It was found that if the embryo was removed, no enzymes (particularly amylases) appeared. However, if very low concentrations of gibberellic acid (as low as 10^{-10} M) were added to the seed, the production of digestive enzymes proceeded. β-Amylase was found to be *activated* as a result of gibberellin action, but several lines of evidence suggest that α-amylase and perhaps also proteases are synthesized de novo as a result of gibberellin action.

Varner found that ^{14}C-amino acids are incorporated into the enzymic protein of α-amylase, showing new synthesis of the amino acid chain. When α-amylase was induced in the presence of $H_2{}^{18}O$ instead of normal $H_2{}^{16}O$, the enzyme protein was more dense, as determined after careful centrifugation through a dense layer of cesium chloride solution. This shows that the enzyme had been synthesized de novo from amino acids that contained ^{18}O, which had thus been derived from the hydrolysis of other preexisting seed proteins. All these points suggest that gibberellins act at the molecular genetic level, derepressing the genes responsible for the synthesis of α-amylase. It has been shown that the embryo does indeed provide the gibberellin required to initiate the activation or synthesis of various enzymes that will provide for its own nutrition during and after germination. Furthermore, one of the earliest metabolic events in the embryo on germination is the elaboration of auxin in the coleoptile. This is not only affects coleoptile growth (see Chapter 18) but also induces the formation of vascular tissue along which the gibberellic acid moves and by which the nutrients released from the endosperm may be translocated to the embryo. Thus the embryo, by the synthesis of hormones, controls the mobilization of its own nutrition and thus its growth.

Seeds that store fats, such as castor beans, squash, or pumpkin, have an interesting metabolic sequence for the mobilization and conversion of their fats to sugar, which can be translocated to the growing embryo. The fats are first oxidized to acetyl-CoA by the β-oxidation pathway (Chapter 6, page 122). The acetyl-CoA so formed enters small microbodies called glyoxysomes. These small subcellular organelles surrounded by a single membrane were first discovered by the American (originally British) plant physiologist H. Beevers in 1966. They contain the enzymic machinery of the glyoxylate cycle (Chapter 6, page 125), which has the net function of converting two molecules of acetyl-CoA to one of succinate. The succinate is further converted to oxaloacetic acid, which is decarboxylated by carboxykinase enzyme to make phosphoenolpyruvate (PEP),

$$\text{oxaloacetate} + \text{ATP} \longrightarrow \text{PEP} + \text{ADP} + CO_2$$

The PEP so formed is converted to sugar by reactions that essentially accomplish the reversal of glycolysis. One difference is that FDP is converted to F-6-P by a phosphatase instead of by reverse action of the phosphofructokinase which operates in glycolysis. Phosphatase has an equilibrium that strongly favors the production of F-6-P, and this tends to drive the sequence toward the production of sugars. Reducing power and ATP needed to drive reverse glycolysis are presumably derived from the oxidations of the glyoxylate cycle, from the conversion of succinate to oxaloacetic acid, and from the β-oxidation of fats, which produce NADH. This in turn can feed electrons into the electron transport chain, generating ATP.

Seedling Nutrition. The digestion of reserves in the embryo or cotyledon results in the production of amino acids, amides, sugars, nucleotides, and organic acids. It was originally thought that all these soluble products were mobilized and transported to the growing embryo where they were reassembled into new proteins, carbohydrates, nucleoproteins, and lipids. However, largely as the result of the work of the great Russian physiologist D. Prjanishnikoff, it has become clear that only certain compounds are transported, and the mobilization of nitrogen in particular requires a considerable amount of metabolism. Amino acids that result from the breakdown of storage proteins are mostly deaminated, and the resulting organic acids may be used either for respiration or as carbon skeletons for the formation of translocation compounds. These are usually the amides, glutamine or asparagine (glutamine seems to be the more usual), although some other amino acids are extensively translocated in certain plants (for example, homoserine in pea). The carbon skeletons for amide synthesis appear to come rather from sugar than from protein-derived amino acids. Prjanishnikoff pointed out that amide synthesis requires the presence of sugars as well as soluble nitrogen compounds. Energy for synthesis of transport compounds presumably comes from the respiration of some of the organic acids derived from amino acid degradation.

The nitrogen transport compounds are reworked in the growing embryo, again using carbon skeletons derived from transported sugars, to form the amino acids required for protein synthesis and growth. A summary of these ideas is presented schematically in Figure 17-8.

Certain amino acids may also be translocated as such in some seedlings. The Canadian physiologist Ann Oaks has shown that leucine is so translocated to the root tips of corn seedlings, where it tends to regulate, via its concentration, the de novo synthesis of leucine in the root tip. It is interesting to note that leucine supplied from outside does not have this regulating capacity. Evidently leucine supplied in the normal transport route is accessible to the site of its synthesis, whereas that supplied externally is not. This sort of compartmentation of amino acids has been shown by Steward and his coworkers to be a common phenomenon in plants. It may provide an extra level of control in the metabolism of plants by permitting storage and metabolic activity not associated with protein synthesis to go on without interfering (by mass action or through molecular control mechanisms) with the synthesis of amino acids destined for protein formation during growth and development.

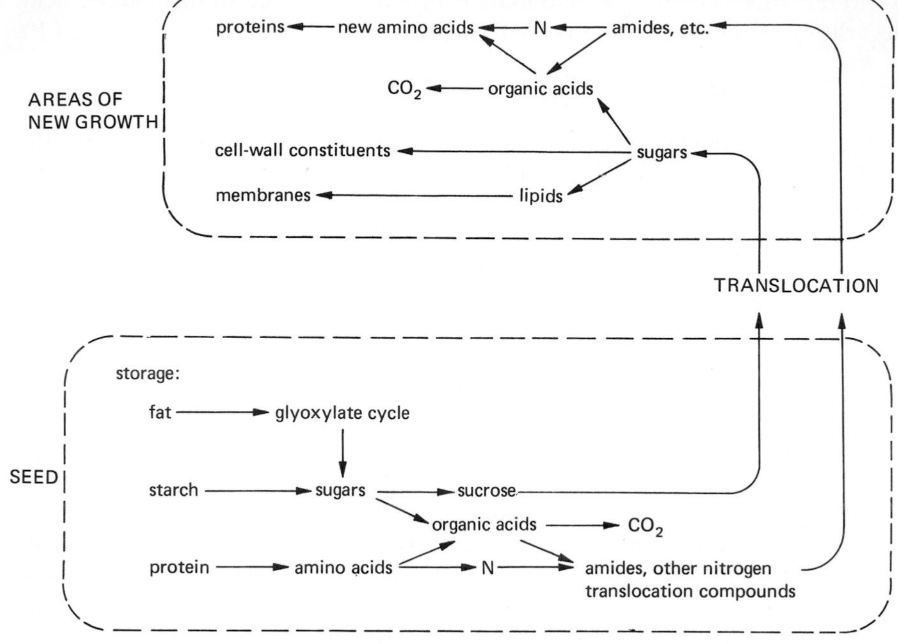

Figure 17-8. Mobilization of nutritional resources in a germinating seed.

Additional Reading

See list for Chapter 16 (page 406).

Black, M.: *Control Processes in Germination and Dormancy.* Oxford University Press, London, 1972.

Crane, J. C.: Growth substances in fruit setting and development. *Ann. Rev. Plant Physiol.*, **15:**203–26 (1964).

Hansen, E.: Postharvest physiology of fruits. *Ann. Rev. Plant Physiol.*, **17:**459–80 (1966).

Rosen, W. G.: Ultrastructure and physiology of pollen. *Ann. Rev. Plant Physiol.*, **19:**435–62 (1968).

Salisbury, F. B.: *The Flowering Process.* Pergamon Press, New York, 1963.

Schwabe, W. W.: Physiology of vegetative reproduction and flowering. In F. C. Steward (ed.): *Plant Physiology: A Treatise,* Vol. VIA. Academic Press, New York, 1971, pp. 233–411.

18 Patterns of Growth

Seedling Growth

Following germination, the root meristem of the embryo is activated and grows rapidly, beginning the development of the primary root. Shortly thereafter, the main meristem of the aerial part of the plant begins to grow. In some plants the cotyledons are carried upward with the growth of the hypocotyl; in others the cotyledons remain in the ground and only the epicotyl grows above ground (see Chapter 4, page 73).

The delicate meristem of the root, which is pushed through the soil by the expansion of cells behind the meristem, is protected from damage by the root cap. The apical meristem has a different sort of protection. In monocotyledons it is protected by the **coleoptile,** the first seedling leaf, which envelops the growing shoot until it reaches the surface of the soil. When the coleoptile reaches the soil surface its growth stops, and the shoot inside pushes through its tip. In many dicotyledons the stem is recurved near the tip, forming the **plumule hook.** Thus the bent portion of the stem, rather than the meristem and its developing leaflets, is pushed first through the soil. When the plumule hook reaches the surface of the soil it straightens out and the cotyledons or primary leaves unfold, as shown in Figure 18-1.

Photomorphogenesis. The above mentioned behavior patterns are evidently responses to the perception of light by the coleoptile or the plumule on reaching the soil surface, since the coleoptile continues to grow and the plumule hook does not unfold if the seedling is grown in darkness. This response appears to be similar to the light response in seed germination noted in the preceding chapter. The tip of the coleoptile and the plumule itself are apparently the sensitive tissues.

The sensing mechanism is through the absorption of light by the pigment phytochrome, and the mechanism of its action will be examined in detail in Chapter 20. Very low intensities of red light are perceived, but light of shorter wavelength does not seem to be so effective. However, certain seedlings that normally do not form a plumule hook (for example, Grand Rapids lettuce) do so if illuminated with weak red light, and this effect is reversed by strong far-red or blue light. The most active wavelengths are all characteristically absorbed by one or another of the forms of the pigment phytochrome. Thus, the nature of the response to illumination seems to depend upon the tissue, although the sensing mechanism is the same in each case.

Figure 18-1. Photograph of bean seedlings germinated for 8 days in darkness (**A**) or in light (**B**). Note the plumule hook in A that has straightened out in B.

Seedlings grown in darkness are characteristically **etiolated;** that is, they are without chlorophyll, are much elongated, their vascular and support tissues are poorly developed, the leaves usually do not grow, and the plumule hook unfolds slowly, if at all. This condition is shown in Figure 18-2. All these growth anomalies can be prevented by illuminating the plant with very dim light, at levels that are inadequate for the photosynthetic nourishment of the seedling. Further, the action spectrum of the light that prevents all of these anomalies indicates that they are all either entirely or at least partially mediated by phytochrome.

Light, or its absence, is thus one of the most important environmental factors that trigger various developmental steps in the seedling and in the growing plant. Not only the presence (or absence) of light is important but also its quality and sometimes its quantity and periodicity, and some of these responses are much affected by temperature as well. The whole subject of periodicity and the measurement of time will be discussed in Chapter 20.

A number of the responses normally elicited by light can be simulated by the application of suitable hormones—usually gibberellins or auxins. Reactions of various plants to different hormones are not all the same, and frequently one hormone will cause opposite results in different plants. We shall consider the details of some of these responses in the next two chapters. At this point it is sufficient to stress the fact that the whole train of biochemical and genetic responses that result in development are set in motion, either directly or through the agency of growth substances, following the perception of light and other environmental stimuli by the plant.

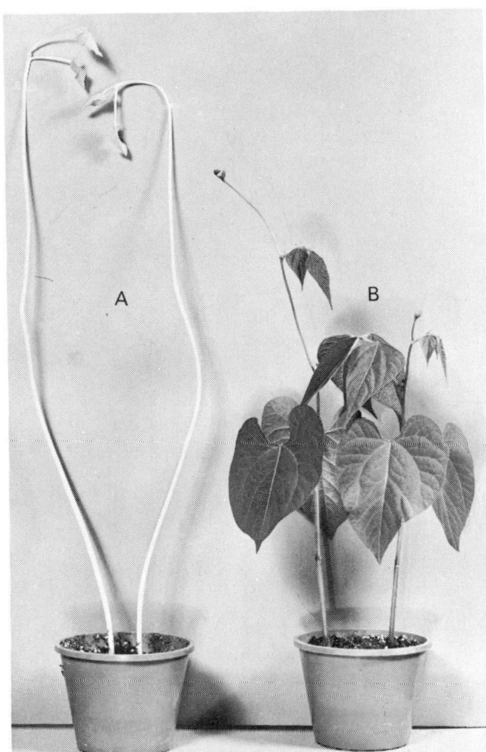

Figure 18-2. Photograph of bean plants grown for 22 days in darkness (**A**) or light (**B**). Note the elongated thin stem, small leaves, absence of color and presence of the plumule hook in the etiolated seedling (**A**).

Initiation of Organs in Tissue Cultures

Much work has been done on the growth of calluses of tissue in culture and on the factors that cause them to undertake organized growth. Some of the experiments were considered in the previous chapters in the discussion of totipotency and the control of growth and morphogenesis. We are here concerned with the factors that cause certain tissues to become roots or leaves or stems or flowers.

It will be remembered, for example, that roots form on the basal end of a cut willow stem and shoots at the apical end (Figure 16-18, page 401). This results from the basipetal distribution of auxin in the twig. The question that must be asked is, does the initiation of roots and shoots require qualitatively different stimuli? That is, are there root-forming or root-initiating stimuli that differ in kind from shoot-forming or shoot-initiating stimuli? Experiments by the American physiologists F. Skoog and C. O. Miller show that varying the relative concentration of two different growth substances, IAA and kinetin, results in the production of either roots or shoots in tobacco callus cultures as shown in Figure 18-3 (see also Figure 16-13, page 394). These results clearly indicate that there is no need to postulate specific organ-forming substances. The release of information leading to the formation of either roots or shoots is clearly accomplished by the same agencies—IAA and cytokinins—but in different relative concentrations. More IAA is needed to stimulate root growth, as was also shown by the willow twig experiment, but more cytokinin is needed for shoot growth.

The experiments of Skoog and his coworkers also show that the callus grows much larger in the presence of high IAA concentration, regardless of the kinetin concentration

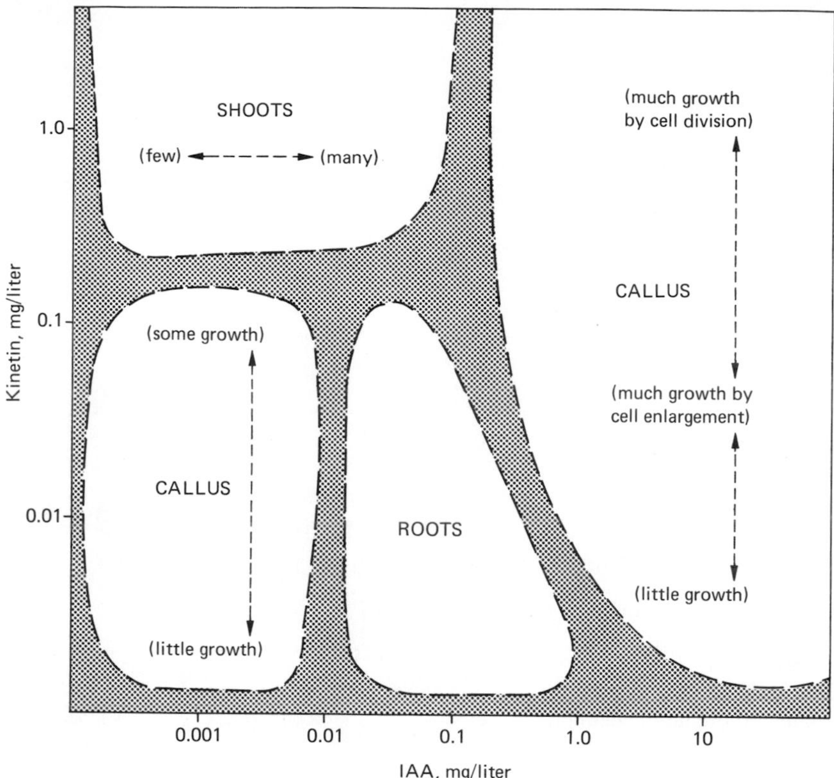

Figure 18-3. Interactions of IAA and kinetin on growth and development of tobacco callus. [From the work of F. Skoog *et al.*]

(Figure 18-3). However there is a qualitative difference in growth. At low kinetin concentration, the callus is watery and composed of large cells that have grown by cell expansion. In the presence of 1 mg/liter of kinetin, the calluses are dense and composed of large numbers of small, meristematic cells.

Similarly, it has been shown by F. C. Steward and his group that the addition of different stimuli (fractions isolated from the liquid endosperm of horsechestnut or coconut, either alone or together with natural or synthetic growth substances) will cause various plant tissue cultures to grow in different ways: either by cell division or by cell enlargement. Thus there are clearly no specific cell-division or cell-enlargement substances. Once again, the responses are evoked by the appropriate relative concentrations of naturally occurring growth substances or their substitutes. Not all tissues are equally responsive in the laboratory. It seems likely that this is not a matter of a missing or unknown growth factor but that the precise combination of growth factors and circumstances has not yet been discovered.

The complexity of this situation may be illustrated by references to work on plant tumors induced by the tumor-forming bacterium *Agrobacterium tumefaciens*. Cells of many plants which have been damaged or irritated are susceptible to infection by the bacterium. After infection, tumors appear that may ultimately become free of the infecting organism, which dies off, and sterile secondary tumors may appear at other places in the plant. These tumors may be cultured, and under normal conditions they do not revert to normal growth. If the causative organism is killed by heat shock after only a short time, however, the tumors may revert partially to organoid forms of growth; apparently the capacity for normal growth is not completely lost but is merely masked.

An interesting system has recently been developed by the American physiologists W. R. Sharp and J. E. Gunckel using two species of tobacco, *Nicotiana glauca* and *N. langsdorffii,* and a hybrid between the two. A range of different types of tumors arise spontaneously, or following wounding, or as the result of infection by various strains of *A. tumefaciens.* Differences among these tumor types include their degree of autonomy (their requirement for specific organic nutrients or growth factors), their ability to organize to the extent of producing organs, organlike structures, or whole new plants, and their tendency to produce cancerous growth in the intact plant. Illustrations of these tumor types are shown in Figure 18-4. Each tumor type is stable and grows in vitro, but a number of them can be converted into other forms (for example, from amorphous to organoid, or from organoid callus to intact plant) by the addition of organic nitrogen, vitamins, the hormone IAA, or by heat shock (Figure 18-4, numbers 12 and 13). Thus these systems can be changed from one stable state to another. Since they are interconvertible, they all contain the same genetic information. This system thus forms a valuable model that is being used for the study of the metabolic patterns in various types of tumors and the switch mechanisms that convert one into another.

One of the great challenges put to plant physiologists by modern agricultural demands is the need for cloning unusually (or unexpectedly) desirable specimens of a cultivated plant. For example, a mutant cereal plant may be shorter and sturdier, or produce more grains, than all the other plants in a field. This propensity may be difficult to isolate or recover by breeding experiments and certainly takes time. Tissue-culture techniques now permit growing masses of new plantlets from isolated cells or cell cultures from such a plant, thus greatly shortening the time needed to bring an advantageous new cultivar to testing and production. However, many cell cultures are difficult to grow. Some form only shoots but no roots; others form only roots. Many research programs for studying the development of plants are now active all over the world, particularly in China, India, Viet Nam, and other developing countries. Research workers, particularly in China, have developed ingenious ways of initiating and promoting shoot and root growth from callus using chemical and hormone treatments or physiological shock.

Root Growth

Terminal Meristem. Root growth is simple compared with stem growth. Since the root meristem produces only one axis, it does not branch; thus the root has no nodes. There is some question about the size of the meristem itself. Some experiments suggest that only one or very few cells give rise ultimately to all the cells of the root. The American physiologist R. T. Brumfield treated root tips with x-rays, causing chromosomal anomalies in certain cells that could be detected by microscopic examination. All of those cells derived from a "marked" initial would themselves be marked, so that the distribution of the progeny of specific initials could be followed. The root was then allowed to grow for a time, following which it was carefully examined. Brumfield's experiments suggest that only a few—perhaps only three—initial cells are responsible ultimately for all the cells in the root.

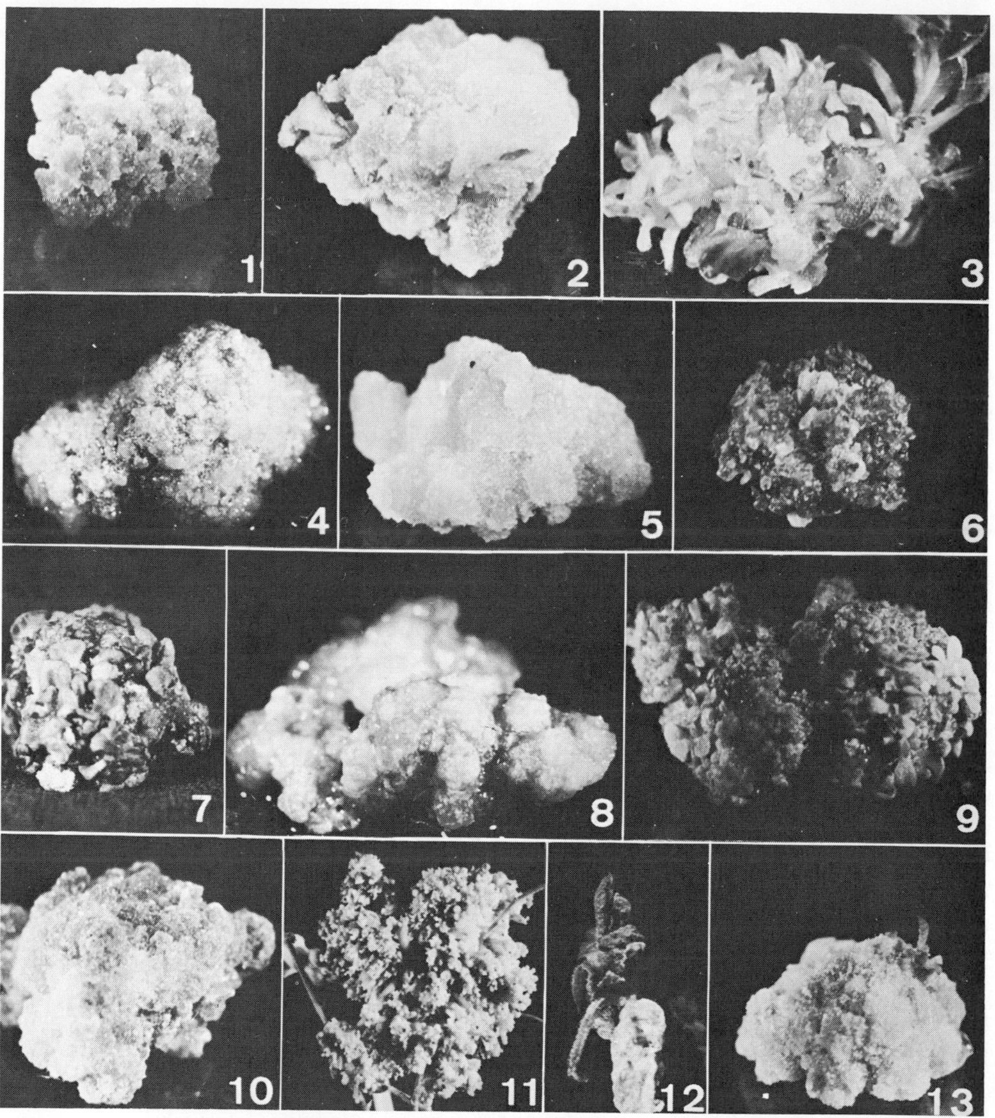

Figure 18-4. Explants of *Nicotiana glauca*, *N. langsdorffii*, and the hybrid between them after 4 weeks' culture.
(**1**) *N. glauca* pith callus (amorphous, light green, friable). (**2**) *N. glauca* crown gall (strain B-6) tumor (amorphous, white, hard). (**3**) *N. glauca* crown gall (strain T-37) tumor (organoid, green-white, soft). (**4**) *N. langsdorffii* pith callus (amorphous, light green, hard). (**5**) *N. langsdorffii* crown gall B-6 tumor (amorphous, white, friable). (**6**) *N. langsdorffii* crown gall T-37 tumor (organoid, green-white, soft). (**7**) Hybrid pith callus (organoid, green-white soft). (**8**) Hybrid crown gall B-6 tumor (amorphous, yellow-white, hard). (**9**) Hybrid crown gall T-37 tumor (organoid, green-white, soft). (**10**) Hybrid seed tumor (amorphous, light green, hard). (**11**) Hybrid spontaneous tumor (organoid, green-white, hard). (**12**) *N. glauca* crown gall B-6 tumor, which has been transformed from amorphous to organoid form by heat shock. (**13**) *N. langsdorffii* crown gall B-6 tumor in early stage of transformation from amorphous to organoid form after heat shock. [From W. R. Sharp and J. E. Gunckel: *Plant Physiol.*, **44:**1073–79 (1969). Used with permission. Photograph courtesy J. E. Gunckel.]

Later experiments by the British cytologist F. A. L. Clowes, however, show a different pattern. Roots were treated with thymidine labeled with the radioactive isotope of hydrogen, tritium (^{3}H). Thymidine is utilized in the synthesis of DNA but not RNA. Thus cells synthesizing DNA in preparation for cell division take up ^{3}H-thymidine and can be identified by the process of tissue autoradiography. In this technique, a thin section or a squash preparation of the tissue is placed on a microscope slide and covered with a photographic emulsion sensitive to the radiations (β particles) emitted by tritium. After a period of time the film (usually still on the slide) is developed. The localization of tritium within the tissue and within the cells can be seen by the darkening of the emulsion, as is shown in Figure 18-5.

Figure 18-5. An autoradiograph of a root tip section of *Sinapis* labeled with thymidine-^{3}H, showing the location of cells that are actively dividing. The dividing cells are synthesizing DNA, and have incorporated labeled thymidine, which shows in the radio-autograph. Note the quiescent center, where no cell divisions are taking place. [From F. A. L. Clowes and B. E. Juniper: *Plant Cells*. Blackwell Scientific Publications, Oxford and Edinburgh, 1968. Used with permission.]

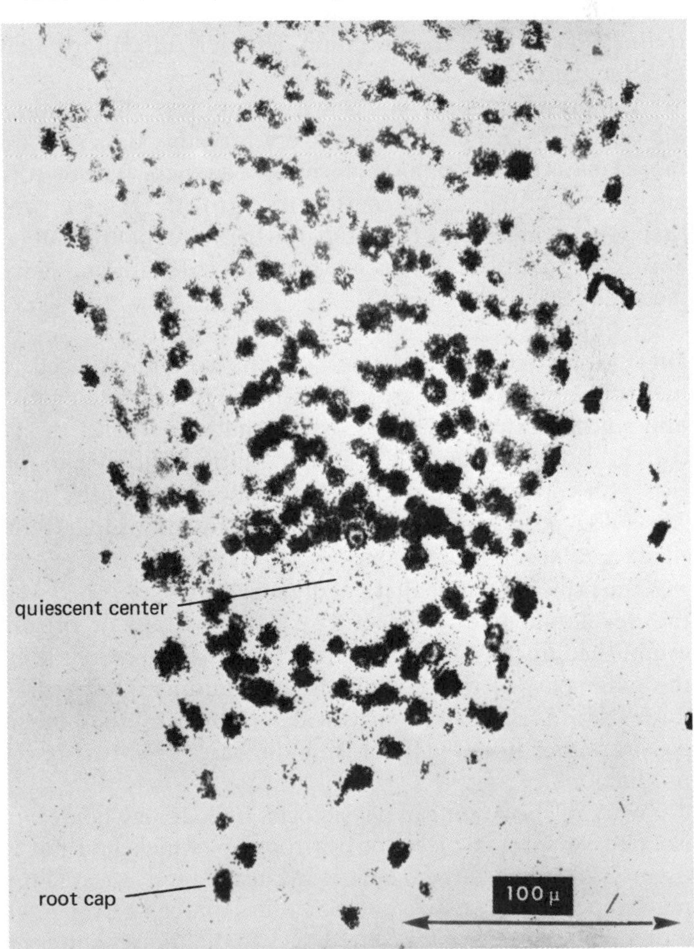

Clowes's experiments show the existence of a **quiescent center,** consisting of a number of cells that do not divide, located just behind the top of the root (Figure 18-5). These are surrounded by a group or layer of actively dividing cells, which give rise to columns of cells forming the tissues of the root. The tip of the root is protected by the root cap; this grows from a meristem close to the apical surface of the root, new cells being continuously produced in the center to replace cells at the outer surface that are abraded or sloughed off as the root grows.

Control of Root Growth. Root growth appears to be normally under the control of auxin concentration, but a good deal of uncertainty exists about this. Many years ago the American physiologist P. R. White succeeded in culturing tomato root tips in a relatively simple medium containing salts, sugars, and brewers yeast extract. The roots could apparently grow indefinitely. Thus if they needed a supply of IAA, they must have synthesized it themselves. Yet no conclusive evidence has been found for the natural synthesis of IAA in roots.

Usually IAA is required only in very small amounts, and the difficulty may be really detection and analysis. It is apparent that under normal circumstances IAA is transported from the shoots of plants to their roots. Although IAA is required for root elongation, its concentration in roots of intact plants is normally well above the extremely low optimal concentration. Thus it is difficult to determine an auxin requirement in attached roots.

The presence of cytokinins is necessary for cell division in roots. As in other tissues, the type and pace of growth are very probably dependent not only on the presence of these factors but upon the balance between them. The Swedish physiologist H. Burström has suggested that auxin controls root growth through two separate effects. He found that auxin accelerates very early growth of the root tip but inhibits later expansion. This apparent duality of action may be due to changing concentrations of other growth factors such as cytokinins.

Differentiation of Tissues. As the columns of cells that are generated by the meristem at the root tip are left behind by the advancing tip, they begin to elongate and differentiate into the various tissues characteristic of the mature root: the epidermis, cortex, and stele. The stele itself is organized into xylem, forming a star-shaped central axis with columns of phloem between the points of the star, encircled by the pericycle. The xylem pattern may be **diarch** (two-pointed), **triarch** (three), **tetrarch** (four), **pentarch** (five), and so on (see Chapter 4, page 79).

The rather precise pattern of organization of tissues raises two questions of interest in the development of the root: (1) Is the pattern or organization of developing cells established by the meristem (that is, from the tip) or by the already differentiated cells in the older part of the root? (2) Is the stimulus for specific differentiation transmitted directly from cell to cell, or is it determined by the emission of hormones or growth factors, either from the tip or from the base, which act in cells far from the site of their production?

Work of the American physiologist J. G. Torrey bears on both of these questions. He isolated very small pieces of pea root tip, which had not yet begun to form vascular tissue, and cultured them in vitro. Many of these isolated tips formed the normal triach pattern of vascular tissue, but sometimes diarch or even monarch patterns resulted. The monarch or diarch roots reverted back to the triarch pattern with time as they grew. This shows that the pattern is established in or through the agency of the tip and not as the result of the influence of mature tissue. Some of the small root tips were sufficiently disturbed by excision to change this pattern, but the altered pattern was not maintained and reverted to normal as the root grew.

In complementary experiments Torrey removed the tip from roots and allowed the decapitated root to regenerate a new tip. The pattern of vascular tissue laid down by the new root tip did not always match that of the older tissue, and, if auxin (10^{-5} *M*IAA) was added to the culture medium, hexarch vascular tissue was formed. This strongly suggests that hormones or growth substances, possibly (but not necessarily) including IAA, are involved in establishing the pattern of vascular tissue. The size of the meristem, which is probably affected by growth factors, appears to influence the pattern of vascularization either directly or through its production of hormones.

The pattern of secondary vascular tissue follows the pattern set by primary vascularization, but its development is not automatic. It has been noted that secondary vascular tissue does not develop in detached roots in culture, regardless of how old they are or how long they grow. The roots of many plants, particularly those like the radish that normally develop large roots, thicken best under short-days conditions. This suggests that factors from the leaves (or at least, resulting from the perception by leaves of short days and their reaction to this stimulus) are responsible for the development of secondary cambium and secondary vascular tissue in roots. Experiments have been performed in which the addition of IAA at high concentration (10^{-5} *M*) together with sucrose solution causes initiation of secondary cambium in some roots and stimulates it to divide with the formation of secondary xylem and phloem.

Certain roots also require small amounts of cytokinins and a cyclic hexitol (myo-inositol) for successful secondary growth. These points illustrate again the necessity for a precise balance of several nutritional and stimulatory substances to achieve a specific kind of growth. They strongly suggest (but do not prove) that some or many of these factors are derived from the shoot. Apart from the carbohydrates for nutrition and probably several growth factors, roots are dependent on the shoot for B vitamins (thiamin and nicotinic acid) which they require for growth, and possibly for certain amino acids as well.

Lateral Roots. Branching occurs by a simple mechanism in roots. The meristem does not divide or branch, but some distance back from it, at a region where vascular differentiation is well under way, new meristems arise at more or less regular intervals in the pericycle. These areas are usually close to or opposite the points of the xylem star, so that triarch roots usually have three rows of secondary roots, tetrarch have four rows, and so on. The new meristem differentiates and grows outward through the cortex, and its vascular elements integrate or connect with the vascular system of the main axis by the differentiation of cells behind it (see Figure 4-9, page 80).

Available evidence indicates that the initiation of branch roots is under the control of auxin, presumably supplied from above, and very probably in balance with the concentration of cytokinins that are produced at the root tip. Thus branch root formation is inhibited near the root tip, where the auxin concentration is lowest and the cytokinin concentration is higher. Further up the root, higher auxin and lower cytokinin concentrations prevail, and a more favorable balance of these two regulators results in the initiation of branch roots. Branch roots are frequently initiated at very regular intervals, and this, together with their relationship with the pattern of protoxylem in the main axis, strongly supports the concept that gradients of translocated substances are important in their initiation. Experiments have shown that branch root formation can indeed be controlled by the addition of auxin.

Shoot Growth

Terminal Meristem. The apical meristem of the stem produces not only the stem but also the leaves, branches, and other appendages of the shoot. Unlike the root, all the branches and appendages arise at the surface of the apical meristem, as bumps or outgrowths from it, as shown in Figure 18-6. The apical meristem is thus a very much more complex structure than the root meristem. However, like the root meristem, it is very small, usually no more than a fraction of a millimeter in diameter. Leaf primordia are formed at regular intervals and the stem is divided by **nodes** at the points of leaf insertion. **Internodes** are usually very short at first. Stem elongation takes place after the formation of leaves. In certain plants the internodes never do elongate, resulting in a characteristic bulb or rosette habit of growth.

The organization of cell division in the meristem has suggested the **tunica-corpus** theory (see Chapter 4, page 74). The outer layer or layers of cells (tunica) divide by anticlinal division (at right angles to the surface) and form primarily the epidermis of the

Figure 18-6. Scanning electron micrograph of the shoot apex of Swedish ivy (*Plectranthus australis*) ×250. Two pairs of opposite leaf primordia may be seen surrounding the meristem. The older (outside) leaves have been dissected away to reveal the apical meristem. (Original photograph kindly supplied by Dr. R. L. Peterson, University of Guelph, Ontario.)

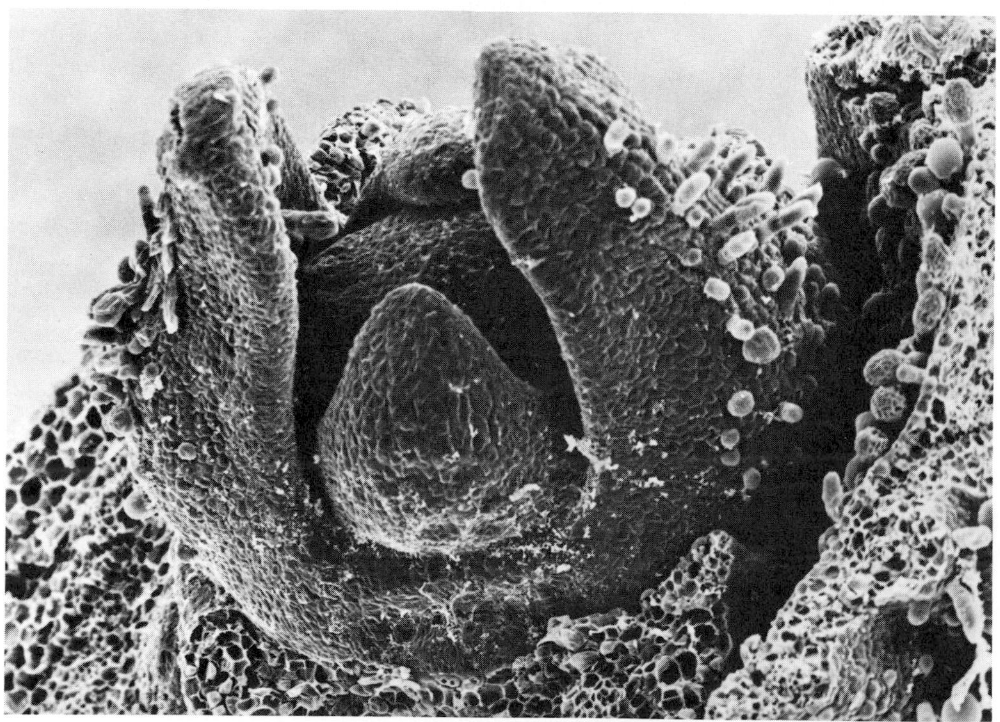

organ derived from the meristem. The cells of the corpus, lying inside the tunica, divide initially in an apparently random orientation, and later in an oriented way, to produce rows of cells that ultimately form the majority of the internal tissue of the new stem and organ. The formation of primordia that will give rise to buds, new shoots, or leaves involves both tissues. Increased periclinal divisions of the corpus causes the projection of bumps that form the new meristem; increased anticlincal divisions of the tunica permit it to push out and enlarge to accommodate the increased growth of the corpus below. We cannot now say which process comes first, but it seems possible that the corpus enlarges first and the tunica development follows.

Stem Growth. The elongation of the shoot is a complex phenomenon very much under the influence of environmental factors and controlled by hormones or growth factors. As one might expect, many different patterns of growth occur, not only among different plants but also among the different internodes of the same plant. It seems unlikely that there are qualitative differences in response among different parts of the plant, but rather quantitative differences in the levels of growth-regulating factors.

Auxins have long been known to affect shoot elongation. Coleoptiles are strongly reactive to IAA, as has been described previously (Chapter 16, page 389), and there is a strong positive correlation between the rate of elongation of the stem of some plants and the amount of IAA they contain, as is shown in Figure 18-7. In fact, the growth of excised pea stem sections is used as an assay system for IAA. Intact plants do not usually react to applied auxin by increased growth, but this is probably because they already contain optimal concentrations of IAA supplied by the shoot apex. If the terminal bud is removed, growth stops. If IAA is applied, it starts again. Thus it appears that IAA is necessary for normal elongation of stem tissue.

Gibberellin also affects stem growth, but in different ways. Whereas auxin causes elongation primarily by causing cell elongation, gibberellins stimulate cell division as well as elongation. Figure 18-8 shows the response of the growing tip of the water pimpernel (*Samolus parviflorus*) to gibberellic acid application; the great increase in cell divisions just below the apex is evident. This particular plant has a rosette form of growth; the application of gibberellic acid causes it to grow tall. It will be remembered that a number of genetically dwarf plants will grow as tall as their normal counterparts with the addition of gibberellic acid. In rosette plants, as in genetic bushy or dwarf plants, the endogenous production of gibberellic acid is very low. As with IAA, there is a good correlation between the amount of gibberellic acid present in the stem and the

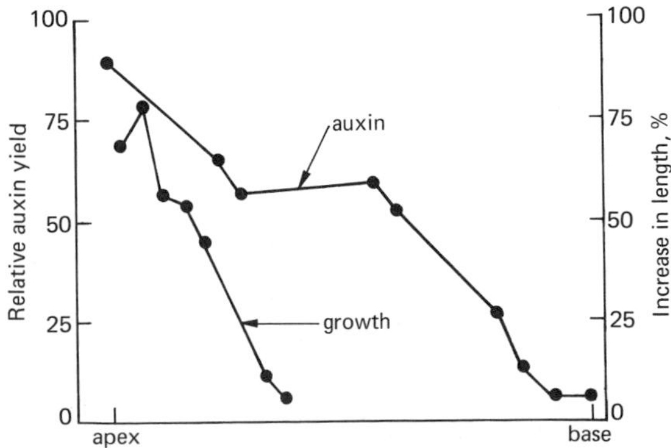

Figure 18-7. A comparison of the yield of diffusible auxin and growth in the epicotyl of Alaska peas (*Pisum sativum*). [Redrawn with permission from T. K. Scott and W. R. Briggs: Auxin relationships in the alaska pea (*Pisum sativum*). *Am. J. Bot.*, **47**:492–99 (1960).]

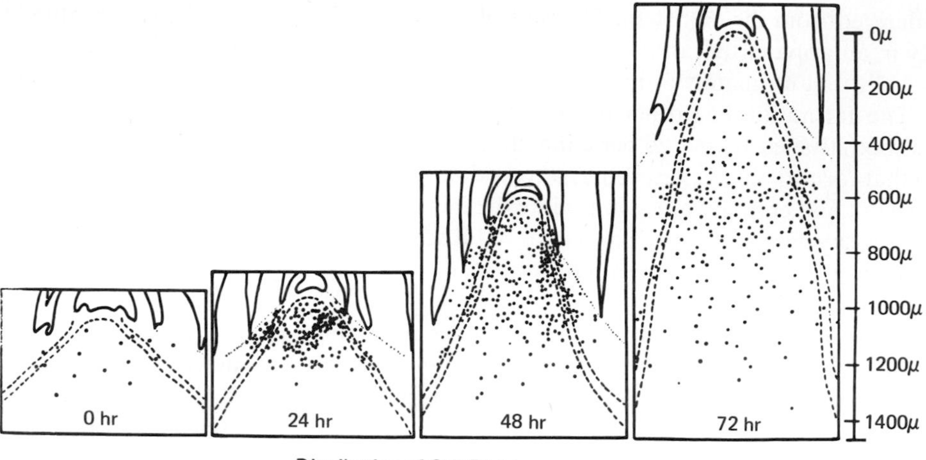

Figure 18-8. Number and position of mitotic figures, indicated by dots, in the median 64μ of the apex of *Samolus parviflorus*, following application of gibberellic acid (25 μg gibberellic acid were applied at 0, 24, and 48 hr). [From R. M. Sachs, C. F. Bretz, and A. Lang: *Am. J. Bot.*, **46:**376–84 (1959). Used with permission.]

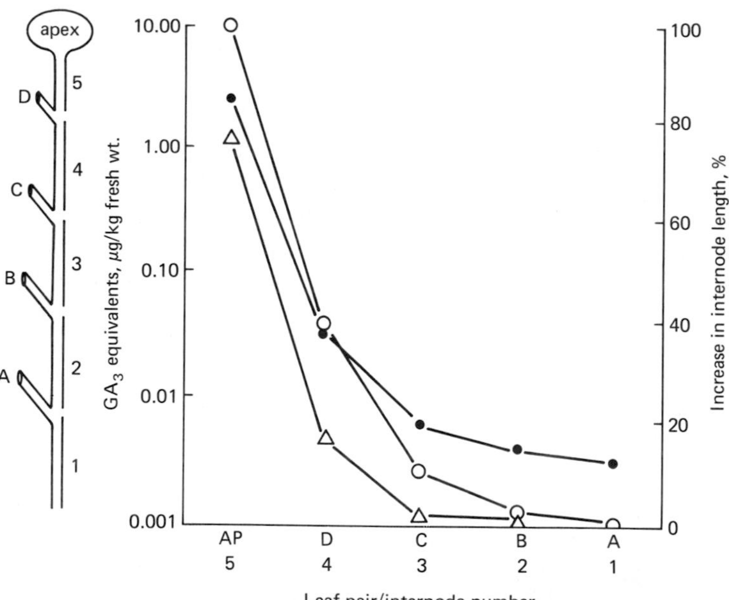

Figure 18-9. Relationship between the amounts of diffusible gibberellic acid-like (GA-like) substances obtained from leaf pairs A to D, apical buds, and internodes 1 to 4 of mature sunflower plants and the growth rates of comparable internodes. ○—○, diffusible GA-like material from apical buds and leaf pairs D to A. △—△, GA-like material diffused from internodes 5 to 1. ●—●, percentage increase in internode length. [From R. L. Jones and I. D. J. Phillips: *Plant Physiol.*, **41:**1381–86 (1966). Used with permission.]

rate of growth, as shown in Figure 18-9. We may conclude that gibberellic acid is normally involved in the growth of the stems.

Since IAA and gibberellic acid both affect stem growth, the question of their interaction naturally arises. Intact plant stems usually react to gibberellic acid but not to IAA. On the other hand, excised stem sections that react to IAA are usually unaffected by giberellic acid. When IAA and giberellic acid are added simultaneously to isolated stem sections, however, the reaction has been shown to be much greater than even from IAA treatment alone. Evidently the presence of auxin (endogenous in the intact stem, added to the stem sections) is necessary for the gibberellin to exert its full effect. Also, it seems likely that cytokinins supplied by the roots are important in stimulating growth and differentiation in shoots. Once again, the importance of the balance between different growth substances must be stressed. However, the interaction of various growth substances in the control of stem growth is still far from being worked out.

Leaf Primordia. Bumps or elevations appear on the periphery of the meristem in a regular pattern. When leaves are borne in pairs (opposite leaf arrangement), each pair usually appears at right angles to the preceding pair. However, when they are borne singly, their arrangment may be a little more complex. The angular distance between one primordium and the next determines the angle between successive leaves on the stem.

The arrangement of leaves on the stem is called the **phyllotaxy** of the plant. Leaves that are directly above one another are in one **orthostichy** in the stem. The description of the phyllotaxy of a plant is best achieved by following a spiral line through the leaf primordia in the order in which they appear, as in Figure 18-10. Then the phyllotaxy is described by the ratio of the number of spiral turns between two leaves in the same orthostichy to the number of leaf primordia through which the line passes. Thus, a plant with alternating leaves in two orthostichies would have a phyllotaxy of $\frac{1}{2}$, and if in three orthostichies, $\frac{1}{3}$. A phyllotactic system of $\frac{1}{4}$, is seldom seen; normally the next in the series is $\frac{2}{5}$ (five rows of leaves, two complete spirals between leaves directly above one another). The series $\frac{1}{2}, \frac{1}{3}, \frac{2}{5}, \frac{3}{8}, \frac{5}{13}, \frac{8}{21}, \ldots$ is called the Fibonacci series. In this series both the numerator and the denominator of each term represents the sum of the numerators or denominators, respectively, of the two preceding terms. The angle between successive leaves in the same spiral approaches a limit of $137° 30' 28''$, and at this point an endless series of leaves would be produced such that no two were precisely above one another.

Some advantage may result in terms of efficiency of light-trapping if leaves are arranged so that they do not overlap; however it is unlikely that the precision of plant growth is sufficient to make any difference beyond phyllotaxies of $\frac{5}{13}$ or $\frac{8}{21}$. Indeed, it is very difficult to determine precisely what is the phyllotaxy of plants with a more complex spiral arrangement. No matter how complex the phyllotaxy is, however, the organization of leaves is so precise that it has been used to measure nonlinear time in development. This is done by using as a time unit the **plastochron,** the time interval between the initiation of successive leaves. Numerous attempts are being made to analyze phyllotaxy mathematically, and it seems likely that these attempts will eventually provide some insight into the causes of the highly specific arrangements that are found in different plants. However, the underlying cause for the precise arrangement is still unclear.

The more likely explanations relate to the organization of the meristem. One possibility is that some positive signal is required to activate the formation of leaf

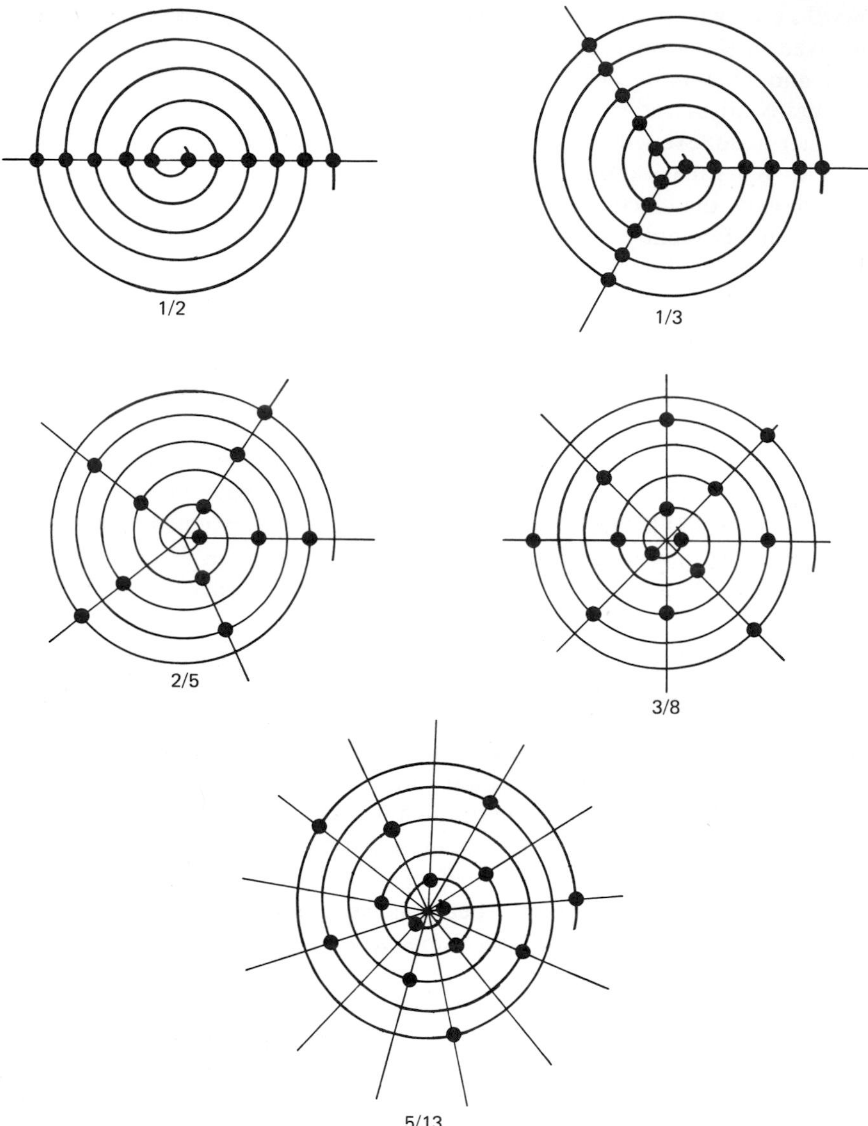

Figure 18-10. Patterns of phyllotaxy.

primordia at specific times, in specific places. Alternatively, areas not destined to form primordia must be suppressed. The presence of a spiralling "leaf-initiating stimulus" has been proposed on theoretical grounds, but this idea is not supported by evidence. As we have seen, the concept of organ-forming substances is not strongly supported by fact. Alternative theories depend on the idea of available space. The arrangement of leaves according to higher terms in the Fibonacci series results in maximum spacing between primordia, and it has been suggested that primordia develop only when a certain minimum spaces is available. This might be physical space. For example, a certain number of uncommitted cells might be necessary before primordial initiation can take place, or tensile stresses that develop in the meristem as the result of the development of earlier primordia might play a part. Alternatively, available space might be defined in terms of the influence of inhibitors or promotors, or the interaction of such substances, produced by older primordia already present.

It is not possible at present to decide which of these possibilities is correct. The British physiologists R. and M. Snow did surgical experiments on the apex of the white lupine (*Lupinus albus*) which indicated that the isolation of the areas in which primordia were expected to appear (using radial cuts in the apex) resulted in no primordia developing if the isolated area was too small. This suggests that the physical size of available space is important.

On the other hand the British morphologist C. W. Wardlaw, using fern apices, showed that a primordium in the apex isolated by surgical cuts grew more rapidly than normal, which suggested that inhibitory substances are normally produced by nearby primordia. He also showed that if this space where a new primordium was expected was isolated by surgery, a bud primordium would form instead of a leaf primordium. This suggests that, as in the callus tissue of tobacco, interaction of different stimuli, perhaps from different areas in the apex, are involved in the determination of the newly developing primordia. Unfortunately the surgical technique has severe limitations because wound-hormones involved in regeneration may be formed, and aberrant metabolism may result from these or from compounds produced in the actual metabolism of the regenerating tissue. As a result, it is not possible to arrive at a firm conclusion regarding factors which cause the organization of the apical meristem.

Differentiation. Provascular tissue, initially procambial, begins to differentiate very close behind the meristem, but the strands of vascular tissue do not originate from the central dome of the meristem. Instead they form columns below the leaf or branch primordia, and indeed they usually bypass the central meristem, passing up into the developing appendages. In dicotyledons, vascular tissue forms a cylinder. The projections or spurs of vascular tissue into leaves or branches are called **leaf** or **branch traces.** The gap left where the leaf trace branches from the central vascular cylinder is called a **leaf gap.** Since the vascular tissue of the main cylinder closes together above each leaf trace, the leaf gap appears as a small "window" in the cylinder (see Figure 18-11).

The provascular tissue does not all form at the same rate or in the same pattern. Cells of the primary phloem mature acropetally—that is, toward the top—and the differentiation of phloem follows the leaf trace from the stem up the axis of the leaf. Xylem has a different pattern. It matures first at about the point where the leaf joins the main axis of the stem, then up and down. The cells of the protoxylem, phloem, and cambium begin differentiation by elongation. The cambial cells retain their meristematic characteristics (thin walls, small vacuole, and so on), whereas the xylem and phloem cells undergo characteristic changes as they develop into mature conducting tissue (see Chapter 4, page 75). As in the root, the questions must be asked: Are certain cells destined to become xylem, phloem, cortex, pith, and so on, or are they directed to do so as the result of external forces? If so, what factors or forces influence their differentiation and where do they originate (in the meristem or from previously differentiated tissue)?

A number of experimental attacks have been made on these questions. Wardlaw and his associates did surgical experiments in which the central portion of the meristem was isolated from its leaf primordia and their associated vascular tissue. The meristem continued to grow and produced new leaf primordia, which developed procambial strands that were not associated with the preexisting vascular tissue. Experiments by the

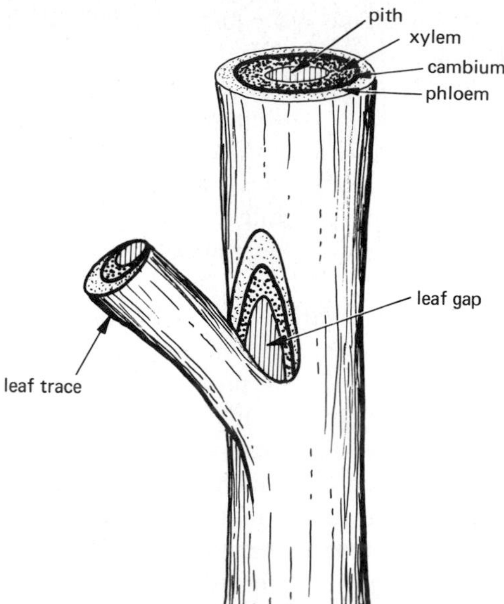

Figure 18-11. A diagram of the anatomy of the vascular tissue in a dicot stem showing a leaf trace and a leaf gap.

French physiologist G. Camus showed that buds would spontaneously arise on cultured callus tissue of chicory or dandelion, and these had organized vascular tissue. Furthermore, he found that, if a small bud was implanted into a piece of callus, vascular tissue connected with that in the bud was induced in the callus tissue below the bud. These experiments support the conclusion that, as in roots, the agent affecting tissue differentiation comes from the meristematic tissue at the stem tip and does not arise in the stem below. This concept, and particularly the experiments of Camus, rule out the possibility of "predetermination" of cells. Clearly, any cells, even those in a callus, can be programmed to develop into vascular tissue.

The next question that needs to be answered is that of the nature of the agency by which the meristem controls the various tissues below it. The American physiologist W. P. Jacobs found a very close relationship between auxin production in leaves and xylem regeneration in a healing wound in the stem. He also found a close correlation in time between the rate of IAA production and the rate of xylem-element formation in young leaves. His investigations led him to suggest that xylem formation is in some way affected by the double influence of auxin moving down the stem and nutrient substances, derived from the photosynthesis of mature leaves, moving up.

The American physiologist R. H. Wetmore and his associates, notably J. P. Rier, showed that buds grafted onto callus tissue of lilac stem would induce vascular tissue in the callus, in a manner similar to Camus' experiments with chicory. Then they found that a drop of IAA solution instead of a bud graft would also induce vascular tissue in the callus. Moreover, the presence of a sugar was found to be necessary for this induction, and the concentration of sugar affected the type of tissue formed. Low sugar concentration (1.5–2.5 percent) caused xylem formation; high concentration (3–4 percent) caused phloem formation; and intermediate concentrations usually resulted in xylem and phloem with cambium in between. The vascular tissue usually differentiated in nodules more or less evenly spaced in a circle in the callus tissue below the applied

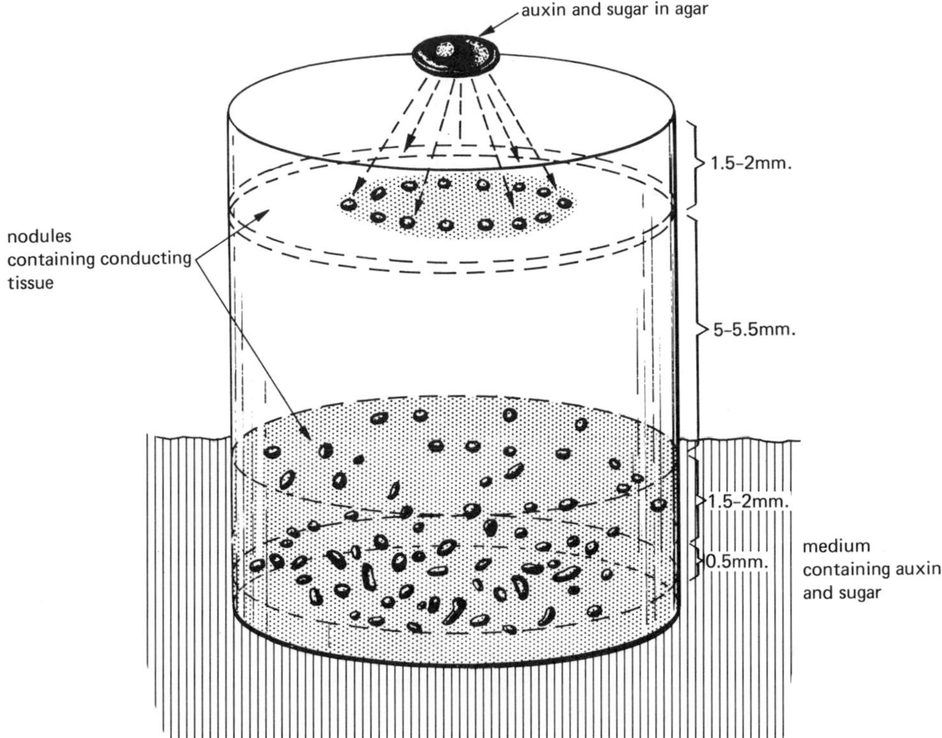

Figure 18-12. Stereogram of a piece of callus of *Syringa vulgaris*. To a small area in the top of the callus an auxin and a sugar (in the experiment 0.1 mg/liter NAA and 3% sucrose) are applied in 1% agar, and the same concentrations were used in the medium. Note the position of the circle of nodules 1.5 to 2 mm from the top, each of which showed tracheids toward the center and sieve elements toward the periphery (not shown in stereogram). Below the medium are found nodules, which may be spherical or irregular shaped or even resemble short strands in haphazard distribution. Characteristically, these nodules show xylem toward the medium and phloem away from it. The diameter of the upper circle of nodules increases with the concentration of auxin within the physiological range. The relative amounts of xylem or phloem in nodules, whether in the circle or below the medium, depend upon the concentration of sugar; lower concentrations of sucrose (1–2.5%) favor xylem formation, higher concentrations (3.5% upward) favor phloem, and middle concentrations encourage both xylem and phloem. This callus was killed for study 35 days after application of the auxin-sugar agar. [From R. H. Wetmore and J. P. Rier: *Am. J. Bot.*, **50:**418–30 (1963). Used with permission. Photograph courtesy R. H. Wetmore.]

auxin and sugar, as shown in Figure 18-12. The diameter of the ring of vascular tissue was proportional to the auxin concentration applied, and the orientation of tissues in nodules was such that the xylem was toward the center and the phloem toward the outside.

These results indicate that the differentiation of vascular tissues, as well as the diameter of the vascular cylinder, controlled by an interaction of IAA and nutrient sugars. Thus the specific agent of control is derived from the meristem, but the nutrient that interacts with it must come from below. It seems probable that the effects of sugar concentration are associated with the normal sugar concentration in vascular tissue: high in the phloem, low in the xylem. Thus phloem is induced by preexisting phloem and xylem by preexisting xylem, as determined by the sugar concentrations in them.

The differention of secondary tissue requires the formation of new cambium. Experiments by J. G. Torrey and R. S. Loomis on radish roots grown in culture showed that not only are IAA and sucrose necessary but also a cytokinin and a cyclitol such as myoinositol. The sugar concentration was not critical, and the cyclitol was not absolutely essential, but strongly beneficial. However, the requirement for auxin and cytokinin was absolute. The stimulus to cell division is believed to require both growth substances present together at optimal concentrations. Subsequent differentiation of cells and tissues may result from the expression of genetic information already programmed, as modified by the position of given cells in the tissue. Thus the hormones are required to trigger the event, but subsequent stages follow automatically.

The foregoing discussion suggests that the orientation and development of vascular tissues are under the external control of surrounding tissues and the gradients imposed by them. However, experiments by the Canadian developmentalist T. A. Steeves and his associates indicate that cambium from mature stems may possess a high degree of autonomy. Small square panels of bark were detached from stems of a number of trees and grafted back in place but reoriented at 90° or 180° from their original position. The cambium of the patches continued to function normally, producing new secondary xylem and phloem, but in its own orientation, and not aligned with the orientation of the stem. This demonstrates that whatever external factors initiated the development of vascular tissues, once formed, they retain a high degree of intrinsic autonomy and resist realignment by the gradients or flow patterns of the stem into which they are grafted.

Growth of Leaves

Leaves start as dome-shaped or bumplike primordia on the apical meristem and grow initially in a nearly cylindrical form. After a short time lateral meristems develop that grow sideways, giving the leaf its flattened form. The lateral meristems grow more or less rapidly at different positions along the leaf margin, thus giving the characteristic shape to leaves. The rapidity of growth might relate to nutrition because growth is often greatest opposite the ends of major veins. Alternatively, growth factors could be readily translocated by the veins, and act most strongly at their ends to stimulate growth.

Much variation is possible, and besides normal leaves, the primordia may develop into bracts, scales, floral parts, or any of the many structures developed from modified leaves, such as spines or tendrils. In many plants the shape and form of the leaves are quite firmly fixed, and little variation is found among them. However, a number of plants are **heterophyllous;** that is, they have several different forms of leaves. Some graduate from one type to another, whereas others change abruptly and often dramatically, as shown in Figure 18-13. Heterophylly is not uncommon in water plants such as *Potomogeton,* where one type of leaf is borne underwater and a quite distinctive type on the aerial part of the plant. The inference is that a number of factors, both internal and external, can modify the expression of the genetic information that determines leaf form. However, the constancy of leaf form on any one part of the plant, or under any given set of circumstances, indicates that the program, once started, is not readily interrupted by factors from outside.

Figure 18-13. Photo of *Acacia salicinae* showing heterophylly. Leaves produced when the plant was young have a typical acacialike leaflet arrangement at the tip. Leaves produced on the upper parts of the plant are straplike.

Various factors affect leaf shape. These include the periodicity, intensity, and quality of light. The perception of morphogenetic light (as opposed to photosynthetic light) is probably via the pigment phytochrome, which will be discussed in detail in Chapter 19. Other factors that affect leaf shape are oxygen and carbon dioxide concentrations. These may be the important factors that control the shape of leaves in heterophyllous plants which grow both above and below water. The concentrations of carbon dioxide and oxygen also control the length of the petioles of floating leaves such as water lily. It is not known how these stimuli operate. However, F. Went has shown that IAA affects linear growth in leaves, particularly the length of veins, hence the leaf area and some aspects of its shape. This is supported by the fact that heterophyllous plants usually show a progression of leaf forms up the axis, suggesting the influence of a gradient of growth substances in the stem (Figure 18-14).

In fact, the influence of a gradient of growth substances may arise in the primordia. As the shoot grows, the apical meristem develops or matures, and it may enlarge considerably. Primordia formed on a larger or more mature meristem will themselves be larger, will have a more rapid initial development, and will respond to their genetic information differently than will those formed earlier or on a smaller or less mature meristem. The size of the apex probably affects the relative amounts of cytokinin and auxin present, thus affecting the cell division and elongation pattern of the leaf. These changes in the apex may in some plants be responsible for the ultimate changeover from the vegetative to the flowering state (see Chapter 20).

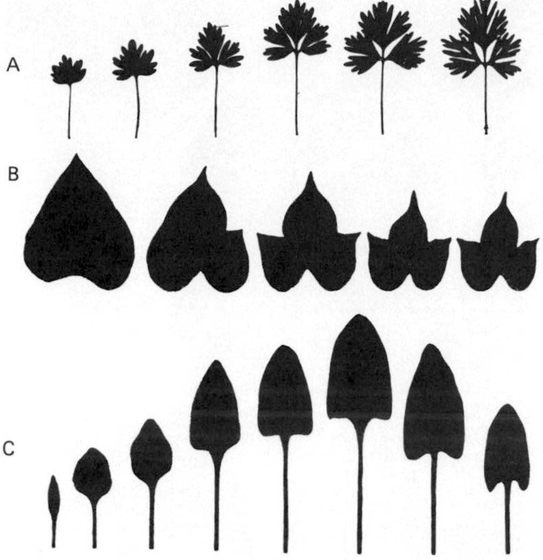

Figure 18-14. Examples of heterophyllic development.
(**A**) First six leaves of *Delphinium ajacis;* (**B**) first five leaves of *Ipomoea caerulea;* (**C**) first eight leaves of sugar-beet. [From E. Ashby: *New Phytol.,* **47:**153–76 (1948). Used with permission.]

Floral Development

The mechanism of floral initiation will be considered in detail in Chapter 20. Floral initiation is a dramatic event involving a total changeover of the character and developmental pattern of the meristem. The stimuli that lead to the induction of flowering are many and varied and may be either internal or external (that is, age or developmental stage of the plant, as opposed to periodicity of light, temperature, and so on).

There is much discussion in the literature about **florigens,** or flower-forming substances, that act in the formation of flowers. In fact, all steps in the flowering process are preprogrammed in the totipotent cells of the meristem. All that is needed is a trigger or a release that sets these cells on the way in the program for flowering. Once this path has been entered, the process, like most other developmental programs, is irreversible and automatic. The capacity to flower is inherent, like the capacity to form leaves. What is required is the signal (one or several growth substances already known, and perhaps others yet to be discovered) for a changeover to take place from one program to another.

Figure 18-15. (opposite) Floral buds of *Aquilegia formosa* cultured on agar containing artificial media.
A. Recently explanted bud at the dimpled-carpel stage, viewed from above, also showing staminodia and stamens; approx. ×75. **B.** Same bud as A after 27 days; sepals were removed when bud was dissected, but part of one sepal remains at lower left; two petals at front and right; approx. ×35. **C.** Recently explanted bud at the grooved-carpel stage, showing staminodia and stamens; approx. ×75. **D.** Same bud as C after 19 days; stamens and staminodia have aborted; sepals and petals were removed when bud was dissected; approx. ×30. **E.** Freshly dissected bud at the elongating-carpel stage, sepals removed; approx. ×55. **F.** Freshly dissected bud with young erect carpels, maturing stamens and stamodia, young petals; sepals were removed for visibility, front stamens removed to expose carpels; approx. ×20. **G.** Freshly dissected bud with mature carpels, staminodia, and stamens; petals are mature except for the spur; sepals were removed; approx. ×20. [From S. S. Tepfer, R. I. Greyson, W. R. Craig, and J. L. Hindman: *Am. J. Bot.,* **50:**1035–45 (1963). Used with permission. Photograph courtesy Dr. S. S. Tepfer.)

A
C
B
D
E
F
G

There seems to be a possibility that certain quiescent cells in the meristem that are "carried along" at the apex (rather like the quiescent center in the root) become activated and participate in flowering. The American cytologist L. F. Randolph exposed corn seeds to strong ionizing radiation, then germinated them and grew plants. Certain anomalies were detected in the leaves as a result of radiation damage to cells in the embryo. However, other and different anomalies were found in the flowers. This suggests that some cells initially present in the seeds, which were not concerned with vegetative growth but were carried in the shoot growing point, were later activated and became involved in flowering.

Wetmore observed that the central part of the meristem in certain plants contains large, quiescent cells. Toward the end of floral induction, however, these become activated and start dividing into small cells. In other words, cells that are latent in the vegetatively growing shoot become activated after floral induction. If this is correct, and the cells are specifically concerned with floral development, the stimulus to flowering becomes the stimulus that "turns on" these cells. However, this situation does not really involve any new principle. Such "floral" cells must have arisen initially from divisions of the totipotent zygote, and there has merely been an increase in the separation, in time and in space, between the programs for vegetative and for floral development.

That the initiation and development of flowering depend upon the balance of hormones or growth factors is shown by experiments with buds of *Aquilegia* (columbine). These can be cultured in vitro and will start, but, as may be seen in Figure 18-15, they cannot be made to complete, the formation of flowers. For unknown reasons the presence of sepals inhibits further development; if the sepals are removed, floral development continues. Some development occurs on a mineral medium fortified with

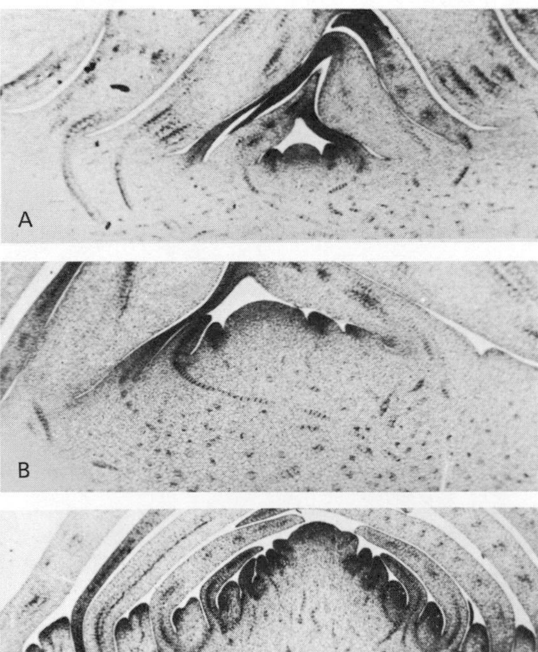

Figure 18-16. Development of the shoot apex of the banana plant (*Musa acuminata* var. Gros Michel) from (**A**) the vegetative state to (**C**) the floral state. Note how the meristem is enlarged and raised in (**B**). Floral primordia are visible in C. [From W. G. Barker and F. C. Steward: *Ann. Bot.*, **26:**413–23 (1962). Photographs courtesy Prof. Steward.]

coconut milk and vitamins. However, for maximum development additions of IAA, kinetin, and gibberellic acid are necessary.

Many experiments and observations have demonstrated the foliar nature of floral parts. However, the changeover from the vegetative to the floral state results in a substantial change in meristem organization, as shown in Figure 18-16. Primordia of the various organs—sepals, petals, stamens, and carpels—form in rapid succession on the enlarged apex. The opening of the flower is accomplished by differential cell enlargement on different surfaces of the petals. This may be a rapid response, and many flowers open and close regularly in response to light or other stimuli. Pollination induces far-reaching changes throughout the whole flower, as a result of hormones or growth factors produced by the developing endosperm and ovary. As a consequence, petals die and absciss, and fruit formation begins.

Throughout this account two main points have reappeared many times: we know very little about how development is programmed, or how the program is made into a physical entity. It is clear, however, that developmental responses, the initiation, growth, or development of organs, or the triggering of new developmental sequences, are seldom if ever brought about by one morphologically active substance working alone. What is required is a balance between two, three, or even more different substances, having different kinds of activities, different sources, and different physical and physiological properties, all acting together on specific cells at specific locations within the organism. Much more understanding of development will doubtless come from a clearer knowledge of the precise action of each of these morphogenetic factors, in their various possible permutations and combinations, on cells in every different site and situation in the organisms.

Additional Reading

See list for Chapter 16 (p. 406). Accounts of plant structure and development will be found in books on anatomy and morphology.

Clowes, F. A. L.: *Morphogenesis of the Shoot Apex.* Oxford University Press, London, 1972.

Maksymowych, R.: *Analysis of Leaf Development.* Cambridge University Press, Cambridge, England, 1973.

Northcote, D. H.: *Differentiation in Higher Plants.* Oxford University Press, London, 1974.

Sachs, R. M.: Stem elongation. *Ann. Rev. Plant Physiol.*, **16:**53–72 (1965).

Street, H. E.: The physiology of root growth. *Ann. Rev. Plant Physiol.*, **17:**315–44 (1966).

Wain, R. L., and **C. H. Fawcett:** Chemical regulation of plant growth; and **F. W. Went** and **L. O. Sheps:** Environmental factors in regulation of growth and development. Both in F. C. Steward (ed.): *Plant Physiology: A Treatise,* Vol. VA. Academic Press, New York, 1969.

Williams, R. F.: *The Shoot Apex and Leaf Growth.* Cambridge University Press, Cambridge, England, 1975.

19 Organization in Space

Direction of Growth

As we have seen, plants respond to a number of external or environmental stimuli as well as to the genetic instructions that they carry. Much of the orientation of a plant in space results from its reaction to directional stimuli, particularly light and gravity. Plants may react by **growth movements,** which are plastic or irreversible changes resulting from growth. Alternatively, they may react by **reversible movements,** which are elastic changes usually resulting from turgor changes in certain cells. Further, the reaction may relate to the direction of the stimulus (that is, be in the same direction, or opposite, or at a specific angle to it). Such a reaction is called a **tropic** response. Examples include geotropism (response to gravity), phototropism (light), thigmotropism (touch), hydrotropism (water). Responses that are unrelated to the direction of the stimulus are called **nastic** responses and include epinasty (bending down), hyponasty (bending up), nyctinasty (sleep movements or rhythmic raising and lowering of leaves), seismonasty (response to mechanical shock), and the reactions of various types of traps in carnivorous plants. We shall first examine tropisms.

Tropic Responses

GEOTROPISM. Plants may grow up (negative geotropism—opposite to the direction of gravitational pull) or down (positive geotropism), horizontally (at right angles to gravity, **diageotropism**) or at some other fixed angle with respect to the vertical **plagiotropism).** Plants can evidently sense gravity and have a mechanism for responding to it. Gravity is not perceived throughout the whole plant. Root caps appear to be the perceptive area for roots; if the caps are removed, no geotropic response occurs. Similarly, the stem tip is essential for geotropic response in stems. Many experiments have been done using the natural gravity of earth; gravity can be experimentally varied by the use of centrifuges, which may increase it, or by the **klinostat** (Figure 19-1). This is a device for rotating a plant about its long axis, usually while it is lying on its side, so as to neutralize the directional effect of gravity.

It can be concluded that the response to gravity is inductive, because the response can be separated in time from the sensing of gravity. If a plant is placed horizontally for

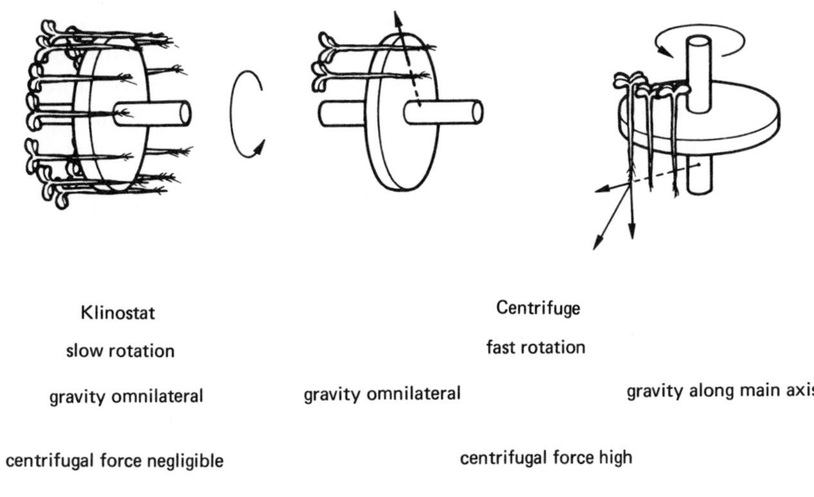

Figure 19-1. Diagrams showing the klinostat and centrifuge techniques for studying geotropism. [From L. J. Audus: Geotropism. In M. B. Wilkins (ed.): *The Physiology of Plant Growth and Development*. McGraw-Hill, London, 1969, p. 214. Used with permission.]

a short time and then replaced vertically, it will subsequently bend as if it were still on its side. Later, it will return to the vertical again. Othei factors, such as light and temperature, also affect the response. For example, the stolon (underground stem) of certain grasses is normally diageotropic (grows horizontally) if the leaves receive light. However, if the leaves are placed in continuous darkness, the stolon becomes negatively geotropic and grows upward. This response ensures that the plant will spread by stolons horizontally along the surface of very uneven ground and prevents the plant from accidentally growing down into the ground or into the sides of hummocks.

Perception of Gravity. The fact that perception of the stimulus and reaction to it may be separated in time means that there are two separate processes involved: perception and reaction. Gravity might be perceived by a cell in two ways—by the perception of the pressure differential across the cell, or as result of the differential distribution of lighter or heavier particles in the cell resulting from their sinking or floating in the cytoplasm. Measurement of pressure differential seems very unlikely because the pressure exerted by the weight of the cell contents is many orders of magnitude less than that exerted by the turgor of the cell. Further, even the fluctuations of the internal pressure resulting from the changing water status of the cell would greatly exceed the pressures due to the weight of the cell contents. Hence the existence of a mechanism that could measure the pressure differential across a cell and detect its lateral displacement with high precision seems extremely doubtful.

The much more attractive hypothesis is that gravity is perceived by means of **statoliths.** Statoliths are small bodies of high specific gravity that sink to the bottom of the cell. Sedimenting bodies in the cytoplasm include the nucleus, dictyosomes, mitochondria, and starch grains (or more properly, the **amyloplasts,** because starch grains occur in plastids and not free in cells). Alternatively, lighter floating bodies would include the vacuole and fat globules.

A large body of evidence now indicates that starch grains or amyloplasts are normally the statoliths in cells that perceive gravity. First, they are the only particles that are heavy enough to sink fast enough for the known perception time. Many tissues can react after a stimulus lasting only $\frac{1}{2}$–1 min (the duration of the stimulus is called the **presentation time**). Second, starch grains are present in nearly all geotropically sensitive tissues. In the early twentieth century the German physiologists C. Zollikoffer and E. Stahl experimentally starved roots and noted that as the starch disappeared from the root caps during starvation so did their sensitivity to gravity.

A few tissues do not lose geosensitivity on starvation. However, it has been noted that, although starch grains in most tissues are easily removed by starvation, statolith starch is very hard to eliminate in this way. There is a striking parallel between the rate of sinking of starch grains and the time required for the perception of gravitational stimulus, as is shown in Figure 19-2. Microscopic examination of gravitationally stimulated cells shows that the starch grains inside them do indeed sink to the bottom, as is shown in Figure 19-3. The American physiologist A. C. Leopold, whose work is shown in Figure 19-3, has experimented with a corn mutant that has fewer and smaller amyloplasts than the wild type. As would be expected if starch grains are statoliths, this mutant shows slower and weaker responses to gravitational stimuli. This work, together with studies on geotropism in decapitated root tips and during root-cap regeneration done by British physiologists B. E. Juniper and P. W. Barlow, demonstrates quite clearly that starch amyloplasts are normally the statoliths through which plant cells perceive gravitational fields.

As mentioned previously, certain tissues do not lose their sensitivity to gravity even after starvation. The American physiologist K. V. Thimann found that, in wheat coleoptiles, treatment with gibberellic acid resulted in the loss of all starch, but the

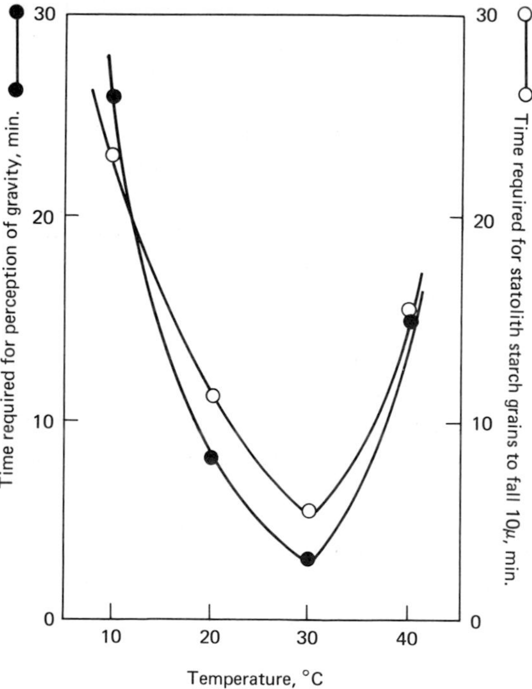

Figure 19-2. The effect of temperature on the rate of sinking of statolith starch grains compared with the time required for perception of a gravitational stimulus. [Recalculated from data of Lillian E. Hawker: *Ann. Bot.*, **47:**503 (1933).]

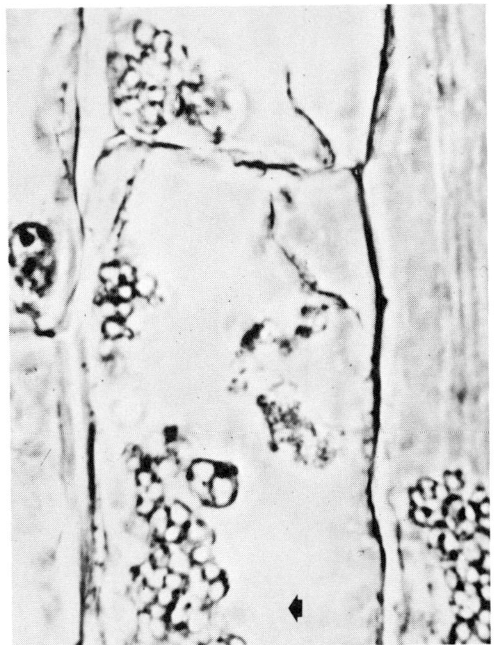
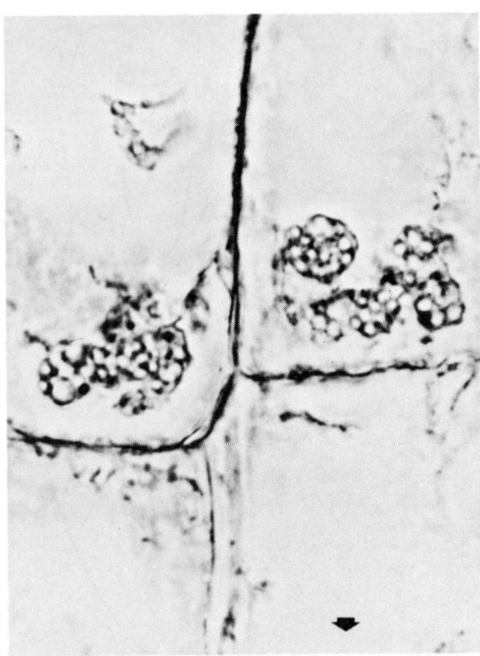

Figure 19-3. Statoliths in cells of corn (*Zea mays*) coleoptiles. The cells on the right were "upright," those on the left were lying on their side. The arrows show the direction of the gravitational field. [Photograph courtesy Professor A. C. Leopold.]

coleoptiles were still geosensitive. In this case, perhaps some other subcellular particles than amyloplasts act, either temporarily or normally, as statoliths. It has been suggested that dictyosomes or mitochondria may also act in this capacity. However, the evidence mentioned earlier makes it likely that amyloplasts normally function as statoliths in most geosensitive tissues.

Mechanism of Response to Gravity. The physiologist F. Went, now working in America, early discovered that extension growth is promoted by auxin. Curvature of a stem is brought about by differences in rates of growth between one side and the other. Hence, N. Cholodny and F. Went suggested that responses in which a stem or leaf undergoes curvature result from asymmetrical distribution of auxin in the tissue. Furthermore, roots appear to respond to auxin in the opposite way from stems; that is, added auxin decreases the growth of roots instead of increasing it. Thus, Cholodny and Went invoked a common mechanism for geotropic responses based on an increase in the auxin content of the lower side of the gravitationally stimulated tissue, which would work for both positively geotropic roots and negatively geotropic stems. The increased auxin content on the lower side would lead to increased growth on that side of the stem, causing it to bend upward, and to decreased growth on the side of a root, causing it to bend downward. This mechanism is illustrated in Figure 19-4.

Substantial evidence supports this mechanism for shoots, but not for roots. It is possible to measure the existence of an auxin gradient across gravitationally stimulated stems, but not in roots. Instead, root caps have been found to be the source of a growth-inhibiting substance. If part of a root cap is removed, the root bends toward the

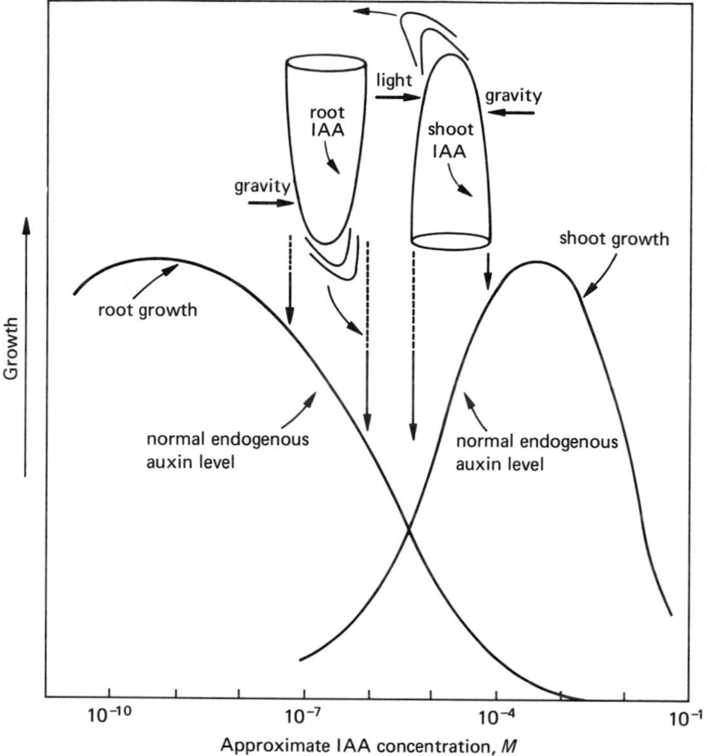

Figure 19-4. Diagram to show how a common mechanism of lateral auxin redistribution might mediate geotropic and phototropic response in shoots and the opposite geotropic response in roots.

side having the remaining portion of the cap attached. This suggests that the cap produces growth inhibitors. Furthermore, removal of the root cap causes a brief increase in root growth, showing it to be the source of inhibitors rather than growth-promoting substances. Abscisic acid has been found in root caps, and it is now considered possible (but not proved) that geotropic growth stimulus is mediated by the asymmetric production of ABA in the geotropically stimulated root cap.

The situation with stems is somewhat different. In experiments with sunflower hypocotyls, the German physiologists L. Brauner and A. Hager placed tissue in the dark to decrease its auxin content. After this treatment, the hypocotyls became insensitive to gravity. However, if after gravitational stimulus some auxin was added to the hypocotyls, they then underwent the normal responses, bending away from the direction of the gravitational field. These workers found that hypocotyls would not react to a gravitational stimulus if they were cooled. However, if the hypocotyls were warmed after the period of stimulus, they would begin to react up to 12 hr after stimulation. Oxygen was found to be necessary for the reaction, showing that metabolic reactions were involved in the response.

Recently, using radioactive IAA, Leopold and his colleagues have demonstrated that auxins are transported laterally under the influence of a gravitational field in corn coleoptiles, as is shown in Figure 19-5. It would thus appear likely that the gravitational

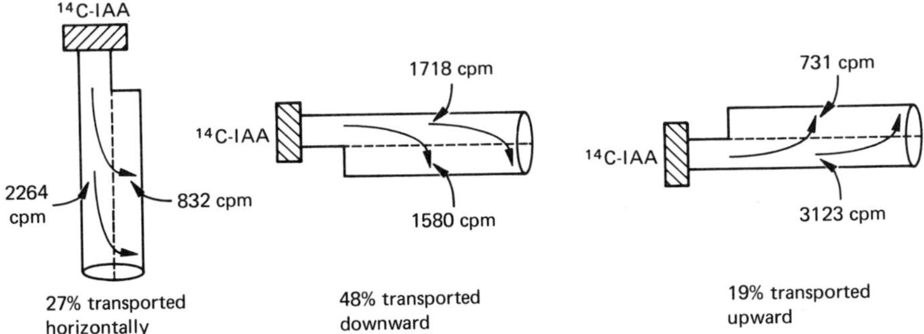

Figure 19-5. Experiment to demonstrate lateral transport of auxin in corn coleoptile sections under the stimulation of gravity. [From data of R. Hertel, R. K. de la Fuente, and A. C. Leopold: Geotropism and the lateral transport of auxin. *Planta,* **88:**204–14 (1969).]

response is indeed mediated by asymmetric lateral distribution of IAA in the tissue resulting from the gravitational polarization of sensitive cells.

The question remains, how does a gravitational stimulus, through the behavior of statoliths, affect the redistribution of hormones, either ABA or IAA, in the tissue? It seems likely that the polar distribution of auxin in stems results both from lateral transport and from increased production of IAA on the lower side of the stimulated tissue. It has been suggested that the higher concentration of cellular organelles at the lower side of the cell under the influence of gravity reduces the efficiency of their metabolism by decreasing their access to metabolites and causing a build-up of metabolic products. Thus the slower metabolism at the lower side of the cell may in some ways bring about an increased synthesis or a preferential transport of IAA. Gravitationally stimulated tissues have been noted to develop an electrical potential perpendicular to the field of gravity. However, tissues in which an IAA gradient is artificially induced have also been found to develop such a potential; it therefore seems likely that the electrical potential is the result of an IAA gradient rather than its cause.

The situation in roots has been critically examined recently by Juniper, who has proposed a hypothetical model in which statolith movement results in lateral redistribution of hormone. His model depends on the ability of statoliths to block the plasmodesmatal connections between cells, perhaps by pressing flaps of endoplasmic reticulum across the plasmodesmatal openings. As can be seen in Figure 19-6, when the cells are tipped on edge, the statoliths tend to tumble into a corner. As a result, diagonal free passages are available for the transport of inhibitors or stimulators, permitting them to move in a lateral direction. It must be emphasized, however, that this model is hypothetical at present. Proof still awaits further experimentation.

Thus, at present, we do not really know how the action of statoliths is used to cause stem or root bending. However, it seems very likely that a lateral bias is applied to the existing polarized (up or down) movements of hormones. Such polarized transport must be strongly maintained; remember that, if a stem is placed upside down, it still retains its original basipetal orientation, and it does not respond to a reversed gravitational stimulus by "changing ends."

PHOTOTROPISM. The understanding of the mechanism of phototropic reaction goes back to Went's experiments leading to the discovery of auxin (Chapter 16, page 389). It was

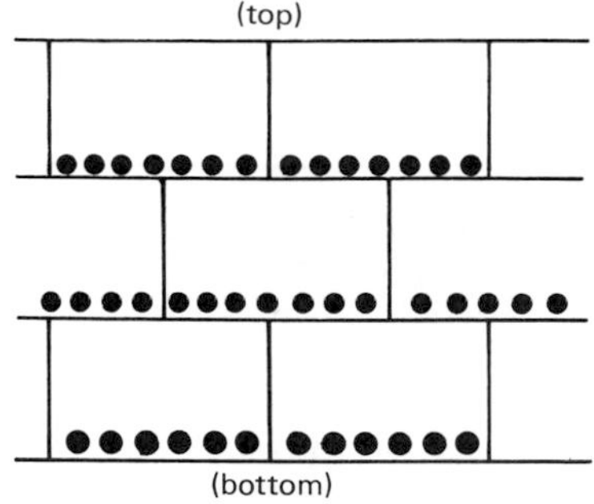

A. Tissue in normal position.

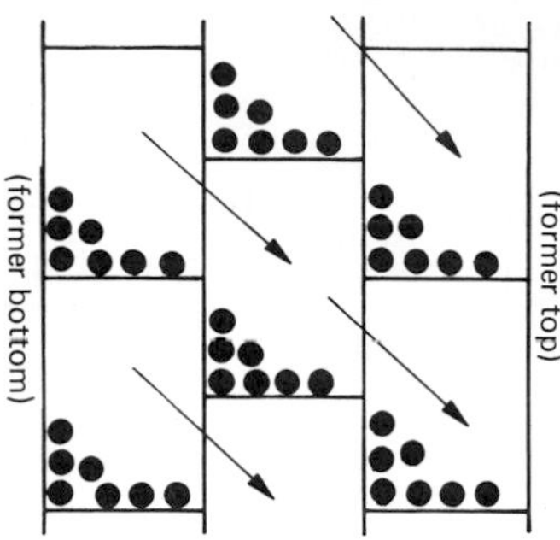

B. Tissues tipped on one side (rotated 90°).

Figure 19-6. Diagram to show a hypothetical model for statolith action. (**A**) Statoliths (black balls) occlude plasmodesmata when they lie against the cell wall, perhaps by pressing flaps of endoplasmic reticulum against the cell wall. When the tissue is disoriented, as shown in (**B**), they tumble across the new lower wall of the cells, thus permitting (or forcing) lateral movement of substances as shown by the arrows. Growth substances could move to the lower side as a result of an upward translocation pattern (as shown in B) or to the upper side if the main translocation pattern was down in the normal orientation of the tissue.

found that, if a coleoptile is illuminated from one side, an asymmetric distribution of auxin occurs so that auxin accumulates on the dark side of the coleoptile. The higher auxin on the dark side causes it to elongate more than the illuminated side, and the asymmetric growth causes the coleoptile to bend toward the light. The scheme of Went and Cholodny shown in Figure 19-4 illustrates how the reaction might work.

Earlier the asymmetrical distribution of auxin was thought to result from a combination of three different mechanisms—photodestruction of auxin on the illuminated side, increased synthesis of auxin on the dark side, and lateral transport of auxin from the illuminated to the dark side. However, there is now abundant evidence that photodestruction of auxin does not occur, and the work of the American physiologist W. R. Briggs showed clearly the lateral transport of auxin is the important mechanism. Recent experiments by Leopold and his group demonstrate the fact that lateral transport of applied ^{14}C-IAA occurs away from the lighted side of a corn coleoptile toward the darkened side, as shown in Figure 19-7. Apparently the normal lateral movement of auxin is not prevented by the presence of light, but its movement in the direction away from light is greatly stimulated.

The phototropic response of leafy shoots has been shown to depend on unequal illumination of leaves facing toward or away from the light. Consequently it has been suggested that unequal synthesis and export of auxin take place as a result of the unequal illumination of the leaves. According to this view more auxin would be exported from a darkened leaf than from an illuminated one, resulting in increased stem growth below the darkened leaf.

Leopold's group showed that, as would be expected, stem growth was greater when the cotyledons of sunflower seedlings were darkened. When one cotyledon was darkened, the side of the hypocotyl below the darkened cotyledon grew more than the side below the illuminated one, resulting in a curvature of the stem, as shown in Figure 19-8. Further, when diffusible auxin was extracted from below the cotyledons of seedlings, the

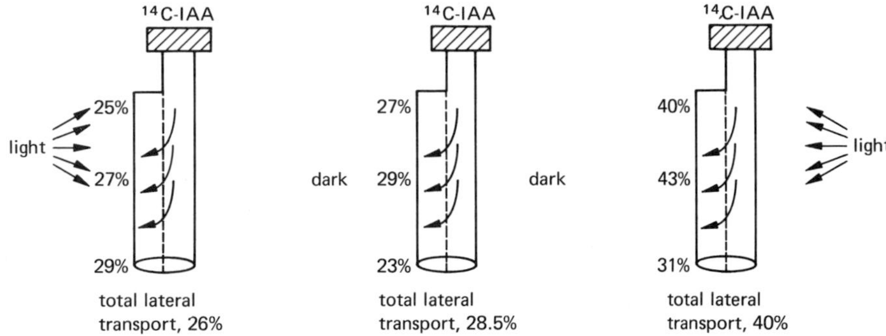

Figure 19-7. Experiment to show the effect of light on lateral transport of IAA in corn coleoptile sections. Lateral transport is increased away from light and slightly inhibited toward light. [From data of R. K. de la Fuente and A. C. Leopold: Lateral movement of auxin in phototropism. *Plant Physiol.*, **43:**1031–36 (1968).]

Avena coleoptile curvature test showed much stronger positive results if the seedlings were in light than in darkness (Figure 19-9). The results of these experiments agree with the well-known effects of auxin on elongation of stem tissues. However, there have been recent suggestions that gibberellic acid might act as well as (or instead of) IAA.

The work of Leopold and others shows that the cotyledons are necessary for phototropic response—the stem tissue will not react if the cotyledons are removed. On the other hand, recent experiments in Holland by J. Bruisma and others suggest that the illuminated cotyledon produces an inhibitory substance, rather than the darkened cotyledon producing a growth promoter. The conclusion will have to await further research.

Some questions must be asked about the exact location of the photoreceptor for phototropic light. Seedlings that have one or both cotyledons darkened by covering with black paper will still show positive phototropism, regardless of the orientation of the light and the darkened cotyledon(s). Covering cotyledons reduces the intensity of the response, but not its direction. However, covering the hypocotyl prevents phototropic

Figure 19-8. Effect of shading one or both cotyledons of sunflower seedlings on extension growth of the hypocotyl during a 24-hr period. [Redrawn with permission from S. Lam and A. C. Leopold: Role of leaves in phototropism. *Plant Physiol.*, **41:**847–51 (1966).]

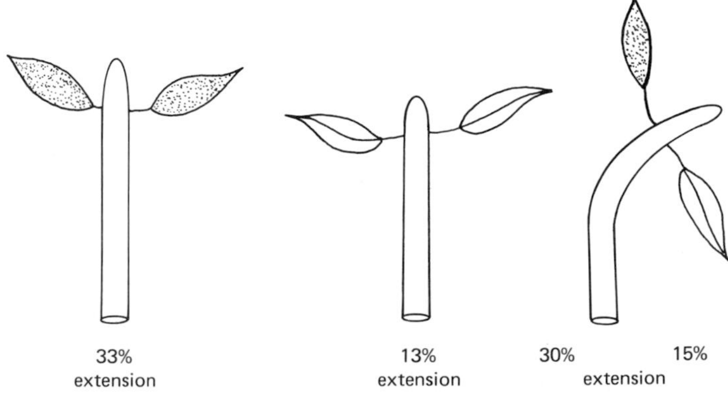

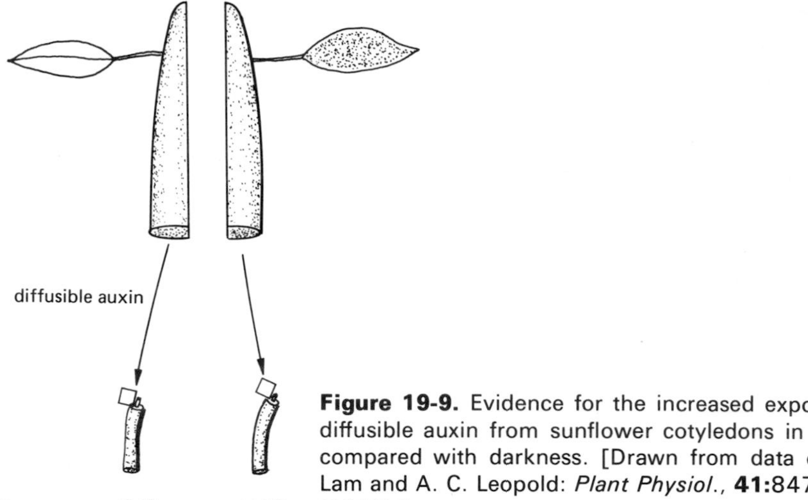

Figure 19-9. Evidence for the increased export of diffusible auxin from sunflower cotyledons in light compared with darkness. [Drawn from data of S. Lam and A. C. Leopold: *Plant Physiol.*, **41**:847–51 (1966).]

response, even if the cotyledons are exposed. This suggests that the illuminated cotyledons are acting primarily as the source of growth-regulating substances. The lateral distribution of IAA (or other stimuli) that imposes asymmetric growth appears to be imposed through the action of the photoreceptor in the hypocotyl.

Some workers have thought, on the basis of economy of operation, that the geotropic and phototropic stimuli should be mediated by the same mechanism operating in opposite directions. This argument is parallel to the idea (inherent in the Cholodny-Went hypothesis) that the same mechanism operating in opposite directions should control both roots and shoots. Although economy of operation may be a sound biological principle, it is nevertheless very dangerous to apply it in this way. Geotropic and phototropic responses, and the growth patterns of roots and shoots, may be quite similar. However, they probably evolved at widely separated times, and are as likely to be mediated by different mechanisms as by the same ones.

Phototropic Light Perception. The mechanism of perception of light that gives rise to phototropism is still a matter of question. It was early discovered that the most effective light for phototropic response is short wave; red light is ineffective. The action spectrum of phototropism, when contrasted with the absorption spectra of a representative carotene and flavin, shows resemblances to both but is identical to neither (Figure 19-10). The pigment that is responsible may be present in extremely small amounts and may therefore be difficult to detect. Certain mutant organisms that have less than 20 percent of the normal amount of carotene are still phototropic, but this does not rule out the participation of a particular carotenoid or a specific fraction of the plant's total carotenes. At the present time, the identity of the pigment is unknown.

Assuming that a specific pigment or pigment association is responsible for receiving light, the mechanism for translating the perception into reduced auxin production and export is an equally difficult question. As in geotropism, a transverse electrical potential is set up in phototropically stimulated plants, but the potential appears to follow the induction of auxin imbalance and may be caused by it. The situation is, in fact, much more complex than the foregoing description might suggest. Phototropic light is meas-

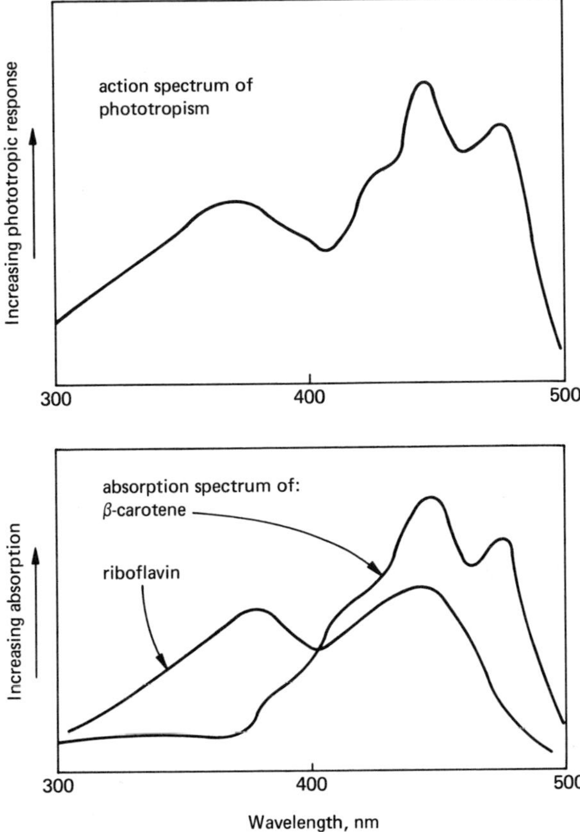

Figure 19-10. Action spectrum of phototropism compared with absorption spectra of a flavin and a carotene. Adapted from several sources.

ured cumulatively by plants. That is, within limits, prolonged weak light produces the same effects as brief strong light. However, the reaction to increasing stimulus is not simply a continuously increasing response. After a certain amount of light has been received, the tissue begins to respond less and less, and finally a negative phototropism takes place. Subsequently, with even more light, a second positive response may occur, but it shows certain differences in character from the first response. A clear interpretation of these effects is lacking. It is possible that the stimulation of the phototropic pigment may in some way lead to changes in cell permeability that result in increased transport of auxin, but this is not known.

Thigmotropism. The reaction of a plant part such as a tendril to a touch stimulus is called thigmotropism if the reaction is aligned with the stimulus, or **thigmonasty** if it is not directional. Tendrils are apparently able to distinguish surfaces, since they respond much more effectively to rough or textured than to smooth or soft surfaces. The response is rapid and may partly involve differential changes in turgor, producing differential expansion or contraction of cells on opposite sides of the organ. However, some differential growth is also involved, and many thigmotropic responses are permanent or growth movements. Rapid responses in tendrils are probably accomplished by movement of electrolytes or salts.

Recent experiments show that rapid changes in ATP and inorganic phosphate content take place upon touch stimulation of pea tendrils. It appears that either rapid changes take place in the permeability of membranes, which lead to water movement, or else ATP-stimulated active transport of ions occurs with the same result. Auxin is known to affect tendril coiling, and unilaterally applied auxin causes tendrils to bend. However, this may not be directly related to the normal response. How the touch response is perceived is not known.

Other Tropisms. Plants react by aligning themselves through growth with various other stimuli. Roots are sometimes said to grow toward regions of high water content in the soil. Such **hydrotropism,** if it occurs, might be directly mediated by a mechanism that permits roots to detect and react to differences in water concentration. A more likely alternative, however, is that roots simply grow faster in regions of higher water content.

Certain vines, particularly of tropical origin, climb up the trunks of trees. When their seedlings germinate, they grow directly toward potential support trees; they do not find hosts by haphazard growth or random searching. Apparently they do this by growing toward the darkest sector of their horizon, which is provided by the darkly colored trunks of large trees. This phenomenon is called **skototropism** (dark-seeking) by the two American scientists, D. R. Strong and T. S. Ray, who first reported it in 1975. After becoming established on its host and starting to grow, the seedling becomes positively phototropic, which ensures adequate photosynthetic nutrition.

Roots, shoots, and coleoptiles tend to align during growth in electrical fields. This again raises the question of whether asymmetric distribution of auxins results from the development of an electrical potential gradient or is the cause of it. Experiments in phototropism and geotropism suggest that the gradient is a result of auxin asymmetry. However, it is possible that an externally applied electrical potential gradient may result in polarized auxin movement, so that asymmetric auxin distribution results, and directional or tropic growth takes place. Insufficient study has been done for critical analyses of these growth responses.

Shape

Correlative Effects. The shape of an organ or an organism results from the fact that it grows in different directions at different rates. The development of a plant is usually closely correlated with its growth. For example, tillers do not begin to develop on wheat plants until the main shoot has almost completed its growth; and the axillary buds of many dicotyledonous plants, such as bean or tomato, do not begin to grow until two or three or more new internodes have been added to the main stem above them. In other words, the development of different parts of a plant is correlated or interrelated. Such **correlative effects** may operate in time, as in the case of apical dominance (which will be discussed in greater detail later in this section). Other correlative effects operate in space; they relate the sizes that different parts of the plant attain. Others relate the rate of growth of plant parts. In **allometric** growth, different plant parts may have different growth rates, but these are closely related so that the ratio of growth of one part to that of another is constant. Most such correlations are effected by hormones or by the establishment of gradients or patterns of nutrient distribution.

Shape is often established by a field or gradient of a morphologically active substance, but this does not seem to be true in all plants. A simple mechanism of this sort may determine the shape of an anatomically simple tissue like the fruiting body of a fungus, but the complex organization of the tissues in a higher plant precludes this except under limited circumstances. Instead, different parts of the plant engage in what is essentially indeterminate growth, but they are interrelated and controlled by correlative effects to produce a characteristic shape or form. An example is the shape of leaves, which is to a considerable extent determined by the pattern of vascularization. Thus the outline of a maple or an oak leaf seems to be generated through the action or interaction of growth hormones (and perhaps also nutrients) transported through the major veins. The final shape of the leaf is thus strongly affected by the pattern of vascular tissue laid down in the embryonic leaf. However, the hard core of the problem still remains: how the genetic information is translated into a specific anatomical pattern in the embryonic leaf is not known.

OTHER FACTORS. In addition to correlative effects, physical limitations may enforce specific shapes. The flat tops of many trees, for example, may be the result of the incapacity of the leaders to grow above a certain height because of water transport problems. Low or otherwise distorted plants grow in exposed places, prevented by adverse environmental factors from achieving their normal shape.

Temperature has a powerful effect on both the rate and the total extent of growth in a plant or organ. This is illustrated by the curves for oat coleoptiles shown in Figure 19-11, derived from the early data of the German physiologist E. Vogt. It may be seen

Figure 19-11. The effect of temperature on the growth rates, duration of growth, and final size of oat (*Avena sativa*) coleoptiles. [Recalculated data of E. Vogt: *Z. Bot.*, **7:**193–270 (1915).]

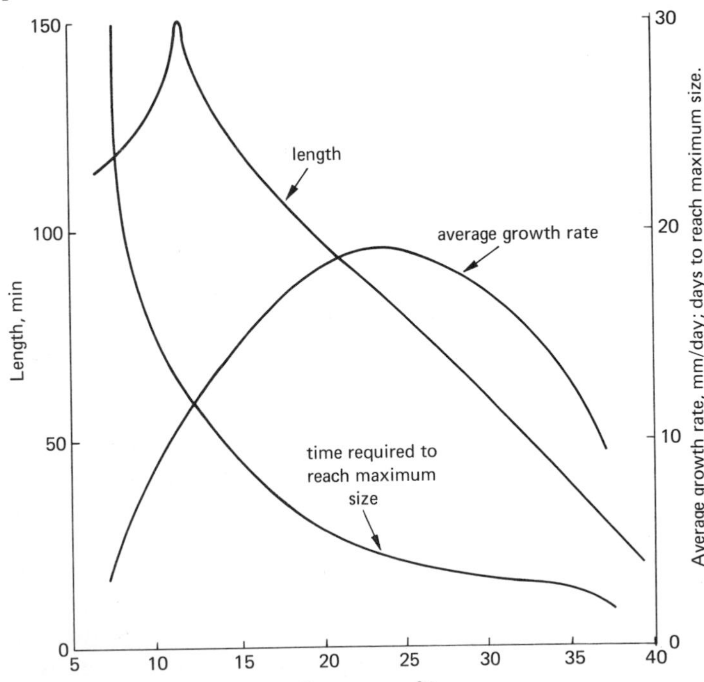

that the ultimate length, the time during which growth took place, and the average growth rate of the coleoptiles all varied in different and complex ways with temperature.

Other environmental factors that affect growth include light (the effect of reduced light, etiolation, is discussed in Chapter 18, page 428). One of the characteristic results of reduced light is increased growth in length. The availability of water exerts a profound effect on growth and the shape and form of plants. Not only does lack of water, or its oversupply, affect the size of the plant but the shape and expression of xerophytic characteristics may be dramatically changed. It has been observed that gorse plants (*Ulex* sp) lose their characteristic spiny appearance when grown on well-watered soil and become almost unrecognizable as gorse. It has been shown for several species that water shortage causes a dramatic increase in the amount of the dormancy-inducing hormone, abscisic acid (see Chapter 22). This may be related to the development of xeromorphic characteristics by some plants.

Apical Dominance. One of the most important and well-studied correlative effects is apical dominance. Early German physiologists considered that it might be due to a "struggle for existance" among shoots of the plant, with the main central shoot "winning." By this theory, apical dominance is established by the direction of nutrient flow, and this has become known as the **nutritive theory.** Later, physiological experiments by K. V. Thimann and F. Skoog, and the beautifully executed surgical experiments of R. and M. Snow, showed that apical dominance was caused by auxin diffusing from the apical bud which inhibited the growth of the lateral shoots. The removal of the apex releases the lateral buds from apical dominance, but this is reestablished if auxin is applied to the decapitated stump. This sequence is illustrated in Figure 19-12.

Experimental difficulties with the theory that auxin alone causes apical dominance were resolved to a considerable extent by the discovery that auxin and cytokinins interact. Increased cytokinin allows lateral buds to be released from apical dominance, in spite of the presence of auxin. This effect is seen in its extreme form in the witch's broom disease of coniferous trees (see page 619), where excess cytokinins produced by the pathogenic organism cause the development of many lateral and adventitious buds. The concept of the interaction of two growth substances provides a way around the principal difficulty of the auxin theory, namely, that the apical bud is not inhibited by its own supply of auxin.

An alternative theory that relates to the known capacity of hormones to affect the translocation of nutrients (see Chapter 21, page 523) is essentially a combination of the nutritive theory and the hormone theory. According to this **nutrient-diversion** theory, apical dominance is maintained because the nutritional traffic up the shoot is directed to the apical bud and not to the lateral shoots, owing to the auxin gradient that results from the production of auxin in the apex. Thus, leaves and any shoots that escape from apical dominance will, once they have begun to grow and produce auxin, automatically be assured of a supply of nutrients. The application of cytokinin would cause the start of cell division and the production of auxin, so that release from dominance would automatically follow.

Apical dominance functions to a greater or lesser extent in most plants. In its extreme form it results in the columnar or excurrent growth habit characteristic of coniferous trees. Many plants characterized by bushy forms or dense growth of lateral branches have very little apical dominance. Apical dominance also affects the form of roots, resulting in tap root development, or fibrous roots in its absence. Recently it has been

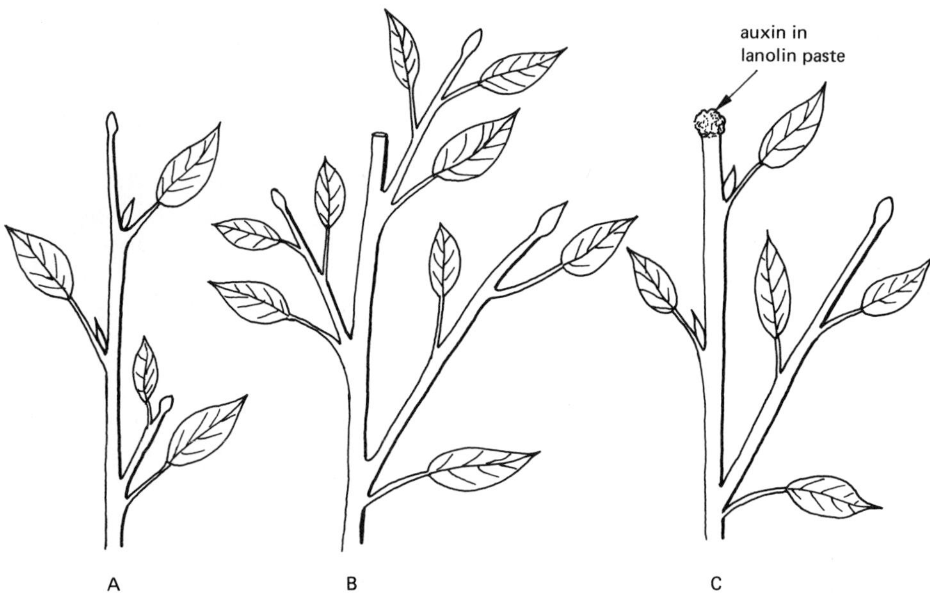

Figure 19-12. Apical dominance.
A. Apical dominance prevents the two upper lateral buds from growing. The lowest one has escaped from the dominance of the apex, and has started growing. **B.** The apex is removed and all the buds have started growing. **C.** The apex is removed and replaced by lanolin paste containing auxin. Lateral bud growth is prevented (except in the lowest bud, which had already broken dormancy and started growing). [After P. M. Ray: *The Living Plant*. Holt, Rinehart and Winston, New York, 1965.]

suggested that the auxin effect is in fact mediated by ethylene. If this is so, it does not alter the overall picture but merely inserts another link in the chain of reactions that correlate growth in different parts of the plant. The mechanism by which auxin, or ethylene, causes either the nutritionally influenced or direct suppression of lateral buds is still not known.

Nastic Responses

Nastic responses are movements made in response to stimuli but are not oriented in a direction related to the direction or vector of the stimulus. They may be growth movements, which are plastic and hence permanent, or they may be **variation movements,** which are reversible. A number of characteristic plant movements take place. Many of them, like the response of the sensitive plant, *Mimosa,* are rapid, vigorous, and very striking. Others seem to be related in a direct way with endogenous rhythmic pace-setters or biological clocks that help to time or pace the events of development. These will be discussed in detail in the following chapter. Several such movement patterns, although apparently not of great importance in the life of the plant, have been extensively studied because it is hoped that an understanding of them will eventually reveal something of the mechanisms that regulate development.

Variation movements usually involve movement of water. An important and typical variation movement is the opening and closing of stomata, described in detail in

A

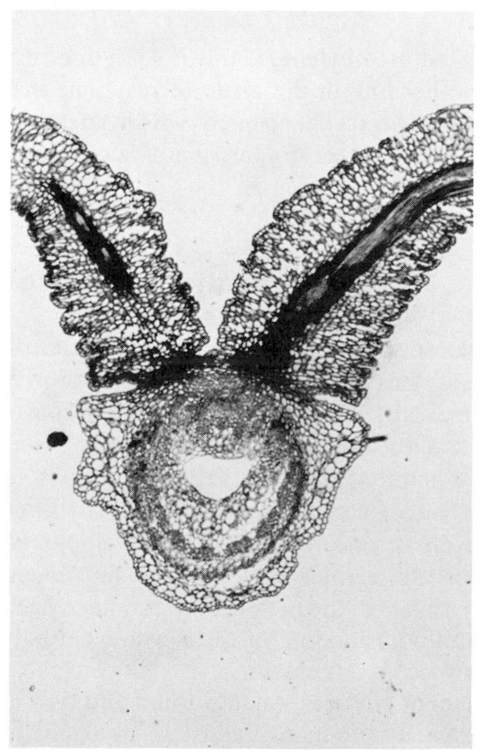
B

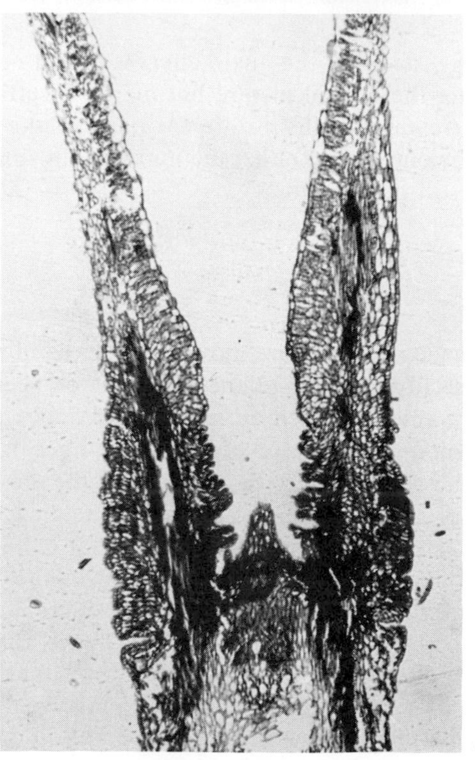
C

Figure 19-13 (opposite). Illustration of *Albizzia* pulvinules. **A.** A scanning electron micrograph of the columns of accordionlike motor cells. **B.** A cross section of the pulvinule showing turgid motor cells (in cross section) at the base of each leaflet. **C.** Similar tissue to (B), but with the motor cells flaccid and the leaflets closed. [Electron micrograph and photographs by Dr. Ruth L. Satter. Original photographs kindly supplied by Professor Arthur W. Galston, Yale University, New Haven, Conn.]

Chapter 14. Movements of leaves, leaflets, and even of small branches are often brought about by a special organ called the **pulvinus,** shown in Figure 19-13A and B. This bulb-shaped mass of parenchymatous cells is located at the base of the leaf, leaflet, or branch. Water moves rapidly into or out of the **motor cells** (Figure 19-13C), which are located on opposite sides of the pulvinus, and the resulting rapid expansion or contraction of the opposite sides of the pulvinus cause the leaf or branch to move up or down. Water appears to move in the pulvinus as the result of osmotic potentials. These are probably set up in the same manner as in guard cells: by the very rapid active transport of potassium ions moved under the influence of ATP-driven transport mechanisms. A variety of different kinds of stimuli may activate the transport of potassium and, hence, the movement of the organs. Some of these will be described now. We shall also examine diurnal rhythmic movements in the next chapter, in connection with the problem of endogenous rhythms or "clocks" that affect flowering.

Epinasty. Epinasty is the bending downward that commonly occurs in petioles and permits the leaves to assume a position in which their tips incline toward the ground rather than up in the air. This does not seem to be a gravitational response because plants show epinasty regardless of their orientation with the gravitational field or even when placed in a klinostat. It appears to result from different amounts of auxin being transported into the upper and lower sides of the petiole from the leaf, which causes differential growth resulting in the bending of the petiole. Many developmental responses (for example, the unfolding of flowers, the unrolling of fern fronds) are epinastic responses. This is a common response to excess auxin or to ethylene treatment (see Figures 16-14 and 16-19, pages 395 and 402).

The reverse effect, called **hyponasty,** may also occur. It can be induced by gibberellic acid application.

Thermonasty. Some plants, such as tulips and crocuses, show repeated opening and closing movements of the flowers in response to temperature changes. The response is quite sensitive and has been noted following a temperature change of only a fraction of a degree. In spite of their reversibility, these thermonastic movements are permanent growth movements resulting from a growth differential between the upper and lower tissues of the petals. The mechanism is unknown.

An interesting temperature indicator is the *Rhododendron* bush. Its evergreen leaves undergo a pronounced thermonastic response in winter, the leaves hanging almost vertically down close to the stem as temperatures approach $-15°$ C (0° F), and standing out nearly horizontally at about 0° C (32° F). The response may involve changing water tension in the petiole, but this is not known.

Nyctinasty. The leaves of many plants undergo "sleep movements," a rhythmic raising of leaves in the morning and lowering or folding of them in the evening, called nyctinasty. This phenomenon has been much studied in bean plants, which exhibit strong nyctinastic movements. Careful analyses with an electron microprobe (see page 337) have enabled the American physiologist A. W. Galston and his colleagues, particularly Dr. Ruth Satter, to detect a substantial movement of potassium ions from the upper to the lower side of the pulvinus and back. The movement of potassium ions causes large changes in the osmotic potential of the motor cells of the pulvini, causing the rise and fall of leaves. Analysis showed that sugars could not be involved because they only accounted for a small fraction of the osmotically active substances in the motor cells.

It has been suggested that auxins are involved in this response. Apparently leaves produce larger amounts of IAA by day, and this is transferred primarily to the lower side of the petiole. Potassium ions move to the area of high auxin, water enters the lower side of the pulvinus, and the leaf rises. The auxin transport is reduced at night, and the reverse reaction takes place. Auxins added to the upper or lower side of the pulvinus cause falling or rising of the leaf, respectively.

However, the picture is not so simple as this. Plant physiologists have long been fascinated by the fact that sleep movements of many leaves continue regularly for a period of days even when the plant is placed under constant conditions, as is shown in Figure 19-14. This means that nyctinasty in intact plants is phased or timed by an internal rhythm. This rhythm can be reset by changing the normal pattern of light and darkness or by short periods of illumination with far-red light. The latter effect is reversed by red light, and this indicates that the response is mediated by the pigment phytochrome. This pigment is implicated in a number of plant responses to light, including flowering, and it is discussed in detail in the following chapter.

Experiments with isolated pulvini show that they are reactive to light and also undergo potassium redistribution and turgor reactions as do pulvini in intact leaves. This means that the pulvini themselves are not only photoreceptive but contain all the apparatus for reacting to the light stimulus as well as the energy sources to carry out the reaction. As will be considered in the next chapter, it seems likely that the phytochrome response is mediated through a combination of various membrane effects. It affects the permeability of membranes and is also capable of controlling or modulating the activities of membrane-bound enzymes, such as ATPase, that mediate active transport of ions. Through these and related actions phytochrome can cause or affect the redistribution of ions, particularly potassium, and so cause osmotic turgor changes and movement.

Seismonasty. Seismonasty means response to shaking. A variety of plants, of which the sensitive plant, *Mimosa pudica* is the best-known example, respond to a touch or blow by folding their leaflets and lowering their leaves. Their responses are very fast—they may start only 0.1 sec after stimulation and be completed in a few seconds. These plants respond to a variety of stimuli besides touch or mechanical shock, including heat and electrical and chemical stimulation. Another peculiarity is the propagation of the stimulus. Not only the leaf or leaflet that is stimulated but much or all of the plant may react. The reaction spreads up or down the plant very rapidly, at rates up to 40—50 cm/sec.

When a sensitive plant reacts to touch or shock, the pulvini undergo two sorts of reactions: in the leaflets, the upper sides shrink so that the leaflets close upward; in the petioles, the lower sides contract so that the whole leaf drops. In either case, the reaction follows rapid ejection or loss of water from the motor cells into intercellular spaces. How

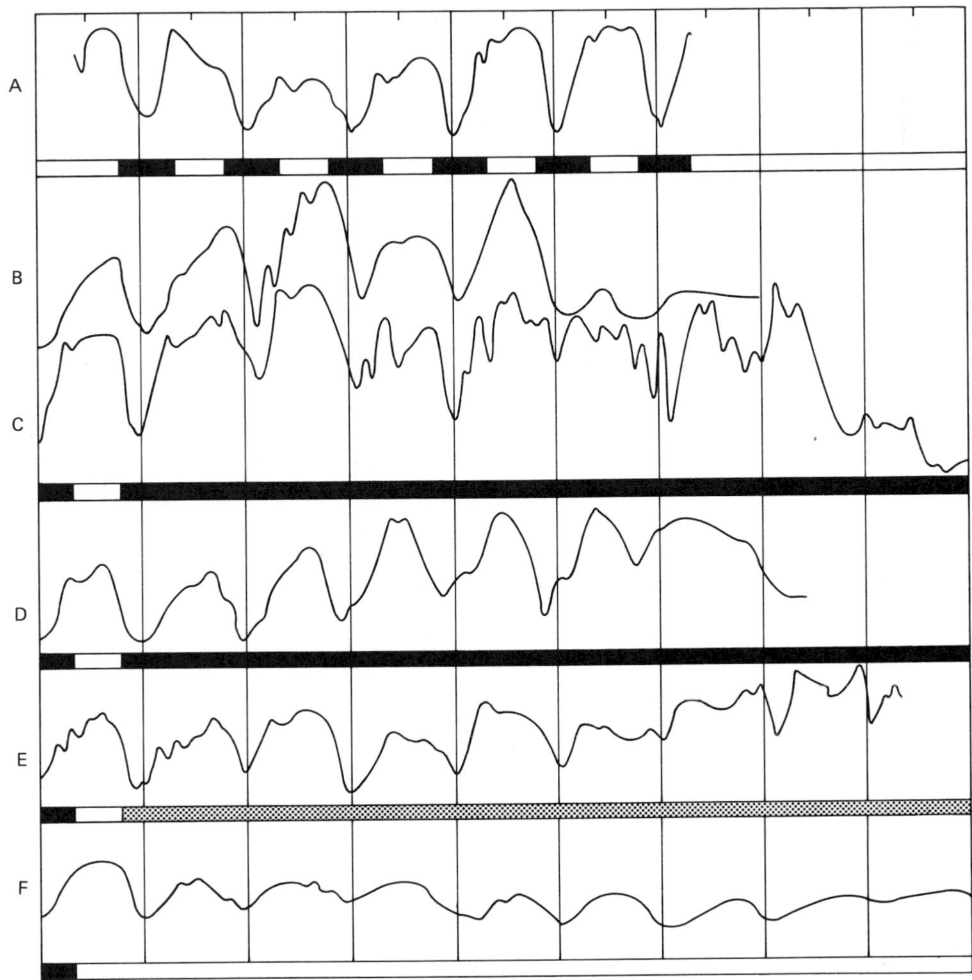

Figure 19-14. Nyctinastic leaf movement tracings of *Coleus* under different light conditions. Vertical lines are 24 hr apart and indicate 2400 eastern standard time. Each recording is a typical representative from a group of five plants.
A. Under light-dark cycles of 12 hr light (empty bars) and 12 hr dark (black bars). **B.** Plant entrained under 12 hr light and 12 hr dark then released to constant darkness. **C.** Plant released to constant darkness; sucrose was added to the pot. **D.** Plant released to constant dim light of 10 ft-c. **E.** Plant released to constant dim light of 30 ft-c. **F.** Plant released to constant bright light of 1300 ft-c. [From Ruth Halaban: *Plant Physiol.*, **43:**1883–86 (1968). Used with permission.]

the water is expelled is unclear. One suggestion is that rapid permeability changes lead to leakage of water. Another theory rests on the presence of very small vacuoles that have been noted in the cytoplasm of turgid pulvinus cells; these disappear after pulvinus reactions. This leads to the thought that the water loss may be by ejection of the contents of these vacuoles, a sort of reverse pinocytosis. However, calculations show that there is not room at the cell surface for sufficient of these small vesicles to discharge enough water in the time required. An early suggestion that pulvinus action depends on the hydration and dehydration of proteins has recently been revived. It is possible that cell diameter is changed by contraction and expansion of cell colloids upon the addition or removal of water. The small vacuoles that have been observed may be involved in the storage of water used in this process. Energy for the process probably comes from ATP, which decreases sharply in amount during movement, and increases again during recovery.

The transfer of the stimulus still remains interesting. The Indian physiologist J. C. Bose considered that sensitive plants transmitted the stimulus by means of a nervous system. This idea was not widely credited, but recent research on the propagation of action potentials makes it once again attractive. The action potential is apparently transmitted through the xylem tissue of the petioles and stem. As in geotropism and phototropism, hormones also stimulate an electrical potential in the tissues of *Mimosa*. However, the nature of the potential is different and, in this case at least, it appears more likely that the potential stimulates hormone production and not the other way around.

Some confusion still exists because early experiments by the Italian physiologist U. Ricca showed that stimulation could clearly be transported through specialized cells in the phloem by a chemical substance, presumably a hormone. However, known transport rates in *Mimosa* are not fast enough to account for such rapid movement of a hormone. Another suggestion is that the stimulus may be propagated through successive loss of turgor in specialized cells. It may be that the final explanation will involve all types of mechanisms, interrelated in some way. It seems, however it works, that the reaction of the sensitive plant is the nearest analogy that has evolved in plants to muscle and nerves in animals. However, this analogy should not be pushed too far!

Traps. A number of insectivorous plants are equipped with traps that react rapidly enough to catch live insects. These are interesting in that they combine the rapid movements of the sensitive plant with a special trigger device to set off the trap. The aquatic bladderwort, *Utricularia* sp., has small spherical bladders with an ingenious trapdoor arrangement. As a small insect or other organism swims to the bladder, it touches a triggerlike hair and the door of the bladder opens rapidly inward. The insect is swept inside by the movement of the door and by water entering through the doorway of the trap, whose inside is under pressure. An illustration of how this trap works is shown in Figure 19-15.

A more widely studied trap is the leaf of the Venus fly trap, *Dionaea* sp., shown in Figure 19-16. This flat leaf has a fringe of hairs at its edge and two or three small triggerlike hairs on the surface of the leaf. If one of the triggers is touched twice, or two of them in succession, the leaf closes rapidly around the midrib until the hairs at the edge intermesh, trapping any insect that is walking on the leaf surface and holding it until it is digested. The requirement for two successive trigger actions prevents the unnecessary springing of an empty trap by a passing insect or an object falling or brushing against the leaf.

Another group of carnivorous plants is the sundews (genus *Drosera*). They have leaves covered with sticky tentacles that trap small insects. Rapid movements of tentacles stimulated by the struggle of captured insects serve to hold them and force them to the center of the leaf. Slower movements fold the leaf around the insect and bring the tips of tentacles, which secrete digestive enzymes, into contact with the insect.

The operation of all these traps suggests nervelike signals or action potentials. Such action potentials were in fact demonstrated by Charles Darwin and others in the preceding century, but attention has recently been called to them again by the work of

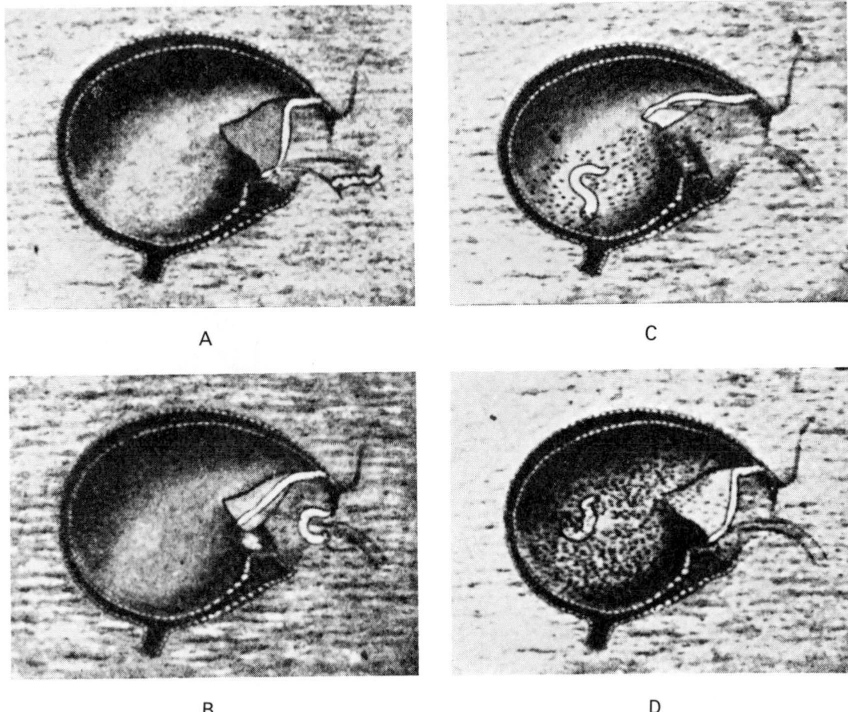

Figure 19-15. The operation of a *Utricularia* trap.
In (**A**) the insect touches a trigger hair and trips the trap, which opens and engulfs the insect (**B,C**). The trap is then reset (**D**).

the American physiologist S. E. Williams. He points out that the rapid, tropistic movements initiated by mechanical stimuli are mediated by action potentials produced by the trigger hairs or tentacles. The nature of the mechanism requiring two stimuli for the *Dionaea* trap is unknown. The possession of neuroid action potentials makes these plants very interesting from an evolutionary point of view, since they must clearly have evolved entirely independently from those of animals. The slow, nastic movements (such as the leaf and tentacle bending in *Drosera*) appear to be mediated by a hormonelike or chemical stimulus.

Rapid Leaf Movements. In addition to sleep movements, epinasty, and other such responses, plant parts are in almost constant motion. Time-lapse cinematography shows a continuous twisting and fluttering of the leaves. Bean leaves have been shown to undergo rotation movements sufficient to raise or lower the edges through 2 cm. These movements are periodic, with a cycle time slightly under 1 hr, and only occur during daytime when the leaf is not in the sleep position. In addition, the leaf waves up and down through an angle of about 10°, also at a period of about 1 hr.

These movements are superimposed upon the normal nyctinastic movements and are shown in Figure 19-17. These rapid movements are not synchronized among adjacent plants and, thus, are presumably not entrained to an external environmental fluctuation. It has been suggested that these movements are caused by rhythmic fluctuations in the auxin content or in its transport from leaves, but evidence is lacking.

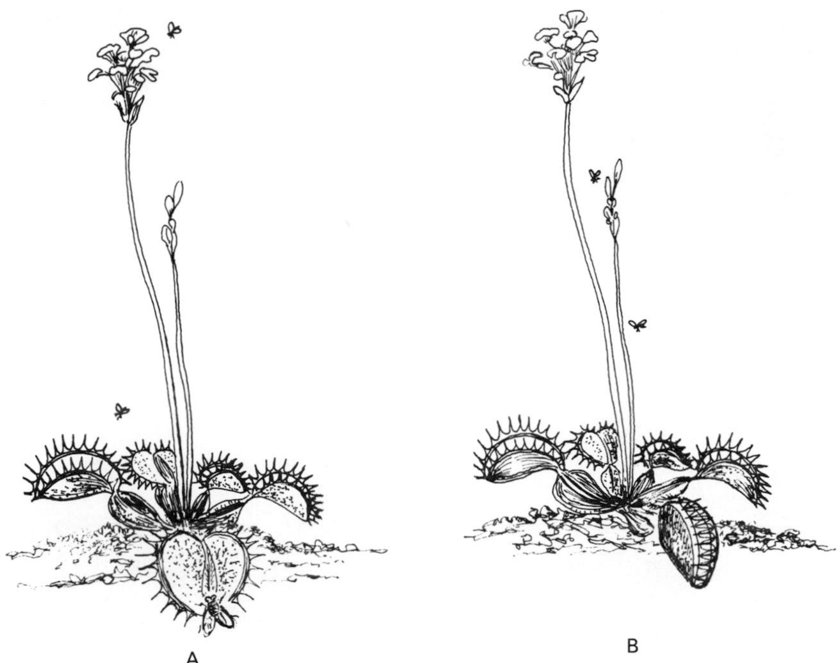

Figure 19-16. The Venus fly trap (*Dionaea muscipula*).
The trigger hairs are visible on the leaf (**A**) where the fly is landing. In (**B**) the trap is sprung and the insect is caught. [From W. H. Muller: *Botany,* 3rd ed. Macmillan Publishing Co., Inc., New York, 1974. Used with permission.]

Nutation. Plants are usually thought of as growing more or less "straight up." However time-lapse cinematography shows that the stem tip describes a continuous tight spiral or weaves from side to side as it grows. Such movements are called nutation movements. The amplitude of nutation varies from almost nothing to as large as 5 ft, observed by Charles Darwin for *Ceropegia gardnerii,* a member of the milkweed family. The rate of nutation varies from one cycle per day up to about one per hour and is

Figure 19-17. Movement of primary leaves of two separate *Phaseolus angularis* plants grown simultaneously.
(**a**) Blade rotation in millimeters as recorded by measuring the distance between the lateral margins of the leaf as it appeared on the projected image. (**b**) Upward and downward movement in angular degrees. [From D. K. Alford and T. W. Tibbitts: *Plant Physiol.,* **47:**68–70 (1971). Used with permission. Photograph courtesy Dr. T. W. Tibbitts.]

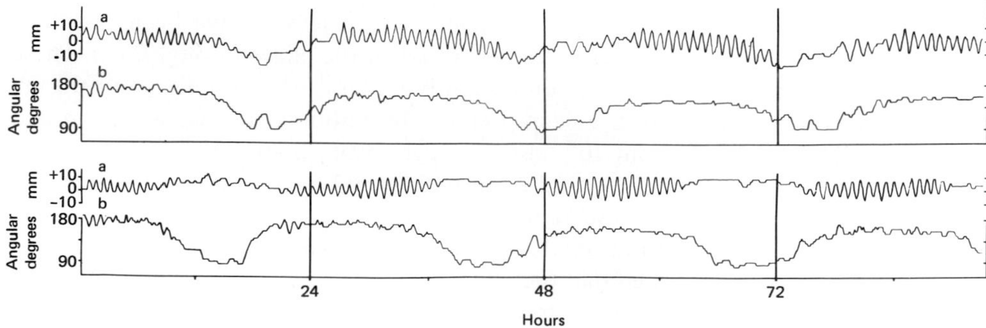

temperature sensitive. Many tendrils wave about in a surprising fashion; perhaps this increases their chance of making contact with a potential support.

Nutation is a growth movement and is caused by uneven growth in opposite sides of the stem. It has been suggested that it is caused by an unstable equilibrium at the stem tip, resulting in some manner in an oscillation in the production of growth substances. Alternatively, it may be a "hunting" oscillation around the vertical in a geotropic response.

Growth is not a smooth unbroken process but takes place by a series of discrete steps. Thus there are brief periods of extension followed by periods of fixation, during which the cells that previously extended undergo cell wall thickening. The periodicity of this process is set by an internal oscillator that appears to be temperature sensitive, unlike the internal oscillators involved in flowering and the daily rhythmic processes. Side-to-side nutation movements may possibly result from the fact that opposite sides of the stem are out of phase with one another. In circumnutation, the oscillations presumably circle the growing tip.

Additional Reading

See list at end of Chapter 16 (page 406).

Downes, R. J., and **H. Hellmers:** *Environment and the Experimental Control of Plant Growth.* Academic Press, London, 1975.

Goldsmith, M. H. M.: The polar transport of auxin. *Ann. Rev, Plant Physiol.,* **28:**439–78 (1977).

Juniper, B. E.: Geotropism. *Ann. Rev, Plant Physiol.,* **27:**385–406 (1976).

Phillips, I. J. D.: Apical dominance. *Ann. Rev, Plant Physiol.,* **26:**341–67 (1975).

Sibaoka, T.: Physiology of rapid movements in higher plants. *Ann. Rev, Plant Physiol.,* **20:**165–84 (1971).

Torrey, J. G.: Root hormones and plant growth. *Ann. Rev. Plant Physiol.,* **27:**435–59 (1976).

Williams, S. E.: Comparative sensory physiology of the Droseraceae—the evolution of a plant sensory system. *Proc. Am. Phil. Soc.,* **120:**187–204 (1976).

20 Organization in Time

Introduction

The Importance of Timing. The importance of timing is revealed by the two following concepts: It would not be good for a plant to flower before it developed sufficient leaves and roots to support the nutritional development of the fruit. And, it would be decidedly bad for a plant to flower so late in the season that the fruit and seed development could not be completed before the onset of winter.

The first concept illustrates the importance of the correlation of events into proper sequence so that development is an orderly and not a haphazard process. This correlation results largely from the fact that development is organized in a linear way at the genetic level. Many developmental processes rely on the completion of previous steps for their "priming" or initiation, or else are directly related (often by hormone messages) with the progress of parallel processes. This represents the timing that results from built-in programming, as was discussed in Chapters 17 and 18.

The second concept points out the importance of absolute time measurement, particularly to plants living in a climate in which seasons of good weather alternate with periods during which growth is impossible. Annual plants must live their entire life cycle between winters, and the time of flowering is most important to their survival. Biennial plants endure the winter in the form of an underground storage organ and flower in their second year. Timing of the formation of the winter form and flowering are both of importance here. Perennial plants such as trees must enter dormancy before the onset of winter; furthermore, lest they emerge from dormancy too soon (for example, during a particularly warm January thaw), they must possess some mechanism for timing the dormant period. Finally, perennial plants too must flower at an appropriate time of the year for fruit and seed development. The seeds of many plants need some timing device to prevent their premature germination at a time (for example, during a fine spell of weather in late fall) when they would be unable to develop to a suitable stage to endure winter.

Thus every type of plant has some need for mechanisms that measure time at one stage or another of its life. The most widespread and probably the most studied aspect of timing is flowering. In this chapter we shall examine biological timing and biological clocks, primarily in relation to flowering. In the course of this discussion, reference will be made to the timing of certain rhythmic processes, such as nyctinasty, described in the previous chapter. Certain timing processes related to dormancy will be discussed in Chapter 22.

Ways to Measure Time. Timing devices of all sorts fall into two categories: the cumulative or **hourglass,** and the rhythmic or **oscillator.** Cumulative devices include the hourglass, the crude candle-clock of King Alfred, the flowing-water clock of the ancient Greeks, and any device by which time is measured as the interval required for complete operation of some mechanism or reaction that goes at essentially constant speed. Chemical analogs can easily be devised. This sort of clock is normally affected by temperature, thus it is not usually very accurate except under rather constant conditions.

The rhythmic clock depends on the regular oscillation of some system or object of fixed period, such as the pendulum or balance wheel in a clock. However, unless the pendulum is very long, its period is short, and the clock requires "gearing" if it is to be used to measure long intervals of time. Biological or chemical analogs are not so easy to devise. It is possible, however, for a cyclic enzymic reaction system in unstable equilibrium to "hunt," that is, the concentrations of intermediates may oscillate about their equilibrium points; thus such a system could constitute a pacesetter for rhythmic timing. A hypothetical mechanism for an oscillator with a relatively long period might be devised on the basis of a pair of reactions whose operation is mutually exclusive (for example, because the substrate of each inhibits the other). Such a model would be temperature sensitive because it depends on the rates of chemical reactions. It is very difficult to devise a temperature-insensitive model for a biological timer. Nevertheless, they do seem to exist.

How Biological Clocks Might Work

Hourglass. The simplest hourglass timer is the life of a plant. The early German physiologist G. Klebs developed the concept **ripeness to flower,** which expresses the idea that a certain stage of development must be reached before the flowering process can be initiated. The tomato plant, for example, usually does not flower until five leaf nodes have developed on the main stem. An alternative mechanism is the diurnally reset chemical hourglass. For example, the pigment **phytochrome** is known to exist in at least two forms that may be interconverted by illuminating with light of specific wavelength. Such a system could constitute the basis for a time-measuring device that would measure the length of day.

Oscillator. Numerous examples are known of events that have a rhythmic oscillation. These include the opening and closing of flowers, the movement of leaves, growth rates, rates of various metabolic processes, and many more. One of the characteristics of these rhythmic processes is that they continue in an artificial environment in which all rhythmic fluctuations (such as light or temperature) have been eliminated. This indicates that there is an intrinsic rhythmic process or oscillation within the plant.

Under constant conditions the intrinsic rhythm is approximately 24 hr, usually somewhere between 21 and 27 hr. Since it is not exactly 24 hr, the rhythmic oscillation is called a **circadian** (approximately daily) rather than a diurnal rhythm. Such rhythms are **free running**—that is, they have a natural period that does not need to be pulsed or restarted at each cycle. Furthermore, they are usually **entrained** to the normal 24-hr

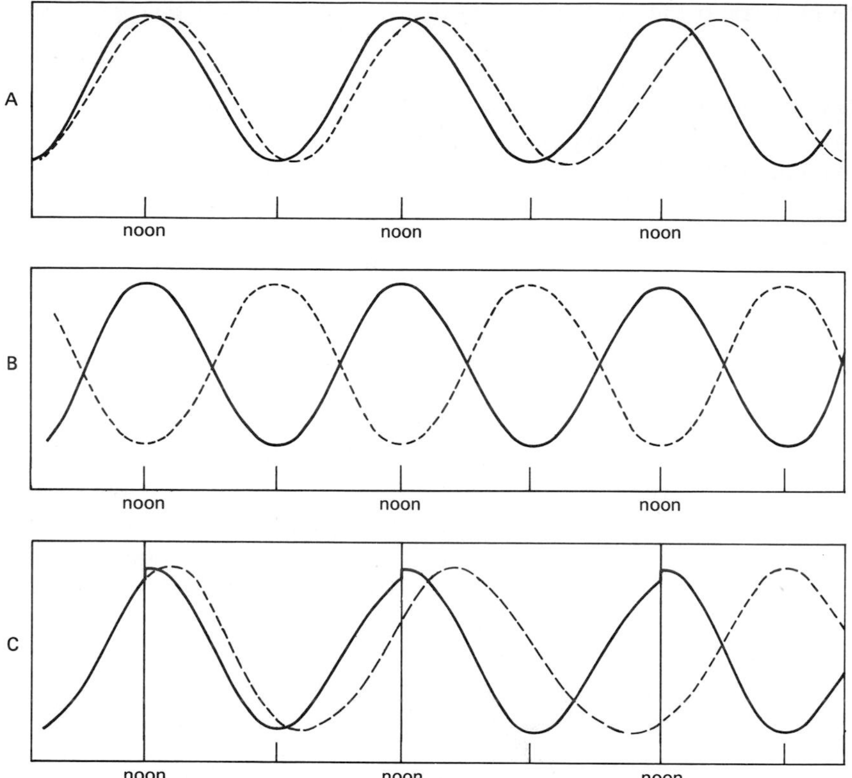

Figure 20-1. Oscillators.
A. Two circadian oscillators that are almost in phase. **B.** Two circadian oscillators that are out of phase. **C.** A circadian oscillator with a period of 28 hr is entrained and reset at noon each day. If it were allowed to run free, it would behave as shown by the dotted line.

cycle of day or night. That is to say, they are reset each day or night, or by the transition from one to the other. Such rhythms are usually essentially temperature insensitive. They can be **rephased** (that is, reset to oscillate in the same period, but peaking at different times of day) in various ways. These ideas are illustrated in Figure 20-1.

We shall later examine additional evidence which strengthens the view that some intrinsic cyclic timing device exists in plants with a period of approximately 24 hr. Its nature is unknown. Nor is it known whether the basic oscillation is circadian or relatively rapid with a mechanism for amplifying or "gearing down" its period. The latter possibility definitely exists, however. Many rhythmic oscillations of metabolic processes or pools of reactants have been reported with periods ranging from less than a minute to several hours. Several of these have been thought to be part of, or a reflection of, a basic oscillating timer responsible for circadian rhythms. Proof of this, however, is elusive.

Interactions. Certain periodic phenomena seem to indicate that there is more than one intrinsic oscillator in some plants. These appear to operate at different frequencies, and the interaction or phasing of two oscillators may time certain events, as illustrated in Figure 20-2A. The movements of certain tidal algae appear to obey a lunar as well as a diurnal clock. Their periodic behavior continues in a constant environment, so the

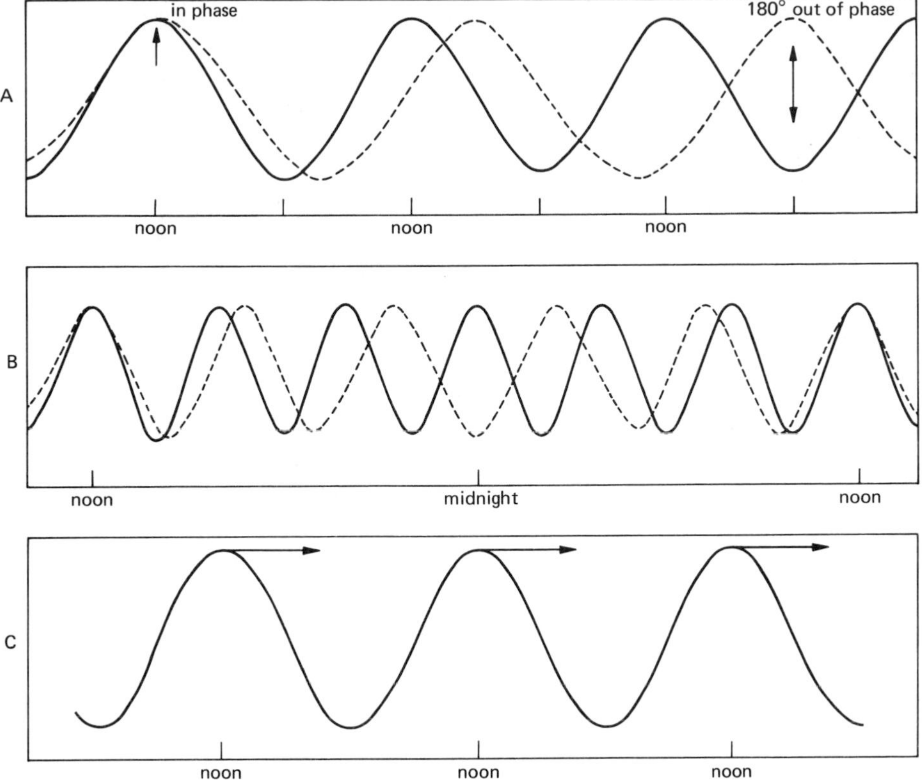

Figure 20-2. Interactions of rhythmic processes.
A. Two circadian rhythms of slightly different period that go into and out of phase. **B.** An example of how a 24-hr periodicity might be established by two rhythmic processes that come into phase once every 24 hr. **C.** How an hourglass timer (represented by the arrows) might be reset each day by a diurnal rhythmic process.

rhythms do not result directly from lunar or diurnal phenomena but from some internal rhythmic devices. It is possible that circadian rhythms might be the result of two or more slightly out-of-phase rapid rhythmic processes, so timed that their beat frequency (frequency of going in and out of phase) is about 12 or 24 hr, as is shown in Figure 20-2B. An interesting concept is that these long-period oscillations may result from the interactions of several short-period oscillators (perhaps of a biochemical nature). Conclusive evidence is lacking, however.

It is possible for hourglass and oscillating timing systems to interact, as is shown in Figure 20-2C. The oscillator would provide the absolute timing for some repeatable event, while the hourglass would time the duration of the event. In such a system, the oscillator, being temperature insensitive, would function as an effective time setter. The hourglass, being temperature dependent, would automatically compensate for variations in environmental conditions. Such an interaction might be advantageous for timing or controlling periodic metabolic events or events that require a minimum concentration of some reactant for successful completion.

Extrinsic Rhythms. Numerous subtle geophysical factors undergo rhythmic cycles of intensity or polarization. The possibility exists that plants can perceive and be timed by one or another of such factors as cosmic rays, weak electromagnetic radiation, geomagnetism, or solar radiation fluxes (the result of sunspot activity—the sun's period of rotation is about 27 days). Many ingenious experiments have been designed to counteract or eliminate the effects of such factors. Plants have been tested for their ability to measure time in specially constructed boxes with applied magnetic fields to neutralize that of the earth, at the bottom of deep mines where they would be free from the influence of radiation from outside the earth, in jet airplanes flying over the surface of the earth, in space capsules, and even on a turntable at the South Pole rotating counterclockwise at one cycle per day, so that they were stationary with respect to the earth's rotation.

Time measurement in plants continued in all these circumstances. So far, no conclusive evidence has been found that plants are entrained or pulsed by any of these extrinsic natural rhythms, except for the major diurnal rhythm of night and day, which is presumably responsible for entraining the circadian rhythm.

Experiments on time perception are very difficult—one set was ruined because the plants, in a constant dark environment, were watered at the same time each day by the light of a faint red safelight. Diurnal changes of noise level, electromagnetic radiation (for example, generated by fluorescent lights in an adjacent room), mechanical vibration (people walking or machinery operating) take place in a laboratory so that even in a supposedly constant environment, plants may be influenced by some unsuspected rhythmic timing device that may help to set or maintain internal rhythms.

Timing of Flowering

Photoperiodism and Vernalization. The two most important timing mechanisms for flowering both measure the progress of the season. One is **photoperiodism,** a mechanism that enables plants to respond to day length so that they flower at a specific time of the year as determined by the length of the day. The other is a cold requirement that many plants have, without which they will not flower. Many such plants, or their seeds, can be artificially cold treated so that they will flower in one year without the normally required overwintering period. This treatment is called **vernalization.** The term (it means "spring-ization") comes from the fact that winter varieties of cereal (wheat or rye), which must be planted in the fall and allowed to overwinter in the ground, can be made to behave like spring varieties by cold treatment, so that they flower the same year that they are sown.

The remainder of this chapter will be devoted to a study of photoperiodism and vernalization, with the objective of determining, so far as we are able, how these mechanisms work and what sort of timing devices (hourglass or oscillator) are involved in time measurement.

The Discovery of Photoperiodism. In 1920, following the earlier work of the French physiologist J. Tournois, W. W. Garner and H. A. Allard, working at the United States Department of Agriculture Research Laboratories at Beltsville, Maryland, made the critical observation that led to the discovery of time measurement by organisms. They noted that a certain tobacco variety, Maryland Mammoth, grew well vegetatively in summer, but only flowered in winter in the greenhouse. They also noted that soybeans planted at various times during the spring all flowered at about the same time in late

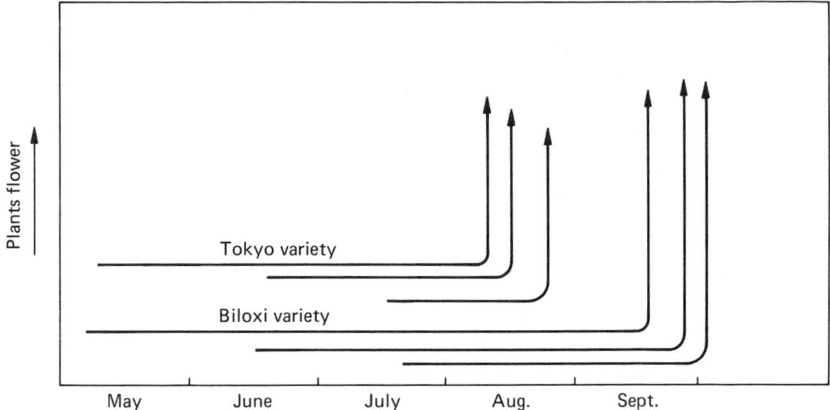

Figure 20-3. Diagram representing an experiment of Garner and Allard in which soybeans planted at different times all came into flower at about the same time in late summer.

summer, regardless of how long they had been growing. A diagram illustrating these experiments is shown in Figure 20-3.

Garner and Allard tested a number of possible variables that might have affected flowering, including temperature and light quality, and discovered that the critical factor was the length of the day. The Maryland Mammoth tobacco plants would not flower if the length of the day (or of artificial illumination in a growth chamber) exceeded 14 hr. Different varieties of soybeans similarly had critical day lengths—during midsummer the days were too long, and flowering did not take place. After September 1 the day length was short enough, and flowering did take place.

Plants like Maryland Mammoth tobacco that only flower when the day length is shorter than some critical maximum are called **short-day plants.** Later Garner and Allard showed that there are **long-day plants,** which will only flower when the day length is longer than a certain critical minimum, and **day-neutral plants,** which are unaffected by day length. Examples of these types are shown in Figure 20-4. It should be noted that the actual critical day length of a short-day plant is not necessarily very short. The important point is that if it is exceeded the plant will not flower. Similarly, long-day plants may have quite a short critical day, but it must be exceeded for flowering to occur.

Most plants have now been classified according to their photoperiodic requirements; a partial list is shown in Table 20-1. However, there are a great many variations and intermediate conditions. Some plants have an *absolute* requirement for short or long days, whereas in others flowering is merely promoted or inhibited. A few plants require special day lengths only early in their development; later they become day neutral. A few species are known that flower only in intermediate day lengths, not in long or short days. A few plants show interrelations with other factors, such as temperature, and the poinsettia (*Euphorbia pulcherrima*) and the morning glory (*Ipomoea purpurea*) are short-day plants at high temperature and long-day plants at low temperatures. Some plants are very closely tied to the seasons. **Long-short-day plants** flower only in the fall, during the short days that follow the long days of midsummer. **Short-long-day plants** flower only during the long days preceding midsummer, following the short days of early spring.

Photoperiodism is clearly a mechanism that regulates flowering time to a particular

Figure 20-4. Photoperiodism of plants: (**1**) Short-day species (*Perilla nankinensis*); (**2**) day-neutral species (*Nicotiana tabacum*); (**3**) long-day species (*Rudbeckia bicolor*).
Absence of flowering in red perilla under long-day (LD) conditions and in *Rudbeckia* under short-day (SD) conditions are adaptations to survival under unfavorable environmental conditions: summer drought in the case of the short-day species, winter frosts in the case of the long-day species. [From M. Kh. Chailakhyan. Reproduced by permission of the National Research Council of Canada from the *Can. J. Bot.,* **39:**1817–41 (1961).]

season of the year. The peculiar adaptability of this process is illustrated by the fact that in a number of widespread species, whose range extends for large distances north and south, the critical day length varies with the latitude at which the plant is growing. The further north, the longer the critical day, regardless of whether the plant is a short- or long-day plant. This is in agreement with the fact that long-day plants are more commonly found in high latitudes, whereas short-day plants tend to be tropical or subtropical.

Vernalization is evidently a phenomenon that occurs at high latitudes where there are cold periods in winter. Therefore, a vernalization requirement is much more frequently found in long- than in short-day plants. Other factors, such as light intensity, light quality, mineral supply, moisture, relative humidity, and the like, may interact in various ways to modify the expression of long- or short-day characteristics in many plants.

Night Interruptions and Dark Measurement. It was initially thought that plants measured the duration of daylight. However, experiments by the American physiologists K. C. Hamner and J. Bonner showed that, in fact, it is the length of night that is critical. They found that the cocklebur (*Xanthium strumarium*), a short-day plant, would flower when the dark period in its daily cycle exceeded 9 hr regardless of the day length, as shown in Figure 20-5.

Table 20-1. A partial list of short-day, long-day, and day-neutral plants

Short-day	*Long-day*	*Day-neutral*
	Monocots	
	Barley (*Hordeum vulgare*)	Bluegrass (*Poa annua*)
	Bentgrass (*Agrostis palustris*)	Corn (*Zea mays*)
Rice (*Oryza sativa*)	Bromegrass (*Bromus inermis*)	
	Canary grass (*Phalaris arundinacea*)	
	Oats (*Avena sativa*)	
	Orchard grass (*Dactylis glomerata*)	
	Rye grass (*Lolium spp.*)	
	Timothy (*Phleum spp.*)	
	Wheat grass (*Agropyron smithii*)	
	Wheat (*Triticum aestivum*)	
	Dicots	
Bryophyllum (*Bryophyllum pinnatum*)	Cabbage (*Brassica spp.*)	Balsam (*Impatiens balsamina*)
Chrysanthemum (*Chrysanthemum spp.*)	Clover (*Trifolium pratense*)	Bean (*Phaseolus spp.*)
Cocklebur (*Xanthium strumarium*)	Coneflower (*Rudbeckia bicolor*)	Buckwheat (*Fagopyrum tataricum*)
Cosmos (*Cosmos sulphureus*)	Dill (*Anethum graveolens*)	Cotton (*Gossypium hirsutum*)
Goosefoot (*Chenopodium rubrum*)	Henbane (*Hyoscyamus niger*)	Cucumber (*Cucumis sativus*)
Japanese morning glory (*Pharbitis nil*)	Hibiscus (*Hibiscus syriacus*)	Holly (*Ilex aquifolium*)
Kalanchoë (*Kalanchoë blossfeldiana*)	Mustard (*Sinapis alba*)	Jerusalem artichoke (*Helianthus tuberosus*)
Morning glory (*Ipomoea purpurea*)	Petunia (*Petunia sp.*)	Potato (*Solanum tuberosum*)
Poinsettia (*Euphorbia pulcherrima*)	Radish (*Raphanus sativus*)	Rhododendron (*Rhododendron spp.*)
*Strawberry (*Fragaria chiloensis*)	Spinach (*Spinacea oleracea*)	*Strawberry (*Fragaria chiloensis*)
*Tobacco (Maryland Mammoth) (*Nicotiana tabacum*)	Swiss chard (*Beta vulgaris*)	*Tobacco (*Nicotiana tabacum*)
Violet (*Viola papilionacea*)		Tomato (*Lycopersicon esculentum*)

*Different varieties of these plants have different day-length requirements.

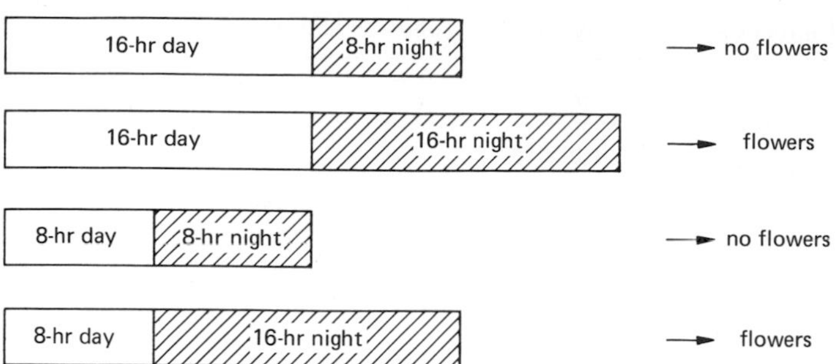

Figure 20-5. Diagram to show the effects of long and short days or nights on flowering of the cocklebur (*Xanthium strumarium*).

Later, experiments of Hamner and his associates showed that a brief interruption of the dark period with light would nullify the effect of a long night, showing that the plant is indeed measuring the duration of uninterrupted darkness. This effect works equally well for long- or short-day plants; however, a period of darkness during the daytime has no effect on flowering. Clearly it is the length of uninterrupted dark, not light, that is important. The experiments that show this are summarized in Figure 20-6. It should be noted that the light intensity of the flash that interrupts the dark period need not be very great; bright moonlight is sufficient in some plants, and certain short-day plants avoid the effects of moonlight by folding their leaves at night so that they are parallel with, instead of at right angles to, the incident moonlight. Certain other short-day plants require a critical period of low-intensity light (comparable to moonlight) instead of a dark period.

Sites of Perception. A question of early interest was, what part of the plant perceives the length of darkness? It was soon found that the leaves, and only the leaves, were the receptive organs. If all the leaves are removed from a plant, it becomes insensitive to photoperiod. If one leaf is left on the plant, the plant responds to the treatment that leaf receives, regardless of the treatment that is given to the remainder of the plant. Experiments by the Russian physiologist M. Kh. Chailakhyan* illustrating this point are shown in Figure 20-7.

*Also transliterated M. C. Cajlakjan or Cajlakhjan.

Figure 20-6. Diagram to show the effects on long- and short-day plants of a dark interruption of a long day and of a light interruption of a long night.

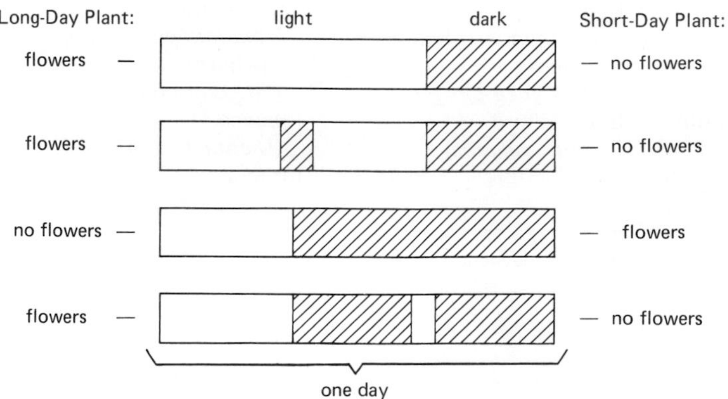

A

B

Figure 20-7. The role of the leaf in photoperiodism.

A. A short-day species, red perilla. On the left (the leaf under long-day conditions), the shoot vegetates; on the right (the leaf under short-day conditions), the shoot is flowering and setting fruits.

B. A long-day species, *Rudbeckia*. On the left (the leaf under long-day conditions), the shoot is flowering; on the right (the leaf under short-day conditions), no flowering shoots and only small leaves are formed.

[From M. Kh. Chailakhyan. Reproduced by permission of the National Research Council of Canada from the *Can. J. Bot.*, **39:**1817–41 (1961).]

The most important observation is that if one leaf on a plant is kept at the correct day length, the plant will flower, regardless of the conditions surrounding the rest of the plant. (In passing, it should be noted that the presence of other leaves on the wrong day length does reduce the sensitivity of the system.) Thus it can be concluded that the leaf is the perceptive organ for photoperiodism. Further, these experiments indicate that the result of perception is the creation of a stimulus that must leave the leaf and travel to the meristem to initiate flowering. We shall consider the nature of the mechanism of perception before going on to look at the nature of the stimulus and its action in initiating flowering.

Phytochrome. The discovery that a flash of light in the middle of a long dark period prevents flowering in short-day plants opened a fruitful avenue of research. The flash of light must be absorbed; this means that there must be a pigment involved. The further discovery that a longer period of low illumination would work about equally as well as a short period of bright illumination made possible an analysis of the action spectrum of the pigment.

Such an analysis was made by the American physiologists S. B. Hendricks and H. A. Borthwick, who, like Garner and Allard, worked at the United States Department of Agriculture in Beltsville, Maryland. Leaves of a number of plants were arranged in the light of a very large spectrograph so that each leaf was illuminated by light of different wavelength, as shown in Figure 20-8. The plants (soybean, a short-day plant) were exposed to appropriate short days, but the night period was interrupted by a flash of light of various wavelengths. The wavelengths that were most active in preventing flowering were in the red region of the spectrum, as shown in Figure 20-9. Unlike photosynthesis, however, short wavelength (blue) light was not effective. Also, there was a rapid drop-off of efficiency in the far-red region.

Now Hendricks and Borthwick returned to some earlier work on the effect of light on lettuce-seed germination done by the American physiologists L. H. Flint and E. D. McAllister. These workers found that germination was promoted by red light, not by blue or far-red, as shown in Figure 20-10. Hendricks and Borthwick tried experimentally to determine whether the inhibitory far-red light would reverse the promoting effect of red light and vice versa, with the results shown in Table 20-2. Then they tried a similar experiment on the flowering of a short-day plant. Interrupting the long night with flashes of red or far-red light would inhibit or permit flowering, respectively. Further, a flash of far-red light following a flash of red light would nullify the effect of the red flash, and the plant would flower as if it had received no light. The results of these experiments are shown in Figure 20-11.

These experiments show that a light-receptive pigment exists and suggested to Hendricks and Borthwick that it has two forms. One form absorbs red light, and in doing so it is converted to the other form. The other form absorbs far-red light, and as a result it is converted back to the first form. The red-absorbing form has been called P_r, and the far-red-absorbing form has been called P_{fr}. The relationship between them was initially thought to be

$$P_r \underset{\text{far-red light}}{\overset{\text{red light}}{\rightleftharpoons}} P_{fr}$$

The experiments that confirmed this hypothesis were done with the assistance of W. Butler, using a special spectrophotometer of extreme sensitivity. Experiments were conducted using etiolated tissue to minimize the interference of chlorophylls and other

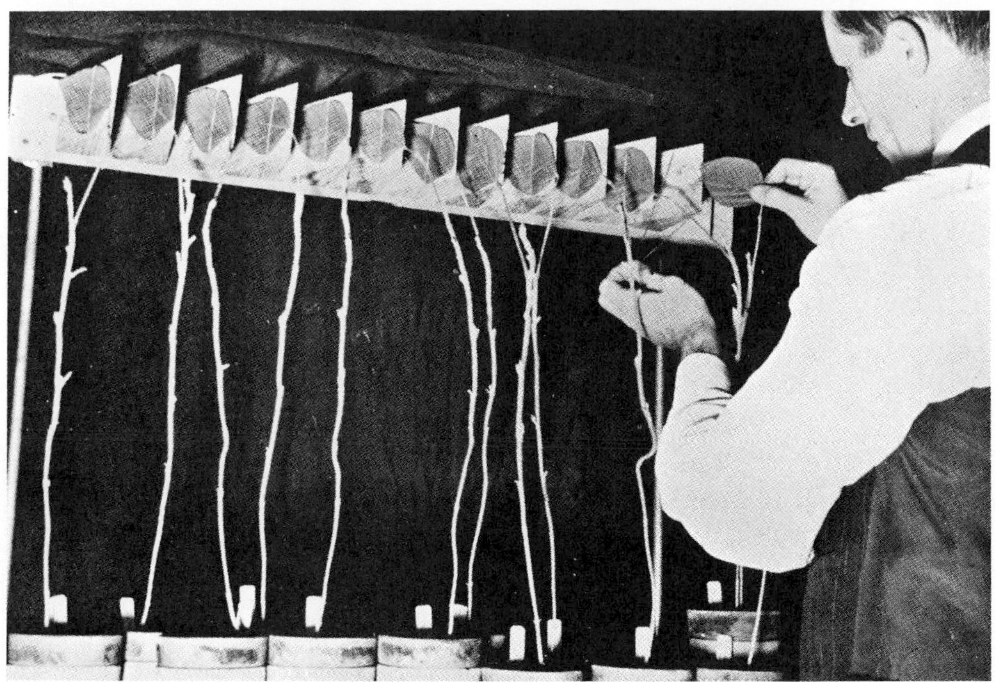

Figure 20-8. Soybean plants arranged in the beam of a spectrograph for irradiation with various wavelengths of light. [From S. B. Hendricks and H. A. Borthwick: Photoperiodism in plants. *Proc. 1st. Int. Photobiol. Congr.*, 1954, pp. 23–35. Used with permission.]

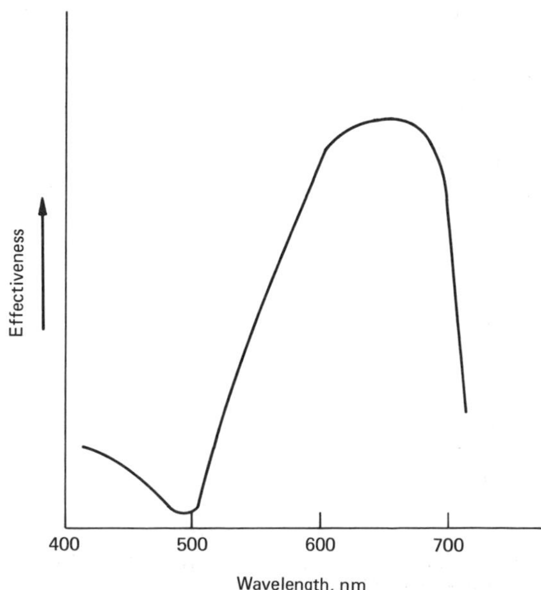

Figure 20-9. The action spectrum of photoperiodism.

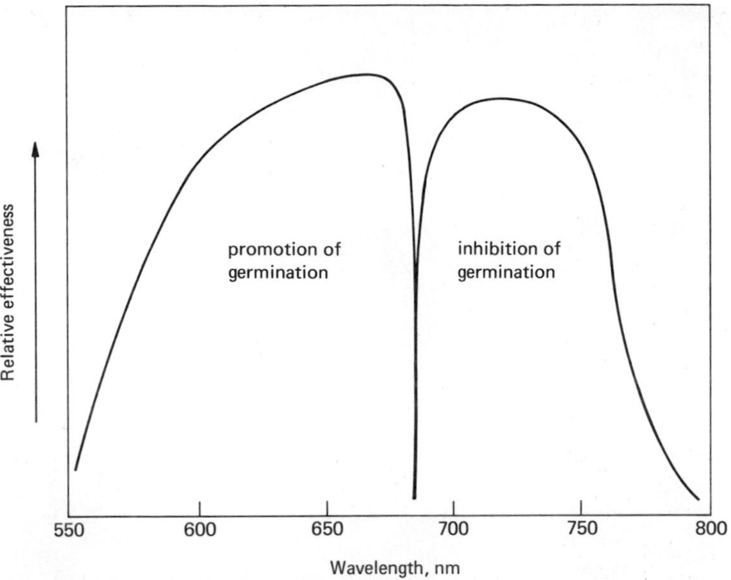

Figure 20-10. The effect of various wavelengths of light in promoting or suppressing germination of Grand Rapids lettuce seeds. [Curves drawn from data of H. A. Borthwick, et al., *Proc. Nat. Acad. Sci.*, **38:**662–66 (1952).]

pigments. When the tissue was irradiated with red light (660 nm), the optical density or absorption of light in the far-red region (730 nm) increased while the absorption of red light decreased. Similarly, when the plant was irradiated with far-red light, the absorption of far-red light decreased while the absorption in the red region increased. Clearly, the red-absorbing pigment was converted to the far-red-absorbing form by red light and vice versa.

The isolation of the pigment involved was very difficult because it is present in extremely small amounts. However, a group of scientists at the Beltsville laboratories were finally able to identify and partially purify a pigment that has the absorption properties detected in vivo. This pigment, called **phytochrome,** has since been further

Table 20-2. Germination of Grand Rapids lettuce seeds after exposure to red (R) or far-red (FR) light in sequence

Light	*Germination, %*
None	8.5
R	98
R, FR	54
R, FR, R	100
R, FR, R, FR	43
R, FR, R, FR, R	99
R, FR, R, FR, R, FR	54
R, FR, R, FR, R, FR, R	98

SOURCE: Adapted from H. A. Borthwick, et al.: *Proc. Nat. Acad. Sci.*, **38:**662–66 (1952).

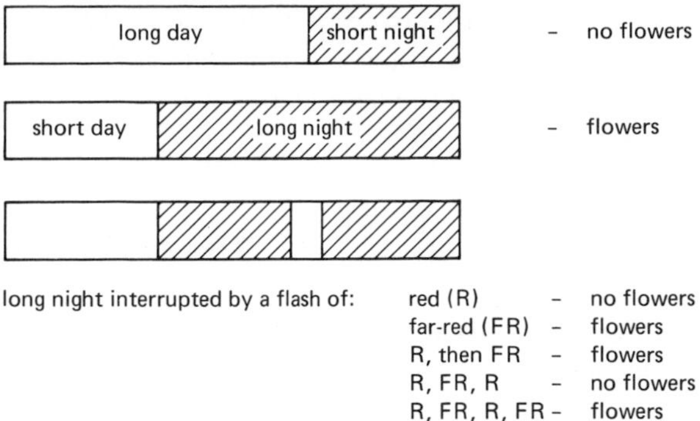

Figure 20-11. The effect of a light flash of red (R) or far-red (FR) light or of sequences of R and FR during the inductive night of a short-day plant.

purified and characterized. It is a glycoprotein–pigment complex with a molecular weight of approximately 125,000. It contains an open-chain tetrapyrrole molecule, similar to the phycobilin pigments (see Chapter 7, page 154). Considerable difficulties were initially encountered in measuring the molecular weight of phytochrome, using techniques involving its sedimentation or equilibrium flotation in a centrifugal field. These were resolved when it was found that most preparations were contaminated with traces of protease enzymes that degraded the phytochrome molecule and gave highly variable results. Its chemistry is very complex, and there seem to be several intermediate steps between P_r and P_{fr}. P_{fr} is somewhat unstable, and in vivo it appears to undergo slow decomposition and possibly also a reversal back to P_r in the dark.

The absorpiton spectra of P_r and P_{fr} are shown in Figure 20-12. Two characteristics need to be mentioned. Both forms of the pigment absorb blue light, and blue light is evidently biologically active in the system, although to a much lesser extent than far-red light and only affecting certain events. The second characteristic is that the absorption

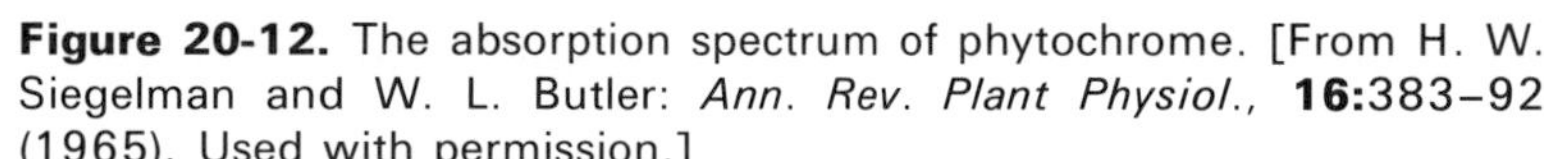

Figure 20-12. The absorption spectrum of phytochrome. [From H. W. Siegelman and W. L. Butler: *Ann. Rev. Plant Physiol.*, **16:**383–92 (1965). Used with permission.]

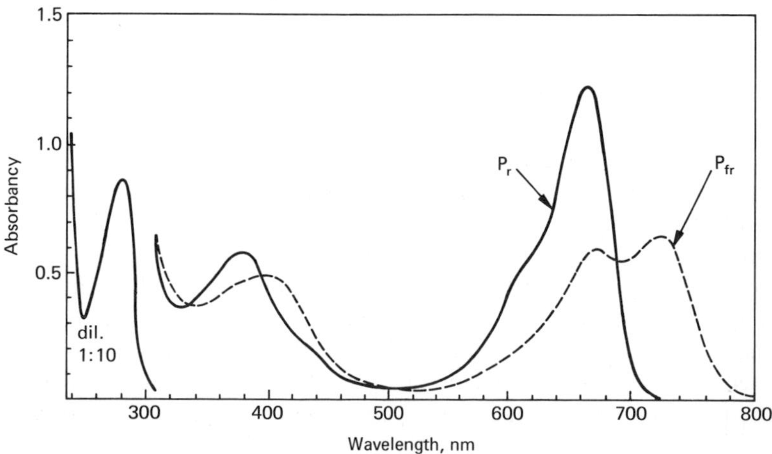

spectra of P_r and P_{fr} overlap in the 650–690 nm region, so that red light irradiation (or daylight) does not result in a complete conversion of P_r to P_{fr} but produces a mixture of the two pigments. These factors will be important in the discussion of phytochrome action.

The Mechanism of Phytochrome Action

The Range of Reactions Mediated by Phytochrome. Phytochrome appears to be involved in a wide variety of responses. The criterion for a phytochrome-controlled response is that it is turned on and off, or affected in opposite directions, by alternating flashes of red and far-red light. Many such responses are also affected by blue light. The action spectrum of a response can be determined by using flashes of monochromatic light of different wavelengths, and this can be compared with the action spectrum of known phytochrome responses and with the absorption spectrum of phytochrome.

Although the flowering stimulus is perceived only by the leaf, other reactions that appear to be mediated by phytochrome are perceived by various parts of the plant. A brief survey of some of these reactions, many of which have been mentioned in the previous chapter, will help to clarify the arguments about how phytochrome works in flowering. Besides flowering, phytochrome is involved in many other phenomena, such as growth (stem elongation), the orientation of leaves and stems to the direction of light (phototropism), sleep movements of leaves (nyctinasty), and the orientation of chloroplasts within leaves or cells. It has been considered possible that the same mechanism of action of phytochrome controls all these things (although this is by no means proven).

The first characteristic is the extreme rapidity of some of these mechanisms. Leaf movements can begin within 5 min of illumination, chloroplast movement within 10 min. Leaf unrolling, which is affected by red and far-red light, is sensitized in 1 min (that is, a 1-min illumination is sufficient so that unrolling will take place at some time afterward). The second characteristic is that phytochrome seems to be located throughout the plant. The sleep movements of leaves are sensitized by light perceived by the pulvinus; the unfolding of the plumule hook of many seedlings is sensitized by light perceived by the epicotyl; germination of lettuce seeds is mediated by light perceived by the growing tip of the embryo. Even roots, though not normally exposed to light, have certain typical phytochrome responses.

Cellular Location of Phytochrome. The precise location of phytochrome in cells has been possible from several different kinds of experiments. Work on the movements of chloroplasts in the alga *Mougeotia* by the German physiologist W. Haupt is of interest. When the cell is illuminated with red light, the large, flat chloroplast turns to become oriented at right angles with the light beam. Haupt found that a microbeam of red light only 3 μ in diameter would cause this reaction. He was able to show that the chloroplast would move if the beam was shone on any part of the cell, including the extreme edge, even well away from the chloroplast and the nucleus. Haupt concluded that the phytochrome is located in the outer membrane or plasmalemma of the cell. He was able to show, using polarized light, that the pigment molecules are oriented in a specific way in the plasmalemma and that the direction of orientation changes 90° when the pigment is converted from P_r to P_{fr} and vice versa. This, of course, does not prove that phytochrome is similarly located in all cells, but it does give some clues to a possible mechanism of action of the pigment.

In other experiments the subcellular organelles have been separated by centrifugation, and the distribution of phytochrome among them and in the soluble phases of the cytoplasm measured. P_{fr} particularly has been found associated with various membrane systems of cells and organelles. P_r may be soluble in the cell, but on conversion to P_{fr} (by absorption of red light) it quickly becomes bound to membranes.

Some Attempted Explanations of Phytochrome Action. The original suggestion for the action of phytochrome is as follows:

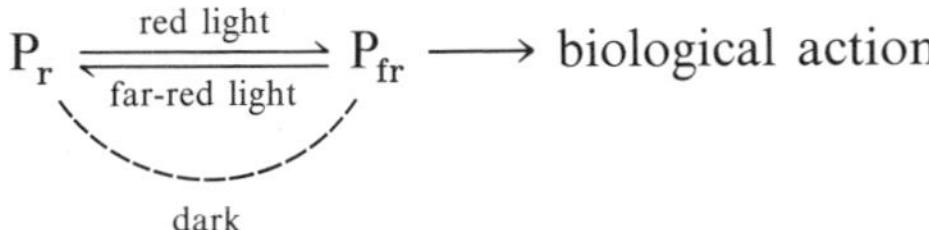

However, the nature of the biological action, the intermediates or processes of the reactions, and the ways in which the products of the initial reaction might operate are all as yet unclear. The German physiologist H. Mohr has suggested that some reaction occurs that results in changes at the genetic level—the activation of a gene or genes or the synthesis of enzymes. The induction of the enzyme phenylalanineammonia lyase (which produces cinnamic acid, required in lignin synthesis) by white or red light, is a good example. The effects of red and far-red light can be prevented by the addition of inhibitors of protein or RNA formation, such as actinomycin D, puromycin, or chloramphenicol. But this does not prove that the primary action of phytochrome is at the genetic level, since the RNA or protein-synthesizing reactions could be secondary. Further, the rapidity of some of the reactions (for example, leaf movement, chloroplast rotation, and so on) is such that enzyme induction seems to be out of the question.

It appears that electrical potentials may be established in tissues reacting to phytochrome stimulation. However, such reactions usually involve auxin, and in every response where auxins are involved, even when they are artificially applied, an electrical potential results. It seems quite likely that the potential results from the presence of the auxins and not the other way around. It therefore seems unlikely that phytochrome action results from a direct generation of electrical potential.

Active and Inactive Phytochrome. Further problems with the simple model of phytochrome action have come from measurements of the total amounts of phytochrome in plants after illumination with red or far-red light. First, it was found that only etiolated seedlings contain large amounts of phytochrome; normal tissues have such minute amounts that it can seldom be detected by present techniques, even in tissues that show the red-far-red response. Second, in some tissues the complete conversion of P_r and P_{fr} seems not to result in morphogenetic response, and some responses occur with only partial interconversion of the pigment forms. Third, it appears that the interconversion of P_r and P_{fr} may not be as simple as it was first thought. There seems no question that P_r is converted to P_{fr} under the influence of red light. However, the reverse reaction is less clear. Spectrophotometric measurements suggest that P_{fr} may be broken down by far-red light, rather than being converted back to P_r. This idea follows from the discovery that increases in P_r do not seem to be related quantitatively with decreases in

P_{fr}. Also, P_{fr} appears to be converted to some other unknown derivatives that are photochemically inactive. Finally, certain paradoxes have been reported that are difficult to explain. In the *Pisum* paradox, response is obtained to far-red light in the absence of detectable P_{fr}. By contrast, in the *Zea* paradox, red-light responses are saturated by an amount of light sufficient to convert only a small fraction of the P_r and P_{fr}.

These facts have led the American physiologists W. R. Briggs and W. S. Hillman independently to propose that most of the phytochrome, as detected spectrophotometrically, is present in an inactive form. It may undergo photoconversion, but only a small, perhaps specially oriented, "active" fraction could have any morphogenetic consequences from doing so. This idea is strengthened by the fact that several different physical forms of phytochrome have been found in plants; all of these are capable of a greater or lesser degree of photochemical reactions, but they differ in subtle ways.

Some Recent Ideas. As a result of this evidence, several revisions of the simple theory of phytochrome operation have been suggested. At this time we do not know exactly how it does work. The following modification of the original scheme by the German physiologist K. M. Hartmann

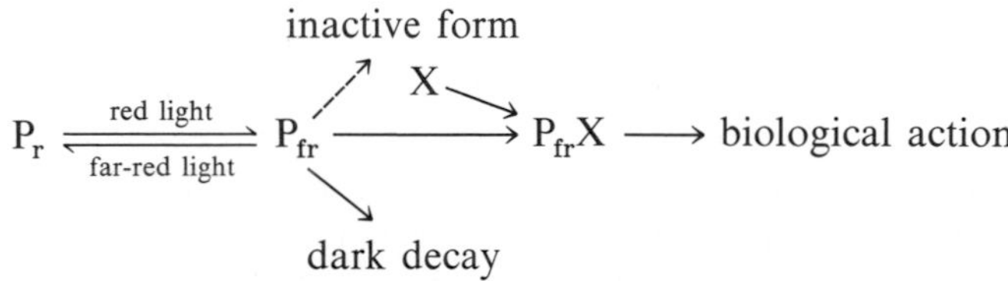

suggests that some unknown derivative of P_{fr} is the biologically active form. It is possible that this derivative could be a membrane-bound form of phytochrome. According to this scheme, the synthesis of the derivative would not depend on the relative amount of P_r and P_{fr} present but on the absolute amount of P_{fr} *being formed* from P_r. This is because P_{fr} can undergo two or more reactions, only one of which leads to biological action. This scheme explains several experimental situations where a balanced mixture of red and far-red light, or light of intermediate wavelength, has the greatest biological effect because it maintains an optimal concentration of P_{fr} in the tissue. Bothwick and Hendricks have elaborated a more complex model with several alternative reaction paths that P_{fr} or $P_{fr}X$ could follow, which would explain a number of the apparently contradictory observations that have been made. However, a clear picture has not yet emerged.

A hypothesis that relates the membrane location of phytochrome to its activity has been put forward by the British physiologist H. Smith. His model, represented in Figure 20-13, accounts for the fact that the pigment appears to be membrane-bound and that many of its activities seem to be associated with sudden and major changes in the permeability of membranes. The observed effects of poisons that might inhibit RNA and protein synthesis may be interpreted on the basis of secondary processes following (but not directly associated with) reactions of phytochrome.

A recent and interesting observation by the American physiologist M. J. Jaffe suggests how phytochrome might effect changes in membrane permeability. Jaffe and his associates found that the animal neurohumor acetylcholine (ACh) could mimic the effects of red light and that ACh mediates a variety of phytochrome responses in roots that seem to involve cell membranes. (ACh acts on membranes in animals.) This

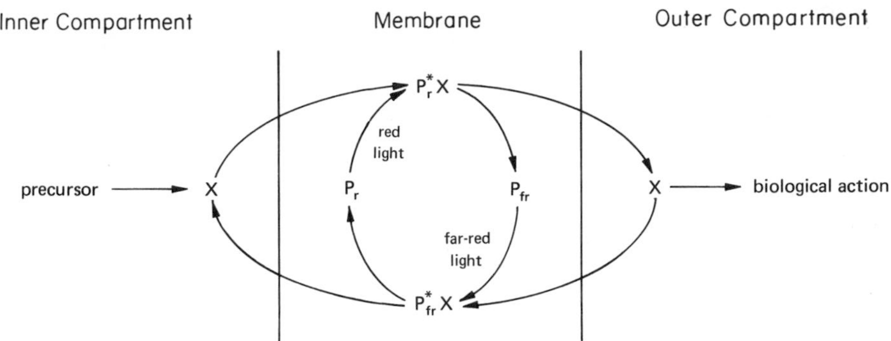

Figure 20-13. A speculative model of phytochrome action proposed by H. Smith. [Redrawn with permission from *Nature,* **227:**665–68 (1970).]
P_r^* and P_{fr}^* represent intermediate photoproducts of P_r and P_{fr}. The scheme should not be taken to indicate that P_r or P_{fr} moves within the membrane but rather that binding sites for X are anisotropic, leading to binding of X on one side of the membrane and its release on the other.

suggests that illumination with red light results in the synthesis of ACh, which affects both cell and mitochondrial membranes, increasing ion transport, oxygen uptake, and ATP utilization. Illumination with far-red light is thought either to induce destruction of ACh or to prevent its further synthesis. It must be emphasized that this hypothesis is also speculative; ACh has not been found in many situations where phytochrome-mediated reactions are known to take place, and ACh has not as yet been widely tested as a morphogenetic agent. However, the hypothesis provides an interesting lead that is well worth following.

Many analyses of events controlled by phytochrome have led to the idea that P_{fr} (or $P_{fr}X$) may somehow, in the words of the Australian physiologist L. T. Evans, "control the flow of reactive substances into several linked and competing synthetic pathways." Some reactions may require high P_{fr}, some low. In addition, the P_{fr} requirements may be related to requirements or pool sizes of intermediates, the supply of photosynthetic carbon, and other similar considerations.

Thus the difference between short-day and long-day plants may be related to the relative rates of P_{fr} and P_r interconversion. Short-day plants appear to require the maintenance of high levels of P_{fr} for a considerable period in the dark, because flowering is inhibited by a flash of far-red light late in the night. One the other hand, flowering in long-day plants is enhanced by far-red light shortly after darkness. It appears that there may be processes requiring high P_{fr} and processes requiring low P_{fr}. In short-day plants, the high P_{fr} processes require a long time to complete in order for induction to take place. In long-day plants it is the low P_{fr} processes that are critical. However, the nature of the processes is not known, and clearly we cannot explain the difference between short-day and long-day plants on the basis of phytochrome action alone.

There are currently three main hypotheses of phytochrome action: (1) gene activation, (2) enzyme activation, and (3) modulation of membrane properties. The activation of genes, or even of enzymes, is quite possible and may account for some phytochrome responses. It does not explain the many rapid responses, however. Phytochrome responses do affect membrane permeability, and many of the rapid responses depend on fast transport of ions to set up osmotic responses. Thus phytochrome (or P_{fr}, the active form) may affect the activity of membrane-bound enzymes (including ion-transporting ATPases) or of other regulating molecules (particularly gibberellic acid). The answers to these questions are not yet clear, but a massive research attack is now underway on this problem, and we may hope that it will soon be solved.

High Energy Reactions. A number of morphogenetic responses require illumination with a much higher intensity of light than is required for the interconversion of P_r and P_{fr}. Some of these have been found to have an action spectrum that closely resembles photosynthesis, and the high energy light requirement can be replaced by supplying sugars. Thus after a long dark period, punctuated with short intervals of light to prevent induction, a short-day plant would not flower when given an inductive dark period unless it was also given either a prolonged light interval or a supply of sugar. Evidently energy-requiring metabolism is necessary in the process of floral induction.

Other high energy requirements appear to be related to the basic requirement for phytochrome action: the maintenance of small but critical concentration of P_{fr} for long periods of time. The optimal light spectrum of these high energy requirements is usually such as to cause destruction of P_{fr}, but also to cause its continuous formation from P_r. This has been interpreted on the basis of the scheme shown in Figure 20-13 as being the continuous operation of the cyclic interconversion of P_r and P_{fr} to maintain a steady supply of the unknown intermediate X at the key reaction site.

However, it must be stressed again that the mechanism shown in Figure 20-13 is hypothetical and that the true mode of P_r and P_{fr} action is not yet known. Many laboratories around the world are working on this key problem in plant physiology.

The Relationship Between Flowering and Rapid Responses. As was described in the preceding chapter, a number of rapid movements in the plant, including such phenomena as nyctinasty, leaf rolling, and the chloroplast movements in the *Mougeotia* experiments mentioned above, are all mediated through the phytochrome system. These reactions are characterized by extremely short inductive periods, from a few minutes to less than a minute. Yet the flowering response is much slower and takes at least several hours. Further, all the rapid responses are perceived by the reacting tissue; however, in flowering the receptor is the leaf and the reactor is the stem tip. Still another difference is the permanence of the flowering response and the ephemeral nature or ready reversibility of movement responses. These differences are probably advantageous in the life of the plant.

However, these differences also point out the possibility that flowering and rapid responses may not necessarily be mediated by the same mechanism. There seems to be no doubt that perception takes place through the pigment phytochrome. In rapid movement, perception is quickly followed by a response including membrane phenomena and energy metabolism and resulting in ion fluxes that change turgor and cause movement. The change is ephemeral and can be rapidly reversed. In flowering, however, membrane phenomena and ion fluxes need not be invoked (although they may, in fact, be involved). The changes take place more slowly and are permanent. As a result of the change in the leaf, metabolism takes place that leads to the formation of some stimulus which promotes flowering and is translocated to the growing tip, where the final response takes place. The two processes may be compared as follows:

Flowering	Rapid movements
1. Perception (phytochrome)	1. Perception (phytochrome)
2. Reaction	2. Reaction
3. Transformation (to create permanence)	3. Response (ion flux)
4. Reaction (formation of flowering stimulus)	
5. Translocation	
6. Response (flowering)	

Although there may be similarities in the initial reaction mechanism that leads to transformation in flowering and to response in rapid movement, this need not be so. Evidently induction of flowering is a more complex sequence of events. The coincidence of the perception mechanism may be no more than a coincidence. Little evidence is available on which to make a final judgment at this time.

Floral Induction

Induction and Floral Development. Induction does not happen all at once. The fact that certain plants can be induced to flower by one critical night (long or short as the case may be) suggests that the process is an all-or-nothing one. However, the American physiologist F. B. Salisbury showed that the transformation of the vegetative apex to flowering depends upon the intensity of the stimulus. That is, one inductive photoperiod results in the vegetative apices of a few plants developing slowly; several inductive photoperiods result in more plants developing more rapidly.

Such a quantitative response can be measured. Salisbury describes eight stages in the development of staminate flowers of cocklebur (*Xanthium*), as shown in Figure 20-14. It is thus possible, by summing and averaging the values given by the developmental stages of a number of plants, to get a quantitative measurement of the average degree of floral response. Other methods of quantifying floral response (depending on the plant in question) include measuring the number of flowers or buds per plant, the height of a bolting shoot, or the time required for flowers to appear. Using the technique of scoring floral apices as shown in Figure 20-14, Salisbury showed that floral development of *Xanthium* takes place more rapidly when the inducing treatment is more vigorous, as shown in Figure 20-15.

This phenomenon, and the fact that different plants require different numbers of inductive photoperiods for complete flowering, indicate that floral induction involves the creation of some more or less permanent change in the induced plant, resulting in a continuously applied stimulus to flower. If the inducing stimulus is very weak, some plants may revert to the vegetative form after a short period of flowering. This indicates that the induction can be reversed, and then the stimulus to flowering ceases to operate.

Perception and Translocation of Floral Stimulus. The fact established earlier, that the leaf is the site of perception of the light signals that induce or prevent flowering, implies that something must be translocated from the leaf to the growing tip where

Figure 20-14. Stages of development of the staminate inflorescence primordium of *Xanthium*. The degree of pubescence varies somewhat, depending on the experimental conditions. [From F. B. Salisbury: *Plant Physiol.*, **30:**327–34 (1955). Used with permission.]

flowers will form. Chailakhyan has pointed out that the flowering response requires four steps: (1) the perception of the stimulus, (2) the transformation of the perceiving organ (to some new pattern of metabolism), (3) the translocation of a resulting stimulator, and (4) a response in the growing tip that results in flowering. The perception is through phytochrome, as we have seen. Transformation is some change in metabolism mediated by phytochrome or a derivative of it. The following questions must now be asked: How

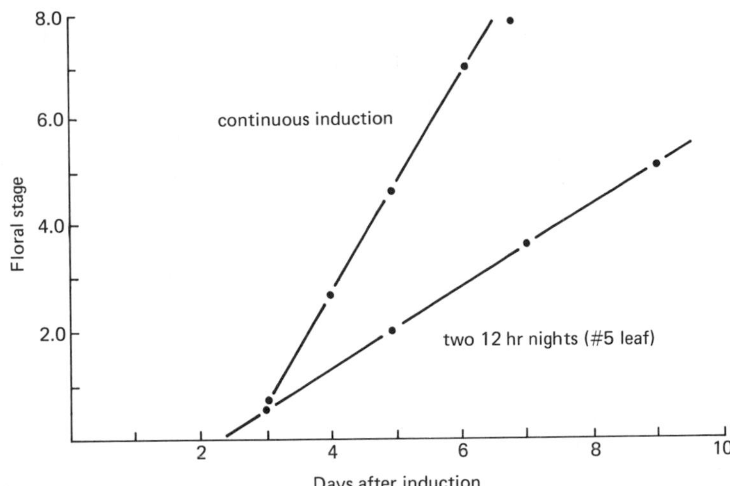

Figure 20-15. The rate of floral bud development. Continuous induction was begun June 3, 1953. Two 12-hr nights were given June 6 and 7, 1953. [From F. B. Salisbury: *Plant Physiol.*, **30:**327–34 (1955). Used with permission.]

does transformation take place? What is translocated (or is the translocation of some inhibiting substances interrupted)? How does it act to cause flowering?

If a stimulus is translocated, then it should be possible to measure its rate of translocation. One technique is to remove leaves from a plant at intervals after an inductive night. Salisbury found that if leaves are removed immediately after the inductive night, no flowering occurs. After several hours, however, removal of the leaves no longer prevents flowering. The stimulus, whatever it is, has already moved out of the leaf and can initiate flowering.

Experiments were conducted by Evans to measure the rates of translocation of the flowering stimulus. He found that the rates of movement in a long-day and a short-day plant are quite different, and in each case those rates are different from the rate of translocation of photosynthate, as shown in Table 20-3. This suggests that perhaps a different substance is translocated in long-day and short-day plants. Evidently the translocation of these substances, like that of IAA, takes place via different pathways than the movement of photoassimilated carbon. Flower-inducing substances are presumably translocated in the phloem, however, because grafting experiments show that a graft union must occur. Also, girdling prevents transmission of flowering stimulus.

Table 20-3. Rates of translocation of floral stimulus and photosynthate in rye grass (*Lolium,* a long-day plant) and Japanese morning glory (*Pharbitis nil,* a short-day plant)

Rate of translocation, cm/hr	*Rye grass*	*Morning glory*
Floral stimulus	2	24–33
Photosynthate	77–105	33–37

SOURCE: Data from L. T. Evans: *Ann. Rev. Plant Physiol.*, **22:**365–94 (1971).

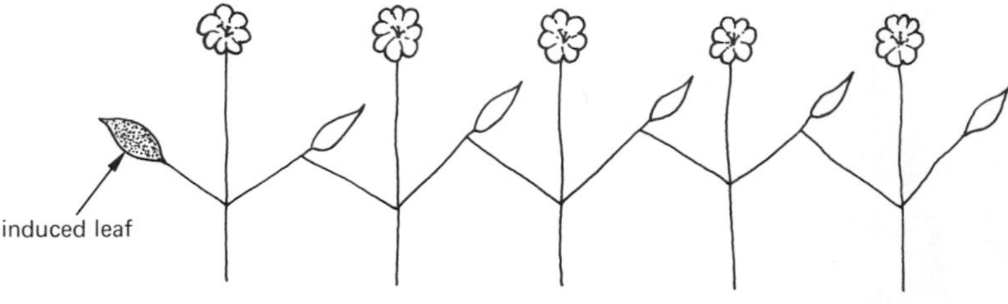

Figure 20-16. Diagram to illustrate Chailakhyan's grafting experiment, which demonstrates the translocation or transmission of the floral stimulus from one induced leaf to five grafted plants.

Translocation of the floral stimulus is further supported by experiments of Chailakhyan in which as many as five cocklebur plants were grafted together, as shown in Figure 20-16. If one leaf on one plant is induced by placing it in the correct photoperiod (short days) all the plants will flower, even if they are all maintained in long days, which would normally keep them in the vegetative state. Further, an induced leaf can be grafted to a noninduced plant maintained under the wrong photoperiod, and the plant will flower. This experiment has even been successfully performed using a leaf that was induced by the correct photoperiod while detached from the plant. Evidently the induced leaf is capable of exporting something to the uninduced plant that causes flowering.

The question of the transformation of the induced leaf is illuminated by some interesting grafting experiments by the Dutch physiologist J. Zeevart, now working in America. He found that induced leaves of the short-day plant *Perilla* could be grafted onto vegetative plants after periods of up to 3 months and still cause them to flower, showing that the induced state is essentially permanent. In *Perilla,* even parts of leaves may be induced (by exposing one half of the leaf to short day), and only the induced part of the leaf will cause flowering in a vegetative plant. This indicates that the induced state may involve a permanent transformation and not merely the formation of a diffusible substance. The state of induction confers the capacity of exporting a floral stimulus.

An interesting question is the difference between short-day and long-day plants: Are there special long-day and short-day substances? Many grafting experiments have shown that, if a long-day plant and a short-day plant are grafted together, the induction of one will cause the other to flower. Clearly, flowering results in both short-day and long-day plants after induction by either short-day or long-day leaves. Thus the floral stimulus must be the same, or two different substances must produce the same results.

The Dutch physiologist S. J. Wellensiek has examined the plant *Silene armeria.* Flowering in this plant can be induced by a variety of conditions: long days, vernalization, short days at high temperature, or by gibberellic acid. Any induced plant, no matter how induced, can act to induce other plants grafted to it. All these points suggest the possibility that, following initial induction (however it is brought about), some secondary reaction takes place that leads to a more universal flowering hormone:

$$\left.\begin{array}{l}\text{long days} \searrow \\ \text{short days} \longrightarrow \\ \text{vernalization} \nearrow\end{array}\right. \text{induction} \longrightarrow \text{translocatable flowering hormone}$$

This is not the only possible explanation; it is also possible that different substances can act in the same way or that a group of substances acting together might be involved. The difference between long-day and short-day plants might then be a matter of the normal concentration of one or another of the required compounds. These possibilities will be examined later in the chapter (page 499).

The general term **florigen** has been coined for flower-stimulating substances or floral hormones. It must be stressed that, at this time, florigen is a concept, not a substance. A number of experiments have been performed in which induced plants or leaves have been extracted and the extracts tested for florigenic activity. A few such tests have been positive, as shown in Figure 20-17, suggesting that a florigen substance may indeed exist. However, it has not been possible to isolate any substance that acts as if it were a florigen; that is, a substance which, when applied to the leaf of a plant, will be translocated to the apex where it will stimulate a previously uninduced meristem to flower. It may be found that florigen is a group of compounds, perhaps not even related. There is some evidence, as we shall see, that some of the presently known growth substances have florigenic properties, although none of them have all the required characteristics to be considered *the* florigen.

Recent experiments by the American plant physiologist W. L. Wardell show that DNA extracted from stems of florally induced tobacco will cause flowering in noninduced buds. Heat-denatured DNA is more effective, but DNAse-treated extracts are inactive. Extracts of uninduced shoots are also inactive. The nature of the stimulus leading to the formation of the active principle in the extract is not known.

Inhibitors. Certain experiments indicate the existence of flowering inhibitors, substances that act in the opposite way from florigen. In some plants flowering is suppressed somewhat if one or more leaves are kept in the noninduced state. Since floral initiation is often accomplished by only one induced leaf, the negative effect of a noninduced leaf suggests an inhibitor. In some plants, such as *Hyoscyamus niger* (henbane, a long-day plant), the plant flowers if all the leaves are cut off regardless of the day length. Again, this suggests that uninduced leaves exert a repressing influence over the growing tip, preventing flowering.

Experiments by the American physiologist G. D. Fratianne with the parasitic dodder plant also suggest the presence of a long-day inhibitor. When dodder grows on a short-day soybean host, it flowers only in short days. If the host is held in long days, flowering is inhibited. However, if the host plant, in long days, is defoliated, then the inhibition is removed. When two short-day soybean plants are linked together by dodder, flowering in one (kept in short days) is inhibited when the other is placed in long days. These experiments indicate a transmissible or translocatable inhibitor produced by short-day plants in long days. However, it is not known whether the phenomenon is general.

It is also possible that the inhibitor affects growth, rather than flowering, so that the effect on flowering may be a secondary one. The fact that in grafting experiments done on *Perilla* or *Xanthium* or other plants, a single leaf, or part of one, can cause flowering in a whole plant under noninducing conditions suggests strongly that inhibitors are not involved in these plants. If they were, the stimulating effect of so small an inducer as a part of one leaf would probably be overcome by inhibitory effects from all the noninduced leaves on the host plant.

Growth Substances. The difficulty of isolating a florigen raises the question of whether a special florigen substance exists or whether it is really the expression of the

A

B

Figure 20-17 (opposite). Effect of extractions from Mammoth tobacco leaves on the growth (**A**) and flowering (**B**) of *Rudbeckia*.
(A1) Gibberellin 0.02%; (A2) extract from the leaves of a long-day plant; (A3) extract from the leaves of a short-day plant; (A4) control, water. (B1) Extract from the leaves of a long-day plant; (B2) extract from the leaves of a short-day plant; (B3) control, water. [From M. Kh. Chailakhyan. Reproduced by permission of the National Research Council of Canada from the *Can. J. Bot.*, **39:**1817–41 (1961).]

effects of one or a combination of the known growth hormones? A striking flowering response is the effect of auxin on pineapple—auxin is used commercially to ensure synchronized flowering of plantation crops. However, pineapple seems to be an isolated case; in most other plants the addition of IAA inhibits flowering rather than stimulating it. There is evidence that cytokinins are involved in flowering, but their effects seem to be limited to a very few species of plants.

Gibberellic acid, on the other hand, has strong effects on bolting and flowering in many plants. It was found to replace both vernalization and photoperiodic induction in many long-day plants, as shown in Figures 20-18 and 20-19. A number of experiments have shown that natural gibberellins extracted from induced long-day plants will cause flowering in uninduced plants. An examination of the recent literature shows that gibberellic acid causes flowering in many, but not all, long-day plants (*Lolium,* ryegrass, is an exception). Virtually all rosette plants that have been tested respond to gibberellic acid. However, gibberellic acid causes flowering in only a few short-day plants such as *Impatiens balsamina;* in most it is ineffective.

Two alternative explanations are possible. (1) Gibberellic acid is effective mainly in long-day plants and only occasionally in short-day plants; therefore, although it may be essential in all plants, it is probably already present in sufficient amounts in most short-day plants. Alternatively, (2) gibberellic acid is essential in the stimulation of

Figure 20-18. The effect of gibberellin on a long-day plant, *Samolus parviflora,* kept in short days. Left to right: plants received 0, 2, 5, 10, and 20 μg gibberellin daily. [From A. Lang: *Proc. Nat. Acad. Sci.*, **43:**713 (1957). Used with permission. Original photograph courtesy Dr. A. Lang.]

Figure 20-19. Effect of gibberellin on the growth and development of *Rudbeckia* under short-day conditions: on the left, a plant treated with gibberellin solution (450 μg during $1\frac{1}{2}$ months) has formed the stem and flowered; on the right the control plant at rosette stage. [From M. Kh. Chailakhyan. Reproduced by permission of the National Research Council of Canada from the *Can. J. Bot.*, **39:**1817–41 (1961).]

growth leading to bolting, which is a prerequisite to flowering in many long-day plants. In either event, it seems clear that although gibberellic acid may be part of a complex floral stimulus for many, and perhaps all, plants it is not in itself the florigen. Further, no combination of the known growth hormones appears able to function as a floral stimulus in a general way in all plants.

At present, the experimental evidence seems to require a florigen, but attempts to extract and isolate it have been singularly unsuccessful. The resolution of this question remains one of the major problems facing plant physiologists.

Anthesin. The term *anthesin,* coined by the Russian physiologist N. K. Cholodny, has been used by Chailakhyan to denote a hypothetical floral-initiating hormone that works with gibberellin to produce florigenic activity. He suggested that gibberellin is nearly always present in sufficient amounts for flowering in short-day plants (regardless of the day length), since its addition seldom causes short-day plants to flower. Long-day plants, however, lack sufficient gibberellin, so it must be added either artificially, or naturally by

Figure 20-20. The anthesin hypothesis, as elaborated by M. Kh. Chailakhyan. Short days promote anthesin formation; long days promote gibberellin formation. Short-days plants have gibberellin; long-day plants have anthesin; day-neutral plants have both. Both are needed for flowering.

	Short days		Long days	
Short-day plants	gibberellin + anthesin	= flowering	gibberellin only	= no flowering
Day-neutral plants	gibberellin + anthesin	= flowering	gibberellin + anthesin	= flowering
Long-day plants	anthesin only	= no flowering	gibberellin + anthesin	= flowering

inductive long days, for flowering to take place. However, short-day plants, although they contain gibberellin, do not flower except in short days. Chailakhyan therefore suggested that another flowering factor, anthesin, is necessary in addition to gibberellin to cause flowering. According to his theory, anthesin is formed in short-day plants, as a result of inductive short days, but is probably present in sufficient quantities in long-day plants regardless of day length (like gibberellin in short-day plants). Day-neutral plants have sufficient quantities of both anthesin and gibberellin to flower without induction. This concept is summarized in Figure 20-20.

Unfortunately this concept requires another hypothetical substance that has not yet been isolated, so it does not improve matters over the simple florigen concept. Chailakhyan has found that methanolic extracts of long-day plants show some weak florigenic activity, but no identifiable compound has been found. Chailakhyan points out that not only is the presence of both gibberellin and anthesin necessary for flowering, but also the apex of the plant must be receptive, or in the right condition, for these hormones to act. Many plants have a juvenile period during which they cannot be induced to flower. It may last for a few days in some rapidly maturing species or many years in some trees. In certain plants the duration of juvenility is dependent on the developmental stage rather than on time. For example, morning glory (*Pharbitis*) normally flowers as soon as three to four leaves have formed, whereas *Eupatorium adenophorum* does not flower until 30 to 40 leaves have formed. Nutritional factors are also involved, and photosynthesis is essential for floral induction.

Changes at the Shoot Apex. Changes in the conversion of the vegetative apex into the flowering state have been described in Chapter 18 (page 448). The morphological changes leading to the production of floral primordia are preceded by changes in the meristematic activity of the growing tip. Some cells that were previously quiescent may become mitotically active, and the mitotic index of the whole meristematic area increases. Flower induction can be shown to activate RNA and DNA synthesis. It seems possible, therefore, that the floral stimulus (or some metabolite produced as a result of its action) depresses a polycistronic operon so that the floral genes become operational. An alternative possibility is that floral parts are produced without the activation of new genes but as a result of changes in the activity of the meristem. Wardlaw showed (see Chapter 18, page 441) that the nature of a primordium—whether leaf or bud—may depend on the available space and organization of the meristem. It is possible therefore that florigen acts by causing the meristem to enlarge and change its activity in such a way that the resultant meristem geometry automatically permits the development of floral parts instead of leaf or bud primordia.

At present, we do not have evidence to distinguish between these possibilities. The fact that induced stem tips of sugarcane can be cultured for as long as 6 months and still flower supports the former suggestion, since evidently some permanent or stable change takes place in the meristem that can be expressed much later. It seems unlikely that a specific meristem geometry could be so permanent. It is not known what changes taking place in the meristem result in its conversion from the juvenile (noninducible) state to the mature state, capable of flowering. There is some indication that this may be the result of a slow increase in diameter with time, but the meristem of some plants is capable of induction and conversion to the flowering state while still very young and small, so this suggestion is not generally supported.

Phytochrome as an Hourglass Timer. We must now return to the question with which we opened this discussion: the nature of the mechanism that times flowering. The phytochrome system appears to provide the means by which plants can measure the length of night by the hourglass method, since P_r is converted to P_{fr} in the day, the P_{fr} then either reverts to P_r or else decays at a slow rate during darkness while new P_r is being formed.

A number of objections have been raised against this simple and apparently elegant theory. The first is that the night-length measurement of many plants is nearly temperature insensitive, but the interconversions of P_r and P_{fr}, being chemical reactions, are dependent on temperature. The second is that the night decay of P_{fr} in many short-day plants is complete in a few hours—a fraction of the minimum inductive dark period. This was shown by Salisbury in the following experiment. The inductive night of a cockleburr plant was interrupted with a flash of red light, which would convert P_r into P_{fr}, thus prevent flowering. If, however, the red flash was followed by a far-red flash, then P_{fr} would be reconverted to P_r. However, if the flash of far-red was separated from the red flash by as little as 35 min, it was no longer effective, showing that the P_{fr} effect was complete within that time. The time was longer or shorter in different plants, but always less than the length of an inductive night.

Finally, it was noted that the *effectiveness* of light flashes in preventing flowering of the short-day Biloxi soybean which was grown in continuous darkness varied in a peculiar manner. Light flashes inhibited flowering during the "night" period but did not prevent flowering during the "day," even though the plant was kept in continuous darkness. Thus somehow the plant can distinguish between the times of day and night, even though it is in continuous darkness. Evidently it has some sort of timing mechanism that cannot be an hourglass type, because this type of timer can only measure one interval without being reset. An oscillating timer must be postulated. As we shall see, the timing of flowering is almost certainly accomplished by an interrelation between an oscillating timer and a resettable hourglass timer involving phytochrome.

Rhythmic Processes

Circadian Rhythms. The German physiologist E. Bünning investigated sleep movements of bean leaves (see Chapter 19, page 465) and found that their behavior correlated with certain photoperiodic events that had been observed. First, they had a

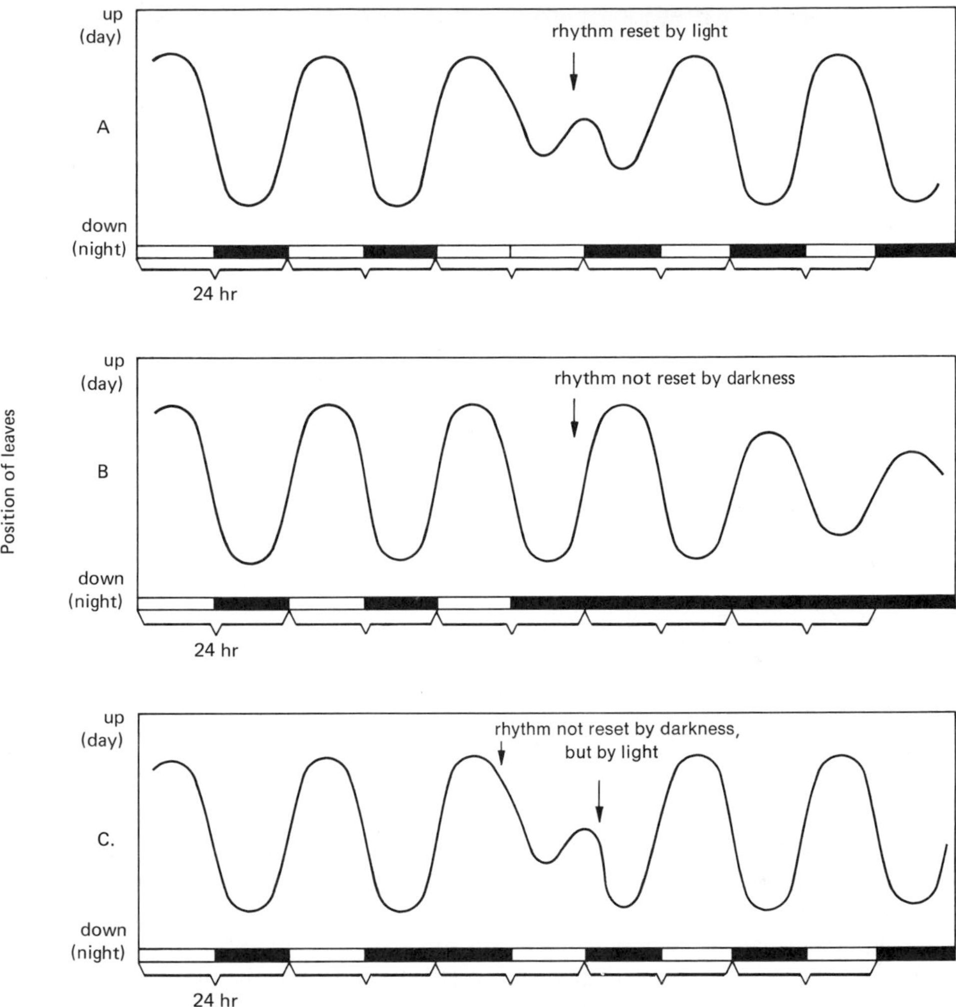

Figure 20-21. Rhythmic movements of leaves, showing how the rhythm may be reset by light (**A** and **C**), not by a dark interval (**C**) or continuous darkness (**B**).

circadian rhythm that persisted even in constant conditions. Second, the rhythm could be reset by a flash of light during the dark period, but not by a dark interval during the light period, as shown in Figure 20-21. If leaves were put in continuous darkness, the rhythm could be reset by a flash during the time that would have been night had the plant still been in its natural environment, but not by a flash during the time that would have been day. Bünning suggested that the circadian rhythm had a **photophile** phase, corresponding with daylight hours, and a **scotophile** phase, corresponding with the hours of darkness (*photos,* light, *scotos,* dark). He postulated that the response to light in photoperiodism relates to these phases in the plant's endogenous diurnal rhythm.

The American physiologist K. C. Hamner tested this hypothesis experimentally. It will be remembered that a flash of light during the inductive dark period will prevent a short-day plant from flowering. Hamner put short-day Biloxi soybeans into a 72-hr dark period, which would normally induce fairly extensive flowering in the plant, though

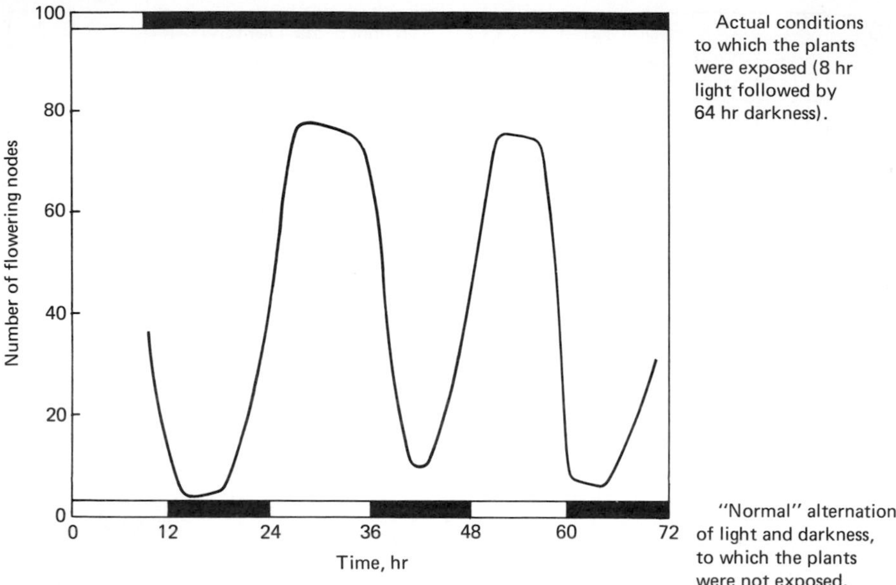

Figure 20-22. Response of Biloxi soybean to 4-hr flashes of light during a prolonged dark period. The plants flower when given a flash during the photophile phase but are inhibited when the flash occurs in the scotophile phase. [Redrawn from data of K. C. Hamner: In L. T. Evans (ed.): *Environmental Control of Plant Growth*. Academic Press, New York, 1963. Used with permission.]

perhaps not as much as three normal days consisting of 12-hr light and 12-hr darkness. He then interrupted the prolonged dark period with 4-hr flashes of light at various times, using different plants for each treatment. The plan of the experiment and the results that Hamner obtained are shown in Figure 20-22. Light flashes during the time when the plant would normally be experiencing night—the scotophile phase—inhibited flowering; however, if the light flashes were given during the period when the plant would normally be experiencing day—the photophile period—they greatly enhanced flowering. This clearly indicates that the day-length response, which appears to be mediated by phytochrome, is interlocked with a diurnal rhythm that persists for at least several cycles under constant conditions.

Many additional studies clearly link the phytochrome response with an endogenous circadian rhythm. The effect of red and far-red flashes varies depending on the point in the endogenous rhythm at which they are given. In the long-day rye grass, *Lolium temulentum,* red light inhibits and far-red promotes flowering if given during the early part of the dark period; later on the situation is reversed. In the short-day plant, *Chenopodium rubrum,* additional far-red light inhibits flowering in the day, whereas, as we have seen, red light inhibits flowering during the night. These results show that the phytochrome response is closely linked with, and in fact dependent upon, an endogenous circadian rhythm in the plant.

The converse does not hold true—that is, the measurement of time by the endogenous rhythm is not affected by phytochrome interconversion. Salisbury has shown that several factors that clearly affect phytochrome interconversion do not alter the timing mechanism. Thus, if flowering is inhibited 50 percent by low continuous light, which considerably decreases the dark loss of P_{fr}, time measurement (as determined by the effect of a flash of light at different times) is unaffected.

It must be remembered that the rhythm is not in fact a true diurnal one, that is, an accurately set oscillator with a 12- or 24-hr period. Instead, natural free-running circadian rhythms of anywhere from 20 to 27 hr are usually found. However, the rhythm is reset each day, usually by a dawn or dusk signal, or both, and is sufficiently accurate to measure time during the subsequent 24-hr period.

Circadian Rhythms and Photoperiodism. It appears that the flowering response and such rapid movements as the sleep movements of leaves depend both upon an internal oscillator and upon the phytochrome system. It is therefore necessary to consider how these two timing mechanisms might interact and whether both systems are indeed universally present. Unfortunately very little clear evidence is available.

The phytochrome system seems to be quite general, and responses attributable to it have been detected in most plants studied. The oscillating timer is also widespread—a "biological clock" has been detected in many organisms. It seems to be involved not only in simple time measurement but also in such activities as navigation of birds and other migratory animals. Some plants do not show any rhythmic responses or other evidence of a biological clock. This does not mean that they lack the capacity for time measurement, however. One may conclude that both the hourglass and the oscillating timer are in all likelihood very widely distributed even among primitive organisms.

How the timing mechanisms interact is far from clear. As suggested earlier, the nature of the phytochrome response appears to depend to some extent upon the phase of the circadian oscillator. That the circadian rhythm is set by light responses mediated by phytochrome is also clear. It is possible to maintain a plant on a **skeleton photoperiod**—that is, the plant is given a flash of light at the beginning and at the end of the "day." The flashes correspond to dawn and dusk, respectively, and entrain the circadian rhythm effectively. Red light is effective; if followed by a flash of far-red, the effect is nullified. If the "dusk" flash is delayed, the whole rhythm will be reset; but if it is delayed beyond some critical time (usually somewhere near halfway through the "night"), then the rhythm is suddenly rephased by 180°. That is, the plant behaves as if the flash is dawn instead of dusk, and the rhythm is reset accordingly. Thus the rhythm is clearly set and entrained by phytochrome-mediated light-dark changes, and the responses in the plant to the state of the phytochrome interconversion are equally dependent on the diurnal rhythm, at least in some plants.

A certain amount of evidence has been found that suggests not all plants require or respond to rhythmic processes in flowering. In some plants (for example, soybean) flowering can be inhibited by light treatments that do not affect the rhythmic movements of leaves. However, here the effect of the light flash may be on that part of floral induction for which the phytochrome response is essential, even though the diurnal rhythm is not affected or reset. It has been suggested that rhythmicity in flowering is a matter of the responsiveness of the shoot apex to the floral stimulus generated in the leaves. Doubtless some light will be shed on this fascinating question when more is known about the nature of the rhythmic response.

The Nature of the Oscillating Timer. The nature of the oscillator that sets rhythmic timing is still very mysterious. It is relatively temperature insensitive, and its period is not affected by respiratory inhibitors (although they may prevent the manifestation of the rhythm in the plant). In some plants, the period of the circadian rhythm is lengthened by treatment with heavy water, D_2O, suggesting chemical steps. There is some evidence that protein synthesis is involved, since inhibition of protein synthesis appears to abolish rhythmicity, but the data are difficult to interpret.

The site of the oscillator has been investigated using the giant algal cell, *Acetabularia,* which shows a number of pronounced rhythmic patterns of behavior. Nuclear transplants between cells that are in opposite phases of their diurnal rhythms indicate that the nucleus is responsible for setting the phase, because the rhythm of photosynthetic oxygen evolution in the resulting transplant is that of the nucleus, not of the receptor cell. Yet, the rhythm continues undiminished for 40 days in an enucleate cell and can be abolished and reset in enucleate cells as well as in intact ones. Thus, though the oscillator is set by the nucleus, it appears to reside in the cytoplasm.

The transcription inhibitor actinomycin D interferes with the cycle after 2 weeks, suggesting the existence of long-lived mRNA involved in setting or maintaining the cycle. However, the inhibitor chloramphenicol, which interferes with translation, affects the cycle after 48 hr, suggesting a proteinaceous mechanism that needs to be continuously renewed. However, it is known that the photosynthetic rhythm in certain organisms is the result of rhythmic variations in photosynthetic enzymes, so the *Acetabularia* results may only reflect the capacity of an enzyme to react to the oscillator. The question of what causes the enzyme to vary rhythmically still cannot be answered.

Many suggestions have been made as to the nature of the oscillator. The possibility that a cyclic feedback mechanism oscillates or "hunts" around a null point has been suggested. Various metabolic systems have periodic oscillations that range from a few minutes to an hour or more. The possibility has been suggested that a population of oscillators could, by their phasing, engender a 12-hr or circadian rhythm. This is one of the more attractive hypotheses for "gearing" rapid oscillations to produce a cycle of long period (analogous to the gears that transform the rapid oscillation of the flywheel of a watch into the hourly or daily sweep of the hands). But the details of such a mechanism, though they could be guessed at, have not been demonstrated.

The existence of endogenous rhythms seems incontrovertible. Workers in Hamner's laboratory grew pinto beans from seed in continuous unvarying light and showed that the plants had strong circadian rhythm of leaf movements for up to 4 weeks. The intensity, but not the period, of the rhythm depended on the light intensity at which the plants were grown. Moreover, while both primary leaves on one plant showed rhythms of about the same period, the two leaves were often out of phase with each other. Thus the oscillators were clearly related to and presumably located in the leaves, not in other parts of the plant. Moreover, they could not have been entrained or generated by some unsuspected environmental rhythmic event. Even cultures of cells or tissues from plants that show strong rhythmic behavior have been shown by the Canadian physiologist B. G. Cumming and his coworkers to have endogenous rhythms in their metabolism. This suggests that rhythmicity is a cellular rather than a tissue phenomenon. The resolution of this and the related problems of the phytochrome effect in flowering will be among the great breakthroughs which are yet to be made in plant physiology.

Vernalization

Cold Induction. Many plants require a period of cold treatment sometime during their development before they can flower. Many cereals exhibit this requirement, and wheat provides a good example. Some varieties of wheat, known as spring wheat, are planted in the spring and flower in one season. Other varieties, called winter wheat, must be planted in the fall. These sprout, then lie in the ground all winter, and flower next year. The winter cold treatment is essential; if they do not get it, either they do not flower or their flowering is much reduced. Russian physiologists experimented with this requirement, because winter wheat is often a better producer than comparable spring varieties. However, the winter is too severe in many parts of Russia for winter wheat to survive. The Russian geneticist T. D. Lysenko found that artificial cold treatment of seeds of winter wheat that are beginning to germinate would permit them to behave like spring wheat when planted in the spring. This process is called vernalization (vernal, springlike).

Many plants require vernalization. These include so-called winter annuals such as winter cereals, most biennials, and a certain number of perennials. Many biennials will remain in the vegetative state for a period of years if protected from cold in winter. Perennials that require vernalization must receive cold treatment each winter in order to flower the following year. Vernalization thus represents essentially another type of hourglass timer with a long duration, which must be reset annually.

Vernalization may be absolute, as in the case of many biennials that cannot flower without it. However, many of the winter annuals, such as winter wheat and Petkus rye (*Secale cereale*), have a qualitative response to vernalization. In these plants the flowering response is improved by increasing the length of vernalization. Full vernalization requires up to about 50 days of treatment between −2° and about +12° C, as shown in Figures 20-23 and 20-24. However, some flowering will occur after a long period without any cold treatment.

Interactions with Other Factors. Chailakhyan pointed out that only temperate plants that undergo winter can be expected to have a vernalization requirement, and these are likely to be long-day plants. Short-day plants are more usually subtropical (with the exception of a few like *Chrysanthemum* that flower late in fall). As a result, the majority of plants requiring vernalization are also long-day or day-neutral plants.

There are a number of odd interactions. In Petkus rye, the vernalization requirement can to some extent be replaced by short-day treatments. However, after the plant is vernalized, it requires long days as well for flowering. Similarly, the henbane (*Hyoscyamus niger*) requires vernalization when in the rosette stage; bolting and flowering will then occur in long days only.

The interrelations of flowering response and vernalization may not necessarily be complex; in fact, there may be no direct interconnection of these processes at all. However, at present neither process is well enough understood to say with certainty.

Site of Perception of Cold Stimulus. Seeds of Petkus rye (*Secale cereale*) that have imbibed some water are capable of being vernalized. The British physiologists F. G. Gregory and O. N. Purvis showed, by culturing excised embryos, that the embryo by itself can be vernalized, and even cultured apices are capable of vernalization. However, not all plants can be vernalized as seeds. Many require the exposure of vegetative parts to cold, and there is considerable variety in which part must be exposed. Those biennials that overwinter in the vegetative state respond to cold treatment of vegetative parts of the plant, including the leaves. However, in *Chrysanthemum*, the stem tip is the perceptive part of the plant. If the leaves are cold treated while the stem tip is kept warm, no

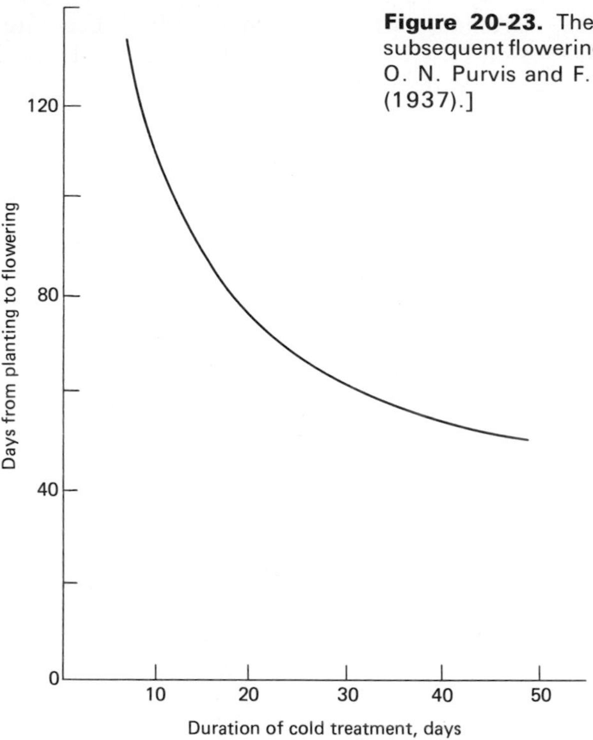

Figure 20-23. The effect of length of vernalization on subsequent flowering of Petkus rye. [Redrawn from data of O. N. Purvis and F. G. Gregory: *Ann. Bot.* (*N.S.*), **1:**569 (1937).]

vernalization occurs. Other plants are less specialized. The Dutch physiologist S. J. Wellensiek has shown that leaves and even isolated roots of *Lunaria biennis* are capable of being vernalized, so that plants that regenerate from the vernalized parts are induced and will flower. Thus there is no clear picture at the moment of where specifically the receptor or perceptive mechanism for vernalization is located. It appears to be present in different parts of different plants.

Vernalin and Gibberellins. Vernalization appears to be a complex process and may in fact be several processes. In some plants, like Petkus rye, vernalization can take place

Figure 20-24. The effect of the temperature during a 6-week period of vernalization on subsequent flowering of Petkus rye. [From H. Hansel: *Ann. Bot.*, **17:**417–32 (1953). Used with permission.]

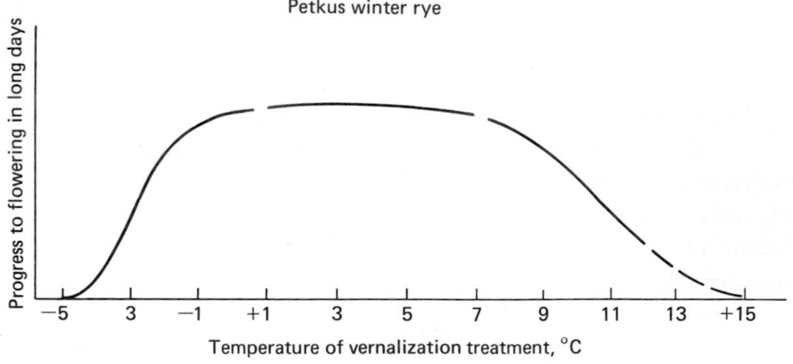

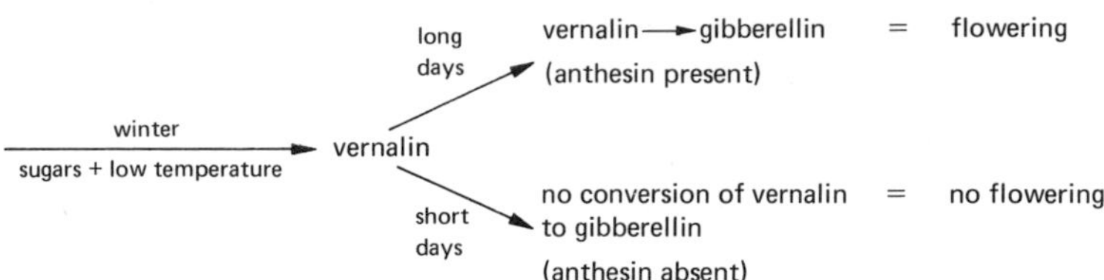

Figure 20-25. The role of gibberellin, vernalin, and anthesin in vernalization of long-day plants, as suggested by M. Kh. Chailakhyan. Note that, since anthesin is considered to be present in long days, the addition of gibberellin will cause flowering of unvernalized plants.

in the seed, and all the tissue derived from the vernalized meristem in a direct cell line is induced. In others, like *Chrysanthemum*, only the meristem can be vernalized, and, as in Petkus rye, only tissue derived from a vernalized meristem is vernalized. In still other plants, such as the biennial henbane (*Hyoscyamus niger*), the stimulus can be translocated via a graft union, so that an unvernalized plant can be induced.

The latter result led the German physiologist G. Melchers to postulate a substance responsible for transmitting the vernalizing stimulus which is called **vernalin.** This substance is hypothetical, however, and it has not been possible to isolate it. It was later shown by A. Lang, now in the United States, that the application of gibberellin would replace vernalization in many cold-requiring plants. However, gibberellin is not vernalin, because the response to it is different from the vernalization response. In rosette plants treated with gibberellic acid, the stem first elongates and produces a vegetative shoot; later, flower buds appear. In true bolting resulting from vernalization, the flower buds appear before stem elongation. Moreover, gibberellic acid does not induce many plants.

Chailakhyan has extended his anthesin hypothesis to account for vernalization in long-day plants as follows. He suggested that a vernalin is produced in low temperatures, which in long days is converted to gibberellin. This, with the anthesin already present in long-day plants, causes flowering (see pages 498–499). In short days the vernalin is not converted to gibberellin, and flowering does not occur. However, unvernalized long-day plants can be induced to flower in long days by the addition of gibberellin because they already contain anthesin. In short-day plants, the addition of gibberellin has no effect because no anthesin is present. Chailakhyan's suggestion is illustrated in Figure 20-25. Unfortunately there is still no firm evidence either for vernalin or anthesin. It seems likely that hormones or growth substances are involved in flower induction, but how they are related to the vernalization process is not known.

The Nature of the Vernalization Process. Vernalization is aerobic because it does not take place when plants or seeds are cooled in an atmosphere of nitrogen. It is a cumulative process because plants become gradually more and more effectively vernalized with time up to as long as about 2 months. It can be reversed in its early stages by heat treatment. All these points suggest a chemical process. However, vernalization must be an unusual chemical process, because it has a negative coefficient of temperature—that is, it goes faster (or better) at lower temperatures.

The British physiologist O. N. Purvis has attempted to resolve this paradox. She suggested that there are two reactions involved, in which one has a much higher temperature coefficient than the other, as shown in Figure 20-26. It can be seen that, if

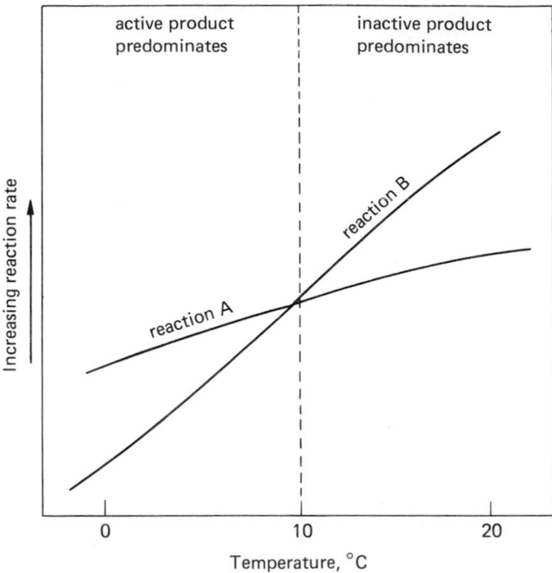

Figure 20-26. A diagram of the hypothesis of O. N. Purvis to explain the apparent negative temperature coefficient of the vernalization process.

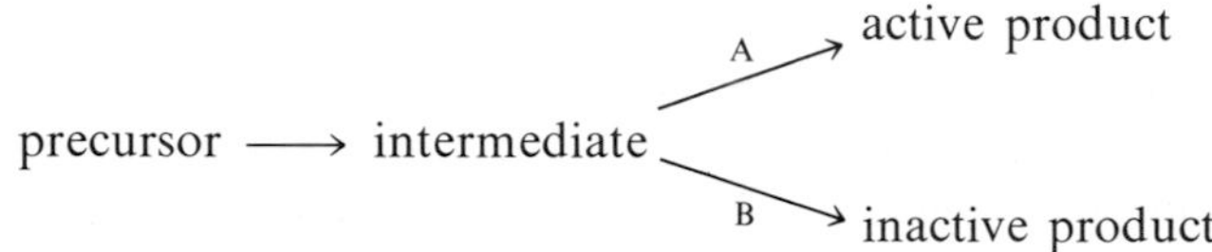

the Q_{10} of reaction B is higher than that of A, then at high temperature an inactive product will predominate, whereas at lower temperature the active products will override. Particularly if the initial reaction preceding the intermediate compound is slow, or if the inactive product is in fact inhibitory, the reaction will appear to have a negative temperature coefficient. This is only a hypothetical explanation, however, and is not supported by direct evidence.

It has been noted that the induced state in such plants as Petkus rye can be spread to many daughter cells without apparent dilution (that is, loss of potency). Experiment has shown that laterals growing on laterals growing on laterals growing on laterals growing on an originally vernalized main stem are all fully induced (in this experiment lateral production is stimulated by removal of the shoot apex). This has led to the suggestion that the flowering stimulus generated by vernalization is self-reproducing, either as a gene or group of genes turned on by vernalization or as a self-replicating organelle, activated by vernalization, that multiplies during cell division. Neither of these ideas, however, has been proven.

Evidently the nontransmissible stimulus of rye grass or *Chrysanthemum* is different from the hormonelike, graft-transmissible stimulus of henbane. This points clearly to the existence of a dual effect; possibly the two effects are unrelated. This resembles the situation in flower induction by day-length effects, where the induction of leaves is clearly a separate phenomenon from the stimulus of flowering in the stem tip. Perhaps the flowering stimuli in both cases are related; equally, there may be similarities in photoperiodic induction and vernalization. Some unifying principle is lacking at present, however, and the processes remain nearly as puzzling as they were when they were first discovered.

Summary: Flowering and Floral Induction

Flowering and floral induction constitute a complex subject, and it seems worthwhile to summarize briefly some of the information presented in this chapter.

The genetic program for flowering is present in the cells of the apex (and of the whole plant), but it is not expressed until the right time. This is determined in two important ways. Some plants flower when they are "ripe to flower" (they have grown to a sufficient size or stage of development), whereas others have devices that determine when the correct season for flowering has arrived. The season is determined by two important requirements: the correct day length (photoperiodism) and a cold requirement (vernalization).

Photoperiodism is a mechanism whereby the interval of darkness between periods of illumination is measured. This is done through the agency of the pigment, phytochrome, which is changed from the red-absorbing form P_r, to the biologically active far-red-absorbing form, P_{fr}, by the absorption of light. In darkness P_{fr} either decomposes or is changed to other forms, and P_r is regenerated. This apparently forms the basis of an hourglass timer that is reset daily. However, how this works is not clearly known or understood at present. The photoperiodic response is linked with an endogenous circadian rhythm, the cause or mechanism of which is also unknown. As a result, phytochrome-mediated reactions (floral induction, leaf movements, nyctinasty, and so on) are also timed on a 24-hr cycle and only occur at specific times of the day or night.

Photoperiodic plants are usually either long-day or short-day plants (other more complex reactions occur); that is, they flower when the night is either shorter (long day) or longer (short day) than some critical length. The photoperiod is measured by the leaves; when a critical, or inductive, dark period (or a sufficient number of critical dark periods) has been perceived, the leaf becomes induced, and a floral stimulus of unknown nature (sometimes called a florigen) diffuses or is translocated to the apex. The apex then becomes induced and will flower.

The difference between long-day and short-day plants is not clear. They may react differently to the P_r–P_{fr} system but produce the same (that is, universal) flowering stimulus. Alternatively, there may be two different types of flowering stimulus for long- or short-day plants. A third possibility is that gibberellin is produced in long days and a hypothetical substance, anthesin, in short days, Long-day plants would, in this scheme, contain anthesin but lack gibberellin; short-day plants would contain gibberellin but lack anthesin. The flowering of the day-neutral plants would depend on the developmental stage of the plant (ripeness to flower), which would permit sufficient production of both hormones.

Vernalization is essentially a cold requirement that induces cells, in either the floral apex or those from which it will develop, so that they can flower when other conditions (day length, temperature, ripeness to flower, and so on) are correct. Many perennials and most biennials have this requirement. The stimulus may be perceived in the shoots or leaves of different plants. Once tissue is vernalized, the induction is essentially permanent. Cells derived from vernalized cells are themselves fully induced. The nature of the induction or of the induced state is unknown. Nor is it known what relationship induction by vernalization bears to photoperiodic induction, if any.

Additional Reading

See list for Chapter 16 (p. 406).

Briggs, W. R., and **H. V. Rice:** Phytochrome: chemical and physical properties and mechanisms of action. *Ann. Rev. Plant Physiol.*, **23:**293–334 (1972).

Chailakhyan, M. Kh.: Internal factors of plant flowering. *Ann. Rev. Plant Physiol.*, **19:**1–36 (1968).

Cumming, B. G., and **E. Wagner:** Rhythmic processes in plants. *Ann. Rev. Plant Physiol.*, **19:**381–416 (1968).

Evans, L. T.: Flower induction and the florigen concept. *Ann. Rev. Plant Physiol.*, **22:**365–94 (1971).

Evans, L. T.: *Daylength and the Flowering of Plants.* Benjamin, Menlo Park, Calif., 1975.

Hillman, W. S.: *The Physiology of Flowering.* Holt, Rinehart and Winston, New York, 1962.

Marme, D.: Phytochrome: membranes as possible sites of primary action. *Ann. Rev. Plant Physiol.*, **28:**173–98 (1977).

Salisbury, F. B.: *The Flowering Process.* Macmillan Publishing Co., Inc., New York, 1963.

Sweeney, B. M.: Biological clocks in plants. *Ann. Rev. Plant Physiol.*, **14:**411–40 (1963).

Zeevaart, J. A. D.: Physiology of flower formation. *Ann. Rev. Plant Physiol.*, **27:**321–48 (1976).

21 Patterns of Nutrition in Development

Photosynthesis and Nutrition

In most green plants the main source of nutrition and energy is photosynthesis. Photosynthesis takes place primarily in leaves or other specifically adapted organs. Products of photosynthesis are largely used in other parts of the plant, however, so they must be translocated to places where they will be utilized. Photosynthesis also takes place to a lesser extent in various other parts of the plant, particularly in green stems, floral bracts, and parts of the fruit. The extent to which the photosynthesis of such organs contributes to the overall nutrition of the plant may be surprisingly great. This raises the question of the overall efficiency of the plant: Does the plant in fact need all the photosynthate that it manufactures? Does photosynthesis go on merely because the photosynthetic apparatus is there? Are there any control mechanisms to regulate the use of energy absorbed by the plant? Is there any advantage to mechanisms that might reduce or control the overall efficiency of use of absorbed energy? These questions will be considered as we examine the patterns and control of nutritional traffic in developing and mature plants.

It seems very likely that the majority of plants produce more photosynthate than they require for growth and reproduction. Biennials and perennials normally produce and store substantial amounts of reduced carbon that serve as a source of energy for the following year's growth. In addition, a considerable amount of carbohydrate or other storage carbon is deposited in seeds. Probably plants living under climatic extremes that reduce photosynthesis (periods of inclement weather, low temperature, drought, and so on) are short of carbon and depend upon maximum photosynthetic efficiency for survival. Nonetheless, there is substantial evidence that most plants carry on more photosynthesis than is required even for the production of seeds or storage materials. Some of the excess is undoubtedly lost during leaf fall, and much of it is consumed in the increased respiration that accompanies senescence. Whether or not mechanisms exist that cause a reduction in photosynthesis when storage sinks are full is not yet clear; we shall examine the evidence of such feedback control mechanisms later in the chapter.

The growing or mature plant usually has several or many leaves distributed over it, at different levels on the stem, that have various quite specific vascular relationships with other leaves and other parts of the plant. As a result, different leaves stand in different physical relationships to the shoot, to the root, and to each other. Translocation takes place most readily between plant parts that have more or less direct vascular connection.

Thus the carbon nutrition of different heterotrophic parts of the plant depends on the photosynthetic activity of different leaves.

In addition, certain metabolic events such as the reduction of nitrate or fixation of nitrogen take place largely in the roots of some plants; these events require carbon nutrition not only to provide the needed energy and reducing potential but also to provide carbon skeletons to detoxify ammonia and form the organic nitrogen compounds of translocation. Amino acids produced in the root must be translocated to other parts of the plant. Some other amino acids are made in leaves as the result of photosynthetic activity; these also must be distributed throughout the plant. Different parts of the plant require different types of carbon compounds for metabolism. Some of these may be made from sugars or other primary products of photosynthesis at the site of metabolism. Others may themselves be primary products of photosynthesis, translocated as such to the sites of metabolism. Thus there is a complex traffic of various types of nutrients being translocated both up and down the plant, among various leaves, and between them and the roots and shoot or shoots of the plant.

The questions we shall consider concern not only the rate of photosynthesis but the possibility of mechanisms that control the nature of the products of photosynthesis, the amounts exported from leaves, the direction of translocation, and the metabolic activity of various parts of the plant as they relate to the needs of other parts. What this amounts to is the regulation of the nutritional traffic of the plant in such a way as to interrelate the metabolic capabilities of various organs with the requirements of others. We shall also consider the development of metabolic capacity by the parts of the plant as they grow, and the ways in which their nutritional requirements are met until they achieve their own metabolic competence.

The Onset of Photosynthesis in Seedlings

During germination the seed reserves begin to be metabolized and translocated toward the growing tip, some of them as immediate breakdown products of storage compounds and others after some further metabolism in storage tissue. This process provides nourishment for the seedling until it becomes capable of supporting itself on external nutrients. The root begins to absorb inorganic nutrients at an early stage, in most seedlings almost as soon as the radicle begins to penetrate the soil. However carbon autotrophy does not begin until much later, and the seedling subsists on carbon reserves in the endosperm or cotyledons until the first leaves are approaching maturity. In some plants the first leaves are in fact the cotyledons, modified into photosynthetic organs. In others the cotyledons do not emerge or are shed essentially unmodified, and new primary leaves are the first functional photosynthetic organs on the plant.

Most plant leaves turn green before they are fully grown; many are green at a very early stage of development. However, they usually have little or no photosynthetic activity until nearly fully expanded. Experiments with $^{14}CO_2$ have shown that CO_2 is absorbed only slowly at first, and much of the carbon absorbed is used to make the structural and enzymic proteins and other components required for the photosynthetic apparatus. Only when the leaf approaches full size does a high rate of photosynthesis and the massive production of carbohydrates begin.

The synthesis of chlorophyll precedes the onset of photosynthetic carbon fixation by periods ranging from hours to days. This may be either because all of the necessary

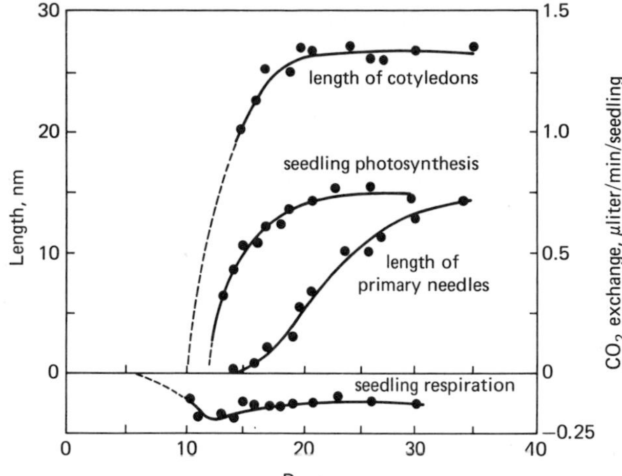

Figure 21-1. Growth, photosynthesis, and respiration of red pine seedlings. [Drawn from data of S. Sasaki and T. T. Kozlowski: *Ann. Bot.*, **33:**473–82 (1969).]

enzymes for photosynthesis have not been formed or because some essential part of the mechanism, although formed, is present in an inactive state. Following the initiation of chlorophyll formation a certain amount of light energy may be utilized in light-driven carboxylation reactions leading mainly to the formation of organic and amino acids These in turn arc used to build up the structural and enzymic machinery of the complete photosynthetic process. Thus the first step in the autotrophic nutrition of photosynthetic organs is the formation of the main autotrophic process itself. That other, catabolic, enzymes are formed previously is clear from the fact that the respiration of seedlings begins as soon as germination starts and reaches a high rate in cotyledons or young leaves before photosynthesis begins. The results of experiments by the American physiologist T. Kozlowski and his associates, shown in Figure 21-1, illustrate these events.

In spite of the fact that photosynthesis does not begin immediately on the emergence of the hypocotyl, it soon begins to contribute substantially to the development of the new seedling. If the newly developing photosynthetic tissue is excised, or if there is insufficient light for effective photosynthesis (but enough to prevent etiolation), seedling development is strongly retarded. Kozlowski has shown that the rate of growth and differentiation in the stages of germination following the onset of photosynthesis is correlated with the rate of photosynthesis in red pine seedlings, even though substantial seed reserves are still being metabolized. Thus it appears that the development of the seedling becomes dependent on photosynthetic carbon production as soon as photosynthesis itself develops.

Patterns of Nutrition in The Mature Plant

Patterns of Assimilation. There is a strong diurnal pattern of photosynthesis related to light intensity and to the diurnal variation in the water status of a plant. Many plants appear to respond only to these environmental stimuli. In such plants the rate of photosynthesis increases and decreases with the intensity of sunlight, and declines somewhat over noon and early afternoon due to increased internal water stress. However, the rates of photosynthesis in other plants clearly deviate from the expected values based on these considerations.

Photosynthesis of many algae placed under constant conditions continues to show diurnal periodicity for several days, rising to a peak around noon and falling at night. This periodicity in photosynthesis appears to depend on changes in the activity of such photosynthetic enzymes as ribulose bisphosphate carboxylase, which in turn may be related to diurnal cycles of activity discussed in the preceding chapter. In some plants the rate of photosynthesis declines somewhat after prolonged periods, suggesting that the size of the photosynthetic sinks may exert an effect on the process of photosynthesis, but in a great many plants this is not so. In some leaves like tobacco, photosynthesis may continue until the chloroplasts are disrupted by massive accumulation of starch.

Various other processes than photosynthesis also show strong diurnal variation. In many plants, translocation of photosynthate takes place mainly during photosynthesis or shortly after the process stops. This means that energy sources for nutrient uptake, nitrate reduction, and so on, are available in larger amounts in the daytime. Further, water uptake is likely to be greater in the daytime so that the intake of minerals is probably greater then. We can infer that the mineral nutrition of the plant is therefore subject to diurnal periodicity. Similarly, energy for synthetic processes, being derived ultimately from photosynthesis, is much more readily available in the daytime. This agrees with observations that most of the protein-synthesizing activity of a plant takes place during the day.

In addition to diurnal variations in photosynthesis, several seasonal variations occur. Many coniferous trees photosynthesize only slowly during winter, even if brought into artificially warm and bright conditions in a greenhouse. Evidently winter dormancy includes also a loss or inactivation of some of the enzymes of photosynthesis, although the chlorophyll content is maintained during winter.

All the leaves on a plant at any time do not have the same photosynthetic capacity. As may be seen in Figure 21-2, the rate of photosynthesis, which is initially high in young leaves of a poplar tree, increases up to the seventh leaf and then decreases in the

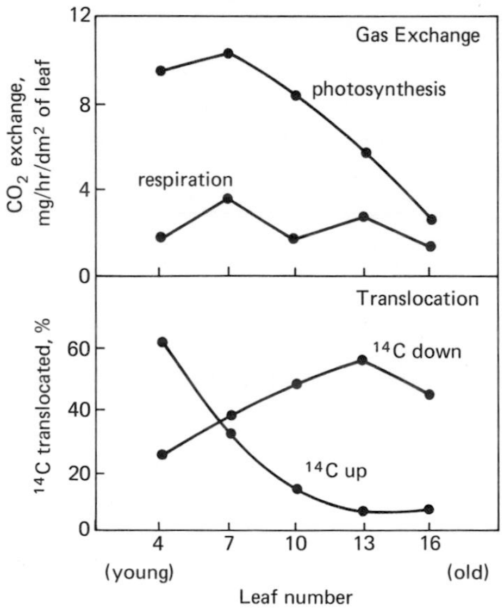

Figure 21-2. The effect of leaf age on net photosynthesis, respiration, and the translocation of ^{14}C following photosynthesis in $^{14}CO_2$. [Drawn from data of P. R. Larson and J. C. Gordon: *Am. J. Bot.*, **56:**1058–66 (1969).]

Table 21-1. Rates of CO_2 absorption in light (2500 ft-c) by wheat leaves at different stages in the development of the plant

	CO_2 *uptake, mg/(dm²)(hr)*			
Leaf no.	*Ear emergence*	*Flowering*	*Early grain filling*	*Mid grain filling*
2	14.0	—	—	—
3	14.7	—	—	—
4	18.6	14.4	10.6	8.6
5	22.1	11.0	16.6	12.7
6	24.3	18.3	16.9	16.2

SOURCE: Data of H. M. Rawson and G. Hofstra: *Aust. J. Biol. Sci.*, **22:**321–32 (1969). Used with permission.

older leaves. On the other hand, the rates of photosynthesis of wheat leaves all increase continuously with age during the formation and filling of the ear, as shown in Table 21-1. Rates of photosynthesis in bean leaves increase slowly with age but show sudden and dramatic increases, which may be short-lived, when axial buds break dormancy and begin to grow (Figure 21-3).

Thus the photosynthetic capacity of the whole plant is the result of the various individual capacities of its leaves. Because these may vary in different ways at the same time, the resultant for the whole plant is complex and perhaps not of great significance. However, in general the rate of photosynthesis seems to be realted to the metabolic activities, and hence requirements, of the plant. This seems to apply both to individual leaves, in relation to the activities of limited regions of the plant, and to the whole plant. Possible mechanisms governing this relationship will be discussed at the end of this chapter.

Patterns of Export from Leaves. The photosynthetic contribution of the leaf to the whole plant changes continuously with its age; that is, the amount of photosynthate exported from a leaf varies from time to time. Moreover, not only does the amount vary

Figure 21-3. The effect of buds breaking dormancy (arrows) on the rate of CO_2 uptake in photosynthesis by a leaf on a bean plant. [Data of W. B. Levin and R. G. S. Bidwell.]

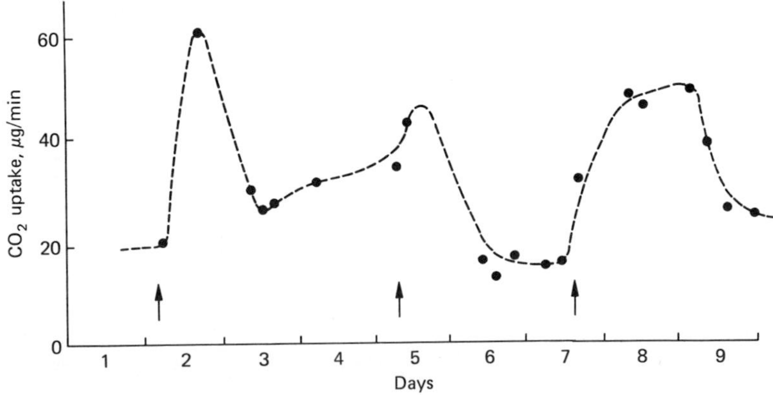

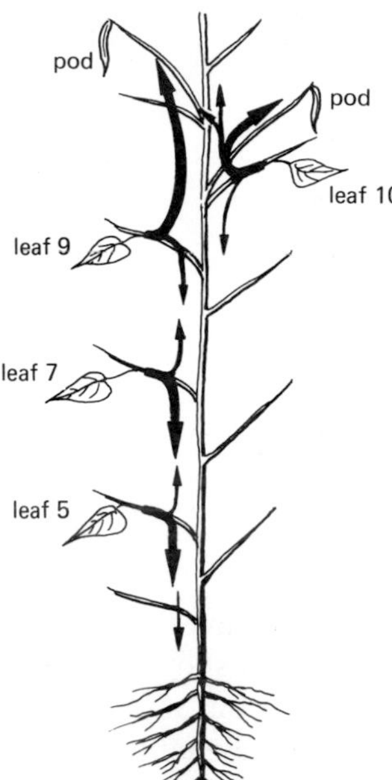

Figure 21-4. Export of photosynthetically fixed carbon from leaves of a pea plant. [Data of A. J. Link. Redrawn from C. A. Swanson: In F. C. Steward (ed.): *Plant Physiology: A Treatise*, Vol. II. Academic Press, New York, 1959, p. 511.]

but the direction in which photosynthate is translocated also changes as the leaf ages. Thus young leaves export principally toward the shoot apex and the younger, still growing regions of the apex, both leaves and shoots. Older leaves export primarily downward toward the lower stem and root. This characteristic is shown diagramatically for a pea plant in Figure 21-4. The same leaf will export mainly upward when it is young, and, as it ages, the direction of export will change toward the root. This pattern of behavior is illustrated by data for wheat leaves given in Table 21-2.

Phyllotaxis (the spatial relationship of leaves to one another along the axis of the stem, see Chapter 18) appears to have an important bearing on the translocation of photosynthate, as shown in Figure 21-5. The relationship here seems to derive from the fact that leaves which are situated one above another are more directly connected by vascular strands, and this suggests that export takes place primarily along the most direct vascular route. However, if the mature leaves opposite a leaf that is treated with ^{14}C are removed, the remaining young and growing leaves must rely on the opposite

Table 21-2. Export of ^{14}C from leaves of a 35-day-old wheat plant that have fixed $^{14}CO_2$ in light

Exporting leaf no.	*^{14}C exported, % ^{14}C fixed*	*Distribution of exported ^{14}C, % total exported ^{14}C in*				
		Roots	*Tillers*	*Leaves*	*Stem*	*Ears*
2	55	47	36	6.9	7.3	1.1
3	61	44	46	3.1	6.5	0.7
4	54	24	57	2.0	15	0.7
5	52	13	35	1.5	48	2.1
6	65	1.5	1.5	0.3	40	57

SOURCE: Recalculated from data of H. M. Rawson and G. Hofstra: *Aust. J. Biol. Sci.*, **22**:321–31 (1969). Used with permission.

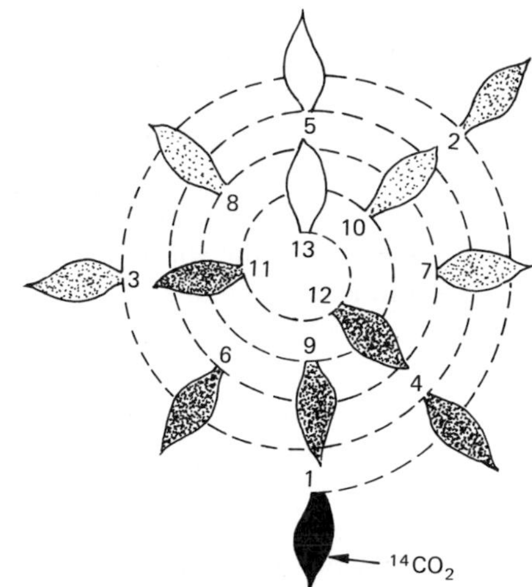

Figure 21-5. Phyllotaxis and distribution of ^{14}C in an 81-day-old tobacco plant. The treated leaf, supplied with $^{14}CO_2$ in light, is shown black. The density of shading indicates the relative amount of ^{14}C recovered. [Redrawn from M. Shiroya, G. R. Lister, C. D. Nelson, and G. Krotkov: *Can. J. Bot.*, **39:**855–64 (1961).]

mature leaves to supply them with carbon. Under these conditions, much more ^{14}C-photosynthate is exported to leaves on the opposite side of the plant. An experiment illustrating this point is shown in Figure 21-6. Thus the direct route is merely a preferred route and not the only possible pathway.

In the course of extended work with field peas, J. S. Pate in Australia has shown that photosynthetic carbon from upper leaves goes directly to the shoot. Carbon is converted there into shoot protein, entering a group of amino acids that are characteristically formed in the shoot, including glycine, alanine, serine, valine, and the aromatic amino acids. Carbon from lower leaves, on the contrary, is transported downward to the root and is there converted into a different group of amino acids, mainly asparagine,

Figure 21-6. Distribution of radioactivity in leaves of a sugar beet plant 1 week after the supply of $^{14}CO_2$ to one leaf.

A. The plant was intact, and only the young leaves on the same side of the plant as the fed leaf received ^{14}C.

B. The expanded leaves on the side opposite the fed leaf were removed. As a result, carbon from the fed leaf has moved to young leaves on the opposite side of the plant as well. [Redrawn from K. Joy: *J. Exp. Bot.*, **15:**485–94 (1964).]

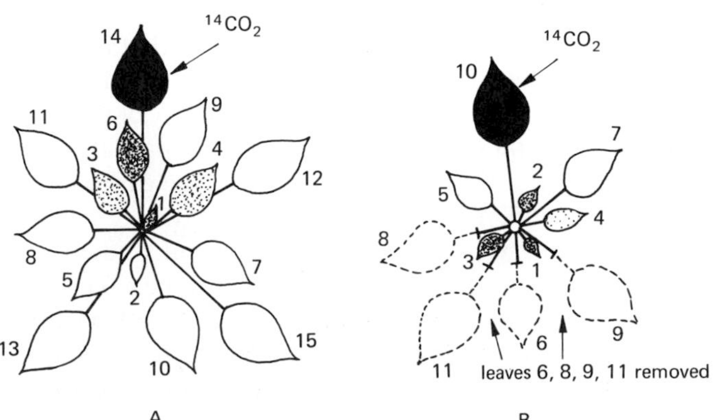

glutamine, aspartic acid, glutamic acid, threonine, lysine, arginine, and proline. These are then transported up the stem to the growing tip. Thus the whole range of amino acids required for protein synthesis is made available for the growth of the stem tip; but some are derived directly from carbon exported by the upper leaves, whereas others come indirectly via root metabolism from carbon exported by the lower leaves.

Inorganic ions also move about the plant during its growth. Early in the spring the mineral content of the soil is likely to be quite high. However, the requirements of a newly established annual plant may not be nearly as high initially as they will be later when it reaches larger size and enters the reproductive stage. At this stage, because of the active depletion of the soil by the growth of plants and by microorganisms, the supply of nutrients may be much restricted. Hence the plant must rely on stored nutrients acquired earlier in the year, and on nutrients released by senescing or dying parts of the plant.

Some ions are immobile and cannot be thus redistributed. A shortage of these ions characteristically produces deficiency symptoms in young leaves of the plant (for example, sulfur, calcium, iron, manganese, boron, copper, zinc). Other ions, which are mobile, such as nitrogen, potassium, phosphorus, and magnesium, are redistributed from old to young tissue, and their deficiencies thus affect older leaves first. It appears likely that when large amounts of nitrogen, potassium, and phosphorus are removed from older leaves as new young leaves appear, the osmotic balance of cells in the older leaves is maintained by the retention of simple sugars produced in photosynthesis. A mature plant in this stage of its growth is usually copiously supplied with reduced carbon and can easily afford this seeming waste of sugar, whereas it cannot afford to retain inorganic salts in roles that are readily filled by simple organic molecules.

Fruit Formation. Several recent studies have assisted in determining the source of carbon for the nutrition of developing fruits and seeds. It was early discovered that only a small percentage of the leaves of a plant are normally required for the carbon nutrition of the fruit. Tomato plants will bear a normal crop of fruit even when stripped of all leaves except those nearest to the fruiting branch, providing that the plants are well supplied with inorganic nutrients, which might otherwise have to be derived from older leaves. In two types of apples it was found that about 15 or 30 leaves were sufficient to support maximum growth of one fruit. In lemon trees, as in apple, only nearby leaves or those produced at about the same time as the fruit and on the same branch translocated carbon to the fruit. More distant leaves or those on lateral branches translocated primarily to the root.

Pate and his coworkers have determined that in the field pea the leaf adjacent to the fruit supplies two thirds of the carbon for the seeds. The diagram shown in Figure 21-7, drawn from their data, shows that the majority of the carbon in the seeds comes from the adjacent leaf and stipules, but a substantial part also comes from photosynthesis of the developing pod. The pod refixes much of the carbon dioxide produced by seed respiration, an effective form of carbon conservation.

Developing wheat plants have also been studied from this point of view. Rates of photosynthesis of the flag leaf (adjacent to the ear) vary in response to the demands of the ear. Although the ear itself supplies 20–30 percent of the total carbon required (refixing the carbon dioxide respired by the seed, as in peas), photosynthesis of the flag leaf alone is at all times sufficient to supply the total needs of the developing ear.

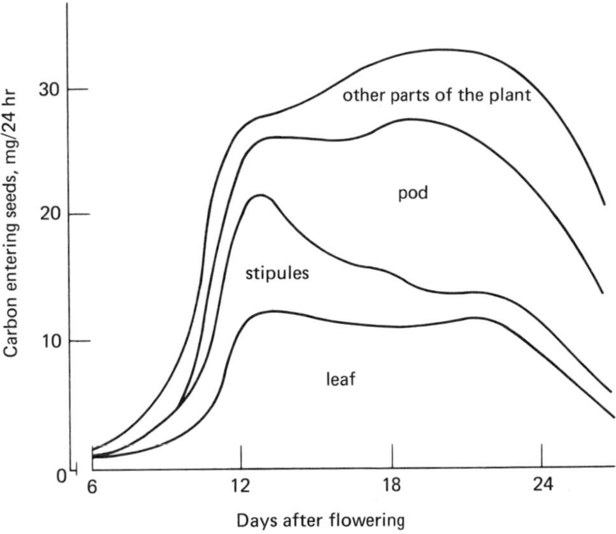

Figure 21-7. Sources of carbon for developing seeds at a single fruiting node of a field pea plant. The estimates include the pod, stipules, and leaf at the fruiting node. Carbon not derived from organs at the fruiting node (including other leaves) is indicated as coming from "other parts of the plant." [Redrawn from data of A. M. Flinn and J. S. Pate: *J. Exp. Bot.*, **21**:71–82 (1970).]

Photosynthate from other parts of the plant is transported primarily to the roots and the developing tillers. Sugars and polysaccharides are normally stored in the stem of cereals, and normally from 5–10 percent of the carbon in the ear is derived from stored sugar in the upper stem internode. Sugars from the lower internodes are used mainly for the nutrition of tillers.

Wood Formation. The use of radiocarbon has made it much easier to study the deposition of wood in trees as it relates to the photosynthesis of individual leaves or groups of leaves. This problem is important because the pattern of wood deposition, which might be affected by the distribution of leaves in a tree, is important in determining the quality or suitability of trees for lumber. Initially experiments were conducted by defoliation or prolonged darkening of specific shoots or groups of leaves, then measuring growth increments by measuring ring diameters in the branches and stems above and below the treated leaves. With isotope techniques, it is possible to supply $^{14}CO_2$ to a group of leaves, then measure precisely where the photosynthetically fixed carbon went. As in fruit trees and herbaceous plants, the majority of the carbon in the wood is derived from nearby leaves, and there is a strong tendency for transport to be directly up and down on one side of the stem; much less lateral transport takes place. Thus a tree that has developed an asymmetric crown for any reason (for example, disease, pruning, shading) will develop an asymmetric stem, with consequent changes in the properties and value of the lumber that can be derived from it.

Nutritional Traffic Control

Control of Translocation. We concluded in Chapter 13 that the larger amount of material being translocated in the phloem probably moves by a mass-transfer mechanism (see Figure 13-7, page 314). This and related mechanisms can be controlled at three points: by the rate of loading, by the rate of unloading, or by the operation of transcellular transport mechanisms at branch points or wherever solutes must pass from one intact cell to another. The loading site (leaves in photosynthate transport) might be controlled by the amount of material in transit—that is, the concentration gradient against which loading takes place. The site of unloading, or metabolic "sinks" toward which translocate is moving, could also be controlled by the demand for materials. Both these mechanisms require mass action or feedback control of the loading mechanism and, if photosynthesis is affected by demand, of photosynthesis too. Control along the translocation path, or control of loading by requirement of a distant sink, may also be possible through specific hormone action. All of these possibilities have been suggested, and we shall consider experimental evidence that bears on this question.

Movement of Nutrients Toward Sinks. Much of the information we have reviewed suggests that nutrients are moved toward sinks or to places where, because of high metabolic activity, a shortage of metabolites exists. When roots are cut off, downward translocation decreases. When growth occurs or increases in rate, translocation to the growing regions intensifies. Translocation to the most actively growing or metabolizing region of the plant exceeds that to less active regions. Factors that affect growth also influence translocation to the affected region. For example, nitrogen deficiency, which restricts growth mainly in leaves and fruit, is associated with decreased translocation to leaves and fruit and increased translocation to the roots.

Increasing the demand for photosynthate from a leaf causes an increase in its rate of export. In the experiment with tomato plants shown in Figure 21-8, darkening the plant or defoliating it increased the amount of carbon exported from a single leaf that was allowed to fix $^{14}CO_2$ in light. The increase was largely accounted for in an increase in the nutritional flow to the fruiting truss. Application of the synthetic hormone naphthylacetic acid, when sprayed on the fruiting truss, also caused increased export, but in this case the increase was largely exported to roots and stems.

All these considerations suggest that the existence and size of a sink in some way control translocation toward it. Indeed, the idea of the sink directly affecting translocation is so firmly established that the concept of "strong" and "weak" sinks is widely current in plant physiological literature. This implies that the sink is filled *because it is there,* and that the amount of material moving to it is proportional to its emptiness. In fact, a good deal of evidence suggests that the movement of the nutrients is affected not only by the avidity of the sink but also by the activity of the source.

If the roots of a plant are excised or chilled, the rate of photosynthesis of its leaves decreases. In experiments with excised leaves that were allowed to grow roots, the rate of photosynthesis of the leaf was found to be directly proportional to the amount of root, that is, proportional to the sink size, as shown in Figure 21-9. In apple trees, the rate of photosynthesis and export is greater when many fruits are present than when most of the fruits are removed. The fact of photosynthetic compensation—that is, when some leaves are darkened or removed from a plant, the rate of photosynthesis of the remaining

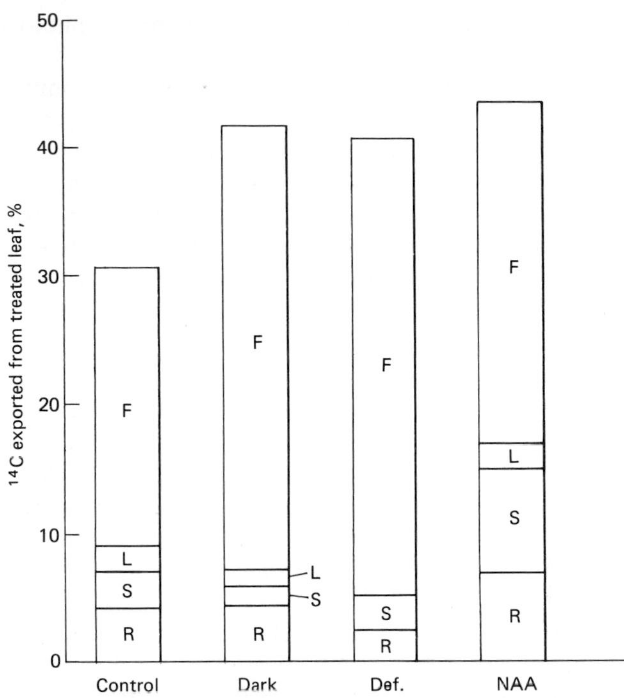

Figure 21-8. The effect of darkening or removing all leaves except the fed leaf (Def.) on the export of ^{14}C from a leaf on a tomato plant allowed to fix $^{14}CO_2$ in light. In one test, the synthetic hormone naphthylacetic acid (NAA) was sprayed on the fruit truss prior to $^{14}CO_2$ feeding. *Code:* F, fruit truss; L, leaves; S, stem; R, roots. [Redrawn from data of A. Khan and G. R. Sagar: *Ann. Bot.,* **33:**753–62 (1969).]

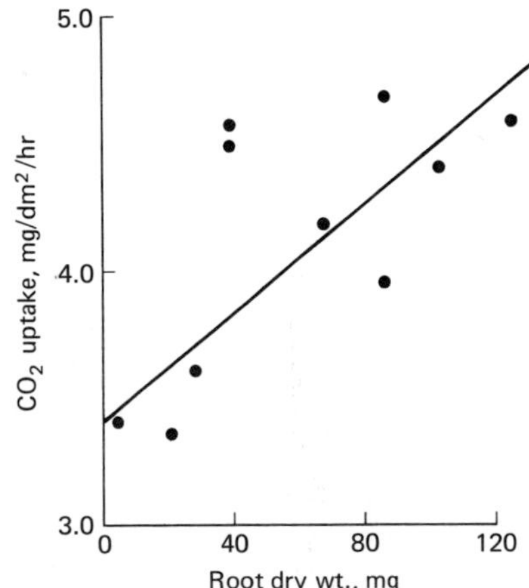

Figure 21-9. Relationship between the rate of photosynthesis of a detached bean leaf and the dry weight of roots that developed on its petiole. [Redrawn from E. C. Humphries and G. N. Thorne: *Ann. Bot.,* **28:**391–400 (1964).]

leaves increases—suggests that the demand for photosynthate, or the size of the sinks, controls photosynthesis.

Careful examination of this idea shows that it implies certain conditions. It is rather like saying that air enters an evacuated bottle because of the vacuum. Actually, air enters the evacuated bottle because there is a pressure differential from outside to inside, down which air moves. Air will move only when there is a direct connection between the evacuated space and outside, and only then will the size and the degree of emptiness of the bottle influence the movement of air. The inference from the idea of a strong or a weak sink is that there is a direct connection between the sink and the source, establishing a gradient down which substances move by diffusion. This would imply, if there are several competing sinks in different parts of a plant, that all translocation systems are equally accessible to the source. However, then all sinks would necessarily be equally accessible to all sources, and one would expect that every leaf on a plant would translocate about the same proportion of its photosynthate to each available sink. As we have seen, this does not occur. Instead, specific leaves clearly supply specific regions of the plant almost exclusively. This fact alone throws some doubt on the concept that sinks directly affect traffic flow.

In addition, certain data we have examined suggest that transport takes place via the most direct route to the nearest available sink, rather than to the strongest one. This follows from the phyllotaxic relationship found in bean and tobacco and in the distribution pattern of photosynthetically fixed carbon in the woody stem of pine trees. Such evidence suggests that the size of the sinks does not control the direction of nutritional traffic. Further, under certain conditions, the creation of a "strong" sink does not necessarily increase traffic flow. If pine trees are grown under very low light, there is inadequate sugar production and roots suffer from a shortage of substrates for growth and metabolism. In spite of this strong sink, little or none of the available carbohydrate is translocated from the leaves; it is used for leaf respiration instead.

Developing fruit appear to constitute a paradox: their nutrient content and concentration are very high, yet nutrients move toward them. Thus, they appear to be sinks that are loaded against a concentration gradient. An important point here is that the loading and unloading of the translocation stream are done by active transport; metabolic energy is used to move nutrients from the translocation stream into the fruit tissue. Consequently, the concentration gradient is probably downward, except across the sites of active transport. However, even a downward concentration gradient may not be necessary, according to some of the theories of phloem transport (see Chapter 13). Evidently the "strength" of a sink does not depend upon how full it is. Some factor other than the size or strength of sinks must be involved in the direction of nutritional traffic.

Apical Dominance and Nutrition. We may examine this question further by considering the role played by nutrient supply in apical dominance. In 1900 the German physiologist K. Goebel suggested that apical dominance resulted from competition for nutrients among parts of the plant. He noted that a leaf would grow to a much greater size if other leaves were cut off, as a result of reduced competition for available nutrients. But it soon became clear that this explanation was not sufficient, and the discovery followed that auxin, derived from the actively growing shoot tip, suppresses growth in lateral shoots (Chapter 19).

However, there are difficulties with the theory that auxin acts directly, and the two points of view were later reconciled by F. Went in his **nutrient diversion theory.** Went

Table 21-3. The effect of hormones, applied in lanolin paste to the cut surface of bean stems, on the accumulation of ^{32}P in adjacent tissue

	^{32}P accumulation; control = 1
Control	1
Gibberellic acid	1
Kinetin	1
Auxin	20
Auxin + gibberellic acid	35
Auxin + kinetin	35
Auxin + gibberellic acid + kinetin	80

SOURCE: Recalculated from data of A. K. Seth and P. F. Wareing: *J. Exp. Bot.*, **18:** 65–77 (1967).

suggested that auxin affects the direction of translocation, resulting in a movement of nutrients toward the site of auxin synthesis, so that the main shoot is well supplied and lateral shoots are maintained in a dormant condition through lack of nutrition. There is evidence to support this idea: if the stem of a bean plant is decapitated and auxin applied to the cut end, the rate of upward translocation increases, as is shown in Table 21-3. This increase takes place in spite of the fact that, since the tip is cut off, there is no sink.

Additional support for this concept comes from the fact that, in certain dormant tissues, dormancy can be overcome by improving the nutritional status of the tissue. Thus the mitotic frequency of the cotyledonary buds of sunflower can be raised by the addition of sucrose alone. Light, which causes an increase in the available sugar through photosynthesis, has the same effect. The point is that the breaking of dormancy is mediated by increased nutrition, however it is brought about. In growing shoots, the presence of an auxin gradient seems to ensure adequate nutrition of the apex.

Hormonal Control of Translocation. The above discussion suggests that hormones may be involved in the direction of metabolic traffic. Experimental verification has been obtained for a number of individual compounds applied to the plant, including sugars and amino acids, as well as for various naturally occurring compounds, including different hormones. Hormonal involvement has also been demonstrated for ^{14}C-labeled products of photosynthesis after the supply of $^{14}CO_2$, as may be seen in Figure 21-10, which shows an experiment with soybean plants. The natural auxin most directly involved in apical dominance, IAA, is frequently the most active hormone in affecting transport.

The question of how auxin controls translocation is an open one at present. Several possibilities may be considered: (1) Auxin acts to create a "metabolic sink" at the point of its application or synthesis. (2) Auxin operates along the transport path—that is, its action is somehow integrated with the transport mechnisms along the vascular tissue. (3) Auxin is known to affect the synthesis of vascular tissue so that, in young or developing (though not in mature) tissue, it could act by establishing transport pathways. (4) Auxin may operate at the source of translocation, that is, at the loading site of the translocation stream rather than at the sink.

The first two possibilities are supported by experiments of P. F. Wareing's group in Wales, who studied the effects of applying IAA and the auxin inhibitor triiodobenzoic

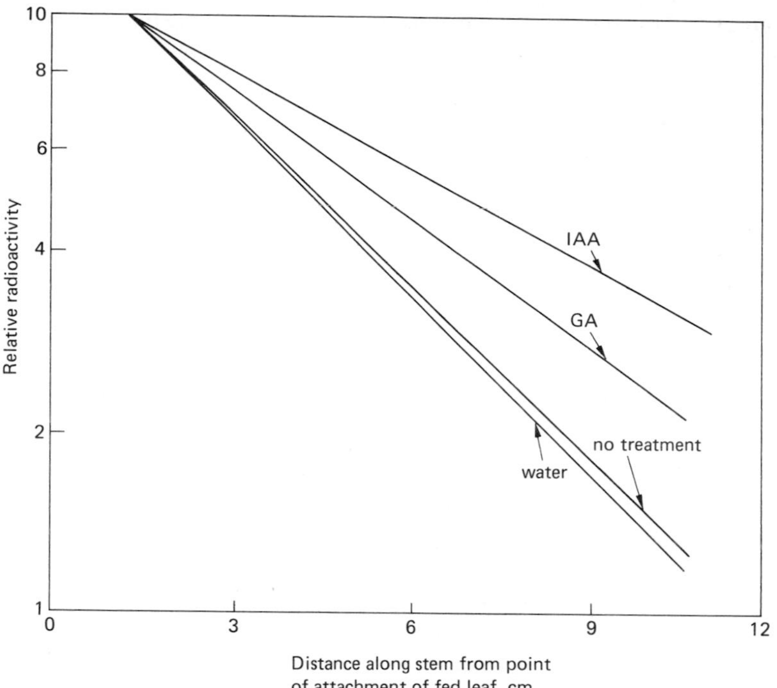

Figure 21-10. The effect of 5 ppm indoleacetic acid (IAA) or 50 ppm gibberellic acid (GA) applied to the cut stem of the apex of a soybean plant on the translocation of ^{14}C from a leaf allowed to fix $^{14}CO_2$ in light. [Redrawn from data of Hew Choy-Sin: M. A. Thesis, Queen's University, Kingston, 1965.]

acid (TIBA) to the stems of translocating plants. The mechanism whereby IAA affects unloading or intercellular transport may relate to the recent discovery that sugar transport across cell membranes probably involves a proton pump. Such pumps, which are also involved in cell wall elongation (see Chapter 23), are thought to be stimulated by IAA.

Such mechanisms, however, would not easily explain the fact that different compounds may be translocated in different directions at the same time. Thus when a mixture of amino acids and amides was supplied to a soybean plant, the Canadian physiologist C. D. Nelson and his associates found that amino acids were translocated toward younger, actively metabolizing tissues, while at the same time the amides glutamine and asparagine were translocated to older parts of the plant. This pattern appears to be related to the need for amino acids in the protein synthesis of growing tissue and the use of amides for nitrogen storage, which takes place primarily in older tissue. However, these requirements do not explain the selectivity of translocation, and it is difficult to see how auxin acting along the path of translocation or at the sink (or sinks) could exert such a selective effect at the source of translocation.

The third possibility (that auxin affects the formation of transport tissue) might be important in young or developing tissues and thus might be a factor in apical dominance. However, this possibility could not account for the effects of auxin on mature tissue, nor for the selective translocation of different substances.

The fourth possibility, that auxin affects nutritional traffic by affecting the site of synthesis or loading of the material being translocated, remains an interesting alternative. The mechanisms by which this could happen have not been worked out, but several suggestions have been offered. Cells appear to contain metabolic compartments; that is, areas or spaces (perhaps coincidental with organelles such as mitochondria or the vacuole, or with regions defined by portions of endoplasmic reticulum) in which different pools of metabolites and different reaction sequences are separated or isolated from each other. These may be important in preventing "crosstalk" or interference between different metabolic sequences that employ the same intermediates, which might otherwise interfere with the control of individual pathways. It appears likely that some degree of metabolic control occurs through the capacity of the cell to control or organize the sequestering of compounds in such compartments.

The accessibility of compounds to the loading sites for transport or translocation may thus be regulated by mechanisms that control the permeability of membranes through which substance must diffuse. Auxins are thought to be effective in changing the permeability of membranes to a variety of substances and might possibly operate in this way. Alternatively, auxin might be directly involved, for example, in activation of transport enzymes responsible for loading the translocation stream, or the processes responsible for the production or accumulation of the compounds being translocated. However they act, it appears likely that the hormones do not control or affect the *process* of translocation, but rather the nutritional *traffic* that occurs as a result of translocation.

Mention should be made at this point of the effects of cytokinins on nutrient mobilization, although the subject will be discussed in depth in Chapter 22. In 1957 it was discovered by A. Richardson and A. Lang in America that cytokinins applied to leaves retard their senescence. Numerous experiments, particularly in the laboratory of K. Mothes and his associates in Germany, have shown that nutrients from other parts of the plant do move to the site of applied cytokinin. Further, cytokinin-treated areas do not lose their nutrients but tend to retain them. This is illustrated in Figure 21-11. The mechanism of this action of cytokinin is not understood (see also Chapter 22, page 550).

Hormone Control of Photosynthesis. Many speculations have been made about possible control mechanisms for photosynthesis. End product control does not seem likely, because the rate of photosynthesis is not clearly related to the duration of the preceding period of photosynthesis (that is, to the amount of photosynthate already present). Furthermore, endogenous rhythms proceed regardless of the nutritional state of the plant. The rate of photosynthesis will decline according to a preestablished cycle even if the plant has just previously been starved of light or carbon dioxide.

Most of the products of photosynthesis are biochemically inaccessible to the site of photosynthetic reactions, either because they are rapidly converted to insoluble storage forms or because they are immediately translocated away from the site of their synthesis. However, it is known that ribulose bisphosphate carboxylase is inhibited in vitro by the product of its reaction, PGA, and also by citrate. PGA would act as an immediate, short-range control factor. Citrate, if it has access to the enzyme in vivo (a point which is by no means certain), would act as a very long-range or distant feedback control when the Krebs cycle becomes overloaded to the point that citrate pileup or storage takes place. Whatever the mechanism, it is now generally accepted that the rate of photosynthesis is not strongly affected by the amount of photosynthate in the leaf. The only exception seems to be when starch grains, after prolonged exposure to light, may swell

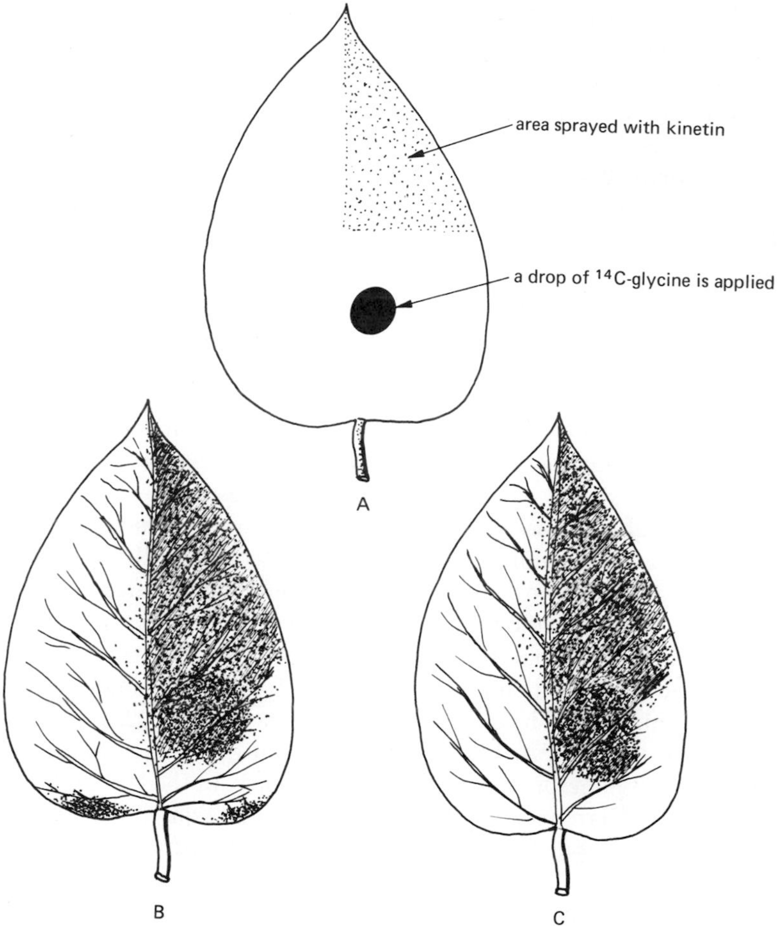

Figure 21-11. Mobilization and retention of a nutrient (glycine in this experiment) under the influence of a cytokinin, kinetin.
(**A**) Kinetin-sprayed leaf. After 55 hr the ^{14}C-glycine has moved to the kinetin-sprayed area in (**B**) light or (**C**) darkness. It has not moved appreciably to other parts of the leaf. [Redrawn from K. Mothes and L. Engelbrecht: *Phytochemistry,* **1**:58–62 (1961).]

up to such a size that they physically interfere with the diffusion of CO_2 or the operation of chloroplasts.

Several workers have considered the alternative possibility that the rate of photosynthesis in a leaf might respond to a signal generated in a distant metabolic sink, and it has been found that certain hormones do increase the rate of photosynthesis in leaves. In one study it was discovered that the rate of photosynthesis in an attached bean leaf responds dramatically to the breaking of dormancy of a nearby bud, as shown in Figure 21-12. It was shown that IAA could elicit the same response in a leaf if it was sprayed on an adjacent leaf (Figure 21-13).

The stimulating effect of auxin on photosynthesis was paralleled by an effect on the translocation of photosynthate. As shown in Table 21-4, when the receiver leaf was in a condition in which it required carbon (that is, it was kept in darkness or was a growing,

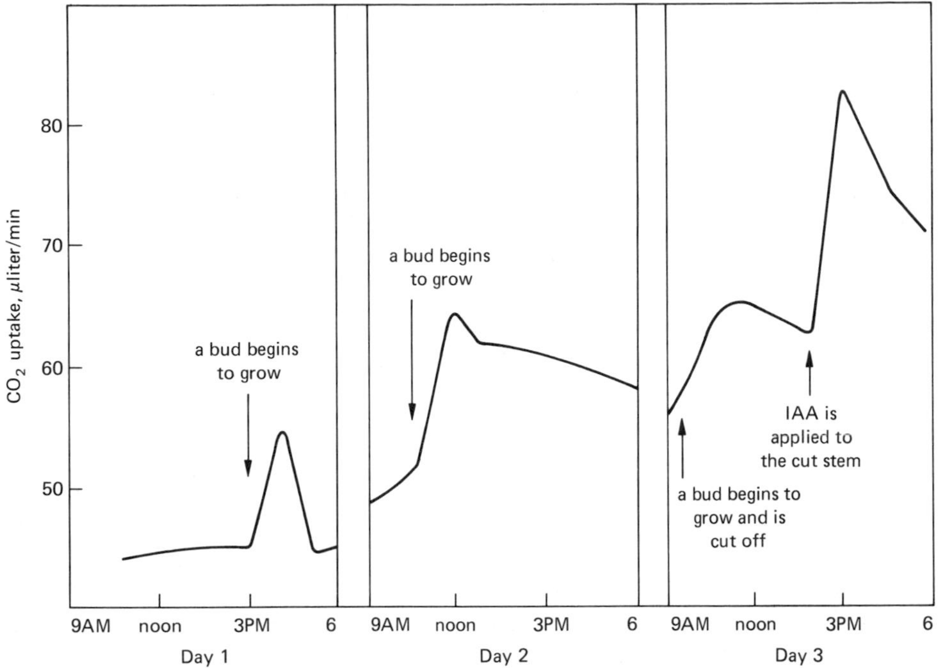

Figure 21-12. The effect of buds breaking dormancy and of the application of 15 ppm indoleacetic acid (IAA) to the cut stump of a developing bud on the rate of photosynthesis of a bean leaf. [Data of W. B. Levin and R. G. S. Bidwell.]

developing leaf), the addition of IAA to the feeder leaf caused an increased export of photosynthate. Applying IAA to the receiver leaf caused an increased import of photosynthate under all conditions. It was shown by experiment that ^{14}C-labeled IAA could move quickly enough within the plant to mediate the responses.

There has been some suggestion, though the evidence is not yet sufficient for critical judgment, that hormones may affect the relative amounts of photosynthetic end products (that is, sucrose, starch, triose phosphate, glycolate, and so on). This might in turn affect the amount of carbon that would be chemically or physically accessible to various metabolic activites and to translocation. Mechanisms like the one described above might partially explain the establishment of complex patterns of nutritional traffic in the plant.

Figure 21-13. The effect of spraying a leaflet with IAA solution and of applying IAA solution to the cut petiole of the leaflet on the photosynthesis of adjacent leaflets of a bean plant. [Data of W. B. Levin and R. G. S. Bidwell.]

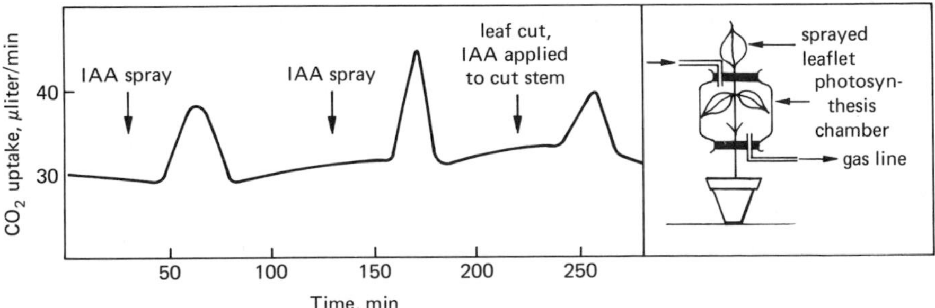

Table 21-4. The effect of sprayed IAA solution (15 mg/liter) on the translocation of recently fixed $^{14}CO_2$ in bean leaves

Condition of fed leaf	*Condition of receiver leaf*	*Amount of ^{14}C translocated to receiver leaf, % of control*
Light	Light	100 (control)
Light + IAA	Light	60–120
Light + IAA	Light (a young leaf)	200
Light	Dark	130
Light + IAA	Dark	300
Light	Light + IAA	125–400*

SOURCE: Data of W. B. Levin and R. G. S. Bidwell.

*Dependent on the ages and relative positions of the leaves.

The mechanism whereby auxin stimulates photosynthesis and the export of photosynthate is not clear, however. Auxin has been found to stimulate carbon dioxide fixation and photosynthetic phosphorylation in isolated chloroplasts. This therefore appears to be a case of genuine hormone action—that is, the stimulation of a primary process by a translocated agent that does not itself take part in the reaction and originates in a distant tissue.

It is quite clear that other mechanisms are also involved in the control of photosynthesis. The rhythmic patterns that have been observed and much of the ontogenetic patterns of photosynthesis in a leaf are related to the amount of important photosynthetic enzymes present, particularly the main carboxylation enzyme ribulose bisphosphate carboxylase. How these are controlled and how the controls relate to the developmental pattern of the plant is not yet known.

Additional Reading

See list for Chapter 16 (p. 406).

Nelson, C. D.: Effect of climate on the distribution and translocation of assimilates. In L. T. Evans (ed): *Environmental Control of Plant Growth.* Academic Press, New York, 1963.

Oaks, A., and **R. G. S. Bidwell:** Compartmentation of intermediary metabolites. *Ann. Rev. Plant Physiol.,* **21:**43–66 (1970).

Peel, A. J.: *Transport of Nutrients in Plants.* John Wiley & Sons, New York, 1974.

Preiss, J., and **J. Kosuge:** Regulation of enzyme activity in photosynthetic systems. *Ann. Rev. Plant Physiol.,* **21:**433–66 (1970).

Wardlaw, I. F., and **J. B. Passioura** (eds.): *Transport and Transfer Processes in Plants.* Academic Press, New York, 1976.

22 Dormancy, Senescence, and Death

Dormancy

Most plants are exposed to seasonal periods of very inclement weather during which they would be damaged or killed if some protective or defense mechanism did not exist. The most common safeguard against freezing cold or extreme dry heat is dormancy. Dormancy can be defined as a state of suspended growth and metabolism. It may be enforced by unfavorable conditions, but dormant tissues often fail to grow even when placed in ideal conditions. This indicates that dormancy can be imposed from within and controlled by mechanisms in the tissue.

Dormancy takes many forms. Many seeds are initially (or become) dormant so that they will not germinate for a period of time after shedding, or until they have experienced a period of cold. Trees shed their leaves and are thus relieved of the danger of desiccation when the air is cold and dry and the soil frozen. In addition, their shoot tips form elaborate and protective buds which are essentially water- and gasproof. In many herbaceous plants the aerial parts die back and the plant overwinters or survives a drought period as a dormant underground bulb, corm, or tuber. Mechanisms that prevent damage due to water loss are largely mechanical and easily understood, but the reasons why dormant tissues are more frost resistant than actively growing tissues are not so clear (see Chapter 28, page 644). However, certain desiccated dormant tissues such as seeds are able to withstand very low temperatures. Even those tissues that do not become much desiccated during dormancy are capable of withstanding lower temperatures than are actively metabolizing tissues.

Dormancy is a defense mechanism against winter frost or summer drought and is a necessary part of the lives of many plants. Dormancy must occur at the right time (that is, before killing intensity of adverse conditions). It must last for a sufficient time, and it must be relieved or broken when the conditions are right for resumption of growth. It follows that dormancy must be rather precisely controlled. Different plants and different parts of plants become dormant in different ways. But each example presents the same basic problem, one that has aroused the interest of physiologists who study this fascinating aspect of plant life; how do plants shut down their metabolic processes in preparation for winter, and how do they start them up again in spring, but not before?

It may be seen that there are, in fact, four basic questions that must be asked about dormancy of a particular plant or organ. The first is, what are the environmental signals that start the process, and how are they perceived? The second concerns the perception of signals that bring about the cessation of dormancy and the resumption of normal metabolism and growth. The third concerns the duration of dormancy: a timing mechanism, presumably of the hourglass type, appears to be necessary to ensure that the plant does not accidentally reawaken during an unnatural period of warm weather in January. Evidently such a timer needs to be variable with the latitude at which the plant is living—periods of winter dormancy may need to vary from several months near the poles to a comparatively short time in southern temperate zones. However, the timer must be quite precisely fixed at any given latitude. The fourth question deals with the nature of dormancy and the mechanisms for bringing about the dormant condition. Dormancy is not just an inactivation of metabolism, but frequently involves the development of specialized organs (for example, bud scales) or substances (for example, gummy waterproofing materials). Many complex events may be associated with dormancy, such as the senescence and abscission of leaves in trees. Dormancy is evidently a programmed developmental event that requires specialized synthetic metabolism in addition to turning off metabolic activities.

Causes of Dormancy

Environmental Factors. The most important factor in inducing dormancy appears to be photoperiod. Short days induce dormancy in many woody plants. The photoperiod is perceived by the leaves, but the apex or buds are the main initial responding parts of the plant. This means that, as in the flowering response, the leaves must be induced to make an inhibitory substance or hormone that is translocated to the buds. Inhibitory substances have been separated chromatographically from leaves of such trees as *Acer* (maple) that were exposed to short days. These substances inhibit the growth of test plants. The inhibition can be relieved by treatment with long days or by the application of gibberellic acid.

Cold itself does not appear to be necessary in the induction of dormancy, and dormancy may not be induced even by short days if the temperature is too low for strong and active metabolism. In fact, cold appears to be the most important prerequisite for breaking dormancy. Temperature may also be important in timing dormancy. The duration of dormancy is probably dependent on an hourglass timer controlled by the rate of decay or breakdown of an inhibitor. Thus the timer is probably temperature regulated and automatically compensates by slowing down in colder temperatures and speeding up in warm weather.

Moisture, or a lack of it, appears to be important in initiating dormancy in some plants, particularly those that rely on dormancy to survive hot dry spells. Again, it appears that the leaves are the perceptive organs, but dormancy of the shoots results. A shortage of nutrients, particularly nitrogen, also appears to trigger dormancy in some plants. However, dormancy apparently does not result from a slowdown of metabolism resulting from nutrient deficiency. On the contrary, slowed metabolism is a result, not a cause, of dormancy.

Abscisic Acid. For a considerable time it was known that some substance formed in the leaves in short days was translocated to the shoots where it inhibited growth and metabolism and caused dormancy. Extracts of leaves of *Betula pubescens* kept on short days were found by the British physiologist P. F. Wareing and his colleagues to contain

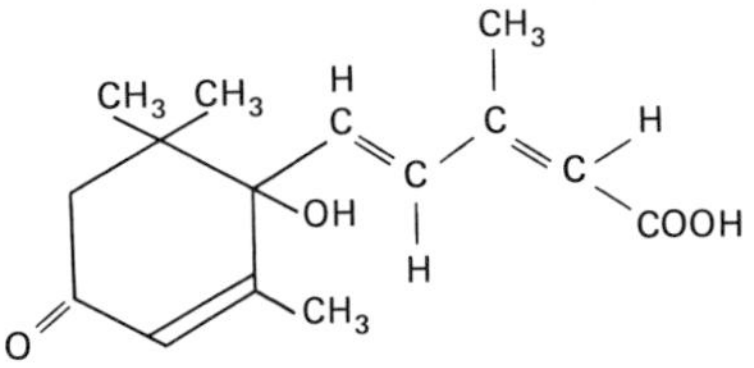

Figure 22-1. Formula of abscisic acid (ABA).

substances that strongly inhibited the elongation of *Avena* coleoptiles. Wareing and his coworkers found that the production of inhibitors actually precedes the onset of dormancy. This strongly suggests that the inhibitors are involved in inducing dormancy.

In 1963, working with *Acer pseudoplatanus,* Wareing and his colleagues were able to isolate a substance having the physiological properties of the dormancy-inducing inhibitor, which they called **dormin.** Meanwhile a group of American physiologists under F. T. Addicott, who were working on senescence and abscission of leaves, isolated a senescence-inducing substance which they called **abscisin II.** Abscisin II was, in fact, by a remarkable coincidence isolated only days before dormin. The two compounds were soon found to have the same structure, which is shown in Figure 22-1, and by agreement they are now called **abscisic acid** (ABA).

ABA is now known to occur very widely in many parts of plants of every type. It is clearly involved in the production and maintenance of dormancy. It is present in dormant tissue, or tissues entering dormancy, and its concentration decreases in tissues treated so as to break dormancy (that is, during chilling of seeds). Even more convincing are the various symptoms of dormancy and senescence that can be induced by the application of ABA.

1. Induction of dormancy.
2. Maintenance of dormancy (Figure 22-2).

Figure 22-2. Bud break in white ash cuttings is reduced or prevented by treatment with abscisic acid (ABA) for 22 days. Treatments (from left to right): control, 0.4, 2, and 10 ppm ABA. [From C. H. A. Little and D. C. Eidt: *Nature,* **220:**498–99 (1968). Used with permission. Photograph courtesy Dr. C. H. A. Little.]

Table 22-1. The effect of abscisic acid (ABA) compared with short days in reducing the growth of blackcurrant shoots

Growth after 30 days, cm	*Long days*	*Short days*
Without ABA	22	4
With 5 ppm ABA	14	—
With 25 ppm ABA	9	—

SOURCE: Adapted from data of H. M. El-Antably, P. F. Wareing, and J. Hillman: *Planta,* **73:**74–90 (1967).

3. Inhibition of germination (Table 22-1).
4. Inhibition of gibberellic acid-induced enzyme synthesis in seeds.
5. Inhibition of flowering.
6. Abortion of buds.
7. Abortion of fruits.
8. Senescence of leaves.
9. Acceleration of abscission.
10. Formation of terminal buds on shoots.
11. Formation of bud scales.
12. Reduction of cell division.
13. Induction of biochemical changes leading to senescence and leaf abscission.

In addition to these effects, ABA appears to be a general inhibitor of growth. It has been found to slow the extension growth of coleoptiles and expanding leaves and the growth of such diverse systems as seedlings, embryos, cultured tissue and shoots (Table 22-2), which are undergoing both division and maturation of cells. Quite large amounts of ABA, added frequently, are needed to bring about some of these effects, such as unseasonable induction of dormancy; that is, an appropriate concentration must be present and maintained in the tissue for ABA to act. However, ABA appears to be nontoxic; the plant reacts in a "natural" way, and the reaction is reversed if ABA treatment is stopped. ABA is rapidly translocated in the plant, and the transient nature of its effects shows that it is readily inactivated. These points all strongly support the natural role of ABA as an important regulating substance and possibly as a true hormone in plants.

Table 22-2. The inhibition of seed germination in tall fescue grass by abscisic acid (ABA)

ABA concentration, ppm	*Percent germination after (days)*		
	6	*10*	*18*
0	12	76	85
1	8	42	78
5	0	7	26
10	0	2	7

SOURCE: Adapted from data of D. C. Sumner and J. L. Lyon: *Planta,* **75:**28–32 (1967).

Table 22-3. The effects of GA and ABA on the germination of seeds of the ash, *Fraxinus ornus*

GA	*ABA*	*Germination, %*
−	−	76
+	−	81
−	+	7
+	+	63

SOURCE: Adapted from data of E. Sondheimer and E. C. Galson: *Plant Physiol.*, **41**:1397–98 (1966).

Interaction of ABA with Other Growth Substances. It has been noted that quite large amounts of ABA must be applied to achieve effective results. Further, the application of ABA must be sustained if the effect is to be maintained; when ABA treatment is stopped, both growth and active metabolism are resumed. These facts suggest that some growth-promoting substance antagonizes the effects of ABA. Many experiments have now shown that gibberellic acid (GA) does indeed exert such an effect. In some situations GA largely overcomes the effect of applied ABA, as shown in Table 22-3 for germinating ash seeds and Figure 22-3 for bud break. When dormant material, such as lettuce seeds kept in darkness, is given extra ABA, even quite high concentrations of added GA do not overcome inhibition (Table 22-4). However, in this situation kinetin also plays a role, and the addition of kinetin counteracts the ABA effect, permitting GA to stimulate germination. The effects of kinetin and GA have also been studied in

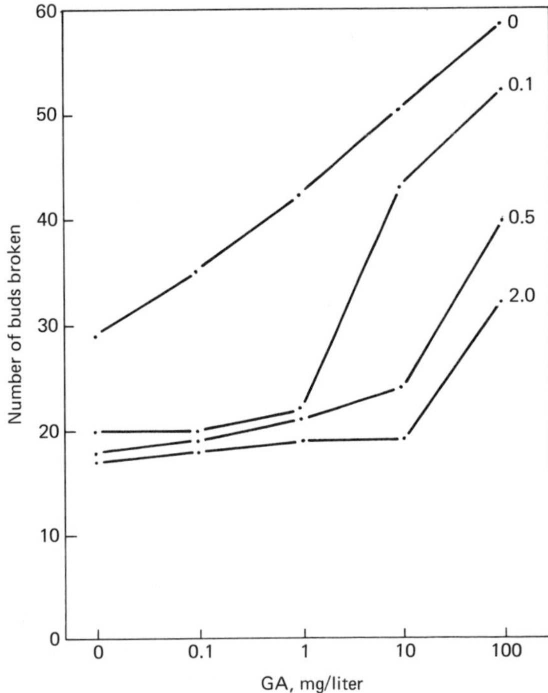

Figure 22-3. Effect of gibberellic acid and inhibitor (ABA) on bud break in isolated segments of birch stem. The numbers opposite the curves represent relative inhibitor concentrations. [From C. F. Eagles and P. F. Wareing: The role of growth substances in the regulation of bud dormancy. *Physiol. Plant.*, **17**:697–709 (1964).]

Table 22-4. The effects of ABA and kinetin on the gibberellic acid (GA) stimulation of germination in lettuce seeds kept in darkness

GA concentration, mM	Germination, %			
	No addition	*+ Kinetin, 0.05 mM*	*+ ABA, 0.04 mM*	*+ ABA and kinetin*
0	10	15	0	0
0.05	21	27	0	17
0.5	66	69	0	57
5	95	97	0	73

SOURCE: Adapted from data of A. A. Khan: *Plant Physiol.,* **43**:1463–65 (1968).

germinating barley seeds. Both the growth of the coleoptile and the induction of α-amylase synthesis are inhibited by ABA, but the inhibition is reversed by the addition of either GA, kinetin, or benzyladenine (another synthetic cytokinin), as shown in Figure 22-4.

The interrelationships of GA and ABA are of great interest. It will be remembered that GA can cause long-day plants to bolt and flower (see Chapter 20). ABA has been found to reverse this action. Furthermore, although ABA can also cause flowering in some short-day plants, ABA is clearly not in any way like Chailakhyan's anthesin, nor is it simply an antagonist of GA. In many situations the two hormones appear to cause different or opposite effects, but they do not always antagonize each other in isolated effects.

These interrelationships led to the postulation of an interesting mechanism for the interaction of GA and ABA by the American physiologists A. W. Galston and P. J. Davies. They pointed out that both GA and ABA are terpene compounds, composed of isoprenoid units and derived from mevalonic acid (Chapter 9, page 237). If they are formed from a common precursor, it appears possible that external or environmental conditions somehow control the pathway at the branch point. Dormancy is commonly associated with short days and breaking of dormancy with long days. Since long and short days are perceived by the phytochrome mechanism (possibly in conjunction with a circadian rhythm process—see Chapter 20, page 503), it seems possible that phytochrome is the controlling agent. The scheme proposed by Galston and Davies is represented in Figure 22-5. This scheme should be compared with suggested mechanisms governing other phenomena such as vernalization (Chapter 20, page 508, Figure 20-26).

The American physiologist A. A. Khan has developed a model for the hormonal control of seed dormancy and germination involving three components—GA, cytokinins, and inhibitors (including ABA). The essentials of this scheme, shown in Figure 22-6, are that gibberellin is necessary for germination and that its absence inevitably results in dormancy, regardless of the presence or absence of an inhibitor. The cytokinin effect is to block the effect of the inhibitor, so that the promoting effect of gibberellin can be realized. This model and the data upon which it is based suggest that dormancy may be caused equally by the presence of an inhibitor, such as ABA, or by the absence of a gibberellin. The model is not necessarily universal, however, and there may be additional or subsidiary mechanisms or yet other growth hormones.

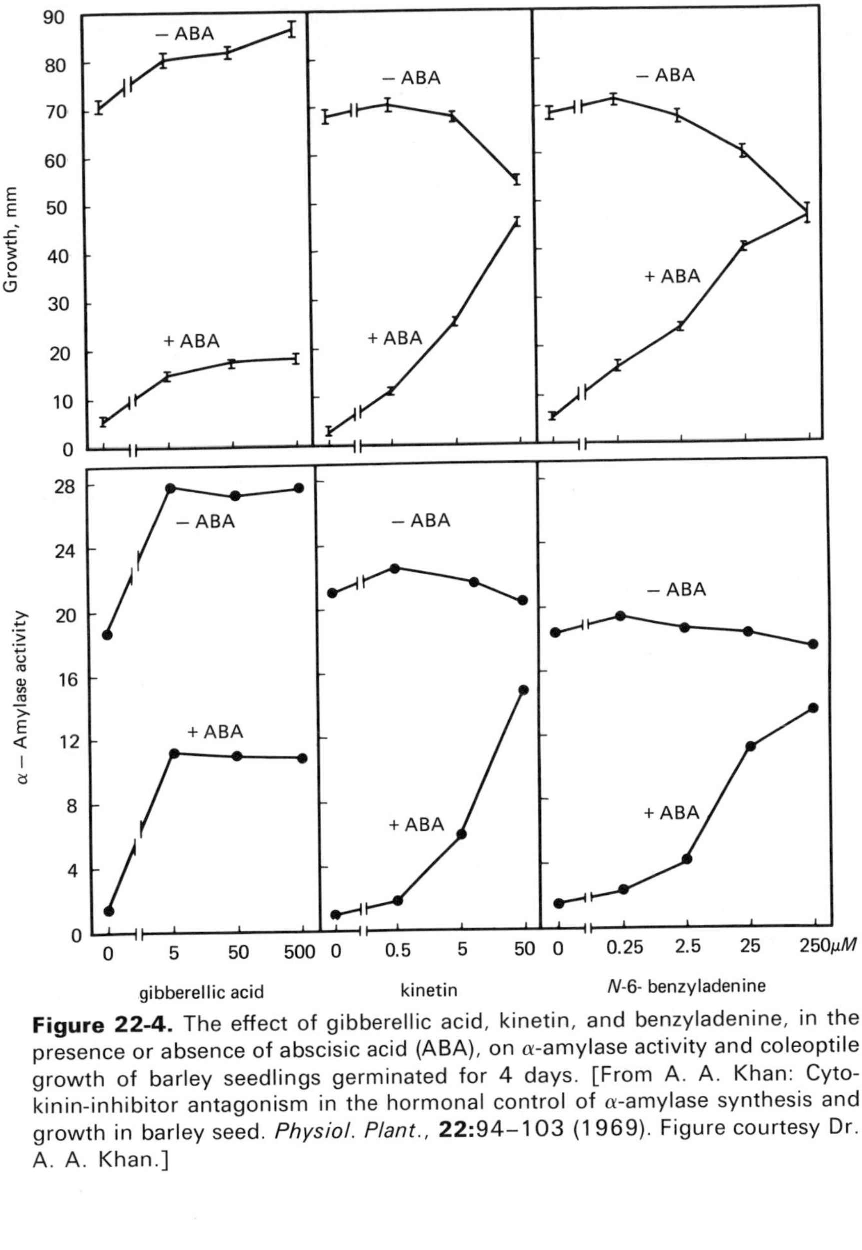

Figure 22-4. The effect of gibberellic acid, kinetin, and benzyladenine, in the presence or absence of abscisic acid (ABA), on α-amylase activity and coleoptile growth of barley seedlings germinated for 4 days. [From A. A. Khan: Cytokinin-inhibitor antagonism in the hormonal control of α-amylase synthesis and growth in barley seed. *Physiol. Plant.*, **22:**94–103 (1969). Figure courtesy Dr. A. A. Khan.]

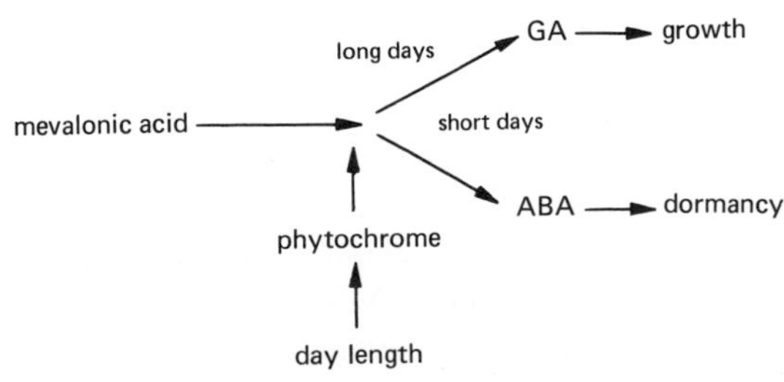

Figure 22-5. Diagram of the mechanism suggested by A. W. Galston and P. J. Davies for the control of dormancy by day length.

Figure 22-6. A model for the hormonal mechanism of seed dormancy and germination. [Adapted from A. A. Kahn: *Science*, **171**:853–59 (1971).]

Gibberellin	Cytokinin	Inhibitor	
+	−	−	Germination
+	+	−	Germination
+	+	+	Germination
+	−	+	Dormancy
−	−	−	Dormancy
−	−	+	Dormancy
−	+	−	Dormancy
−	+	+	Dormancy

Seed Dormancy

Types of Seed Dormancy. Seed dormancy is critically important for the survival of plants, and many different mechanisms have evolved that achieve the same end: an enforced period of low metabolic activity, low water content, and zero growth during which the seed is very hardy to the rigors of cold and drought. In addition, mechanisms have evolved that prevent the seed from germinating immediately after shedding, even when other environmental conditions are right. This requirement, which implies a further period of development after the seeds are shed and before they can germinate, is called **after-ripening.** The main mechanisms that cause seed dormancy or prolong dormancy by preventing germination are as follows:

1. Environmental factors.
   a. Light requirement for germination—positive or negative.
   b. High temperatures.
   c. The absence of water.
2. Internal factors.
   a. Seed coat—prevention of gas exchange.
   b. Seed coat—mechanical effects.
   c. Embryo immaturity.
   d. Low ethylene concentration.
   e. Presence of inhibitors.
   f. Absence of growth promotors.
3. Timing mechanisms.
   a. After-ripening.
   b. Disappearance of inhibitors.
   c. Synthesis of growth promotors.

Light Requirement. The light requirement for germination of many seeds is presumably a mechanism that prevents the germination of small seeds too deep underground, where they would exhaust their reserves before reaching the surface and becoming autotrophic. Many seeds will not germinate under a forest canopy, because insufficient light reaches the forest floor to stimulate germination. However, after a fire

Table 22-5. The effect of day length on the germination at 20° C of birch seeds (*Betula pubescens*) that have not been chilled

Length of photoperiod, hr	*Germination, %*
2	38
4	41
8	32
12	53
16	70
20	89

SOURCE: Adapted from data of M. Black and P. F. Wareing: *Physiol. Plant.*, **8**:300–16 (1955).

many seeds germinate at once, and forest regeneration begins. This mechanism prevents overcrowding in a mature forest and ensures rapid regrowth after a disaster.

The requirement for light to break dormancy is low; 50–200 ft-c for 1 sec may be sufficient, but lower intensities for longer periods of time serve the purpose. As we have discussed earlier (Chapter 20), the data of Flint and McAlister enabled Hendricks and Borthwick to demonstrate that the light requirement is perceived through the pigment phytochrome, red light being active in promoting germination and far-red light inhibiting it. A possible mechanism for the effect, a phytochrome-mediated switch between GA and ABA synthesis, was presented in Figure 22-5. It is now believed that phytochrome becomes membrane bound on illumination and there mediates or modulates the action of GA, so that the mechanism of control (whatever its exact nature) is probably associated with cellular membranes.

Not all seeds require light for germination; some are unaffected and a few are inhibited by light. Blue light at quite high intensity has some effect on the germination of certain seeds, but it is not clear whether this is mediated through the absorption of blue light by phytochrome or by some other pigment. Certain seeds show clear requirements for short or long days. Seeds of birch (*Betula pubescens*) show an improvement in germination in long days, as is shown in Table 22-5. Light and temperature requirements are interrelated: under conditions of alternating temperature, some normally light-requiring seeds will germinate in darkness. Certain chemical treatments (such diverse compounds as potassium nitrate, thiourea, and GA) will eliminate light requirement in some seeds. The light requirement of some seeds may be located in parts other than the embryo, because isolated embryos will germinate in darkness.

Temperature. Low temperature treatment is an essential prelude to germination in many seeds, and high temperature may be inhibitory at the time of germination. The low-temperature chilling requirement is frequently met artificially by the process of **stratification**—the seeds are layered in trays in cold moist air for a period of up to several weeks or months. Temperatures between 0 and 10° C are most effective, as shown in Figure 22-7. The chilling requirement is variously located in the embryo or the seed coat, sometimes in both. The chilling requirement of apple seeds, for example, is much longer for intact seeds than for seeds with the coats removed, or for isolated embryos (Table 22-6).

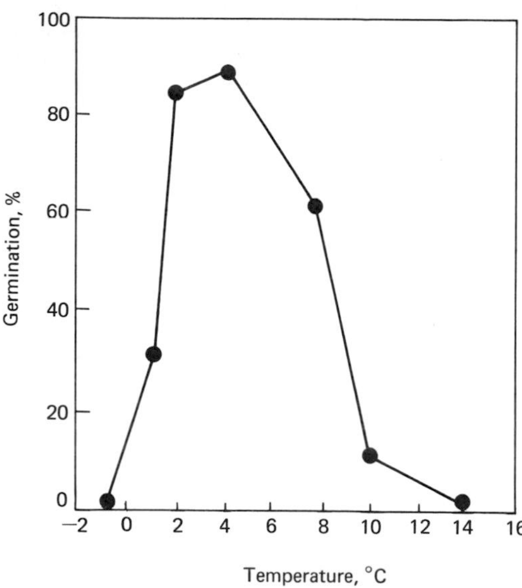

Figure 22-7. Stratification temperature and germination of apple seeds. [Adapted from H. Scharder: *Z. Pflanz.*, **34**:421–44 (1955).]

The period of low temperature appears to be necessary for the breakdown of ABA present in the seeds. In certain seeds (for example, peach and maple) the seed coat is important in causing dormancy. Stratification causes a marked decrease in the measurable amounts of ABA present in the seed coats of these seeds. Low temperature is also necessary for the activation of gibberellin synthesis. Actually, the gibberellin may not be formed until later, under the influence of warmer temperature, but gibberellin synthesis will not take place unless the seed has experienced a period of low temperature, as shown in Figure 22-8. This provides an effective double safeguard mechanism to prevent premature germination—ABA is presumably present initially and ensures that the seed will be initially dormant, and the requirement of a cold period before the synthesis of the growth-promoting gibberellin ensures that winter will be past and spring, with its warm temperatures, already underway before germination can take place.

Red light and GA were found to have a synergistic effect—that is, the combination of both factors stimulates germination more than the sum of the two separately. It has been suggested that phytochrome, in addition to (or perhaps instead of) its possible role in promoting gibberellin synthesis, may also, through its effect on the permeability of

Table 22-6. Stratification requirement of apple seeds

	Stratification time (days) at 3° C required for	
	50% germination	*80% germination*
Intact seeds	64	78
Seeds with coats removed	43	52
Embryos with seed coats and endosperm removed	30	39

SOURCE: Adapted from data of T. Visser: *Proc. Kon. Ned. Akad. van Wetensch. Amsterdam,* Ser. C, **57**:175–85 (1954).

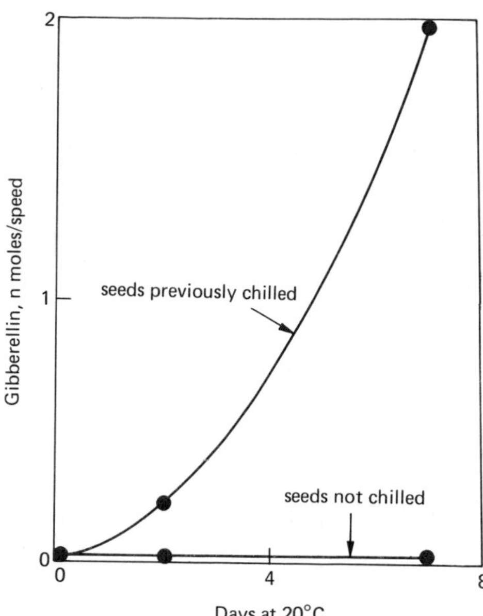

Figure 22-8. The effect of prior chilling for 1 month at 5°C on the subsequent synthesis of GA at normal temperatures in hazel seeds (*Corylus avellana*). [Adapted from data of T. D. Ross and W. J. Bradbeer: *Nature,* **220:**85–86 (1968).]

membranes, facilitate the mobilization of gibberellin to reaction sites. This idea may help to explain the peculiar interactions of light and temperature effects described above.

Seed Coat Effects. In some seeds dormancy is imposed by the presence of the seed coat; if this is removed, the seed germinates. Two possible types of mechanisms could be involved, one biochemical or physiological and the other purely mechanical.

The seed coat is nearly impervious to the diffusion of gases, and the embryo may be maintained in the dormant condition by a lack of oxygen. This might work in several ways. Oxygen is necessary for metabolism. Although it seems likely that metabolism is reduced in seeds as a result of dormancy, not the other way round, certain experiments do suggest that dormancy is at least maintained, if not caused, by lack of oxygen. Wareing's group found that seeds of Birch (*Betula pubescens*), which would not germinate when intact, would do so if the seed coat were scratched or broken. Moreover, added oxygen greatly stimulated germination of such damaged seeds. Evidently the embryos themselves were not dormant; they would germinate perfectly if isolated from the seed. Somehow, dormancy was related to the presence of the seed coat. Oxygen might be necessary for reactions in the breakdown of inhibitors causing dormancy or for the synthesis of growth promotors which would break dormancy. An alternative possibility is that the seed coat might prevent the leaching out of a diffusible inhibitor. Also, in some seeds, all or part of the mechanism that perceives the signal initiating the breaking of dormancy is housed in the seed coat. The inference is that dormancy may be induced by the seed coat.

The second, or mechanical, alternative has been investigated by Y. Esashi and A. C. Leopold using seeds of *Xanthium pennsylvanicum* (also called *Xanthium strumarium*), the cockleburr. This plant produces two kinds of seeds in each fruit—large, nondormant ones and small, dormant ones. The investigators used the specially designed apparatus shown in Figure 22-9. The results, summarized in Table 22-7, show that neither type of seed

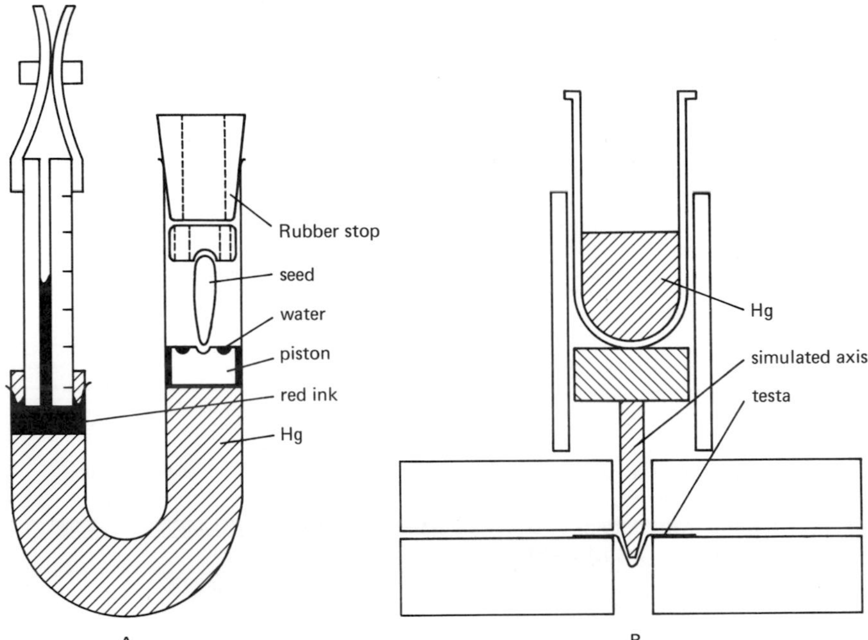

Figure 22-9. Apparatus used for determining (**A**) the thrust of germinating *Xanthium* seeds, and (**B**) the force necessary to rupture the testa.

A. The growing seed depressed the piston and pushed the red ink up in a calibrated tube.

B. Intact pieces of testa from a seed were tested using a piece of soft pencil lead to simulate the seedling axis. [From Y. Esashi and A. C. Leopold: Physical forces in dormancy and germination of *Xanthium* seeds. *Plant Physiol.,* **43:**871–76 (1968). Used with permission.]

generates enough force to rupture the testa during imbibition. During growth, however, the large, nondormant seeds generate sufficient force to break it, whereas the smaller, dormant seeds do not. This shows, for *Xanthium* at least, that the long-held opinion that the embryo must generate sufficient force during germination to rupture the seed coat is correct. Moreover, it is clear that the forces generated by imbibition alone are not sufficient—active growth is needed also.

Table 22-7. Forces required for the rupture of the testas of large (nondormant) and small (dormant) seeds of *Xanthium pennsylvanicum,* and the forces actually developed by the seeds

	Large seed	*Small seed*
Force required to rupture testa (mean of 5 tests), g	67	56
Force developed by germinating seeds during imbibition, g	30	20
Force developed by germinating seeds during active enlargement, g	84	41

SOURCE: Adapted from data of Y. Esashi and A. C. Leopold: *Plant Physiol.,* **43:** 871–76 (1968).

Other Factors. We have examined the interplay of growth-promoting and -inhibiting substances in the previous section. Numerous seeds become capable of germination if they are extensively leached or washed with water, suggesting the presence of soluble diffusible inhibitors that can be washed out. Another substance that may be important in the germination of some seeds is ethylene. A number of workers have shown that ethylene is produced by seeds during the germination period and that seed dormancy may be broken by ethylene treatment. Esashi and Leopold concluded that dormancy of clover seeds is regulated by ethylene produced by the seed. When the seed is lying on the surface of the ground, the ethylene dissipates; if the seed is surrounded by soil, the ethylene concentration becomes high enough for growth to take place. This regulatory mechanism safeguards the seed from germinating when unsuitably located and encourages germination when the seed is buried in the soil.

Some seeds are shed before the embryo has developed sufficiently for it to start growing. In this case, the seed may be dormant for a period because of the immaturity of the embryo. After-ripening mechanisms such as embryo immaturity constitute timing devices that prevent germination for a more or less fixed period of time, even if external conditions appear suitable. This allows time for externally imposed dormancy (resulting from shorter days or reduced temperatures) to become operative. This mechanism for delaying germination would provide a considerable ecological advantage to species that rely upon bird or animal vectors or such natural accidents as floods or washouts for their dispersal.

Dormancy of Vegetative Organs

Day Length and Dormancy. Many plant parts besides seeds become dormant. In woody perennial plants, not only the buds but the entire plant becomes dormant in winter. The meristematic activity of the cambium declines to zero, and metabolism in stem and root tissues falls to a very low level. Various storage organs of biennials or herbaceous perennials, such as tubers, bulbs, corms, and the like, overwinter in the dormant condition. The little water plant, *Lemna* (duckweed), forms a special overwintering dormant bud called a **turion,** which sinks to the bottom.

All these dormant systems seem to share essentially the same mechanism of dormancy—short days somehow promote the synthesis of ABA, through the agency of a phytochrome-mediated system. Wareing and coworkers have extensively investigated the production of growth-promoting (IAA, GA) and -inhibiting (ABA) substances in buds, particularly of the birch, *Betula pubescens*. They found that plants of this species become dormant and form resting buds if exposed to short days. Then, if the whole plant is exposed to long days, the buds swell and resume growth. If the plants are defoliated and only the buds are exposed to long days, they will grow. However, if the leaves are kept in short days the buds will not grow, even if they themselves are exposed to long days.

These effects indicate an increased production of inhibitor in leaves that are kept in short days, which overrides the effect of growth-promoting substances produced in the buds in long days. The relative amounts of bound and free ABA may be more important here than the absolute amounts of inhibitor. Results of analyses of beech buds (*Fagus sylvatica*), illustrated in Figure 22-10, show that most of the ABA is in the free form during the onset of dormancy in October and November, but is exchanged for or

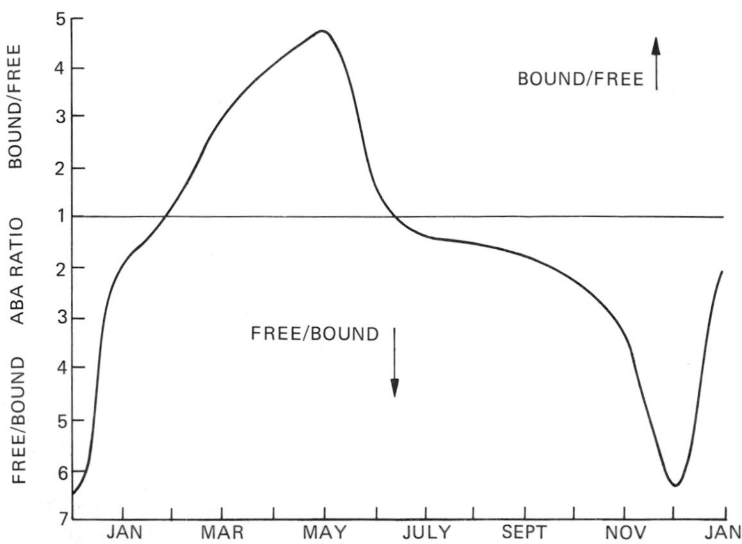

Figure 22-10. Ratios of bound and free abscisic acid (ABA) in buds of beech (*Fagus sylvatica*) during the year. [Redrawn from data of S. T. C. Wright: *J. Exp. Bot.,* **26:**161–74 (1975).]

converted to the bound form during the winter and early spring. It is likely that ABA is produced in leaves in the fall and moves to the shoots. Conversely, growth-stimulating substances (largely GA) are formed in the spring and disappear in the fall and early winter.

Other Factors. The day-length mechanism is clearly not the only one that functions to bring about dormancy. Several trees are relatively insensitive to day length (including some fruit trees such as apple, pear, and plum). Although the seedlings and young trees of many species are clearly influenced by short days in the fall, the older trees stop growing about midsummer, when day length is maximal. In this case it seems likely that shortage of nutrients or changes in the balance of hormonal constituents brought about by some other external or internal factor are of importance. However, it has been pointed out by Wareing that this early cessation of growth is not necessarily true dormancy, which may follow in the fall as a result of short days as in seedlings.

Leopold's group has pointed out that the dormant condition is considered to be one in which the metabolic machinery of the cells lies idle because of a repressed condition of the nucleic acid system. This can happen either because of a failure of the synthetic process resulting from a lack of some critical substance or intermediate or because of a programmed synthesis of substances that cause the cessation of metabolic activity. In experiments with *Begonia* tubers, these workers found that inhibitors of protein synthesis and nucleic acid metabolism actually prevent the onset of dormancy, instead of causing dormancy, as shown in Table 22-8. This indicates that entry into dormancy is an active metabolic process, requiring both nucleic acid and protein synthesis, not simply a passive process such as the cessation of metabolism resulting from inadequate nutrition.

Interacting Factors. The Swedish physiologist A. Vegis has proposed a general mechanism of dormancy based on the interactions between temperature requirements for metabolism and growth and the availability of oxygen as affected by seed coats or

Table 22-8. Prevention of the onset of dormancy in *Begonia* tubers by inhibitors of nucleic acid and protein synthesis (inhibitors were applied during a 20-day period in red light at 15° C, which normally induces dormancy)

Inhibitor	*Inhibits*	*Concentration, mM*	*Sprouting after 70 days, %*	
			Treated	*Control*
5-Fluorouracil	Transcription	10	87	26
2-Thiouracil	Transcription	10	83	11
8-Azaguanine	Transcription	3	86	4
8-Azaadenine	Transcription	5	91	4
5-Fluorodeoxyuridine	Transcription	3	87	4
Canavanine	Translation	1	90	26
Ethionine	Translation	1	69	26
Puromycin	Translation	1	77	26
p-Fluorophenylalanine	Translation	2	67	26
Cycloheximide	Translation	3	87	4

SOURCE: Y. Esashi and A. C. Leopold: Regulation of the onset of dormancy in tubers of *Begonia evansiana*. *Plant Physiol.*, **44:**1200–1202 (1969).

bud scales. Vegis suggests that the combination of high temperature and low oxygen is a cause of dormancy, and so is low temperature. Thus buds and seeds are formed in summer, and the combination of high temperatures and low oxygen (inside bud scales or seed coats) leads to their dormancy. This condition continues through the low temperatures of winter, which also start the processes that lead to the breaking of dormancy. The actual formation of winter buds is a response to short days; their dormancy follows only after their formation. This concept can be modified to account for the high-temperature dormancy of summer-dormant seeds or buds, as shown in Figure 22-11.

An objection to this concept is that dormancy does appear to be caused, as we have seen, by factors from outside buds arising in the leaves. Further, the early stages of the production of dormant buds, including the formation of bud scales, could not be caused by the exclusion of oxygen by the bud scales themselves. Vegis, however, has suggested that the formation of the bud is caused by short days and the presence of ABA from leaves, but that true dormancy is brought about by reduced oxygen and the narrowing temperature limits for growth and metabolism as shown in Figure 22-11.

The effect of oxygen in breaking dormancy at first seems rather straightforward, and it might be thought that increased oxygen merely serves to speed up respiratory and metabolic activities. However, the true situation does not seem to be as simple as this. Certain respiratory inhibitors, such as azide, malonate, or cyanide (see Chapter 6), have been found experimentally actually to increase germination of dormant seeds, although they normally depress oxygen uptake and respiration. This suggests the possibility that some oxidation reaction not connected with respiration is essential for breaking dormancy. The fact that respiratory inhibitors increase germination suggests that the respiratory enzymes may compete with this unknown oxidation reaction for available oxygen. The oxidative reaction, whatever it is, may be connected with the destruction or removal of growth inhibitors such as ABA. Alternatively, the inhibitor effects may relate to the observation that cyanide-insensitive respiration (page 131) is usually more pronounced in tissues (such as seeds) that are just breaking dormancy. Regardless of the exact mechanism, it now seems apparent that dormancy may be regulated, in part at

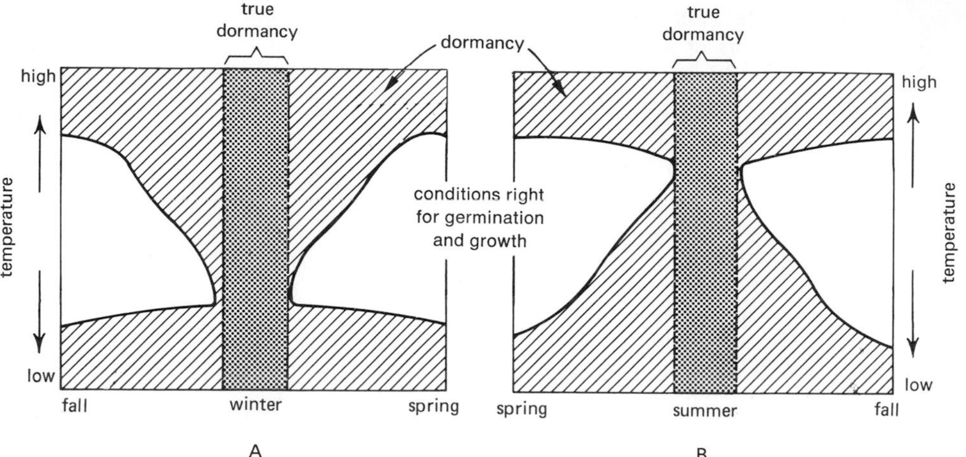

Figure 22-11. Diagram to show seasonal changes in dormancy as related to temperature, following the suggestion of Vegis.
Northern species having a cold temperature dormancy behave as in (**A**); species which become dormant during hot seasons behave as in (**B**); intermediate patterns also occur. [Adapted from A. Vegis: Dormancy in higher plants. *Ann. Rev. Plant Physiol.*, **15:**185–215 (1964).]

least, by the failure of integration of metabolic systems, including respiratory pathways. These are presumably controlled and integrated through some aspects of hormone metabolism.

The foregoing discussion emphasizes the fact that dormancy is not a simple process, and the entry into dormancy may be mediated by very different factors from those that maintain it. Similarly (and perhaps because of this), the breaking of dormancy is also a complex process.

Breaking Dormancy. Several mechanisms appear to act in releasing buds, seeds, and other tissues from dormancy. Perhaps the most important is cold. This seems strange, since dormancy usually takes place during cold weather and may be considered a cold-avoidance mechanism. However, cold is the most obvious attribute of winter, and mechanisms have evolved that permit plants to measure its duration and intensity. Many seeds require cold, as we have seen, in order to overcome the dormancy that sets in soon after they are ripe. Many trees require from 250 to 1000 hr of cold before dormancy can be broken. If a dormant tree is taken into a greenhouse, it will not start growing again unless (or until) it has had the required period of chilling. In certain districts in the southern United States there is sometimes difficulty in fruit orchards because the winter is not cold enough to cause the buds to break in spring; sometimes hormone treatment (for example, GA) is necessary to initiate spring growth.

Cold treatment is not the only requirement for breaking dormancy. Resumption of growth requires warm temperature and, in many species, long days. Thus, if a plant has received its full cold requirement by midwinter, it still remains dormant, either until the occurrence of warm weather or lengthening days or both. A high-temperature shock may break dormancy early. A standard technique for forcing flowers and woody shrubs (such as *Forsythia*) is to immerse them in warm water, 30–35° C, for several hours.

As we have seen, the application of exogenous GA reverses the effect caused by ABA, either preventing or breaking the dormancy induced by this inhibitor. Of more interest, since it appears to be a natural physiological mechanism, is the effect of

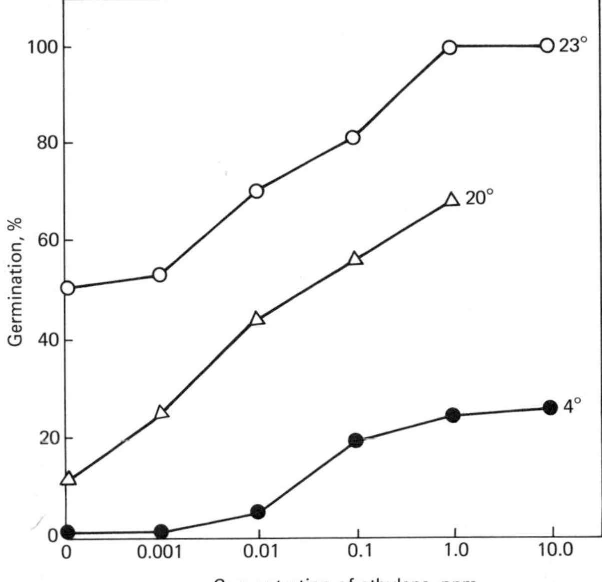

Figure 22-12. The effect of ethylene on the germination of clover seeds (*Trifolium subterraneum* var. Dinninup) at different temperatures. [From Y. Esashi and A. C. Leopold: Dormancy regulation in subterranean clover seeds by ethylene. *Plant Physiol.*, **44:**1470–72 (1969). Used with permission.]

ethylene. Seeds of some small plants such as clover have been shown to break dormancy in response to extremely low ethylene concentrations, as shown in Figure 22-12. Imbibed seeds have been shown to produce ethylene. It seems likely that this represents a mechanism which ensures that the seed will germinate only when it is completely covered with soil. Then the ethylene it produces cannot escape but increases to a sufficient concentration to cause the breaking of dormancy and to permit germination.

The pregermination cold treatment that some seeds require to break dormancy is called stratification (see Chapter 17, page 422). During stratification, several changes occur in the major classes of hormones, as shown in Figure 22-13. ABA, initially very

Figure 22-13. Changes in growth factors in sugar maple seeds (*Acer saccharum*) during stratification at 5°C. [Adapted from data of D. P. Webb, J. van Staden, and P. F. Wareing: *J. Exp. Bot.*, **24:**105–17 (1973).]

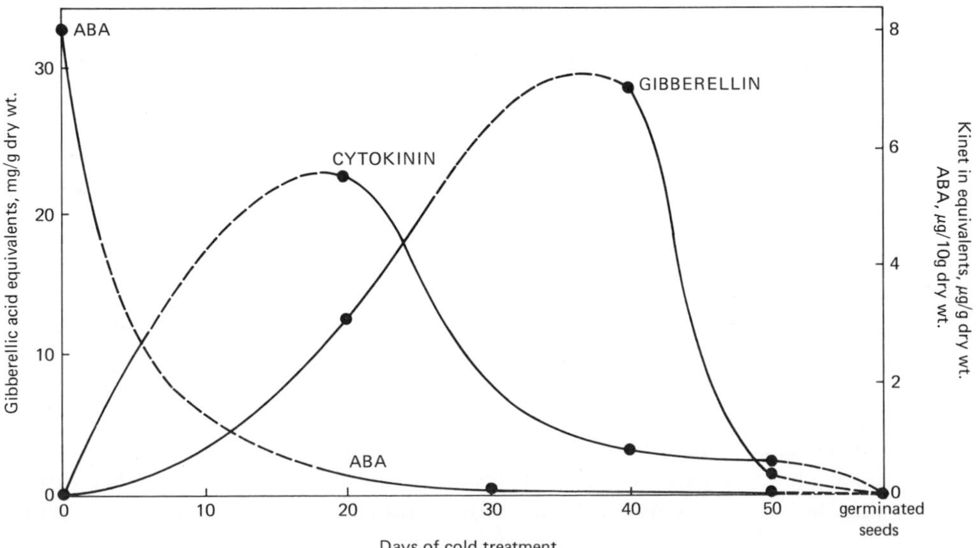

high, declines rapidly. Cytokinins increase, then decrease again as the gibberellins increase. Finally, at the time of germination, all the hormones fall to a low level. It would seem that the interplay of the effects of all these hormones is important in the processes that occur during the breaking of dormancy.

Senescence and Death

Patterns of Aging and Death. Plants and their parts develop continuously from germination until death. The latter part of the developmental process, which leads from maturity to the ultimate complete loss of organization and function, is termed **senescence.** It is a characteristic of plant behavior that senescence is not simply a running down of the life processes but is a highly ordered and programmed process or series of processes. Plants do not "go to pieces" as they age. They age, as they grow and develop, in an orderly fashion.

Plants, according to their habit of growth, senesce in many different ways. The whole plant may senesce and die at one time, as occurs in many annuals after flowering is complete. Alternatively, there may be a progressive senescence of parts as the whole plant ages—some parts (usually those nearest the tips of shoots and roots) remaining active and in the juvenile stage while the older parts (particularly older leaves) senesce and die. Third, there may be a simultaneous or sequential senescence of a part of the plant (the top of an overwintering biennial or perennial, the leaves of a deciduous tree), while the rest of the plant remains alive. Finally, during the process of tissue maturation certain cells, such as xylem vessels and tracheids or sclerenchyma tissue, may senesce and die although the plant as a whole is in a state of vigorous growth.

Patterns of senescence appear to differ in quite important ways, both in their causes and in the nature of the senescence processes, as well as in their degree of reversibility. Some types of senescence appear to be closely correlated with developmental events in the whole plant. For example, senescence in **monocarpic** plants (those that flower only once and then die) is closely related to the process of flowering and the growth of fruits. If flower or fruits are removed, senescence may be postponed; if the plant is maintained under conditions unfavorable to flowering (for example, in the wrong day length), senescence of the whole plant may be postponed for many years. Some monocarpic plants do indeed live for many years before flowering (for example, the "century" plant, or *Agave*), but when they have produced fruit, they die. But some perennial plants, such as the bristle-cone pine (*Pinus aristata*) of the white mountains of California, may live to be 5000 years old.

Thus, on the one hand, the senescence that follows fruit production or the senescence of flower parts after fertilization is apparently irreversible and is an inevitable consequence of flowering or fruiting. It is thus not under the control of the plant in the sense that it might be influenced by internal or external stimuli. On the other hand, the rapid senescence of a detached leaf can be reversed and the leaf rejuvenated by the application of a cytokinin or by allowing the leaf to root. Similarly, older leaves on plants like bean or tobacco senesce as the plant grows. This senescence can be reversed if the top of the plant is cut off.

These observations suggest that there may be more than one process or causal agent of senescence characteristic of different situations or different tissues. This further suggests that senescence is not just the running down of tissues or cells but a programmed part of development. Senescence is often of great advantage to plants, which would be in serious difficulties if it did not occur. Loss of leaves in deciduous trees is an essential part of winter cold avoidance. Older leaves in herbaceous plants often senesce and die, and their nutrient contents are withdrawn for the nutrition of the growing parts of the plant. Senescence and death are essential to the functioning of xylem and sclerenchyma cells. These functions are so important that it is unlikely that they are left to chance.

Metabolic Aspects of Senescence. At the cellular level, senescence appears to be tightly controlled, although the mechanisms of control are not known. Senescing cells undergo reduction of their structure, and much of the membranous subcellular inclusions are disrupted. It has been suggested that the vacuole acts as a lysosome, secreting hydrolytic enzymes which digest cellular material that is no longer needed. Evidently some sort of breakdown of the tonoplast takes place, and hydrolytic enzymes are liberated into the cytoplasm. However, the situation is not so simple because the internal structure of chloroplasts and mitochondria is also reduced, and this seems to happen before their external membranes are breached. Therefore, it seems likely that degradative processes are initiated, or synthetic processes eliminated, in organelles as well as in cells. Possibly the same signal that causes senescence in cells is also perceived by their organelles, causing them to senesce simultaneously.

Conspicuous changes take place in the metabolism and contents of senescing organs. Decreases in DNA, RNA, proteins, inorganic ions, and various organic nutrients have been observed. Pronounced changes take place in the pace of certain metabolic reactions. Some of the patterns of metabolism of senescing bean leaves are shown in Figure 22-14. Photosynthesis decreases somewhat before senescence starts, and chlorophyll breakdown does not occur until much later. This is probably due to the reduced demand for photosynthate, which, as we have seen (Chapter 21, page 513), appears to control the rate of photosynthesis to some extent. However, a major decline in photosynthesis occurs as senescence begins and as proteins and chlorophylls decrease. Shortly thereafter a respiratory climacteric is seen. Soluble nitrogen increases briefly as protein breakdown occurs, but these compounds are rapidly translocated away. When death occurs, much of the valuable nutrients in the leaf have been salvaged and translocated to still growing parts of the plant.

The losses in many major polymeric substances in senescing leaves (for example, RNA, DNA, protein) suggest that degradative activity is greatly accelerated during senescence. The Canadian physiologist R. A. Fletcher and his group have investigated the amounts of hydrolytic enzymes in bean and radish leaves. Contrary to expectation, they found that the amounts of these degradative enzymes decreased sharply during senescence, as shown in Table 22-9. Moreover, if senescence was reversed by the application of cytokinins, then the amounts of degradative enzymes went up instead of down. This means that the losses during senescence are due to decreased synthesis rather than to increased degradation. Many polymeric compounds, such as proteins and RNA, are turning over in an actively metabolizing leaf; that is, they are continuously being broken down and resynthesized. Turnover decreases and stops in senescing leaves. The implication is that the synthetic part of the turnover cycle stops or decreases faster than the degradative part, even though the latter may also decrease in rate.

Nutritional Competition in Senescence. The German physiologist H. Molisch suggested in the 1920s that senescence might be caused by nutritional deficiencies.

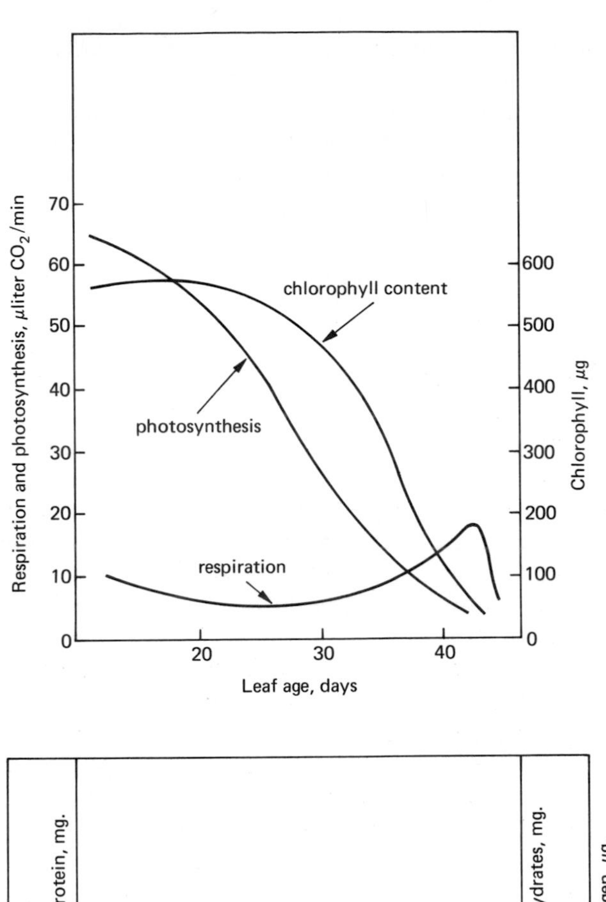

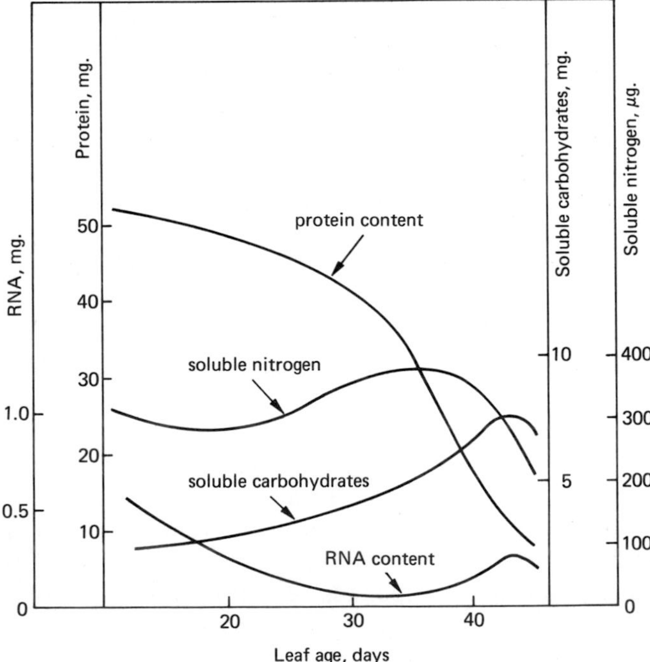

Figure 22-14. Some changes that occur in senescing leaves of bean (*Phaseolus vulgaris*). [Data from laboratory exercises and the work of associates in the author's laboratory.]

Table 22-9. Loss of degradative enzymes in senescing leaves

Days after treatment		*Bean leaf* RNA % of control	*Bean leaf* RNAase % of control	*Radish leaf* Chlorophyll % of control	*Radish leaf* Chlorophyllase % of control
0	Control	100	100	100	100
4	Senescent	—	—	66	29
4	Cytokinin treated	—	—	93	91
8	Senescent	—	—	30	0
8	Cytokinin treated	—	—	78	65
9	Senescent	51	36	—	—
9	Cytokinin treated	94	60	—	—

SOURCE: Adapted from data of D. R. Phillips, R. F. Horton, and R. A. Fletcher: *Physiol. Plant.*, **22**:1050–54 (1969).

Different parts of the plant compete for nutrition, and fruit or growing tips, for example, might form stronger sinks for translocation and thus accumulate so much of the available nutrient that older leaves would be starved. He pointed out that, if the fruit, seeds, or growing tip is removed from plants, senescence of other parts such as leaves is greatly delayed. This theory can be accommodated to later ideas about hormone direction of translocation if it is postulated that starvation of older leaves is due to the direction of translocation by hormones produced in the growing tips or developing fruits. The effect of cytokinins, which delay or prevent senescence when applied to a leaf, might be to cause cell division, possibly accompanied by IAA production, which would then act to direct translocation toward the area of its production.

This view is extremely simple, hence attractive. Unfortunately, a number of observations make it rather unlikely to be correct. For example, in dioecious plants those bearing male flowers, which set no fruit and require no additional nutrition, nevertheless undergo senescence just as do female plants which bear fruit. If annual plants such as *Xanthium* are prevented from flowering by continuous short days, they eventually become senescent without ever flowering. Further, senescence in detached leaves can be reversed and the leaf rejuvenated by the application of kinetin, but senescence cannot be so reversed in attached leaves. Finally, it is not possible to retard senescence in annual plants that have flowered or fruited even by massive application of nutrient fertilizers. If nutritional competition is a major cause of senescence, the opposite effects would be expected in each of the above examples.

Effects of Growth Factors. A clue to the causes of senescence came from the observation that if a detached, senescing leaf begins to form roots, then senescence is reversed. This suggests that roots produce something that is translocated to leaves and prevents or reverses senescence. Scientists working in both America and Germany discovered that cytokinins applied to leaves would reverse senescence in the area of the leaf to which they were applied, as shown in Figure 22-15. Later work showed that roots do indeed produce cytokinins, and it has now become reasonably certain that the antisenescence hormone translocated from roots is in fact a cytokinin or a group of cytokinins.

Figure 22-15. Detached tobacco leaves treated with chloramphenicol, which enhances senescence.
Six days before the picture was taken, the right half of each left-hand leaf was sprayed with water; the right half of right-hand leaves was sprayed with a solution of kinetin. The kinetin-sprayed area has retained its green color. [From L. Engelbrecht and K. Nogai: Zur Frage der Kinetinwirkung gegenüber der Stoffwechselhemmung durch Chloramphenikol. *Flora,* **154:**267–78 (1964). Used with permission. From an original photograph courtesy Dr. Engelbrecht.]

The mechanism of action of cytokinin is not wholly clear, but there are indications from Mothes' experiments. He found that when a drop of kinetin or other cytokinin is placed on a leaf, various organic and inorganic nutrients are mobilized in peripheral areas of the leaf and move into the kinetin-treated area. It is not clear whether the increased nutrition is the immediate cause of rejuvenation or the cytokinin causes some other events to take place that result in both rejuvenation and mobilization of nutrients.

The British physiologist A. C. Chibnall early found that detached leaves without roots invariably became senescent, even if cultured on a complete nutrient solution; therefore, it would appear that rejuvenation is not merely the result of cytokinin-induced mobilization of nutrients. However, cytokinins are known to cause cell division and to enhance many metabolic processes, including protein, DNA, and RNA synthesis. The work of J. Cherry suggests that cytokinins may protect RNA from degradation (see Chapter 23) and thus permit continuation of protein synthesis even in the presence of RNAse activity. This aspect of cytokinin behavior is therefore more probably responsible for its rejuvenating effect. IAA production often occurs in actively metabolizing cells, and this may be responsible for the directed translocation of nutrients (see Chapter 21, page 522).

The question might be asked: Why do certain (that is, older) leaves on an intact plant senesce, whereas younger leaves do not? Both apparently have the same access to the root system. The answer may lie in the fact that nutritional traffic in the plant tends to be strongly directed toward younger and more actively growing parts of the plant (see Chapter 21). This direction of translocation is possibly the result of more vigorous auxin production in rapidly growing tissue; auxins have been shown to increase translocation

Figure 22-16. Retardation of senescence of the primary leaf of a bean plant treated with 30 mg/liter benzyladenine (right) compared with one treated with water (left). [From R. A. Fletcher: Retardation of leaf senescence by benzyladenine in intact bean plants. *Planta* (*Berlin*), **89:**1–8 (1969). Used with permission. Photograph courtesy Dr. R. A. Fletcher.]

toward the site of their application or production. The movement of cytokinins from the roots may be affected in the same way; thus, senescence may be the result of the starvation of older leaves, not only of nutrients but of cytokinins also. The addition of the cytokinin benzyladenine has been shown to prevent senescence in older leaves of bean plants, as shown in Figure 22-16. This suggests that the cause of senescence is indeed a lack of cytokinins.

As usual, there are complications. Not all plants respond to the same hormones. Cytokinins appear to be more effective in many herbaceous plants. Gibberellins are effective in retarding senescence of the dandelion (*Taraxacum officinale*) and the ash (*Fraxinus*), and endogenous gibberellin levels fall progressively during leaf senescence. Auxins (IAA and 2,4-D) have been found to retard senescence in certain trees, although they cannot always be shown to have this effect in all plants. Ethylene strongly promotes senescence in many tissues; it appears to be physiologically involved in ripening fruits, in which its concentration may build up to effective physiological levels.

There is strong evidence to suggest that ethylene is intimately connected with aging. It has a strong "phytogerontological" effect if applied externally. The American physiologists A. D. Hanson and H. Kende have shown that ethylene is derived largely or exclusively from the amino acid methionine. Methionine is also involved in methylation reactions, which keep its concentration low in young, actively growing plants. As activity decreases the methionine content builds up greatly, and ethylene synthesis takes place. The inference is that the ethylene may be an integral part of the aging process, but does not cause it. The ultimate cause of aging is still unknown.

Abscission. Abscission of leaves and fruit is one of the more obvious characteristics of senescence. Leaves do not fall simply because they are dead. A zone of cell division, the **abscission zone,** develops near the base of the leaf so that numerous crosswalls form at right angles to the long axis of the petiole, as shown in Figure 22-17. Then pectinases and cellulases are induced in these cells of the abscission zone as shown in Figure 22-18. These dissolve the lamellae of the crosswalls of these cells, so that the petiole breaks off. The vascular connections break and usually become plugged by the formation of **tyloses** (deposits of gummy substances) and layers of corky cells. Thus, at least two important

Figure 22-17. Drawing of the abscission zone showing the line of separation. Cell separation takes place across the abscission zone, leaving intact cells at the abscission interface. [Drawing by Mrs. A. B. Addicott. Original print courtesy Dr. F. T. Addicott.]

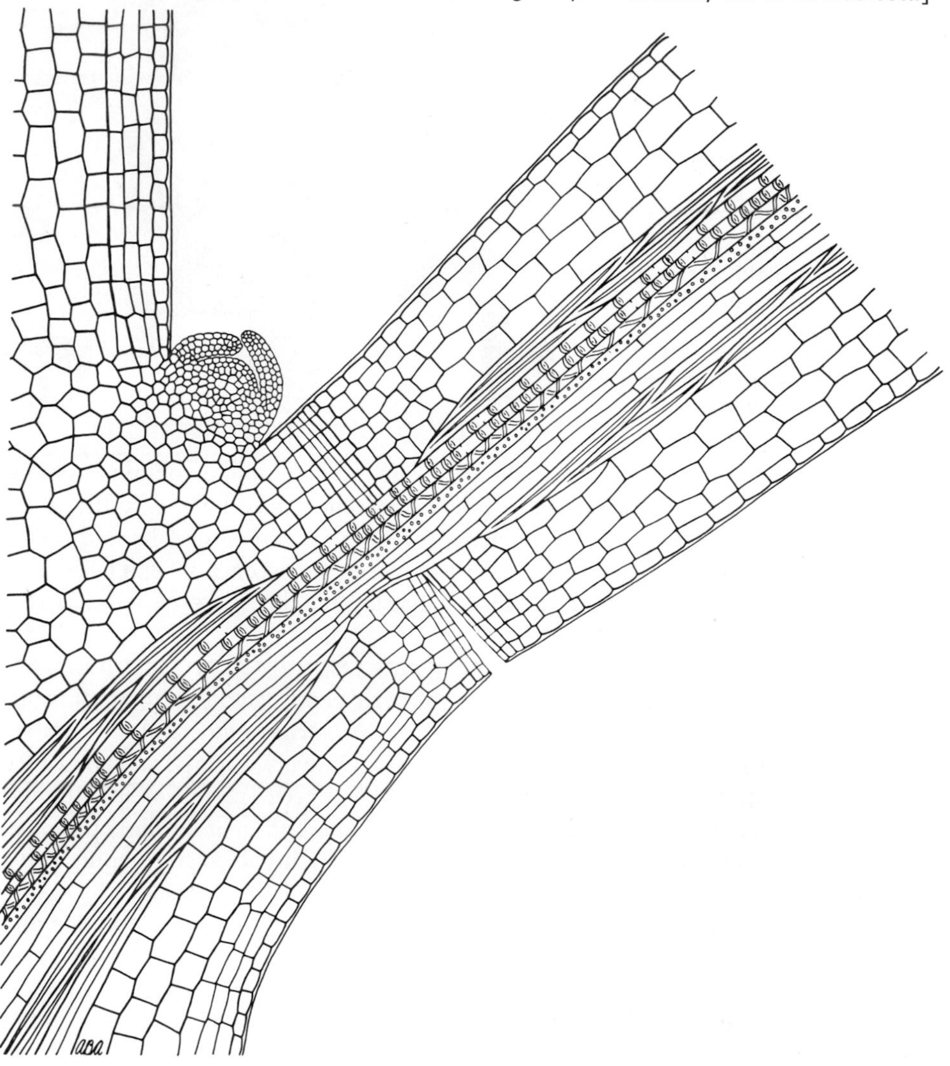

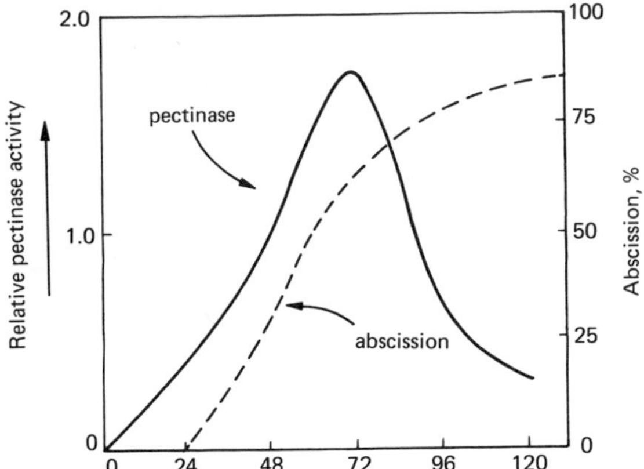

Figure 22-18. Pectinase activity and abscission of bean petiole explants following removal of the leaf blade. [From data of H. W. Mussell and D. J. Morré: *Plant Physiol.*, **43:** 1545–59 (1968).]

events are involved in abscission: cell division and the induction of hydrolases. Both of these are processes of active metabolism and must therefore be a programmed part of the development of the plant.

The beginning of abscission seems to follow the cessation of growth and active metabolism in the leaf. However, the formation of the abscission layer is not in itself senescence; it is a resumption or continuation of the processes of development that constitute senescence. The protein and RNA contents of the abscission zone increase during abscission, and the formation of the abscission zone can be inhibited by various metabolic and protein synthesis inhibitors. It seems clear, however, that abscission only occurs when the leaf blade is senescent, and the nutrition required for the metabolism of the abscission zone appears to come largely from the nutrients released from the senescing cells of the leaf.

Causes of abscission involve several interlocking events. It seems possible that certain growth-inhibiting substances may be involved. One of the discoverers of ABA was searching for a senescence factor in cotton leaves, and the ABA content of ripening fruit is known to increase sharply about the time of fruit development. ABA clearly stimulates abscission in cotton petiole explants, as does GA (Figure 22-19). However, this

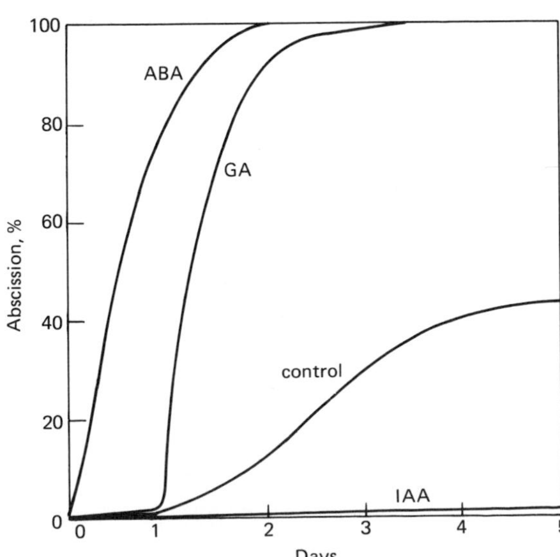

Figure 22-19. The influence of ABA, GA, and IAA (as the trimethylsilyl derivatives) on the abscission of cotton petiole explants. [Redrawn from data of L. A. Davis, D. E. Heinz, and F. T. Addicott: *Plant Physiol.*, **43:** 1389–94 (1968).]

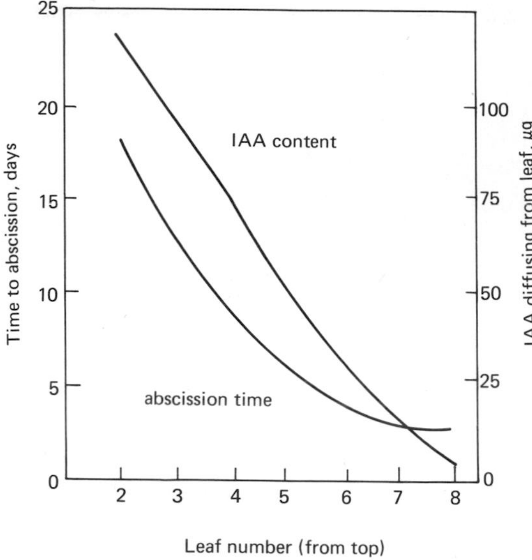

Figure 22-20. With increasing age, *Coleus* leaves show a decline in the time required for the development of abscission upon being debladed and an associated decline in diffusible auxin content. Leaf age is taken to be equivalent to the position of the leaf on the plant, leaf number 1 being nearest the plant apex. [Adapted from data of R. M. Meyers: *Bot. Gaz.*, **102:**323–38 (1940).]

Figure 22-21. The progressive decline in breaking force at the abscission zone with time after excision.
The curve for control plants is compared with progress curves for bean petiole explants under 0.1 and 1000 ppm ethylene. [From R. K. dela Fuente and A. C. Leopold: Kinetics of abscission in the bean petiole explant. *Plant Physiol.*, **44:**251–54 (1969). Used with permission.]

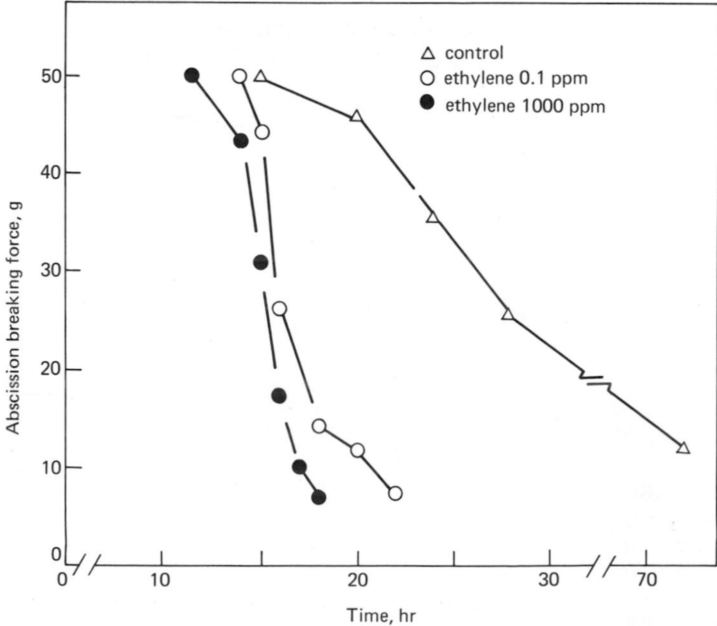

cannot be the only factor involved because ABA inhibits the RNA synthesis and growth required for abscission. Senescence is probably largely related to the decrease of cytokinin supply; abscission may follow automatically upon senescence. This view is supported by the fact that, as leaves age, their auxin production decreases, as is shown in Figure 22-20. Addition of auxin to the petiole or blade of a senescing leaf prevents the formulation of the abscission layer, and thus inhibits abscission. Therefore, abscission may be a preprogrammed series of events that can only start when the auxin supply from the leaf reaches a specific low level. This happens when the leaf ages and begins to senesce.

Ethylene appears to play a role in leaf abscission. When petioles are excised, an abscission layer forms within about 3 days and the force required to break the petiole decreases abruptly. The addition of ethylene greatly accelerates this process, as is shown in Figure 22-21. If the ethylene is removed, the formation of the abscission layer proceeds at the original (control) rate (Figure 22-22).

It has been suggested that ethylene has a twofold effect: a gerontological action, which causes or accelerates senescence in the leaf, and a stimulation of the induction of cell wall degrading enzymes in the abscission zone. It is the latter effect that is counteracted by IAA, as shown in Table 22-10. Although IAA treatment has been found to prevent abscission, it actually stimulates the rate of abscission when it is applied late (after senescence has begun). This may relate to an IAA-stimulated formation of ethylene. Some leaves are known to produce a short burst of IAA late in senescence,

Figure 22-22. Changes in decline of the abscission breaking force following the introduction of 1 ppm ethylene (after 26 hr), and removal of ethylene by evacuation (29 hr after excision). Arrows indicate times of ethylene addition or evacuation. Vertical lines indicate standard errors. [From R. K. dela Fuente and A. C. Leopold: Kinetics of abscission in the bean petiole explant. *Plant Physiol.*, **44**:251–54 (1969). Used with permission.]

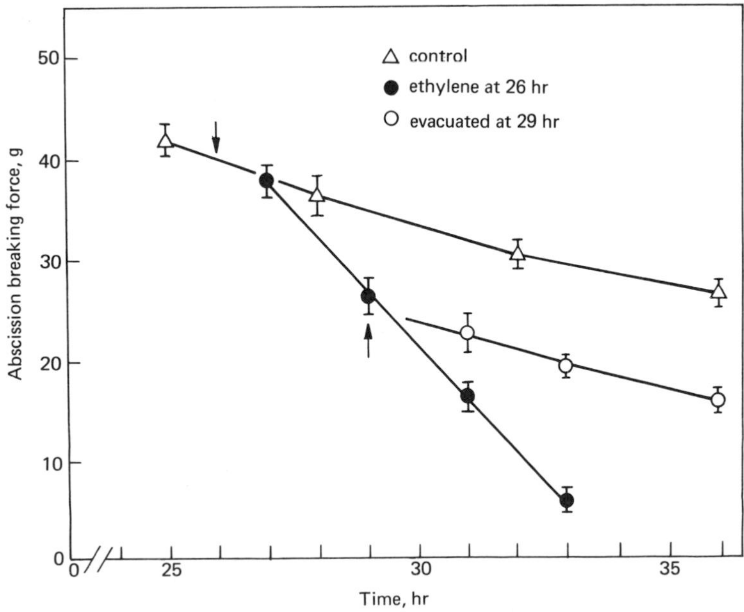

Table 22-10. Increase in cellulase activity in bean leaf abscission zones as affected by ethylene or the synthetic auxin, 2,4,5-trichlorophenoxyacetic acid (2,5,4-T)

	Activity, % of control
Control	100
+ Ethylene	166
+ 2,4,5-T	23

SOURCE: Adapted from data of R. F. Horton and D. J. Osborne: *Nature,* **214:**1086–88 (1967).

shortly before death. At this stage, the IAA would probably act to stimulate abscission rather than to prevent it.

The overall picture for abscission thus may be summarized as follows: Growth decreases, caused partly by ABA production resulting from short days and partly by decreased supply of cytokinin and nutritional factors resulting from decreased IAA production. This is presumably the first stage of senescence. As a result of decreased growth, IAA production decreases sharply, and the formation of the abscission layer begins. A further consequence of decreased IAA production may be greatly reduced translocation to the leaf; for this reason, and perhaps others, senescence progresses rapidly. A frequent consequence of progressive senescence is ethylene production; this stimulates the production of cell wall degrading enzymes in the abscission zone, and abscission takes place. By this time senescence has proceeded to the point where the leaf is dead, or nearly so, and most of the nutrients that can be mobilized have been translocated out. Following abscission, corky cells and tyloses seal off the wound left by abscission.

Additional Reading

See list for Chapter 16 (p. 406).

Addicott, F. T., and **J. L. Lyon:** Physiology of abscisic acid and related substances. *Ann. Rev. Plant Physiol.,* **20:**139–64 (1969).

Taylorson, R. B., and **S. B. Hendricks:** Dormancy in seeds. *Ann. Rev. Plant Physiol.,* **28:**331–54 (1977).

Villiers, T. A.: *Dormancy and the Survival of Plants.* Edward Arnold Ltd., London, 1975.

Wareing, P. F., and **P. F. Saunders;** Hormones and dormancy. *Ann. Rev. Plant Physiol.,* **22:**261–88 (1971).

Woolhouse, H. W.: *Ageing Processes in Higher Plants.* Oxford University Press, London, 1972.

23 Action of Hormones and Growth Substances

Introduction

We have considered the behavior of plants in a sequence that approximates a "life history" of plants in the preceding six chapters. The common thread which runs through every phase of plant development and behavior is control by hormones or growth substances. We have seen something of the many different processes and behavioral patterns that are influenced by each hormone. Now we shall summarize the information that is available about the actual mechanism or mechanisms whereby these hormones exert their manifold effects.

One of the most bewildering aspects of plant physiology is the fact that a small group of chemically quite simple compounds, the plant hormones, exerts such an extraordinary variety of effects in such a large number of different situations. Some of the effects are so widely different (for example, the effects of IAA on cell walls, water uptake, growth, metabolism, translocation, photosynthesis, or cell expansion) as to suggest a variety of different mechanisms of action for the hormones. Yet the principle of Occam, "Entities are not to be multiplied unnecessarily," is still very attractive. If there is a single site or mode of action of a hormone, it will serve as a unifying concept that will aid in bringing order out of the chaos of hormone action.

For this reason much of the recent work in hormone biochemistry has been directed toward finding a single master control effect for each hormone. The most natural place to look for such an effect is at the genetic level—the switching on or off, or the selection among, or the modification of various programs of development written into the genetic complement of the cell. In this view, the hormones are looked upon as selecting, activating, or modifying whole programs rather than as individual keys operating many individual steps in a variety of separate and independent pathways. It is possible that hormones act in both ways. Recent evidence does indicate, however, that there are basic and generalized levels of hormone action that clearly override in importance any specific action on individual enzymes or reactions of metabolism.

Auxins

Synthesis, Movement, and Inactivation. Control by hormones may be achieved through the operation of the hormone in specific or general ways or through the establishment of polarized concentration gradients in tissues. Gradients are developed by the localized synthesis of the hormone, by its movement or translocation, and by its destruction. Growth seems to be a prerequisite of IAA synthesis, and the IAA seems to be produced mainly in growing tips, expanding leaves, and such meristematically active tissues. There is a question about roots; some experiments on root growth suggest that auxin is the mediating agent in the control of root morphology by the apex. Yet the amount of auxin present in roots is vanishingly small, and no direct evidence of its production there has been obtained. It seems more likely that the auxin present in roots is translocated there from the shoots. However, this problem has not been completely resolved as yet. Biochemical reactions leading to the synthesis of IAA and its derivatives are described in Chapter 9, page 240. The main pathways of IAA formation are summarized in Figure 23-1.

The polar translocation of auxin accounts for much of its specificity of action. Reasons for polar transport are not yet clear but may be related to electrical or ionic gradients and the differential permeability of auxin in the ionic form or as the free acid. Auxin moves quite slowly, no faster than would be expected for diffusion from cell to cell across the cell walls. Not all auxin movement is basipetal; numerous recent experiments have shown that applied radioactive IAA will move acropetally with considerable velocity. Unfortunately it is not possible to determine how much auxin moves in this way, or what proportion of the total mobile pool of auxins is involved. It seems possible that auxin might move simultaneously in opposite directions, but in

Figure 23-1. Pathways leading to IAA formation in plants, and the structures of some important indole compounds.

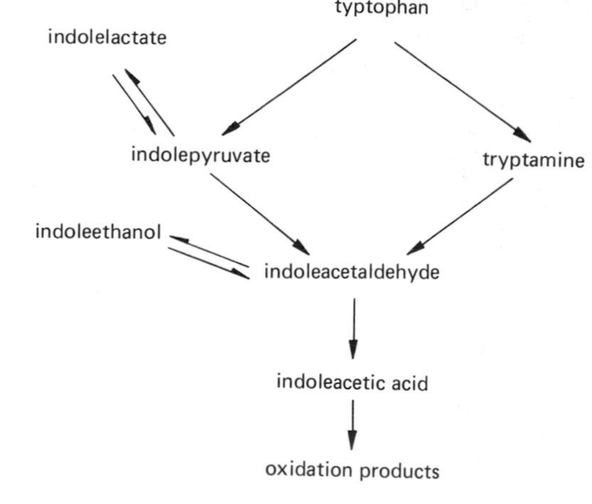

different tissues or in different columns of cells. A. C. Leopold has noted that many conditions or factors that influence the transport of IAA also influence its growth effects to the same extent. This suggests that transport and growth-regulating mechanisms are closely linked and may perhaps have a common primary process.

Auxins are almost continuously produced by some tissues in the plant; however, they do not accumulate in large amounts. This means that some process or processes of inactivation or destruction must take place. In fact, auxin inactivation is an important part of the system by which growth control and correlation are achieved, because the auxin concentration at a given site is proportional not only to its rate of delivery but also to its rate of destruction.

One of the important distinctions between natural auxins like IAA and some of the synthetic herbicidal ones like 2,4-dichlorophenoxyacetic acid (2,4-D) or 2,4,5-trichlorobenzoic acid (2,4,5-T) is that the synthetic compounds are more stable. Perhaps, because they are unnatural compounds, there are few enzyme systems that attack them easily. Thus they accumulate more readily in the plant and poison it. One of the difficulties of experiments with auxins like IAA is the fact that added auxin is very rapidly inactivated in most tissues, and it is often very difficult to maintain unnaturally high concentrations of IAA experimentally in the tissue.

Numerous bacteria degrade IAA; this is another experimental hazard. Areas where IAA is frequently handled, particularly nonsterile places like greenhouses, tend to develop a substantial population of microorganisms that oxidize IAA. This makes sterile conditions a most important experimental consideration. A small amount of contaminating bacteria can destroy a very large amount (relative to the amounts usually used) of IAA in a short time.

Several patterns of IAA destruction or inactivation are known. It may be irreversibly oxidized to the compound methylene oxindole by a widely distributed enzyme (Figure 23-1). This compound was originally thought to be inactive, but it has now been shown to act as a growth inhibitor in yeast, bacteria, and certain higher plant systems. However, it is not yet clear whether methylene oxindole is normally involved in the regulation of growth. IAA may be destroyed by peroxidase enzymes or by oxidases that interact directly with molecular oxygen. It may also be decarboxylated enzymatically.

Another important way in which IAA can be rapidly inactivated is by its conjugation with a variety of chemicals normally available in the cell. The Canadian physiologist W. A. Andreae discovered that IAA rapidly conjugates in roots with aspartic acid, making the peptide derivative aspartyl-IAA (Figure 23-2). IAA can evidently conjugate with a number of other compounds, including other amino acids, certain sugars and polysaccharides, and proteins. It is not clear whether the IAA so conjugated can be readily released, but many conjugates are considered to be only temporarily inactive, so that their formation really constitutes storage rather than inactivation. However,

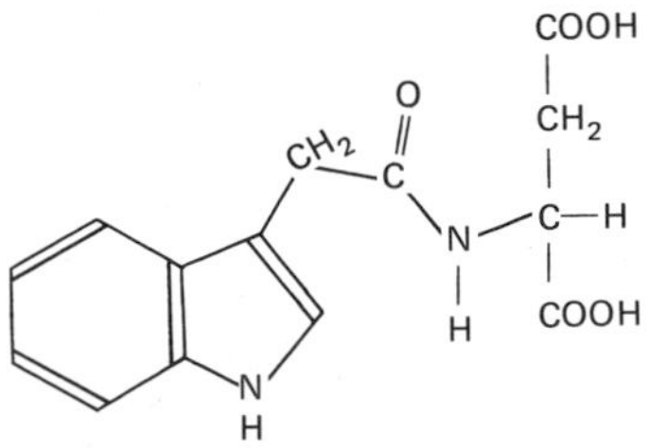

Figure 23-2. Indoleacetyl aspartic acid, the conjugation product of IAA and aspartic acid.

regardless of their permanency, the conjugates do serve to reduce the IAA concentration in tissues.

Some questions arise about the total amounts of IAA that are present naturally in tissues. Not all the IAA is extracted from tissues with equal ease, and it seems likely that much of the IAA present in tissues may be there in storage forms or as conjugates. It has been suggested that IAA is transported in the form of a conjugate, but this view is not widely held. Recent experimental work shows that auxins are strongly compartmented in tissues, and that the regulation of their synthesis and destruction, as well as of their action, may be concerned with this compartmentation.

IAA and Ethylene Formation. IAA has been shown to stimulate ethylene formation in many situations, and ethylene itself causes epinasty and many of the formative effects that are also attributed to IAA. As a result, for a time it was suspected that the primary action of IAA was to cause the formation of ethylene. However, it has been noted that ethylene has opposite effects to IAA in certain situations. For example, the effect of ethylene and IAA on senescence and abscission are quite different. Further, it is doubtful if sufficient amounts of ethylene are produced, particularly in stem tissues where it would diffuse rapidly away, to achieve an effective concentration.

The situation in roots is different. The American physiologists A. V. Chadwick and S. P. Burg have shown that IAA does stimulate ethylene formation in roots and that the auxin effects in roots (particularly geotropism) may well be attributable to ethylene rather than directly to IAA. Ethylene inhibits root growth, as does IAA, and the inhibition is largely removed by carbon dioxide. Carbon dioxide stimulation of the growth of several systems affected by auxins, including the coleoptiles of grass seeds, has been observed, and carbon dioxide is thought to be a competitive inhibitor of ethylene. Many of the effects which had previously been attributed to auxin are caused by the application of ethylene alone. Since roots are enclosed in soil, ethylene produced there would not diffuse away as rapidly, and so effective concentrations could build up. It thus seems likely that ethylene is important in mediating the auxin effect in roots. This subject will be further considered later in the chapter, in the section dealing with ethylene.

IAA Effects on Specific Enzymes. IAA has been reported to cause a direct (possibly allosteric) enhancement of the citrate-condensing enzyme in corn cotyledons, and it inhibits pyruvate and α-ketoglutarate decarboxylation in the Krebs cycle of pea roots. Thus there is some evidence that IAA may directly affect the operation of certain key enzymes in the energy metabolism of cells. However, these effects are by no means clear. It has been observed that certain phosphorylation reactions are affected by IAA in intact cells but not in cell-free extracts or brei. This suggests that the integrity of the whole cell is necessary for IAA to affect the metabolic reactions. If this is so, it is unlikely that the effect is directly on the enzymic reaction. It is more likely to be on some basic aspect of cell metabolism or cell structure (for example, membrane permeability) that occurs only in intact cells. It has been suggested that IAA has an effect on RNA polymerase, but this might also be a secondary effect mediated by some other hormone (both gibberellins and cytokinins have been implicated) under IAA control.

An interesting point, which relates to the synthesis of auxin and its control, is that IAA strongly inhibits the enzyme indoleethanol oxidase. This is an important enzyme on one of the main pathways of IAA synthesis (see Figure 23-1), which seems to be the major one in certain plants. This discovery by the American physiologist W. K. Purves indicates that the synthesis of auxin may be under feedback control, at least in plants (for example, cucumber) in which this is a main pathway of auxin synthesis.

Auxins and Translocation. The effects of IAA and other auxins on the translocation of metabolites are not at present understood. As described in Chapter 21, the application of IAA or synthetic auxins seems to cause the formation of an artificial "sink" to which recent photosynthate or applied labeled nutrients are moved. It has been suggested that some of the growth effects of auxins are the result of increased nutrition and metabolism caused by auxin direction of food and cytokinin transport. However, as will be discussed in the next section, many of the important actions of auxin, such as the stimulation of cell elongation which causes tropic responses, are very rapid and direct and in no way related to nutritional phenomena.

This does not mean that there are necessarily two or more different mechanisms of auxin action. There may well be some primary mechanism whereby auxin exerts its many different effects. But if so, this mechanism must operate at the level where it can influence a great many reactions. The translocation effects are examples of true hormone action that operate over long range and seem to correlate growth and metabolism in various parts of the plant. However, the primary action of the hormone, as distinct from the fact of its operation at a distance, may well be the same in the different types of response: namely, the stimulation or activation of metabolic reactions, whether those that affect nutritional flow by causing the loading of specific translocation streams or those that affect cell elongation. These effects might operate through a genetic mechanism, or on some general facet of cell metabolism such as membrane activity or permeability.

Cell Wall Effects. Long ago it was found that IAA stimulates cell enlargement, and early theories were developed around this facet of its effect on cells. Cell walls contain crisscrossed layers of cellulose fibrils and are normally quite rigid. Thus for a cell to grow, there must be a mechanism for relaxing or loosening the cellulose fibrils and for allowing new ones to be added. Many theories of the action of auxin have been developed that depend on its acting to release the crosslinking bonds which hold the microfibrils together, allowing them to slip over each other. Thus, auxin is thought to cause the walls to become plastic, and then osmotic water intake causes the cell to swell like a balloon. The Dutch physiologist A. N. J. Heyn in 1931 tackled this problem by fixing *Avena* (oat) coleoptiles horizontally and applying weights to the tip, as shown in Figure 23-3. That the IAA effect on growth is directly related to its effect on cell wall plasticity is illustrated by the work of the American physiologist R. Cleland, shown in Figure 23-4.

Recently the American physiologist P. M. Ray and his coworkers have developed microscopic techniques for measuring growth or extension of such tissues as coleoptiles on a minute-to-minute basis. They observed that the IAA effect on elongation begins after a lag period of only 8–12 min. More recently it has been found that with certain auxins (the methyl esters of IAA), and even with IAA itself used under the correct conditions, the lag period can be reduced to 1 min or less. It has been argued that this necessarily means that IAA acts directly on cell walls rather than on genetic or biochemical mechanisms that in turn affect enzymes. For example, low pH has also been found to induce cell wall plasticity and extension growth for a short time. Cleland and

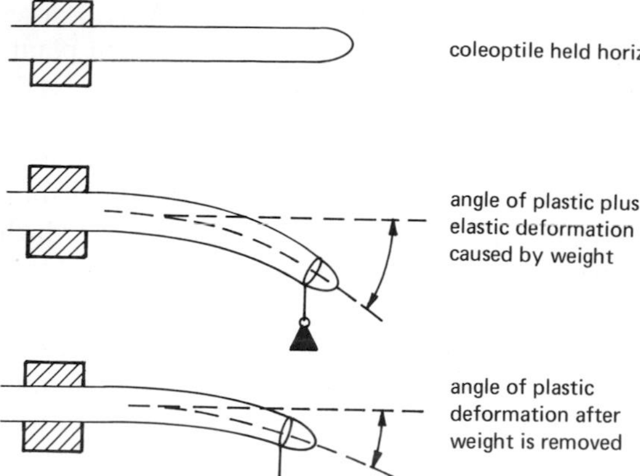

Figure 23-3. A method of measuring elastic and plastic deformation. Plasmolysed coleoptiles are used so that the results are not complicated by the effects of turgor.

his associates, particularly D. L. Rayle, have shown that this effect can also be made to work in cells that have been disrupted by freezing and thawing. They suggested that cell wall plasticity is related to the breaking of acid-labile bonds, not to synthetic reactions (since no synthesis occurs in frozen and thawed cells), and that IAA works by facilitating the release of hydrogen ions.

This concept has been called the "acid growth theory." It depends on observations with many tissues that rapid changes in proton (H^+) concentration follow the application of IAA within minutes. It has been suggested that IAA stimulates a proton/potassium ATPase, a transport mechanism powered by ATP that exchanges H^+ and K^+ ions across a membrane (see Chapter 12, page 298). A considerable body of evidence supports this theory, but it cannot yet be taken as proved.

Although this hypothesis goes far to explain the effects of IAA on cell extension, it is probably not the only mode of IAA action. Evidently it is not concerned with the many effects of IAA on RNA and protein synthesis that have been demonstrated, which will be considered in the next section.

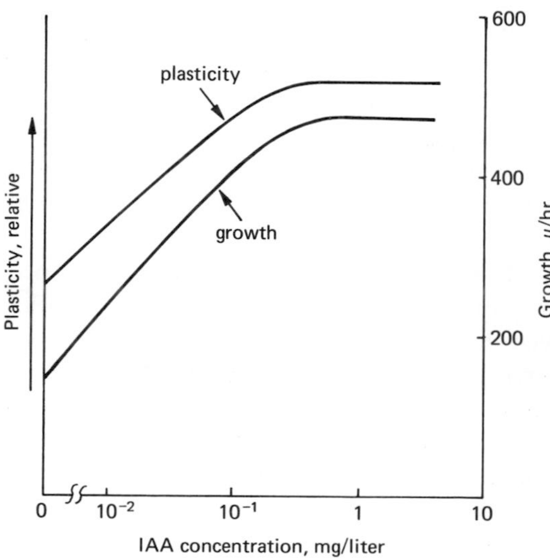

Figure 23-4. A comparison of the effects of IAA on growth and cell wall plasticity in *Avena* coleoptiles. [Adapted from R. Cleland: *Ann. N.Y. Acad. Sci.*, **144:**3–18 (1967).]

Effects on RNA and Protein Synthesis. F. Skoog and his coworkers first noted in 1953 that auxin-increased growth of tissue cultures was associated with increased RNA and DNA synthesis. In particular, it has been observed in many systems that increased RNA synthesis (ribosomal RNA as well as tRNA and mRNA) accompanies growth, indicating that continuous protein synthesis is necessary during growth. This evidence led to the suggestion by the American physiologist J. L. Key that hormones may affect growth by stimulating RNA synthesis and thus the necessary protein synthesis that must accompany growth.

Experiments show that inhibitors such as actinomycin D and cycloheximide, which prevent RNA synthesis and protein formation, respectively, not only inhibit growth by cell elongation but prevent hormone function as well, as shown in Figure 23-5. Key and his colleagues have investigated this problem in depth. It is now well established that concentrations of auxin that promote growth enhance RNA and protein synthesis, whereas inhibitory concentrations reduce their synthesis. If growth is artificially suppressed, as by putting the tissue in osmotic solution strong enough to prevent water uptake, RNA synthesis still takes place. This shows that the RNA synthesis is a primary event, not just the consequence of increased growth. All types of RNA synthesis (mRNA, tRNA, and particularly ribosomal RNA) appear to be enhanced under the influence of auxin.

Several attempts have been made to explain IAA action on RNA synthesis. The Australian physiologist K. T. Glasziou and his coworkers have noted an increase in sugarcane invertase on treatment with IAA. They concluded that the effect of IAA was to stabilize mRNA, not to increase its amount, so that transcription was increased. The effect was quite specific for invertase, however, and did not operate for a peroxidase, which they also investigated. The American physiologist D. J. Armstrong has suggested on theoretical grounds that IAA functions as a signal for polypeptide chain initiation. IAA may act like certain amino acids that are essential for tRNA synthesis in bacteria.

Figure 23-5. Parallel inhibition by actinomycin D of auxin-induced growth and RNA synthesis in soybean hypocotyls. [From J. L. Key, N. M. Barnett, and C. Y. Lin: RNA and protein biosynthesis and the regulation of cell elongation by auxin. *Ann. N.Y. Acad. Sci.*, **144**:49–62 (1967). Used with permission.]

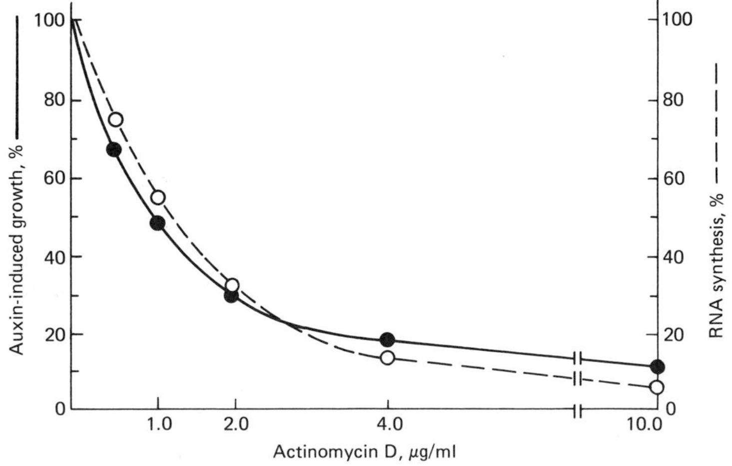

Key and his coworkers, on the basis of their work on IAA synthesis and that of other workers on in vitro RNA synthesis, suggested that auxin may work at the level of a regulator or activator of transcription.

Clearly the action of auxins cannot be explained precisely at present. The evidence suggests that there may be more than one primary site of reaction. One of these, responsible for the extremely fast reactions such as the initial phase of extension growth (and perhaps the transmission of stimuli as in *Mimosa*), may well act directly on the cell wall or plasma membrane as suggested by the acid growth theory. The second, responsible for continuation of growth and the synthesis of proteins that must accompany it, appears to act at the level of RNA and protein synthesis. We should expect that modern biochemical techniques will elucidate these mechanisms in the near future. Even then, however, the whole problem of the regulation of hormones themselves during the development of plants will still have to be worked out.

Structure and Activity. Regardless of exactly how or where auxins exert their effects, they must complex or react in some way with some entity in the cell in order to modulate the cell's chemical activity. Certain known facts about the structures that confer auxin activity on a chemical have to be considered. Initially both a ring structure and a side chain with a carboxyl group were thought to be necessary, as in IAA (Figure 23-1). However, certain of the synthetic auxins lack a ring structure (for example, carboxymethylthiocarbamate, Figure 23-6), and some auxins lack the carboxyl group (for example, indoleethanol). It has been suggested that there are two points of attachment of the auxin molecule to its substrate, the carboxyl group and a partial positive

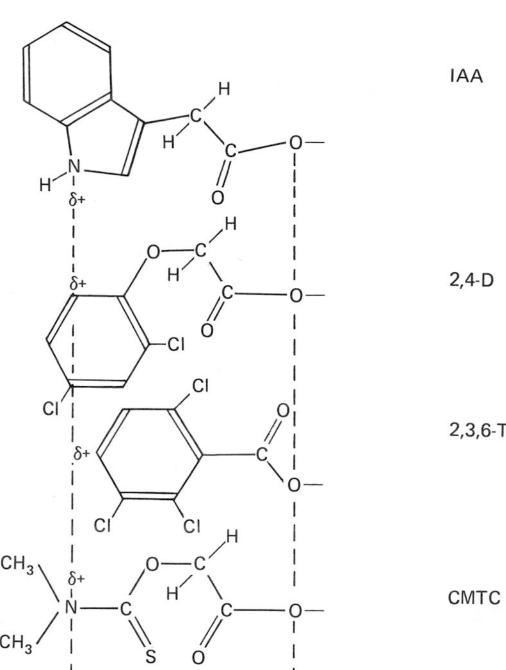

Figure 23-6. Indoleacetic acid 2,4-dichlorophenoxyacetic acid (2,4-D), 2,3,6-trichlorobenzoic acid (2,3,6-T), and carboxymethylthiocarbamate (CMTC); four active auxins, showing the charge-distance relationship between the negative charge on the carboxyl and a fractional positive charge on the nucleus. [From K. V. Thimann: *Ann. Rev. Plant Physiol.*, **14:**1–18 (1963). Used with permission.]

charge on the ring or some other portion of the molecule. These two groups must be separated by a distance of 5.5 Å, as shown in Figure 23-6, in order for the compound to have auxin activity.

Earlier it was thought that covalent bonding occurred at the point of attachment. The known ability of auxin to form peptides (Figure 23-2) suggested that peptide bonding with a protein receptor might be important. However, recent studies with chemical derivatives of auxins, with certain compounds called **antiauxins** (which resemble auxins, but block their action), and with inactive stereoisomers of the active naturally occurring or synthetic auxins show that this view is not correct. It now seems more probable that covalent bonding merely inactivates the auxin.

It now appears that activity results from two, three, or many points of interaction of the auxin and its substrate through weak bonds, van der Waals forces, electrostatic attraction, hydrogen bonds, or the formation of charge-transfer complexes. The very precise relationship of activity and the position of halogen substitutions in the ring of a number of synthetic auxins clearly relates to the steric requirements of multipoint attachment. For example, 2,4-D is an extremely powerful auxin, whereas 2,6-D is inactive. Evidently the substitution at position 6 renders the molecule unable to combine properly with its substrate.

One of the interesting relationships of structure to function is seen in the indole acid series of auxins. IAA is active, indolepropionic acid (IPA) is relatively inactive, indolebutryic acid (IBA) is strongly active, indolepentanoic acid is inactive, and so on. Those acids with an even number of carbons in the side chain are active; an uneven number of carbons in the side chain confers inactivity. The probable explanation is that the side chains are oxidized two carbons at a time by the β-oxidation cycle (Chapter 6, page 122). Thus even-numbered side chains will be converted to IAA, which is the active auxin, but odd-numbered side chains will not. The activity of some other naturally occurring compounds such as indoleacetonitrile and indoleacetaldehyde is due to their conversion in the plant to IAA.

Receptors and Binding Sites. Good progress has been made recently by animal physiologists in defining the target sites or receptor sites for hormones. The problem with plant hormones seems to be much more difficult, and although some data are accumulating, not much progress has been made to date. Part of the trouble is that hormones react with many sites that have nothing to do with their hormonal activity, and thus create real problems for the scientist. Accumulation of kinetic data is also difficult, because of the low concentrations of hormones, their reactivity, and their instability in vivo.

Nevertheless, some advances have been made, and a real breakthrough in the field seems likely in the near future. Plant physiologists in European countries and in America have recently shown that the binding sites of maize coleoptiles are located on the membrane of the rough endoplasmic reticulum. It has been suggested by P. N. Ray, at Stanford University in California, that primary auxin action may occur at endoplasmic reticulum membranes, where it could facilitate the transfer of hydrogen ions (as suggested in the acid growth theory, page 562). Auxins acting in this way might also affect the delivery of secretory proteins and material for cell wall synthesis from the endoplasmic reticulum–Golgi system described on page 45. It is too early yet, however, to elaborate a general principle of auxin action on the basis of these results.

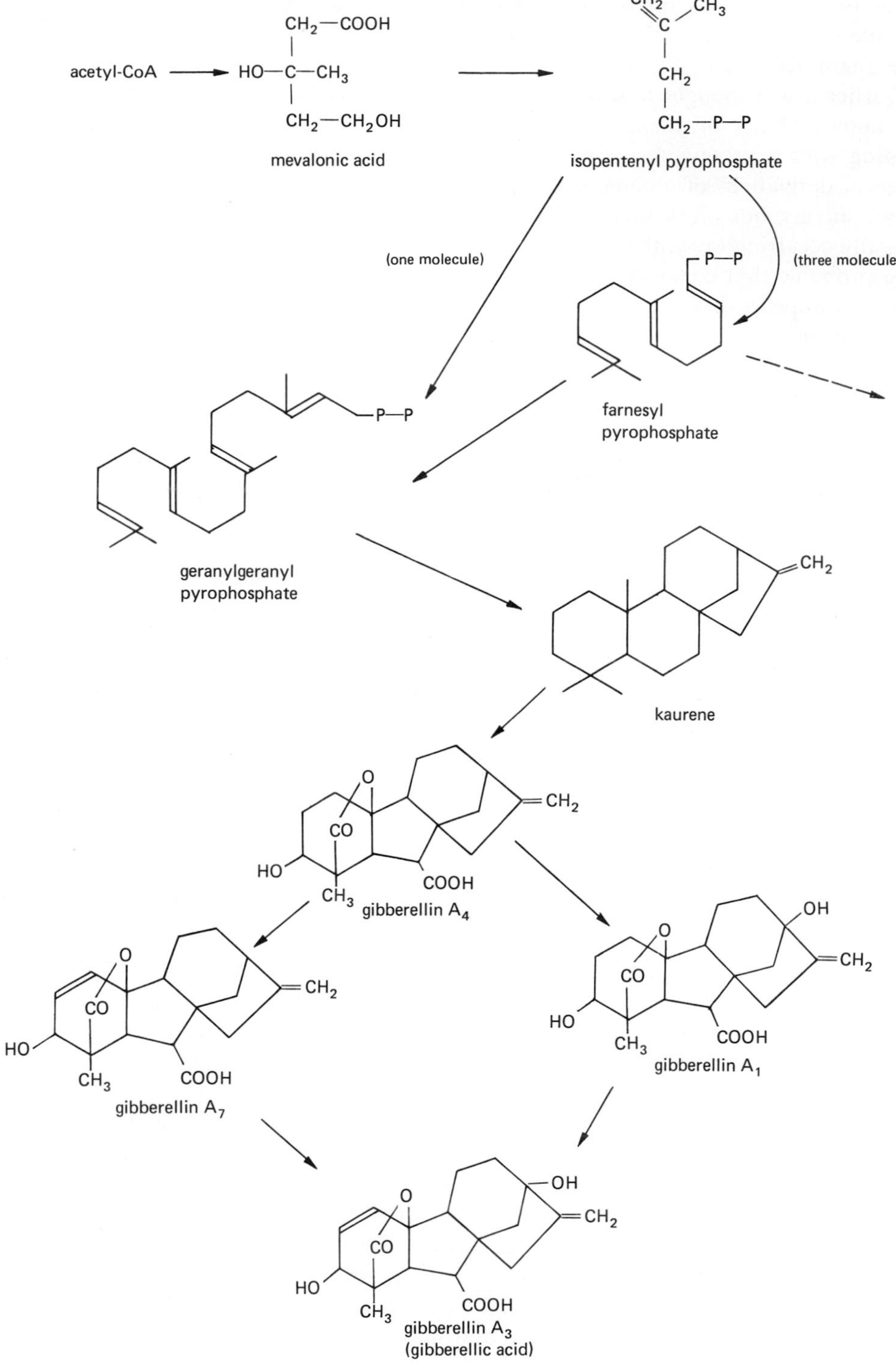

Figure 23-7. Outline of the biosynthetic pathway of some gibberellins.

Gibberellins

Synthesis and Distribution. There are over 40 gibberellins now known; all have the same basic ring structure derived from the isoprenoid synthetic pathway (Figure 23-7, and see also Chapter 9, page 236, and Figures 9-3 and 9-4). The distribution of gibberellins is rather specific, although many tissues contain two or three or several of the known gibberellins. Further, tissues differ in their reactivity to different gibberellins. A compound that causes rapid growth in dwarf corn, for example, may be inactive in promoting the growth of pea internodes and vice versa. The literature on the variety of gibberellins and their distribution is very large, but it does not seem possible at this time to make any clear generalizations that might lead to a greater understanding of gibberellins and their activity. It has been suggested that many of the various gibberellins are in fact intermediates on the pathway of synthesis of one or another active form. Alternative views are possible, as we shall see later.

Gibberellins seem to be synthesized in many parts of the plant, but more especially in actively growing areas such as embryos or meristematic or developing tissues. The relationship between the age of tissue and its gibberellin content is shown in Figure 23-8.

Figure 23-8. Relationship between the amounts of diffusible GA-like substances obtained from leaves and internodes of mature sunflower plants. [Adapted from R. L. Jones and I. D. J. Phillips: *Plant Physiol.*, **41**:1381–86 (1966).]

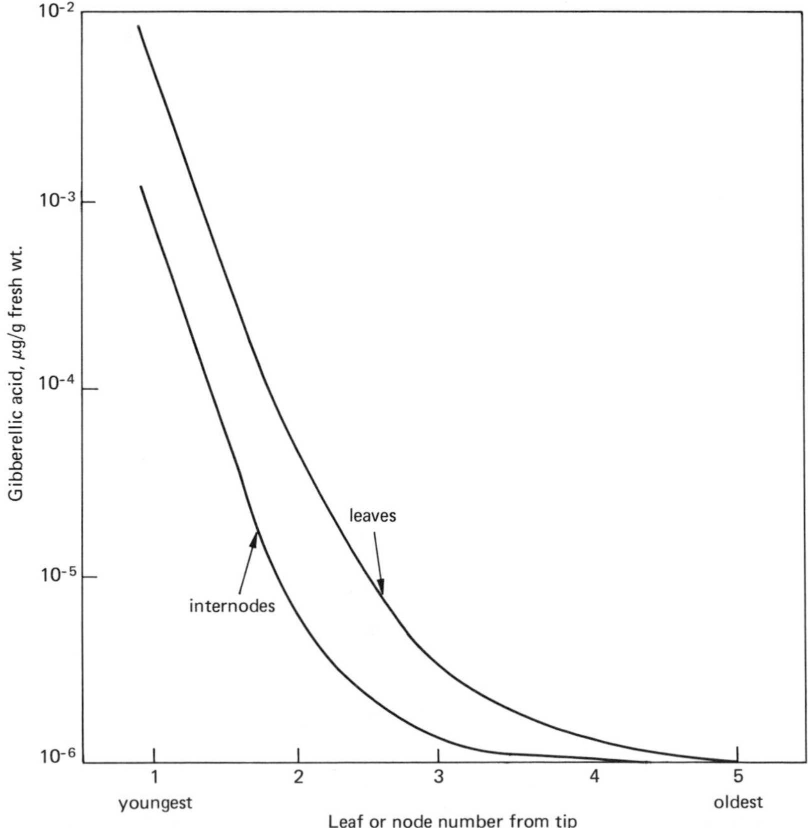

Gibberellins are transported readily about the plant, apparently moving passively in the transport stream of either the phloem or the xylem. A considerable portion of the plant's gibberellins may be bound or compartmented and inactive at any given time. The rapid production of gibberellin that occurs in germinating seeds is suspected to be, in large part, a release of bound gibberellin whose synthesis took place much earlier, perhaps during or soon after the cold treatment that is often necessary prior to germination of seeds. The synthesis of gibberellins is subject to feedback control, the oxidation of kaurene being inhibited by gibberellins.

A number of synthetic compounds are known that prevent gibberellin action, including such growth retardants as **AMO-1618, CCC,** and **phosphon-D.** These compounds appear to act primarily by inhibiting gibberellic acid (GA) synthesis, as well as in other ways. Little is known of the natural inactivation of gibberellins. Abscisic acid (ABA) antagonizes GA effects, and the two substances may have a common precursor (see Figure 23-7 and Figure 22-5, page 535). If this is so, it seems unlikely that they would be present simultaneously in high concentrations in one tissue at one time, so their direct interaction may not be very significant.

ELONGATION. GA produces effects similar to those of IAA on cell and stem elongation, but the effects are not identical. GA works on a number of tissues upon which IAA is ineffective or inhibitory and vice versa. In some tissues, if IAA is first applied, GA has little effect. However, if GA is applied first, then IAA has a greater than usual effect in causing elongation. This suggests that gibberellin action takes place at some site that precedes auxin action in the reaction sequence leading to stimulated growth in length. Because at least some IAA effects seem to involve protein synthesis, GA may act at the level of enzyme induction or DNA transcription (see below).

Lang has shown that GA stimulates cell division at the stem apex as well as stem elongation. However, this may be a subsidiary effect; the effect on cell elongation appears to be the main one. Nucleic acid metabolism is clearly involved; GA speeds up RNA synthesis. In fact, GA increases the synthesis of RNA in isolated nuclei. It does not seem unreasonable to assume that this is related to the mechanism of its action.

FLOWERING. GA applied to many nonvernalized rosette plants will cause bolting and subsequent flower initiation (Chapter 20, page 497). It is not clear yet whether this is a true florigenic activity. Chailakhyan has suggested that GA is one of the two main hormones that together constitute florigen, the flowering stimulus. However, the effect of vernalization is not the same as that of GA; in the latter, flower initiation takes place after extension growth, not before, as is the case in vernalization. It has been suggested that flower initiation is merely a consequence of GA-caused rapid growth and is not really a direct effect of GA. The answer to this question is not apparent yet.

Another interesting effect of gibberellins is that they overcome juvenility in young trees and permit them to flower years before they would normally come to maturity. This phenomenon, which has been extensively investigated by the Canadian physiologist R. P. Pharis, has proved extremely useful in tree breeding by dramatically shortening the generation time in certain forest species.

ENZYME SYNTHESIS. One of the most dramatic effects of GA is its induction of hydrolytic enzymes in the endosperm of germinating barley seeds (Chapter 16, page

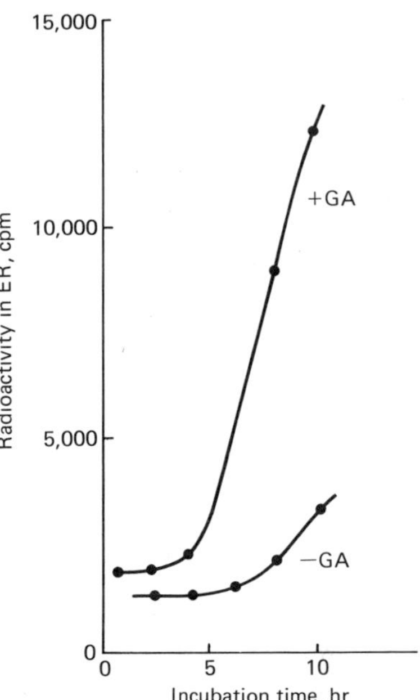

Figure 23-9. Stimulating effect of GA_3 on the incorporation of ^{14}C-choline into endoplasmic reticulum (*ER*) of barley aleurone layers.
Duplicate samples of 40 aleurone layers from barley seeds were treated for various times in buffer or buffer + 1 μM GA, then transferred to the solution of labeled choline for 30 min. The ER was isolated and counted. [Redrawn from W. H. Evans and J. E. Varner: Hormone-controlled synthesis of endoplasmic reticulum in barley aleurone cells: *Proc. Nat. Acad. Sci.*, **68:**1631–33 (1971). Used with permission.]

398). In this system new enzyme proteins are formed, and the GA treatment also stimulates a substantial synthesis of new mRNA. Thus GA acts to uncover or derepress specific genes, which then cause the synthesis of these enzymes. It is assumed, but not as yet proved, that GA acts on the DNA of the nucleus.

Other enzymes are known to be affected by GA in quite different situations. Endoplasmic reticulum formation in the barley aleurone cells has recently been shown to be stimulated four- to eightfold by the addition of GA_3 as shown in Figure 23-9. Glasziou has shown that GA_3 causes an increase in invertase synthesis (though not of peroxidase) in exactly the same manner as IAA. These effects point to a wide variety of action of gibberellins in inducing enzyme synthesis and other developmental steps.

Mechanism of Action. Gibberellins are closely related to steroids, many of which have strong hormonal effects. Indeed, gibberellins have ecdysonelike effects in insects (**ecdysone** is the molting hormone of insects). Extracts of insects, but not ecdysone itself, have weak gibberellinlike effects in plants. The steroids have very specific effects in derepressing genes and thus activating specific enzymes. The large number and wide variety of the chemical forms of steroids are thus probably related to the number and specificity of the molecular sites at which they must react.

It seems possible, then, that the gibberellins act in plants in much the same way. If this view is correct, the large number, the complex distribution, and the specificity of action of the gibberellins may reflect the number and variety of reactions that they control. According to this view, the many known gibberellins are not merely intermediates on the pathways of synthesis of one or a few hormones but are themselves all active or potentially active hormones. This view is hypothetical; the exact mode of action of the gibberellins is not known. We do know that they act to derepress genes and to stimulate RNA synthesis. It seems probable that they are loosely bound to the reaction sites by weak forces, in a manner similar to the auxins.

Cytokinins

Distribution. The discovery of cytokinins stems from the observation that adenine-containing compounds can modify the expression of growth. This led to the discovery of 6-furfurylaminopurine, or **kinetin,** by F. Skoog and C. O. Miller, in a hydrolyzate of herring sperm DNA. All the known cytokinins, both natural and synthetic, are derivatives of adenine (Figure 23-10). This chemical relationship of the cytokinins to a naturally occurring purine which occurs in DNA and RNA caused great excitement among plant biochemists and physiologists when cytokinins were discovered. Strangely, however, the resulting high intensity of research has not provided any clearer insight into the mode of cytokinin action than we have of the other classes of hormone, the gibberellins and auxins.

Cytokinins do not appear to move about the plant as readily as gibberellins or auxins. However, there is unequivocal evidence that cytokinins are made in the roots and translocated to the shoots and leaves of plants. This stems from the discovery by the German physiologist L. E. Engelbrecht that tobacco plants from which the roots were removed senesce rapidly; the presence of roots or the addition of kinetin prevents senescence. Experiments with many different tissue systems have amply confirmed this phenomenon. The hormone appears to be translocated via the xylem; there is some evidence that cytokinin is moved toward the source of auxin, as are other nutrient substances and photosynthetically fixed carbon. However, it should be noted that many

Figure 23-10. Structures of some cytokinins.
Kinetin was initially isolated from animal DNA and does not occur naturally in plants. Benzyladenine is synthetic. Zeatin and IPA are formed naturally in plants.

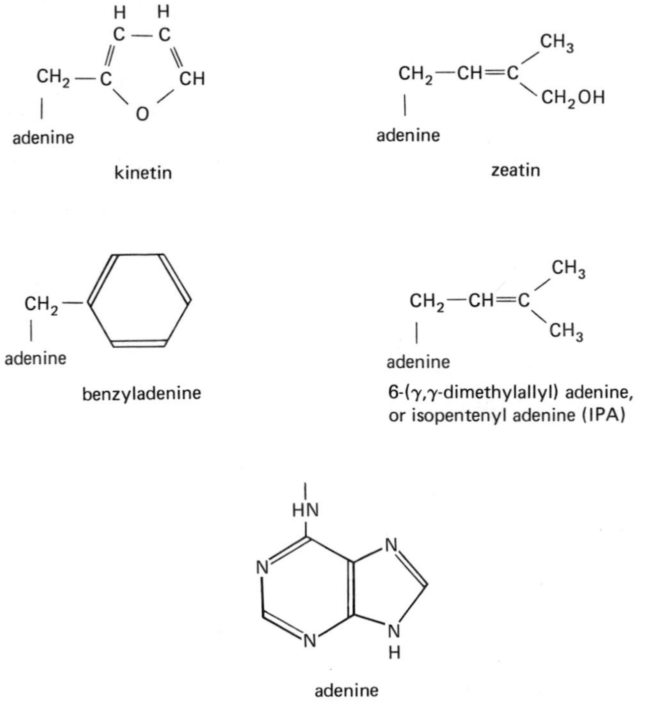

H H
C—C
CH_2—C CH
O
adenine

kinetin

CH_3
CH_2—CH=C
CH_2OH
adenine

zeatin

CH_2—
adenine

benzyladenine

CH_3
CH_2—CH=C
CH_3
adenine

6-(γ,γ-dimethylallyl) adenine, or isopentenyl adenine (IPA)

HN
N N
N N
H

adenine

experiments have shown that when cytokinin is applied to a leaf or tissue, it does not move but stays where it was applied.

Effects. Among the effects of cytokinins are organ formation in cultured tissues, cell enlargement and division, the prevention of senescence, and the induction of flowering under certain circumstances. The Indian morphologist B. M. Johri has shown that cultured endosperm of mistletoe will develop shoots if treated with as little as 8 ppm cytokinin. The work of Skoog and his colleagues has shown that IAA and cytokinins together stimulate organ formation in tobacco tissue culture, the relative amounts of the two hormones determining what sorts of organs will be formed (see Chapter 18, page 429). Bud formation in intact shoots that follows cytokinin application may result from the synthesis of auxins, which seems to be caused by cytokinin application.

The interaction of auxins and cytokinins has other effects as well. IAA induces elongation of pea stem internodes; the addition of cytokinin does not inhibit cell enlargement but induces sideways rather than lengthwise enlargement. The enlargement of mature cells under the influence of cytokinin is also well known. Figure 23-11 depicts a radish seedling, one of whose cotyledons has grown very large as a result of cytokinin-induced cell enlargement. Cytokinins also promote cell division, and the effect occurs with concentrations as low as 5×10^{-11} *M*. The fact that cytokinins act to release buds from apical dominance, as shown in Figure 23-12, may be related to their effects on cell division. This, in turn, may be related to their ability to stimulate auxin production in cells.

Figure 23-11. A radish seedling.
The cotyledon on the right was painted with a solution of the synthetic cytokinin, 6-benzylamino-9-(tetrahydropyran-2-yl) purine (100 mg/liter). The stimulation of cotyledon development is caused mainly by promotion of cell enlargement. [From D. S. Letham: Cytokinins and their relation to other phytohormones. *BioScience,* **19:**309–16 (1969). Used with permission. Redrawn from photograph courtesy Dr. Letham.]

Figure 23-12. A *Nicotiana glauca* plant. Zeatin was applied as a lanolin paste (0.5%) to the lateral bud at the base of the petiole bearing the wire loop. This released the bud with apical dominance. Zeatin paste was then applied over the surface of the resulting lateral shoot. A very thick vigorous shoot with developing lateral buds was formed. The photograph was taken 34 days after the start of the experiment. [From D. S. Letham: Cytokinins and their relation to other phytohormones. *BioScience,* ***19:***309–16 (1969). Used with permission. Photograph courtesy Dr. Letham.]

One other cytokinin effect should be mentioned. IAA can stimulate flowering in a few plants, notably pineapple, and GA causes flowering in unvernalized long-day plants. Similarly, a few instances are known in which cytokinins can induce flowering in short-day plants. M. Kh. Chailakhyan found that cultured shoots of *Perilla* could be induced to form flower primordia by the application of kinetin, and the Indian physiologists S. C. Maheshwari and R. Venkataraman have found that zeatin application will cause plants of the minute duckweed *Wolffia* to flower. Conversely, kinetin inhibits the formation of flower buds in cultured tobacco stem segments.

Prevention of Senescence. The effectiveness of cytokinins in preventing senescence when applied to isolated leaves or leaf discs was discovered early and has been the basis of several bioassays. The overall effect is probably the result of at least two rather distinct cytokinin actions. Cytokinins are known to prevent the formation of hydrolytic enzymes such as nucleases and proteases, thus they interfere with the breakdown of polymers. This prevents the degradative changes which are among the main characteristics of senescence. The other factor that acts to prevent senescence is that cytokinins cause immobilization of nutrients or their transport to cytokinin-treated areas. This phenomenon was investigated by the German physiologists K. Mothes and L. Engelbrecht. They showed that cytokinins not only prevent the loss of nutrients but cause the transport of endogenous or applied nutrients to the site of applied cytokinin.

Figure 23-13. Radioautographs of *Nicotiana rustica* leaves to which a radioactive spot of DL-aminoisobutyric acid-^{14}C has been applied.
A. Young leaf showing almost no spreading of radioactivity because of high natural retention.
B. Mature leaf; radioactivity spreads primarily to the petiole and to the veins.
C. Mature leaf of which the left half has been supplied with kinetin; radioactivity spreads only in the kinetin-treated half.
D. Mature leaf of which the right half has been supplied with kinetin; radioactivity does not spread because of the increased retention. [From K. Mothes: The role of kinetin in plant regulation. *Régulateurs Naturels de la Croissance Végétale*, No. 123, Centre National de la Recherche Scientifique, 1963. Used with permission.]

A typical experiment is shown in Figure 23-13. Younger leaves already contain some cytokinin and tend to lose less of their applied ^{14}C-amino acids than do older leaves. Added cytokinin causes radioactivity to move from its site of application to the cytokinin-treated area. How this transport phenomenon is mediated is not understood. One suggestion is that cytokinin may stimulate the formation of transport protein molecules involved in the active movement of metabolites. Another is that cytokinins stimulate cell division, which in turn causes IAA formation. The IAA so formed may then act to promote translocation to its site of synthesis. Neither theory appears to be strongly supported by hard evidence.

Enzyme Formation. Cytokinins have been shown to stimulate the formation of enzymes in several situations. The German physiologist J. Feierabend has shown that kinetin speeds the formation of photosynthetic enzymes in rye seedlings. However, the effect is nonspecific and quantitative rather than qualitative, and it does not seem likely that this is a specific cytokinin interaction at the gene or enzyme formation level. On the other hand it has been noted that the de novo synthesis of certain enzymes, including tyramine methylpherase in barley seedlings, is stimulated by the addition of kinetin. Similarly alkaloid synthesis is stimulated in lupine roots. Like GA, cytokinin will also stimulate α-amylase formation in germinating barley seedlings, but the effect is not so pronounced.

It appears that cytokinins are not directly involved at the genetic level in any of these effects. That is, the cytokinins are not themselves active as gene derepressors. Their effect seems rather to be on synthetic processes, and many of their side effects may be due to their ability to cause auxin formation in previously quiescent tissue.

Cytokinins as Constituents of RNA. A great deal of excitement followed the discovery in Skoog's laboratory that cytokinins are a normal constituent of RNA. Analysis has revealed that certain tRNAs in many organisms, including plants, yeast, bacteria, and animals, in fact contain a molecule of a cytokinin in the position adjacent to the anticodon, the point where the tRNA is encoded to the mRNA on the ribosome. This structure is shown in Figure 23-14. The cytokinin most frequently found in tRNA to date is isopentenyl adenine (IPA, Figure 23-10), but others, including zeatin (also shown in Figure 23-10), have been detected. The amount of cytokinin in the tRNA is very small, averaging only one cytokinin per 20 tRNA molecules. Only certain tRNAs have a cytokinin component. The one for serine has, as is shown in Figure 23-14, but the tRNAs for alanine, tyrosine, and phenylalanine have not.

Initially it was thought that the presence of cytokinins in tRNA was closely linked with their mode of homone action and would explain it. However, more recent experiments now throw very strong doubts on this idea. When ^{14}C-labeled cytokinins are added to tissues that need cytokinins for growth, they are not incorporated into the tRNA intact. Instead, it has become clear that the characteristic cytokinin side chain is

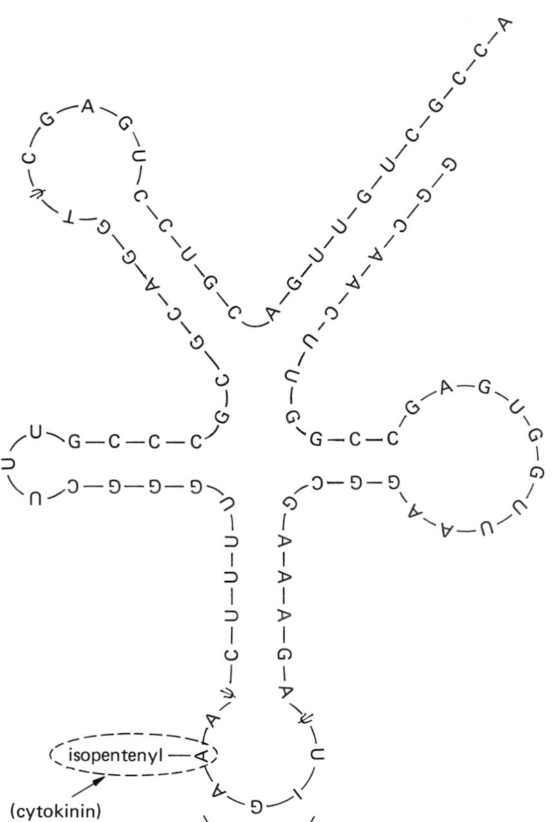

Figure 23-14. Structure of serine tRNA. The cytokinin IPA is next to the anticodon. [Redrawn from A. W. Galston and P. J. Davies: *Science,* **163:**1288 (1969). Copyright 1963 by the American Association for the Advancement of Science.]

attached to an adenine residue already present in the tRNA. It has been suggested that, although cytokinins are necessary for the function of certain tRNAs, this has no relation to their hormone activity. This view is further enhanced by the fact that the zeatin isolated from corn seed RNA is a different isomer from the one found free in the seed, which is presumably responsible for cytokinin activity. Also it has been found that a cytokinin acting on bud formation in moss is only loosely attached. It can be washed out, and bud development is thus arrested. This shows that the cytokinin activity does not reside in molecules of cytokinin that are covalently bonded to the site of action but in those that are only loosely bound to the target.

Cytokinins may be present as breakdown products of tRNA. There may be no connection at all between their role in tRNA and their role as hormones. A final possibility is that their hormone activity is related to their presence in RNA but not a direct consequence of it. One explanation based on this concept will be considered next.

Cytokinin Action. There appear to be rather stringent structural requirements for cytokinin activity. Most known cytokinins have an adenine nucleus with the purine ring intact and with N^6 substituents of moderate size. The only exception to date is diphenylurea and some of its derivatives, originally found in coconut milk by E. Shantz and F. C. Steward. However, these may be a different class of growth substances. The cytokinin is most active if the side chain contains about five carbons, and activity is increased by the presence of a benzene ring or unsaturations in the side chain (see Figure 23-10). There appear to be specific requirements for planarity of the structure, the dimensions of the molecule, and the location of polar groups with respect to the main nucleus. Just how these structural features are related to activity is not known. Presumably, as in the structures of the auxins, these features are important in the formation of weak bonds which bind the cytokinin to its site of action.

The location of cytokinin action is not known yet. Cytokinins have been found to bind to ribosomes, and measurements indicate that rather precisely one molecule of cytokinin binds to each ribosome. Cytokinins act characteristically on enucleate *Acetabularia* cells, so they probably act on cytoplasmic ribosomes. However, the Indian physiologists A. Datta and S. P. Sen have shown that cytokinins will stimulate RNA synthesis in isolated nuclei; presumably, therefore, cytokinins can act on nuclear ribosomes as well. If indeed the hormone action of cytokinin requires their binding to ribosomes, they may act on chloroplast ribosomes also, since they can influence the synthesis of protein molecules within the chloroplast.

The American physiologists J. E. Fox and J. H. Cherry have independently suggested a possible mechanism of action of cytokinins that relates to the presence of cytokinins in tRNAs. They suggest that tRNAs lacking the isopentenyl side chain on the adenine next to the anticodon are not active. Addition of the isopentenyl side chain would thus activate the RNA, but a nuclease is present that can deactivate it by hydrolyzing the side chain. The soluble cytokinins act to protect the tRNA by complexing with this enzyme and inhibiting its action, thus permitting protein synthesis to take place. It must be stressed that this is a hypothetical mechanism; other ideas are possible. However, it seems likely that the solution to this problem must account for the known presence of cytokinins in tRNA, and also for the fact that they have frequently been shown to facilitate and increase the rate of both RNA and protein synthesis. The recent observation that cytokinins regulate polyribosome levels through an effect on protein synthesis at the translational level would be consistent with this mode of action.

Abscisic Acid

Abscisic Acid Effects. Abscisic acid (ABA) is a growth inhibitor, and its primary action seems to be to inhibit gibberellin action and to promote dormancy. One stimulating effect has been noted—like GA, ABA causes an increase in the production of invertase in sugarcane. This effect seems to operate at the level of the translation of the mRNA, at the point of the synthesis of the enzyme protein. However, other stimulatory effects of GA are opposed by ABA. ABA inhibits the GA stimulation of synthesis of endoplasmic reticulum and α-amylase in barley seeds. The effect of ABA seems to be rather specific for α-amylase, inhibiting this enzyme, while the synthesis of others continues. This suggests that ABA specifically inhibits the translation of mRNA for this enzyme but not for others.

The effects of ABA on dormancy and senescence are paralleled by its influence on protein and RNA synthesis in general. It therefore seems likely that much of its dormancy-inducing action is due to these effects.

ABA has an interesting effect on flowering responses: like GA, it initiates flowering in some uninduced plants. But whereas GA causes flowering in long-day plants, the ABA effect is on short-day plants. This may be related to the postulated phytochrome-controlled mechanism whereby an isoprenoid precursor is converted either into GA or into ABA under the influence of long or short days (see Figure 22-5, page 535, and Figure 23-7).

ABA is also the agent that mediates stomatal closure under the effect of drought (see Chapter 14, page 332). The effects of certain pathogenic fungi that cause wilting of infected plants is due to the production by the pathogen of ABA-antagonizing substances that prevent stomatal closure. It is an interesting question how one substance, ABA, mediates both short-term responses such as stomatal closure and long-term effects such as senescence or dormancy. The answer to this question may relate to the compartmenting of hormones or to the modification of the action of one hormone by another. Clearly, more work is needed on this interesting subject.

Abscisic Acid Action. The mechanism of action of ABA appears thus to follow its effect on translation of RNA. ABA does inhibit RNA synthesis, but this may be a secondary effect. If translation is reduced, the synthesis of tRNA normally falls off. ABA does not appear to affect the derepression of DNA but, in situations where mRNA synthesis is not taking place, ABA still inhibits protein synthesis. This is consistent with an effect of ABA at the ribosome level, on translation and protein synthesis, but not at the nuclear level, where mRNA is being formed. As yet no mechanism for the operation of ABA is known; further experimental work is needed before even hypothetical mechanisms can by advanced. Like other hormones, the interaction of ABA with its target site is probably by weak forces and not by covalent bonding.

Ethylene

Ethylene Effects. A real problem in the study of ethylene is the difficulty of separating its effects from those of auxins. It is now clear that IAA causes ethylene production in tissues, and many of the supposed IAA effects are really secondary effects caused by ethylene produced as a result of IAA stimulation. For example, the IAA effect on pineapple flowering may be due to the ethylene so produced. It was early noted that, if pineapples are placed on their side, they flower. It has recently been suggested that the geotropic response is mediated by IAA-induced ethylene formation. Thus the flowering of geotropically stimulated pineapples may be a side effect of the ethylene produced in response to IAA accumulation on the lower side of the plant.

Experiments have shown that the IAA effect on roots is different from that of ethylene, showing that they cause separate reactions. However, application of an auxin causes ethylene production in roots, thus the IAA and ethylene effects are difficult to study independently. In some plants, ethylene production has been held responsible for the characteristic optima of the IAA stimulation of tissues (Chapter 19, page 454, and Figure 16-8). IAA may continue to have a stimulatory effect by itself at very high concentrations, but the inhibitory effect of ethylene, produced as a result of IAA action, overcomes this stimulation and eventually causes inhibition when a critical level of IAA is reached.

Ethylene has a very wide range of effects, from strongly stimulatory to powerfully inhibitory. Ethylene is usually classed as an inhibitory hormone. However, though its full range of action is not known as yet, its known regulatory activities are so varied that it defies easy classification. Its effects on fruit ripening and leaf abscission appear to be due to the stimulation of synthetic processes required for the development of senescent characteristics or the formation of the abscission zone. Thus the inhibitory effects of ethylene may be in large part due to a truly stimulatory effect, operating on degradative processes. The inhibition of the ethylene effect (for example, by carbon dioxide) appears to slow down the production of degradative enzymes.

Mechanism of Action. Several of the known ethylene effects have similar saturation levels, and the same concentration (0.1–0.2 ppm) is required for half-maximum response. This has suggested to the American physiologists S. P. and A. E. Burg that there is a common reaction site for several of the important ethylene effects. Carbon dioxide competitively inhibits ethylene action in many of its responses, including those that follow the application of IAA (for example, geotropism in roots). It seems likely that carbon dioxide and ethylene react at a common binding site. What sort of reaction is inhibited by carbon dioxide, or how, is not clear. Presumably it is an enzyme reaction.

The location of the primary ethylene reaction site is not known. Ethylene is very lipid soluble, so it may be associated with the lipid portion of cell membranes. It has been found that ethylene stimulates the excretion of α-amylase from aleurone cells of barley seeds, although it does not stimulate its production. Ethylene does not seem to have a very pronounced effect on any biochemical reactions, but it may affect membrane permeability or possibly stimulate the activity of permease systems.

Other Substances That Influence Growth

A number of substances not now considered as hormones nevertheless influence plant growth and development, some profoundly. The possibility that the phytochrome effects might be mediated through the agency of acetylcholine has been mentioned (Chapter 20, page 488). Most of the naturally occurring cholinergic drugs are of plant origin (for example, pilocarpine, muscarine, nicotine, *d*-tubocurarine, atropine, eserine, solanine, scopolamine, and arecoline). It is within the realm of possibility that these and other

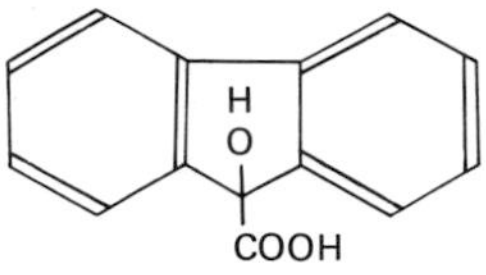

H
O
COOH

flurenol
(9-hydroxyfluorene-9-carboxylic acid)

H
O
Cl
COOH

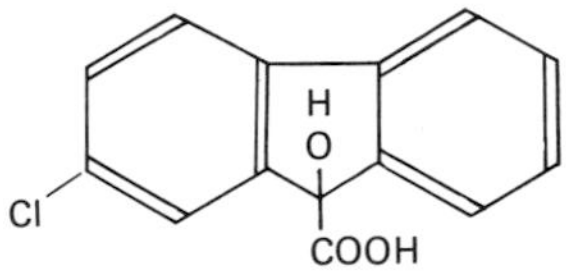

chlorflurenol
(2-chloro-9-hydroxyfluorene-9-carboxylic acid)

Figure 23-15. Structure of two morphactins. Compare their structure with that of gibberellins shown in Figure 23-7.

physiologically active compounds (for example, alkaloids, terpenes, possibly tannins, and so on) may play some role in the regulation of growth or the mediation of effects which are already known.

Much has been written recently about 3,5-cyclic AMP, which has powerful regulating capacities in animal metabolism. So far, it has not been convincingly demonstrated in higher plants. When applied to plants it may elicit a variety of effects, but there is little evidence to suggest that it is a naturally occurring growth regulator in plants.

An interesting new group of synthetic growth regulators has recently been manufactured, largely in the laboratory of the German chemist G. Schneider. These compounds, called **morphactins,** have a fluorene ring with various substitutions on the fluorene ring structure (Figure 23-15). These compounds are inhibitory in specific ways and work over a very wide range of concentrations—much wider than presently known

Figure 23-16. Comparison of the range of active concentration of two morphactins with three other synthetic growth substances.
Width of bar is proportional to growth-regulating effect. The substances were applied twice in the form of droplets (0.02 ml per plant) to seedlings of *Galium aparine*. [From data of G. Schneider: *Ann. Rev. Plant Physiol.*, **21**:499 (1970).]

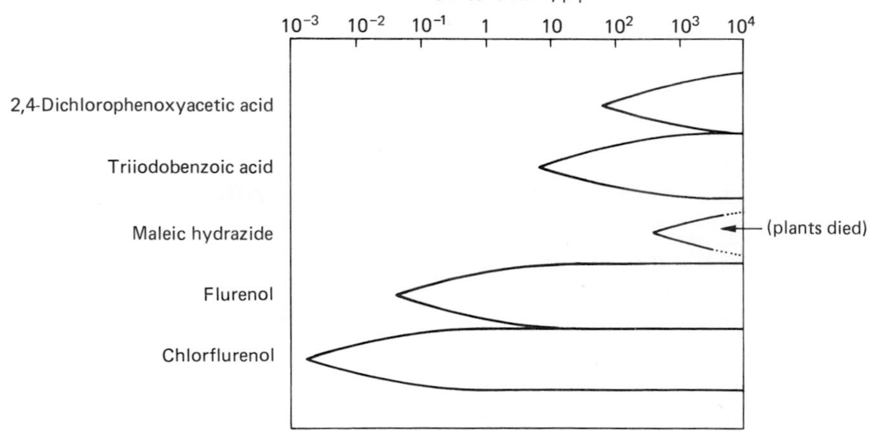

growth regulators or specific herbicides (Figure 23-16). They appear to inhibit the transport of IAA, thus cause a number of growth anomalies including the abolition of tropic responses and apical dominance. Morphactins also inhibit seed germination, and they appear to antagonize GA in that they prevent bolting of vernalized rosette plants. This class of compounds appear to be very interesting for the control of plant growth because they are not very toxic, they exhibit a high degree of selectivity, and they are not too persistent in the soil.

Interaction of Hormones

The Canadian development biologist R. H. Hall, in his article on cytokinins cited at the end of this chapter, suggests that scientists should put more effort to examining the totality of development and worry less about the causes of specific events. The key to developmental patterns and events is the whole of the organism, its previous experience, and the sum of all the factors that influence each other as well as the future behavior of the organism itself. In a study that illustrates this principle the American physiologist P. V. Ammirato examined the interacting effects of ABA, the cytokinin zeatin, and GA on suspension cultures of cells from caraway (*Carum carvi*) stems that were induced to undergo embryogenesis. The results are shown in Figure 23-17.

In the absence of added hormones, small, aberrant embryos developed; only a few were normal (Figure 23-17A). If zeatin or GA was added, the embryos grew faster but were quite abnormal (Figures 23-17B and C). ABA caused many of the embryos to appear normal, but they were quite small (Figure 23-17D). ABA reduced the proportion of deformities when added with zeatin or GA (Figure 23-17E and F). GA and zeatin together caused massive deformities (Figures 23-17G), but all three factors together led to the formation of a large number of normal embryos (Figure 23-17H). Clearly each hormone by itself causes some degree of disruptive abnormality or retardation, yet the correct balance of these factors leads to the orderly development of normal embryos. It is the balance or interaction of the hormones, rather than a summation of their individual actions, that must provide the key to development.

Summary of Hormone Actions

Auxins

Organ formation (interacts with cytokinins)
Tissue organization (interacts with other factors)
Stimulation of cell division (interacts with cytokinin)
Cell elongation } through stimulation of proton secretion
Cell wall relaxation }
RNA and protein synthesis
Direction of translocation
Enzyme effects
Ethylene production
Tropic and nastic responses (perhaps sometimes due to ethylene)
Apical dominance
Prevents abscission

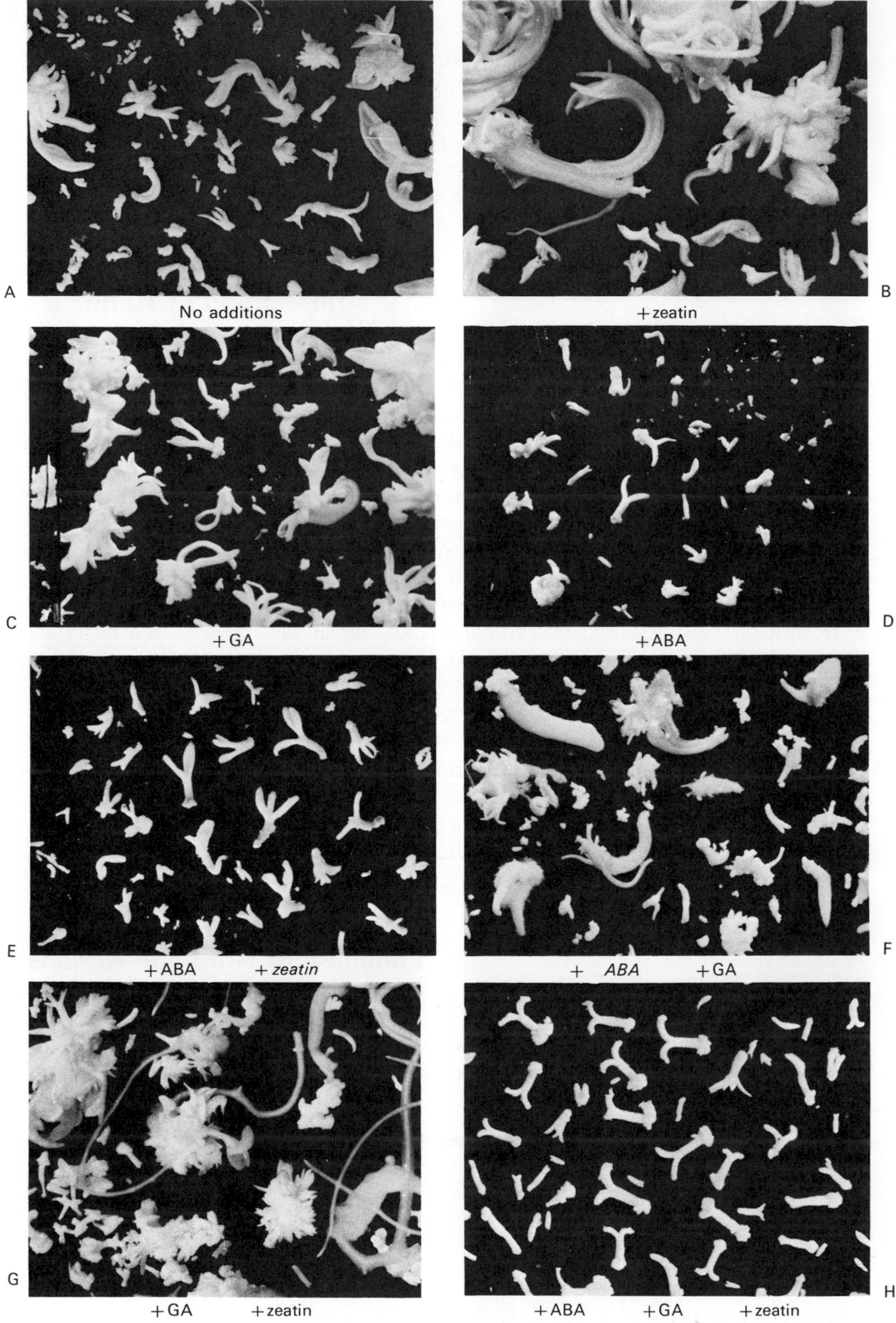
A
No additions
B
+zeatin
C
+GA
D
+ABA
E
+ABA +*zeatin*
F
+ *ABA* +GA
G
+GA +zeatin
H
+ABA +GA +zeatin

Figure 23-17 (opposite). Development of somatic embryos from caraway cells as affected by the addition of hormones. **A.** No additions. **B.** 10^{-6} M zeatin. **C.** 10^{-6} M GA_3. **D.** 10^{-6} M ABA. **E.** 10^{-7} M ABA + zeatin. **F.** 10^{-6} M ABA + GA_3 **G.** 10^{-7} M GA + zeatin. **H.** 10^{-7} M ABA + GA_3 + zeatin. [From P. V. Ammirato: Hormonal control of somatic embryo development from cultured cells of caraway. *Plant Physiol.,* **59:**579–86 (1977). Used with permission. Photographs kindly supplied by Dr. Ammirato.]

Gibberellins

Cell elongation (not the same site as auxins)
Cell division
Enzyme induction
Flowering (long-day plants)
Overcomes dormancy (antagonizes ABA)
Inhibits organ formation
Precocious flowering of trees

Cytokinins

Cell division (induction as well as promotion—interacts with auxins)
Cell enlargement
Organ formation (interacts with auxins)
Overcomes dormancy
Releases apical dominance
Prevents senescence
Mobilization of nutrients
Regulates polyribosomes

Abscisic Acid

Dormancy
Gibberellin antagonist
Flowering (short-day plants)
Abscission
Stomatal closure
Control of embryo development (with cytokinins and GA)

Ethylene

(Auxin causes its production)
Epinasty
Geotropism
Ripening of fruits
Senescence
Abscission

Hypothetical Compounds That Cause Flowering

Florigen
Vernalin
Anthesin

Additional Reading

See list for Chapter 16 (page 406).

Abeles, F. B.: Biosynthesis and mechanism of action of ethylene. *Ann. Rev. Plant Physiol.*, **23:**259–92(1972).

Davies, P. J.: Current theories on the mode of action of auxin. *Botan. Rev.*, **39:**139–71(1973).

Evans, M. L.: Rapid responses to plant hormones. *Ann. Rev. Plant Physiol.*, **25:**195–223(1974).

Goldsmith, M. H. M.: The polar transport of auxin. *Ann. Rev. Plant Physiol.*, **28:**439–78(1978).

Hall, R. H.: Cytokinins as a probe of developmental processes. *Ann. Rev. Plant Physiol.*, **24:**415–44(1973).

Jones, R. L.: Gibberellins: their physiological role. *Ann. Rev. Plant Physiol.*, **24:**571–98(1973).

Kende, H., and **G. Gardner:** Hormone binding in plants. *Ann. Rev. Plant Physiol.*, **27:**267–90(1976).

Key, J. L.: Hormones and nucleic acid metabolism. *Ann. Rev. Plant Physiol.*, **20:**449–74(1969).

Marme, D.: Phytochrome: membranes as possible sites of primary action. *Ann. Rev. Plant Physiol.*, **28:**173–98(1977).

Milborrow, B. V.: The Chemistry and physiology of abscisic acid. *Ann. Rev. Plant Physiol.*, **25:**259–307(1974).

Phillips, I. J. D.: *The Biochemistry and Physiology of Plant Growth Hormones.* McGraw-Hill Book Co., New York, 1971.

Schneider, E. A., and **F. Wightman:** Metabolism of auxin in higher plants. *Ann. Rev. Plant Physiol.*, **25:**487–513(1974).

Schopfer, P.: Phytochrome control of enzymes. *Ann. Rev. Plant Physiol.*, **28:**223–52(1977).

Sheldrake, A. R.: The production of hormones in higher plants. *Biol. Rev.*, **48:**509–59(1973).

Steward, F. C. (ed.): *Plant Physiology: A Treatise,* Vol. VIB. Academic Press, New York, 1972.

Thimann, K. V.: *Hormone Action in the Whole Life of Plants.* The University of Massachusetts Press, Amherst, 1977.

Torrey, J. G.: Root hormones and plant growth. *Ann. Rev. Plant Physiol.*, **27:**435–59(1976).

SECTION V

Physiology of Special Organisms

24 Tree Physiology

Special Characteristics of Trees

Trees have long been regarded by nonbotanists as the apex of plant development because of their obvious characteristics of size, longevity, and widespread distribution. Botanically, trees differ from other plants only in degree, not in kind. But as a result of their large size, slow development, and long maturity (all interrelated factors), they have reached a high degree of specialization in certain directions. Special solutions have evolved for the particular problems associated with their characteristic habit of growth. Much of the physiology of trees is common to all plants, and certain of the problems of plant physiology best exemplified by trees have been examined in earlier chapters. In this chapter we shall consider only certain aspects of plant physiology that are peculiar to trees, and which have not been treated elsewhere in this book.

Assimilation

The processes of photosynthesis and respiration in trees are like those of other plants. However, their overall relationship in the plant is complicated by the large mass of nonphotosynthetic tissue in trees. This means that, compared with herbaceous plants, the photosynthetic tissue of trees must support a relatively larger mass of nonproductive, respiring tissue. Consequently the relative effects of temperature, water supply, and other factors on photosynthesis and respiration are of greater significance in trees. The relative effects of temperature on photosynthesis and respiration in the leaves of a typical boreal forest tree are shown diagrammatically in Figure 24-1. Clearly, although the rate of *gross* photosynthesis is close to maximal over the range 0–40° C, the great increase in respiration at higher temperatures has a marked effect on *net* photosynthesis, which decreases abruptly above 20° C.

The picture is further complicated by the fact that the greater mass of respiring tissue of a tree is in the root, stem, and branches. The most active tissue in the root and stem is the cambium, which is roughly porportional in size to a square function of the height of the tree. The mass of metabolically inactive wood is, however, a cubic function of the height of the tree. Thus during early growth, when height and diameter are increasing rapidly, a tree becomes more efficient in terms of the ratio of photosynthesis to respira-

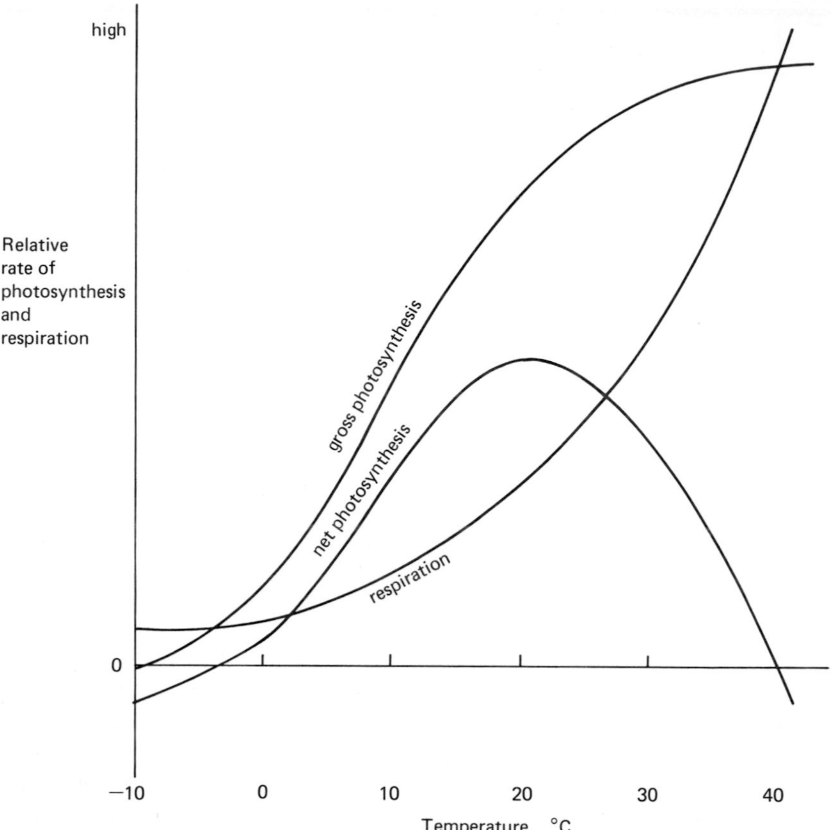

Figure 24-1. Diagram illustrating the relative rates of photosynthesis and respiration in leaves of a boreal forest tree.

tion. This difference is reflected in the data for 10- and 40-year-old trees shown in Table 24-1. As the tree gets older, the proportion of nonphotosynthetic tissues increases without concomitant increase in leaf area. Thus the proportion of photosynthetically fixed carbon available for growth decreases, as is shown by the differences between 40- and 90-year-old trees shown in Table 24-1. Therefore, the tree's sensitivity to environ-

Table 24-1. Loss and retention of carbon in beech trees

Percentage of total carbon fixed in photosynthesis lost by	*Age of tree, years*		
	10	*40*	*90*
Leaf respiration	22	19	22
Root, stem, branch respiration	21	22	26
Leaf loss	16	12	15
Root and branch loss	4	5	6
Percentage of total carbon utilized for growth	37	42	31

SOURCE: Data recalculated from C. M. Möller, M. D. Müller, and J. Nielsen: *Dan. J. Forest.*, **21**:327 (1954).

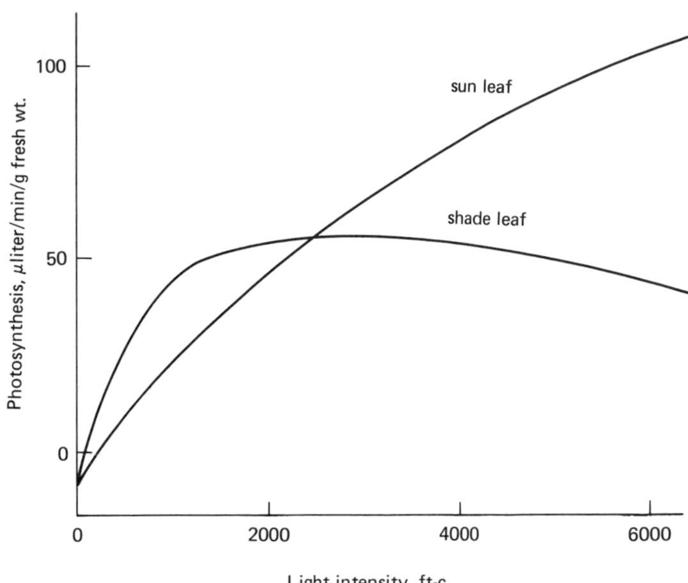

Figure 24-2. Photosynthesis of sun and shade leaves of a maple tree as affected by light intensity. [Data from plant-physiology student lab reports, University of Toronto.]

mental conditions that differentially affect photosynthesis and respiration varies with its size and age.

Another consequence of the larger size of trees is that their upper leaves shade the lower ones to a much greater extent than usually occurs in herbaceous plants. Thus whereas many species of herbaceous plants are specialized for shade tolerance or as sun plants, the leaves in an individual tree may show wide variation in their degree of anatomical (see Chapter 14, page 326) and physiological specialization (see Figure 24-2). For example, lower leaves may reach their maximum photosynthetic rate at only 30 percent of full sunlight, whereas leaves in the crown may continue to increase in photosynthetic rate up to the light intensity of full summer sunlight.

Recently scientists have been showing real concern over the rising atmospheric levels of CO_2. This trend started in the nineteenth century because of the increasing industrial use of fossil fuels, but it has accelerated dramatically in recent decades. The air balance of CO_2 is largely maintained by photosynthesis, and it is now becoming apparent that forests are the largest and most active biological sink for CO_2. Much of the recent rise in atmospheric CO_2 is probably due to the nineteenth century destruction of the world's great forests. The consequence of this act of civilization on the weather patterns of the world will probably be severe and possibly calamitous. Photosynthesis is important to people as well as to trees!

Wood Formation

Wood is not homogenous but contains numerous gradients and discontinuities of cell type. The most obvious of these are the annual rings. **Spring wood** or **earlywood** is made up of larger, thin-walled cells, and these grade into **summer wood** or **latewood** consisting of thicker walled, smaller cells. In ring-porous hardwoods, the vessels may form a conspicuous layer in the early spring wood. Since the quality of wood for commercial purposes depends upon its density and strength, which in turn depend on wall thickness, cell diameter, and cell length (beside the inherent mechanical properties of the raw material itself), tree physiologists have carefully examined the factors governing the deposition of wood cells.

Hormones. Cambial activity and growth are regulated by hormones. If IAA is applied to a tree, earlywood is produced adjacent to the site of application. The auxin antagonist 2,3,5-triiodobenzoic acid (TIBA) has the opposite effect; its application results in the formation of latewood. Hormones produced in the shoots as they break dormancy in the spring appear to initiate cambial activity as well as regulate the kind of wood cell produced. Auxins diffuse or are transported from the apex of shoots, so the auxin concentration tends to be higher near the top of the tree and lower near the base. As a result, latewood formation commences at the base of the tree and spreads upward as the season progresses.

The relationship between gibberellic acid and auxin appears to govern the relative amounts of xylem and phloem made. Increased auxin and reduced gibberellic acid tend to result in xylem cell production, whereas a relative increase in gibberellic acid compared with auxin results in phloem cell production.

Photoperiod. Photoperiodism affects the pattern of growth of the leaves and shoots, and this affects wood formation because of the different levels of hormone production associated with different rates and kinds of growth. Spring wood production corresponds with extension growth of buds and shoots, which is photoperiodically regulated. As this ceases, following natural or artificial changes in photoperiod, summer wood formation begins. The response apparently depends on the foliage, which perceives and reacts to photoperiod, not on the trunk or buds of the tree.

Water. Water supply affects leaf growth, and this, through the above-mentioned hormone effect, regulates wood type. Generally, rapid leaf growth is associated with spring wood formation. In periods of drought, when growth decreases, summer wood is formed. Severe drought followed by a period of ample water supply may cause the formation of **false rings,** in which a graduation occurs from earlywood to latewood to earlywood to latewood during one season.

Temperature. Temperature seems to influence primarily cell length. In gymnosperms, increased temperature causes increased cell length. However, this is apparently not related to photosynthesis, because night temperature has even more influence than day temperature. The temperature effect appears to operate directly upon cells of the trunk, not through the leaves or apical meristems.

Assimilation. Assimilation is important because it provides the carbon from which the wood cells are made. Groups of leaves tend to supply carbon to some specific areas of wood in the stem or branch beneath them, so that partially defoliated twigs develop asymmetrically. However, the most important relationship here is that wall thickness is directly related to net assimilation. Thus, although the type of wood cells (early or late) is determined by factors governing hormone production in leaves, the cell length is affected by temperature, and wall thickness by net assimilation or photosynthetic rate.

Reaction Wood and Orientation Movement. The trunks of most trees are rather precisely oriented vertically with respect to the Earth's gravitational field. If a tree is reoriented artificially, several things happen. The apical bud will make the usual geotropic response by growing upward. The main stem or branch will move somewhat toward the vertical, although the movement may not be more than a few degrees. This is called **orientation movement.** At the same time a characteristically different type of new wood is formed on the upper side (in angiosperms) or lower side (in gymnosperms) of the stem. This is known as **reaction wood.**

Reaction wood forms under the influence of gravity. In angiosperms, more cambial activity occurs on the upper side of stems placed off the vertical, resulting in the formation of so-called **tension wood.** This has fewer and small vessels and, generally, more cellulose and less lignin than normal wood. In gymnosperms, more cambial activity occurs on the lower side making **compression wood.** In compression wood, the tracheids appear more rounded in section and have more lignin and less cellulose in their walls. Orientation movement and the formation of reaction wood are illustrated in Figure 24-3.

The fact that reaction wood forms under the influence of gravity and not as the direct result of tension or compression of the wood, as was previously thought, is illustrated by the experiment shown in Figure 23-4. If the trunk of a sapling is bent into a loop, reaction wood forms on the upper side of the loop only (if the tree is an angiosperm; in gymnosperms the reaction wood forms on the lower side). If tension caused the formation of the reaction wood, it would form entirely around the loop. Instead, the system responds to gravity, and the reaction wood forms on the outside of the loop at the top and on the inside of the loop at the bottom, as illustrated.

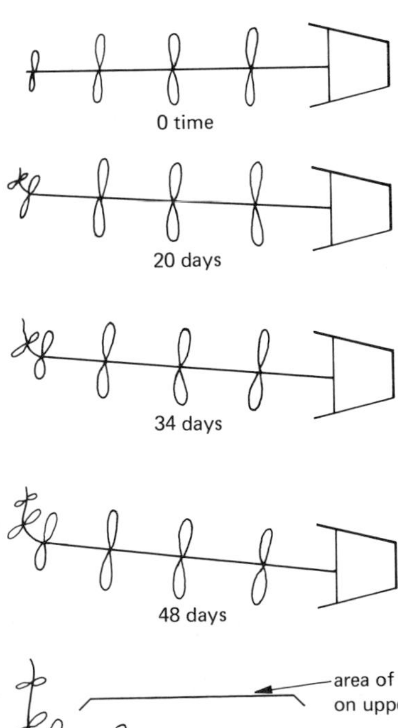

Figure 24-3. Orientation movement of a seedling of *Tristania conferata* placed horizontally. [Redrawn and adapted from A. B. Wardrop: Reaction anatomy of arborescent angiosperms. In M. H. Zimmermann (ed.): *The Formation of Wood in Forest Trees*. Academic Press, New York, 1964, p. 406.]

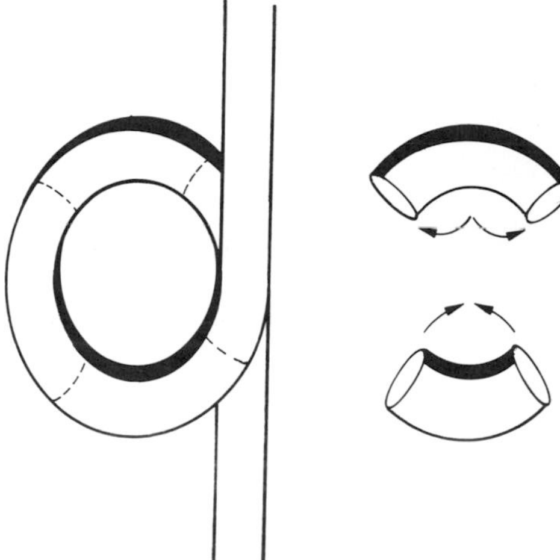

Figure 24-4. Location of reaction wood in an angiosperm tree trunk bent into a circle, and direction of bending when the circle was cut into pieces as shown. [Redrawn from A. B. Wardrop: Reaction anatomy of arborescent angiosperms. In M. H. Zimmermann (ed.): *The Formation of Wood in Forest Trees*. Academic Press, New York, 1964, p. 407.]

The experiment in Figure 24-4 also shows that the formation of reaction wood is responsible for orientation movements, which tend to reorient stems that have bent from the perpendicular. If the circle of wood is cut into pieces as illustrated, the pieces bend in different directions relative to the tree trunk but in the same direction (upward) relative to the gravitational field. The differences in radial growth in the disoriented branch resulting from reaction wood formation create an elliptical cross section, with excess wood forming on the upper side in angiosperms. This results in internal tensions in the branch that cause it to bend upward, toward the side in which the tension wood is formed. In gymnosperms the reaction is opposite, but the formation of compression wood on the lower side has the same result, bending the branch upward.

The mechanism of reaction wood formation is thought to be mediated by auxin, which tends to move to the lower side of branches by the normal gravitational response (see Chapter 19, page 454). Higher concentrations of auxin on the lower side of a stem appear to result in the formation of compression wood in gymnosperms. The situation in angiosperms is less clear. In some earlier experiments, added auxin seemed to stimulate the formation of reaction wood. Also, when auxin is applied to the lower side of a horizontal stem, cambial activity increases, not on the lower side immediately next to the point of application, but on the upper side farther away. This suggested that auxin moves to the upper side of the angiosperm branch or trunk and there exerts its influence to cause increased cambial activity resulting in tension wood formation. However, several workers have reported finding more auxin on the lower side of horizontally placed angiosperm stems. Moreover, recent experiments have shown that added auxin actually prevents the formation of reaction wood in inclined stems of poplar trees, and the application of the antiauxin TIBA causes tension wood formation in upright stems of white elm. It thus appears more likely that the formation of tension wood on the upper side of angiosperm stems is associated with a decrease in the normal supply or concentration of auxin.

The mechanism by which decreased or increased auxin mediates the formation of tension or compression wood is not clear. It has been suggested by the American physiologist K. V. Thimann that the relationship between peroxidase enzymes, which oxidize auxin and are also involved in the formation of lignin precursors, and auxin itself may regulate reaction wood formation. Auxin inhibits polyphenol polymerization, hence lignin formation, but is itself oxidized by peroxidase. However, no clear relationship between the distribution of polyphenol oxidase activity and reaction wood formation has been shown. Other factors may also be involved.

Form

Crown Shape. Trees take many characteristic shapes. Some result from strong apical dominance, as in pines, others because of a strict angle of branching, as in spruce or poplar (plagiotropism, see below), and still others because of a characteristic determinate growth of many apices resulting in a crown, as in beech or oak. The mechanisms of apical dominance and plagiotropism are dealt with elsewhere (pages 462 and 397). The characteristic crown shape may well be the result of physical as well as physiological factors. The following argument has been suggested: As a tree grows, the crown retains a nearly constant area and, hence, a nearly constant rate of photosynthesis. However, the trunk, which has substantial respiratory activity and thus consumes sugars produced in photosynthesis, increases continuously in size as the tree ages. As a result, less sugar is available to the roots which grow less vigorously and absorb less water and nutrients. This results in decreased leader growth, decreased apical dominance, more lateral growth, and a characteristic flat top. However, it is equally possible that upward growth may be limited by the internal water potential, resulting in decreased growth of the highest leaders where the water potential is lowest.

Plagiotropism. The tendency of branches to maintain a constant specific angle throughout growth is called plagiotropism. This may be related to the balance of reaction wood formation in the branches. Plagiotropism appears to be controlled by the leader because, if the leader or a branch at a crotch is removed, the adjacent branch usually reorients itself by orientation movement and geotropic response. However, once plagiotropism is established in some species of trees, it is apparently retained without further influence from the leader, since the branch continues its characteristic angle of growth even after the leader has grown a great distance beyond. In certain species plagiotropism is so firmly established that a rooted branch will continue to grow at the angle established in the tree from which it was removed. How such an environmentally determined control is impressed semipermanently or permanently upon the genome of the branch apex is not clear.

Consequences of Perennial Growth

The three most obvious consequences of perennial growth are

1. The maintenance of a large mass of nonliving tissue.
2. The need for storage and dormancy capabilities for winter.
3. The need for redistribution of nutrients prior to leaf fall in deciduous trees to avoid excessive annual loss.

Metabolism of Perennial Tissue. The metabolic activities of tissues in a tree trunk are shown in Table 24-2. Although the bark and cambium tissues are metabolically very

Table 24-2. Heat-labile respiration of various tissues in the trunks of black ash trees

Tissue	*Respiration, mm^3 O_2/(hr)(g fresh wt.)*
Phloem	87.6
Cambium	181
Sapwood	16.6
Heartwood	0.3

SOURCE: Data recalculated from R. H. Goodwin and D. R. Goddard: *Am. J. Bot.,* **27:**234 (1940).

active, the sapwood is much less so and the heartwood is almost without detectable metabolic activity. Thus support for the metabolism of a trunk is largely required only in the narrow band of tissue around the surface of the trunk, which does not increase each year in proportion to the mass but only to the area of the trunk. If this were not so and the whole trunk were metabolically active, the growth of a tree beyond a small size would be impossible.

DORMANCY. Dormancy and storage problems of trees have received much attention. The storage of carbon reserves appears to be largely regulated by temperature. Many trees store considerable amounts of lipid, whereas others store starch or both. The low temperatures encountered by the stem tend to induce fat storage; roots, which do not encounter such extremes of temperature, store mainly starch. In the spring, rising temperatures cause mobilization of the storage reserves. Prior to the start of spring growth, large amounts of sucrose may be synthesized from stored compounds and released into the xylem. Simple changes of temperature are not sufficient for this reaction; what is required is an alternation of warm days and freezing nights. Sap flow in maple trees, for example, continues only as long as these conditions prevail.

The condition of dormancy in trees and its relationship to survival has been much studied. Different parts of trees undergo different degrees of dormancy at different times caused by different stimuli. Thus the dormancy of a tree involves a number of separate and unconnected phenomena. The buds of many trees have a winter chilling requirement for dormancy, whereas the roots and perhaps also the cambium do not. During the winter many metabolic and developmental processes go on in the tree, and dormancy is by no means complete. Also, the dormancy of various parts of the tree seems to be at least partially independent, and different parts of the tree may enter or break dormancy independently. It seems likely that different mechanisms, different regulators, and presumably different genes may be involved in dormancy of different parts of the plant.

Patterns of dormancy are closely related to the environmental pressures that the tree experiences. Short days and decreasing temperatures trigger the cessation of growth, the fall of leaves, and the onset of dormancy. Dormancy is broken by the reversal of these trends, that is, by increasing temperatures and lengthening days. However, mechanisms have evolved to safeguard the tree from the danger of premature breaking of dormancy by unseasonable warm spells in winter. Dormancy requires a prolonged period of chilling, the temperature requirement varies with the species but is usually close to +5° C. Temperatures much above or below are ineffective. Thereafter the tree can be awakened from dormancy by warm treatment. However, the tree must be exposed to a minimum time, of the order of 300 hr, at temperatures near 25° C before new growth starts. These mechanisms prevent premature growth during "Indian summer" or even during prolonged January or February thaws. The mechanisms rather precisely fit the environmental requirements. Trees of the same species growing in subtropical or northern temperate areas behave entirely differently when treated with short days and low temperatures—those of the northern type drop their leaves and enter dormancy, whereas those of the southern type continue to grow, because they have no mechanism for the induction of dormancy.

Nutrient Salvage Prior to Leaf Fall. In autumn a large proportion of the nitrogen and mineral content of leaves is translocated back into the twigs before abscission occurs. The minerals lost by the leaves are not rain leached but are actually translocated, and they may ultimately be stored during the winter in the trunk or roots. The mechanism governing nutrient export from the leaves is unknown. Presumably, as the leaves senesce, they become less and less able to retain organic and mineral solutes against the competition of more physiologically active trunk and root tissue. The mechanisms whereby more actively growing or metabolizing tissues compete successfully with less active or senescing organs for available nutrients, based on the production of auxins and cytokinins, are described in detail in Chapters 21 and 22. However, specially programmed senescence of leaves, part of the overall developmental pattern of the tree, is required to initiate the degradative processes that make leaf nutrients (particularly nitrogen) available for export.

Communities of Trees

A forest is, at the very least, a close association of individual trees. However, the association among individuals may be so close that the whole forest grows and develops as if it were one organism, and it may even senesce as one. For example, a forest is said to **stagnate** when mutual shading and the mass of individuals, compared with the photosynthesizing area, become so great that net growth eventually ceases, respiration in the forest as a whole becoming equal to or greater than photosynthesis. At this point, unless selective cutting of the forest is carried out, productivity will be seriously reduced.

Many forests consist of associations that are, in fact, closer than merely collections of individuals. The formation of root grafts, shown in Figure 24-5, quite commonly occurs between two or many trees, and in a typical white pine forest over 50 percent of the trees may be so linked together. That these are viable living grafts is shown by the fact that a cut stump so linked will survive for many years, its root system maintained by and essentially becoming part of that of surrounding trees. An experimental attempt at thinning such a root-grafted forest with poisons proved disastrous because all the trees connected to the poisoned tree by grafts also died. Experiments have shown that dyes, radioactive organic and inorganic materials, and even fungal spores can be transmitted from tree to tree through root grafts.

An interesting results of this tendency to form extensive root grafts is that a large group of trees may become so closely associated that, in essence, they form one individual. Older or small trees die, and their roots are captured and survive as part of the system. Larger trees control the growth and development of smaller ones through the effects of shade, nutrient competition, and translocated hormones. Thus whole sections

A

B

Figure 24-5. A. A living stump showing graft connections with the living tree on the left which has taken over the root system of the stump. **B.** Root grafts between two living trees. [From F. H. Bormann: In T. Kozlowski (ed.): *Tree Growth*. Ronald Press Company, New York, 1962, pp. 237–46. Used with permission.]

of forest, not simply individual trees, form a closely controlled physiological unit, living, growing, and behaving as one organism.

Additional Reading

Cote, W. A., Jr. (ed.): *Cellular Ultrastructure of Woody Plants*. Syracuse University Press, Syracuse, N.Y., 1965.

Kramer, P. J., and **T. T. Kozlowski:** *Physiology of Trees*. McGraw Hill, New York, 1960.

Kozlowski, T. T. (ed.): *Tree Growth*. Ronald Press, New York, 1962.

Kozlowski, T. T., and **T. Keller:** Food relations of woody plants. *Bot. Rev.*, **32:**293–382 (1966).

Perry, T. O.: Dormancy of trees in winter. *Science,* **171:**29–36 (1971).

Zimmermann, M. H. (ed.): *The Formation of Wood in Forest Trees*. Academic Press, New York, 1964.

Zimmermann, M. H., and **C. L. Brown:** *Trees—Structure and Function*. Springer-Verlag, New York, 1971.

25 Physiology of Marine Algae

Introduction

Most textbooks on plant physiology concentrate largely on land plants and only mention algae where they have been extensively used for experimental purposes (for example, *Chlorella* and *Scenedesmus* in photosynthesis research). Moreover, the physiological problems that relate to the marine habitat and the solutions that have evolved to these problems are seldom discussed in any detail. It is now becoming apparent that a large part (perhaps 50 percent) of the world's biomass is present in the form of marine algae, and this constitutes an important fraction of the ecological resources of the planet. Not only does a large part of the world's photosynthesis take place in the sea, but an extremely large resource of potential food and raw material, which we may need to exploit in the near future, exists in the sea.

It therefore seems appropriate to include in this chapter a brief discussion of some of the important factors of algal physiology and biochemistry that distinguish them from land plants and permit their adaptation to various kinds of marine habitat. No attempt will be made to cover those aspects of algal physiology that are similar to higher plants discussed elsewhere in the book, and of necessity this discussion is far from exhaustive.

Productivity of Marine Algae

Natural Food Chains. An astonishing fact that has recently become apparent is that marine algae are among the most productive plants on earth. Calculations show that beds of kelp and similar brown seaweeds in the North Atlantic are more productive, on the basis of growth per unit mass, than even a tropical rain forest or a highly cultivated agricultural crop growing under ideal conditions. The data in Figure 25-1 give some idea of the high net productivity of some marine forms compared with plants of familiar terrestrial ecosystems.

Many of the large marine algae die or decompose at the tips as they grow from the base; others die back substantially or entirely in the winter. It used to be thought that the material derived from the death of seaweeds was simply decomposed and "wasted" in the marine ecosystem, because the polysaccharides that make up the greater part of seaweeds are mostly indigestible by animals. However, it is now clear that most of this

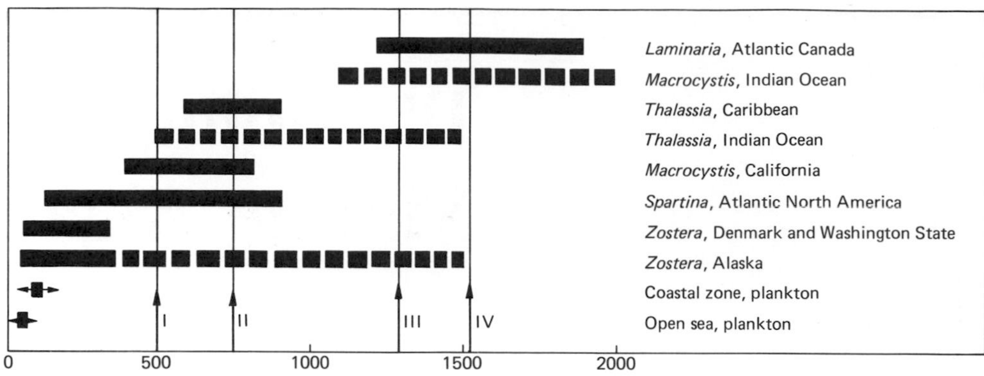

Figure 25-1. Productivity of marine macrophyte systems compared with phytoplankton and four medium-productivity land systems.
I. Medium-aged oak-pine forest, New York. **II.** Young pine plantation, England. **III.** Mature rain forest, Puerto Rico. **IV.** Alfalfa field, intensively managed, United States. [Adapted from K. H. Mann: Macrophyte production and detritus food chains in coastal waters. *Proceedings IBP-UNESCO Symposium on Detritus and Its Ecological Role in Aquatic Ecosystems*. Pallanza, Italy, May, 1972. Data courtesy Dr. Mann.]

massive primary production enters the marine animal food chain by an unusual and interesting link.

Pieces of dead seaweed become infested with marine bacteria, which can digest the polysaccharides. These are eaten by small animals, which assimilate the bacterial protein and excrete the remaining seaweed material. This becomes reinfested, is again consumed by animals, and the whole process is repeated until the "indigestible" polysaccharide is consumed. The small animals that feed in this way are food for fish and other larger animals. Thus the great productivity of marine ecosystems is due to a large extent to the very high productivity of large seaweeds.

Economic Use of Algae. Many algae are edible and form a useful supplementary food source for people or as animal feed, particularly in eastern countries. Also, because of their gelatinous nature and resistance to bacterial degradation, several algal polysaccharides are much in demand for world industry. These include agar, carrageenin, laminarin, and alginate, all of which are excellent stabilizing or gelling agents and are used in the food, cosmetic, and chemical industries. Exploitation of the natural resources of algae throughout the world has led to shortages, and the possibility of onshore tank cultivation is now being explored. Figures 25-2 and 25-3 show a typical commercial seaweed cultivation in the sea and an experimental tank-cultivation research station.

Preliminary tests have been disappointing because it has been very difficult to realize the growth rates in tanks that our understanding of algal productivity in nature would lead us to expect. This has led to a more useful analysis of the reasons why natural marine algal systems are so productive. The marine environment is much more stable than the terrestrial one. It contains huge quantities of carbon in the form of bicarbonate, and other nutrients are continuously available, though in seasonally varying amounts. Temperatures are lower than in terrestrial environments, but extremes (particularly of cold) do not occur. Some interesting discoveries are now being made about the ways in which marine algae exploit their peculiar environment.

Figure 25-2. The author examining kelp (*Laminaria japonica*) grown in commercial aquaculture beds by members of a commune near Tsing Tao on the China Sea. [Photograph courtesy Dr. T. K. Tseng, Academia Sinica Institute of Oceanology, Tsing Tao, Shan Tung Province, People's Republic of China.]

Physiological Adaptations of Marine Algae

Photosynthesis. CO_2 in air has a concentration of 330 ppm. Free, dissolved CO_2 in the sea has about the same concentration, which is equivalent to about 10 μM CO_2. However, the sea also contains bicarbonate (HCO_3^-), and its concentration is 200 times greater, about 2 mM. The substrate of RuBPcase is CO_2 (see page 347), and for a long time it was supposed that CO_2 was absorbed for photosynthesis by algae. However, from time to time reports appeared suggesting that algae can absorb HCO_3^-, and this has now been shown to be correct. Data in Figure 25-4 show that a normal curve is obtained when photosynthesis is plotted against HCO_3^- in the water rather than against CO_2 supplied. The normal high levels of growth are only realized when the air supply is greatly enriched with CO_2. This is because the plants absorb $HCO_3^-$from the water faster than it can be replenished from air. It emphasizes that these plants can make full use of the high levels of carbon in their all but inexhaustible environment, but they do depend on water movement from currents and wave action to replenish continuously their growth medium. A further interesting consequence of the high concentration of their available source of CO_2 is that algae do not normally photorespire and have lower levels of the associated glycolate metabolism (see Chapter 15, page 352).

Seasonal Growth. Perennial marine algae tend to become dormant during seasons of adverse conditions, in that their physiological mechanisms slow down considerably and

A

B

Figure 25-3. (**A**) An experimental tank cultivation site for *Chondrus crispus* (irish moss) in Nova Scotia. The plants (normally sessile on coastal rocks) are agitated by bubbling air and grow in the form of free-floating balls (**B**) [Photographs (**A**) R. G. S. Bidwell. (**B**) J. Crosby, courtesy National Research Council of Canada.]

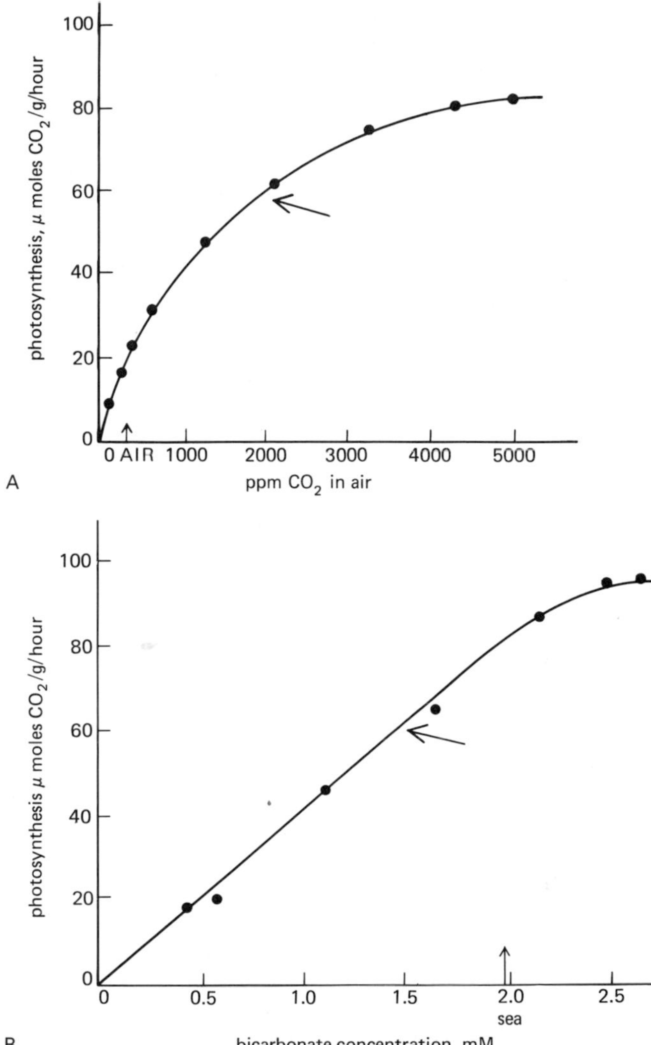

Figure 25-4. Photosynthesis of *Fucus vesiculosus* plotted against the CO_2 content of air bubbling through the seawater medium (**A**), and against the actual HCO_3^- content of the medium (**B**). Compare these curves with a normal photosynthesis/CO_2 curve of a land plant (Figure 7-24, page 189). The level of photosynthesis needed to sustain observed growth is shown by an arrow. [Data of R. G. S. Bidwell and J. McLachlan, National Research Council of Canda, Halifax, Nova Scotia. Used with permission.]

growth essentially ceases. However the characteristics of the marine environment include a dramatic increase in the concentration of inorganic nutrients and vitamins early in the season, usually before conditions of temperature and illumination become optimal. Since these dissolved nutrients are usually in short supply and therefore normally limit growth, their early utilization is extremely important for the survival of the plants.

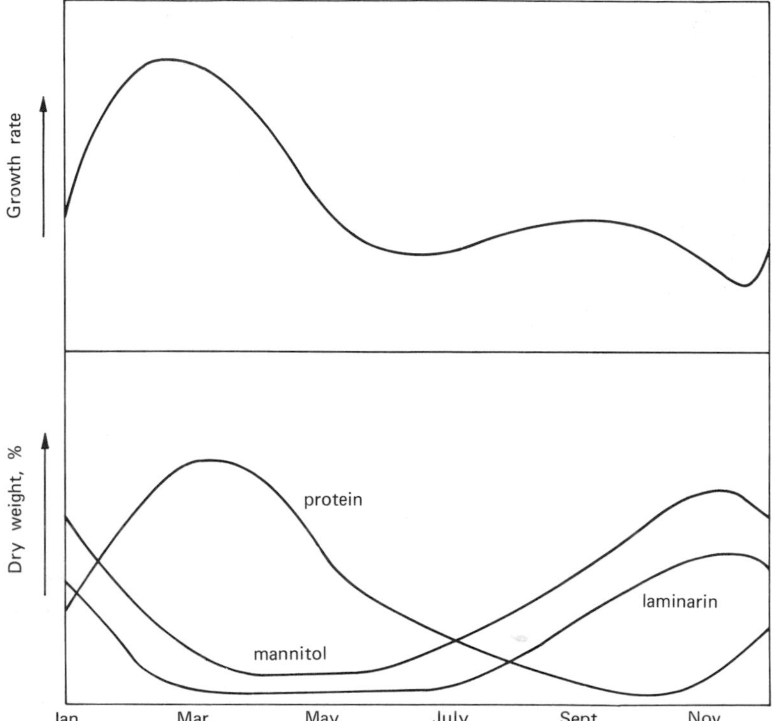

Figure 25-5. Diagram illustrating the growth cycle and the content of carbohydrate and protein of a typical brown marine alga from the North Atlantic.

Algae have developed the capacity to absorb and store a number of nutrients as well as carbohydrates so that growth can commence very early in spring at the expense of stored compounds, instead of waiting for conditions suitable for rapid photosynthesis, which do not occur until later. The annual growth cycle of a typical brown alga is represented in Figure 25-5, together with the seasonal variation in its protein content and principal storage carbohydrates. The major period of growth in late winter is characterized by the synthesis of new protein that accompanies development and fruition. Later in the year the storage carbohydrates, which were converted to proteins during the early growth, are replenished by photosynthesis. Not all seaweeds exhibit the same periodicity or restrict the period of their growth to this particular time of year, but many are highly seasonal in their growth and development patterns.

Uptake of Nutrients. Photosynthesis of marine algae continues even at very low temperatures and light intensities, and plants continue to fix carbon and assimilate nutrients under the ice in the Arctic winter. Marine algae, particularly unicellular algae that have relatively lower food reserves, are highly dependent on the various nutrients they can scavenge from the sea. In fact, many planktonic species normally live far below the limit of light penetration and survive heterotrophically on dissolved substances present in the ocean. Under conditions of active photosynthesis, marine algae tend to be rather "leaky," and a large amount of their assimilated carbon is liberated into the ocean in various organic forms. Frequently over 25 percent of assimilated carbon is leaked out in this manner, and up to 80 percent may be lost under certain conditions. Compounds leaked out in large amounts include sugars, sugar alcohols, peptides, amino acids, glycerol, glycolic acid, and some other more elaborate molecules. Because of this and because of the wide variety of organic and inorganic compounds that enter the sea from land runoff, a rich diversity of nutrients is available for the heterotrophic nutrition of phytoplankton.

The work of the Canadian physiologist J. A. Hellebust has shown that the uptake of many organic and inorganic nutrients is mediated by enzymic transport mechanisms that tend to be very specific for individual compounds. However, the supply of dissolved compounds varies greatly in composition and amount from time to time, as do the needs of the organisms. The continuous maintenance of a large number of specific uptake systems would immobilize large amounts of carbon and nitrogen in specialized enzymic proteins, and nitrogen in particular is often in short supply. Perhaps because of this the main uptake systems are not permanently present in marine algae but are inducible and may be destroyed when conditions change so that they are no longer required.

Thus, the uptake of glucose by phytoplankton diatoms is induced, not by the presence of glucose but by darkness, and the uptake mechanism is destroyed in light. Presumably photosynthesis supplies sufficient sugar-carbon in light to satisfy the organism's need. Similarly, certain organisms have inducible uptake systems for amino acids and organic acids; the systems are induced by the presence of the substance and lost again in its absence. Usually these systems are also inducible only in darkness; they are not needed in light. Forms living in the open ocean normally lack uptake systems for sugars, amino acids, and organic acids; these compounds seldom occur in their environment in significant amounts. On the other hand, they do usually have efficient uptake systems for vitamins and inorganic nutrients. Estuarine forms and organisms found in tidal pools or near the shoreline usually have the capacity to induce transport systems for various organic substances as required.

The uptake of nitrate is similarily related to its presence and distribution. In a single widely distributed diatom species, the open ocean form has a transport system with a very high affinity for nitrate, forms living in coastal waters have a lower affinity, and estuarine forms living in water normally high in nitrate have the lowest affinity for nitrate.

There are strong ecological advantages in having this sort of inducible or adjustable uptake system. The plants are safeguarded against possible poisoning by the uptake of unrequired compounds or metabolic analogs. They are not required to maintain a large amount of protein carbon and nitrogen tied up in the form of specialized proteinaceous uptake systems that may not be needed. Finally, absorbing nutrients takes energy. The uncontrolled absorption of unrequired carbon compounds would be a great waste of energy, besides creating a storage problem.

Reactions to Environmental Factors

LIGHT. Plants that live in the sea are in a different situation from land plants because of the absorption of certain wavelengths of light by water. Water is less transparent to red light than to blue and becomes quite opaque to certain wavelengths in the infrared region. As a broad generalization it can be said that red algae, whose pigments absorb more in the blue region, tend to live in the greatest depths of water (up to 600 ft). Brown algae tend to occur at intermediate depths, and green algae (absorbing more in the red part of the spectrum) are found nearest to the surface.

There seems to be a real significance to this generalization, since the action spectrum of red algal photosynthesis indicates that much of the light absorption is done by the red accessory pigments. The chlorophyll a present is still involved, however, because the energy trapped by phycoerythrin is transferred to chlorophyll for the generation of reducing power. Exceptions to the generalization are frequent; a number of red algae exist in shallow water. However, the deep-water adaptation does not preclude their living in shallow water. The reverse exception, deep-living green algae, is less common. This relationship of the red and brown pigments to depth is not absolute. Water is quite transparent to light of certain wavelengths in the infrared spectrum, and some of the less common chlorophylls have absorption bands that correspond to these "windows" of transparency.

One of the important characteristics of the marine habitat is low light intensity. In particular, water under the Arctic ice or in circumpolar seas receives extremely low light intensities. Temperatures are not extreme, never going below 4° C, and many algae are physiologically adapted to grow most rapidly at low temperature. Thus algae may flourish or grow under what appear to be impossibly dim conditions. Summer and late summer photosynthesis, taking place when nitrogen and other nutrients are at their lowest level, results in storage of large amounts of polysaccharides. The major part of new growth and development takes place in winter, when more nutrients are available, utilizing the energy and carbon of previously stored polysaccharides. In this way the season of major growth may precede (or follow) the season of maximum photosynthesis by several months, allowing time for the completion of the complex life cycle and the establishment of new plants during the short northern summer.

Algae have motile spores and gametes, and many are themselves motile. As one might expect of organisms existing in a medium where movement is possible, various types of phototactic responses are common. Phytoplankton may show rhythmic positive and negative phototaxy relating to their photosynthetic requirement. Zoospores of some algae such as *Ectocarpus* are positively phototactic, whereas those of others, for example, *Fucus,* are negative. *Fucus* zoospores are capable of orientation or movement along the plane of light polarization; presumably, therefore, they contain a mechanism, probably related to the semicrystalline structure of the polysaccharides, that can perceive and react to polarized light (see page 399).

The phototactic responses of *Acetabularia* gametes and zygospores is helpful in producing bacteria-free cultures. The gametes are positively phototactic, and can be made to swim down a tube of sterile water toward light, thus leaving behind nonmotile contaminating organisms. After gamete union, the zygotes become negatively phototactic and can be made to swim away from the light, thus freeing them from contamination by positively phototactic organisms that might have been present with the gametes.

Temperature. Tolerance of variation or extremes in temperature is related to the ability of seaweeds to survive on intertidal zones and is doubtless related to the ability to survive desiccation (see below). The requirement of deep shade that is quite common among littoral and sublittoral red algae is probably due to lack of heat tolerance.

A more important temperature relationship is that of photosynthesis and respiration. Respiration appears to increase in many algae as temperature rises, whereas photosynthesis rises more slowly and begins to decrease again as some critical temperature (often as low as 10–15° C) is passed. As a result, the light compensation point varies directly with temperature, hence the photosynthetic efficiency per unit of light is much greater at low temperatures. This fact may permit surprisingly high rates of net photosynthesis under what would appear to be extremely poor conditions as, for example, in the cold water and dim light under Arctic ice.

Desiccation. Sublittoral algae are not normally exposed to drying and have very little resistance to it. Intertidal algae, however, may spend a large proportion of their time under intensely drying conditions, exposed in sunlight on bare rock between tides for periods that may range up to nearly 12 hr. Not only are the plants resistant to drying but some of them, as certain species of *Fucus,* are difficult to culture unless they are periodically exposed.

Several mechanisms are involved in the resistance to desiccation. The cells of resistant forms are small, with dense, sometimes elastic cell walls that tend to expand and shrink with the protoplast so that plasmolysis does not occur. Many have a low internal osmotic potential and either resist or survive plasmolysis. Some cells can accumulate salt against a gradient, and thus lower their osmotic potential. The presence of many hygroscopic sugar derivatives (for example, polyhydric alcohols) and mucilaginous compounds (for example, alginate and the sulfated polysaccharides) doubtless assists in keeping exposed seaweeds from excessive drying. The drying time of algal thalli is related directly to their normal habitat. Deep-water or tide-pool forms usually dry quickly on exposure, but intertidal forms may take many hours to lose up to one third or one half of their water, even under extreme conditions.

pH. The pH of water in the open seas is quite constant at 7.9–8.0, but in isolated tidal pools it may vary over a wide range. As bicarbonate ions are absorbed in photosynthesis, the pH tends to rise sharply, presumably because of anion exchange for hydroxyl ions. Intertidal forms that are exposed to the sun may cease photosynthesis after only a few minutes, even while still moist, because the pH of the remaining water film on the thalli rises quickly to 9 or over. At this high pH the concentration of bicarbonate ions drops to a low level and only carbonate remains, which is unavailable for photosynthesis. Many tide-pool or littoral algae survive wide variations of pH, but they do not grow or carry on their normal physiological functions except at or near the normal pH of sea water.

Salinity and Osmotic Potential. Seaweed distribution is limited by salinity in areas where fresh water mixes in substantial amounts. However, very wide variations in salinity and in the osmotic potential of water occur in tide pools, particularly those replenished only by spring tides or storm waves. Mechanisms of protection against the deleterious effects of such changes have evolved in organisms that thrive in such situations. The Canadian phycologist J. S. Craigie has shown that the tide-pool flagellate, *Monochrysis lutheri,* can synthesize the compound **cyclohexanetetrol,** shown in Figure 25-6, in large quantities in the cells as a response to increased salinity. This

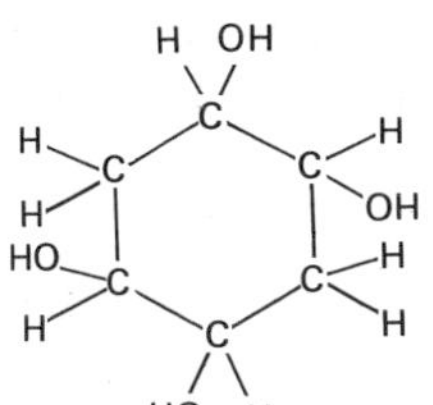

Figure 25-6. Cyclohexanetetrol, the osmotic "ballast" of the tidal flagellate alga, *Monochrysis lutheri.*

compound can be expelled rapidly into the medium in response to a decreased salinity. The reaction time for the formation of this "osmotic ballast" is 4 hr or less, and it can be dumped in only 10 min. Floridoside (a glycerol galactoside) apparently plays the same role in some red algae and golden-brown unicellular algae, being formed and stored when salinity is high and rapidly excreted when salinity decreases. Green flagellates such as *Dunaliella* and brown algae like *Fucus* use photosynthetic glycerol or mannitol, respectively, in the same way. Such mechanisms enable organisms to live in environments inaccessible to others, against which they might not be able to compete in a more stable or hospitable environment.

Wave Action. Multicellular, littoral algae must be extremely tough and flexible to withstand the constant battering of wave action. This may perhaps be a reason why so many different types of polysaccharides are found in seaweeds. The distribution of alginate types, examined by the Norwegian phycologist A. Haug, suggests that this polysaccharide is involved in resistance to wave action. Alginate containing a high proportion of guluronic acid residues is firmer, stiffer, and forms a tougher gel than alginate made up largely of mannuronic acid residues. The content of guluronic acid is much higher in older plants, in the stipes or stalks of such surf-surviving algae as *Laminaria* and *Fucus,* and in plants that are present in rough-water areas. Conversely the alginate from the younger more flexible fronds, from fruiting bodies, and from plants living in calmer water is much higher in mannuronic acid.

Similarly, the sulfate content of fucoidin is related to its toughness and water retaining capacity. *Fucus* plants high up the beach, which are much exposed and more likely to be battered by wave action, have a higher proportion of sulfate in their fucoidin than do plants of the same species living lower in the tidal zone.

Peculiarities of Algal Metabolism and Biochemistry

Chemotaxonomy. Many efforts have been made to correlate similarities in chemical constituents or biochemical pathways with taxonomic groupings, but usually without conspicuous success. The main problem seems to be that major biochemical pathways are essentially common to all plants and probably evolved early in the development of plants, before most of the modern taxa had appeared. The distribution of other less meaningful pathways or compounds is often sporadic and evidently unrelated to taxonomic groupings. However the algae fall into several extremely widely separated groups, whose relationship is very remote. One, the Chlorophyta (green algae), contains members that are essentially similar to the probable ancestral forms that evolved into the higher plants. The Cyanophyta (blue-green algae) are procaryotic and may be related to bacteria. A number of characteristics of some of the Rhodophyta (red algae) are very similar to those of certain fungi, and a relationship may exist. However, regardless of the reality of these relationships, the differences between the various taxa are real and diagnostic.

Pigments. The major pigments of the algal taxa are shown in Table 25-1. All the algae contain chlorophyll a and β-carotene, but the distribution of the remainder of the pigments is quite specific. Chlorophyta alone among the algae contain the accessory

Table 25-1. Major pigments of the algae

Pigment	*Chlorophyta* (*green algae*)	*Phaeophyta* (*brown algae*)	*Cryptophyceae** (*cryptoflagellates*)	*Dinophyceae** (*dinoflagellates*)	*Chrysophyta** (*golden-brown algae*)	*Xanthophyta** (*yellow-green algae*)	*Bacillariophyceae** (*diatoms*)	*Rhodophyta* (*red algae*)	*Cyanophyta* (*blue-green algae*)
Chlorophyll a	++	++	++	++	++	++	++	++	++
Chlorophyll b	++	0	0	0	0	0	0	0	0
Chlorophyll c	0	+	+	+	+	0	+	0	0
Chlorophyll d	0	0	0	0	0	0	0	+	0
Chlorophyll e	0	0	0	0	0	+	0	0	0
α-Carotene	+	+	+	+	0	0	+	+	0
β-Carotene	++	++	++	++	++	++	++	++	++
ε-Carotene	0	0	0	0	0	0	+	0	0
Lutein	++	+	++	+	+	+	0	++	+
Fucoxanthin	0	++	++	++	+	+	++	0	0
Myxoxanthin	0	0	0	0	0	0	0	0	++
Phycoerythrin	0	0	0	0	0	0	0	++	0
Phycocyanin	0	0	0	0	0	0	0	0	++

CODE: ++ = large amount; + = present; 0 = absent.

*All species in these taxa are unicellular or form only small colonies.

pigment chlorophyll b, like higher plants. Phaeophyta (brown algae) contain the brown pigment **fucoxanthin,** Rhodophyta contain the red-colored **phycoerythrin,** and the other groups contain additional specific pigments. The less common chlorophylls c, d, and e are distributed among the algae in an interesting fashion, as shown in Table 25-1.

Small Molecules. The products of photosynthesis of algae are frequently different from those of higher plants. Sucrose is commonly formed by green algae, as in higher plants, and also to a lesser extent in blue-greens. Brown algae produce larger amounts of the sugar alcohol mannitol, as do some diatoms, dinoflagelletes, and related types. Red algae characteristically produce the compound **floridoside,** a glycoside of galactose and glycerol.

It might be thought that the biochemical pathways producing such end products as mannitol and floridoside would be different from those in higher plants that produce sucrose. However, it appears probable that the photosynthetic pathways of all algae are basically similar, and the differences occur only in the final stages. For example, mannitol is very likely formed by reduction of fructose-1-phosphate, which is produced in brown algae as in higher plants. The two parts of floridoside could be derived from triose phosphate and hexose phosphate, and the synthesis of floridoside may thus be considered essentially analogous to the formation of sucrose in higher plants, but using different substrates. Other more unusual products of photosynthesis, such as cyclic alcohols and benzene derivatives, may require more extensive metabolism. However, the primary processes of carbon fixation appear to be similar in both land and marine plants.

Storage Compounds. The cell walls of most algae are quite different from those of land plants and contain a number of different substances. In addition, the reserve polysaccharides of many algae are unlike those of the higher plants, which commonly store starch or occasionally the polyfructosan inulin. Some algae store mainly oil instead of polysaccharides, and large parts of the natural petroleum deposits of the world may have been derived from oil stored by phytoplankton sinking to the ocean floor and being held there in sediments.

Recently the work of the American physiologist A. A. Benson has shown that huge amounts of highly unsaturated fats and waxes (see page 230 for a description of these compounds) are produced by the photosynthesis of both free-floating and attached marine algae. Most such algae are food for animals, or live in close association with them (for example, in corals or in the gills of clams and other molluscs). These fats and waxes thus constitute a major resource in the world's food chain. Also, being liquid at low temperatures, they make an effective antifreeze for polar organisms, besides being efficient storage forms for carbon and energy. It is a matter of real concern to plant physiologists that these fascinating compounds, which constitute such a large proportion of the world's photosynthetic production, have been so little studied.

In addition to the common polysaccharides made up of glucose, such as laminarin and cellulose, a wide variety of polysaccharides are known from various green, brown, and red algae, which contain fucose, galactose, arabinose, xylose, rhamnose, uronic acids, glycerol, erythronic acids, and even amino acids in various combinations. Many of these polysaccharides contain sulfated residues. Table 25-2 shows the major types of polysaccharides appearing in the important groups of algae. Many of these polysaccharides are very important to industry as stabilizing agents (for example, alginate, car-

Table 25-2. Common marine algae polysaccharides

Organism	*Storage*	*Cell wall*
Green algae	Starch (glucose)	Mannan (mannose) Cellulose (glucose)
Brown algae	Laminarin (glucose)	Alginic acid (guluronic acid + mannuronic acid) Fucoidin (fucose, sulphate) Cellulose (glucose)
Red algae	Floridean starch (glucose, more branched than starch)	Cellulose (glucose) Carrageenin and agar (both galactose, sulphate)
Blue-green algae	Starch (much branched)	Pectin (glucuronic acid)
Diatoms	Fats	Pectin (glucuronic acid) Chitan (acetamidodeoxy-glucose)

rageenin, agar-agar), but they are generally indigestible by animals and hence not directly useful (without major reprocessing) as food. However, the mass and productivity of seaweeds are such that greater economic use of them will undoubtedly be made in the future.

Calcareous Algae. Frequently coral reefs are discussed as if they consisted entirely of animal (coral) skeletal remains. In fact, the major proportion of many coral reefs is made up of plant remains. Many red algae deposit substantial amounts of calcium carbonate (essentially limestone) on their external surfaces. These may live to a depth of 600 ft in the tropical seas and make up the bulk of many of the constantly changing coral atolls of the south seas. Their form is adjusted to the degree of wave action, since they tend to be brittle. Heavier deposits of carbonate occur in sites with less wave action. Tests with radioactive carbon show that up to half of the carbon absorbed by these algae may be deposited as calcium carbonate, the remainder being reduced photosynthetically. The majority of calcareous algae are tropical red algae. However a number of temperate reds as well as a few green algae, such as the marine siphonaceous alga *Acetabularia* and the fresh water *Chara,* also tend to form calcareous deposits.

Antifouling. One of the problems facing large marine organisms is the presence of smaller saprophytic or parasitic forms that grow on them. Nevertheless many thallose seaweeds, growing under conditions which ought to promote the growth of parasitic forms, are surprisingly clean. The reason appears to be the release of a variety of compounds—some of them phenolic and some colored (**gelbstoff,** a yellow-brown substance found in tidal pools). Advantage was taken of this fact in early times when extracts of certain brown seaweeds were made into antifouling substances for the bottoms of ships. The ecological relationships of marine algae are probably much dependent on the presence of these and similar antibiotic compounds, but little is yet known of their chemistry or occurrence.

Pheromones. It has long been supposed that marine forms with free-swimming gametes have some sort of sex attractant to increase the chance of gametic union. Sex attractants of some green algae have been postulated, and a hormone called **sirenine** has

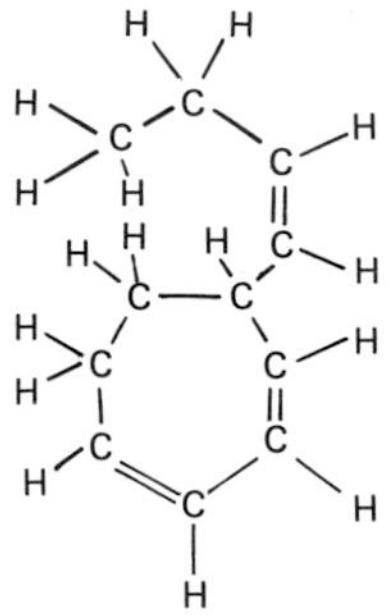

Figure 25-7. *Ectocarpus* sirenine, the sex attractant (gamone) of *Ectocarpus siliculosus* female gametes.

been found to be produced by the gametes of the water mold *Allomyces*. Recently a similar type of **gamone** from the brown alga *Ectocarpus siliculosus* has been isolated by a group of German scientists under the leadership of D. G. Müller. The compound, a butene substituted cycloheptadiene (Figure 25-7), is called *Ectocarpus* sirenine. It is produced in minute amounts by the female gametophytes and strongly attracts the smaller free-swimming male gametophytes. After fusion of a male gamete with the female, the production of sirenine stops and the remaining male gametes disperse. No clue is presently available concerning the mechanism of action of such attractants, more of which will undoubtedly be discovered in the future.

Additional Reading

Boney, A. D. (ed.): *A Biology of Marine Algae*. John Wiley & Sons, Toronto, 1970.

Dawson, E. Y.: *Marine Biology*. Holt, Rinehart and Winston Inc., New York, 1966.

Hill, M. N. (ed.): *The Sea*. Interscience, New York, 1963.

Oppenheimer, C. H. (ed.): *Symposium on Marine Microbiology*. C. Thomas, Springfield Ill., 1963.

Percival, E., and **R. H. McDowell:** *Chemistry and Enzymology of Marine Algae Polysaccharides*. Academic Press, New York, 1967.

Proceedings of the International Seaweed Symposia.

Reports of the Norwegian Institute of Seaweed Research, Trondheim.

Stewart, W. D. P.: *Algal Physiology and Biochemistry*. University of California Press, Berkeley, 1974.

26 Parasites and Disease

Introduction

Organisms form associations of many types. One may simply be physically supported by another. Alternatively the tissues of one organism may be closely associated with or invaded by those of another, and considerable exchange of materials may take place. If no harm ensues to either partner from such an association, or if mutual benefit occurs, the association is called **symbiosis.** If the presence of one organism is detrimental to the health and well being of the other, then the association is called **parasitism.** The result of parasitism is **disease.** Diseases may also be produced as a result of other agents or inadequacies, such as the presence of toxins, deficiencies in nutrition, and the like. Examples may be found of every possible gradation between mutually beneficial symbiosis and deadly parasitic disease.

Specificity is important in the study of parasitism and symbiosis. It may range from very low to very high and may relate to the host, or the invader, or both. It is one of the most interesting and least clearly understood aspects of physical relationship among organisms. Specificity may be extremely precise, so that very small differences in the genetic makeup of a species may result in large differences in the infectivity of a pathogen. A single gene difference in the host may confer susceptibility or resistance. Much genetic work has been done in breeding resistant varieties of agriculturally valuable plants, but there is very little precise knowledge of the exact mechanisms of specificity. These may include specific inhibitory substances, the production of required substrates, a balance of nutrients, the presence of growth stimulants or retardants, or physical factors.

The mechanisms of disease are manifold. There may be the production of toxic substances, mechanical, physiological, or biochemical disruption of the host metabolism, nutritional debilitation of the host, or disruption of the host's normal life cycle. Much empirical work has been done in plant pathology to find ways of combatting disease and to produce disease-resistant plants. It seems probable that even greater advances will be possible when the physiological mechanisms of disease and disease resistance are clearly understood. The brief account in this chapter is intended to outline some of the aspects of disease that may bear on the physiology and biochemistry of parasitism and to describe some to the known physiological responses of plants to infection and disease. The physiology of pathogenicity is a field where many problems and techniques are clearly defined. Plant disease is an old field of study, but there is an ever-increasing need for its further development in this hungry world.

Infection

Organisms of Disease. Plants are parasitized by a large number of microbes, both bacteria and fungi, and also by viruses. Recently it has been shown that a number of plant diseases may be caused by mycoplasma, small organisms that resemble very small, primitively organized bacteria. Fungal diseases are common in gymnosperms and angiosperms; bacterial and viral diseases are widespread among angiosperms. Few plants are free of parasites. However some so-called living fossils such as the ginkgo appear to have outlived their pathogens and are susceptible to very few diseases.

In addition to microbial diseases, plants are also parasitized by insects and worms, which live or secrete their eggs in the plant body. Many higher plants are saprophytic or epiphytic, living on dead plant material or supported by a living host, and a few have become clearly parasitic, deriving much or all their substance from the host plant. Lower plants are involved in several epiphytic and symbiotic situations, but they do not appear to have become prominent as parasites. It is probable in any event that they would have become unrecognizable as plants by degenerative evolution if they had adopted the parasitic habit.

Resistance. Plants are extremely resistant to microbial infection; resistance is the rule rather than the exception. Healthy tissue is usually free of microorganisms or infested only with nonpathogenic microflora. Most lesions heal without infection and the invading microorganisms die. Intact plants are protected by the cuticle, which cannot be digested by most pathogens. Many plants produce or contain substances that inhibit the growth of microbes, such as the cyanide-containing exudates of flax roots and the poisonous glycosides of oat leaves. Toxic phenols are often produced on wounding, and organic acids that inhibit bacterial growth are present in epidermal hairs that, being easily broken, would otherwise provide an invasion route. Many plants contain phenolic compounds that are highly inhibitory to pathogenic fungi, such as chlorogenic acid, which inhibits scab growth on potato, and catechol and protocatechuic acid, which protect onions from smudge disease (See Figure 26-1, also Chapter 9, page 241).

Resistance to insect attack is often provided by specific repellents such as terpenes, essential oils, mustard oils, or complex glycosides, or by deterents such as poisonous alkaloids. In certain plants, such as reed canary grass (*Phalaris arundinacea*), intense lignification takes place at sites of attempted penetration by pathogens, preventing infection. This may be related to the same pathways of metabolism that produce phenolic antifungal compounds. Mechanistic, tactile, or chemical stimuli are important in insect attack. For example, certain caterpillars that eat the edges of leaves can only attack holly leaves if the prickles are first trimmed off. It seems unlikely that the nutritional value of the host plays any part in the resistance or susceptibility, since this would require the ability of the pathogen to prejudge the nutritional value of the prospective host.

Some plants react to certain kinds of infections by isolating the pathogen. Thus, infected leaves may rapidly form an abscission layer and drop off, preventing the spread of the disease to healthy tissue. Some tissues posses "hypersensitive" cells that very

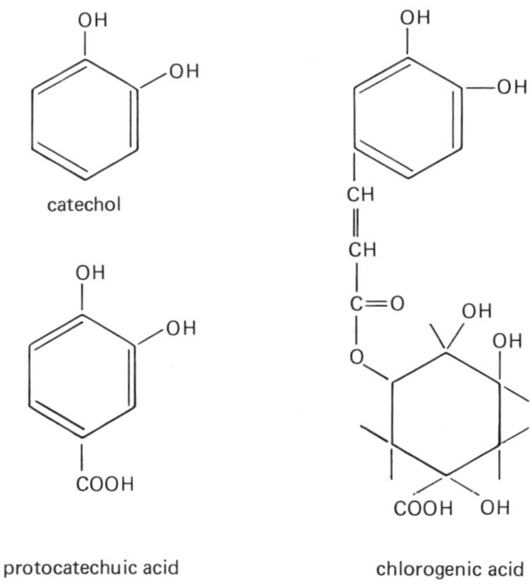

catechol

protocatechuic acid

chlorogenic acid

Figure 26-1. Phenolic substances conferring disease resistance.

readily succumb to infection and die. However, these cells are then "walled off" from the remaining healthy tissue, and so the spread of the pathogen is prevented.

Resistance to pathogens is frequently under very precise genetic control, so that resistance may depend upon a single gene. The nature of such resistance is not known, however, and is much affected by such factors as the nutritional status and age (both chronological and physiological) of the plant.

Immunity. Immunity may be inherent and based on factors providing resistance. Alternatively, mechanisms may exist that seem to protect the plant from the spread of infection. Some infections spread rapidly, whereas others are self-limiting. Plants may be protected from or immunized to virulent pathogens through infection by an avirulent strain or by the application of killed or inactivated bacteria. This is essentially similar to the immunization of animals by the injection of an avirulent or inactivated pathogen and suggests that plants may also have an immune reaction. Such is indeed the case. Certain compounds are produced in plants, only during or following pathogenic invasion, that are toxic to the pathogens. These are called **phytoalexins** (alexin means to ward off or guard against, hence compounds that guard the plant and ward off disease).

Phytoalexins are usually chemically similar to some normal component of the plant and do not require major changes in metabolism for their production. Several have been discovered, and the structures of a few are known. They are frequently phenolic derivatives, such as **pisatin** and **phaseollin,** flavones from peas, and beans respectively, **ipomoeamarone,** an abnormal sesquiterpenoid from sweet potatoes, and **orchinol,** a phenanthrene from orchids (see Figure 26-2). Some of them, however (like safynol and wyerone acid), are basically highly unsaturated straight-chain carbon compounds.

Susceptibility of various pea cultivars to the pathogenic fungus *Ascochyta pisi* is directly correlated with the amount of pisatin formed following infection. Conversely, the degree of pathogenicity of various strains of the fungus to a single variety of pea is inversely related to the amount of pisatin formed. That is, the more virulent strains do not activate the resistance mechanisms as strongly as do the less virulent ones. The nature of the mechanism whereby phytoalexin formation is stimulated is not known. It

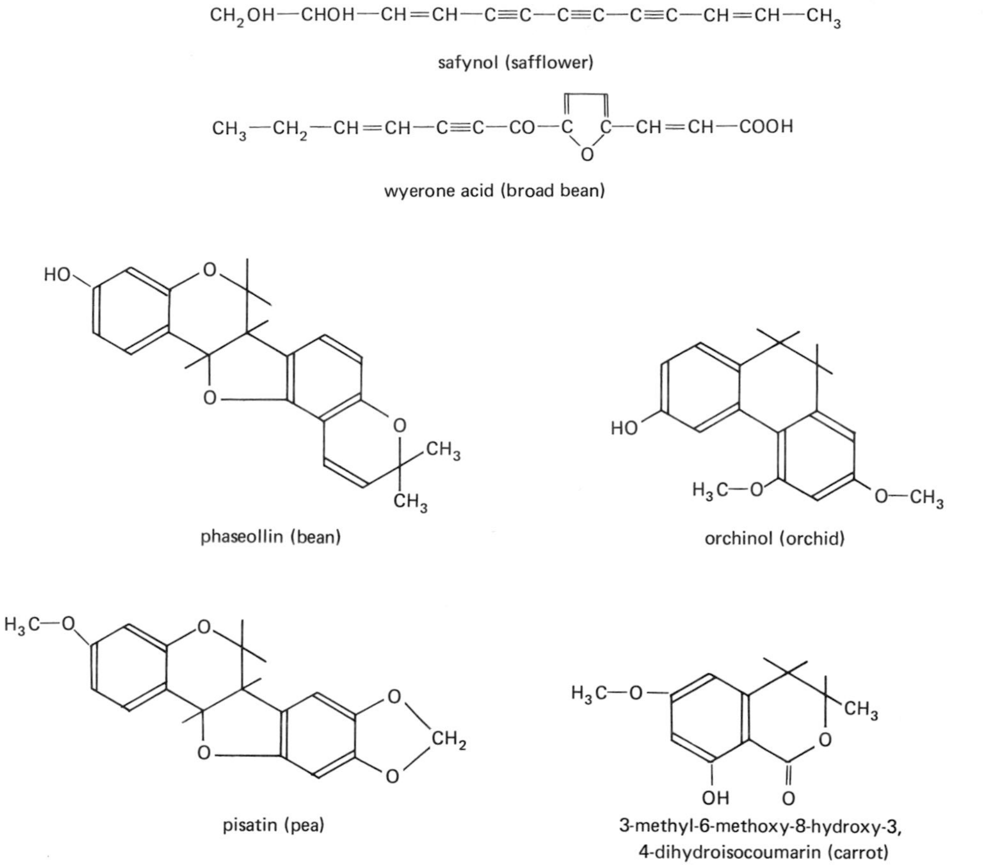

$$CH_2OH{-}CHOH{-}CH{=}CH{-}C{\equiv}C{-}C{\equiv}C{-}C{\equiv}C{-}CH{=}CH{-}CH_3$$

safynol (safflower)

$$CH_3{-}CH_2{-}CH{=}CH{-}C{\equiv}C{-}CO{-}C_4H_2O{-}CH{=}CH{-}COOH$$

wyerone acid (broad bean)

phaseollin (bean)

orchinol (orchid)

pisatin (pea)

3-methyl-6-methoxy-8-hydroxy-3, 4-dihydroisocoumarin (carrot)

Figure 26-2. Structures of some phytoalexins.

may be artificially triggered by the application of certain inorganic chemicals or growth factors such as 2,4-*D*.

An interesting protective mechanism is the formation in some plants of terpenoid substances that act as antagonists to insect hormones. Juvenile hormone, a steroid hormone that controls insect development, is made in insects from plant-produced precursors. Some plants also contain a chemical that prevents the formation or action of juvenile hormone, thus preventing insect maturation and reducing subsequent infection. A number of similar mechanisms are known.

Stimulus to Infection. The stimulus to infection may be active or passive. There is often a substantial interaction between the root of the host and soil microbes. Contact may be necessary for spore germination of certain fungal pathogens. Some fungi are attracted to hosts by substances released from the roots. Cellophane-coated roots are attractive, so the substances in question must be diffusible. High nutritional status of the host often increases the likelihood of infection, and it has been suggested that the nutrients (carbohydrates, amino acids, or organic acids) which are exuded by the roots may serve as attractants. Potassium deficiency is accompanied in some higher plants by lowered resistance to infection of the roots by fungi; this may be related to the increased sugar and amino acid content of potassium-deficient plants, which results from reduced polymer synthesis.

Insects may be attracted by specific substances secreted by plants, such as essential oils or mustard oils. These are secondary substances, however, and are not nutrients or primary metabolites.

Invasion. The physiological situation that permits a pathogen to enter and become established is poorly understood. Wounds and natural openings, such as stomata, are common ports of entry, as are thin-walled or delicate structures, such as leaf or root hairs. High humidity increases the chance of infection by bacteria but not by viruses. Unlike most fungi and some bacteria, the entry of viruses is a passive process and often occurs via minute wounds in individual cells that are later healed or through injection by insects. The invasion of a plant is influenced by its physiological status. Any factor that increases the carbohydrate content of a leaf, such as high light or low temperature, decreases the chance of viral infection, whereas conditions that favor decreased carbohydrate content render the plant more susceptible. The establishment of a successful invasion also requires the precise interplay of nutritional factors of the host and requirements of the pathogen.

Interest is now being focused on chemical factors present on cell surfaces either of the host or of the parasite that permit "recognition." Usually, in the absence of recognition the pathogen cannot bind to the host or subsequently penetrate. Such factors may be proteins that can selectively bind to both host and parasite cells. A good example is the **lectins,** glycoproteins occurring in soybeans, that have strong and specific interactions with the surface lipopolysaccharides of certain strains of host-specific N-fixing *Rhizobium* (see Chapter 8, page 193).

Toxins. The symptoms of disease are often (but not always) the result of toxic substances produced by the pathogen. The symptoms of several diseases can be simulated in plants by applying extracts of cultured pathogens. The symptoms of a disease may be manifest far from the site of infection, showing that toxins may be diffusible or transportable substances. Several toxins have been shown to be analogs of required metabolites that act by inhibiting important reactions in the host metabolism.

One such toxin, which has been well studied, is produced by *Pseudomonas tabaci* and causes the **wildfire disease** of tobacco. The toxin is a dipeptide that acts to inhibit the enzyme glutamine synthetase (see Chapter 8, page 211), thus causing a toxic buildup of ammonia. It is presumably a glutamine analog, as shown in Figure 26-3. Another specific toxin is fusaric acid, also shown in Figure 26-3, which seems to act by chelating iron. Other toxic substances include the polysaccharide slimes produced by certain bacteria that damage the plant by blocking its water transport system. Some resistant plants do not respond to the application of toxins, which suggests that susceptible plants may be conditioned in some way by the effect of the toxin to permit the establishment and spread of the pathogen.

Growth Substances. Some pathogenic microorganisms and insects produce substances that are identical with natural plant hormones; they would thus produce symptoms characteristic of severe hormone inbalance. However, the same symptoms may also be produced when the host plant, as the result of invasion by a pathogen, is itself stimulated to produce an excess amount of growth substance. In this case it is

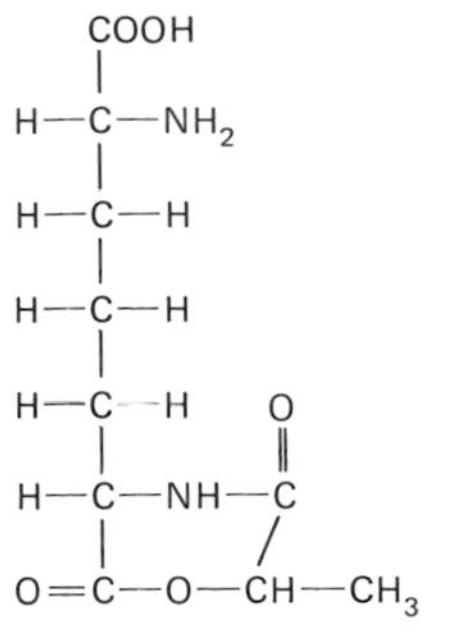

```
     COOH                      COOH
      |                         |
  H—C—NH2                  H—C—NH2
      |                         |
  H—C—H                     H—C—H
      |                         |
  H—C—H                     H—C—H
      |                         |
  H—C—H    O              O=C—NH2
      |     ‖
  H—C—NH—C
      |     /
  O=C—O—CH—CH3
```

wildfire toxin of *Pseudomonas tabaci*

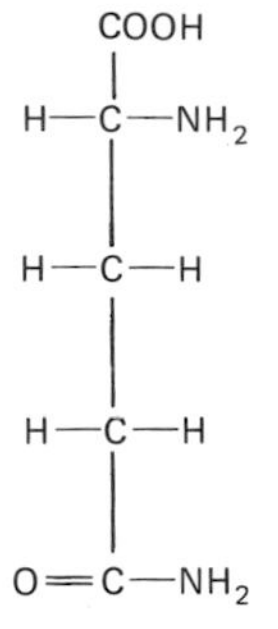

glutamine

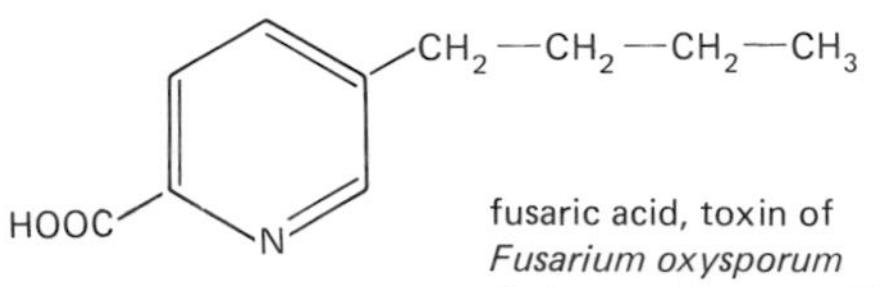

```
        /\ CH2—CH2—CH2—CH3
       |  |
HOOC— \N/
```

fusaric acid, toxin of *Fusarium oxysporum* that causes vascular wilt

Figure 26-3. Fungal toxins. (Glutamine is not a toxin, but is shown for comparison with wildfire toxin.)

extremely difficult or impossible to determine the source of the "toxin" causing the characteristic symptoms. In many cases the overproduction of hormones is the result of both host and parasite metabolism.

Examples of this kind of pathogenic effect are the production of IAA by crown-gall bacteria (*Agrobacterium tumefacians,* see Chapter 18, page 430), of gibberellins by *Gibberella fujikuroi* causing "foolish seedling" disease of rice (Chapter 16, page 391), or of cytokinins by many pathogenic organisms. Some of these produce the same effects as applied cytokinins to senescing leaves: surrounding tissue ages, but the infected tissue remains green due to the cytokinin effect (see Chapter 22, page 550). Some organisms produce ethylene or ABA, which causes aging or senescence of the host tissue.

Physiological Responses to Parisitism

Respiration. The most common reaction to infection is a large increase in respiration, up to ten times that of healthy tissue. Although some of the increase may be due to the respiration of the parasite, most of it is due to the host. The increase might be partly due to the breakdown of barriers that separate substrates and enzymes or to the respiration of sugars and other soluble compounds, which are characteristically mobilized to the site of infection.

Respiration appears to become uncoupled in some infections; that is, the oxidation of carbon compounds is no longer associated with the normal production of ATP. When this happens the temperature of infected tissues may rise considerably, as also happens in tissues that are treated with dinitrophenol, an uncoupler of respiratory phosphorylation. Narcotics or agents that suppress respiration or the action of oxidative enzymes are known to increase susceptibility, but this may be due to suppression of the synthesis of poisonous compounds of defense, such as oxidized phenols, or alternatively to the reduction of the host's ability to remove compounds mobilized as a result of infection.

Infection affects a number of enzymes. Some are increased in activity, others decreased, and no distinct pattern emerges. The Pasteur effect (page 125) may be abolished, and a switch from the Embden-Meyerhoff-Parnass pathway of glycolysis to the pentose-phosphate shunt may occur. The respiration of affected tissue sometimes becomes more cyanide resistant, and the proportion of respiration occurring via copper-containing oxidases, instead of the normal iron-containing cytochromes system, may increase substantially. The significance of these changes is not clear.

Photosynthesis. Photosynthesis is often reduced in infected plants, sometimes due to loss of chlorophyll or disorganization of chloroplasts, sometimes for obscure reasons. The rate of photosynthesis may increase substantially during early stages of infection; this may be a response to increased growth substances synthesized at the site of infection either by the host or by the parasite.

Nitrogen Metabolism. An important result of disease is the use of large amounts of host nitrogen to make parasite protein. Some virus multiplication appears to require concomitant protein breakdown in the host; compounds liberated from host protein breakdown are used to synthesize the viral protein. Some infections stimulate tumorous growth in the host, diverting the host's nitrogen supply to the nourishment of the tumor or causing the synthesis of new proteins in the host that were absent from the healthy tissue. Other disease organisms may secrete digestive enzymes, which destroy the host's proteins. Conditions that reduce a tissue's capacity for protein synthesis, such as detaching leaves, frequently make the tissue more susceptible to parasitic attack. Soluble nitrogen usually increases in diseased tissues, presumably as the result of protein degradation. The amides behave in a characteristic way, glutamine decreasing and asparagine increasing dramatically. Ribonucleic acid content characteristically increases in diseased tissues.

Translocation. Many plant diseases are characterized by wilting or "damping off" caused by interference with water transport. Root infection by bacteria, fungi, or nematode worms greatly reduces the plant's ability to absorb water. Infections by bacteria or fungi may block the xylem of the stem mechanically by triggering the deposition of masses of gum, slime, or mucilage derived either from the host or the parasite, as in Dutch elm disease. Even virus infections may trigger this sort of reaction. The mass of microbial infection may be so great as to block the xylem vessels, or these may be blocked by the formation of **tyloses** (expansions of parenchyma cells into the lumen of vessels). Finally, a common response to disease is increased transpiration. Any or all of these factors may so upset the plant's water balance as to cause temporary or permanent wilting.

Disease seems to have a strong effect on translocation and mobilization of solutes. Commonly there is a great increase in the concentration of soluble metabolites at the site of infection. This may be caused in part by breakdown of host tissue, by increased import of compounds from other parts of the plant, and by decreased outward transport. Photosynthate formed from $^{14}CO_2$ in rust-infected wheat leaves remains in the leaf to a much greater extent than in healthy leaves. Concentrations of phosphate up to 1000 times that of normal tissues have been found in diseased leaves. High concentrations of certain substances may be maintained in leaves even after the pathogen has been experimentally killed, perhaps as the result of aberrations of the host metabolism which are secondary to the infection.

The pathogen itself may travel in the host's conducting tissue. Many pathogens entering leaves or roots or through the surface of the stem appear to find their way readily to the phloem and spread throughout the plant by this route. Penetration of the xylem frequently occurs following wounding.

Growth Substances and Morphological Response. Pathogenicity is frequently accompanied by great increases in auxins, particularly IAA. Many of the symptoms of disease, such as tumor formation, epinasty, changed growth rate and form, aberrations in cell wall formation, and so on, can be mimicked by, or are clearly the result of, increases in the concentration of IAA, ethylene, cytokinins, or gibberellins. Many bacterial pathogens have the capacity to synthesize ethylene, which is doubtless responsible for the characteristic epinasty of some bacterial diseases.

It is usually difficult or impossible to determine whether the excess auxin is made by the host as a reaction to infection, or by the pathogen, or if it is due to a loss of auxinase activity in the host. Investigations with radioactive precursors of IAA suggest that both host and parasite may contribute to the increased IAA that causes epinasty in potatoes infected with the Granville wilt disease (*Pseudomonas solanacearum*). The gall-forming bacteria, *Agrobacterium tumefaciens,* is capable of synthesizing IAA, but tumors caused by the infection and the associated imbalance of IAA metabolism continue long after the pathogen has died off (see Figure 26-4).

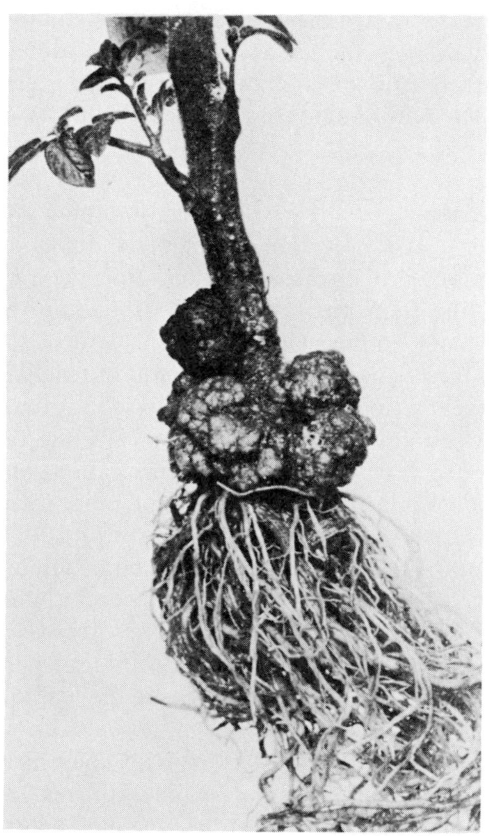

Figure 26-4. Crown gall on tomato. [From C. Chupp and A. F. Sherf: *Vegetable Diseases and Their Control.* Ronald Press Company, New York, 1960, p. 35. Used with permission.]

Certain virus diseases are accompanied by increased IAA formation, and the disease symptoms can be simulated by spraying tissues with IAA. Since viruses cannot make IAA, the increase must be the result of stimulated host metabolism. IAA oxidase activity may be depressed, but this is not the result of a general inhibition of enzymes because polyphenol oxidase and ascorbic acid oxidase may increase. The fact that viruses also frequently induce premature aging in plants has led some workers to link the phenomena of disease and of aging in terms of disturbances in hormone metabolism. It has been shown that rust resistance in wheat is associated with a higher IAA oxidizing activity of the tissue, which prevents the characteristic rise in IAA on infection or reduces it to a normal level more rapidly.

Increased IAA concentration may be an important factor in infection by softening cell walls and inhibiting secondary wall formation, thus improving conditions for the growth of the pathogen. It may also be partly responsible for the increased translocation of solutes. Root infection by nematode worms is often accompanied by the formation of giant cells derived from the breakdown of cell walls between adjacent cells. The development and maintenance of giant cells require the continued presence of nematodes, which evidently supply some material or substance whose continuous presence is necessary for the maintenance of giant cells. The presence of these giant cells provides living room for the nematodes, which pierce surrounding cells, often without killing them, to suck their required nutrients.

Some nematodes cause galls; extracts of these worms also cause galls, suggesting that they secrete gall-forming substances. Alternatively, it has also been suggested that the proteases that nematodes secrete may release large amounts either of protein-bound IAA or of the IAA precursor, tryptophan. However although all nematodes secrete proteases, not all make galls. Furthermore, the existence of protein-bound IAA has been questioned.

Gall formation is also characteristic of several insects, such as aphids, *Diptera, Lepidoptera, Hymenoptera* (wasps and saw flies), which lay eggs in plants. Gall formation may be triggered by IAA or some other growth substance secreted by the ovipositor, but this is not always the case. The fact that different insects produce highly characteristic galls, in terms of size, shape, and color, indicates that the physiology of gall formation is more complex than the mere introduction or formation of a growth substance at the point of invasion (see Figure 26-5). Aphid infestation of trees or woody shrubs may produce changes in wood anatomy very similar to those caused by IAA application, but again it cannot be proved that the aphids produced the IAA or its precursors.

The result of parasitism of trees or shrubs by members of a number of higher plant genera, loosely called mistletoe, is very frequently a marked aberration in growth called "witch's broom" (Figure 26-6), which appears to be caused by a disturbance in the balance of growth hormones in the host. There is usually either marked hypertrophy of tissue or the formation of a dense mass of short branches that have the appearance of a broom. The mistletoe feeds through haustoria that penetrate into the stem tissues of the host. The nature and source of growth hormone imbalance that produces the characteristic symptoms are not known. Overproduction of cytokinins has been suggested as the immediate cause.

Another parasite that attacks a number of important crop plants is dodder (*Cuscuta* sp.). It also invades the tissues of the host by means of haustoria (Figure 26-7) that penetrate to the phloem of the host. There the tips of the haustoria form into hand-

Figure 26-5. Galls on plants. Different insects cause different kinds of gall, even on the same plant.
Compare (**A**) the Goldenrod Bunch Gall and (**B**) the Goldenrod Ball Gall; caused by different species of fly. Also note the galls (**C**) on the upper surface (parenchyma cells), and (**D**) lower surface (vascular tissue) of oak leaves, caused by a mite and a wasp. Each gall is a massive outgrowth of host-tissue cells; (**E**) the Mossy Rose Gall (caused by a wasp) consists of a mass of bractlike tissue.

Figure 26-6. Witches brooms on black spruce. The stunting of the main stem and dense proliferation of branches are caused by infection with mistletoe.

shaped structures that grip the sieve tubes with many fingers (Figure 26-8). Host tissues above the site of infection usually die, presumably from nutritional deficiency.

Certain infections appear to interfere with growth hormone metabolism to the extent that they cause abscission of leaves. Normally abscission is prevented by excess IAA and favored by the reduction of IAA or the production of abscisic acid. How infections may affect these hormones is not known, but the mechanism acts as a protection against the spread of disease by causing premature abscission of infected leaves.

A few characteristic symptoms are caused by interference with the normal gibberellin metabolism of plants. Certain virus infections may reduce the content of natural gibberellins; the resulting stunting of growth can be overcome by the application of gibberellic acid spray. The best-known gibberellin disease is the **bakanae disease** (literally "foolish seedling") of rice, in which rapid extreme elongation of seedlings occurs following infection by the fungus *Gibberella fujikuroi,* which produces gibberellin. This is one of the few diseases whose symptoms are easily explained and clearly understood.

Responses to Environment. Plants infected with a disease that causes wilt or affects water uptake and transport are much more susceptible to drought. A few less obvious responses have been noted, which may be physiological rather than mechanical. Resistance to extremes of temperature is sometimes affected by disease. Rust-infected wheat, for example, is more susceptible to frost injury but becomes more resistant to heat. Such responses might be due to changes in the plant's water balance or some other more basic physiological mechanisms.

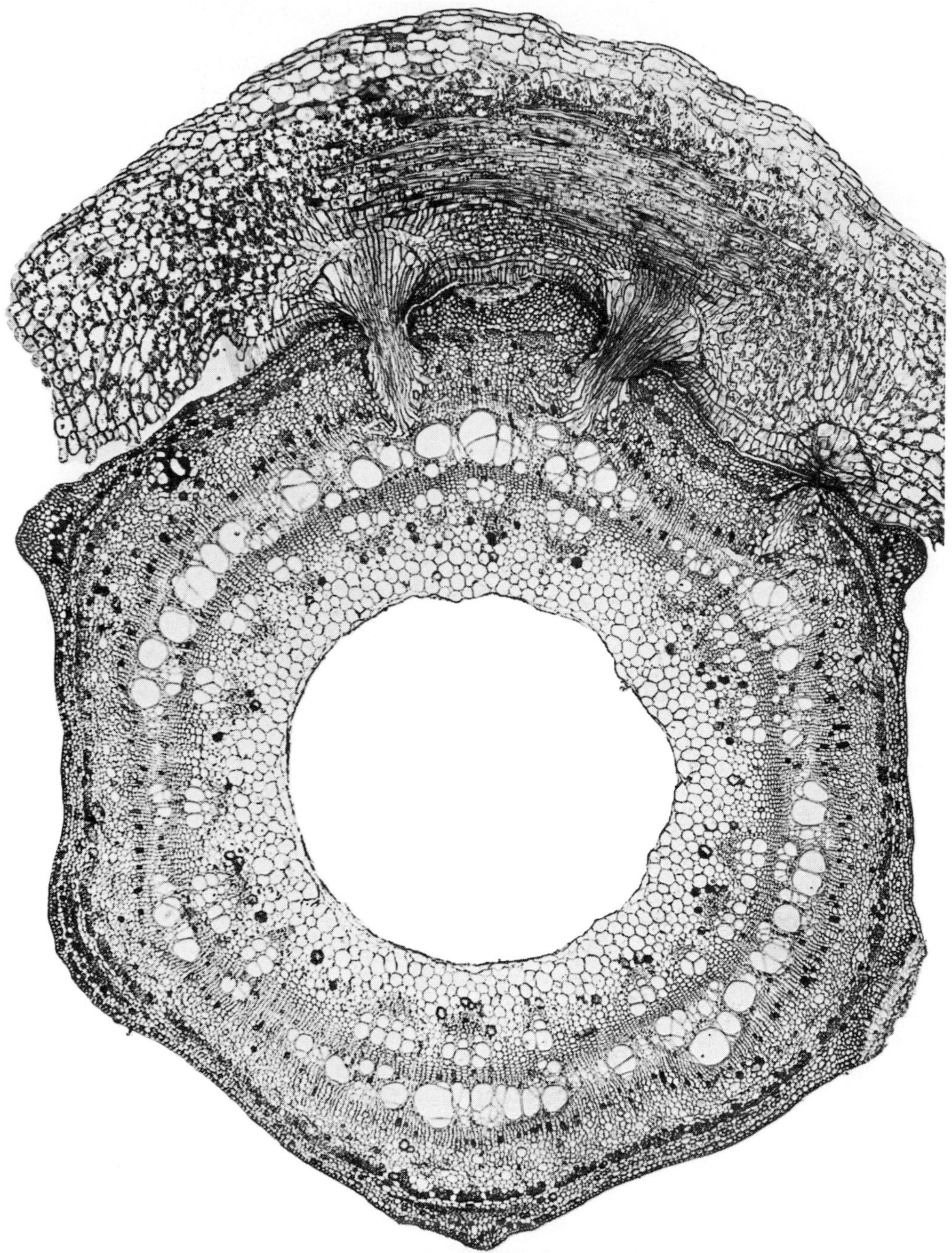

Figure 26-7. Micrograph showing haustoria of the dodder invading a plant stalk. [© 1960 by H. J. W. Uitgeversmaatschappij N. V., Amsterdam. English Text © 1960 by George G. Harrap & Co., Ltd. Reprinted from *Plant Marvels in Miniature* by C. Postmas by permission of the John Day Company, Inc., Publisher.]

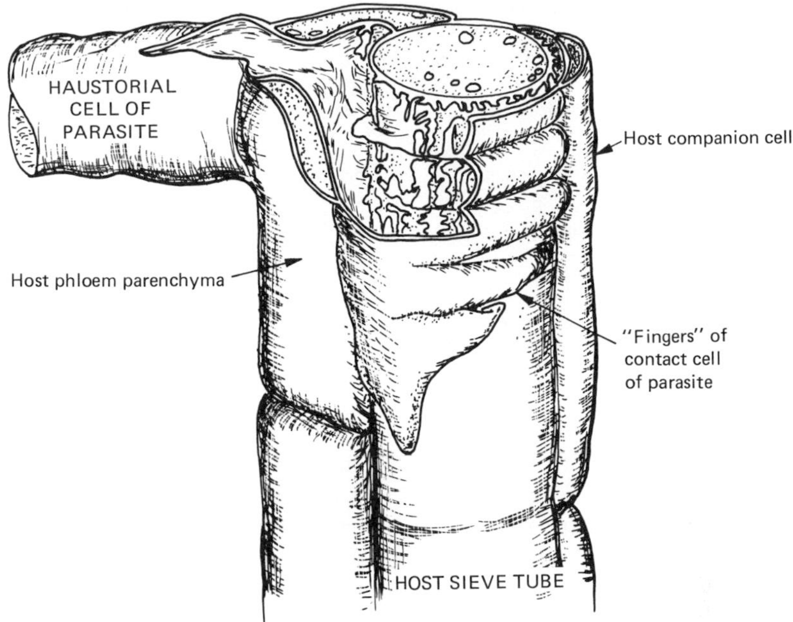

Figure 26-8. Drawing of a haustorium of dodder making contact with a sieve tube in the host phloem. [Redrawn from I. Dörr. *Protoplasma,* **75:**167–84 (1972).]

INJURY. Diseased plants may be mechanically injured by parasites or disease organisms. Blockage of transport vessels or even rupture of tissue by masses of infecting microorganisms may occur. Nutritional disorders follow shortages of various nutrient elements, which are diverted to support the metabolism of the parasite. This is particularly evident in the case of infection by parasitic higher plants like vines or such parasites as dodder (*Cuscuta* spp.). It has been determined that from 75 to 90 percent of ^{14}C-sucrose applied to leaves of parasitized *Vicia faba* is transported to the parasite *Cuscuta reflexa* in a period of 24 hr. The removal of the buds and growing tips of the parasite did not prevent the transport of sugars to it, suggesting that the mechanism causing diversion of the sugars from the host acted at the point of infection rather than as the result of translocation to sinks generated by the actively growing tips of the parasite. Mechanical strangulation may follow infestation of a tree or shrub by certain vines (for example, bittersweet, *Celastrus scandens*). Trees may succumb to the effects of shading by parasites such as spanish moss (*Tillandsia usneoides*), which does not seem to cause any serious physiological or pathogenic disorders in the host.

Host-Parasite Interaction

Much thought and considerable experimentation have gone into the question: What aspect of host-parasite interaction determines the high degree of specificity of most infections? The Canadian physiologist M. Shaw has pointed out that several levels of interaction between the host and the parasite are possible, as shown in Figure 26-9.

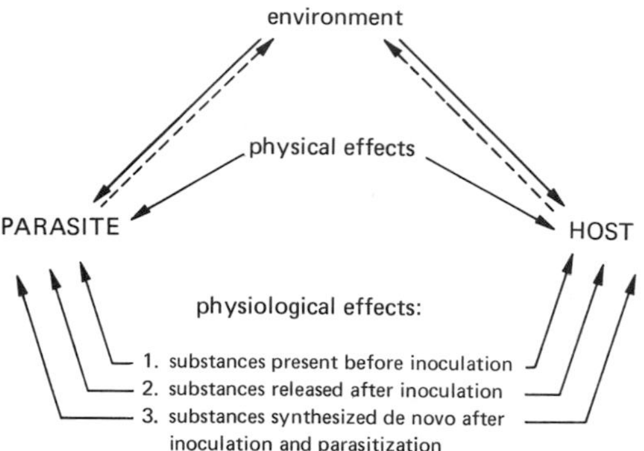

Figure 26-9. Summary of host-parasite interactions. [Modified from M. Shaw: *Can. J. Bot.*, **45:**1205–20 (1967).]

Clearly, the key to understanding and eliminating disease lies somewhere in this difficult and complex area of the physiology and biochemistry of these interrelationships.

Recently Shaw has shown that new proteins are made in rust-infected wheat leaves that are characteristic only of the infection but not of either the uninfected host or the parasite in pure culture. This suggests that "hybrid" functional proteins or enzymes may occur in the infected tissue that are necessary to the establishment or continuation of the host-parasite relationship. This in turn indicates that there must be an exchange of material between the host and parasite. There might, as in viral infections, be an exchange of DNA comprising whole cystrons or sections of cystrons. Alternatively an exchange of mRNA, or of some small molecules that could act as specific inducers to activate enzyme synthesis, could take place. It has been found that at least one disease organism, *Cronartium* sp., which causes pine blister-rust, will grow in culture when separated from the host tissue by a sheet of cellophane. This means that the exchange, if it takes place, must be of small molecules capable of diffusing through cellophane.

Whatever the exact mechanism of its establishment, it is evident that the host-parasite relationship should be considered as a new system, different from either the host or the parasite alone. The way is now beginning to open to the understanding of this sort of system. From understanding can come control. The elimination of disease by the control of the disease (that is, the host-parasite interaction itself), rather than by the attempted genocide of disease-causing organisms, would be a giant step forward for the science of agriculture.

Additional Reading

Articles in *Annual Review of Entomology and Phytopathology.*

Bollard, E. G., and **R. E. F. Matthews:** The physiology of parasite disease. In F. C. Steward (ed.): *Plant Physiology: A Treatise,* Vol. IVB. Academic Press, New York, 1966, pp. 417–550.

Chupp, C., and **A. F. Sherf:** *Vegetable Diseases and their Control.* The Ronald Press, New York, 1960.

Goodman, R. N., Z. Király, and **M. Zaitlin:** *The Biochemistry and Physiology of Infectious Plant Diseases.* D. Van Nostrand Co., Inc., Princeton, N.J., 1967.

Wood, R. K. S.: *Disease in Higher Plants.* Oxford University Press, London, 1974.

Yarwood, C. E.: Responses to parasites. *Ann. Rev. Plant Physiol.,* **18:**419–38 (1967).

27 Symbiosis

Types of Symbiosis

Symbiosis may be considered a form of mutual parasitism in which either one or both partners benefit from the association, but neither suffers from it. A great many different levels of association are possible, from a casual loose aggregation of different species to the close, precise, and permanent relationship of root nodules or lichens. All symbioses have one thing in common: the symbiosis permits one or both partners to better withstand the rigors of the environment.

Some associations have a clear host-invader relation, in which the host, though suffering no harm from the presence of the invader, does not benefit from it. This can be considered an extremely successful form of parasitism, in that the parasite does not prejudice itself by harming its host. Other associations are clearly of mutual benefit, each partner providing something that the other needs or can use. Indeed, the partnership may provide conditions or chemical substances that would be unavailable to either partner alone. It has been suggested that the presence of chloroplasts and mitochondria in the cells of eucaryotic plants evolved from symbiotic relationships between early heterotrophic eucaryotic cells and primitive procaryotic algae or bacterialike invaders.

Symbiotic associations may be **facultative,** where the partners can live either alone or in association, or **obligate,** where either one or both of the partners are unable to survive independently. All different levels of association are possible. **Epiphytes** merely grow or are supported on their host, such as ferns and some tropical orchids on tree trunks. An intimate association of the cells of both partners is achieved in lichens, which develop the relationship to the point of penetration of algal cells by fungal haustoria in some species. The closest association occurs when one parasite dwells inside the tissue or cells of the other, as in **mycorrhiza** (fungal-roots) and nitrogen-fixing bacteria (see page 192) or the not infrequent animal-plant association in which algal cells live inside the bodies of protozoa or coelenterates.

Specificity of the association varies from very low, as in the case of some (but not all) epiphytic plants, to very high, as in the case of nitrogen-fixing bacteria. This specificity is presumed to be due to the capacity of proteins such as lectins in one member of the symbiosis to recognize binding sites in the cell surface of the other member (see Chapter 26, page 613). The basis of specificity is very probably the same as that for parasitic infection. The epiphytic red alga *Polysiphonia lanosa* grows only on the nodes of the brown alga *Ascophyllum nodosum,* although adult plants are capable of surviving and photosynthesizing for long periods apart from the host. Presumably the mechanism of this extreme specificity is related to some requirement for the germination of the epiphyte spores, which is only met in that precise site on that particular host where the epiphyte grows. Other specificities may be related to precise nutritional requirements or balance, to hormone requirements, to the presence of specific metabolites, attractants, or repellents, to physical characteristics of the host plant, or to combinations of many factors.

The following sections briefly examine some characteristic symbiotic associations whose physiology has been investigated in some depth.

Associations

An interesting chemical association is found in rice paddies where hydrogen sulfide (H_2S) may accumulate to toxic levels in the anoxic rhizosphere. The bacterium *Beggiotoa* obtains its energy by oxidizing H_2S and is a common inhabitant of flooded rice paddies. The rice survives because the toxic H_2S is removed, and the bacterium flourishes because of its need for the O_2 produced by the rice seedlings.

Recently interest has been focused on "plant defense guilds." These are facultative associations of plants that are functionally interdependent with respect to herbivores. In some of these one member of the association supports a parasite of a serious pest. For example, California wild grape is attacked by a leafhopper that is in turn controlled by an egg parasite. The egg parasite requires wild blackberry for its alternate host. Grape and blackberry plants often grow in close association, and the grape varieties in the association delay their spring leaf production until the egg parasites have developed sufficiently on the earlier blackberry leaves to control the leafhoppers.

Other associations include toxic or repellent members that may sometimes even be mimicked by the more edible members. Protection may be by chemical or physical means, for example, by spines, or even by concealment. Many other mutually helpful associations are possible; these probably represent rather loose symbiotic systems that have not persisted long enough for close interdependence of species to have evolved.

Mycorrhiza

Mycorrhiza are small roots or root hairs of many species, mostly trees, that become infected with fungi and form a long-lived association in which the fungus dwells in or on the cells of the root. A mantle or sheath of fungal hyphae may surround the root, acting essentially like a sponge and replacing root hairs, which do not or cannot grow. Some morphological changes occur, particularly the formation of short, many branched roots with club-shaped swellings on the tips, as shown in Figure 27-1. These may result from auxin action; they can be simulated by the application of IAA, and many mycobionts are known to form IAA. Many fungi are involved (they may be members of the Basidiomycetes, Hymenomycetes, or Gastromycetes), and some of the associations formed are not very specific. Certain common forest mushrooms are the fruiting bodies of mycorrhizal fungi. Characteristically, fungi involved usually do not secrete cellulase or protease.

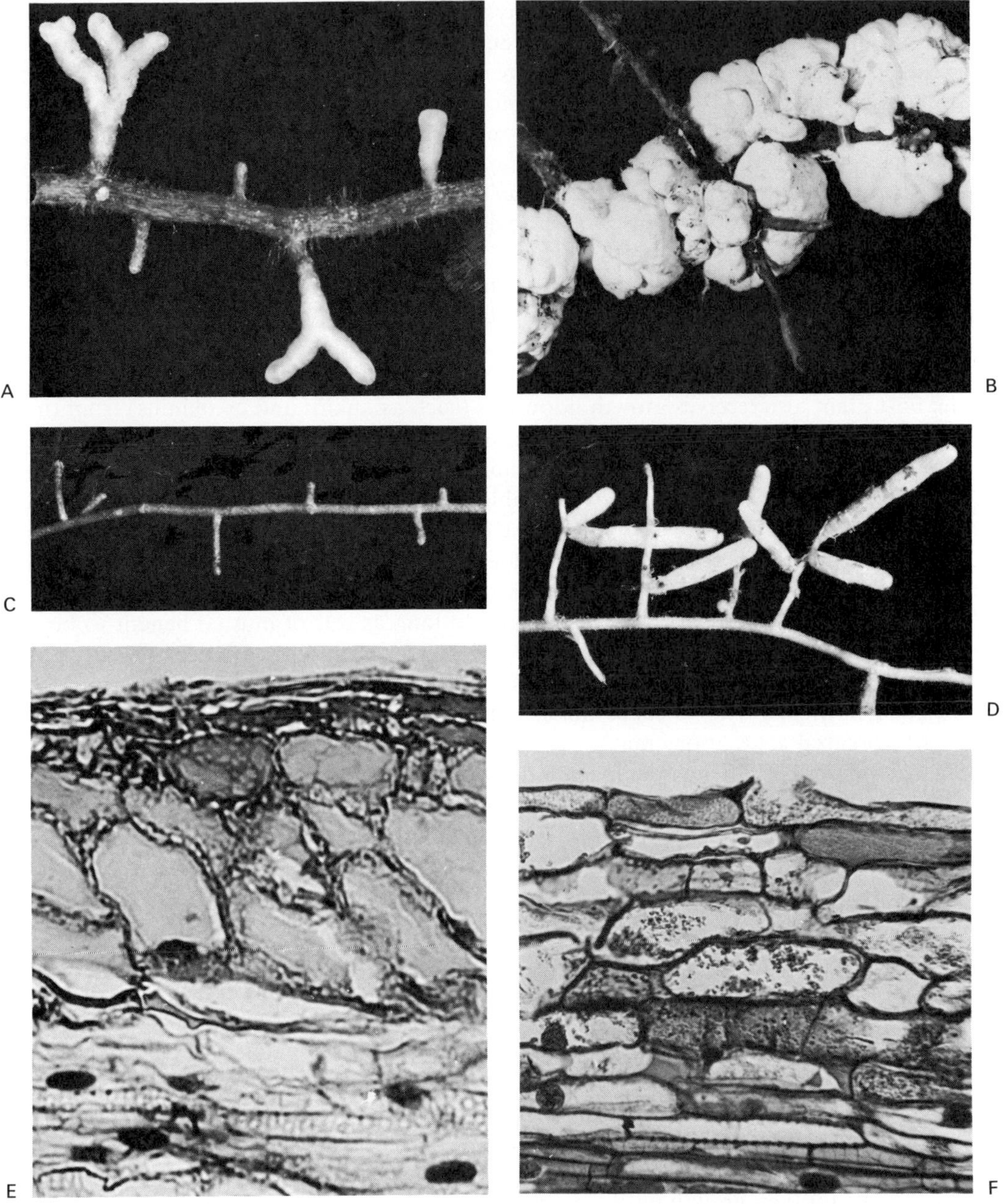

Figure 27-1. Mycorrhizal roots.
A. Two uninfected and three mycorrhizal short roots of pine (*Pinus strobus*) (×10). **B.** Short roots of pine (*Pinus strobus*) converted into tubercle or nodule mycorrhizae (×5). **C** and **D.** Sterile and mycorrhizal roots of yellow birch (*Betula alleghaniensis*) (×10). **E** and **F.** Longitudinal sections of *P. strobus* short roots showing epidermis, cortex, and part of the stele. Note that the corticle cells have enlarged transversely in the infected root (E), and the intercellular spaces are filled with a network of fungae hyphae (×500). [Photographs courtesy Dr. V. Slankis, Canadian Ministry of Natural Resources, Forest Research Branch, Maple, Ontario.]

Some unknown factors seem to be involved in the establishment of mycorrhizal associations. Although the fungi will grow on a complex defined medium, their growth is greatly increased if some tree root pieces are cultured with them. Some exudate from the root apparently controls or may even prevent entry of the fungus. Tomato roots prevent the entry of fungi, and in pine roots the infection is confined to specific parts of the rootlets.

Mycorrhizal symbiosis is not one sided. The fungi absorb sugars from the host, and other factors such as B vitamins, α-keto acids, and amino acids. On the other hand, the absorption of minerals by roots is greatly increased by the presence of mycorrhiza, possibly because of changes in the permeability of the root cells. The presence of mycorrhiza is necessary for the normal growth of many trees. Although it was previously thought that mycorrhiza function only as nutrient-supplying organs for the host plant, it is now apparent, particularly from the work of the Canadian physiologist V. Slankis, that the host plant also receives growth hormones (such as auxins and cytokinins) from the symbiotic fungus. These excess hormones profoundly affect not only the external appearance (Figure 27-1A to D) but also the internal morphology (Figure 27-1E and F) of the roots. They may well be responsible for the more efficient mobilization and transport of nutrients in the host plant (see Chapter 21, page 523). Also, the more intense green color and the increased drought and temperature resistance of mycorrhizal plants may be ascribed to the increased hormone concentrations in the host plant.

The association appears to continue only so long as it is of mutual benefit to both partners. Factors that reduce the capacity of the tree to supply nutrients (for example, excessive shade) or that reduce the tree's need for absorption, such as oversupply of water or nutrients, tend to cause the association to break up and disappear.

Mycorrhizal associations may be complex and extensive, involving more than one macrosymbiont. The heterotrophic chlorophyll-lacking plants of the genus *Monotropa* (pine sap, Indian pipe), which exist on the floor of temperate coniferous forests, were originally thought to be saprophytic. They are now known to obtain nutrition from neighboring trees by the passage of salts and carbohydrates via mycorrhizal hyphae, which are associated with both their own roots and those of nearby trees.

Orchids

Most green orchids and all saprophytic ones form a close, obligatory association with a fungus, usually of the genus *Rhizoctonia*. The association is essentially mycorrhizal, with the fungus penetrating the roots and root cells of the orchid. However, the association goes beyond that of tree mycorrhiza in several ways.

Orchid seeds are minute and incapable of germination unless provided with a supply of nutrients for their growth until they become autotrophic. In nature, many such seeds will not germinate until they are infected by a fungus, which presumably provides the stimulus to germinate. Subsequently a flow of carbohydrates and nutrients is maintained from the fungus to the orchid, which thus grows essentially as a parasite of the fungus. However, when the orchid becomes green and autotrophic or develops roots or rhizoids permitting saprophytic nutrition, the roles are reversed. Now the fungus receives its nutrition from the orchid, and the plant that was formerly a parasite on the fungus becomes parasitized by it.

The development of some orchid tubers is similarly dependent on infection by fungi, which supply the initial stimulus to grow and the nutrition for growth and later become parasitic on the autotrophic plant. This sort of association differs from other symbioses in that, although bidirectional movement of nutrients occurs to the benefit of both partners, it is separated in time. However, it is an effective and successful association.

Some saprophytic orchids such as *Corallorhiza* and *Neottia* have a very close association with their mycorrhiza that includes extensive fungal cell destruction. The mycobionts live and appear to thrive in the outer cortex of the host roots or rhizoids, performing an essential duty to the symbiosis by absorbing nutrients and water. However, the fungus penetrates deeply into the inner layers of the cortex and is there lysed or digested by the host tissue. It is not clear whether this is a reaction to excessive activity amounting to pathogenicity by the fungal partner or merely a method for the nutrition of the orchid. In spite of the similarity to a disease reaction, this symbiosis is obligatory and remarkably stable.

Lichens

Lichen Associations. Lichens form one of the most successful symbiotic relationships. The composite plant has an entirely separate and different existence from either of the partners, and its morphology is often so different that either partner cultured separately is unrecognizable. Several different groups of fungi are found in lichens, Ascomycetes, Basidiomycetes, and Deuteromycetes being the most common. The algal partner is usually a single-celled or filamentous Chlorophyceae, although Cyanophyceae are not uncommon. Filamentous growth does not occur in algae that are associated in lichens, however. The algal partner is usually present as single cells or in small clumps of cells (Figure 27-2). The fungus is the dominant partner, controlling the morphology and fruiting of the lichen.

The association is one of convenience only and breaks down due to death or uneven growth of the partners if nutrients, water, or other factors get out of balance. The lichen association is formed easily. "Trial marriages" may occur when the germinating fungal spores encounter algae. If the algae are capable of supporting or withstanding the demands of the fungus (which may include haustorial invasion of the algal cells), a permanent association is formed. Additional forms may also be involved. Some lichens contain nitrogen-fixing bacteria as well as algae, and are thus very independent.

One of the more obvious characteristics of lichen metabolism is that it is rather slow. Rates of carbon fixation and respiration are only a fraction of those of higher plants on a weight basis, and the growth rate of some lichens is so slow that it is measured in millimeters per year. In fact, the slow (and quite predictable) growth of lichens has been used as a historical dating technique, for example to establish the age or period of most recent disturbance of tombstones. Lichenologists have also used lichen growth rates and the size of existing plants to establish the rates of advance or retreat of glaciers.

Lichens can survive in extremely inhospitable places, on bare exposed rocks and in poorly lighted crevices, and are the first invaders in many ecological successions (see Figure 27-3). They are able to do this partly because their nutritional requirements are very low, as a result of their low growth rate. However, their extremely fast rate of water loss and absorption is probably more important as a mechanism of survival. When conditions become unfavorable, lichens dry out very fast and thus are able to withstand much greater extremes of heat than most tissues. When conditions again become favorable for growth, they absorb water rapidly and quickly resume their capacity for metabolism.

Metabolic Interactions. It thus appears that the algae are "protected" within the thallus of the lichen, and provided with water and mineral nutrients as they are available. In return, the alga provides the carbon nutrition for both partners. There is

Figure 27-2. Diagrams of cross sections of the three main types of lichens: **A,** Crustose; **B,** foliose; **C,** fruticose. [Reprinted by permission of the publisher, from Vernon Ahmadjian: *The Lichen Symbiosis*. Ginn and Company, Boston, copyright 1967.]

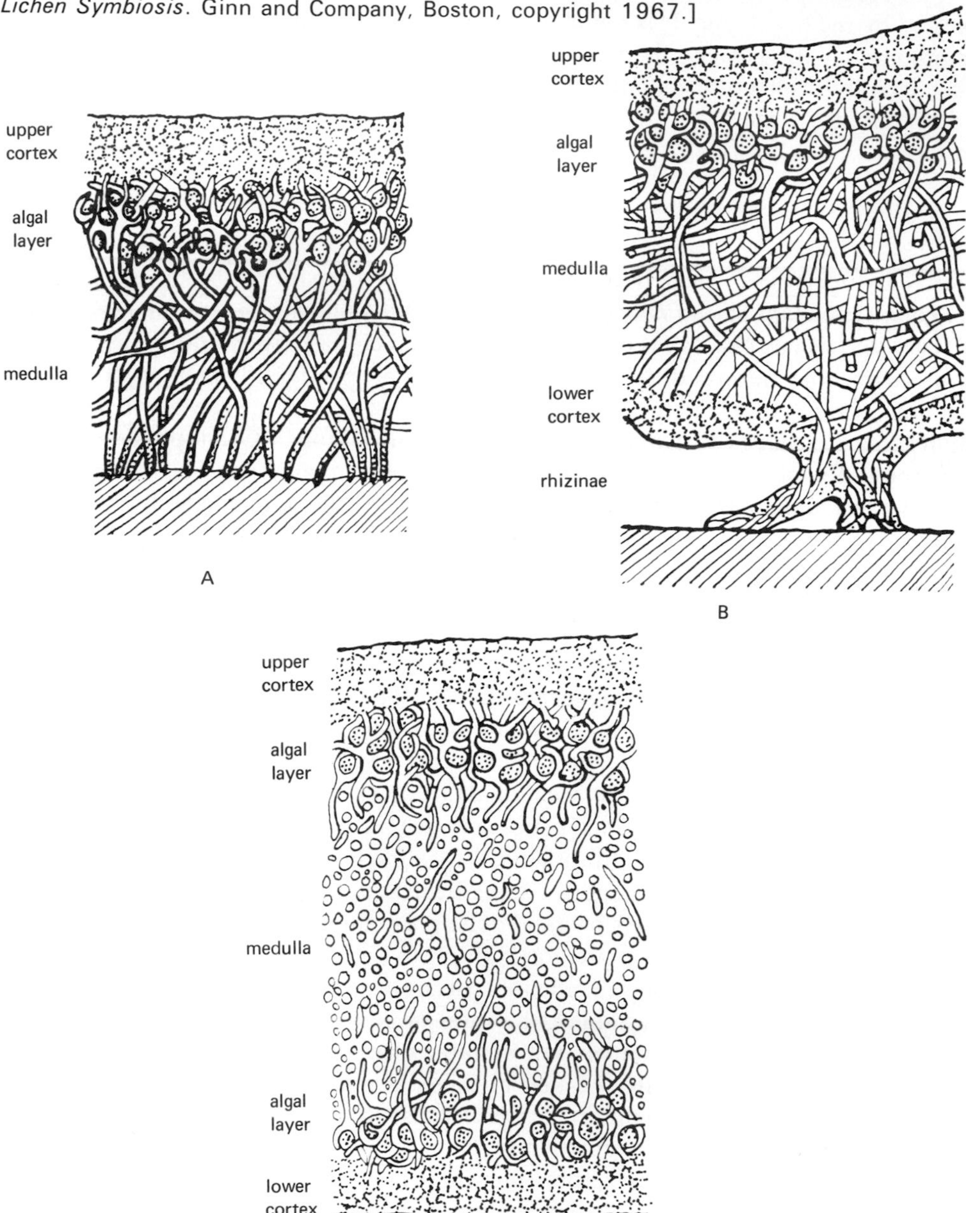

Figure 27-3. Lichens growing on (**A** and **D**) granite rock, (**B**) the side of a stone building, and (**C**) a tree trunk. They get little or no nutrient from their substrate. [Photographs (A) and (B), from the collection of the late Dr. R. Beschel, Queen's University; (C) courtesy G. Bidwell; (D) courtesy Dr. W. Maas, National Research Council of Canada, Halifax, N.S.]

some evidence that substances from the fungus, possibly auxins or ascorbic acid, stimulate the photosynthesis of the alga. Algae from lichens characteristically tend to retain much of their fixed carbon in the soluble form, rather than converting it to starch or other reserves.

When algae enter symbiosis, certain metabolic changes occur. Surplus soluble carbohydrates are produced in much larger amounts, and some new carbohydrates may be produced that were not produced by the algae during free growth. Changes in the rate of carbohydrate production may be related to the fact that growth of the alga is greatly reduced, although its rate of metabolism may not be. Products of photosynthesis of symbiotic algae may include much larger proportions of polyhydric alcohols, characteristic of lichens, than are found in the free-living form. Carbon from the symbiotic alga is unquestionably used for the nutrition of the whole mass of the lichen.

When lichens are supplied with $^{14}CO_2$ in light, radioactive carbon fixed by photosynthesis in the alga moves rapidly to all parts of the lichen, even where no algal cells are present. The fungal partner can often be cultured well on simple sugars or disaccharides, so the contribution of the alga is probably limited to carbohydrate food. There may in addition be some metabolism that is characteristic only of the association, but not of either partner alone. Certain diphenolic compounds called **depsides** are produced in lichens, which are not found in either the alga or the fungus. The fungal component may be able to produce the monophenolic precursors, but only the complete association is able to produce depsides.

The photosynthesis of algae in fungal association is much affected by the fungus, and the export of photosynthate from algal cells is dramatically affected by the association. In one experiment, $^{14}CO_2$ was supplied to the *Trebouxia* (algal) symbiont of the lichen *Xanthoria aureola*. Export of photosynthate from the algal cells declined from 40 percent of fixed carbon in the lichen to only 2 percent in the isolated cells, which made direct use of a much higher proportion of fixed carbon (Table 27-1). It appears that the fungal component of the symbiosis affects or controls the export for its own use of food manufactured by the algae.

The rate of metabolism of the algal component also speeds up during periods of rapid growth or activity (such as fruiting) of the fungus. A possible mechanism has been suggested by the American physiologist V. Ahmadjian. As the fungal metabolism increases, it produces larger quantities of urea from arginine and other sources. This is hydrolyzed by the enzyme urease, commonly found in lichens, to produce ammonia and

Table 27-1. Export of photosynthetically fixed ^{14}C from the algal symbiont of a lichen.

	Amount of fixed $^{14}CO_2$, %		
	In lichen association	*Algae freshly isolated*	*Algae cultured free-living*
Exported	40	8	2
Retained in soluble phase of algae or respired	58	72	48
Incorporated into algal insoluble phase	2	20	50

(Data of T. G. A. Green, recalculated from D. C. Smith: *The Lichen Symbiosis*. Oxford University Press, London, 1973.)

carbon dioxide, which stimulates the alga to increase photosynthesis. The alga is thus called upon to produce more carbohydrate to nourish the fungus at times of increased need. Other mechanisms based on hormonal interaction have been suggested, although without clear supporting evidence, and it is possible that several such mechanisms of interaction exist.

Water Relations. The algal partner in lichens is sensitive to light, and may be bleached or else may grow too fast for the continuation of the association if it receives too much light. An interesting mechanism has developed for regulating the light reaching the phycobiont, based on the ability of lichens to react rapidly to moisture conditions.

The cortex of the fungus acts as a light screen. When the sun shines brightly, it dries out rapidly. Conversely, in shade it absorbs moisture from the surroundings. When the fungal cortex is dry, much less light passes through than when it is wet, due to changes in light scattering and reflection. Also, cortical cells contract when dry, and the increased density of cell walls acts as a light filter. Finally, the algal cells shrink as they dry, so absorbing less light per cell. Thus, within the wide limits often met by lichens growing in exposed conditions, the comparative activities of the fungal and algal partners are kept in the proper relationship. This prevents the destruction of the partnership which follows imbalance in this relationship due to excessive growth or metabolism by either symbiont.

Pigments. Control of photosynthesis and growth of the phycobiont is further achieved by the synthesis of a number of characteristic pigments. Lichens are noted for the production of intense orange, red, yellow, brown, or even black pigments when growing in exposed situations, whereas members of the same species growing in the shade are gray or white. The deposition of pigments undoubtedly helps screen the sensitive phycobiont from excess light. The mechanism whereby such pigment formation is stimulated is not clear. Lichens growing partly in shade and partly in light form a precise demarcation of pigmented and nonpigmented areas. This shows that not only are the pigments not transported but the stimulus for their formation is precisely delineated by the site of illumination, and does not extend beyond it. The mechanism of light perception for this reaction is unknown but presumably resides in the algal symbiont.

Algae-Invertebrate Symbiosis

Many symbiotic associations are known between algae of various kinds and various invertebrates. Some of the invertebrates are quite large and immobile, including the giant clam and other molluscs, sea anemonies and hydras. Others are small, including coelenterates, flatworms, and protozoans. The mobile forms often seek illuminated areas where photosynthesis is optimal, and some anemonies seek shady places or withdraw their tentacles if illuminated with very high (above optimal) light intensities. Algae (or, on occasions, chloroplasts) are often located inside the animal cells and frequently divide synchronously with the host cells so that the number of algal cells per host cell is constant.

The algae photosynthesize and transmit specific products of photosynthesis to their hosts. If $^{14}CO_2$ is supplied, a range of radioactive products is produced inside the algal

Table 27-2. Products of photosynthesis in $^{14}CO_2$ by a zooxanthelline alga isolated from symbiotic association in a coelenterate dinoflagellate, as affected by incubation with homogenate of host cells.

	Products in cells	*Products transferred to medium*
	(relative size of chromatographic spot)	
Organic phosphates	+++	+
Glucose	++++	+
Glycerol	+++	++++
Lipid	+++	–
Glutamate	++++	–
Alanine	+++	++
Succinate	+++	++
Glycolate	–	+

(Recalculated from data of R. K. Trench in D. C. Smith: *Symbiosis of Algae with Invertebrates*. Oxford University Press, London, 1973.)

cells. Little material leaks out of the algal cells if they are cultivated outside the animal host. However, if a homogenate of animal host cells is added, there is an immediate release of selected compounds to the medium (Table 27-2). This suggests that the animals influence the export of metabolites by the algal cells. Clearly the algae do "feed" their hosts, and do so on command from the host cells, as happens in lichens. Some of the algae (though not all) are also nitrogen fixers, which would make them specially valuable to their animal hosts.

The obvious advantages of this sort of symbiotic association, and the variety and large number of them that are known, have been considered as positive evidence for the evolution of chloroplasts from associations of algae with heterotrophic host cells (see page 57). Of course, these present-day associations really do not constitute proof of the endosymbiont origin of eukaryotic organelles. However, it is easy to speculate that environmental pressures from nutritionally deteriorating environments could have favored the development of such associations.

Nitrogen-Fixing Symbiosis

The well-known association of *Rhizobium* bacteria with the roots of legumes is described in Chapter 8 (page 192). Many other symbiotic associations are known, and several of them involve interesting developmental modifications of the host tissue that provides more efficient expression of the combined capabilities of the symbiosis. In several associations, such as those between algae and ferns, liverworts or some higher plants, little obvious morphological modification takes place. In others, as in legumes and in the association between alder and an Actinomycete, nodules form on the roots in which the invading symbiont lives.

One of the more interesting modifications is in the symbiosis between the roots of the primitive Cycad *Macrozamia riedlei* and a species of the blue-green alga *Anabaena* shown in Figure 27-4. The infected roots become swollen at the tip and reverse their normal geotropic response so that they grow upward toward the surface of the ground where the algal symbionts would presumably intercept sufficient light for photosynthesis.

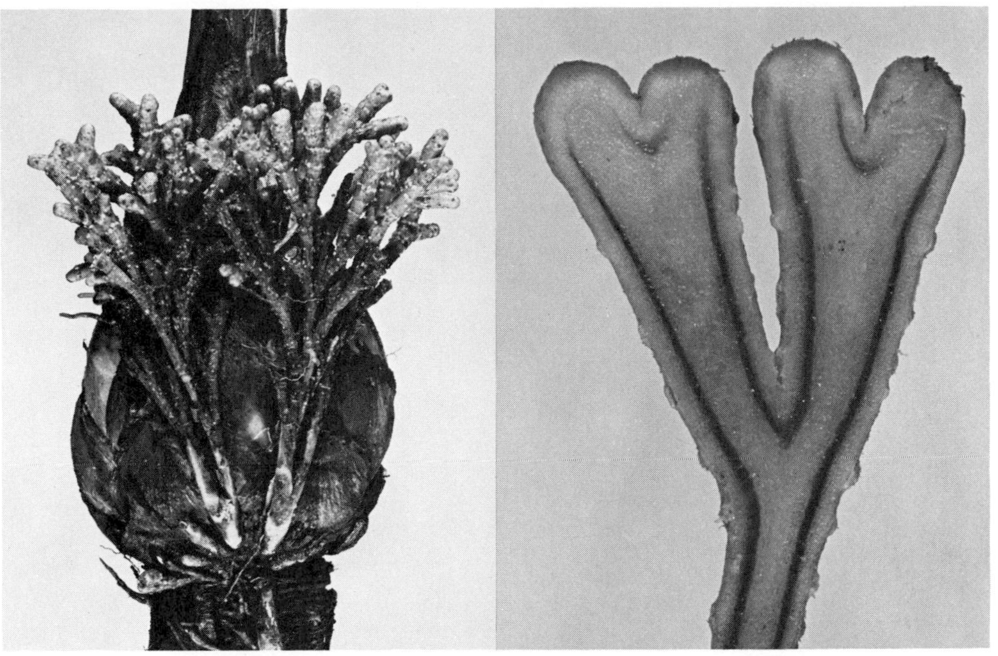

Figure 27-4. Roots of the cycad *Macrozamia riedlei* infected with the nitrogen-fixing blue-green alga *Anabaena* sp. Infected roots have lost their normal positive geotropism and grow upward (**A**), and become swollen and club shaped (**B**). (A originally published in J. S. Pate: Transport in symbiotic systems fixing nitrogen. In V. Luttge and M. G. Pitman (eds.): *Encyclopedia of Plant Physiology*. New Series, Vol. **2B,** pp. 278–303, Springer Verlag, Berlin, 1976. Used with permission. Photographs kindly supplied by Prof. J. S. Pate.)

The mechanism underlying this modification of root growth has not been investigated but might be interesting in the study of geotropism.

Additional Reading

Ahmadjian, V.: *The Lichen Symbiosis*. Blaisdell Co., Waltham, Mass., 1967.
Atsatt, P. R., and **D. J. O'Dowd:** Plant defence guilds. *Science,* **193:**24–9 (1976)
Scott, G. D.: *Plant Symbiosis*. Edward Arnold Ltd., London, 1969.
Smith, D. C.: *Symbiosis of Algae with Invertebrates*. Oxford University Press, London, 1973.
Smith, D. C.: *The Lichen Symbiosis*. Oxford University Press, London, 1973.
Smith, D., L. Muscatine, and **D. Lewis:** Carbohydrate movement from autotrophs to heterotrophs in parasitic and mutualistic symbiosis. *Biol.* Rev., **44:**17–90 (1969).

SECTION VI

Physiology of Plant Distribution and Communities

28 Physiology of Plants Under Stress

Introduction

Normal physiology is maintained under ideal environmental conditions. However, plants seldom exist under just the right conditions. Usually something is lacking; often several factors are far from ideal. Because of the fact of competition, plants often live at the limit of their capability to survive one or more adverse conditions. This creates considerable stress in the organism, which reacts by various biochemical and physiological mechanisms to overcome, avoid, or neutralize the stress.

A considerable amount of research has been done on plants under stress for two reasons. The first is that insight into physiological mechanisms may often be gained by studying plants whose mechanisms are affected by specific types of stress, and the mechanism of the plant's reaction to stress may be worth studying for itself. Second, in agricultural situations plants are frequently under stress of one sort or another, and their ability to withstand such stress is of extreme economic importance. In addition, many agricultural problems stem from the fact that good land exists in areas that have difficult or unfavorable climatic conditions—winter cold, early frost, long periods of drought, and so on. Less than 10 percent of the land surface of Earth is suitable for cultivation. There is a real need to create crops that can resist or tolerate the stresses of these extremely unfavorable conditions and make use of what might otherwise be unusable land. It is easier to breed plants that are more resistant or tolerant to stress if we understand the mechanisms of stress tolerance and resistance. Such plants may, in the long run, be more effective than high-yield varieties for sustaining the "green revolution" in developing countries that have tropical or otherwise stressful climates.

Effects of Stress

Stress of any sort is essentially analogous to an applied force; the organism must yield to some extent. The reaction to stress may be **elastic,** that is, after the stress is removed the organism returns to its former state. Alternatively the reaction may be **plastic,** the organism remaining deformed or changing in some way as a result of the stress. In either case, if the stress is too great, something must break. The organism becomes irretrievably damaged and dies.

Stress may have a direct effect on the organism, immediately observable. This may or may not be accompanied by **conditioning effects.** Many plants become more resistant to stress after exposure to sublethal doses of the stress, a process called **hardening.** Winter grain, for example, can survive low temperatures in winter after prolonged exposure to progressively lower temperatures during the fall. Exposure to similarly low temperatures in summertime would quickly kill the plants because they are not then hardy.

In certain plants stress may produce effects that are carried over one or more generations and behave as if they were inherited factors. The American physiologist H. Highkin found that pea plants grown under abnormally low temperatures became smaller from generation to generation, becoming much dwarfed after eight generations. If seeds from dwarf plants were grown under normal conditions, they nevertheless produced dwarf plants. Only after eight generations under normal conditions were normal plants again produced. Furthermore, crosses between dwarf and normal plants produced intermediate offspring.

The genetic basis of stress resistance is only now beginning to receive adequate study. Genetic adaptation can be achieved in two ways: by evolving a genotype that confers resistance (this may be a major process involving many genes) or by developing a set of genes that are capable of producing various phenotypes adapted to different environments as required. It may be difficult, in fact, to differentiate between plants that have developed resistance as the result of these two mechanisms. However, plants developed by the second mechanism, that is, with adaptability built into their genetic makeup, would be more versatile in agriculture than plants developed for a specific condition.

It is worth noting that the initial developmental program with which a plant starts its life may be varied by environmental stress without any genotypic response. Due to environmental effects on the metabolism, translocation, and growth of the parent plants, the composition of their seeds may subsequently be affected. Thus, the experience of the parents may be transmitted to their offspring without the intervention of genetic mechanisms of any sort.

The reactions of plants to environmental stress are complex and involve many kinds of physiological response, from simple chemical or biochemical direct responses, through complex hormonal or developmental responses, to inheritable effects that appear to be genetic in character. The inference is that the study of stress response or stress resistance is as complex as all of plant physiology. Indeed, the study of these aspects of plant behavior, which are fundamental to agriculture and to the study of plant ecology, can really be said to be the most important emerging branch of plant physiology.

Types of Stress

The major kinds of stress to which plants are exposed are the environmental stresses resulting from extremes of climate: drought, heat, cold, and frost. Certain other stresses result from the geographic or physical location of plants and their proximity to one another: shade, radiation levels, soil deficiencies or excesses (including minerals and water, Chapters 10 and 11), and altitude, which is a complex of many stresses. Other kinds of stress may result from the toxic effects of natural or artificial (for example, industrial) pollution, ionizing radiations, the leaching effects of excessive precipitation, and so forth. We shall focus our attention primarily on the effects of, and adaptations to, the major environmental factors of drought, high temperature, low temperature, and freezing.

The effects of these factors are closely interrelated. Resistance to high temperature may also involve resistance to drought conditions, which often accompany it. Resistance to freezing appears to be largely interconnected with resistance to dehydration of tissues. The development of hardiness to one factor often confers some degree of hardiness to other stresses. As a result, the study of stress resistance has been difficult and slow, and no far-reaching general theories are widely held. We shall consider special studies of resistance to specific stresses, but it must be continuously borne in mind that stress resistance is a complex and multifaceted phenomenon; all the details must fit into any general theory before it can be accepted.

Stress Resistance: Avoidance and Tolerance

Resistance to stress is not a simple phenomenon, nor is there merely one mechanism for the resistance to any single kind of stress. Two broad types of stress resistance are avoidance and tolerance. It should be noted that the terms avoidance and tolerance do not imply any kind of active capacity on the part of the plant to determine its own fate; they are merely convenient terms to describe different types of reaction mechanisms.

Avoidance is usually based on a mechanism whereby an internal environment is created within the plant such that its cells are not under stress, even though the external environment may be very stressful. Examples are a leaf that avoids high temperature by transpiring, thus maintaining a lower internal temperature; or a cactus plant that avoids drought by intense conservation of its internal water, so that it does not suffer from drought internally.

Tolerance, on the other hand, is the capacity to endure the stress; to survive or even to function normally under internal as well as external conditions of extreme stress. Examples are certain mosses, which can endure extreme desiccation in times of drought but revive on rehydration, or algae and bacteria, which can live and thrive in hot springs, functioning at temperatures that would kill other organisms.

Both types of resistance have developed for most stress situations, and both types may be present in the same plant. Stress avoidance does not necessarily involve specialized physiology but merely mechanistic or morphological devices that enable the plant to escape from the effects of environmental extremes. As such, this type of stress resistance is not as interesting to the physiologist as stress tolerance. Tolerance implies the development of special physiological mechanisms that enable the organism to survive under conditions which would be inhibitory or lethal to nonhardy species or individuals. We shall consider the mechanisms of tolerance to the more important forms of environmentl stress in this chapter. We shall then continue in the next chapter to consider how stress-hardiness is concerned in the distribution of plants throughout the world, and how it impinges on agricultural practice.

Measurement of Hardiness

Measuring hardiness is extremely difficult because tolerance of stress varies greatly, not only with the individual, race, or species of plant but also according to the previous history of the individual. Thus, plants can often develop or acquire hardiness by exposure to sublethal stress, but sometimes it develops as the result of other apparently unrelated experiences such as changes in day length. In addition, the rate of application of a stress as well as its duration are vital interacting factors that determine the magnitude of effect of the stress.

Most attempts at measuring hardiness have been made by determining the degree of stress, arbitrarily applied (but under carefully controlled conditions) that kills 50 percent of the experimental population. This does not measure the spread of resistance (for example, what percentage of the population would survive a 5 or 10 percent greater or lesser stress). However, simulating natural conditions in the laboratory is usually difficult or impossible or simply too time consuming. Thus, this form of stress measurement is not particularly useful to agricultural scientists. In addition, the natural processes of hardening need to be considered when measuring stress resistance in crop plants. As a result, farmers are more apt to consider the extremes of stress (duration as well as intensity) that a given crop can endure without serious damage (perhaps with less than 10 percent mortality). Since the duration and intensity of stress are often linked together in complex ways (usually, the longer duration the lower intensity that can be resisted), the best possible measurements of hardiness are only rough guidelines to hardiness in the field.

Drought

Drought Avoidance and Tolerance. Drought is probably one of the most common stresses that plants must endure. Many mechanisms of drought avoidance have developed. Annual plants survive periods of drought as seeds; desert plants may live their entire life cycle during a brief period following a rain. Many plants have developed special means for absorbing water efficiently or for retaining it strongly (cuticle, stomatal modifications, and so on, see Chapter 14). Such plants survive drought because their inner tissues are protected from being subjected to too high a stress. There are some species of desert plants, like the cactus *Opuntia,* which survive and even continue to metabolize for months under the most extreme conditions, as for example in a desiccator surrounded by the strongest available water absorbers. However, these plants are merely retaining water. If the drought conditions are sufficiently extreme, or continue long enough, they will lose water in spite of their protective mechanisms. Then, if their tolerance is not high, as is often the case, they must succumb.

Consequences of Dehydration. Drought tolerance mechanisms are not yet completely understood. The consequences of drying out are complex for living protoplasm. Drought often accompanies the problem of excess heat, which causes a number of characteristic lesions leading to disintegration and death. These will be considered in the next section (page 642).

The first direct consequence of dehydration is probably the loss of water molecules that act as protective layers about colloidal micells, on membranes, and on (as well as in) the complex convolutions of the tertiary structure of proteins. Water molecules act not only as a solvent for chemicals but as spacers that help to keep complex fluids in a stable configuration. When they are removed, charged particles or surfaces move closer together. Not only do solutions become more concentrated but the reactive colloidal surfaces approach each other to the point where they coalesce and denature. The increased concentration of cell sap and intercellular fluids causes a great decrease in the water potential of the fluids, which further subjects the protoplasm to stress by an increased tendency for water loss. Other effects of concentration may occur—the imbalance of biochemical processes caused by unnaturally high concentrations of metabolites may contribute to molecular disruption. In addition, the high concentration of certain solutes may effectively "salt out" proteins. The same result may follow changes in cell pH caused by the concentration of ionized acidic or basic solutes.

Mechanisms of Drought Tolerance. Because drought has such varied effects, it is not surprising that several different mechanisms for drought tolerance seem to have developed. All land plants presumably have some degree of drought tolerance. In most plants this seems to be conferred by the presence of hydrophilic substances in the protoplasm. These may be complex, high molecular weight substances, like proteins themselves, or certain carbohydrates, such as the alginic acid and other colloidal polysaccharides of many marine seaweeds. Low molecular weight compounds can have a twofold effect. Some may be strongly hydrophilic, such as the polyhydric alcohols which commonly occur in littoral seaweeds. These plants are subject to severe drying stresses between tides and, being unprotected by a cuticle, must rely on internal devices to retain water. Even if they are not specifically hydrophilic, low molecular weight substances such as sugar are often elaborated in times of drought, perhaps because their presence in solution directly lowers the water potential of cell sap, thus helps to retain water.

However, such devices merely result in water conservation and do not help to protect delicate protoplasm from dehydration. Water held tightly by colloids or strong solutions is as strongly withheld from the cytoplasm. Thus some plants having very high sugar concentration, such as sugarcane, are also susceptible to drought, whereas others, such as pine, which contain only low concentrations of sugars and other solutes, are highly resistant. Evidently the important factors of drought resistance lie more deeply in the basic chemistry of the protoplasm.

The American physiologist Y. Vaadia has suggested that drought hardiness may relate to the plant's capacity to bind water to proteins. Such bound water may be present in a configuration approaching the crystalline state of ice, strongly resistant to removal from the tissues. It has been suggested that under the stress of drought certain types of resistant proteins appear, perhaps characterized by a configuration that resists denaturation (that is, does not form internal or intermolecular bonds readily). Drought hardiness would then depend mainly upon the plant's ability to synthesize certain proteins.

Attempts to isolate such proteins or to transfer drought resistance by extracted plant fractions that might contain them have given only equivocal results. The Russian physiologist P. A. Henckel has suggested that resistance is associated with protoplasmic elasticity. However, he points out that most factors conferring or associated with drought resistance, such as smaller cells, high nucleic acid content, lower starch, and higher sugar and colloid content, are built into the plant during its development under the influence of water deficiency. Thus these mechanisms of drought resistance are presumably not involved in the innate tolerance of sudden or unexpected drought.

Metabolic functions in some drought-tolerant plants are relatively undamaged by desiccation, as photosynthesis in the red alga *Porphyra* which recovers immediately on rewetting. In other plants, such as the drought-tolerant mosses and yeast, the important characteristic is an ability to repair or rebuild the drought-damaged photosynthetic or respiratory mechanism, not to maintain it.

Certain plants have developed the capacity to tolerate or survive extraordinary extremes of drought. Some mosses survive drought in an essentially desiccated form but are reactivated on rehydration. Many desert plants, such as the creosote bush (*Larrea divaricata*), can survive for long periods with a water content as low as 30 percent of their total weight, although active growth and metabolism virtually cease under these conditions. Just what physical or chemical property of the protoplasm in these plants permits this behavior is not known. It appears to relate to the capacity of the protoplasm to bind water, which is then retained with extraordinary tenacity by the tissues.

Heat

Limits of Heat Tolerance. Plants vary greatly in their tolerance to heat. Some heat avoidance is possible in such organs as transpiring leaves, but a temperature decrease of no more than a few degrees is possible by transpiration and then only at the expense of a greatly increased water loss. Most plants that survive high temperatures do so because they have internal characteristics which enable them to endure or tolerate heat. Desert plants are usually characterized by high heat tolerance. Members of the genus *Cactus,* which are drought-avoiding plants, can stand up to 60° C, and species of *Atriplex,* characteristic drought-tolerators, can survive temperatures to 50° C. Certain lower plants, algae, fungi, and some bacteria, can withstand even higher temperatures. Organisms that live in volcanic hot springs can endure temperatures close to the boiling point of water. On the other hand, most plants not acclimated or specialized to desert conditions are damaged or killed if kept for any length of time at temperatures in excess of 35–40° C.

Mechanisms of Heat Tolerance. The direct effect of high temperature is to denature and coagulate proteins. However, an important side effect is the increased rate of water loss that accompanies high temperature. Thus, many mechanisms of heat resistance are in fact mechanisms of drought resistance. The frequent correlation exhibited by plants between heat and drought has lead several workers, notably the American physiologist J. Levitt, to point out that any general theory that explains heat tolerance must also explain drought tolerance, and vice versa.

Since different proteins have different degrees of heat stability, it is reasonable to expect that heat tolerance should be associated with the stabilization of more sensitive enzymes in cells. This might be done simply by increased rate of production of the enzymes to counter their increased rate of destruction. Alternatively, existing enzymes might be stabilized by some secondary mechanism, or mechanisms might develop that would enable the organism to manufacture more stable proteins.

Experiments with microorganisms have shown that the addition of simple compounds will often restart growth after it has ceased as temperature is raised. Many organisms can grow at much higher temperatures when supplemented with ascorbic acid or other vitamins. Evidently the systems producing these substances are more sensitive to heat than other metabolic machines. The fact that unsupplemented organisms rapidly recover when the temperature is lowered indicates that the effect is on the metabolic machinery rather than on the genetic material in the cells. Only one or two enzymes appear to be affected initially; as the temperature is raised, the situation becomes more complicated and the organism is progressively more difficult to maintain as more systems are affected. Few comparable experiments have been successful with higher plants. There is some evidence that the addition of adenine may improve heat tolerance in certain plant tissues.

Where experiments have been performed, the results suggest that high-temperature organisms tend to have enzymes that are more temperature stable than their counterparts from temperature-intolerant plants. The heat stability of many enzymes appears to depend to some extent on the temperature at which they are formed. Thus heat tolerance is a condition that may be acquired to some degree. Although this may not be advantageous to organisms exposed to a climate having rapid, wide, but infrequent variations in temperature, it could be decidedly useful to organisms growing in climates where very high temperatures usually occur.

The sum of our present knowledge suggests that heat tolerance is largely the result of the capacity of certain organisms to produce more heat-stable proteins. Their capacity to replace heat-damaged proteins rapidly may also be important. The nature of the heat-stable proteins or the mechanisms whereby proteins can be stablized are not clear.

Low Temperature and Freezing

Chilling and Freezing. Resistance to freezing, like resistance to drought, is a many-sided process complicated by the fact that most plants are capable of hardening to cold, that is, acquiring increased resistance by exposure to low temperature. There may be several different effects of low temperature, relating either to the direct effect of reduced temperature on the life processes of the plant or to the effects of ice formation and freezing. Any one of a number of factors may be the ultimate cause of death by freezing, depending on the plant and on its circumstances.

Tropical plants are usually susceptible to **chilling,** that is, the damaging or lethal effects of low temperatures above freezing. Such plants may be injured by temperatures as mild as 12–13° C and may be killed by temperatures between 0 and 5° C. Evidently freezing is not involved. This effect may result from low-temperature sensitivity of proteins. Most temperate and Arctic plants are not seriously damaged by chilling, however. The problems they face are the effects of freezing of their internal water and the consequent formation of internal ice. This is emphasized by the fact that dehydrated plant material, such as seeds and other dried tissues that can normally survive extreme desiccation, suffers no damage from freezing and thawing. Upon hydration, however, these tissues lose their special resistance to frost damage. Characteristically, actively growing tissues are much more susceptible to frost damage than are dormant ones, some of which may withstand temperatures to −196° C (liquid nitrogen).

Freezing damage may be twofold. Ice crystals themselves may cause mechanical damage, disrupting delicate membranes and cell organization. In addition, the consequence of ice formation is the reduction of the water content of the tissue, eventually causing a drought situation. Water in the intercellular spaces has a high potential, whereas that inside the cytoplasm or vacuole tends to have a lower or more negative value of ψ. Thus ice crystals tend to form initially in intercellular spaces, and continued freezing causes water to leave the protoplasts as the intercellular ice crystals grow. On thawing, plants that withstand freezing well tend to reabsorb the water derived from the melting of these crystals into their protoplasts. In nonhardy plants, the water tends to remain in the intercellular spaces. Further, in nonhardy plants ice crystals form more readily within the protoplasts, where they may do direct mechanical damage.

The sequence of events on freezing has been summarized by the American physiologist P. Mazur:

1. Ice crystals form outside, but not inside, the protoplasts.
2. The solutes within the protoplasts become more concentrated as water is removed. If cooling is very rapid, the protoplasts may freeze; but if cooling is slow, they may only dehydrate.
3. Precipitation or coagulation of the concentrated solutes takes place in the protoplasts. This may cause considerable change in the internal pH if ionized compounds are precipitated.
4. Below the **eutectic** temperature (usually $-35°$ to $-40°$ C) all water in the tissue freezes.
5. With time, small angular crystals of ice are converted to large, spherical ones that have lower surface free energy. The resulting distortion of cell components may do extensive mechanical damage. Slow cooling may cause more injury from the effects of dehydration; during fast cooling the injurious effect of ice crystal formation may be greater.

There is much variation among plants in the actual rates of "slow" or "fast" freezing. The normal field condition is a relatively slow freeze, and damage may occur from the following: concentration and precipitation of solutes, pH change, reduction of cell water, cell shrinkage or plasmolysis, and critical reduction of the spatial separation of sensitive macromolecules. All or any of these factors could cause cell death.

Theories of Freezing Resistance. As might be expected, a number of different theories have been developed to account for the effects of frost and frost resistance. These can be roughly classified as relating to (1) low temperature denaturation of proteins, (2) dehydration effects, (3) electrolyte concentration effects, (4) effects of sugar, (5) steric effects, and (6) ice crystal formation.

It has been suggested that special proteins not susceptible to denaturation or dehydration may be formed in frost-hardy plants or that observed increases in other factors, for example RNA, in frost-hardy plants may retard denaturation. Antidehydration mechanisms may include the formation of special hydrophilic proteins, as suggested by the Canadian physiologist D. Siminovitch, or an increased concentration of electrolytes that would protect tissue water against its removal by ice formation. Frost-hardy tissue has long been observed to have higher sugar concentration than nonhardy tissue, and it has often been suggested that the mechanism of frost-hardiness involves sugars. Unfortunately the parallelism between frost-hardiness and sugar concentration is not complete and, in many plants, developing hardiness is not accompanied by increasing sugar concentration.

The American physiologist P. Steponkus has observed that frost-hardiness may result from placing tissues in sucrose solution, but full hardiness is not achieved unless some cold pretreatment has been given. He has suggested that a dual effect is required: (1) the synthesis of new, specially adapted proteins, stimulated by cold treatments, and (2) these proteins must be of a type that may be further stabilized by the presence of higher concentrations of sugars. Just how the sugar would stabilize proteins is not known.

Levitt has suggested that protein molecules are brought closer together by cell shrinkage resulting from dehydration. S—S bonds are broken, and on regeneration they may not re-form in the correct configuration due to the unnaturally close packing of molecules; such a process would result in protein denaturation. He suggests that frost-resistant plants form proteins with more hydrophilic bonds (thus yielding water less easily) and that hardening is accomplished by the formation of more hydrophilic and fewer hydrophobic bonds in important reactive proteins. An alternative mechanism has come from the work of the American physiologist C. R. Olien, who observed that cell wall polysaccharides extracted from frost-hardy plants tend to prevent ice crystal formation, whereas those extracted from nonhardy plants do not. A third possibility is revealed by the recent work of H. G. Volger and U. Heber in Germany. They demonstrated the presence of certain small protein molecules (molecular weight 10,000–20,000) in frost-hardened leaves that are more than 1000 times as effective in protecting chloroplast membranes from freezing than low molecular weight compounds like sucrose or glycerol. Each of these mechanisms provides interesting and alternative possibilities; the final solution to the problem will probably come with further research.

Frost Hardening. The process of frost hardening is complex and not well understood. In many plants it is closely tied to the effects of photoperiod, and some plants require a series of specific pretreatments or experiences in order to achieve maximum hardiness. Many plants require periods of dormancy or an appropriate photoperiod at the start of the hardening process. This preliminary induction is followed by a requirement for a period of growth, or survival, at reduced (but not freezing) temperature. As a rule, the length and severity of the cold treatment determine the degree of hardiness achieved.

Plants may lose their hardiness at this stage of their development if they are subjected to high temperatures. In many plants hardiness protects them only from low temperatures, not from freezing. These plants will die if they freeze, hardy or not. Many hardy plants are able to withstand lower temperatures (below 0° C) because their freezing point is depressed and they actually freeze at lower temperatures. A few extremely hardy species acquire maximum hardiness only after exposure to very low (below 0° C) temperatures.

Apparently a source of energy and some process of metabolism are required for certain stages of the process of frost hardening because the presence of metabolic inhibitors, or the degree of partial starvation resulting from low light intensity during hardening, retards or prevents the acquisition of frost hardiness. It is not known whether the energy requirement is a primary or specific one or merely part of the normal metabolic requirement that must be met for the reactions of hardening to proceed.

Radiation

Plants may be subject to stress from too much or too little radiation in the form of light. Excessive shade causes starvation, growth anomalies (elongation of internodes, weak, poorly developed stems, poor branching), and general debility. However these latter effects are the direct result of inadequte illumination; the consequences of starvation and of malfunction of the light-mediated growth-controlling mechanisms. Adaptations to light or shade conditions (stress tolerance) work primarily to increase the efficiency of photosynthesis and include changes in leaf area, blade thickness, chlorophyll content, the numbers and orientation of chloroplasts, and the thickness of the palisade layer. These are discussed in Chapter 14 (page 327). Other types of radiation include heat (see page 343) and ionizing radiation effects. Natural radioactivity is seldom high enough to disturb plants, but artificially generated fluxes of very high intensity may be encountered under experimental conditions. Plants have frequently been subjected to radiation of various degrees of intensity for the purpose of studying the effects of the radiation, but we shall not consider these effects here.

A more frequent occurrence, whose consequences are seldom recognized or examined, is the exposure of plants or cells to very high radiation fluxes during physiological experiments using isotopes as tracers. Few comprehensive studies have been conducted; however, high specific activity $^{14}CO_2$ has been found to affect strongly the reactions of photosynthesis under study, and high doses of tritiated water ($^{3}H_2O$) will strongly inhibit or kill metabolizing cells in germinating seeds. It is frequently overlooked in tracer studies that, although only small amounts of isotope are used (microcuries or millicuries), the tissues are in intimate contact with the isotope, and the dose of ionizing radiation per unit of tissue may be very large. The possibility that radiation influences the reactions or processes under study must always be carefully eliminated in tracer experiments.

Soil Conditions

We have discussed the consequences of mineral deficiencies (Chapter 10) that result when minerals are present in the soil in too low concentration or are too tightly bound. In addition, deficiencies or inadequacies in the soil can result from drought or flooding. The consequence of excessive water may often be that soils become anaerobic, and some sensitive plants may suffer from anoxia. This may affect the root's ability to absorb water. Some plants (corn is a good example) may literally drown, dying from lack of water because a shortage of oxygen prevents their roots from absorbing it, even though the roots are immersed in it.

Soils may contain toxic materials or compounds, such as salt, that are injurious in excess. Various mechanisms have developed in plants for the toleration or avoidance of the stress of excessive salt, as described in Chapter 10 (page 268). Avoiders such as the mangrove (also known as a **salt regulator**) do not absorb salt, but have mechanisms for its exclusion from their roots. Tolerators (salt **accumulators**) such as *Atriplex* have cell sap of very low ψ (approximately -200 bars, compared with -20 to -30 bars for normal leaves), thus are able to absorb salt water of high concentration. These plants tolerate high salt concentrations internally and eliminate excesses by secreting it through special glands in their leaves.

Altitude

The effects and stresses of altitude are the complex of all the climatic conditions that are characteristic of high altitudes. Weather conditions are much more violent at high altitude, and the success of plants is more dependent on the surface microclimate, particularly as modified by topographical features.

Radiation is higher at high altitudes. Direct radiation may be high, but diffuse radiation is much greater because of the relative thinness of cloud cover. Violet and ultraviolet radiation is particularly high at high altitude, although the levels of red and infrared radiation are not so much affected. Plants growing at high altitude have presumably developed resistance to the possibly damaging effects of ultraviolet light. A thick epidermis, which is characteristic of alpine plants, appears to act as an ultraviolet filter.

The average temperatures tend to be lower at higher altitudes by a factor of approximately 5° C per thousand meters. This situation is more a matter of a short growing season than because of extremely low temperatures during winter; the lowest temperatures reached are not much affected by altitude. Soil and plant temperatures at high altitude are often rather higher than at low altitude because of the thinness of soil cover and the high levels of radiation. However, high winds may disturb the snow cover in winter, allowing leaves and even the soil to be exposed and to become very cold. Worse, shock fluctuations of temperature occur very frequently. The most traumatic aspect of freezing is the actual transition across the freezing point; this may occur much more frequently at high altitude than at low altitude. Plants growing at high altitudes are usually much more frost-hardy than their counterparts at low altitude; the conditions at high altitude appear to be more conducive to hardening.

Wind and drought are major stresses faced by high altitude plants; nearly all such plants show xerophytic form. The effect of wind is not so much to lower the temperature (high wind merely speeds up the process of temperature change) as to remove water. The frequent shortage of water and the thinness of soil cover make the development of extensive root growth ineffective as a means of drought avoidance. Most high altitude plants have characteristically xerophytic leaves—thick cuticle, small area, sunken stomata, hairs, and so on—and rely on the control of water loss for survival.

One of the most striking adaptations to the stress of high altitude is in the photosynthesis of alpine plants. They tend to have much higher light-saturation values than low altitudes plants; values as high as 7000–10,000 ft-c have been recorded. In addition their efficiency at low carbon dioxide concentrations is greater. This compensates both for the reduced carbon dioxide at high altitudes and for the fact that carbon dioxide penetration in xerophytic leaves is hampered by the mechanisms that prevent water loss. Finally, the process of photosynthesis functions at lower temperature in alpine than in low altitude plants. Temperature optima of 10–12° C or lower are not uncommon in alpine plants, compared with 20–30° C at most low altitude plants.

Pollution

Pollution, only rarely a natural hazard for plants, has increased to the point of crisis over the past two decades. The stresses of pollution are largely chemical and are the result of either direct poisoning by toxic materials or the effects of secondary toxic substances created in the air or in the plant from pollutants. Defense mechanisms as such are the normal resistance of the plants to toxic compounds. Recent attempts to develop pollution-resistant lines of some plants have been moderately successful. However, it is to be hoped that pollution control will make this aspect of practical research unnecessary.

Smog and pollution damage to plants is of two main types. Ozone (O_3) in the atmosphere appears to be responsible for much damage, but apparently not directly. The American physiologist J. T. Middleton, working in California, showed that some unstable or transitory reaction intermediates of ozone and unsaturated hydrocarbons cause the visible pollution damage. Neither oxidized hydrocarbons nor ozone by itself produced the same effects. The damage consists primarily of glazing and bronzing of leaves and the development of chlorotic and necrotic spots. The damage occurs mainly on leaf surfaces having stomata and does not occur if stomata are closed during exposure. The epidermal cells, particularly guard cells, absorb excessive amounts of water and may rupture, while mesophyll cells are dehydrated. The growth and development of plants are not much affected by pollution until lesions develop, but many species do grow better in filtered air than in "normal" unfiltered air. Such diseases as **weather fleck** of tobacco, **black spot** of grapes, and tumors on broccoli have been shown to result from ozone and organic pollutants in the atmosphere (see Figure 28-1C and D). Damage from ozone may be prevented by spraying the plants with abscisic acid

Figure 28-1. Some effects of pollution on plants.
A. PAN smog damage to a tobacco plant. Damage like this occurs up to 75 miles from the source of pollution. **B.** Sulfur dioxide (SO_2) injury to white birch. **C.** Weather fleck injury to tobacco caused by ozone. **D.** Ozone damage to White Cascade Petunia. [Photographs courtesy United States Department of Agriculture.]

A

B

C

Figure 28-2. Sulfur dioxide and ozone pollution damages white pines at relatively low levels causing "chlorotic dwarf" disease. United States Department of Agriculture scientists enclosed several 10-year-old diseased pines in plastic chambers fitted with filters which remove pollutants (**A**). After three years the trees were healthy and vigorous (**B**), but if they were reexposed to unfiltered air, they quickly showed severe chlorotic dwarf symptoms again (**C**). [Photographs courtesy of the United States Department of Agriculture.]

solutions that close stomata. Unfortunately, this also stops photosynthesis and production!

Apart from ozone damage, which appears to be much more severe in mature or fully developed leaves, plants may suffer from the effects of poisoning caused by **peroxacyl nitrates (PAN).** These compounds, which attack primarily young or developing leaves, are formed from unsaturated hydrocarbons together with nitric oxide (NO) or nitrogen dioxide (NO_2) and oxygen under the influence of light or ultraviolet radiations from the sun. Among other effects, respiration is increased and photosynthesis is decreased in leaves damaged by PAN smog (see Figure 28-1A).

Much has been said about the effects of pollutants, particularly surfactants (detergents), on plant life and on photosynthesis in particular in the oceans. Algae are extremely susceptible to this form of pollution, which destroys membrane and thylakoid structure. Even seaside land plants may suffer. Increasing salt damage to Norfolk Island Pines (*Araucaria heterophylla*), grown as ornamentals along the beaches in Adelaide, Australia, has been shown to result from increasing detergent in the sea from municipal sewers. The detergents lower the surface tension of spray that falls on the leaves, and salt absorption is sufficiently increased to damage or kill the foliage.

One of the worst pollutants from industry (particularly mining) and urbanization is sulfur dioxide (SO_2). This gas is especially poisonous to trees, causing chlorosis and dwarfing (Figures 28-1B, 28-2). Huge tracts of forest and many suburban areas throughout the industrialized world are seriously affected by SO_2, or a combination of SO_2 and ozone, which seems to be especially poisonous.

There has been some success in protecting leaves from oxidative damage due to smog by the addition of reducing substances to the plants. Sucrose solutions prevent some damage in pinto beans and spinach leaves. Carbamate spray or drench and potassium ascorbate drench seem to protect some plants from the effects of ozone damage, but the leaves must be thoroughly covered with the drench. Tobacco plants have been protected by growing them under a canopy treated with reducing substance that destroys ozone. Some smog-resistant varieties of plants have been bred. However, preventative measures are expensive and unsatisfactory. The only sensible solution to this problem is the removal or control of sources of pollution. In the long run, this will be cheaper and much more effective, as well as aesthetically more pleasing.

Additional Reading

Articles in the *Annual Review of Plant Physiology*.

Levitt, J.: *Responses of Plants to Environmental Stresses*. Academic Press, New York, 1972.

Sutcliffe, J.: *Plants and Temperature*. Edward Arnold (Publishers) Ltd., London, 1977.

Weiser, C. J.: Cold resistance and injury in woody plants. *Science,* **169:**1269–78 (1970).

Wolstenholme, G. E. W., and **M. O'Connor** (eds.): *Ciba Foundation Symposium on the Frozen Cell*. J. A. Churchill, London, 1970.

Woodell, S. R. J.: *Xerophytes*. Oxford University Press, London, 1973.

Woolhouse, H. W. (ed.): Dormancy and survival. *23rd Symposium of the Society for Experimental Biology*. Cambridge University Press, New York, 1969.

29 Physiological Factors in Plant Distribution

Introduction

The distribution of plants and the physiological factors that underlie the principles of ecology are extremely important subjects; however, they are peripheral to our treatment of basic plant physiology, therefore they are discussed only briefly here. Nonetheless, one of the very important applications of physiological principles is in the study of the relationships of plants to climate, to physical and physiological environmental factors, and to each other. This is the basis of the study of ecology, and this chapter describes some of the more important physiological aspects of plant distribution and ecology.

The plants that live in a specific region can be roughly analyzed in two ways: as the vegetation and as the flora. The **vegetation** is the type or types of plants that live in the region, the sorts of plants that can successfully live within the limitations imposed by the climate and the environment. The **flora,** on the other hand, is the group of actual species of plants that makes up the vegetation of a region. Thus, the vegetation type of an area might be classified as, for example, a deciduous forest. This indicates the sorts of plants that dominate in the region. The flora would list the actual species of deciduous forest trees and associated plants that actually live in the area; those species that are able to compete successfully with each other and coexist to become a more or less stable part of the plant community of the area.

Physiological Factors in Ecology

All those factors resulting in stress to the organism may affect its distribution. Physiological factors may thus be positive or negative; physical environmental factors such as heat or light may be too intense or not intense enough; substances or substrates may be present in excess or in insufficient amounts. The environment is composed of many factors, and these are interrelated. Thus, heat and water supply are interdependent to the extent that excessive heat may be unendurable except in the presence of excessive water. All those factors that grossly affect the environment also affect the type of vegetation that can live successfully in an area. **Climatic** factors include heat and light intensity and periodicity, rainfall, and relative humidity. **Physiographic** factors include soil structure, acidity, and nutrient composition which are often related to the nature of underlying rock, and local climatic conditions caused by the physical contours of the land. **Biological** factors result from the nature of the vegetation established and include shading (competition for light), water competition, nutrient competition, alteration or provision of substrate, plant-animal interactions, antibiotic activities, saprophytic and parasitic relations, and modification of microclimatic conditions such as pH or wind within vegetation masses. All these factors, interacting in bewildering complexity, affect the nature of the vegetation and the composition of the flora that exists in any given time in a given area.

One of the principles of ecology is that a vegetation or flora is seldom static. Since the existence of a flora almost inevitably results in the modification of its environment, it changes conditions to the point where changes in the flora result. This process, which can begin with bare ground or a sterile body of water, is called **succession.** An ecological succession often culminates in a stables **climax,** that is, a flora which maintains itself because it generates conditions best suited for its own reproduction and survival. Under these conditions competition continues unabated, but is restricted to competition among individuals of the successful climax flora, unsuccessful species having been eliminated.

On the other hand, a metastable or unstable condition may prevail if a flora generates conditions favorable for the flora of some previous stage of the succession. This will result in a cyclic situation in which two or more stages in a succession alternate with each other. Regardless of the stages in a succession or the kind of climax that is established, the physiological factors that govern the capacity for survival of each

Table 29-1. Adaptations of plants in response to environmental stress*

Adaptation	*Advantages or stress relief*
Growth rate	
Greater	Seedlings escape shading
Less	Less damage to sensitive young tissue
Final height	
Taller	Competition for light
Shorter	Less wind damage
Leaf size	
Larger	Shade tolerance
Smaller	Sunlight and drought tolerance
Xeromorphism	Survive drought
Changes in periodicity or dormancy	Adapt to climatic periodicity
Cold or frost hardiness	Survive cold
Increased photosynthetic capacity	
Maximum attainable rate	Better survival for tropical, sunplants
Efficiency at low CO_2	Better survival in dense stands, bright light, high temperature
Efficiency at low light	Shade tolerance
Decreased photorespiration	Higher efficiency, particularly in long days or high temperature
Decreased dark respiration	Greater productivity, particularly in short days
Greater or earlier reproduction	Competition
Improved yield	Survival in agriculture
Tolerate noxious environment	Survival in man-dominated ecosystem

*Adaptations may be temporary or permanent in an individual, or they may be inherited, resulting in the formation of ecotypes.

individual or each species (for example, in growth, development, and reproduction) are always the factors that determine the nature of the vegetation and the composition of the flora.

The vegetation and flora are not the only dynamic aspects of an ecosystem. Environmental factors may vary due to outside influence or as the result of ecological succession. In addition, the facts of competition and extremes of environment place the species and individuals of the community under stress, and they may react to stress in various ways. Adaptations resulting from stressful environmental factors may be nonheritable, that is, the reactions of individuals to stress (as discussed in the previous chapter). Alternatively, adaptations may be heritable, resulting in the appearance of distinctive forms or varieties called **ecotypes,** which are better suited for competition or survival under local conditions. A list of some more common types of adaptation is given in Table 29-1.

Any factor, or any combination of factors, may limit the distribution of a plant (that is, limit its capacity either to survive or to compete). Justus von Liebig long ago established the **law of limiting factors,** which states essentially that the final growth of an organism depends upon the amount of that foodstuff which is available to it in minimum quantity. For plants, this includes light, water, and carbon dioxide, as well as mineral nutrients. The law may also be extended to include nonfood factors; it must be recognized that adverse factors or an overdose of some normally desirable factor may equally limit growth. In a complex system governed by an interplay of many factors, the critical concentration (minimum quantity) may well depend upon the concentrations of other factors that are above the range normally regarded as limiting. The analysis of such a complex of environmental factors may thus be extremely difficult.

Factors Affecting Vegetation

Vegetation Types. There are four main types of land vegetation whose distribution is largely dependent on climatic factors. These are **forests, grassland, tundra,** and **desert.** These grade into each other, and certain well-recognized subtypes have been described. Thus **savanna** represents a grassland with scattered or isolated trees. The typical Mediterranean or **sclerophyllous** forest is a rather distinctive type of savanna or scattered forest composed of trees and bushes characterized by tough, leathery, xeromorphic leaves. Many other subsidiary classes of vegetation have been described, but these need not be considered here.

Fresh-water aquatic vegetation (marsh, bog, sublittoral) is largely dependent on local or physiographic factors. Marine vegetation, apart from temperature dependence, is also largely related to local phenomena, such as nutrient supply, water movement, depth, and turbidity, as well as rock substrate.

Historical Factors. Many components of a vegetation may be present as a result of historical geological events. The distribution of plants has been governed in the past by climatic changes, which are often cyclic in nature. Thus a continuous population over large land masses may be fragmented and separated by successive ice ages, and the fragments may be further isolated by crustal movement and continental drift. The vegetation of a particular location is therefore composed of those plants that survive under present conditions and those that were also present at the time of the latest major climatic or geological event or were able to invade subsequently. Plants that could theoretically survive or even dominate the vegetation may be absent because physical barriers prevent them from getting there; other plants less well adapted may dominate because of the absence of more effective competition. This is the reason why introduced species sometimes spread rapidly or explosively in a new country to the detriment of the indigenous flora.

Geographic Factors. Vegetation is affected by geographic factors that modify climate. Large land masses tend to have extremes of climate in the middle, which are much modified by the presence of the sea at the edges. A mountain range along the sea coast usually produces a climate with very high rainfall, as moist air from the ocean rises to cooler altitudes on encountering the mountains. Similarly, the continental side of a great mountain range is likely to be drier than the maritime side. Thus rain forests tend to cloak the seaward side of maritime mountain ranges, yielding to savanna or desert on the other side. Weather patterns, hence vegetation, are much affected by lesser relief patterns of the land, which affect wind velocity and direction, local rainfall, and seasonal temperature. Ocean currents are immensely powerful in regulating or affecting climates and, hence, the distribution not only of the oceanic flora in them but of the land flora on surrounding land masses.

Rainfall. Water is probably the most important factor affecting vegetation. The vegetation patterns of the world can be directly related to rainfall in summer and winter as shown in Table 29-2. Forests, whether tropical or temperate, require water or high relative humidity throughout the year, and thus tend to cover mainly the edges of continents where maritime climates prevail.

Areas with adequate rainfall in spring or summer, but where drought conditions may prevail at other times of the year, are usually covered by grassland of various sorts. Many grasses, besides having exceptionally large and effective root systems, may also become dormant after midsummer. Thus they are able to withstand considerable drought in the growing season, and are able to survive periods of extreme drought that often follow in late summer or fall. Their requirement for winter rainfall is low.

When rainfall is infrequent, low, or very uneven, desert conditions prevail. Deserts vary greatly in their climate and vegetation, as do forests and grassland regions. Cold deserts exist on snow fields in the far north and south and on rocky barrens close to them, as well as on the sands and dry rocky plains of the more commonly recognized

Table 29-2. Relationship between seasonal rainfall and vegetation types

Summer rainfall	*Winter rainfall*	*Vegetation type*
heavy	heavy	**Forest.** Mesophytic to hydrophytic; requires water all year
heavy	light	**Grass savanna.** Dormant in the fall, thus survives fall or winter droughts that kill trees
light	light	**Desert.** Xerophytic plants; species may survive several years of adverse conditions
light	heavy	**Schlerophyllous** (forest savanna). Xerophytic, leathery-leafed trees; typical Mediterranean vegetation; heath

tropical and subtropical deserts of Africa, Australia, or southern United States. Certain areas of the world have low rainfall and relative humidity in the summertime but may have periods of quite high rainfall in the winter. These areas are usually covered by savanna and sclerophyllous forests. Grasses are usually present, and larger vegetation is usually broadleaf evergreen forests of strongly xerophytic character in warm areas or heath and moorland at lower temperatures.

The importance of seasonal distribution of rainfall may be seen in Table 29-3, which shows a correlation of several climatic variables with the vegetation. The transition from forest to grassland in several widely separated (climatically as well as geographically) locations is dependent on the amount of winter rainfall. Below 7–8 in. at high latitudes or about 12 in. at low latitudes, grasslands take over from forests. Only in the hottest regions analyzed, where summer rainfall is also extremely low, does grassland give way to savanna and desert. The existence of extensive grasslands is always correlated with strongly differentiated wet and dry seasons. This is brought out in Table 29-4, which shows seasonal variations in some typical areas of tropical and temperate grassland vegetation.

Relative Humidity. This may be an important independent factor under certain circumstances. RH, rather than the actual amount of rain, is probably the critical factor affecting the dividing line between vegetation types. It is normally quite closely related to rainfall. However, in desert areas or places that have infrequent heavy rains followed by extremely dry periods, the RH may be so low most of the time that only extremely xerophytic plants can survive. Alternatively, it is possible to have what is essentially a rain forest in an area of constant high humidity, such as the California coast, even though rainfall there is not very high.

Temperature. Within the limitations set by water, vegetation is markedly affected by temperature. The limitations of temperature are not quite so great, however. It is seldom, if ever, too hot for vegetation and only rarely too cold. Plants have developed within all the major vegetation types that are adapted to various temperature regimens. Thus, forests flourish from the hottest tropical regions to the far north, until limits are reached, not of cold but of the accompanying low relative humidity, which prevent further growth.

In addition to the direct effects of temperature, its periodicity is important. Temperate and Arctic climates are subject to great seasonal variations in temperature, and components of vegetation must be able to withstand them. Dormancy is one of the commonest mechanisms for avoiding extreme cold temperature, but this requires further mechanisms which ensure that the correct periodicity is observed. Since periodicity of inclement temperatures is closely related to latitude, this component of the climate, rather than the absolute temperature, probably explains the quite strong tendency of vegetation to follow lines of latitude in many parts of the world. Deviations from these lines are usually the result of changing availability of water.

Because of the strong dependence of relative humidity on temperature, water availability and temperature interact strongly in establishing limits to vegetation types. These interactions are presented diagramatically in Figure 29-1. The diversity of vegetation types decreases from bottom (hot) to top (cold), as is evident from the triangular shape of the diagram.

Table 29-3. The correlation between seasonal rainfall, temperature and vegetation in the United States

	*Rain in winter**	*Rain in summer**	*Total rain*	*Months below freezing**	*Temperature of three warmest months*	*Longitude*	*Vegetation*
Approximately along latitude 47° N							
Sault Ste. Marie, Mich.	12.2	13.7	25.9	4	61	84° 21′	Forest
Duluth, Minn.	9.0	20.9	29.9	5	63	92° 06′	Forest
Park Rapids, Minn.	6.4	20.5	26.9	5	65	95° 10′	Forest
Moorhead, Minn.	6.4	18.1	24.5	5	66	96° 44′	Grassland
Grand Forks, N.D.	4.4	15.7	20.1	5	65	97° 05′	Grassland
Oakdale, N.D.	4.7	13.2	17.9	5	65	102° 50′	Grassland
Glendive, Mont.	5.2	10.7	15.9	5	69	104° 30′	Grassland
Great Falls, Mont.	3.5	9.9	13.4	3	66	111° 15′	Grassland
Approximately along latitude 43° N							
Port Huron, Mich.	14.3	16.7	31.0	4	66	82° 26′	Forest
Grand Haven, Mich.	16.0	18.8	34.8	4	67	86° 13′	Forest
Milwaukee, Wis.	12.6	18.4	31.0	4	68	87° 54′	Forest
La Crosse, Wis.	9.2	21.7	30.9	4	71	91° 15′	Forest
Charles City, Iowa	8.2	21.6	29.8	4	71	92° 40′	Forest
Sioux City, Iowa	5.6	19.9	25.5	4	72	96° 24′	Grassland
Rosebud, S.D.	5.5	12.8	18.3	4	71	100° 52′	Grassland
Oelrichs, S.D.	7.2	12.1	19.3	4	70	103° 13′	Grassland
Fort Laramie, Wyo.	4.5	9.7	14.2	3	71	104° 31′	Grassland
Approximately along latitude 40° N							
Pittsburgh, Pa.	16.5	20.3	36.8	1	73	80° 02′	Forest
Columbus, Ohio	18.7	19.5	38.2	1	73	83° 00′	Forest
Indianapolis, Ind.	19.4	22.5	41.9	1	74	86° 10′	Forest
Springfield, Ill.	16.1	21.3	37.4	2	74	89° 39′	Forest
Keokuk, Iowa	12.3	22.8	35.1	3	75	91° 26′	Forest
Corydon, Iowa	10.5	24.8	35.3	3	73	92° 40′	Forest
Atchison, Kan.	10.1	27.0	37.1	3	75	95° 08′	Forest
Concordia, Kan.	6.5	20.3	26.8	2	76	97° 41′	Grassland
Colby, Kan.	3.6	15.2	18.8	3	74	101° 02′	Grassland
Cope, Colo.	3.6	14.8	18.4	3	72	102° 49′	Grassland

Approximately along latitude 37° N							
Cairo, Ill.	21.2	20.4	41.6	0	77	89° 10′	Forest
Olden, Mo.	18.8	22.6	41.4	0	75	91° 54′	Forest
Springfield, Mo.	17.2	26.4	43.6	0	75	93° 18′	Forest
Columbus, Kan.	14.8	29.7	44.5	0	77	94° 50′	Forest
Independence, Kan.	12.5	24.5	37.0	0	79	95° 43′	Grassland
Englewood, Kan.	5.1	15.4	20.5	0	78	99° 58′	Grassland
Viroqua, Kan.	4.2	13.5	17.7	0	76	101° 46′	Grassland
Blaine, Colo.	3.6	11.7	15.3	1	75	102° 15′	Grassland
Approximately along latitude 34° N							
Memphis, Tenn.	23.4	27.4	50.8	0	80	90° 03′	Forest
Little Rock, Ark.	25.5	24.1	49.6	0	79	92° 06′	Forest
Fort Smith, Ark.	18.5	23.3	41.8	0	79	94° 24′	Forest
Oklahoma City, Okla.	11.3	20.4	31.7	0	78	97° 33′	Grassland
Fort Sill, Okla.	10.3	19.8	30.1	0	80	98° 25′	Grassland
Roswell, N.M.	4.4	11.2	15.6	0	77	104° 30′	Grassland
Approximately along latitude 30° N							
New Orleans, La.	25.6	32.0	57.6	0	82	90° 04′	Forest
Lake Charles, La.	26.9	26.4	53.3	0	81	93° 06′	Forest
Houston, Texas	21.2	27.0	48.2	0	82	95° 15′	Forest
San Antonio, Texas	10.7	17.7	28.4	0	82	98° 28′	Scrub
Fort Clark, Texas	7.4	16.0	23.4	0	84	100° 24′	Desert
El Paso, Texas	3.0	6.3	9.3	0	80	106° 30′	Desert
Tucson, Ariz.	4.2	5.6	9.8	0	85	110° 53′	Desert

SOURCE: H. A. Gleason and A. Cronquist: *The Natural Geography of Plants*. Columbia University Press, New York, 1964. Used with permission.

*In this table, "winter" refers to the six coldest months of the year, and "summer" to the six warmest months; "rain" is total precipitation (both rain and snow), measured in inches; temperature is given in degrees Fahrenheit; and "months below freezing" refers to months in which the mean temperature is below 32° F.

Table 29-4. Seasonal rainfall variation in some typical grassland areas (total inches per season)

		Dry season	*Wet season*
Tropical	Loango, W. Africa	1	58
	Minas Geraes, Brazil	11	51
Temperate	Natal, S. Africa	4.6	23.2
	Corrientes, Argentina	15.2	30.8
	Great Falls, Mont.	3.5	9.9
	Colby, Kan.	3.6	15.2

SOURCE: Modified from H. A. Gleason and A. Cronquist: *The Natural Geography of Plants*. Columbia University Press, New York, 1964.

WIND. Wind is seldom important except as a local destructive force. Physical damage done by high winds seldom affects vegetation permanently. However continuous strong winds in exposed places can have a powerful modifying effect on the climate by increasing evaporation and transpiration. Thus strongly xerophytic conditions may be enforced and the vegetation modified thereby. Typical wind effects, such as dwarfing or stunting, are often seen in exposed places such as coastal headlands or Arctic tundra. These are the result both of wind desiccation and of the effect of materials such as salt, dust, or snow carried by the wind.

PERIODICITY AND SEASON LENGTH. These factors are extremely important, particularly in the north and south temperate and subarctic zones, where periods of clement and inclement weather alternate. Plants may grow well during the growing period and may even be capable of surviving the hardships of winter, but if they emerge too early or remain active too late, they cannot survive the climate. Some far northern areas have climatic conditions suitable for vigorous growth of more southern species, but the growing season is not long enough to permit plants to complete their life cycle; thus, they are unable to survive. The onset of dormancy is related to day length as well as

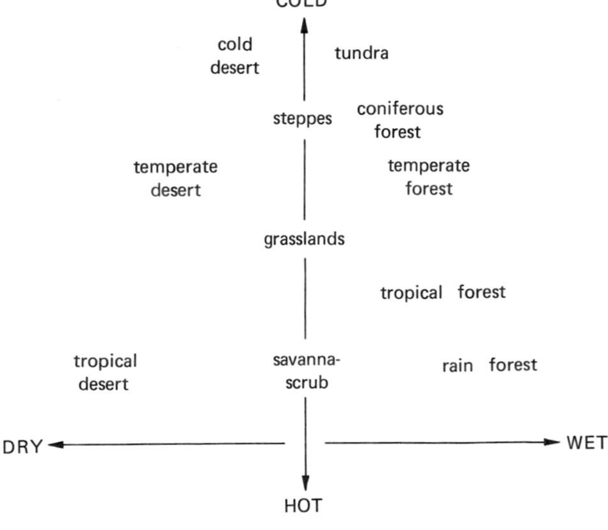

Figure 29-1. Relationship between vegetation, temperature, and rainfall or humidity.

decreasing temperature in many plants. Plants that have adapted to lower latitudes tend to enter dormancy only as days become quite short. When these species are grown in northern latitudes, they are frequently killed by early season frosts, which they could have survived easily if dormancy had been advanced by a few days. Season length, or the number of frost-free days, and similar parameters are sometimes, but not necessarily always, directly correlated with mean temperature isotherms. Although some plants may be limited by temperature, the former parameters are frequently critical in marking the limitation of vegetation types.

Factors Affecting Flora

Climatic. It is possible to draw lines on maps that connect points of equal temperature or temperature limits (isotherms), light intensity, rainfall, and so on. The geographic limits of a species are often found to coincide very closely with one or another of such lines. For example, a rather precise relationship occurs between the low temperatures occurring in January and the northern limit in Europe of a species of madder (*Rubia peregrina*) as shown in Figure 29-2. Other plants may be limited by water or rainfall. An example of this is the distribution of the genus *Stylidium* in Australia, shown in Figure 29-3. Here the limitation is a requirement of 20–30 in. of rainfall per year. The relationship is not quite so precise as in the previous example; evidently some additional factors limit distribution in certain parts of the range.

In fact, the distribution of a plant may be limited by a group of factors that are not necessarily related. The distribution of the sugar maple (*Acer saccharum*) is shown in Figure 29-4, together with several meterological limits. This species appears to be limited by the −40° C mean annual temperature isotherm in the north. To the south, distribution is related to winter warmth, the −10° C mean annual minimum temperature. The western limitation is clearly related to water; the line for the 20-in. annual rainfall in the northwest and 30-in. in the southwest. However, a much closer relationship on the western limit is shown with the line dividing areas where evaporation exceeds precipitation (west of line A-A) from those in which transpiration exceeds evaporation (east of line A-A). The line A-A also marks limits of the forest vegetation to the east and the grassland to the west. This factor is related to both rainfall and temperature, whose complex interactions preclude either one or the other being considered a simple limiting factor.

Thus the climatic limitations of a flora or its components are usually set not by a single characteristic (although this may happen, as in Figure 29-2) but by some limit produced by the interrelationships of two or more factors, in which none of the components by themselves constitute limiting factors. A multivariate analysis of climatic conditions as they affect distribution has been done by the Finnish ecologist V. Hintikka. In Figure 29-5 the distribution of Norway spruce (*Picea abies*) is compared with a "climate curve," a line mapping a complex datum involving rainfall, mean summer temperature, and mean winter temperature. A very close relationship between distribution and the climate curve can be seen over most of the range of this plant. However, even this analysis is not complete. Evidently some other climatic condition exerts a strong influence on the eastern side of the Adriatic sea, where the distribution and the climate curve do not coincide.

Apart from the absolute limits of climatic factors, floras are much affected by the

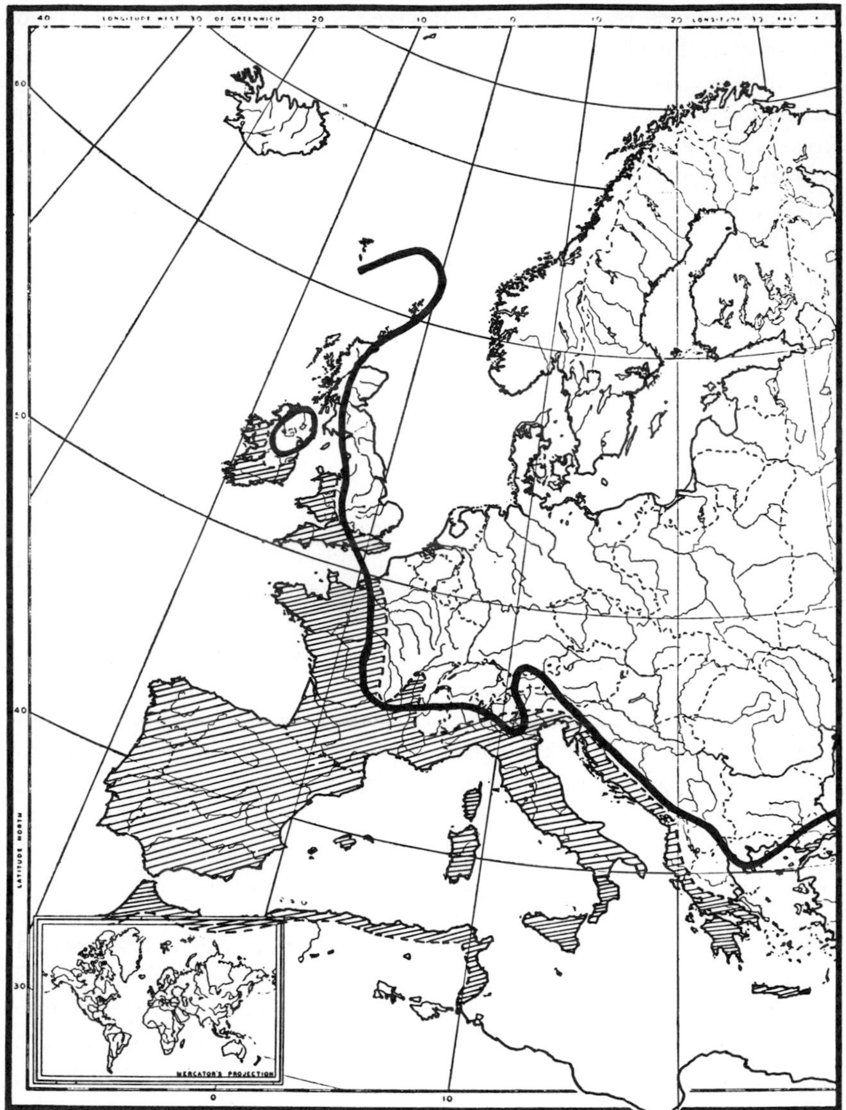

Figure 29-2. The 4.5° January isotherm, a climatic limit for *Rubia peregrina* in Europe. [From P. Dansereau: *Biogeography*. The Ronald Press, New York, 1957. Used with permission. Original figure courtesy Dr. Dansereau.]

stability of the climate or of certain factors in it. Instability is usually accompanied by diversity in the flora, whereas stable climates are characterized by fewer species. As a result of climatic stability those species that are not very well adapted are eliminated by competition, and only the best-adapted species survive in any one niche. When conditions vary greatly, broader limits are set on the range of conditions limiting species, and a less stable and more varied flora results. The limits of the variation in flora are set by the degree of instability of the climate.

Physiographic. Factors of the local climate are affected by the relief and structure of the land, and these affect the pattern of flora in local areas. The physiographic and geological nature of the bedrock upon which the landscape is built are extremely important in determining the conditions of moisture, soil type, structure, and fertility and the extremes of temporary drought, cold, heat, or excess water that may occur. For example, a junction between limestone and either granite or acidic metamorphic sedimentary rocks is usually marked by sharp changes in the flora. This is brought about by several factors.

First, granitic soil is usually loose, sandy, acidic, and very low in many nutrient elements. Limestone soil, on the other hand, tends to be richer, alkaline, and clay. The granite soil, because of its openness, may be more prone to summer drying, whereas the clay soil is more apt to become waterlogged in wet periods. The flora of the two areas is dependent on the capacity of the individual plants to succeed or compete effectively in the special conditions of pH, nutrition, water stress, and so on, characteristic of each

Figure 29-3. Mean annual rainfall in inches as related to the distribution of *Stylidium* in Australia. [From P. Dansereau: *Biogeography*. The Ronald Press, New York, 1957. Used with permission. Original figure courtesy Dr. Dansereau.]

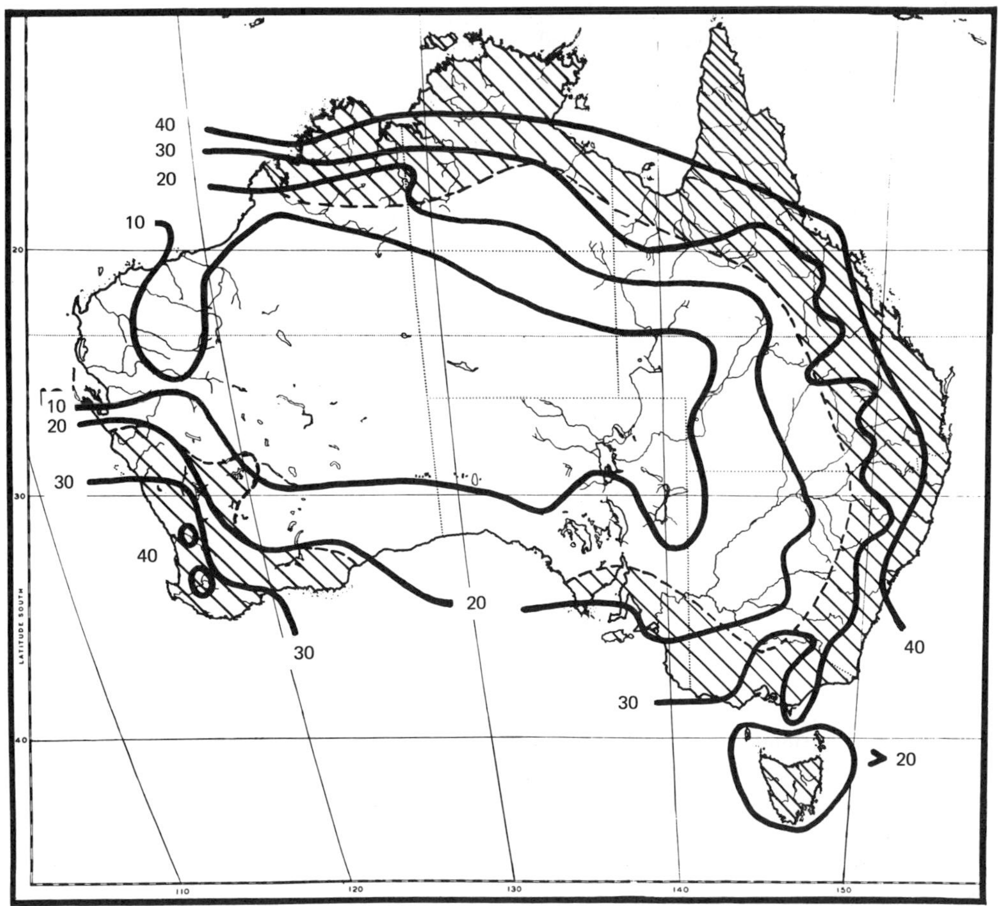

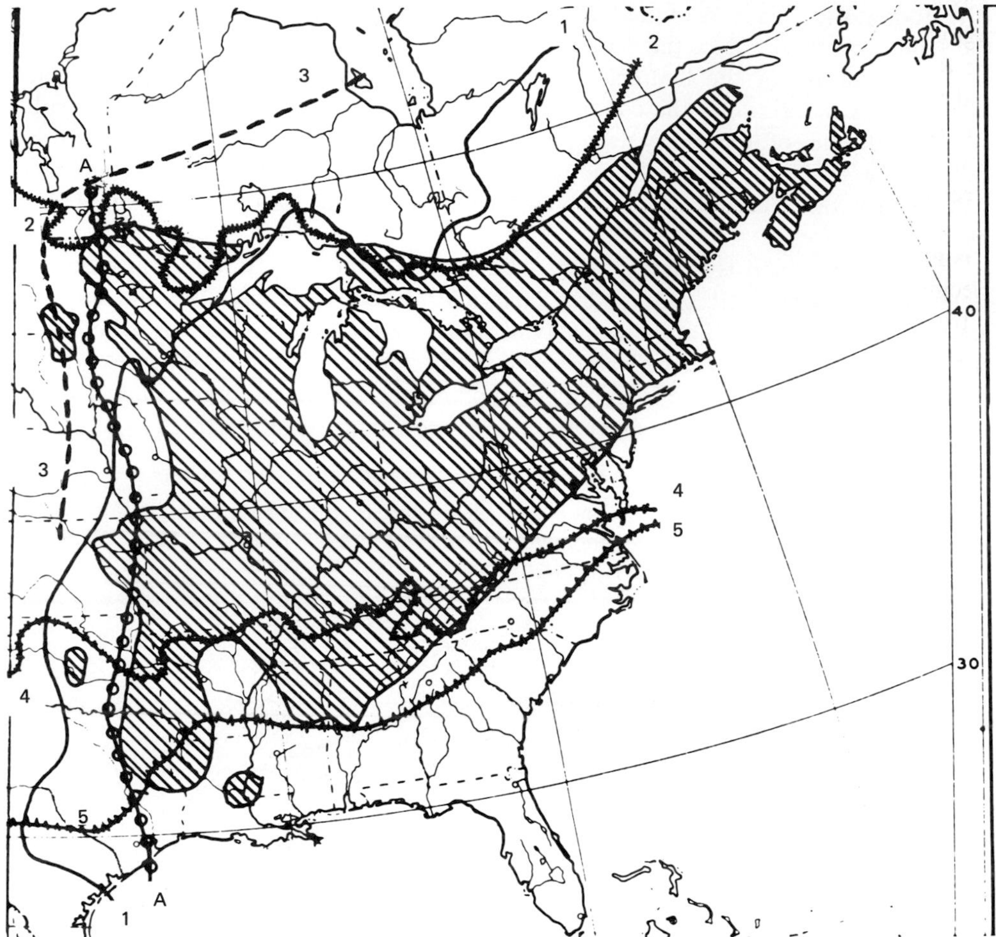

Figure 29-4. Bioclimatic limits of sugar maple (*Acer saccharum*), showing coincidence with meterological elements (shaded area is distribution of sugar maple).
1, 30-in. annual rainfall; 2, −40°C mean annual minimum; 3, 20-in. annual rainfall; 4, 10-in. mean annual snowfall; 5, −10°C mean annual minimum. Line A-A, see text page 659. [From P. Dansereau: *Biogeography*. The Ronald Press, New York, 1957. Used with permission. Modified from an original figure kindly supplied by Dr. Dansereau.]

area. But this pattern is secondary to the main pattern of vegetation, which is imposed by the major climatic factors that were considered above.

Pollution. Within defined areas the flora may be largely altered by the presence of natural or man-made pollution. A visit to Yellowstone National Park will quickly convince one of the powerful effect of pollution, not only man-made but also natural (in this case, the outflow of hot underground water with very high mineral content). Man-made pollution is sometimes more subtle, often more widespread, and can be very serious. Hundreds of square miles of forest in areas around large mining and smelting operations have been killed by sulfur dioxide (SO_2) fumes or so severely modified as to be useless for lumber or recreation. Lichens are very sensitive to pollution and have been drastically reduced in areas where high concentrations of heavy industry occur. Whole regions of central and northern Europe are affected by the industrial fumes generated in the Ruhr. In a study of the lichens of the area around Sudbury, Canada, it was found that the diversity of lichens was closely related to the level of sulfur dioxide in the air generated by the nearby smelters, as shown in Figure 29-6. Clearly, pollution is a powerful destructive force that may ultimately seriously disturb the environment in huge areas of the world if it is not brought under control.

Competition. Competition results from the fact that plants need space to grow. The specific factors for which they compete are mainly nutritional—light, carbon dioxide, nutrient elements, and water. In special circumstances competition may approach the level of physical strife (for example, the strangling effects of certain vines), but normally

Figure 29-5. Distribution of Norway spruce (*Picea abies*) compared with a "climate curve" derived from data on rainfall, mean summer temperature, and mean winter temperature. [From V. Hintikka: *Ann. Bot. Soc. Vanamo,* **34:**5 (1963). Used with permission.]

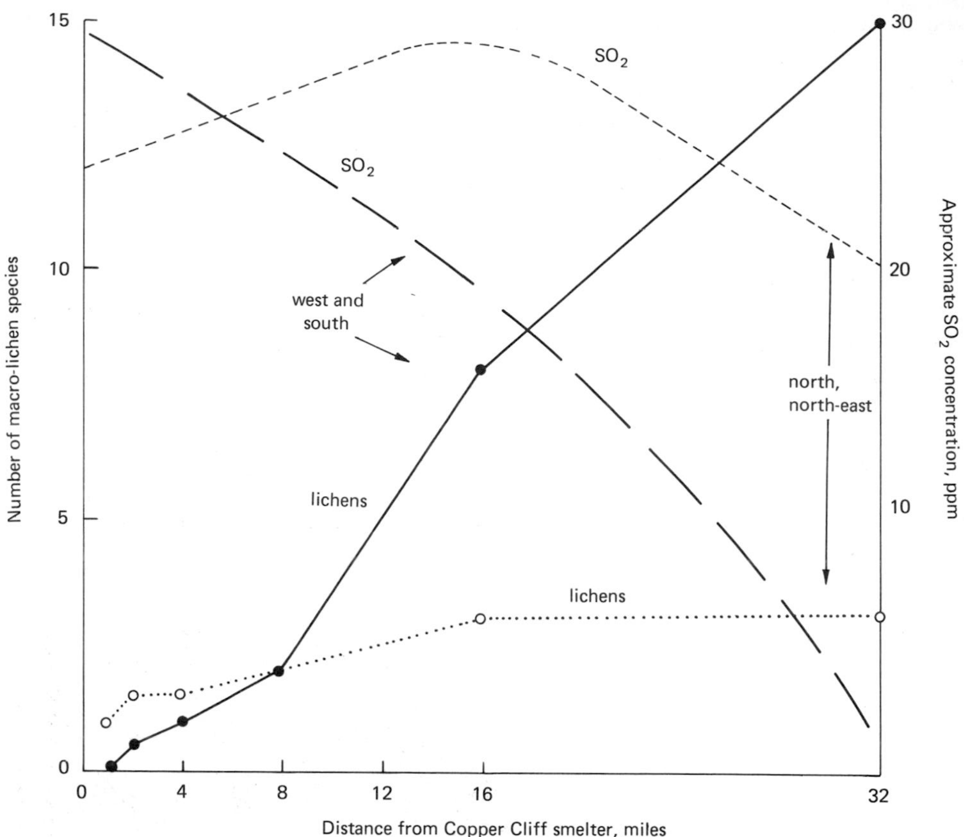

Figure 29-6. Relationship between lichen distribution and pollution. [Adapted from data courtesy D. H. S. Richardson and K. Puckett, Laurentian University, Sudbury, Ontario.]

competition is essentially a passive process. A successful competitor must be able not only to survive but to complete its entire life cycle more quickly or more effectively than other plants under the stress of reduced nutritional or physiological factors.

The physiological characteristics that enable a plant to survive or compete successfully are those that enable the plant to tolerate stresses better and are discussed in Chapter 28. The stresses that usually result from competition are shade, drought, nutrient limitation, and the presence of biotic pollutants—antibiotics produced by one organism that inhibit another. Successful competitors need to be able not only to survive better under conditions of stress but also to create conditions of stress for other plants, by growing taller more rapidly, producing larger and denser leaf canopies, or larger and more effective root systems.

The competition between red maple (*Acer rubrum*) and sugar maple (*Acer saccharum*) in areas that are basically satisfactory for the growth of both species provides an interesting example. Red maple is more resistant to climatic factors and can grow better under normal circumstances in the areas where the ranges of the two species overlap. However, sugar maple is more resistant to shade, and its seeds grow better and more quickly in shaded places than do those of red maple. Thus sugar maple tends to dominate mixed stands. In spite of the fact that it is less hardy, it is a successful competitor because of its greater reproductive capacity in the shaded conditions created by forest growth.

Another example that illustrates the unexpected breadth of competition is that between a tree growing in a lawn and the surrounding grass. Grass may suffer in this association by the obvious competition for light; its habitat may be made much less desirable by leaf fall from coniferous trees; and the tree, by reason of its extensive root system, may subject the grass to severe water and nutrient stress. Indeed, fertilizing a lawn frequently results in an unexpected burst of growth in nearby trees, without noticeably improving the appearance of the lawn!

SUCCESSION. The effect of competition is that a flora is seldom completely stable. One species or a group of species is eliminated through competition by another, until essentially a more or less stable situation is reached. This process is called succession, ultimately reaching a climax of greater or lesser degree of stability. The process starts with the colonization of previously barren ground by **invasion** of hardy species that are capable of surviving the often extreme conditions. Invaders are usually xerophytic because of the tendency of barren ground to dry out very quickly; they are often prolific and rapid reproducers, which enables them to become established during short periods of good conditions.

Following invasion, a succession begins in which new plants become established as each stage of the succession modifies the existing conditions to suit more fastidious plants. Very often conditions are modified so that the existing plants grow better, but in spite of this they are replaced by new species that are even more capable under the new conditions. Such new conditions include the improvement of soil structure for water retention, better release of nutrients from soil particles resulting in higher fertility, greater plant mass with a consequent increase in shading, less extremes of temperature and wind, higher RH values, and antibiotic effects.

Ultimately the succession moves to a climax consisting of species that can successfully complete their life cycles repeatedly under the conditions that have become established and can compete effectively at all stages of growth. Such a climax will be stable. Sometimes a climax flora tends to bring about conditions that prevent or inhibit its own reproduction, and this leads to an unstable condition in which the last stage or stages of the succession repeat constantly. The frequency and degree of periodic reversion of such an unstable climax depend largely on the rate of growth and reproduction of the dominant species it contains, rather than directly on environmental conditions. Instability of a climax may also result from nonclimatic events, such as repeated fire or repeated infestation with insect pests or diseases.

Physiological Mechanisms of Competition

One of the most important mechanisms of competition is the capacity to grow better under adverse conditions. We have mentioned above the sorts of stresses to which plants may be subject as the result of competition. It is important to realize that plants seldom grow under perfect or ideal conditions; even when such conditions exist for a plant, some other species may be able to grow better and thus be able to compete successfully. Thus, many plants that grow under extreme conditions do so only because they grow there better than other plants can. For example, halophytic plants can endure conditions of extreme saltiness that kill nonhalophytic plants. The halophytes, in fact, may grow much better in nonsalty conditions, but they are not able to survive competition with more successful nonhalophytes. They do compete successfully under salty conditions because their capacity to survive this extreme is greater than that of other plants.

Direct competition among plants results in shortages of water and minerals. Under these circumstances plants that have the capacity to produce a larger or more extensive or more efficient root system are likely to compete successfully. Capacity for growth seems to be an inherent function, but its expression is much affected by environmental conditions. Response mechanisms for root growth have not been well studied. It is known that low night temperatures promote root growth of many plants, although it is not clear why. However, an increased sensitivity of such a response would be an excellent mechanism for improving the chances of survival of a species in a temperate zone. Whether such a mechanism is actually involved is not known. Alternative mechanisms allowing effective growth with lower supplies of water and minerals, or lower water loss, would equally improve chances of success in competition.

Probably the most important competition among plants is for the raw materials of photosynthesis—mainly light, but also, in dense leaf canopies, CO_2. Shade tolerance is an important factor in competition, particularly in seedlings and developing plants. As a consequence, a variety of mechanisms have evolved in plants. These can be divided into three broad categories: shade-avoidance mechanisms, mechanisms that increase interception of light or CO_2, and mechanisms that increase efficiency.

Shade avoidance consists mainly of growing fast and rising above the shade-producing competitors. Plants in which the hormone-mediated response to increased stem and internode length at low light is highly developed compete successfully in dense stands. Here the detrimental effect of a thin, weak, elongated stem is minimal; the important requirement is to get a canopy of leaves as high as possible and as quickly as possible.

Increased interception of light may be accomplished by auxin-mediated morphological responses, for example, the development of mechanisms for leaf orientation to light and for the bending of petioles to produce maximum coverage in the leaf mosaic, as shown in Figure 29-7. In addition, all the physiological mechanisms of shade tolerance, including increased chlorophyll, increased number and orientation of chloroplasts, and larger thicker leaves with increased photosynthetic capacity may be important in competition. There is some indication that the leaves of shade-tolerant plants have a greater carboxylation efficiency, in spite of the fact that they have a lower concentration of the carboxylating enzyme, ribulose bisphosphate carboxylase. It is possible that they maintain a higher concentration of the substrate, ribulose bisphosphate.

Improvements in many factors could improve the competitive capacity of the plant. The primary fixation of CO_2 by β-carboxylation as in C_4 plants (see Chapter 7, page 176, and Chapter 15, page 357) appears to be more efficient, particularly at low CO_2 concentrations and high light, than the RuBP primary carboxylation of C_3 plants. This would enable plants having the Hatch and Slack cycle to compete successfully with plants lacking it, particularly in dense stands under conditions of intense illumination. These conditions prevail in grasslands, where light penetration is good, and in tropical areas, where light intensity and temperature are high. Under these conditions, CO_2 concentration may be somewhat reduced and competition for it may be an important

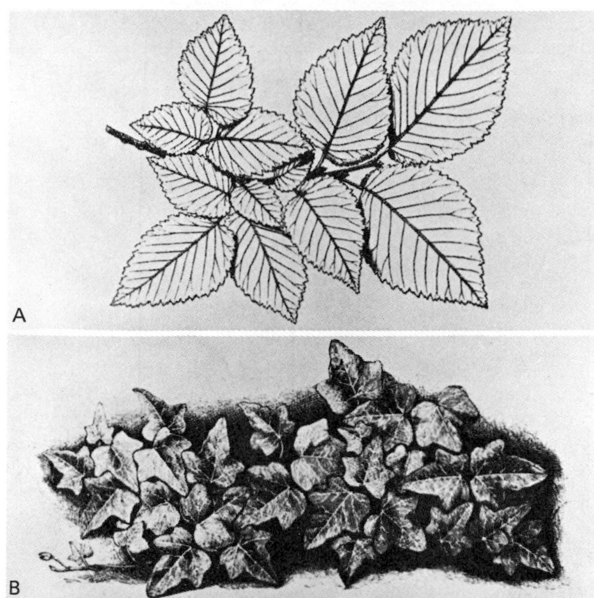

Figure 29-7. Leaf mosaic under shade conditions: (**A**) elm, (**B**) ivy. [From F. W. Went and L. O. Sheps: In F. C. Steward (ed.): *Plant Physiology: A Treatise*, Vol. VA. Academic Press, New York, 1969. Used with permission.]

factor in success of the plant species. The reduction of photorespiration in C_4 plants cuts respiratory losses, particularly at high temperatures. Moreover, C_4 plants can better withstand water stress because photosynthesis continues strongly with reduced stomatal aperture. Therefore, it is not unexpected that C_4 plants are largely (although not exclusively) grasses and plants that inhabit or originate from dryer tropical habitats.

Mechanisms that increase efficiency are largely physiological or biochemical. These include decreased rates of respiration, lower photorespiration, and increased efficiency of photosynthesis at low light or low CO_2 concentration. These overlap somewhat with the mechanisms mentioned in the preceding paragraphs but relate more to the efficiency of use of the raw materials, rather than to the efficiency of their accumulation. Experimental values for the efficiency of photosynthesis vary widely, and there are grounds for the belief that this variable may be under genetic control. Thus, genetically more efficient varieties or species may exist that would be at a competitive advantage under shade conditions. Similarly, the degree of coupling between the respiration rate and the energy requirement or energy charge (see Chapter 5, page 103) seems to vary among plants and tissues. The most effective coupling will mean the highest efficiency and the lowest waste. It is this factor, not simply the overall rate of a process, that is important in the success or failure of plants or species in competition.

Additional Reading

Articles on "Environmental Physiology" in *Annual Reviews of Plant Physiology* and textbooks on plant ecology.

Evans, L. T. (ed.): *Environmental Control of Plant Growth*. Academic Press, New York, 1963.

Kellerman, M. C.: *Plant Geography*. Methuen & Co. Ltd., London, 1975.

Larcher, W.: *Physiological Plant Ecology*. Springer-Verlag, Berlin, 1975.

30 Plants and Man

Introduction

There are many levels of interaction between plants and man in which physiological principles play an important part. Man depends on plants for food, clothing, shelter, the maintenance of the environment, and natural beauty. But plants and the flora depend also on man, because man's activities usually alter the environment, frequently destructively. In many of his activities, man's effect on the environment is not deliberate, particularly in more primitive societies. However, as societies evolve and develop, deliberate attacks on naturally established vegetation become more and more vigorous. Destructive acts include the exploitation or total harvesting of nonrenewing (often nonrenewable) resources, such as the early lumbering activities in North America, or the elimination of much or all of a flora by destructive modification of the environment, such as road building, the practices which led to the creation of the dust-bowl, and so forth.

Some practices of more highly developed civilizations lead to constructive attacks on the environment, including agriculture, the modification of the physiography or terrain to suit the growth of plants (for example, paddying, terracing, irrigation), and the modification of the plants themselves by various means. All the techniques for the practice and improvement of plant cultivation must rely heavily on sound knowledge of the physics of the environment and of the physiology of the plants. It is not essential to know, for example, the mechanisms of cold temperature resistance in order to select for more resistant plants. But if we did know the exact mechanisms, we might be able to select more precisely for some specific property, not now readily apparent, that would by itself or in combination with other factors confer the desired resistance.

Man's Impact on the Landscape

There has been endless philosophical argument as to whether the effects of man on the environment are natural or unnatural. The argument as such is irrelevant here, but it adds perspective to the fact that man does influence plants and the flora in marked ways, by rendering unstable previously stable climaxes, sustaining unstable conditions, or completely eliminating the normal ecological succession. Many of the effects of man are unintentional and the results of other activities, but much of man's effort is spent in the deliberate cultivation of plants for all sorts of reasons, from agriculture to landscape gardening to total environment engineering.

In every situation one of the most important factors is the physiological reactions of the organisms under cultivation. Animals react to their environments substantially through behavioral expression as well as through their basic physiology. Plants react wholly through their physiological responses. It is therefore worthwhile to make a brief summary of the physiological bases upon which plants and man interact.

Levels of Interaction. There are several levels of interaction, depending largely on the state of civilization or culture achieved by man. At the lowest level, in which man acts solely as a **gatherer,** little interaction occurs because the intensity of man's activities is usually low. When the cultural development of man tends toward **hunting,** the resulting effects on vegetation are usually secondary to the changes caused in the fauna. For example, large herds of grazing animals such as buffalo tend to maintain an unstable grass climax that may give way to bush or forest if the animals are killed off. As hunting gives way to **herding,** widespread changes in the vegetation result from the practices of burning and grazing, which maintain areas in grass or prairie that would, if left alone, result in a forest climax.

However, it is not until the level of **agriculture** is reached that large-scale and deliberate changes in the flora are brought about. Domestication of plants leads to vigorous selection of specialized characteristics for yield, hardiness, convenience of harvest, and the like. Modifications in the structure, physiography, and fertility of the environment are made to ameliorate growing conditions. Pure stands of plants are frequently cultivated, and this leads to special problems, such as shortages of certain specific nutrient elements and the development of high concentrations of virulent pathogenic organisms. This in turn leads to further modifications of the plants to lower their nutrient requirements and to raise their resistance to disease.

The final stage of man's cultural development, **industry and urbanization,** leads to further interaction. Plants now become affected by pollution and destruction of the environment, and further selection and cultivation of plants for landscaping take place. Highly specialized plants are developed, with carefully regulated habits of growth and maximum resistance to environmental hardships, for use in landscaping. Lawn grass, ornamental shrubbery, and shade trees are all the result of this level of interaction.

Modification of the Environment. Modifications by man are often drastic because the climax vegetation, being stable, holds and ameliorates the soil. If this is disturbed, the soil may deteriorate or be lost. Frequently, particularly when thin layers of soil are present on a hard rock substrate, soil so damaged or lost cannot be replaced, perhaps for centuries. As a result, the physiography and even the climate of a region may be modified. A typical example of this type of interaction results from various lumbering practices, as illustrated in Figure 30-1. Secondary effects of such activities as lumbering include the creation of much wider fluctuations in local climatic conditions and in the availability of water. When the forest is cut, water runs off more quickly and less is retained. Thus the level of water in rivers, ponds, and the low-lying areas becomes more irregular, and many plants cannot grow because of too much or too deep water in spring and not enough in summer. As a result the fauna dependent on the flora is also adversely affected.

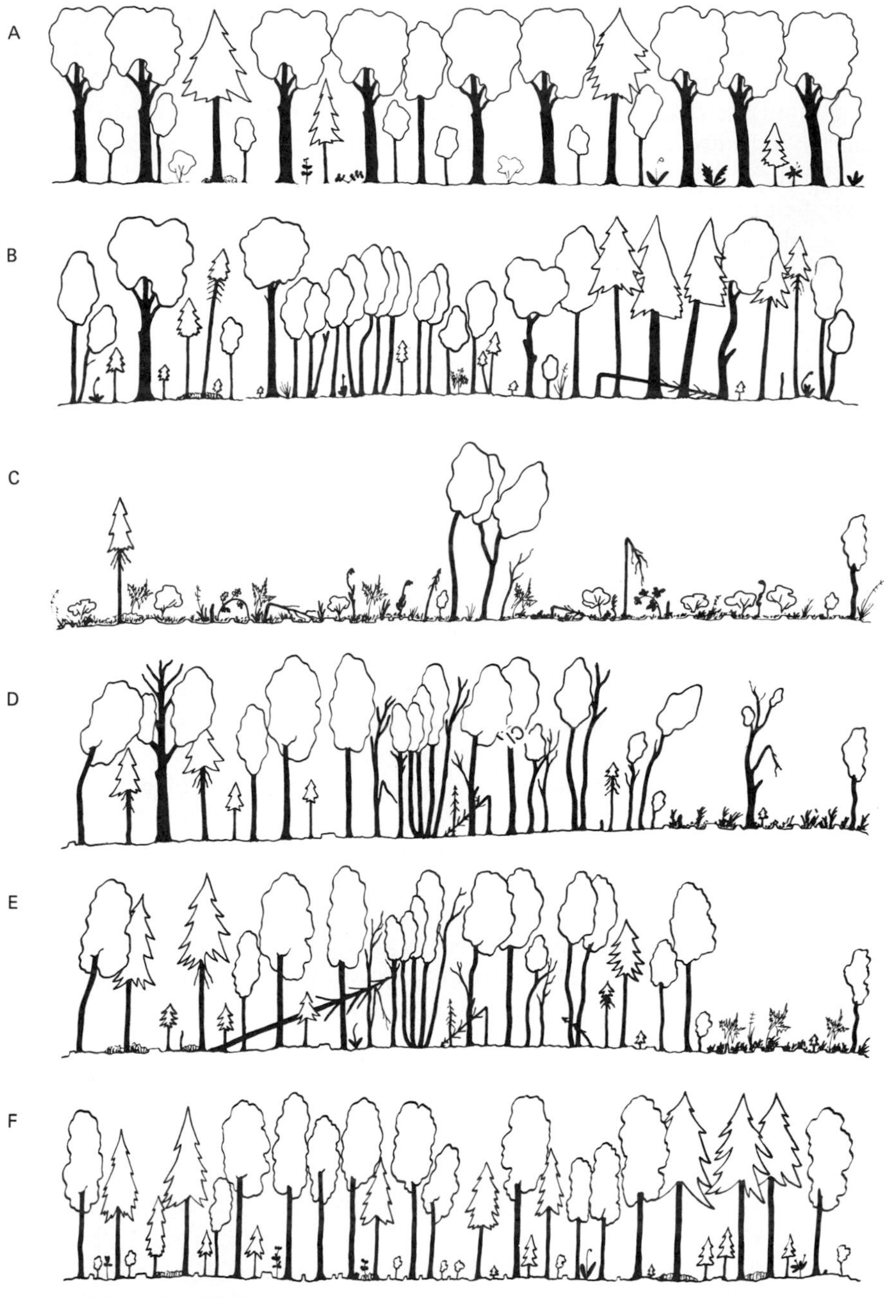

Figure 30-1. Some effects of lumbering in Quebec.
A. Virgin forest. **B.** Forest somewhat modified by man. **C.** 10-15 years after clear-cutting. **D.** The B forest 10-15 years after "highgrading" (removal of best trees). **E.** The B forest modified by "diameter limit" cutting. **F.** The B forest selectively cut. [From P. Dansereau: *Biogeography*. The Ronald Press Co., New York, 1957. Used with permission. Photograph courtesy Dr. Dansereau.]

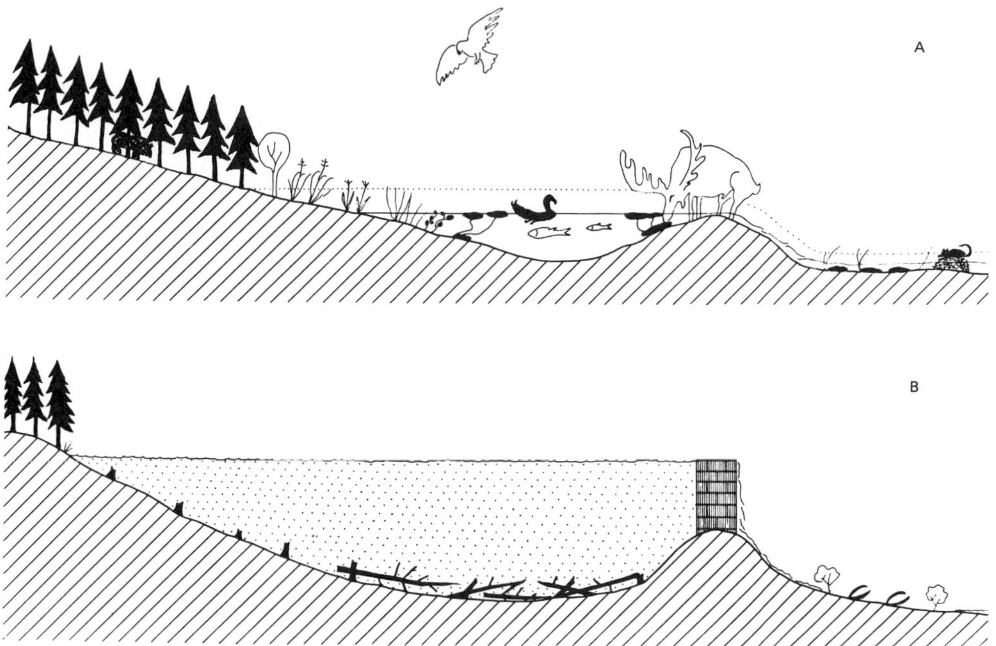

Figure 30-2. Effects of damming on the Upper Peribonka River, Quebec. **A.** Before damming. **B.** After damming. [From P. Dansereau: *Biogeography*. The Ronald Press, New York, 1957. Used with permission. Original figure courtesy Dr. Dansereau.]

The construction of dams frequently has the same effect, as shown in Figure 30-2. Many dams are built in an attempt to rectify the water problems following improper forest management, but they frequently fail because of lack of knowledge of the conditions required for successful growth of vegetation. The dam illustrated in Figure 30-2 floods an area of about 200 square miles of forest and scrub in Quebec, Canada. It creates irregular water flow so that the spring flooding downstream now prevents salmon from spawning in the river. There is too little water in the summer so the aquatic vegetation has died; as a result the aquatic molluscs have vanished, and muskrats which feed on the plants no longer inhabit the area. The original aquatic flora in the lake above the dam has been eliminated because most of the lake is too deep and because fluctuating water levels expose it in summer along the edges. Finally, even the moose that fed on this vegetation have left. What was once a productive forest land teeming with wildlife is now a barren wilderness. Water control must be brought about through vegetation control to maintain the forest soils. It cannot be done by the creation of dams and puddles.

Modification in Agriculture. Agriculture is the deliberate modification of the land for the growth of useful plants. However, this in itself creates problems of disease, pest and weed control, soil and fertility maintenance, and so on. Desirable crop plants are frequently unable to sustain themselves or grow well in areas at or beyond the normal limits of their range, and breeding programs must be undertaken to improve them and extend their useful range. The biggest developments in this area follow an awareness of the physiological attributes that need to be selected and improved by breeding.

This is not always easy or obvious. Much work has been done developing plants for extreme regions, for example, potatoes for mountainous areas or rice for dry or temperate climates. High yield depends substantially on the rates of photosynthesis and photorespiration by day and the rate of respiration by night. Thus, if a region is characterized by low night temperature, a plant having high photosynthesis and high respiration would be a likely choice, since the low night temperature would offset the tendency to high carbon loss at night. However, such a plant would not grow well in a climate characterized by warm nights. Similarly, a high rate of gross photosynthesis is not much use if photorespiration is also high, because the net productivity under these circumstances will be low.

Water is expensive and often difficult to get, so it is desirable also to relate net photosynthetic ability with the capacity to obtain and retain water. Nutrient requirement is another factor that must be related to field situations. The increased value of a high-producing line of a crop plant may be more than offset by its increased requirement for fertilizer or water! This is a real problem for developing countries, where the green revolution has brought increased agricultural production from high-yield crops. These crops require better growing conditions and much higher levels of fertilization. Thus the cost of sustaining high productivity may outstrip the benefits of the higher production, leaving the country worse off than before.

Environment Management. Landscape gardening, or environmental engineering in its broadest sense, is a highly specialized form of agriculture, and its problems are basically the same. Only the "crop" is different—tidiness, beauty, land maintenance, or pollution abatement. (A grove of trees can as effectively screen a factory from the ear as from the eye, and will absorb atmospheric pollution, too, if the plants are hardy.) There is the same requirement for tolerance of extreme conditions. For example, shade tolerance is necessary in lawn grass planted under trees, and tolerance of industrial and urban pollution is even more important. Disease resistance is interrelated. The resistance of long-lived shade or ornamental plants is extremely important and may be closely related to pollution levels.

Productivity and Agriculture

There is a sound physiological basis for most successful agricultural practices. Many such practices were developed empirically, but the growth of plant physiology has made possible the discovery of the scientific basis of many successful "arts" so that their usefulness has been greatly extended. Much physiological research has been deliberately directed toward the improvement of agriculture. For example, several early Russian physiologists worked on the nitrogen metabolism of plants that grow successfully on infertile barrens. The plants they studied were unfit for food themselves, but they hoped, through this study, to unlock the secret of growing profitable crops on barren or unproductive lands. Modern insight into physiological mechanisms of growth and metabolism can now be used in the search for increased quality and quantity of crops and for improving the survival or extending the range of desirable plants.

Use of Growth Factors. Many different agricultural techniques are based on various different aspects of plant physiology. One of the most evident is the application of growth factors and hormones and synthetic chemicals to modify growth and development.

Figure 30-3. Improved fruit set after auxin spray.
The tree in the foreground was sprayed with naphtheleneacetic acid, which prevents fruit drop until harvest time. The large pile of apples at the right fell prematurely from the unsprayed tree. [Courtesy U.S. Department of Agriculture.]

Auxins and ethylene were among the first compounds to be used commercially, and chemicals of this type have now been put to so many uses that to list them all would be impossible. Besides their well-known uses for rooting, the induction of parthenocarpy, the improvement of fruit set (Figure 30-3), and the thinning of fruit to improve yield (Figure 30-4), auxins have been successfully used to improve the growth density and yield of many crops. A partial list of the uses of auxins is given in Table 30-1.

Gibberellins have been very widely used in agriculture. Among their earliest uses was to induce parthenocarpy, which results in seedless fruit, often of larger size (Figure 30-5). Increased harvest and better shape in fruits that form bunches, such as grapes, may be obtained by treating the cluster with GA, which causes elongation of the cluster and thus reduces crowding (Figure 30-6). Many other uses of gibberellins are listed in Table 30-2.

Cytokinins have been used, though not extensively, to prolong the life of plants or plant parts. Growth retardants are often useful to preven unnecessary vegetative growth before seed production in some plants and to control their size for ease of harvesting. Certain retardants, such as maleic hydrazide, are useful in preventing sprouting of root crops such as potatoes during storage. Another technique is the induction of polyploidy by such agents as colchicine, resulting in larger, more abundant fruit (Figure 30-7).

In all these applications it is usually advantageous to use synthetic auxins such as naphthaleneacetic acid, indolebutyric acid, parachlorophenoxyacetic acid, and so on, rather than naturally occurring ones. This is because enzymes exist in most plant tissues

Table 30-1. Some uses for auxins

Compound	*Use or purpose*	*Conc. (amount)*	*Treatment*
Various natural and synthetic auxins	Stimulate rooting of cuttings Promote budding or sprouting Control of cotton boll shedding	100–10,000 ppm	Dust or dip
IAA	Prevention of leaf and fruit drop	10–100 ppm	Dip
NAA	Control of preharvest fruit drop	20–100 ppm	Preharvest spray
2,3,5-Triiodobenzoic acid	Increase yield of soybeans	3–4 oz per acre	Foliar spray
N-arylphthalamic acids	Induce flower formation, increase fruit yield of tomatoes	200 ppm	Spray on young seedlings
Natural and synthetic auxins	Increase yield in potatoes, peas, beans, beets, corn	1–10 oz per acre	Dust or spray during growth
NAA, other synthetic auxins	Control fruit set in many fruits	20–200 ppm	Spray blooms

A

B

Figure 30-4. Chemical thinning to improve fruit set in Golden Delicious apple trees. This variety tends to bear only in alternate years unless thinned. Treatment with synthetic hormones shortly after bloom causes a decrease in the number of flowers and fruits but ensures a more regular burden of fruit year after year and an increase in the average yield.

A. Trees sprayed with 250 ppm Sevin (1-naphthyl N-methylcarbamate) 15 days after bloom the previous year; they have bloomed again this year. The trees without blooms were not treated.

B. Trees at left sprayed the previous year with 34 ppm NAD (naphthaleneacetamide) 17 days after bloom, and then with 500 ppm Sevin 22 days after bloom. The treated trees bear a good crop this year. The untreated trees, which bore a heavy crop last year, bear almost no fruit this year.

[Original photograph courtesy U.S. Department of Agriculture, through the kindness of Dr. M. W. Williams.]

Figure 30-5. The induction of parthenocarpy in Wealthy apples. Gibberellin A_4 treatment (10^{-2} *M*) made the apples on the right grow larger than the untreated ones on the left. [Original photograph courtesy Dr. M. J. Bukovac, Michigan State University, East Lansing, Michigan.]

Figure 30-6. Treatment with gibberellin 3 weeks before bloom caused the flower stalk of these Zinfandel grapes to elongate. This makes the clusters looser and easier to harvest and reduces the chance of rotting on the vine. Gibberellin concentration (left to right): 0, 10, 100, and 1000 ppm. The highest concentration clearly had bad side effects. The recommended concentration for this variety is 10 ppm. [Original photograph courtesy Dr. R. J. Weaver, University of California, Davis, Calif.]

Table 30-2. Agricultural uses of gibberellins

Use or purpose	*Crops*	*Conc./amount*	*Treatment*
		Vegetable crops	
Uniform bolting Increase seed production	Lettuce seed production	5–10 ppm or 1–5 g/acre	Spray at 8–12 leaf stage
Increase stalk length Higher yields	Celery	25–50 ppm or 5–10 g/acre	2–3 weeks before harvest
Reduce chilling requirements Increase yields	Forcing rhubarb	250–500 ppm and 50–100 ml/crown	Beginning of forcing period
Enhance development Shifts harvest earlier	Globe artichoke	25–50 ppm or 5–10 g/acre	2–3 sprays in fall
Break dormancy Uniform crop emergence	Seed potatoes	1/2–1 ppm	Dip or spray seed pieces before planting
Induce staminate flowers on gynoecious types	Hybrid cucumber seed production	500–1500 ppm	Spray at 2–4 leaf stage
		Fruit crops	
Increase berry and cluster size. Loosen clusters	Black Corinth Seedless grapes	2.5–5 ppm	Spray once shortly after bloom
Improve berry size Loosen clusters	Thompson seedless grapes	2.5–20 ppm 20–40 ppm (16–48 g/acre)	Spray at full bloom Spray at fruit setting (Wet cluster area thoroughly)
Induce seedlessness Increase berry size Hasten maturity	Delaware grapes	100 ppm 100 ppm	Dip clusters before bloom Dip at fruit setting
Loosen clusters Reduce berry rot	Compact clustered wine grapes	1–10 ppm depending on variety and 100 gal/acre	Spray 2–3 weeks before bloom
Delay rind senescence Reduce rind disorders	Navel oranges	10 ppm at 500 gal/acre	Spray as marketable color develops
Delay fruit maturity	Lemons	Same as for Navel oranges	Spray prior to loss of green color
Reduce virus yellows Improve fruit quality	Red Tart cherries	15–25 ppm	10–15 days after petal fall
Delay maturity Extend harvest Firmer, larger fruit Less fruit cracking	Sweet cherries	5–10 ppm	3 weeks before normal harvest
Improve fruit set Reduce nonproductive period	Pears (some varieties)	10–20 ppm	Spray at flowering or petal fall
Improve fruit size Cause fruit to set with low seed number	Blueberries, cranberries, seeded grapes, tomatoes	10–50 ppm	Spray at flowering or petal fall
		Other uses and crops	
Increase α-amylase and proteolase. Improve malting properties	Barley malting	1 ppm or 1 g/ton	Add to steeping water
Promote early spring growth	Grass and sod	10–50 ppm or 2–10 g/acre	Spray in early spring ——
Improve flower size and longevity. Accelerate development of tree types	Geranium	5–10 ppm	Spray flower buds at first color Repeat sprays for upper foliage
Promote growth of stalks Increase yield of sugar	Sugarcane	200 ppm	Aerial application in winter, 3 months before harvest

SOURCE: S. H. Wittwer: Growth regulants in agriculture. *Outlook on Agriculture,* **6**:205–17 (1971). Reproduced by permission of Imperial Chemical Industries, Limited.

Figure 30-7. Increased cluster size as a result of colchicine-induced polyploidy in the Loretto grape. Left, diploid. Right, tetraploid. [Courtesy U.S. Department of Agriculture.]

for the rapid detoxification or destruction of natural auxins, whereas the unnatural ones are not so readily attacked and so persist and act much longer.

Among the most useful new growth regulators is N,N-dimethyl-aminosuccinamic acid (Alar-85 or B-9). Sprayed on fruit trees at or soon after the time of blooming, it greatly improves fruit quality and hastens maturation. It also increases the ease of mechanically harvesting the fruit. Some of the uses to which this chemical has been put are listed in Table 30-3. Other new compounds that have undergone extensive trials are 2-chloroethyl-trimethylammonium chloride (CCC or Cycocel) and 2-chloroethylphosphonic acid (Ethrel). The very wide range of effects that these chemicals have on various plants can be seen from Tables 30-4 and 30-5.

Evidently the use of synthetic agents to control the growth and form of plants is a major new front in agricultural technology. Much research is currently devoted to this branch of crop management. Vast numbers of chemicals are being tested on the off chance that they may prove useful, but many of the more profitable studies of naturally occurring chemicals and their synthetic analogs have been directed by physiological considerations. Unusual or interesting aspects of development, the suspected presence of new or unknown agents, and the known or suspected effects of plant hormones and synthetic compounds have all provided leads for this important research.

Chemical control of weeds has developed dramatically since the discovery of 2,4-D, 2,4-5-T, and similar synthetic auxins. The extent of growth of this dynamic branch of plant research is revealed by the fact that a Canadian government publication on chemical weed control lists over 200 chemical agents available in 1974, and a 1977 guide to insecticides, herbicides, and fungicides in the United States lists nearly 300 preparations.

Timing. Crop value and suitability may be greatly improved by careful choice of planting and harvesting time for reasons that are not empirically obvious. For example, early sowing of seeds might reasonably be expected to provide an early maturing crop.

However, an overall deleterious effect may follow early-season exposure to unfavorable conditions, and the seed, lying long in the soil, may fall prey to insect or fungal pests. As another example, silage harvested early in the season usually has a much higher fructosan content than if harvested late. High sugars promote effective fermentation of the silage and prevent the development of bacteria causing proteolysis and spoilage, so early-harvested silage provides much better feed.

Many plants have rather precise day and night temperature requirements for maximum productivity and proper development, and these requirements may change during the growth of the plant. Thus, an exact knowledge of the requirements of the plant and physiographic or climatic data of a region permits the selection of varieties most likely to grow and produce well and of the correct planting time for maximum yield. Such analyses have been done by the American physiologist F. Went and his colleagues. Some data from their experiments are represented in Figure 30-8.

Environmental Control. Other agricultural techniques are designed to take advantage of the physiological adaptability of plants. Fertilizer application is an important

Table 30-3. Agricultural and horticultural uses for N,N-dimethylaminosuccinamic acid (Alar-85 or B-9)

Use or purpose	*Crops*	*Conc./amount*	*Treatment*
Reduce tree vigor—promote flowering Prevent premature fruit drop Enhance fruit quality and storage life Delay bloom, increase fruit set	Apple*	1000–2000 ppm 1000–2000 ppm 4000 ppm (5–15 lb/acre)	Postbloom spray 45–60 days before anticipated harvest Fall application
Induce uniform ripening Hasten maturity Increase color	Peach*	1000–4000 ppm (6 lb/acre)	At beginning of pit hardening
Enhance fruit set	Grape*	2000 ppm	At early bloom
Advance fruit harvest 7 days Improve—and more uniform—color Reduce fruit removal force Firmer and more uniform fruit Increase color in frozen packs and pies	Sour cherry*	2000–4000 ppm 4 lb/100 gal (6–12 lb/acre)	2–3 weeks after bloom
Advance fruit maturity and color Increase soluble solids Fruit removal force reduced Increase resistance to cracking	Sweet cherry*	1000–2000 ppm (3–6 lb/acre)	Two weeks after bloom
Increase yields Higher grades Drought resistance	Peanuts	(1–1½ lb/acre)	Timing not critical
Increase yield, number of tubers	Potatoes	3000–6000 ppm	Apply when tubers are forming
Improve transplants Concentrate fruit set Facilitate mechanical harvest	Tomatoes	2500–5000 ppm	Timing variable
Produce attractive compact, dark-green plants that flower early	Bedding plants (petunia, snapdragon, marigold, salvia, zinnia)	2500–5000 ppm	Spray young plants
Promote rapid footing of stem cuttings of herbaceous and woody plants	Ornamentals	1000–5000 ppm	Momentary dip of cut stems

SOURCE: S. H. Wittwer: Growth regulants in agriculture. *Outlook on Agriculture,* **6**:205–17 (1971). Reproduced by permission of Imperial Chemical Industries, Limited.

*Approved for commercial application.

Table 30-4. Present and projected uses for 2-chloroethyl-trimethylammonium chloride (CCC or Cycocel)

Use or purpose	*Crops*	*Conc./amount*	*Treatment*
Shorten and stiffen the plant Prevent lodging Increase yields Enhance root growth Increase tillering	Wheat, oats, rye in western Europe	1 to $1\frac{1}{2}$ lb/acre spring wheat $1\frac{1}{2}$ to 2 lb/acre winter wheat	Soil applications
Compact symmetrical sturdy plants Early flowering	Poinsettia Chrysanthemum Azaleas and other ornamentals	2000–10 000 ppm	Soil drench
Drought resistance Cold resistance, Salt resistance	Soybeans, cabbage, tomato	2500–5000 ppm	Foliar spray
Growth stimulant	Snapdragons	50–500 ppm	Foliar spray
Increase fruit set Greater fruit size or berry weight	Grape Tomato	100–1000 ppm	Foliar spray
Compact, thick stems adapted for transplanting Earlier flowering	Tomato	10–100 ppm	Soil drench
Increase flowering, boll frequency, and yield	Cotton	25–50 ppm 5 g/acre	Foliar spray 70 days after emergence

SOURCE: S. H. Wittwer: Growth regulants in agriculture. *Outlook on Agriculture,* **6**:205–17 (1971). Reproduced by permission of Imperial Chemical Industries, Limited.

Table 30-5. Potential uses for 2-chloroethylphosphonic acid (Ethrel)

Use or purpose	*Crops*	*Conc./amount*	*Treatment*
Accelerate maturity	Pineapple	1–6 lb/acre	Foliar spray
Induce uniform ripening	Fig	250–500 ppm	Foliar spray
Enhance color Increase yields	Tomato	500–1000 ppm	Spray at mature green fruit stage
Accelerate fruit abscission Facilitate mechanical harvest	Many types of tree- and bush-borne fruit	500–2000 ppm	Spray before harvest
Induce femaleness	Cucumber, squash, and melon	100–250 ppm	Spray at first true leaf stage
Induce flower abscission Thinning fruit	Apple, peach	100–200 ppm	Spray 10 days after bloom
Induce fruitlessness	Ornamentals	1000 ppm	Foliar spray
Induce uniform flowering	Pineapple Bromeliads	1–6 lb/acre 100 ppm	Foliar spray Foliar spray
Stimulate bottom breaks	Rose	2500 ppm	Two sprays on stems
Effect preharvest defoliation	Nursery stock	2000 ppm	Foliar spray
Induce bulbing	Onion	500–10,000 ppm	Spray in early growth
Reduce incidence, severity of lodging Increase tillering	Cereals (wheat, barley, oats, rice, rye) peas	1–2 lb/acre	Spray young crop

SOURCE: S. H. Wittwer: Growth regulants in agriculture. *Outlook on Agriculture,* **6**:205–17 (1971). Reproduced by permission of Imperial Chemical Industries, Limited.

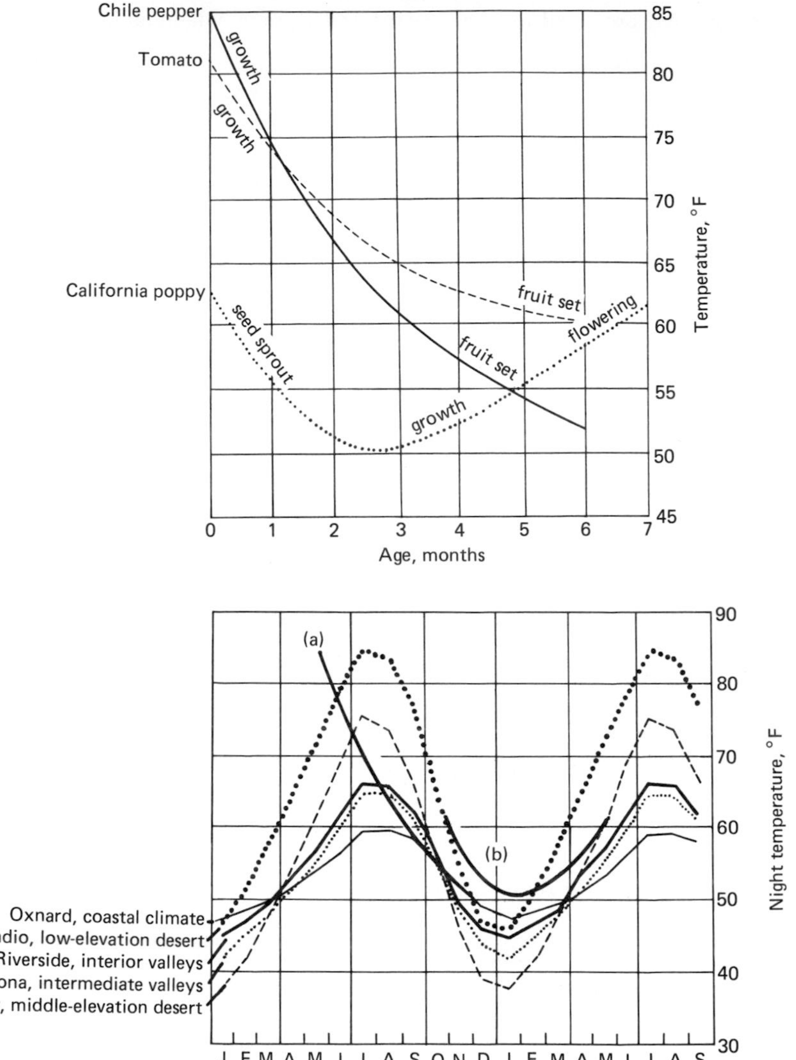

Figure 30-8. Optimum night temperatures for various stages in the life of three annual plants are given by the curves in the diagram at the top. The curves in the diagram at the bottom give the average night temperatures during 1950 and 1951 for four locations in California. By fitting the curves at the top to the curves at the bottom, it is possible to determine when it is best to plant these species. The curves fitted in the lower diagram are for (a) the chile pepper and (b) the California poppy. [From F. W. Went: Climate and agriculture. In J. Janick, R. W. Schery, F. W. Woods, and V. W. Ruttan (eds.): *Plant Agriculture*. W. H. Freeman & Co., San Francisco, 1970. Used with permission.]

example: because of cost and pollution factors, it is desirable to reduce application to a minimum. Here it is not only the amounts that are important but a knowledge of the balance of the nutrients required; a thorough knowledge of soil conditions is conducive to most effective use of native and applied nutrients.

Water control is equally important. In areas of sparse rainfall, careful analyses determine the times during the growing season when water is indispensable or most useful in terms of yield. In some areas, irrigation is so closely controlled that rain may be considered a nuisance, damaging or washing away the soil. Plastic film has been widely used to control water loss from soil, to kill weeds between the rows, and to heat up the soil in the spring to promote early germination and growth of plants.

Great strides have been made in greenhouse technology. This has led to the development of sophisticated installations for producing out-of-season crops of fastidious fruits and vegetables in North America. More important, it has provided large areas of cheaply modified agricultural environment for food production in developing countries. Many agricultural communes in China, for example, have acres of simple greenhouses made essentially of pits or trenches in the ground with low mud-brick parapets and glass or plastic covers, like the ones shown in Figure 30-9. Being underground, little or no heating is required, and rush mats can quickly be unrolled over the glass on unusually cold nights. Huge amounts of fresh produce for the masses of people in China are grown in this way. Their diet is in sharp contrast with the lack of fresh vegetables in winter in more northern and agriculturally less well-managed countries like Russia.

Certain specialized culture techniques are sometimes of advantage. For example fruiting vines such as grapes may be stem-girdled to prevent downward flow of nutrients. This increases nutrient supplies to the fruits and results in a larger crop. Damage to the root stock ensues, however, if the process is not carefully controlled. Hydroponic or sand culture of plants on a large scale has been found feasible under special conditions, for example in the Arctic, where light may be more than adequate in summer, but frost-free periods are too short and the climate too variable for successful out-of-doors growth. There the expense of the technique becomes relatively less important when compared with the cost of flying in fresh vegetables. Similarly, in greenhouse practice it is possible to increase flowering and to time it exactly for special occasions by the use of appropriate day-length treatments. Such techniques are now widely used in horticulture.

All these techniques, as well as such practices as crop rotation, green-crop plough-in, selective fertilization, and the like, are based on plant physiological concepts applied to the requirements of agriculture.

"The Sun's Work in a Cornfield." A recent and very interesting paper with this title by a team of agricultural scientists and plant physiologists directed by E. Lemon at Cornell University (see Additional Reading, page 687) emphasizes the importance of plant physiology as well as computer technology to agriculture. Lemon and his colleagues have created a soil-plant-atmosphere model (SPAM) which is designed "(1) to define, on the scale of the leaf surface in a plant stand, how each leaf (and the soil surface) will respond to a given, immediate climate; (2) to calculate from meteorology what that climate is; (3) to calculate the specific leaf and soil responses to that climate; and (4) to add up, leaf layer by leaf layer (and soil surface), the responses for the whole crop." The essential components of the model are shown in Figure 30-10, together with its predictions. The flow diagram of the model is shown in Figure 30-11.

Although a model such as SPAM is limited in its present usefulness (primarily by limitations in our knowledge of the behavior of plants, that is, by limitations in data input), it can nevertheless provide a surprising amount of information about a crop and its relationship with the environment. From the various leaf and community traits and the external climate, the model can predict the microclimate in a community and at the leaf and soil surfaces. It can also predict the activity of the leaves and plant community in such processes as photosynthesis, respiration, evaporation, transpiration, and heat

A

B

C

Figure 30-9. Glasshouse crop production on a commune near Peking, China. **A** and **B** show simple, inexpensive greenhouses that are partly subterranean and have rush covers for cold nights. **C.** The Chairman of the Revolutionary Committee of the commune demonstrates the high quality of produce from the greenhouses to the author's wife.

exchange. Such models can be used, with caution, as powerful tools to help man order his priorities of plant community traits for whatever outcome he desires, be it for production, nature and water conservation, climate modification, or aesthetic enjoyment.

Adaptation and Development of Plants for Special Needs

The selection or creation of new varieties of plants to suit special conditions or for special purposes is primarily a genetic problem. However, it is necessary, before undertaking genetic surgery, to know precisely what is required. In some cases this is not difficult—for example, ornamental plants need to be as hardy as possible as well as being ornamental. In other cases this is not so easy. Productivity of a plant could be thought of initially as a combination of its capacity for photosynthesis, photorespiration, and dark respiration. However, it is not enough merely to attempt to increase the first and decrease the other two. The nature of the production and its timing must also be considered. A bean plant with high photosynthesis and low respiration that produces many leaves and few fruit offers no advantages!

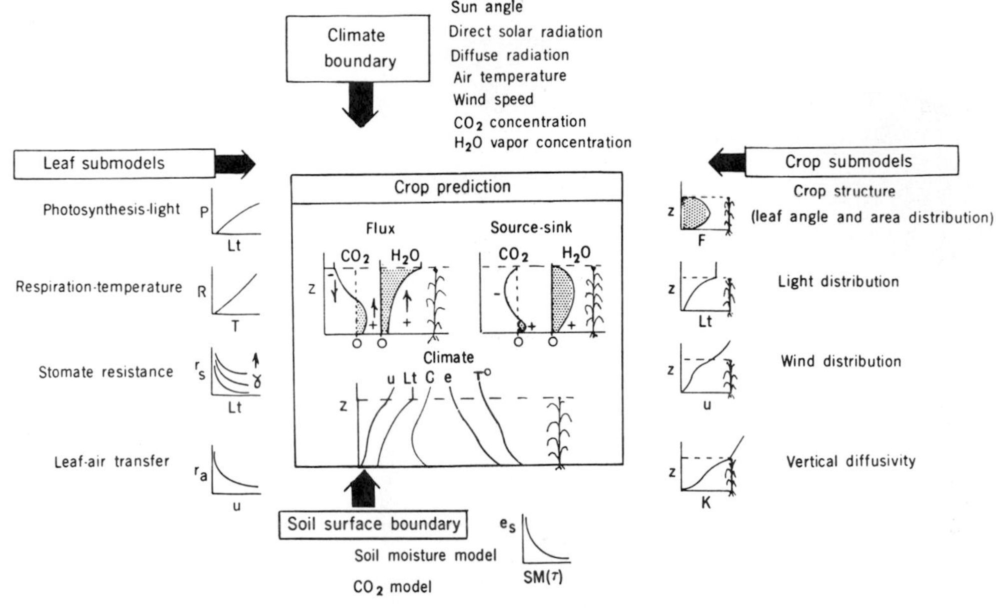

Figure 30-10. Schematic summary of a mathematical soil-plant-atmosphere model (SPAM) giving required inputs, submodels, and representative daytime predictions of climate and community activity (that is, water vapor and carbon dioxide exchange).
Abbreviations: height (z), wind (u), light (Lt), concentration of carbon dioxide (C), water vapor (e), air temperature (T°), surface vapor pressure (e_s), surface soil moisture or water potential SM(τ), photosynthesis (P), respiration (R), leaf temperature (T), stomate resistance (r_s), minimum stomate resistance at high light intensities (γ), gas diffusion resistance (r_a), leaf surface area (F), vertical diffusivity (K), net radiation (R_n), sensible heat (H), latent heat (LE), photochemical energy equivalent (λP), and soil heat storage (S). [From E. Lemon, D. W. Stewart, and R. W. Shawcroft: The sun's work in a cornfield. *Science*, **174**:371–78 (1971). Copyright 1971 by the American Association for the Advancement of Science. Used with permission.]

Sometimes special products are required of plants, such as alkaloids, narcotics, or other drugs and chemicals. As an example, many plant proteins are deficient in the amino acid lysine, which is essential for animal growth. Special genetic lines or mutants of wheat and corn have been produced that have higher than normal lysine content; this greatly increases their food value. In many instances the production of required compounds may be improved by adjusting conditions in specific ways to affect the physiological balance of the plant. For example, radioactive glutamine and asparagine are commercially synthesized by supplying $^{14}CO_2$ to leaves in light. The yields can be greatly increased by the addition of ammonium salts to the leaves' culture fluid and by adjusting the length of light and dark periods.

It seems probable that many more widely useful commercial properties of plants could be improved by selection of the appropriate cultural practice to affect the physiology of the plant in the required way. Profitable investigation might be conducted in such qualities as fiber strength or structure (as in flax, wood), oil content of seeds such as flax, peanut, sunflower, or soybean, essential oils used for flavoring as in mint, or to decrease the yield of toxic compounds (such as the the coumarins in clover that yield the toxic dicoumarol, which causes sweet clover disease of stock animals).

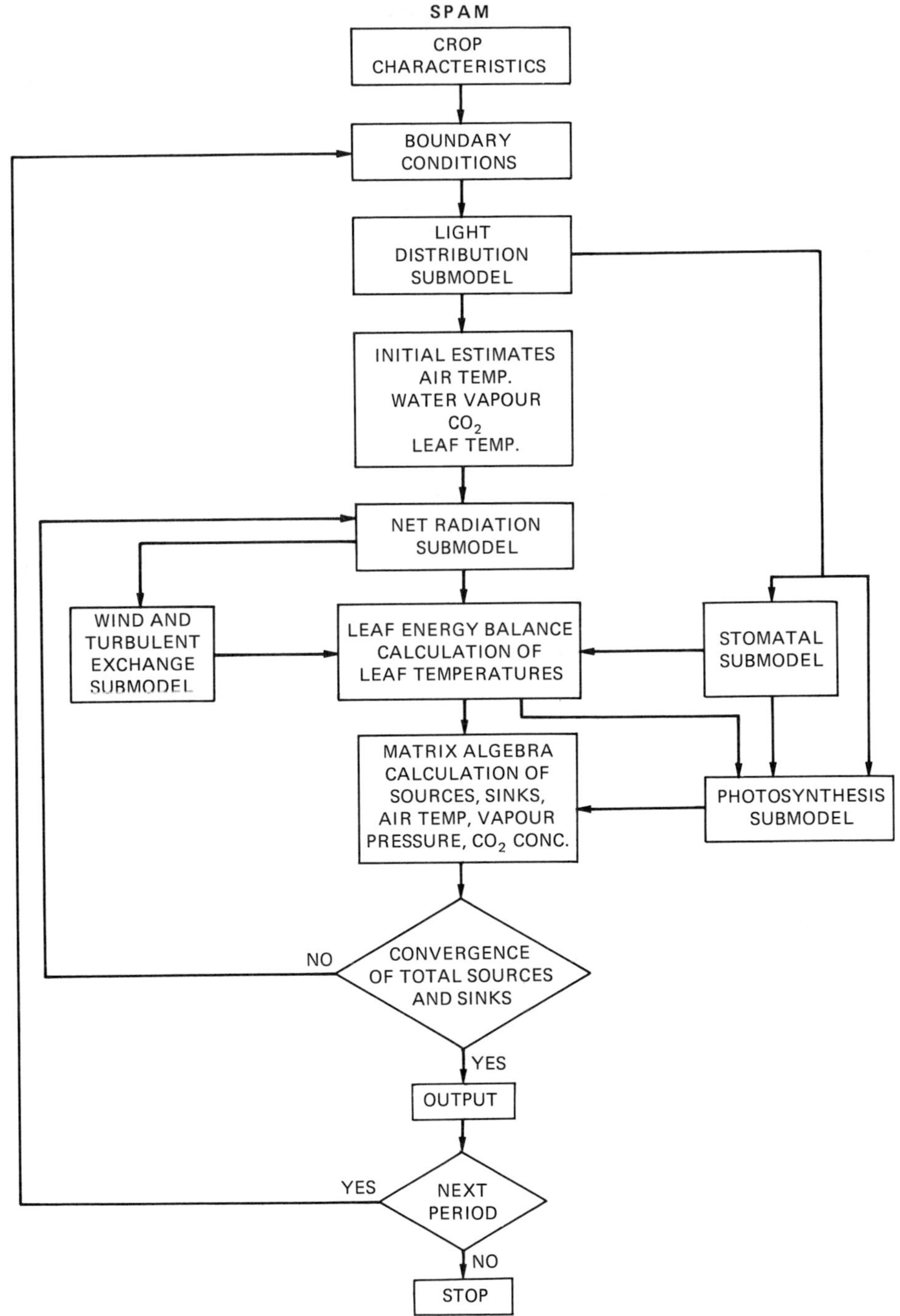

Figure 30-11. The general procedure of SPAM, given as a flow diagram. [From E. Lemon, D. W. Stewart, and R. W. Shawcroft: The sun's work in a cornfield. *Science,* **174:**371–78 (1971). Copyright 1971 by the American Association for the Advancement of Science. Used with permission. Original photograph courtesy Dr. Lemon.]

We hear a great deal these days about the energy crisis. Most people do not realize that from 40 to 90 percent of the food energy from agriculture (plants or farm animals) is derived from oil, not from the sun. Oil is needed to run tractors and farm machinery, to generate energy for fertilizer manufacture, for harvesting, storing, preparing, packaging, and delivering food. Much greater reliance will have to be made in future on our only free source of energy: the sun. M. Calvin, famous for his work on photosynthesis, has recently suggested growing as crops plants that make large amounts of oils, terpenes, or chemicals from which industrial organic compounds or fuels could easily be extracted. J. Levitt, better known for his work on stress in plants, has suggested building small ovens on farms to convert the huge amounts of agricultural waste and useless plants into coke for fuel. Certainly, in these ways, "fuel crops" from plants or regions otherwise useless to agriculture could be made to yield energy directly and cheaply.

Plants and Pollution

Pollution is primarily a man-made problem that damages plants (see Chapter 28). However, the fact that plants are damaged by pollutants infers that the pollutants themselves must interact with the plants and thus be used up in the process. Put in other terms, plants can be effective as absorbers of pollutants from the atmosphere, providing that the level of pollution is not high enough to kill or severely damage them. This fact has been demonstrated by experiments, similar to photosynthesis experiments with $^{14}CO_2$, in which plants have been permitted to absorb such chemicals as carbon monoxide, sulfur dioxide, and ozone from closed chambers. Pollutants are only absorbed when stomata are open; when they are closed (as the result of low water, darkness, or high carbon dioxide), absorption of pollutants ceases.

Because uptake continues for considerable periods of time, it must be concluded that pollutants are metabolized. Except for certain organic pollutants, little is yet known about the pathways of metabolism or the final sinks of absorbed pollutants. Ozone is probably converted to oxygen. Carbon monoxide is known to be actively metabolized by microorganisms and plants, entering into the C_1 pathways, and sulfur dioxide is probably used like sulfate in the plant's normal sulfur metabolism. This latter point is supported by the fact that it is difficult, if not impossible, to grow plants that show characteristic symptoms of sulfur deficiency in large industrial cities. Evidently they get all the sulfur they need from the atmosphere.

A consequence of plant absorption of pollutants is that they purify the atmosphere. In fact, plants are probably of tremendous importance in reducing pollution in the air. The ozone concentration at Los Angeles, where ozone pollution is very severe, decreases from over 150 ppb (parts per billion) to less than 30 ppb some 50 miles downwind. The American ecologist P. E. Waggoner has calculated that a large part of this decrease is due to foliar absorption by the vegetation in the intervening systems. He has concluded that vegetation which survives pollution may really play an important role in improving the polluter's environment.

The Role of the Plant Physiologist

At present the population of the world is increasing almost without control. Only a very few percent of the present world's population are being fed at anything like a satisfactory nutritional level. It seems inevitable that the business of producing and processing food for the world must assume ever-increasing dominance in the near future. Current overproductions in certain more fortunate areas of the world are local phenomena. Factors that lead to overproduction, such as distribution problems and the economics of demand, will hopefully be resolved. Then all the technological and scientific resources of the world must be brought to bear on the problems of increasing production and productivity. Agriculture will necessarily become industrialized, and, through centralization of facilities and direction, will require and use more and more skilled technologists and basic scientists, many of whom must be plant physiologists.

In addition, environmental engineering in all its aspects—beautification, the creation and maintenance of parks and recreational areas, sewage and garbage disposal, environmental regeneration, human sociology and psychiatry in relation to the environment—all these will require basic science and technology. Plant physiologists will be needed at all levels, as technologists, engineers, experts, innovators, in agricultural and environmental sciences. Inevitably, we must also continue the search for deeper basic understanding of how plants work by continuing scientific investigation. Thus, the teaching and practice of plant physiology will hold a valued and assured place in the future.

Additional Reading

Dansereau, P.: *Biogeography.* The Ronald Press Co., New York, 1957.

Evans, L. T. (ed.): *Crop Physiology.* Cambridge University Press, New York, 1975.

Furon, R.: *The Problem of Water—A World Study.* American Elsevier Publishing Co., Inc., New York, 1967.

Janick, J., R., W. Schery, F. W. Woods, and **V. W. Ruttan:** *Plant Science.* W. H. Freeman & Co., San Francisco, 1974.

Lemon, E., D. W. Stewart, and **R. W. Shawcroft:** The Sun's work in a cornfield. *Science,* **174:**371–78 (Oct. 22, 1971).

Milthorpe, F. L., and **J. Moorby:** *An Introduction to Crop Physiology.* Cambridge University Press, New York, 1974.

Page, B. G., and **W. T. Thomson:** *The Insecticide, Herbicide, Fungicide Quick Guide,* Thomson Publications, Fresno, Calif., 1976.

Scientific American Books: *The Biosphere* (1970); *Man and the Ecosphere* (1971). W. H. Freeman & Co., San Francisco.

Weaver, R. J.: *Plant Growth Substances in Agriculture.* W. H. Freeman & Co., San Francisco, 1972.

Wittwer, S. H.: Growth regulants in agriculture. *Outlook on Agriculture,* **6:**205–17 (1971).

Author Index*

* *Italic* numbers refer to figures or tables.

Index of Plant Names*

Plants are listed under their English and (or) Latin names as they appear in the text.

**Italic* numbers refer to figures or tables.

General Index*

*Description and definitions are given in **boldface** numbers. References to figures or tables are in *italics.*